Optical Diagnostics for Thin Film Processing

Optical Diagnostics for Thin Film Processing

Irving P. Herman
Department of Applied Physics
Columbia University
New York, New York

Academic Press
San Diego   London   Boston
New York   Sydney   Tokyo   Toronto

This book is printed on acid-free paper. ∞

Copyright © 1996 by ACADEMIC PRESS

All Rights Reserved.
No part of this publication may be reproduced or transmitted in any form or by any means, electronic or mechanical, including photocopy, recording, or any information storage and retrieval system, without permission in writing from the publisher.

Academic Press, Inc.
525 B Street, Suite 1900, San Diego, California 92101-4495, USA
http://www.apnet.com

Academic Press Limited
24-28 Oval Road, London NW1 7DX, UK
http://www.hbuk.co.uk/ap/

Library of Congress Cataloging-in-Publication Data

Herman, Irving.
   Optical diagnostics for thin film processing / by Irving Herman.
         p.   cm.
   Includes index.
   ISBN 0-12-342070-9 (alk. paper)
   1. Thin films--Surfaces. 2. Spectrum analysis--Industrial applications. I. Title.
QC176.84.S93H47   1995
621.3815'2'0287--dc20   95-16540
                                                           CIP

PRINTED IN THE UNITED STATES OF AMERICA
96 97 98 99 00 01 MM 9 8 7 6 5 4 3 2 1

*In memory of
my mother
Ida*

Contents

Preface   xv
Notations and Symbols   xix

1 Overview of Optical Diagnostics   1
    1.1 Diagnostics vs Sensors vs Transducers   3
    1.2 Attributes of *in Situ* Optical Diagnostics   4
    1.3 Performance Characteristics of Sensors   7
    1.4 The Need for *in Situ* Diagnostics   8
       1.4.1 Probing Fundamental Chemical and Physical Steps   9
       1.4.2 Process Development   9
       1.4.3 *In Situ* Monitoring and Control   10
       1.4.4 Criteria for Implementing Optical Diagnostics   18
       1.4.5 Data Analysis for Diagnostics   19
    1.5 Survey of Optical Probes   21
       1.5.1 Linear Optical Spectroscopies   21
       1.5.2 Nonlinear Optical Spectroscopies   23
       1.5.3 Hybrid Optical/Nonoptical Probes   24
       1.5.4 Probing Bulk vs Surface Properties   24
       1.5.5 Potential Optical Diagnostics of Gas-Phase Species   26
       1.5.6 Focused Reviews of Optical Diagnostics   27
       1.5.7 Applications of Optical Spectroscopies to Thin Film Processing   28
    1.6 Survey of Nonoptical Probes   29
    1.7 Thin Film Processes and Their Diagnostics Needs   34
       1.7.1 Deposition   36

	1.7.2 Plasma- and Ion-Assisted Processing	42
	1.7.3 Rapid Thermal Processing	46
	1.7.4 Photon-Assisted Growth and Processing	47
	1.7.5 Other Techniques	47
	References	50

2 The Properties of Light — 57
  2.1 Propagation — 59
  2.2 Imaging — 61
  2.3 Polarization Properties of Light — 62
  References — 65

3 The Structure of Matter — 67
  3.1 Separation of Electronic and Nuclear Motion — 67
    3.1.1 Nuclear Motion — 68
  3.2 Energy Levels in Atoms and Molecules — 69
    3.2.1 Rotational States — 70
    3.2.2 Vibrational States — 71
    3.2.3 Electronic States — 71
    3.2.4 Population of a State — 73
  3.3 Energy Levels in Solids — 76
    3.3.1 Electronic Structure — 76
    3.3.2 Vibrational Structure (Phonons) — 77
  References — 78

4 Interactions of Light with Matter for Spectroscopy — 81
  4.1 Dipole Moments and Polarization — 81
    4.1.1 Material Parameters for Linear Optical Spectroscopies — 83
  4.2 Quantum Mechanics of the Interaction of Light with Matter — 85
    4.2.1 Absorption and Spontaneous Emission — 86
    4.2.2 Spontaneous Raman Scattering — 90
  4.3 Spectroscopy — 94
    4.3.1 Absorption and Emission Spectroscopies — 94
    4.3.2 Spontaneous Raman Scattering — 100
  4.4 Nonlinear Optical Interactions — 108
  4.5 Heating by the Probing Laser — 112
  Appendix: Converting Units and Equations from the Gaussian CGS to the Rationalized MKSA System — 113
  References — 116

Contents　　　　　　　　　　　　　　　　　　　　　　　　　　ix

5 Diagnostics Equipment and Methods　　　　　　　　　119
　5.1 Optical Components　　　　　　　　　　　　　　　　119
　　　5.1.1 Optical Sources　　　　　　　　　　　　　　　119
　　　5.1.2 Optical Detectors　　　　　　　　　　　　　　124
　　　5.1.3 Components for Spectroscopic Analysis　　　　129
　　　5.1.4 Other Optical Components　　　　　　　　　　137
　5.2 Signal Collection and Analysis　　　　　　　　　　　139
　　　5.2.1 Light Collection, Imaging, and Spatial Mapping　141
　　　5.2.2 Temporal Resolution　　　　　　　　　　　　　151
　　　5.2.3 Signal Processing　　　　　　　　　　　　　　152
　　　References　　　　　　　　　　　　　　　　　　　　154

6 Optical Emission Spectroscopy　　　　　　　　　　　　157
　6.1 Mechanisms for Optical Emission　　　　　　　　　　160
　6.2 Instrumentation　　　　　　　　　　　　　　　　　　162
　6.3 Applications in Processing　　　　　　　　　　　　　166
　　　6.3.1 Plasma Etching　　　　　　　　　　　　　　　167
　　　6.3.2 Plasma-Enhanced CVD　　　　　　　　　　　　186
　　　6.3.3 Sputter Deposition　　　　　　　　　　　　　187
　　　6.3.4 Pulsed-Laser Deposition and Other Laser Processing　193
　　　6.3.5 Thermometry　　　　　　　　　　　　　　　　195
　　　6.3.6 2D and 3D Profiles by Imaging and
　　　　　　Tomographic Reconstruction　　　　　　　　　203
　　　References　　　　　　　　　　　　　　　　　　　　208

7 Laser-Induced Fluorescence　　　　　　　　　　　　　215
　7.1 Experimental Considerations　　　　　　　　　　　　218
　　　7.1.1 Signal Analysis　　　　　　　　　　　　　　　219
　　　7.1.2 Instrumentation　　　　　　　　　　　　　　　223
　7.2 Applications　　　　　　　　　　　　　　　　　　　225
　　　7.2.1 Probing Gas-Phase Species　　　　　　　　　　225
　　　7.2.2 Probing Processes at Surfaces　　　　　　　　　250
　　　References　　　　　　　　　　　　　　　　　　　　257

8 Transmission (Absorption)　　　　　　　　　　　　　　263
　　8.1 Experimental Considerations　　　　　　　　　　　264
　　　8.1.1 Signal Analysis　　　　　　　　　　　　　　　264
　　　8.1.2 Instrumentation　　　　　　　　　　　　　　　268

8.2	Gas-Phase Absorption	272
	8.2.1 Infrared (Vibrational) Absorption	273
	8.2.2 Ultraviolet/Visible (Electronic) Absorption	296
	8.2.3 Laser Magnetic Resonance	308
8.3	Transmission through Adsorbates or Thin Films	309
8.4	Transmission through Substrates for Thermometry	312
	References	320

9 Reflection 327

9.1	Optics of Reflection	328
	9.1.1 Reflection for Simple Structures	330
9.2	Reflectometry, Ellipsometry, and Polarimetry	341
9.3	Optical Dielectric Functions	343
9.4	Reflection at the Interface with a Semiinfinite Material	345
	9.4.1 Thermometry	345
	9.4.2 Monitoring Phase Changes and Annealing	347
	9.4.3 Stress Analysis	350
9.5	Interferometry	352
	9.5.1 Interferometric Metrology	354
	9.5.2 Interferometric Thermometry	368
9.6	Photoreflectance	376
9.7	Surface Infrared Reflectometry	381
	9.7.1 External Reflection Mode: Infrared Reflection–Absorption Spectroscopy	382
	9.7.2 Internal Reflection Mode: Attenuated Total Internal Reflection Spectroscopy	385
9.8	Differential Reflectometry	399
9.9	Surface Photoabsorption and p-Polarized Reflectance Spectroscopies	403
9.10	Reflectance-Difference (Anisotropy) Spectroscopy	413
	9.10.1 Instrumentation	415
	9.10.2 Comparing RDS with SPA and Ellipsometry	417
	9.10.3 Applications	418
9.11	Ellipsometry	425
	9.11.1 Theory and Modeling	427
	9.11.2 Instrumentation	435
	9.11.3 Comparison with Other Reflection Methods	442
	9.11.4 Applications in Real-Time Analysis	443
	Appendix: Terminology	465
	References	466

Contents

10 Interferometry and Photography — 481
    10.1 Interferometry — 481
    10.2 Photography, Imaging, and Microscopy — 489
    References — 491

11 Elastic Scattering and Diffraction from Particles and Nonplanar Surfaces (Scatterometry) — 495
    11.1 Detection of Particles — 496
        11.1.1 Theory — 498
        11.1.2 Particles in Gases — 505
        11.1.3 Particles in Liquids — 515
        11.1.4 Particles on Surfaces — 517
    11.2 Diffraction from Surface Features — 522
        11.2.1 Laser Light Scattering (Scatterometry) from Random Surface Features — 522
        11.2.2 Scatterometry from Periodic Surface Features — 537
    11.3 Speckle Photography and Interferometry — 548
    References — 552

12 Raman Scattering — 559
    12.1 Kinematics and Dynamics of Spontaneous Raman Scattering — 561
    12.2 Instrumentation — 563
    12.3 Thermometry and Density Measurements in Gases — 567
    12.4 Real-Time Raman Probing of Solids and Surfaces — 572
        12.4.1 Surfaces and Interfaces — 572
        12.4.2 Thin Films and Substrates — 578
    References — 587

13 Pyrometry — 591
    13.1 Theoretical and Experimental Considerations — 592
    13.2 Single-Wavelength Pyrometry — 596
        13.2.1 Optical Fiber Thermometry — 604
    13.3 Dual-Wavelength Pyrometry — 608
    13.4 Pyrometric Interferometry — 609
    13.5 Thermal Radiation during Pulsed-Laser Deposition — 614
    References — 614

14 Photoluminescence — 619
    14.1 Experimental Considerations — 622
    14.2 Probing Defects and Damage — 623

	14.3	Thermometry	628
		References	635

15 Spectroscopies Employing Laser Heating 637
    15.1 Laser-Induced Thermal Desorption 637
    15.2 Thermal Wave Optical Spectroscopies 642
           15.2.1 Photoacoustic and Thermal Wave Thermometry 649
        References 651

16 Nonlinear Optical Diagnostics 655
    16.1 Coherent Anti-Stokes Raman Scattering 656
           16.1.1 Theoretical and Experimental Considerations 656
           16.1.2 Attributes and Relative Strengths of CARS 660
           16.1.3 Density Measurements and Thermometry 661
    16.2 Surface Second-Harmonic Generation 665
    16.3 Third-Harmonic Generation in Gases 670
        References 670

17 Optical Electron/Ion Probes 673
    17.1 Photoionization 674
    17.2 Photoemission 681
    17.3 Optogalvanic Spectroscopy 682
        References 685

18 Optical Thermometry 689
    18.1 Temperature 690
    18.2 The Need for Thermometry in Thin Film Processing 692
    18.3 Nonoptical Probes of Temperature 695
    18.4 The Physical Basis of Optical Thermometry 696
           18.4.1 Thermometry of Gases 696
           18.4.2 Thermometry of Wafers 697
    18.5 A Comparison of Optical Thermometry Probes 701
           18.5.1 Gas Temperatures 701
           18.5.2 Wafer Temperatures 702
        Appendix: Representative Citations in Real-Time Optical Thermometry in Thin Film Processing 705
           18A.1 Reviews of Optical Thermometry 705
           18A.2 Citations for Optical Thermometry in Gases 706

Contents

	18A.3 Citations for Optical Thermometry of Wafers	708
	References	713

19 Data Analysis and Process Control 715
    19.1 Data Acquisition and Analysis 716
    19.2 Process Modeling 719
        19.2.1 Statistical Models 720
        19.2.2 Neural Network Training 721
        19.2.3 Applications to Thin Film Processing Development 724
    19.3 Process Control 726
    References 735

Index 739

Preface

Thin film processing techniques are widely used in many types of manufacturing. Clearly, they are central to fabrication in microelectronics, an area that has grown tremendously in recent decades. With the ever-increasing complexity of film fabrication procedures has come the realization that it is important to understand the science underlying many film processing steps and to be able to monitor and control the course of crucial steps. *In situ* diagnostics are needed to achieve either aim.

Optical-based diagnostics probably affect thin film processing activities in more ways than any other type of diagnostic technique. In fact, there has been explosive growth in the use of optical diagnostics in recent years, particularly in their application to manufacturing, and there is every indication that this very strong growth will continue in the future. This book presents the optical spectroscopies and related optical techniques that are used as real-time diagnostics during the processing of thin films. Although it is limited to diagnostics that use photons, which might seem to be unduly restrictive, this book demonstrates the amazingly wide diversity of applications of optical probes. This diversity is a testimony to the power and versatility of optical diagnostics.

This volume helps delineate the role of optical diagnostics in improving thin film processing science and technology by illustrating their uses, ranging from fundamental studies of the physics and chemistry of a process to the ultimate goal of turnkey manufacturing control. The emphasis throughout this volume is on thin film processing of semiconductors and other materials for applications in microelectronics and optoelectronics. Although specific examples of the applications of optical probes are drawn mostly from microelectronics and optoelectronics processing, these methods are applicable to all uses of thin film processing.

This book is broadly targeted to assist all those who want to use optical diagnostics. In particular, it is intended for researchers in thin film processing who have little experience in optical methods and who want to select and implement diagnostics. They may have a specific problem in mind, and may use the book to help solve it by using optical diagnostics. This volume will also serve as a resource handbook for investigators already in the field and for other scientists and engineers who are interested in the variety of existing spectroscopies and their applications. While this book has not been written to serve as a textbook (for instance, no problems have been included), it should prove to be a useful recommended text for graduate courses and seminars in the areas of spectroscopy and processing.

Chapter 1 is a broad overview that discusses the need for real-time diagnostics and the attributes of the spectroscopies that make them useful for diagnostics. Optical and nonoptical process diagnostics are also surveyed, and a brief description of each type of thin film process is presented along with its diagnostics needs. One goal of this chapter is to unify and interrelate the discussions of the different optical diagnostics and their applications that are detailed in subsequent chapters. This is done, in part, in Tables 1.2–1.5 and 1.8–1.10.

The next three chapters survey the physics and chemistry needed to understand the fundamentals of optical diagnostics. Chapter 2 briefly reviews the properties of light, Chapter 3 discusses the structure of matter, and Chapter 4 explores the fundamental interactions of light with matter and how these interactions can be used for optical spectroscopy. These chapters provide a brief yet pithy review of a truly large field. They are clearly not meant to be tutorial in nature. Readers who are not acquainted with this material, at least in an introductory manner, should refer to the textbooks cited in these chapters for a more pedagogical presentation. Most second- and third-year graduate students in applied physics, chemistry, electrical engineering, and materials science doctoral programs should be able to follow this treatment. Space limitations prevent the inclusion of many of the details about spectroscopy needed to understand all of the examples discussed in later chapters. Such details can be found in the original journal articles and in the many excellent texts on optics and spectroscopy. Quantum mechanics is used often in these introductory chapters, but it is rarely (directly) used in the subsequent chapters, which often cite results from these earlier chapters. Chapters 2–4 can be skipped by those lacking a strong technical background with no loss of continuity. All equations in these chapters and the rest of the book are presented in Gaussian CGS units. The appendix to Chapter 4 gives key expressions in Rationalized MKSA (SI) units and describes how to convert between these two sets of units.

The experimental methods and procedures needed to perform *in situ* diagnostics are described in Chapter 5, along with the optical analysis components that are required to conduct these measurements. Subsequent chapters present details specific to the diagnostic described in that chapter.

Chapters 6–17 form the core of this book. Each of these chapters describes the application of one type of optical spectroscopy or technique for real-time diagnostics in thin film processing reactors. The format of each

Preface

chapter is similar: (1) The underlying principles of the spectroscopy are presented in a brief yet rigorous scientific manner. As needed, reference is made to the theory developed in Chapters 2–4. For areas that are too large to summarize completely, standard texts are cited; this should also assist the novice. (2) The experimental implementation of the probe as an *in situ* probe is detailed, often with reference to the overview given in Chapter 5. (3) Applications of the diagnostic in thin film processing are surveyed, along with relevant spectroscopic data for each diagnostic.

Many of the diagnostics presented in Chapters 6–17 can be used to measure the temperature of either the wafer or the gas above it. Such measurements provide insight into the process science and valuable information needed for process control. Whereas specific examples of the use of these probes for optical thermometry probes are presented in each of these chapters, Chapter 18 reviews all of these optical probes of temperatures and compares them. Each of these techniques is briefly described and assessed in this chapter; the appendix to this chapter lists many of the literature references on optical thermometry cited in previous chapters. Chapter 18 also addresses the physical principles that underlie each of these methods.

The mere acquisition of raw optical data is not sufficient for the successful implementation of optical diagnostics for thin film processing applications. In particular, for process control these data must be analyzed rapidly; this reduced data set must be interpreted, which is often accomplished with the help of process models; and decisions must be made about how to best control the process step. This is described in Chapter 19.

I wish to thank D. Aspnes, K. Bachmann, B. Bent, S. Brueck, J. Butler, S. Butler, F. Celii, Y. Chabal, E. Church, R. Collins, V. Donnelly, A. Eckbreth, D. Economou, R. Eryigit, G. Flynn, D. Geohegan, S. George, M. Gross, D. Guidotti, T. Heinz, R. Hicks, G. Higashi, P. Hobbs, E. Irene, S. Leone, H. Litvak, A. Mantz, G. Oehrlein, J. O'Neill, R. Osgood, C. Pickering, H. Ryssel, K. Saenger, K. Saraswat, C. Spanos, B. Shanabrook, M. Taubenblatt, E. Whittaker, J. Woolam, and many others for stimulating discussions and for sending me details about their work. High-quality copies of many of the figures that are reproduced in this volume were supplied by many authors. I thank them for their assistance, as well as for updating me on their recent research efforts.

My warmest thanks go to D. Aspnes, W. Breiland, S. Brueck, V. Donnelly, and R. Gottscho, who helped review the manuscript. Their many insights, suggestions, and corrections helped me improve every aspect and every chapter of this book immensely (including this preface).

Irving P. Herman
New York, New York
November 1995

Notations and Symbols

Acronyms (and Abbreviations) of Optical Terms

ATIR	Attenuated total internal reflection
ATR	Attenuated total reflection
BARS	Brewster-angle reflectance spectroscopy
BBO	β-Barium borate
BOXCARS	Crossed-beam CARS
BRDF	Bidirectional reflection distribution function
CARS	Coherent anti-Stokes Raman scattering
CCD	Charge-coupled device
DR	Differential reflectometry
DSLIF	Doppler-shifted laser-induced fluorescence
DSR	Differential surface reflectometry
ESPI	Electronic-speckle-pattern interferometry
FFT	Fast Fourier transform
FT	Fourier transform
FTIR(S)	Fourier-transform infrared (spectroscopy)
ICCD	Intensified charge-coupled device
IR	Infrared
IR-DLAS	Infrared diode laser absorption spectroscopy
IRRAS (IRAS)	Infrared reflection–absorption spectroscopy (=RAIRS)
IRS	Internal reflection spectroscopy (=ATR)
KDP	Potassium dihydrogen phosphate
LDS	Laser desorption spectroscopy (=LITD)

LIF	Laser-induced fluorescence
LIFE	Laser-induced fluorescence excitation (spectroscopy)
LIMS	Laser ionization mass spectroscopy
LITD	Laser-induced thermal desorption (=LDS)
LLS	Laser light scattering
MCP	Multichannel plate
MIR	Multiple internal reflection
MIRIRS	Multiple internal reflection (MIR) internal reflection spectroscopy (IRS) (=IRS, ATR)
MPI	Multiphoton ionization
NA	Numerical aperture
NSDFS	Near-surface dielectric function spectroscopy
OE	Optical emission
OES	Optical emission spectroscopy
OMA	Optical multichannel analyzer
PAC	Photoactive compound
PAM	Photoacoustic microscopy
PAS	Photoacoustic spectroscopy
PBD	Photothermal beam deflection
PDA	Photodiode array
PDS	Photothermal deflection spectroscopy
PE	Photoemission
PI	Photoionization
PIE	Plasma-induced emission (type of OES)
PL	Photoluminescence
PLE	Photoluminescence excitation (spectroscopy)
PMT	Photomultiplier
POGS	Photoemission optogalvanic spectroscopy
PPE	Photopyroelectric spectroscopy
PR	Photoreflectance
PSD	Power spectral density
PTS	Photothermal spectroscopy
QE	Quantum efficiency
RAIRS	Reflection–absorption infrared spectroscopy (=IRRAS)
RAS	Reflectance anisotropy spectroscopy (=RDS)
RDS	Reflectance-difference spectroscopy (=RAS)
REMPI	Resonant-enhanced multiphoton ionization (also MPI)

SDR	Surface differential reflectometry/Spectroscopic differential reflectometry
SE	Spectroscopic ellipsometry
SFG	Sum frequency generation
SHG	Second-harmonic generation
SIRR	Surface infrared reflectometry
SIRS	Surface infrared spectroscopy
SPA	Surface photoabsorption
SUFSI	Subfeature speckle interferometry
SWE	Single-wavelength ellipsometry
THG	Third-harmonic generation
TIR	Total internal reflection
TIRM	Total internal reflection microscopy
TIS	Total integrated scatter
TPA	Two-photon absorption
TRLIF	Time-resolved laser-induced fluorescence
TRR	Time-resolved reflectivity/reflectometry
TWA	Thermal wave analysis
UV	Ultraviolet
VI	Virtual interface
VSA	Virtual substrate approximation

Acronyms of Nonoptical Terms

ALE	Atomic layer epitaxy
APC	Advanced process control
CBE	Chemical beam epitaxy
CD	Critical dimension
CIM	Computer-integrated manufacturing
CP	Critical point
CVD	Chemical vapor deposition
ECR	Electron cyclotron resonance
FFEBP	Feed-forward error back propagation
FWHM	Full width at half maximum
IIND	Identically, independently, and normally distributed
LPCVD	Low pressure chemical vapor deposition
MBE	Molecular beam epitaxy
MIMO	Multiple-input multiple-output

MOCVD	Metalorganic chemical vapor deposition (also OMCVD)
MOMBE	Metalorganic molecular beam epitaxy
MOVPE	Metalorganic vapor phase epitaxy (also OMVPE)
MS	Mass spectrometry
PECVD	Plasma-enhanced chemical vapor deposition
PLD	Pulsed laser deposition
PMMA	Polymethylmethacrylate
RF	Radio frequency
RGA	Residual gas analyzer (analysis)
RIE	Reactive ion etching
RHEED	Reflection high-energy electron diffraction
RMS	Root-mean square
RSM	Response surface methodology (model)
RTP	Rapid thermal processing
RtR	Run-to-run
SISO	Single-input single-output
SPC	Statistical process control
SWP	Single wafer processing
TDS	Thermal desorption spectroscopy (=TPD)
TOF	Time of flight
TOFMS	Time-of-flight mass spectrometry
TPD	Temperature-programmed desorption (=TDS)
UHV	Ultrahigh vacuum
VPE	Vapor phase epitaxy
XPS	x-Ray photoelectron spectroscopy

Symbols—Greek

α	Absorption coefficient
α'	Absorption coefficient per unit pressure
α''	Euler angle
$\hat{\alpha}$	Electric polarizability
$\tilde{\alpha}$	Linear coefficient of thermal expansion
α_e	Ellipsometry Fourier coefficient
α_s	Seraphim coefficient
α_{sat}	Absorption coefficient with saturation
α_v	Varshni coefficient

β	Phase shift in electric field due to roundtrip
$\hat{\beta}$	Hyperpolarizability
$\tilde{\beta}$	Fractional change of refractive index n with temperature
β_e	Ellipsometry Fourier coefficient
β_s	Seraphim coefficient
β_T	Volume coefficient of thermal expansion ($=3\tilde{\alpha}$)
β_v	Varshni coefficient
γ	Polarizability anisotropy
γ_0	Equilibrium, static γ
γ'	Derived, dynamic γ
$\hat{\gamma}$	Second hyperpolarizability
γ_E	Electric-field beam azimuth
γ_G	Grüneisen parameter
γ_{nonrad}	Nonradiative decay rate
γ_{rad}	Radiative decay (spontaneous emission) rate ($=A_{21}$)
γ_t	Total relaxation rate
γ_{TIR}	TIR decay coefficient
Γ_p	Phonon linewidth
$\Gamma_L^{(\nu)}$	Lorentzian linewidth (FWHM) in Hz
$\Gamma_L^{(\omega)}$	Lorentzian linewidth (FWHM) in rad/sec
$\Gamma_L^{(\mathcal{E})}$	Lorentzian linewidth (FWHM) in energy units
$\Gamma_D^{(\nu)}$	Doppler linewidth (FWHM) in Hz
δ_{ij}	Reflection phase shift
δ_x, δ_y	Electric field phase
Δ	Ellipsometry angle
$\Delta^{(1),(2)}(T)$	Frequency shift in Raman scattering
$\Delta\Omega$	Collected solid angle
ε	Strain
$\tilde{\varepsilon}$	Complex dielectric function
ε' (or ε_1)	Real part of $\tilde{\varepsilon}$
ε'' (or ε_2)	Imaginary part of $\tilde{\varepsilon}$
$\bar{\varepsilon}$	"Extinction" coefficient
$\hat{\varepsilon}$	Emissivity (emittance)
$\eta_{t,o,d}$	Total, optical, and detection efficiency factors ($\eta_t = \eta_o \eta_d$)
η_E	Electric-field ellipticity angle
θ (θ_i)	Angle of incidence (polar angle)
θ_b	Critical-point temperature parameter

θ_B	Brewster's angle, pseudo-Brewster angle
θ_c	Critical angle
$\theta_{m,s}$	Angle of diffraction, scattering
Θ	Surface coverage
λ	Vacuum wavelength
μ	Electric dipole moment
μ_d	Thermal diffusion distance
μ_m	Reduced mass
μ_m	Magnetic permeability
ν	Frequency (Hz)
ν_0	Resonant frequency
$\Delta\nu_L$	Optical source linewidth (Hz)
ρ	"Complex reflectance" ratio ($=r_p/r_s$)
ρ_d	Density
σ	Scattering cross section, absorption cross section
σ	Vertical surface roughness (rms)
σ_{div}	Full divergence angle
$\Sigma, S, \Lambda, L, \Omega, N$	Angular momentum
τ	Relaxation time
τ_{rad}	Radiative decay, spontaneous emission lifetime
ϕ	Azimuthal angle
ϕ_{ex}	Exciton phase
$\Phi(\lambda, T)$	Spectral radiant flux
χ_{ij}	Electric susceptibility ($=\chi' + i\chi''$)
χ_i, χ_r	Polarization state
χ^2_{fit}	Figure-of-merit error estimator
Ψ	Ellipsometry angle
ω	Frequency (radial; radians/sec)
ω_0	Resonant frequency (radial)
ω_e	Vibrational mode constant (also x_e, y_e)
$\omega_{i,s}$	Radial frequency for incident, scattered beams
ω_p	Phonon frequency
ω_v	Vibrational energy of fundamental transition (1–0)
Ω ($\Delta\Omega$)	Solid angle (collected solid angle)

Symbols—Roman (Lowercase)

a	Particle radius
a	Surface roughness parameter

Notations and Symbols

a	Mean polarizability
a_0	Equilibrium, static a
a'	Derived, dynamic a
a_b	Critical-point temperature parameter
a_E	Semimajor axis of electric field
a_n	Elastic scattering coefficient
b_E	Semiminor axis of electric field
$b_{J',J''}$	Placzek–Teller coefficient
b_n	Elastic scattering coefficient
c	Speed of light in vacuum
c_1, c_1', c_2	Radiation constants
d	Beam diameter
d	Groove separation (period) of grating
d	Film thickness
d_c	Critical film thickness
e	Electric charge magnitude
e_E	Beam ellipticity of electric field
$\mathbf{e}_i$	Polarization unit vector
f	Focal length of lens (or frequency of modulation)
$f(\mathcal{E}_i)$	Thermal distribution function
$f_\#$	f Number
f_{12}	Oscillator strength of $1 \leftrightarrow 2$ transition
$f(v',v'')$	Franck–Condon factor
g_i	Degeneracy
$g(\omega, v, \text{ or } \mathcal{E})$	Normalized lineshape
h	Wafer thickness
h	Planck's constant
$\hbar$	$h/2\pi$
k	Imaginary part of refractive index $\tilde{n}$
k_B	Boltzmann's constant
k_f	Force constant
k_n	Pressure broadening coefficient for species n
k_r	Reaction rate constant
$\mathbf{k}$	Wavevector ($2\pi\tilde{n}/\lambda$)
m	Refractive index ratio ($=\tilde{n}_1/\tilde{n}_2$)
m	Diffraction order
m, m_e	Electron mass

m^*, m_i	Effective mass
n	Real part of refractive index $\tilde{n}$ (also atomic quantum number)
$\tilde{n}$	Complex index of refraction, $n + ik$
$n(\omega_p)$	Phonon density
p	Phonon deformation potential
$\mathbf{p}$	Electron momentum
$p\ (\mathbf{p})$	Scattered light component parallel to the plane of incidence
p_i	Partial pressure of gas i
$p_{\text{OES,LIF}, \ldots}$	Emitted power per unit volume
p_{pix}	Pixel size
q	Phonon deformation potential
$q\ (\mathbf{q})$	Propagation wavevector for phonons
$q\ (\mathbf{q})$	Scattered light component normal to the plane of incidence
q_i	Spectrometer image size
r	Phonon deformation potential
r	Reflection coefficient
$\mathbf{r}, r_i$	Electron position vector
t	Time
t_p	Pulse width
v	Vibrational quantum number
v_i	Velocity
v	Scattering volume
w	Beam radius
w_i	Slit width
x	Normalized particle size $(=2\pi a/\lambda)$

Symbols—Roman (Uppercase)

$\mathbf{A}$	Vector potential
A	Aperture diameter
A	Absorbance (absorbed fraction)
A_{21}	Einstein A coefficient $(=\gamma_{\text{rad}})$
A_E	Electric field strength $= \sqrt{a_E^2 + b_E^2}$
B_v	Rotational constant for rigid rotor ($A_v, C_v, D_v,$ and H_v are other rotational constants)

Notations and Symbols

C	Specific heat
$C(v,t)$	Autocovariance function
C_e	Associated ellipsometry parameter
$C_{ext,sca,abs}$	Elastic scattering cross section (extinction, scattering, absorption)
$\mathbf{D}$	Electric displacement vector
D	Depolarization ratio in Raman scattering
D	Linear dispersion (of a spectrometer)
$D_{lens}\ (D)$	Lens diameter
D_t	Thermal diffusion coefficient
$E, \mathbf{E}$	Electric field
$\mathcal{E}$	Energy of a state (e.g., $\mathcal{E}_{elect,vib,rot,total}$)
$\mathcal{E}_{0,1,2}$	Critical-point energy
$\mathcal{E}_{act}$	Activation energy
$\mathcal{E}_b$	Critical-point temperature parameter
$\mathcal{E}_{bg}$	Band gap energy
$\mathcal{E}_{photon}$	Photon energy
F_{pulse}	Pulse fluence
$F_{vv'}(J)$	Vibration–rotation coupling correction factor
$\mathcal{H}$	Hamiltonian
I	Beam intensity (W/cm^2)
I_{sat}	Saturation intensity
J	Rotational quantum number (also K)
$\bar{J}$	Coherency matrix
K	Thermal conduction coefficient
M	Magnification
M_{ij}	Optical propagation matrix
$M(\lambda,T)$	Spectral radiant emittance (or exitance)
$\mathcal{N}$	Total number of species
N	Population density
N_e	Associated ellipsometry parameter
N_i	Density in state i
ΔN	Number of fringes
P	Degree of polarization
$\mathbf{P}$	Electric polarization
$\mathcal{P}_{OES,LIF},\ \ldots$	Collected power
$\mathcal{P}$	Beam power (W)
Q	Normal coordinate for vibrations

xxviii   Notations and Symbols

Symbol	Meaning
$Q_{\alpha\beta}$	Optical "reflection" factor for laser light scattering
Q_i	Normalized elastic scattering cross section ($= C_i/\pi a^2$)
R	Reflectance
R_{12}	Transition rate
$\mathbf{R}, R_i$	Nuclear coordinate
$\overleftrightarrow{\mathbf{R}}$	Raman tensor
$\mathcal{R}$	Beam curvature, wafer curvature
T	Temperature (also T_{rot}, T_{vib}, ...)
T (T_{trans})	Transmittance (T_{trans} in Chapter 13)
$S(J), S(J', J'')$	Hönl–London factor
$\mathcal{S}$	Raman scattering efficiency
$S_{0,1,2,3}$	Stokes parameters
$S_{1,2,3,4}(\theta,\phi)$	Elastic scattering matrix
S_e	Associated ellipsometry parameter
$S_{OES,LIF}, \ldots$	Detected signal intensity
U_{pulse}	Pulse energy
V	Volume
V_{im}	Imaged volume
$W(p,q,t)$	Surface power spectral density ($=$ PSD)
$W(\lambda,T)$	Hemispherical spectral radiant intensity
$Z(x)$	Surface topography
Z_i	Partition function

Physical Constants Used in This Book[a]

	Gaussian CGS units	(Rationalized MKSA units)
Speed of light in vacuum	$c = 2.998 \times 10^{10}$ cm/sec	($= 2.998 \times 10^8$ m/sec)
Planck's constant	$h = 6.626 \times 10^{-27}$ erg sec	($= 6.626 \times 10^{-34}$ J sec)
	$\hbar = 1.055 \times 10^{-27}$ erg sec	($= 1.055 \times 10^{-34}$ J sec)
Electron charge (magnitude)	$e = 4.803 \times 10^{-10}$ esu	($= 1.602 \times 10^{-19}$ C)
Electron mass	$m_e = 9.109 \times 10^{-28}$ g	($= 9.109 \times 10^{-31}$ kg)
Avogadro's number	$N_A = 6.023 \times 10^{23}$	
Boltzmann's constant	$k_B = 1.3807 \times 10^{-16}$ erg/K	($= 1.3807 \times 10^{-23}$ J/K)

[a] Consult the Appendix to Chapter 4 for the parameters and expressions used in converting from Gaussian CGS to rationalized MKSA (SI) units.

Unit Conversions for

Energy:
1 joule (J) = 10^7 erg (=1/4.184 calorie (cal))
1 eV = 1.602×10^{-12} erg
1 eV/atom = 11,600 K = 23.06 kcal/mol = 96.4 KJ/mol

Energy/hc: wavenumber units (cm^{-1}). A 1-μm-wavelength photon has an energy of 1.24 eV, corresponding to 1.0×10^4 cm^{-1}. Also, 1 cm^{-1} corresponds to 30 GHz = 3×10^{10} Hz.

Pressure:
1 atmosphere (standard) = 760 Torr (0°C) = 1.01325×10^5 Pa (pascals)
1 bar = 1.0×10^5 Pa
1 Pa = 1 N/m^2 = 10 dyn/cm^2
1 Torr (0°C) = 133.32 Pa

Consult the Appendix to Chapter 4 for other units conversions.

CHAPTER 1

Overview of Optical Diagnostics

The thin film industry, which has grown rapidly in recent decades, now affects many diverse areas of manufacturing. Thin film processing is essential to the success of the massive semiconductor and microelectronics industry and to the growth of the emerging optoelectronics industry. These techniques are widely employed for fabricating coatings to produce optical elements, to improve machine tools, and to form protective and anticorrosion layers on many components. Along with this growth has come the development of improved and increasingly sophisticated processes that require a deeper fundamental understanding of the science behind the thin film process. This increased sophistication of processing methods and the ever-increasing complexity of new materials and devices have placed new demands on process engineers, thin film equipment vendors, and product manufacturers. These demands have placed a premium on developing and employing diagnostics to understand these methods in the laboratory and to monitor and control them in the fabrication line. In particular, optical diagnostics have been increasingly used in these applications because of their versatility and power. This book addresses these issues by collecting into one volume descriptions of optical techniques for the *in situ* analysis, monitoring, and controlling of thin film processes.

The subject of thin film processing is huge, due in part to the plethora of techniques that are commonly used. Films can be deposited, etched, patterned, doped, oxidized, and annealed. Physical and chemical phenomena involving gases, liquids, surface layers, solids, plasmas, and photons can be significant. Many interactions affect the satisfactory progress of the film process. Complex chemistry can occur in the gas and on the surface, and fluid flow can greatly affect performance, as can the appearance of contaminants and undesirable complications, such as the formation of particles. Moreover, each of these processes must be developed and optimized for each material used in manufacturing. These techniques can be quite complicated whether they are being developed for processing simple elements, such as silicon and diamond, or more complex materials, such as the binary semiconductor gallium arsenide and high-temperature superconductors. Given the stringent fabrication specifications regarding film thickness, composition, purity, and crystallinity, the development and application of optical diagnostics for process elucidation and control are clearly essential.

This chapter explores how optical diagnostics can be used to improve thin film processing science and technology. It also serves as a master guide to the rest of the book, interrelating optical methods and their applications as diagnostics. Sections 1.1–1.3 briefly discuss the attributes and characteristics of diagnostics. Section 1.4 details why *in situ* diagnostics can be essential in thin film processing procedures ranging from investigations of fundamental science to process development and real-time control during manufacturing. In essence, it motivates this entire book. Sections 1.5 and 1.6 respectively survey optical and nonoptical probes. Section 1.7 presents commonly used thin film processing methods and analyzes their diagnostic needs. The tables in this chapter (Tables 1.1–1.10) interrelate diagnostics techniques and their applications, citing as well the sections in which each diagnostic is described.

Before launching into the subject matter in detail, it is best to place the study of optical diagnostics in perspective. Optical spectroscopy describes all types of probing of matter by light. Sometimes the term is used narrowly to refer to fundamental studies in which the goal is to determine the structure of matter. Optical diagnostics is a subset of optical spectroscopy and other optical techniques in which data from more fundamental studies are used to examine processes and materials. In this book the term *optical diagnostics* will refer to the *in situ* examination of process conditions, reactants, and products during a thin film process. *Ex situ* examination of materials, which is a subset of optical diagnostics, is a valuable method for materials characterization. Several spectroscopic techniques are useful both for optical diagnostics and materials characterization.

In situ optical diagnostics can be applied to the science and technology of thin film processing in three ways: (1) determining the fundamental chemical and physical steps in a thin film process, (2) developing the process, and (3) monitoring and control of the process. These three applications are vital to the basic science, applied science/engineering, and manufacturing phases of thin film processing, respectively. In each case, there is a need for nondestructive diagnostics with satisfactory sensitivity, speed, and temporal and spatial resolution.

It is useful to classify diagnostics by where and when the probing occurs, as illustrated in Figure 1.1. In this book, the term *in situ analysis* will refer to any type of analysis with the sample in place within the processing

1.1 Diagnostics vs Sensors vs Transducers

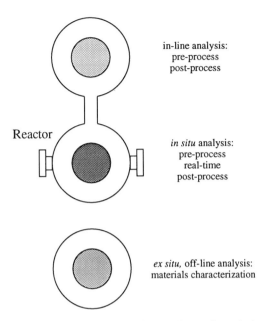

Figure 1.1 Location and time of diagnostic analysis.

chamber. Real-time *in situ* analysis occurs while the thin film process is in progress. Pre-process *in situ* analysis examines the wafer before processing, and post-process *in situ* analysis occurs after the processing has terminated. In-line analysis can occur before or after the process, with the sample moved elsewhere within the reaction chamber or to a chamber interconnected by a vacuum load lock. *Ex situ* analysis is the probing of a sample that has been removed from the reaction chamber, and can be on-line or off-line relative to the processing line. Figure 1.2 diagrams the hierarchy of probe attributes and applications. The emphasis here is on noninvasive optical diagnostics for real-time *in situ* analysis.

1.1 Diagnostics vs Sensors vs Transducers

Several different terms are commonly employed to describe the instruments and systems that are used to probe a process. Because they are often used interchangeably—and consequently incorrectly—it is best to define them here. *Sensors* are sensing devices that provide input signals to a measurement system (Norton, 1989). The purpose of this system is to provide information about a property or condition, the *measurand,* which may be used by a control system to control the value of a given quantity. A *transducer* is a sensing device or instrument that measures a PHYSICAL QUANTITY and provides an electrical signal in response to that measurement. The

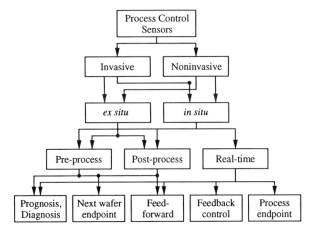

Figure 1.2 Hierarchy of process monitoring and control sensors that can be used during semiconductor manufacturing, as adapted from Moslehi *et al.* (1992). (© 1992 IEEE.)

physical quantity can be a property or condition, such as pressure, sound waves, temperature, optical radiation, heat flux, humidity, flow rates, force, viscosity, density, torque, strain, displacement, velocity, or acceleration. Sensors or transducers of optical radiation are usually called *detectors*. Instruments used for CHEMICAL ANALYSIS, such as mass spectrometers and multichannel spectrometers, are called *analyzers*. Therefore, transducers and analyzers are two different types of sensors (Norton, 1989).

Using this terminology, optical diagnostics are measurement systems that utilize optical sources and/or sensors. The development of optical diagnostics focuses on the measurement method and, at times, the control system as well. It rarely concentrates on the optical sensor because the required optical detectors are usually readily available. A given optical probe can be characterized as a sensor of a specific property or condition; e.g., one that measures temperature is a temperature sensor. It is not uncommon to find the term *sensor* used interchangeably with *diagnostic* or *probe*.

1.2 Attributes of *in Situ* Optical Diagnostics

Only those spectroscopies, be they optical or nonoptical, that can identify species, measure their concentrations, determine film thickness, or monitor process conditions compatibly with a thin film process may be considered suitable for *in situ* real-time diagnostics. Figure 1.3 illustrates the use of optical diagnostics to probe a wafer, any films or adsorbates atop the wafer, and the gas above the wafer.

1.2 Attributes of *in Situ* Optical Diagnostics

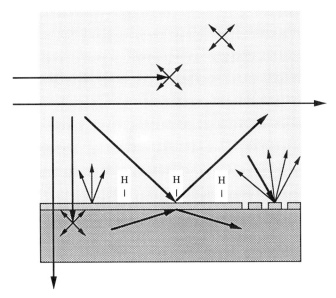

Figure 1.3 Schematic of the use of optical diagnostics for real-time probing of a wafer, the patterned or unpatterned thin film atop it, adsorbates, and the gas above the wafer. The depicted spectroscopies include internal and external reflection probing, transmission through the gas and wafer, optical emission from the gas and the wafer, laser-induced fluorescence and scattering from the gas, photoluminescence and scattering from the wafer, and scatterometry from the patterned film.

One important application of diagnostics is characterizing the properties of the film itself, including its thickness, composition, uniformity, crystallinity, stress, and the like. The measurement of characteristic film dimensions, such as film thicknesses or lateral feature widths (critical dimension, CD), is called *metrology* and is very important in most applications. In film processing by gases, optical diagnostics are desired to probe the gas-phase reactants, intermediates, and final products present during a thin film process. In this sense, probing a species means identifying it and measuring its concentration, and sometimes determining its energy distribution, all with the desired spatial and temporal resolution. Adsorbates on the film surface must also be identified in process science studies, and their density, orientation, and specific binding sites must be determined. Such measurements of species in the gas and on the surface are not needed during manufacturing. Process parameters, such as temperature, can be monitored by using diagnostics both in the gas and on the processed substrate. The measurement of temperature, called *thermometry,* can be essential both in understanding and in controlling a process. A very detailed mapping of species and process parameters may be desired in fundamental and process development studies. Much less detail is required for real-time monitoring during manufacturing, where only a few critical parameters need be measured. Optical spectroscopies are often unique in their ability to probe these process parameters during thin film processing.

Diagnostics should not affect the process. They should not perturb conditions, and must therefore be noninvasive and nonperturbative (Figure 1.2). They should not irreversibly modify the film, and must therefore be nondestructive. These attributes are important in fundamental studies and are absolutely essential during manufacturing. One advantage of most optical spectroscopies is that they are noninvasive and nondestructive.

Optical probes have many desirable features. Specific optical spectroscopies can identify many gas-phase species unambiguously and are often very sensitive, permitting measurements of very low densities. Similarly, they can identify wide-ranging film compositions and can have submonolayer sensitivity. In fact, optical spectroscopies have been developed for virtually each diagnostic need in film processing. Optical probes are usually nondestructive and noninvasive. They can usually probe through gases, unlike electron spectroscopies, and sometimes can probe through liquids and solids. They are also readily adaptable to ultrahigh vacuum (UHV) systems. Optical probes can be devised for real-time monitoring and control, where low cost, small footprint (size), simple access, and robustness are critical. In each of these attributes, an optical probe may be superior, inferior, or complementary to an existing nonoptical probe.

Diagnostics must have the necessary temporal and spatial resolution. Most thin film processes occur in steady state, with processing times on the order of minutes. The speed of the measurement is often not critical in fundamental studies; however, rapid data collection and analysis are essential for real-time monitoring and control. Sometimes the turn-on and turn-off aspects of a steady-state process are significant and need to be investigated. Some types of film processing are transient in nature. For example, rapid thermal processing (RTP) has a characteristic time of several seconds. Also, pulsed lasers are sometimes used in thin film processing, such as deposition by excimer laser ablation (pulsed laser deposition, PLD), where the characteristic interaction times can be in the nanosecond range.

The spatial resolution achievable with optical diagnostics depends on the nature of the spectroscopy, while the spatial resolution that is needed depends on the process and the diagnostic goal. For example, the gas above a substrate during deposition or etching can be probed by propagating a collimated laser with a lateral dimension of ~1 mm through the gas at a given height above the surface. This absorption measurement determines the density of a resonant species averaged over the chamber length, say ~50 cm, for this (variable) height. If the density varies slowly across the wafer or if the laser propagates along a symmetry axis, knowing the measured average density may be very useful. However, better spatial resolution is necessary if density variations are large. One approach is to conduct

1.3 Performance Characteristics of Sensors

a series of absorption measurements across different trajectories, and to use tomographic reconstruction to determine the density profile in three dimensions. A more straightforward solution is to measure the density in a confined three-dimensional volume by light scattering or laser-induced fluorescence, where the laser is focused to a small volume and the signal is collected from that small volume. Spatial resolution down to ~ 1 μm^3 is possible with visible lasers when the collection optics are very near the probed volume, which is achievable in test reactors. However, such fine resolution is rarely needed and is also difficult to achieve in production reactors because they have more limited optical access.

Although the several-millimeter lateral resolution provided by unfocused lasers often gives adequate resolution in characterizing unpatterned surfaces, films, and substrates, it is easy to obtain much better lateral resolution, down to $\sim\lambda$ of the probe light, simply by focusing the laser on the surface. The collection optics must be near the surface to achieve ~ 1-μm resolution. This is common in Raman spectroscopy of films, where such high-resolution analysis is known as Raman microprobe scattering.

The vertical resolution of the probe for a given material is determined by several factors. Some optical probes are very sensitive to the conditions within a few angstroms near the surface because of the nature of the interaction of the light with the material. Such surface specificity is common in many reflection probes, such as attenuated total internal reflection (ATR) and reflectance difference (RDS) spectroscopies. Although reflection ellipsometry senses bulk properties, it is sensitive even to submonolayer-level changes in a film. If the probing light is absorbed by the medium, optical spectroscopies that sense bulk properties, such as ellipsometry and Raman scattering, probe only as far down as the beam is transmitted. For example, in Raman scattering the top $\sim 1/2\alpha$ is probed, where α is the absorption coefficient of the material. Enhanced depth resolution can also be provided by using confocal and Mirau interference microscopies (Kino and Corle, 1990; Kino and Chim, 1990).

Rarely can one spectroscopy satisfy every diagnostic need. Several complementary probes are often needed to obtain all of the desired information. For example, different spectroscopies are needed to probe the gas and the film, or to probe different gas-phase species (Figure 1.3). Sometimes a different diagnostic is needed to measure temperature. When choosing a complete set of diagnostics, it may be best to employ both optical and nonoptical diagnostics.

1.3 Performance Characteristics of Sensors

The performance characteristics of diagnostics that assess the quality of optical probe measurements are the same as those used to evaluate any sensor (Norton, 1989). They characterize the static, dynamic, and environmental features of the sensor itself and, more generally, of the measurement system.

Static characteristics include accuracy, linearity, hysteresis, repeatability, resolution, threshold, and sensitivity. *Accuracy* is the absolute deviation of a measurement from the actual value. *Linearity* between sensor output and measurand is desirable, but not essential as long as there is a *calibration record* or *curve*. For example, the optical constants of an alloy often vary nonlinearly with composition, so they must not be linearly interpolated or extrapolated when interpreting optical measurements. *Hysteresis* is the maximum difference between the output for a given measurand value when the measurand is approached alternately from smaller or larger values; it is rarely a concern in optical measurements. The *repeatability* or *reproducibility* of measurements made consecutively is related to the *precision* of the measurement. The smallest measurable increment in sensor output when a measurand changes is the *resolution*. Related to this is the *threshold*, which is the smallest change in the measurand that produces a measurable change in the output. *Sensitivity* is the ratio of the change in sensor output for a given change in measurand, which is the slope of the calibration curve.

Dynamic characteristics of an optical sensor include the *frequency response*, which is related to the *rise time* or *time constant* of the *transient response, damping, overshooting* and potential *ringing*. Although the fundamental speed of the optical detection process itself rarely limits optical probing of thin film processing, such dynamic features are often rate-limiting in determining the characteristic time for data acquisition of the measurement system, data analysis, and process control.

Environmental features of a sensor describe changes in sensor output due to changes in properties or conditions other than those of the measurand itself. They can be very significant in optical measurements. For example, the interpretation of *in situ* optical measurements of film composition in a deposition chamber is very sensitive to temperature. Consequently, accurate measurements of composition require knowledge of both the composition and temperature dependence of optical properties, such as the index of refraction. Also, transmission and reflection measurements can be sensitive to the transmission losses from the reactor windows, which can change during a run if deposits form on them. Furthermore, small changes in the strain in reactor windows can affect ellipsometric measurements.

1.4 The Need for *in Situ* Diagnostics

A diagnostic is designed to make a specific measurement to achieve a particular goal. These goals are closely linked to the specific film process under examination and to the application for which the probe is used, that

1.4 The Need for *in Situ* Diagnostics

is, whether the probe is being used for a fundamental investigation, process development, or for real-time monitoring and control. The diagnostics requirements of these three applications differ greatly.

1.4.1 *Probing Fundamental Chemical and Physical Steps*

The goal of fundamental investigations of gas-phase processing is to identify the elementary steps that occur in the gas and on the surface and to determine the rate constant of each of these steps; this falls in the realm of chemical kinetics and dynamics. From a chemistry perspective, several questions need to be answered, including: What species decompose in the gas? What intermediates are formed and by what mechanism? How do these intermediates bind to the surface? What chemistry occurs on the surface after these intermediates adsorb on the surface? From a materials perspective, the questions include: How and why does this chemistry affect the composition of the surface and its morphology? Are there defects in the film or on the surface? What is the film crystallinity? Is the film strained? How and why do these steps affect the electrical and optical properties of the film? What fundamental steps can permit process control with monolayer precision? For patterned processing, what physical and chemical steps determine how faithfully the resist pattern is preserved?

1.4.2 *Process Development*

Each of these more fundamental issues meshes closely with a more practical issue in process development, and it is sometimes difficult to distinguish between these two aspects. While the fundamental questions are: What happens and why?, the process development questions are: What happens? Can the experimental parameters be varied to obtain the desired results? and How sensitive is the process to these experimental parameters?

One level of understanding and characterization focuses on the local kinetics. Equally important are more global process issues involving transport, such as: What is the gas flow pattern? and How effective are gas flow and diffusion in bringing fresh reactants to the surface and removing desorbed products? Other important process development questions include: What are the growth or etching rates, and are they uniform across the surface? What is the temperature on the surface and in the gas near the surface, and is it sufficiently uniform? Can the process be modeled well enough to predict changes when process parameters are varied and to correct for errors during processing? Are particulates formed in the gas, and do they deposit on the surface? For patterned processing it is important to ask: How faithfully is the resist pattern preserved?

Sometimes the goal of process development is to establish the process parameters so well that there is no need for real-time monitoring and control. Since attaining tight control of parameters is very time-consuming and expensive, such extensive process development may be cost effective only for very large-scale, high-volume manufacturing operations.

1.4.3 *In Situ Monitoring and Control*

Tight control of process parameters is important, but is not always sufficient. For example, the process parameters in a typical plasma etch system include radiofrequency (RF) power, gas mixture composition, total pressure, flow rate, substrate temperature, load size, reactor wall conditions, and the state of the system based on previous wafer processing. In open-loop control, where there is no real-time feedback from sensors, these parameters are noted and controlled. Because all these features cannot be controlled precisely, especially those dependent on the walls and previous conditions, parameters such as etch rates cannot be controlled precisely solely by monitoring externally controllable parameters. Film processing quality is improved by using *in situ* sensor monitoring for feedback in closed-loop control. Control approaches and strategies are detailed in Section 19.3.

Moslehi *et al.* (1992) and Barna *et al.* (1994) have described the sensor needs for the different steps in manufacturing integrated circuits. In each step, sensors are needed to monitor the states of the equipment, the process, and the wafer. Table 1.1 lists useful *in situ* sensors for process control in several of the thin film processing steps used in the fabrication of semiconductor devices. Such diagnostics needs apply universally to the *in situ* monitoring and control of any thin film process and, with some modifications, of any manufacturing tool. For most film processes, the most important *in situ* measurements are often those of wafer temperature, film thickness, and the endpoint of the process. While real-time monitoring of some parameters is absolutely necessary, for others *in situ* pre-process and post-process analysis suffice. While many of these requirements are served well by conventional nonoptical methods, some are best met with optical probes.

The *equipment state*, which is also known as the run recipe or menu describes input conditions. This includes the reactor set-point temperature; the reactant gas flows and partial pressures for gas processes; applied RF power, substrate bias, and magnetic field strength for some plasma-assisted processes; and optical intensity for lamp-heated and laser-assisted processes. The *process state* describes the process parameters, such as wafer temperature and uniformity; the local concentration of reactants, intermediates, and products; the time to endpoint detection; and local plasma conditions (in RF-excited plasmas). These conditions cannot be predicted by setting the input conditions only, because they also depend on uncontrollable or unidentified perturbations and long-term drifts, such as those due

1.4 The Need for *in Situ* Diagnostics

Table 1.1 Useful *in Situ* Sensors for Process Control in Representative Integrated Circuit Fabrication Steps*

Metal CVD	Wafer temperature (real-time)
	Sheet resistance (endpoint)
	Metal film thickness, surface roughness (post-processing)
Dielectric CVD	Wafer temperature (real-time)
	Film thickness (endpoint)
	Particles (in gas, on wafer; real-time, post-processing)
	Refractive index, film stress (post-processing)
Silicon CVD/epitaxy	Wafer temperature (real-time)
	Film thickness (endpoint)
	Particles (in gas, on wafer; real-time, post-processing)
	Grain size (post-processing) [polycrystalline]
	Sheet resistance, surface reflectance (post-processing)
Plasma etch (anisotropic/isotropic, ashing)	Plasma power, plasma density (real-time)
	Particles (in gas, on wafer; real-time, post-processing)
	Film thickness, plasma emission (endpoint)
	Critical dimension (real-time, post-processing) [anisotropic etch]
	Anisotropy, side-wall angle, overetch (real-time, post-processing)
	Selectivity (post-processing)
Thermal oxidation/nitridation	Wafer temperature (real-time)
	Film thickness, slips, thickness uniformity (post-processing)
Implant activation and drive-in	Wafer temperature, dopant dose and energy (real-time)
	Electrical activation (endpoint)
	Sheet resistance, slips (post-processing)
Resist processing	Resist thickness, uniformity (post-coating, pre-exposure)
	Latent image, image overlay (during exposure)
	Surface condition, plasma emission (dry-develop endpoint)
	Critical dimension (post-processing)
Glass reflow	Wafer temperature (real-time)
	Surface topography (endpoint)
	Film thickness, stress (pre- and post-processing)

* Adapted from Moslehi *et al.* (1992), © 1992 IEEE, and Barna *et al.* (1994).

to prior history (residue buildup on walls, electrode wear, recent changes in gas lines), small leaks, and the like. The *wafer state* describes the actual state of the film, including film thickness and uniformity, film properties (composition, refractive index, resistance, stress), critical dimensions of patterned films, surface roughness, and the density of particles on the surface. Input conditions are controlled by using real-time information from sensors that monitor each of these three states and by employing appropriate process models, as diagrammed in Figure 1.4.

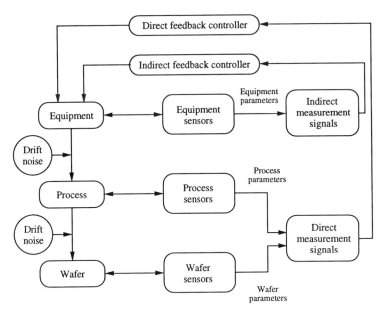

Figure 1.4 Illustration of process control strategies using equipment, process, and wafer sensors, as adapted from Moslehi *et al.* (1992). (© 1992 IEEE.)

Traditionally, statistical process control methods (open-loop control), which chart Run-to-Run trends, have been used to attain and maintain high yields. As part of this control strategy, monitor (or pilot) wafers, which in silicon device processing are often unpatterned silicon wafers that may be bare or overlaid with an oxide, are periodically introduced into tools to test them. Off-line measurements are then made to assess the status of the tool and the process. Disadvantages of this testing procedure include wafer cost and manufacturing down time. A more fundamental weakness of this approach is that the chemistry and many process details can be quite different on a uniform wafer than on a patterned wafer. *In situ* probes obviate the need for such testing procedures. Generally, advanced process control strategies, which utilize *in situ* sensors, have the potential to be superior to statistical process control methods. They are surveyed in Section 1.4.3.1.

1.4.3.1 Hierarchy of Process Control Methodologies

There is a hierarchy in the ways that *in situ* monitoring can be used to control thin film processing.

(1) The lowest level is archiving process conditions during a run. Such records can be examined periodically as part of routine maintenance. They

1.4 The Need for *in Situ* Diagnostics

can also be examined during a particular run for abnormalities if problems are discovered during wafer testing. Process conditions can be corrected in time for subsequent runs, but only after many processed wafers have been wasted. Fast, efficient analysis that identifies abnormal conditions from the complex set of sensor readings would enhance the utility of the collected data. (It has been pointed out that an even lower level of process control is feedback from customers complaining that all the devices shipped to them months ago are failing.)

(2) In the next level, the data are analyzed so fast that the operator can be alerted during the offending process step during that wafer run, or soon thereafter. After the "alarm" has sounded, the operator could decide to stop the run (system shutdown) or to take steps to overcome or sidestep the problem within that wafer run by modifying the run recipe for subsequent steps. The data analysis software must determine the severity of the problem and set thresholds for such alarms, which could lead to shutdown. In traditional open-loop control, alarms have been based on the aberrations in Run-to-Run conditions and results, as determined by statistical control methodology. The sensor readings may indicate a relatively minor problem—the need for routine maintenance, vacuum leaks, need to purge lines, clogged gas flow controllers, a change in wafer heating, etc.—or some more catastrophic failure.

Although the ultimate aim of *in situ* monitoring is real-time control, this less-demanding goal (Run-to-Run control) is still important because the whole process run could be aborted, thereby eliminating subsequent processing steps. This would result in substantial savings in cost and time. Moreover, it is difficult to discover problems in a film after it has been overlaid by other films, although the development of optical probes such as ellipsometry has improved the analysis of such buried layers. This monitoring is particularly important for totally *in situ* processing and for use in interconnected processing reactors, called cluster tools. If real-time analysis is not feasible, even *in situ* or in-line post-processing examination of intermediate layers can be very helpful for this level of control (Figure 1.2).

(3) In the most sophisticated level of real-time monitoring, conditions and wafer parameters are noted so rapidly that tool conditions (temperature, flow rates, etc.) can be controlled to achieve process goals (Figures 1.2 and 1.4). Such closed-loop, real-time control can either implement routine, planned changes or correct unanticipated abnormal conditions or process results. Endpoint detection (Table 1.2) is the goal of many thin film processes, and is used often in plasma etching.

This type of control (regulatory control) requires very rapid collection of the optical signal and, equally important, rapid analysis. For example, the goal in controlling deposition with real-time diagnostics is to measure the wafer state parameters, compare them with the desired material parameters, and then to control the input parameters directly; this is real-time

Table 1.2 Methods for Endpoint Detection

Method	Measuring	Monitoring
Optical		
Optical emission spectroscopy[a,b] (Section 6.3.1.2)	Spectrum of light emitted from discharge and its intensity	Emission from reactive species and/or etch products
Absorption[b] (Chapter 8; Section 8.3 for films)	Transmitted light	Change in concentrations in different molecules.
Optical reflection (Sections 9.4, 9.5, 9.11)	Reflection changes, interference effects,[b,c] or ellipsometry parameter monitoring	Changes in film thickness
Scatterometry[b] (Section 11.2)	Diffracted light	The profile of the surface or of the refractive index distribution.
Pyrometry (thermal imaging) (Section 13.2)	Infrared emission	Changes in emissivity
Photoemission optogalvanic spectroscopy[a] (Section 17.3)	Plasma current	Changes in work function due to changed surface
Nonoptical		
Mass spectrometry	Gas composition	Etch products
Impedance monitoring[a]	Impedance mismatch	Voltage change
Langmuir probe[a]	Changes in electron density or average energy	Current from probe
Pressure	Total pressure	Changes in total pressure

[a] Useful only during plasma etching (Marcoux and Foo, 1981).
[b] Useful for evaluating resist processing.
[c] Interference effects in OES (Section 6.3.1.3), Raman scattering (Section 12.4.2), and pyrometry (Section 13.4) can also be used for endpoint detection.

feedback control (Figure 1.2). Recipes that interrelate equipment, process, and wafer state parameters are needed (Section 19.2). Both routine and emergency control need to be run by software, not by operator intervention. If large errors are detected, the software should be able to mitigate the effect of the error for that particular run, to modify the recipe and endpoint for the next wafer in that tool (next-wafer endpoint, supervisory control), and to adapt to the error for that particular wafer by modifying subsequent process steps (feedforward control). Such *in situ* probes can also be used for equipment diagnostics. These applications of *in situ* monitors are diagrammed in Figure 1.2 and are detailed further in Section 19.3. Collectively, these control strategies are sometimes called advanced process control (Butler, 1995).

1.4.3.2 The Increasing Need for *in Situ* and Real-Time Probes

Improved control strategies are needed to obtain and maintain high yields, even with the increasing sophistication of film processing methods and aims. These include the fabrication of small-critical-dimension, high-density integrated circuits and the growth of complex heterostructures (multiple quantum wells, superlattices, strained layers, δ-doped structures) with varying compositions.

For example, band gaps, band offsets, and strains within semiconductor heterostructures can be controlled by varying the composition of ternary or quaternary alloy semiconductors. Submonolayer monitoring of the surface composition within the time needed for growing a fraction of a monolayer is imperative, so that input conditions can be altered before another monolayer is deposited if the surface composition is found to be incorrect. Otherwise, the error may be uncorrectable. Such problems can arise even when depositing films with constant compositions and can be caused by the transition from start-up to steady-state conditions or for more unexpected or subtle reasons. Tight control is important in other applications. Control of grading of interfaces is important in deposition for heterojunction bipolar transistors (HBTs). Uniform layer thicknesses in resonant tunneling diodes (RTDs) must be controlled to 1 monolayer during deposition for multistate memory (Celii *et al.*, 1993a).

Trends in production methods in microelectronics manufacturing are driven by the need for increased fabrication yields, reduced manufacturing cycle time, improved device reliability and performance, and reduced manufacturing cost per chip (Moslehi *et al.*, 1992; Liehr and Rubloff, 1994). Many of these trends have accelerated the need for *in situ* real-time diagnostics. Integrated circuits are fabricated by a sequence of many steps, performed in different chambers, some of which use wet processing and others dry processing. The wafers must be transferred between these chambers (or tools), sometimes by a cassette transport. While wafers have traditionally been processed by batch systems, where many wafers (~50–100) are often handled simultaneously in hot-wall reactors, there is a strong trend toward single-wafer processing (SWP), where only one wafer is processed in a tool.

Moslehi *et al.* (1992, 1994a) and Masnari (1994) have pointed out that SWP lowers overall factory costs, permits sensor integration for real-time control and computer-integrated manufacturing (CIM), and, when used with real-time sensors and control, decreases the manufacturing cycle time and enhances flexible manufacturing capability. Also, wafer transfer is faster with SWP, the thermal energy requirement (i.e., the thermal budget) is lower and therefore better, and some processes are much more easily implemented in the single-wafer mode, such as rapid thermal processing. *In situ* feedback during each step becomes critical in SWP when several processing steps occur within that one reactor before transport to another tool.

Because the walls are usually cold in SWP, to avoid contamination by the walls, the measurement and control of wafer temperature and temperature uniformity are necessary in single-wafer reactors. (In contrast, temperatures can be measured anywhere in hot-wall reactors because the temperature is uniform.) Thermometry involving direct contact with the wafer, such as with thermocouples, leads to wafer contamination. While the noncontact nature of optical methods is a definite strength, the applicability of optical thermometric techniques to a given process must still be examined carefully (Chapter 18). For example, pyrometry, a widely used method, may not be the best optical probe in some cases because it is sensitive to surface emissivity (Chapter 13). Still, the design of single-wafer tools makes them amenable to diagnostics on the wafer, and in particular to optical diagnostics.

The trend to larger wafer sizes has also accelerated the interest in single-wafer processing because the value of each individual wafer increases with its diameter. While throwing out a 12-inch wafer is a major problem, disposing of a single 4-inch wafer from batch processing is a relatively trivial consideration. Moreover, since the lot size decreases with increasing wafer size, each wafer becomes a larger fraction of the lot size. This drives even further the need for improved real-time process monitoring and control across each individual wafer.

In situ and in-line diagnostics are also important because of the trend to vacuum-integrated cluster tools, where the (single) wafers are transported under vacuum between interconnected systems (McNab, 1990a,b; Bergendahl *et al.*, 1990; Davis *et al.*, 1992; Moslehi *et al.*, 1992). As illustrated in Figure 1.5, several different processing steps can occur sequentially in cluster systems without breaking vacuum, such as cleaning, metal deposition,

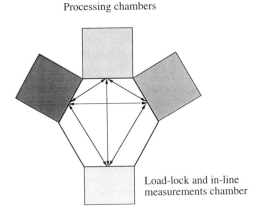

Figure 1.5 Schematic of a cluster tool.

1.4 The Need for *in Situ* Diagnostics

plasma etching, and thermal oxidation. There are usually three to six process modules in a cluster tool. The CMOS process flow described by Moslehi *et al.* (1992) employs about 16 cluster tools, each with three or four process chambers. One chamber could be added to a cluster tool that would be devoted solely to diagnostics—a measurement module. However, since adding a measurement module is expensive, it is better to perform diagnostics inside the transport chamber (the transport region between processing chambers) or in a cool-down chamber. Alternatively, diagnostics can be performed inside a process module. The trend to cluster processing is part of an overall push toward totally *in situ* processing in which more and more steps in microfabrication are performed within a (complex) vacuum chamber. Again, *in situ* diagnostics are essential to the success of this trend in processing. Schneider *et al.* (1993) have discussed the integration of *in situ* metrology sensors into cluster tools, including optical methods, ellipsometry (Section 9.11), thermal wave analysis (Section 15.2), and optical emission spectroscopy (Chapter 6), and nonoptical methods, such as x-ray photoelectron spectroscopy (XPS).

The relative merits of high-volume, general-purpose manufacturing and small-volume, specialty manufacturing need to be evaluated when developing new processes and products, including those involving thin films in semiconductor processing. In high-volume manufacturing the process is highly developed, sometimes relying heavily on *in situ* diagnostics, before constructing the pilot tool and then the final tool. Much time and money are spent developing the process so that it is very tightly controlled, and real-time diagnostics may not be needed to achieve high production yields. In contrast, in specialty manufacturing relatively few wafers are fabricated with any given design. There is relatively little effort devoted to process development and there may not even be a pilot system. Real-time monitoring and control during manufacturing are therefore needed to achieve quality control.

Closely related to specialty production is adaptive or flexible manufacturing, where the manufacturing goals can change with time. Saraswat *et al.* (1994) have described a multifunctional rapid thermal multiprocessing reactor, equipped with sensors, that is capable of flexible process design, through a "virtual factory" of computer-aided design tools, and flexible manufacturing, through this programmable factory. In addition to adaptive design, multiple process steps can occur in this facility, as they demonstrated with the fabrication of an entire MOS capacitor stack by employing three different processes that were performed sequentially: *in situ* cleaning, thermal oxidation, and chemical vapor deposition (CVD) of silicon. Single-wafer systems are particularly suitable for computer-integrated manufacturing, such as that which is needed for adaptive manufacturing, because they can be multifunctional, either alone as in rapid thermal processing chambers or interconnected as in multichamber cluster tools. In contrast, batch processing systems are less suitable for rapid prototyping and adaptive processing, in part because it is not easy to implement *in situ* monitoring for real-time control or to conduct several different types of thin film processing steps in these mass manufacturing systems. Specialty and adaptive manufacturing are expected to become increasingly important in all modes of manufacturing, especially those involving thin film processing, and concomitantly the need for real-time sensor control of such fabrication lines will become increasingly necessary.

The Semiconductor Industry Association (1994) has reviewed current capabilities and projected needs in semiconductor manufacturing, including those for process control.

1.4.4 *Criteria for Implementing Optical Diagnostics*

Although the motivation behind real-time monitoring and control is quite different from that for fundamental and process development studies, the same types of measurements are often needed in each; therefore, the same types of spectroscopies can be employed. For example, the measurement of deposition and etching rates is important for process development and control applications, and optical reflection spectroscopies are often used for each.

Still, these different applications of optical diagnostics have distinctive requirements. While fundamental studies of process mechanisms may be able to employ elaborate and expensive diagnostics that are sometimes "temperamental" or "finicky," practical production-line diagnostics must be simple, reliable, and relatively cheap. (Colloquially, this aim has been summarized as KISS: "Keep it simple, stupid.") Furthermore, the processing chamber is often designed around the diagnostics in fundamental studies, or at least it is designed with diagnostics in mind. In contrast, given historical design of production tools, diagnostics added to a tool must be designed around the existing tool and must not otherwise interfere with operation or the operator.

Any diagnostic used for real-time monitoring and control must be robust and rugged, and require little, if any, maintenance. It must be more reliable and failsafe than the tool itself. The diagnostic must be small (small footprint) and unobtrusive when installed on the tool; i.e., it must not interfere with the process or the tool and must allow routine monitoring without any operator expertise or involvement. The probe must be cost-effective and fast enough for suitable control. The speed is determined in part by the time needed to acquire the spectrum, and in part by analysis. Diagnostics for real-time monitoring may be scaled-down versions of instruments used for laboratory experiments; e.g., they may have fewer features, be geared for a specific wavelength, and so forth. They also may need to be cheaper, as well as more rugged. Often cited "acceptable" costs of sensors are less

1.4 The Need for *in Situ* Diagnostics

than $10,000 (Barna *et al.,* 1994), although more expensive sensors would surely be utilized if they were well developed and critical to the operation of expensive equipment. For example, a $100,000 sensor that controls the operation of a $2,000,000 deposition system would be a welcome investment. The sensor must be available as a complete package that can be installed on the tool (or come already installed on the tool), and then operated in a turnkey fashion.

A diagnostic employed for real-time process monitoring and control must have sufficient sensitivity and satisfactory precision and/or accuracy, and be readily calibrated. Often, only Run-to-Run reproducibility of a condition is required; attaining an absolute value for a parameter is often not important. Sometimes, the diagnostic must be able to adapt to changing conditions (environmental characteristics). For example, data collection and analysis to determine the temperature of a wafer should not be affected by previous processing steps that may have altered the wafer, so that many steps can be conducted in the same chamber without needing to change the sensor or its calibration.

Addition of real-time probes to current tools must involve only minor modifications to the tool that do not affect the process at all. (As important, management and the engineering staff must be convinced of this last point.) Some chambers are not designed for ready optical access to the front or back of a wafer. New generations of tools are being designed to have modular units, to permit upgrading without replacing the entire instrument. The hope is that diagnostics can be integrated into these modules to allow straightforward control in future generations of tools.

1.4.5 *Data Analysis for Diagnostics*

Data analysis and interpretation are as important as the physical implementation of the diagnostic; these are discussed in detail in Chapter 19. The term the chemistry community uses to describe this area is *chemometrics,* which encompasses mathematical and statistical analysis of data (including multivariate analysis), process modeling, and experimental design. The analysis of sensor data is quite different in the three applications of optical diagnostics.

The goal in fundamental measurements is to interpret data in terms of steps of a reaction mechanism involving processes on the surface and in the gas, and to obtain the reaction rate constants or potential energy surface for each step. The experiment and its subsequent interpretation may involve state-to-state resolved steps or thermally averaged processes. Each process may be compared to *ab initio* calculations or to less fundamental, yet detailed and physically reasonable, models.

In process development, one goal is to be able to predict the results for different experimental parameters; another is to find regions in parameter space where the process is optimized and relatively stable to small perturbations in input conditions. The aim is to understand why and how the different features of the process couple to lead to a predictable aspect of the process feature (e.g., the etch rate) and not the "fundamental" goal of why quantum mechanics predicts a certain binding energy. Both physical and empirical models can assist in process development (Section 19.2).

Physical models are based on the chemical and physical steps. From a physical model perspective, one goal of process development is determining which part of the overall process limits the rate of the process. Part of this is determining whether the process is limited by the flux to (or from) the surface or by the rate of reaction, i.e., whether it is transport or kinetically limited. To understand the kinetics of the process, it is necessary to identify the rate-limiting kinetic step and to determine the apparent or overall activation energy $\mathcal{E}_{act}$, which characterizes the overall rate constant of the process k_r in the Arrhenius form $k_r = A \exp(-\mathcal{E}_{act}/k_B T)$ for temperature T. This is analogous to, but much less ambitious than, determining the energy barriers or potential energy surface for each step, which is the goal of more fundamental studies. Detailed process models have been developed for many processes that either couple the chemistry and mass flow aspects of the process, or treat them individually (Section 19.2).

Equipment and process parameters and film properties can be correlated by using an empirical, or "black-box," model that is totally blind to the process chemistry and physics (Section 19.2). Two categories of empirical models are: (1) statistical models, which correlate process parameters and results in a multivariate analysis, often with a polynomial relation as in the response surface method; and (2) neural networks, which permit highly nonlinear correlation between process inputs and outputs. In a procedure called the *statistical design of experiments,* statistical models are also used to characterize a process efficiently by analyzing only a small subset of input parameter space. These empirical models may be used to optimize a process, make it robust (i.e., insensitive to perturbations), or control a process. For closed-loop control, errors must be assessed and evaluated in terms of confidence levels to determine the appropriate action.

In any event, the analysis of the volumes of data provided by the diagnostics must be rapid and efficient. In addition to optimizing a process, physical or empirical models are used to maintain a process under either statistical or advanced process control (Section 19.3). The aim of data interpretation for real-time control is to control and correct process parameters rapidly. A deviation in an experimental parameter, such as temperature or pressure, from the set-point value can be corrected directly, while the control or correction of a film property requires a change in experimental parameters on the basis of how they correlate to that property and to other properties. This can be simple for some cases, e.g., thickness control or endpoint

1.5 Survey of Optical Probes

detection, and much more complex in others, e.g., monitoring film composition, and require detailed modeling.

1.5 Survey of Optical Probes

In each of the optical spectroscopies described in this volume, photons are used to stimulate an interaction and/or emitted photons are detected. Diagnostics employing wavelengths from the ultraviolet to infrared are included in this volume; x-ray and microwave spectroscopies are not covered.

Technically, the term *optical spectroscopy* denotes only those optical techniques that probe a medium as a function of the wavelength of either the beam that excites the medium or the beam that leaves it. As such, elastic scattering from a medium and reflection interferometry at one wavelength are optical methods, but not optical spectroscopies. In this volume, however, the term *optical spectroscopy* will be used in a broader sense that includes all optical methods.

1.5.1 *Linear Optical Spectroscopies*

Many of the optical spectroscopies useful for *in situ* analysis can be classified into one of the four broad categories illustrated in Figure 1.6.

(1) The simplest probe is optical emission spectroscopy (OES) in which light is emitted from the medium without external optical excitation (Figure 1.6a) and is then spectrally analyzed. The two important examples of emission are (a) light emission from atoms, radicals, or molecules in a plasma that have been excited by collisions with high-energy electrons (Chapter 6); and (b) thermal emission (or blackbody radiation), the detection of which is called pyrometry (Chapter 13).

(2) Light that is incident on the medium can be transmitted or reflected (Figure 1.6b). The light transmitted through a gas or solid can be monitored to probe absorption (Chapter 8). Alternatively, collimated light specularly reflected from an interface, such as a gas–solid or solid–solid interface, can be monitored (Chapter 9). The detected and incident beams have the same wavelength in transmission and reflection spectroscopies. The intensity and polarization of the detected light can be monitored, sometimes as the wavelength of the incident light is varied ("spectroscopic" analysis) or as the polarization of the incident light is varied. Resonances in the absorption and reflection spectra can help identify the medium. Also, changes in film thickness or temperature can be monitored by oscillations in the intensity of the detected beam that are due to interference. Several specialized reflection probes have been developed that are useful in *in situ* diagnostics.

22   Chapter 1. Overview of Optical Diagnostics

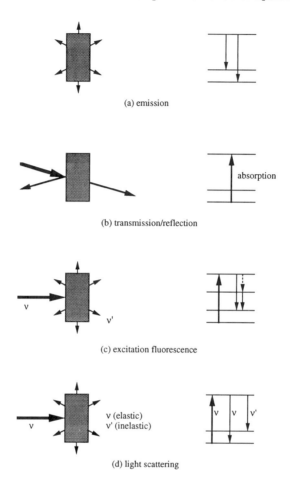

Figure 1.6 Schematic of the optical events and energy level diagrams for the four groups of linear optical diagnostics.

Ellipsometry, which is sensitive to polarization and the angle of incidence, is widely used to determine film thickness and the complex index of refraction of films. Several reflection probes are very sensitive to conditions near a surface or interface, such as attenuated total internal reflection (ATR), surface photoabsorption (SPA), and reflectance difference spectroscopy (RDS).

(3) Incident light can be absorbed by the medium and then be re-emitted at a different (longer) wavelength (Figure 1.6c); this emission is then spectrally analyzed and detected. Emission can occur directly from the state that is excited or from a lower energy state that is populated after this state relaxes. This spectroscopy is commonly called laser-induced

1.5 Survey of Optical Probes

fluorescence (LIF) in gases (Chapter 7) and photoluminescence (PL) in solids (Chapter 14). In these spectroscopies, the wavelength of the incident light can be fixed and the emission analyzed, so the spectrum is similar to an emission spectrum, such as that from OES. Alternatively, the incident wavelength can be varied and the emitted light monitored with very low spectral resolution. With this technique, known as laser-induced fluorescence excitation spectroscopy (LIFE) or photoluminescence excitation (PLE), the emission spectrum is similar to the absorption spectrum.

(4) Incident light can be scattered by the medium, either at the same wavelength (elastic scattering, Chapter 11) or different wavelength (inelastic or Raman scattering, Chapter 12) (Figure 1.6d). Elastic scattering, which is often called laser light scattering (LLS) or scatterometry, falls within the realm of diffraction. Elastic scattering from species much smaller than the wavelength of light λ is often called Rayleigh scattering. Analysis of elastic scattering as a function of scattering angle is useful in analyzing particles in gases, liquids, and on surfaces as well as in determining the random roughness of surfaces and periodic patterns on surfaces. Photons interact with the medium during inelastic scattering, although they are never actually absorbed, as they are in LIF and PL. The scattered beam has frequency components that are shifted both below and above the frequency of the incident light when the photon loses energy to the medium and gains energy from it, respectively; these features are called the Stokes and anti-Stokes lines, respectively. (Only the Stokes process is shown in Figure 1.6d). When these frequency shifts exceed ~ 5 cm^{-1}, inelastic scattering is called Raman scattering. When they are less than ~ 5 cm^{-1}, as from scattering off of sound waves, it is called Brillouin scattering. Generally, only Raman scattering is of interest in thin film diagnostics.

These techniques are called linear optical spectroscopies, because (for the latter three) the detected light intensity is proportional to the incident intensity. The latter three categories of optical probes are also often used in *ex situ* materials characterization. Linear probes are also discussed in the chapter on interferometry and photography (Chapter 10).

1.5.2 *Nonlinear Optical Spectroscopies*

In nonlinear optical spectroscopy, the optical response of the medium is not linear with the incident laser intensity. Nonlinear spectroscopy, which usually requires very high laser intensities, includes saturation effects that are seen in absorption, higher-order scattering processes such as coherent anti-Stokes Raman scattering (CARS), and multiphoton coupling phenomena such as those in second-harmonic generation and multiphoton absorption and ionization. The linear spectroscopies depicted in Figure 1.6 are much more widely used as optical diagnostics and are preferred to nonlinear optical probes, particularly for use in real-time control, because they are simpler. Sometimes, however, only a nonlinear optical spectroscopy can successfully probe a given species in a film process; thus there are some very important applications of nonlinear optics for *in situ* analysis of film processing. Chapter 16 surveys the use of these spectroscopies in diagnostics and highlights the applications of CARS and harmonic generation in diagnostics. Some nonlinear optical methods, such two-photon LIF (Section 7.2.1.2) and multiphoton ionization (Section 17.1), are detailed in the chapters describing the corresponding linear spectroscopy.

1.5.3 *Hybrid Optical/Nonoptical Probes*

The above probes are either photon-in/photon-out or photon-out spectroscopies. Some other probes involve both photons and other particles. For example, the absorption of ultraviolet photons with energy above the work function of a surface to induce electron emission is known as photoemission (PE). Photoemission using ultraviolet and x-ray photons is a widely used surface probe. After an atom or molecule absorbs one or more photons, it can emit photons (LIF) or, if it is excited enough, it can ionize to form an ion and electrons. When resonantly excited, multiphoton ionization (MPI) is known as REMPI, resonant-enhanced multiphoton ionization. These optical/electron probes are described in Chapter 17. Sometimes optical beams are used to heat a medium, and the effect of this heating—e.g., the desorption of adsorbates from a surface (laser-induced thermal desorption, LITD) or the creation of sound waves (thermal wave analysis, photoacoustic and photothermal spectroscopies)—can be probed either by optical or nonoptical spectroscopies. Such diagnostics are discussed in Chapter 15.

1.5.4 *Probing Bulk vs Surface Properties*

Many optical spectroscopies probe three-dimensional regions in a gas or a solid, be it a film on a wafer or the wafer itself. As such, they probe any species in the gas or the bulk solid within the optical beam. In some cases, a spectroscopy can be used to probe the surface itself, i.e., either the few monolayers of the film nearest the interface or adsorbates on the surface. A surface spectroscopy must be very sensitive because relatively few atoms, radicals, or molecules are being probed. Moreover, the probe should be selectively sensitive to the surface and insensitive to the bulk material, so that the contribution from the bulk does not overwhelm that from the surface. Surface-sensitive optical probes are listed in Table 1.3, together with the section in which they are described. Some of these spectroscopies have ~0.01 monolayer sensitivity (Aspnes, 1994a). McGilp (1990, 1995) has coined the term *epioptics* to describe the application of optical spectro-

1.5 Survey of Optical Probes

Table 1.3 Optical Diagnostics of Surfaces

Method	Section
Surface-specific methods	
Surface infrared reflectometry (SIRR) [infrared reflection-absorption spectroscopy (IRRAS), attenuated total internal reflection (ATR) spectroscopy]	9.7
Differential reflectometry (DR, SDR)	9.8
Surface photoabsorption (SPA), p-polarized reflectance spectroscopy[a]	9.9
Reflectance difference (anisotropy) spectroscopy (RDS/RAS)	9.10
Raman scattering [only under certain conditions]	12.4.1
Laser-induced thermal desorption (LITD)	15.1
Surface second harmonic generation	16.2
Surface-sensitive methods (not sensitive to the surface only)	
Transmission through adsorbates	8.3
Ellipsometry	9.11
Laser light scattering (LLS)	11.2.1
Photoluminescence (PL)	14.2
Near-surface sensitive methods (near-surface dielectric function spectroscopy, NSDFS)	
Ellipsometry	9.11

[a] Also known as Brewster-angle reflection spectroscopy (BARS).

scopies to the study of surfaces and interfaces. Reviews of optical surface spectroscopies have been written by Olmstead (1987), McGilp (1990, 1995), and Aspnes (1993, 1994a,b). Olmstead (1987) and McGilp (1995) detail the microscopic theory of the optical response of surfaces. The diagnostic applications of optical surface spectroscopies in probing semiconductor surfaces during growth are noted in Table 2 in McGilp (1995). Whitehouse (1994) has reviewed the optical and nonoptical metrology of surface topography.

Four features of optical spectroscopies can make them valuable surface probes.

(1) An interaction or property may be localized to the surface or interface. For example, a physical effect may average out to zero in the bulk because of symmetry, but may not vanish on the surface because the surface has lower symmetry. This principle makes reflectance difference spectroscopy (RDS/RAS) (Section 9.10) and second harmonic generation (SHG) (Section 16.2) very sensitive to the surface. Closely related to this, Raman scattering is sometimes selectively sensitive to the interfacial region because the selection rules for scattering are relaxed within a confined region *vis-à-vis* those in the bulk or because the atomic structure, and hence phonon frequency, is very different at the interface than in the adjacent bulk (or bulk-like thin film) regions (Section 12.4.1).

(2) The probe can be very sensitive to very small changes in film thickness, the presence of new thin films, or differences in the properties of

different surfaces. Spectroscopic ellipsometry (Section 9.11) is actually a bulk probe that is very sensitive to film thicknesses and compositions, even for multilayers, and is consequently very sensitive to the surface. Since this probe is sensitive to the optical properties near the surface, it is a form of near-surface dielectric function spectroscopy (NSDFS), which is important in epitaxial growth. Spectroscopic ellipsometry is also sensitive to buried interfaces. Differential spectroscopies, such as differential reflectometry (Section 9.8) and surface photoabsorption (SPA) (Section 9.9), compare surfaces relative to reference surfaces or as they evolve in time. Photoluminescence intensities (Section 14.2) often depend on the structure of the surface.

(3) The diagnostic can be sensitive to morphological deviations from a flat interface. Laser light scattering (Section 11.2.1) probes deviations from a perfect interface and can also be used to probe deviations within an otherwise uniform bulk material.

(4) Sometimes the probing light is confined very near the surface. In attenuated total internal reflection spectroscopy (Section 9.7.2), the evanescent wave decays within a wavelength above the surface, so it is sensitive to adsorbates on the surface. The substrate must be transparent to the beam for this probe to be employed.

1.5.5 *Potential Optical Diagnostics of Gas-Phase Species*

Table 1.4 lists the most common optical probes used to identify and determine the density of atoms, diatomics, and triatomics, together with the section in which the diagnostic is discussed. Thorsheim and Butler (1994)

Table 1.4 Probes for Gas-Phase Species Identification and Density Measurements*

Method	Section
OES[a]	6.3
LIF	7.2
Infrared diode laser absorption spectroscopy (IR-DLAS)	8.2.1.1
Fourier transform infrared spectroscopy (FTIRS)	8.2.1.2
Ultraviolet/visible absorption	8.2.2
Interference holography	10.1
Raman scattering	12.3
CARS	16.1
Third harmonic generation (THG)	16.3
REMPI	17.1

* See Table 1.5 for probes of temperature in the gas.
[a] For plasma reactors only.

1.5 Survey of Optical Probes

have listed the diagnostics that can be used to probe ground-state species in diamond CVD reactors such as H, H_2, CH_3, CH_4, C_2H_2, CO, and OH. For example, H atoms can be detected directly by multiphoton LIF, REMPI, and third-harmonic generation, and indirectly by OES and CARS. Table 1.5 lists those optical probes that are useful in determining temperature in a gas. Thermometry is also discussed in Chapter 18.

1.5.6 Focused Reviews of Optical Diagnostics

The reader is referred to several excellent book chapters and books that detail the use of optical, and sometimes nonoptical, diagnostics for specific types of thin film processes. Optical and nonoptical diagnostics during plasma processing have been reviewed in the volume edited by Auciello and Flamm (1989a,b), which includes a thorough review of optical diagnostics by Donnelly (1989); Gottscho and Miller (1984) and Dreyfus et al. (1985) have also reviewed optical diagnostics for plasma processing. Selwyn (1993) has detailed diagnostics for plasma processing and their practical implementation. Breiland and Ho (1993) have reviewed the use of optical and nonopti-

Table 1.5 Optical Diagnostics for Temperature Measurements*

Method	Section
Gas-phase species[a]	
Optical emission spectroscopy (OES)	6.3.5
Laser-induced fluorescence (LIF)	7.2.1
Infrared absorption	8.2.1.3
Ultraviolet/visible absorption	8.2.2.1
Interferometry and photography	10
Raman scattering (rotational)	12.3
Coherent anti-Stokes Raman scattering (CARS)	16.1.3
Resonant-enhanced multiphoton ionization (REMPI)	17.1
Wafers	
Transmission	8.4
Reflectometry at an interface	9.4.1
Reflection interferometry	9.5.2
Photoreflectance	9.6
Ellipsometry	9.11
Interferometry across the wafer	10.1
Imaging	10.2
Optical diffraction interferometry	11.2.2.3
Speckle interferometry	11.3
Raman scattering (phonons)	12.4.2.1
Pyrometry	13.1–13.5
Photoluminescence (PL)	14.3

* See Chapter 18 for an overview of thermometry.
[a] See Thomson scattering for electrons in Chapter 11.

cal diagnostics during chemical vapor deposition (CVD), and Thorsheim and Butler (1994) have reviewed specifically vapor-phase diagnostics in diamond film growth. Aspnes (1993, 1994a,b), Pickering (1994), and Richter (1993) have detailed the application of optical spectroscopies to the study of epitaxial growth; the Aspnes reviews emphasize the use of real-time optical probes for monitoring and controlling epitaxial growth. Geohegan (1994) and Saenger (1993) have surveyed the use of optical probes during pulsed-laser assisted deposition.

Optical spectroscopies that have been developed to study other processes in real time have been very helpful in developing probes for thin film processing. Eckbreth (1996) has written a detailed description of laser diagnostics for thermometry and species measurement during combustion. Macleod (1989) has discussed the use of real-time optical probes during the deposition of optical coatings. The comprehensive review of optical probes of organic thin films and their interfaces by Debe (1987) describes many of the spectroscopies needed for real-time probing of any thin film reaction.

1.5.7 *Applications of Optical Spectroscopies to Thin Film Processing*

The discussion of each diagnostic covered in Chapters 6–17 is divided into three parts: (1) a summary of the theory of the spectroscopic interaction, (2) the experimental methods best suited to implement the spectroscopy as an *in situ* diagnostic, and (3) a survey of applications. The cited applications are intended to serve as representative examples; they most certainly are not a complete review of prior work. The goal is to give a comprehensive understanding of how optical spectroscopies can be used as film diagnostics by way of these illustrative examples, rather than to review the literature. Chapters 2–4 provide the theoretical framework needed to understand these optical spectroscopies. Chapter 5 gives an introduction to the experimental equipment and techniques needed for optical diagnostics.

The discussion in Section 1.7, together with Tables 1.2–1.5 and 1.8–1.10, serves as a guide on how to achieve a specific diagnostic aim with available optical probes. The relative merits of different probes that can be used for the same application are discussed in the individual chapters. Some of the cited focused reviews also compare such competitive diagnostics; for instance, Breiland and Ho (1993) and Thorsheim and Butler (1994) critique the spectroscopies that can be used to probe the gas during CVD. Of particular importance in diagnostics is the use of optical spectroscopy for thermometry. Table 1.5 lists the optical probes used to measure the temperature of solids and gases, along with the section that details each specific application. Chapter 18 reviews the use of optical diagnostics for temperature measurements.

1.6 Survey of Nonoptical Probes

Many nonoptical diagnostics have proved useful in probing fundamental process steps, developing processes, and optimizing process parameters, as well as for real-time monitoring and control. Although some may not be as flexible or nonintrusive as optical diagnostics, others can be superior to optical probes for specific applications, perhaps because they are cheaper, more robust, or are capable of probing features that optical spectroscopies cannot. Breiland and Ho (1993) have contrasted physical probes of CVD, such as mass spectrometry, with optical probes. Physical probes can be simpler to implement in CVD than their optical counterparts, but can also perturb the deposition process. Optical probes are usually nonintrusive, but can be more complex and expensive, and may require a modification of reactor design. At times optical and nonoptical probes are used together to provide complementary information. For example, during plasma processing, sometimes both optical emission spectroscopy (or plasma-induced emission)—an optical probe (Section 6.3)—and residual gas analysis (RGA)—atomic-mass-number-resolution mass spectrometry—are used for real-time monitoring. Each of the general requirements and applications of diagnostics described earlier in this chapter for optical diagnostics also applies to nonoptical probes.

Gas flow into a chamber is often monitored with mass flow controllers and the pressure within the reactor by standard gauges, such as capacitance manometers, ionization gauges, and pirani gauges. These sensors are used both for research efforts and for integrated circuit manufacturing process control (Barna *et al.*, 1994). Optical methods have been developed to monitor the flow of specialty reactants and condensible gases in CVD and atomic flux in molecular beam epitaxy (MBE); these techniques include LIF (Chapter 7), absorption (Chapter 8), and photoionization (Section 17.1).

The use of thermocouples for temperature measurement is common in film studies. But because the thermocouple must be in contact with the wafer, the wafer can become contaminated when measuring temperature this way. Details about nonoptical and optical sensors of temperature can be found in Table 1.5 and Chapter 18.

Mass spectrometry in general, and the more modest implementation of residual gas analysis in particular, can monitor reactants during the process, sense such impurities as water vapor, and detect the process endpoint (Table 1.2). Although time delays can be a concern when RGA is used for endpoint detection, its relatively low cost and the plethora of useful information obtainable with RGA make it an interesting real-time diagnostic. Mass spectrometry, which is sometimes called mass spectroscopy, is also a common tool in fundamental and process development studies, and is sometimes used in conjunction with optical probes. For example, Shadmehr *et al.* (1992) employed mass spectrometry and OES to monitor plasmas for process analysis and control (Section 19.2.3). Differential pumping is used to sample gases from "high-pressure" regions ($>10^{-5}$ Torr), which are common in the reaction chamber and in the effluent from the chamber. In microvolume mass spectrometry, gas is sampled by a capillary during a high-pressure process, such as CVD, and is then analyzed by a differentially pumped mass spectrometer for *in situ* monitoring (Cheek *et al.*, 1993). Breiland and Ho (1993) and Thorsheim and Butler (1994) have detailed the application of mass spectrometry in CVD investigations. Reflection mass spectrometry has been used to monitor fluxes in MBE growth (Tsao *et al.*, 1989; Celii *et al.*, 1993b). Time-of-flight mass spectrometry (TOFMS) has been employed in the study of numerous gas–surface interactions, as well as in the investigation of the plume formed during pulsed laser deposition (Geohegan, 1994). Also, neutrals can be photoionized with a laser and then analyzed by mass spectrometry via a technique called laser ionization mass spectroscopy (LIMS). This can be done by using resonant, nonresonant, or resonant-enhanced multiphoton (REMPI) ionization (Geohegan, 1994; Section 17.1).

Many nonoptical probes have been developed for plasma diagnostics (Auciello and Flamm, 1989a,b), including the Langmuir probe, microwave diagnostics, electrostatic energy analysis, and current/voltage analysis. Some, like current/voltage characterization, are nonintrusive and can be used for control during manufacturing. Others, like the Langmuir probe, are intrusive. Whereas they are widely used in research, such intrusive diagnostics are not suitable for manufacturing control. Ion probes, such as Langmuir probes, can be used to analyze the quasi–steady-state (several kHz to 13.56 MHz) plasmas used in etching and deposition (Hershkowitz, 1989) and the plasma plumes formed in pulsed laser deposition (Saenger, 1993; Geohegan, 1994).

Electrical probes used for *ex situ* analysis are sometimes also used as integrated circuit manufacturing control sensors (Moslehi *et al.*, 1992; Barna *et al.*, 1994). For example, 2- or 4-point probes, which are intrusive, and eddy current measurements, which can be made in line, can be employed to measure sheet resistance of the wafer.

Gas chromatography, which is widely used in the study of all chemical processes, can be used to probe the gas phase during thin film processing. While it clearly identifies products, chromatography cannot be employed for real-time analysis because it requires sample extraction; it is therefore inherently intrusive and requires relatively long times for analysis. Breiland and Ho (1993) have detailed the use of gas chromatography to investigate CVD. Thorsheim and Butler (1994) have reviewed the use of the sample extraction methods of gas chromatography, mass spectrometry, and matrix isolation Fourier transform infrared spectroscopy in studying the CVD of diamond films.

1.6 Survey of Nonoptical Probes

Several acoustic spectroscopies have been applied as diagnostics. These methods probe specific parameters (concentrations, pressure, temperature, and the like) by determining how the speed of acoustic waves depends on these process parameters. For example, Stagg *et al.* (1992) used ultrasonics to monitor the concentrations of metalorganics (MO) in reactant gas streams entering MOCVD reactors. Degertekin *et al.* (1994) measured wafer temperature during rapid thermal processing (RTP) by monitoring the sound speed in the wafer, and Pei *et al.* (1995) used this method to monitor film thickness *in situ*. Section 15.2 discusses those acoustic techniques in which light generates and/or detects the acoustic wave.

Many surface analysis methods have been developed to probe adsorbates on the surface and the first few monolayers of a film. The names, acronyms, and features of nonoptical surface sensitive probes are listed in Table 1.6; Table 1.3 lists surface-sensitive optical probes. The methods listed near the top of Table 1.6, such as AES, XPS, LEED, and HREELS, all probe the surface with charged particles and/or very short-wavelength photons via interactions that are particularly sensitive to the surface. The probes in the middle of the table, STM and AFM, have atomic-level lateral resolution, in addition to atomic-level vertical resolution. Although these two probes are commonly used *ex situ*, they have also been applied *in situ*. The methods at the bottom of the table, including TPD, involve monitoring atoms, radicals, or molecules that desorb or scatter from the surface. Scanning electron microscopy (SEM), sometimes used with energy-dispersive analysis (EDAX), and transmission electron microscopy (TEM) (Table 1.7) are also surface sensitive, and are commonly used *ex situ*. *In situ* x-ray methods include total-reflection x-ray fluorescence spectroscopy (Roberts and Gray, 1995) and x-ray scattering (see references cited in Section 11.2.1). Surface analysis methods have been discussed by Feldman and Mayer (1986) and by Woodruff and Delchar (1994).

All of the surface-sensitive methods listed in Tables 1.3 and 1.6 can be used in real-time during "low-pressure" thin film processing, such as during MBE, although some are difficult to implement during MBE because they block the molecular beams. The surface-sensitive techniques involving charged particles (electrons, ions) usually cannot operate above $\sim 10^{-5}$ Torr, and operate best under ultrahigh vacuum (UHV) conditions. Only methods that involve only photons, including x-rays, can be used at higher pressures. Therefore, charged-particle methods, including AES, XPS, and LEED, cannot be used for *in situ*, real-time monitoring of high-pressure thin film processes, such as CVD and plasma chemical processing, while the surface-sensitive optical methods listed in Table 1.3, such as SIRR (IRAS, ATR) and RDS, can be used. For example, reflection high-energy electron diffraction (RHEED) is often employed to monitor layer-by-layer growth during MBE growth, but it cannot be used under CVD conditions. In contrast, reflectance difference spectroscopy (RDS), described in Section 9.10, can

Table 1.6 Common Nonoptical Surface Science Techniques*

Technique	Probe	Information
Auger electron spectroscopy (AES)	Electrons	Elemental composition, sometimes chemical information
X-ray photoelectron spectroscopy (XPS)[a]	X-rays incident, electrons detected	Elemental composition, chemical environment
Low-energy electron diffraction (LEED)	Electrons	Surface crystallography, adsorbate geometry
Reflection high-energy electron diffraction (RHEED)	Electrons	Surface crystallography
High-resolution electron energy loss spectroscopy (HREELS)	Electrons	Surface vibrations, chemical structure, bonding orientation
Surface extended x-ray absorption fine structure (SEXAFS)	X-rays	Bond lengths and orientations, coordination numbers
Surface x-ray diffraction (crystallography)	X-rays	Surface structure
Scanning tunneling microscopy (STM)	Electron tunneling	Composition, structure
Atomic force microscopy (AFM)	Force deflection	Composition, structure
Secondary ion mass spectroscopy (SIMS)	Cs^+ or O^+ ions	Elemental composition (small concentrations; destructive)
Temperature programmed desorption (TPD)[b,c]	None incident, reactant and product molecules detected	Desorption kinetics, coverages, reaction mechanisms and kinetics, adsorption kinetics (indirectly)
Molecular beam scattering or low-pressure reactant exposure[c]	Reactant molecules	Adsorption kinetics, reaction mechanisms and kinetics, desorption kinetics

* Adapted with permission from J. R. Creighton and J. E. Parmeter, *CRC Critical Reviews of Solid State and Materials Science*, **18**, 175 (1993). Copyright CRC Press, Boca Raton, Florida.

[a] Also known as ESCA (electron spectroscopy for chemical analysis).

[b] Other acronyms include TDS (thermal desorption spectroscopy), TDMS (thermal desorption mass spectrometry), and TPRS (temperature-programmed reaction spectroscopy). When a laser is used to induce thermal desorption, it is sometimes called LITD (laser-induced thermal desorption) or LDS (laser desorption spectroscopy); this is described in Section 15.1.

[c] Products can be monitored by mass spectrometry, laser-induced fluorescence (LIF), or resonant-enhanced multiphoton ionization (REMPI).

monitor layer-by-layer growth under both MBE and CVD conditions. These charged-particle methods are commonly employed in developmental investigations for post-process, in-line analysis after vacuum transfer from the high-pressure process tool to the UHV analysis chamber. Depending on the process, it is possible that weakly bound adsorbates desorb before this in-line analysis. (See Sections 9.11.4 and 15.1.)

1.6 Survey of Nonoptical Probes

Table 1.7 Survey of Common *ex Situ* Thin Film Characterization Methods, Excluding the Common Surface Analysis Methods of Table 1.6*

Method	Probe	Information
Physical analysis		
Scanning electron microscopy (SEM)	Electrons	Surface morphology, crystal quality
Electron microprobe/energy dispersive x-ray analysis (EDAX)	Electrons	Composition
Transmission electron microscopy (TEM)	Electrons	Defect density, interface structure, layer thickness
Rutherford backscattering	MeV He^+ or H^+	Composition, layer thickness, strain (sputtering, destructive)
Sputtering		
with AES	Electrons	Composition depth profile (destructive)
with SIMS	Ions	Elemental depth profile (destructive)
Electron analysis		
Van der Pauw/Hall effect	Electronic transport	Carrier density, mobility, compensation
Capacitance-voltage (C–V)		Carrier density profile
Transient C–V measurements		Deep level characterization
Deep-level transient spectroscopy (DLTS)		Electronic states, defect distribution
Optical analysis		
Optical microscopy (Nomarski-phase contrast)	Photons	Surface morphology
Photoluminescence (PL)	Photons	Material quality, electronic states, defects, interface roughness (see Chapter 14)

* For details see Schroder (1990).

Sometimes wafers are rotated during growth by MBE or CVD to promote uniform growth. This rotation can hinder the application of electron spectroscopies; in most cases it does not affect optical probes. For example, RHEED, which is sensitive to the surface structure, is normally not used with rotating wafers, while the application of laser light scattering (LLS, Section 11.2.1), which is also sensitive to features on the surface, is not inherently affected by rotation. However, LLS is not sensitive to anisotropies on the surface, e.g., those due to misfit dislocations, when the sample is rotating unless synchronous sensitive detection is used. (Such synchronous detection is also possible for RHEED.)

Schroder (1990) has described the many *ex situ* nonoptical and optical tools used for semiconductor material and device characterization. Several are listed in Table 1.7. Other applications of optical methods to the analysis of semiconductor materials and devices can be found in the issue of the

Journal of Selected Topics in Quantum Electronics edited by Fouquet and Merz (1995). Thin film, surface, and interface characterization techniques are detailed in the volume edited by Brundle *et al.* (1992).

1.7 Thin Film Processes and Their Diagnostics Needs

Thin film processing in microelectronics involves the fabrication of devices, their interconnection (sometimes with multiple levels of connections), and larger scale issues, such as depositing protective layers, cooling the wafer, forming electrical connections off the wafer, and mounting chips into larger units (multichip modules), that are collectively called *packaging*. The physics and materials aspects of these processes have been reviewed in books edited by Sze (1983) and McGuire (1988); the process engineering aspects of fabrication have been emphasized by Wolf and Tauber (1986), Elliott (1989), Wolf (1990), Rao (1993), and Middleman and Hochberg (1993). Marcus (1983) has reviewed conventional *ex situ* diagnostics used in VLSI.

Semiconductor device fabrication involves a series of dry and wet thin film processes in which patterning occurs by photolithography. The photolithographic process is central to microelectronics fabrication, as has been detailed by Moreau (1988). In photolithography, a photoresist is deposited on a predeposited film or substrate and is exposed to UV light projected through a mask (Figure 1.7). With positive resists, the exposed regions become more soluble and are removed, while with negative resists the exposed regions become less soluble and the nonexposed regions are removed. This leaves the underlying film exposed in specified regions, where it can be locally removed by etching—a subtractive process, which is fol-

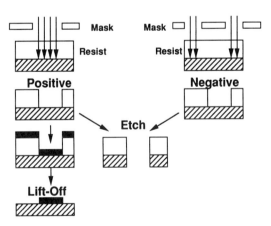

Figure 1.7 Positive- and negative-resist definition in photolithography (from Moreau, 1988).

1.7 Thin Film Processes and Their Diagnostics Needs 35

lowed by resist removal. Alternatively, in patterning by lift-off—an additive process—a film is deposited on top of the patterned resist, remaining on the wafer after resist removal only where the resist was removed during photoresist patterning. Deposition, etching, oxidation, doping, resist processing, and patterning, as well as the diagnostics needed to control these processes, are all important in microlithography. In "resistless" or direct-write lithography, directed, localized photon, ion, or electron beams locally etch or deposit films.

Thin film techniques important in semiconductor processing include deposition, etching, doping, oxidation, and annealing. A thin film process can occur on a wafer that is blank, one that has pre-existing structures or devices, or one that is patterned with resist. Excellent overviews of a wide range of thin film processes are given in the two volumes edited by Vossen and Kern (1978, 1991), the volume edited by Maissel and Glang (1983), and in the series currently edited by Francombe and Vossen (1963–1993). These thin film processes can be promoted by thermal energy, photons, or by energetic charged or neutral particles. Continuous or steady-state sources can be used. Alternatively, pulsed sources can be employed, as when flash lamps are used in rapid thermal processing (RTP) or when pulsed lasers, such as excimer lasers, are used in laser ablation. Such processes can involve dry (gas) or wet (liquid) reactants. Although optical probing of dry processing is emphasized in this volume, the diagnostics of some wet processes, such as wet etching, is also covered.

Table 1.1, adapted from Moslehi *et al.* (1992) and Barna *et al.* (1994), lists several of the thin film processes involved in integrated circuit manufacturing, along with their diagnostics needs. Thermometry (Table 1.5, Chapter 18), metrology (Table 1.8), endpoint control (Table 1.2), and the detection and control of particles (Section 11.1) are very important in most of these processes. In some cases, the measurement of stress and strain is of interest, which can be done by using reflection (Section 9.4.3), Raman scattering (Section 12.4), and photoluminescence (Chapter 14).

Although most of the diagnostics applications described in this book are drawn from the thin film processing of microelectronic and optoelectronic materials, the diagnostics needs of other applications of film processing should also be recognized. The deposition of protective and decorative coatings is very important; for example, diamond coatings can increase the hardness of machine tools and increase thermal conductivity for cooling. Thin film microfabrication techniques developed for microelectronics have also been widely used for nanomachining.

The different types of deposition techniques are now surveyed collectively because they often have common diagnostics requirements. Three film processing techniques—plasma-assisted, rapid thermal, and photon-assisted processing—are then described along with their diagnostics needs;

Table 1.8 Optical Diagnostics for Metrology

Method	Section
Film thicknesses	
Optical emission interferometry	6.3.1.3
Reflection interferometry	9.5.1
Ellipsometry	9.11
Interferometry and photography	10
Laser light scattering interferometry	11.2.1.4
Raman scattering interferometry	12.4.2
Pyrometric interferometry	13.4
Lateral distances on surfaces (CD, critical dimensions)	
Interferometry across wafer	10.1
Imaging	10.2
Scatterometry (etch features, latent images)	11.2.2
Optical diffraction interferometry	11.2.2.3
Speckle photography and interferometry	11.3
Particle sizes	
Elastic scattering	11.1

these techniques are used for deposition, etching, and other types of processing. Then other important thin film processes are briefly examined.

1.7.1 *Deposition*

The deposition of metals, semiconductors, superconductors, and insulators can be classified either as physical vapor deposition (PVD), where no molecular bonds are broken, or as chemical vapor deposition (CVD), where bonds are broken in the molecular precursors. Deposition processes are reviewed in the above-cited overviews and in George (1992). Only gas-phase deposition processes, together with their applications and diagnostics needs, are surveyed here.

Film deposition is widely used in manufacturing integrated circuits for electronics. Semiconductor heterostructures, including strained and unstrained layers, quantum wells, and superlattices, are very common in microelectronics and optoelectronics, and are usually grown by MBE (molecular beam epitaxy) or by epitaxial CVD processes. Metal layers for electronics applications are usually deposited by evaporation and sputtering. Piezoelectric, dielectric, and metal films are also used for microwave acoustics (Krishnaswamy *et al.*, 1993). The piezoelectric films are deposited by several PVD methods, including evaporation and sputtering, and by CVD. The metal films are usually deposited by these PVD methods, and both PVD and CVD are used to deposit the dielectric layers. Ferroelectric films such as $BaTiO_3$, $LiNbO_3$, and $BaMgF_4$, which form a subclass of piezoelectrics, are used in signal processing, electro-optics, and semiconductor memory devices (Francombe, 1993). The deposition of these films by evaporation has been replaced by simpler sputtering methods. Sputtering, thermal evaporation, pulsed-laser ablation deposition, MOCVD (metalorganic CVD), and metalorganic deposition (a liquid-based process, as in photoresist deposition) can be employed to deposit high-T_c superconducting thin films (Dhere, 1992).

Sputtering is the method most widely used to deposit most magnetic films (Fe, Ni, Co) for magnetic recording (Cadieu, 1992) because evaporation becomes complex when three or more elements with different evaporation properties need to be codeposited. Related methods, such as MBE, are not cost-effective because magnetic films usually need to be thicker than 1 μm. Thin-film rare-earth/transition-metal alloy films, such as GdCo films, are deposited by evaporation and sputtering for use in magneto-optic recording (Krusor and Connell, 1991; Cadieu, 1992).

Thermal evaporation is widely used in the fabrication of optical coatings for dielectric mirrors, antireflection coatings, and band-pass filters (Ritter, 1975; Macleod, 1989). Hard coatings, such as stoichiometric and nonstoichiometric nitride, carbide, and carbonitride compounds, are employed to improve the performance of tools, mechanical parts, and components. They are prepared by sputtering and arc evaporation (Musil *et al.,* 1993). Reeber (1993) has reviewed the use of thin film processing technologies for surface engineering of structural ceramics.

1.7.1.1 Physical Vapor Deposition, Including MBE

Evaporation is PVD induced from a source or multiple sources by resistance, induction, electron beam, or laser heating, or by a cathodic arc discharge (Glang, 1983; Deshpandey and Bunshah, 1991). Laser-assisted evaporation includes pulsed-laser deposition (PLD), which is also called laser ablation (Cheung and Sankur, 1988; Saenger, 1993; Geohegan, 1994).

In sputtering, energetic charged particles sputter a target (cathode), sometimes in the presence of a gas (Parsons, 1991). Sputtering chambers are operated with direct current (DC) for conducting (metallic) targets and with radio frequency (RF) excitation for both insulating (dielectrics) and conducting targets. Early diode and triode arrangements have largely been replaced by magnetron sputtering, where the presence of a magnetic field lowers the discharge voltage and increases the deposition rate, even at lower pressures. The various magnetron geometries, including planar, cylindrical (post and hollow), and gun types, can be scaled to deposit very large-area films. In magnetron sputter ion plating, there is enhanced ionization and hence a larger flux of ions to the surface. Magnetron discharges are operated with an argon ambient to deposit the target materials, and with an argon mixture containing a small fraction of reactive gas (N_2, O_2, etc.) to deposit compound films. Reactive sputtering has been reviewed by Westwood (1989). Sputtering can be classified as a type of plasma-assisted deposition. Optical emission spectroscopy (OES) (Section 6.3.3) is a useful real-time probe for monitoring sputtering, as are the optical probes listed in Table 1.9 for real-time diagnostics of MBE and CVD.

Molecular beam epitaxy (MBE) is evaporation conducted under ultrahigh-vacuum (UHV, $\sim 10^{-11}$ Torr) conditions so that impurities can be minimized. In MBE, epitaxy is promoted by cleaning the sample surface, maintaining the substrate at a sufficiently high temperature, and keeping

Table 1.9 Processes and Parameters Probed by Optical Diagnostics during MBE, CVD, and Plasma-Assisted Etching

Feature	Diagnostic locator	Important in MBE	CVD	Plasma etching
In the gas				
Identity and densities of intermediates and products	Table 1.4	—	×	×
Reactant flux to the surface and reflected from the surface	Sections 7.2.1.1 (LIF), 8.2.2.1.3, 8.2.2.1.4 (absorption)	×	—	—
Flow visualization	Table 1.10	—	×	—
Plasma characteristics (electron and ion properties)	Chapters 6 (OES), 7 (LIF), 11 (Thomson), 17 (Optogalvanic)	—	—	×
Temperature of the gas	Table 1.5 and Chapter 18	—	×	×
Particulates	Section 11.1[a]	—	×	×
In and on the wafer				
Surface composition	Table 1.3	×	×	×[b]
Surface morphology and roughness	Section 11.2.1 (LLS)	×	×	×
Particulates on the surface	Section 11.1.4[a]	—	×	×
Film composition	Sections 9.5.1, 9.11 (reflection), 12.4 (Raman), Chapter 14 (PL)	×	×	—
Film thickness	Table 1.8	×	×	×
Strain relaxation in the film	Sections 11.2.1 (LLS) and 12.4 (Raman)	×	×	—
Critical dimensions	Table 1.8	—	—	×
Wafer temperature	Table 1.5 and Chapter 18	×	×	×
General features				
Endpoint detection	Table 1.2	—	—	×

[a] See discussion in Section 1.7.2.

[b] Nondesorbed etch products.

the deposition rate sufficiently slow. Elemental sources are usually used in MBE, and the species hitting the surface are often atoms. For some elements, molecules are evaporated, as in As deposition for GaAs where As_2 mostly impinges on the surface. Reflection high-energy electron diffraction (RHEED, Table 1.6) is often used to monitor crystalline quality and the growth rate. The principles of molecular beam epitaxy have been covered by Herman and Sitter (1989), Chow (1991), Panish and Temkin (1993), and Tsao (1993).

Table 1.9 lists the process parameters that can be probed by optical diagnostics during MBE, and shows where these probes are described in this volume. In addition to probes of the surface and the film, optical monitoring of the flux of reactants to the surface can be important for controlling all types of evaporation processes, including MBE.

1.7.1.2 Chemical Vapor Deposition

In chemical vapor deposition (CVD), the rate-limiting steps in reactant decomposition can occur in the gas (homogeneous processes) or on the surface (heterogeneous processes). Sometimes, both homogeneous and heterogeneous chemical events can be very important. For example, reactant bond breaking can begin in the gas, but further decomposition of surface-adsorbed intermediates may be necessary. Heterogeneous processes are clearly important in epitaxy. (When the goal of a CVD process is epitaxial film growth, it is sometimes called vapor phase epitaxy (VPE).) Chemical vapor deposition is a dynamic process in which fluid flow, heat transfer, and diffusion of the various component gases must be characterized and understood. At lower temperatures, CVD is kinetically controlled, while at higher temperatures it is under transport control.

Chemical vapor deposition is performed in either cold-wall or hot-wall reactors. Energy input in CVD can be supplied by several sources. Resistive, RF-induction, high-intensity lamps (as in rapid thermal processing, Section 1.7.3) and lasers (laser CVD, Section 1.7.4) can be used to heat the substrate in a cold-wall, thermal CVD reactor. Better temperature control can be achieved in hot-walled reactors, where the reactor is situated in a furnace or oven. A laser can also be employed to heat the gas directly (laser CVD, Section 1.7.4). Plasmas supply the energy in plasma-enhanced CVD (PECVD), which is also called plasma-assisted CVD (PACVD) (Section 1.7.2). In addition to heating surfaces and gases, lasers can be used to photodissociate molecular precursors to stimulate deposition; this is also known as photo-CVD (Section 1.7.4).

Historically, thermal CVD has been conducted at atmospheric pressure (atmospheric pressure CVD, APCVD), but now many CVD processes are conducted below atmospheric pressure: 1–100 Torr in reduced-pressure CVD (RPCVD), 10 mTorr–1 Torr in low-pressure CVD (LPCVD), and

<10 mTorr in ultrahigh-vacuum CVD (UHVCVD). Operation at lower pressures leads to better uniformity and suppresses the redeposition of desorbed impurities.

Chemical vapor deposition of compound semiconductors from metalorganic compound precursors is alternatively (and equivalently) called MOCVD (metalorganic chemical vapor deposition), OMCVD (organometallic chemical vapor deposition), OMVPE (organometallic vapor phase epitaxy), or MOVPE (metalorganic vapor phase epitaxy) (Miller and Coleman, 1988). (Although the last two terms specifically denote epitaxy, while the first two do not, the aim of compound semiconductor deposition is nearly always epitaxy, and the four terms are usually used interchangeably.) The nomenclature used in this volume will be that employed in the cited reference. Films grown by CVD on cleaned substrates can be of the same high quality as those grown by MBE. Molecular beam epitaxy of compound semiconductors with molecular sources is an MBE/MOCVD hybrid, which is called *chemical beam epitaxy* (CBE) [when vapor group III and vapor group V sources are used to deposit III–V semiconductors], *gas-source MBE* (GSMBE) [with elemental group III and vapor group V], or *metalorganic MBE* (MOMBE) [with vapor group III and elemental group V] (Houng, 1992).

Chemical vapor deposition has also been used for the atomic layer epitaxy (ALE) of films, in which the goal is precise layer-by-layer growth and consequently fine control of film thickness and excellent thickness uniformity. In the ALE of III–V semiconductors, the two reactants used in MOCVD, such as trimethylgallium and arsine in the growth of GaAs, are alternately pulsed on and off (Creighton, 1994).

The volume edited by Hitchman and Jensen (1993) and the book by Stringfellow (1989) are comprehensive treatments of the science and technology of CVD. Sherman (1987) has surveyed the different CVD methods used in microelectronics. Kern and Ban (1978), Jensen and Kern (1991), Kuech and Jensen (1991), and Creighton and Parmeter (1993) have written review chapters on CVD.

Breiland and Ho (1993) have written an excellent and detailed review on the use of optical and nonoptical diagnostics for fluid mechanics diagnostics, gas-phase species measurements, and the analysis of particles during CVD. They emphasized that fluid mechanics considerations are essential in understanding CVD because carrier-gas flow conditions determine the diffusive and convective mass transport within the reactor, as well as the temperature distribution in the gas phase. Some of these methods are discussed in this volume; for more details refer to their review chapter.

Flow visualization methods developed for applications in aerodynamics have been used to observe flow patterns in CVD (Lauterborn and Vogel, 1984; Merzkirch, 1987; Hesselink, 1988; Yang, 1989; Adrian, 1991; Frey-

1.7 Thin Film Processes and Their Diagnostics Needs 41

muth, 1993). As summarized in Table 1.10, flows can be visualized in three ways.

(1) The flow can be detected by optically monitoring a foreign material, or tracer; this method is best in stationary and incompressible flows. Tracers include added molecules, such as $TiCl_4$ which reacts with water vapor to form TiO_2, and added particles, such as smoke formed by kerosene drops on a hot wire in air. Elastic light scattering (Rayleigh scattering), Raman scattering, and fluorescence are employed to trace the flow of tracer molecules. These and other techniques, such as Doppler velocimetry and speckle photography, are used to track tracer particles.

(2) Since the index of refraction depends on fluid density, compressible flows can be monitored by optical methods, such as shadowgraphy, schlieren, Doppler schlieren, polarization-sensitive absorption and interferometry, Mach-Zehnder interferometry, and interference holography. Interference holography is the best choice for CVD investigations (Breiland and Ho, 1993). Changes in gas temperature can be monitored by using optical path length methods (interferometry) that are sensitive to small changes in the index of refraction. Absorption spectroscopy can also monitor density changes.

(3) The flow can be perturbed by adding energy locally, as by heating or with a discharge, and then monitored by using the above methods.

Table 1.9 lists the process parameters that can be probed by optical diagnostics during CVD, as well as the locations where these probes are described in this volume. This includes an array of probes of the gas phase, the surface and the film, and diagnostics of particulates. Many of these diagnostics are useful during process development. Some are also very important for real-time control; see, for example, Table 1.1.

1.7.1.3 Common and Distinctive Features of Monitoring Deposition

Probes that can sense in real time the composition and morphology of the film surface, together with the thickness, composition, temperature,

Table 1.10 Optical Probes for Flow Visualization

Method	Section
Tracers:	
Laser-induced fluorescence	7.2.1.1
Elastic scattering, including Doppler velocimetry	11.1.2.2
Speckle photography	11.3
Raman scattering	12.3
Interferometry (including interference holography)	10.1
Shadowgraphy and schlieren	10.2

strain, and the crystalline or epitaxial nature of the deposited film, are important during the growth of films and heterostructures by MBE, CVD, and other techniques. Probes used during CVD must neither be perturbed by, nor perturb, the gas above the wafer; this is not a concern in MBE. While electron spectroscopies can be used in real time only during MBE, optical spectroscopies can be employed during deposition by any technique. Gas-phase probes are very important in CVD because they can map the densities of reaction intermediates. They are also important in evaporation-type PVD methods, such as MBE and pulsed laser deposition, where they are used to monitor the flux of species to the surface.

Aspnes (1994a) has divided the common aims of growth diagnostics into the measurement of three types of properties. The primary properties are layer thickness and composition, including the composition near the surface. Secondary properties are film properties such as doping and interface widths and abruptness; these can be controlled by real-time monitoring and control of surface properties. Tertiary parameters describe experimental conditions, such as temperature and the types and densities of reactants. These primary and secondary properties are the wafer-state parameters of Barna *et al.* (1994), while the tertiary parameters fall within their equipment- and process-state classifications, along with other diagnostics measurements, such as the mapping of gas-phase intermediates.

The control of growth in the vertical and horizontal directions is also important in all types of deposition processes. The formation of abrupt interfaces can be very important. This can be accomplished by atomic layer epitaxy, where the goal is to grow one monolayer at a time. Such growth can be limited to a single layer because of the self-limiting features of the growth mechanism. Submonolayer control can also be achieved by using surface-sensitive optical diagnostics (Table 1.3). In selective-area growth, a patterned mask defines two different regions. A given deposition process, such as CVD or sometimes MBE, leads to deposition in one region and not the other (Moon and Huong, 1993). This mask can either prevent or assist growth. A nucleation layer is an example of the latter type of mask (Herman, 1989).

1.7.2 *Plasma- and Ion-Assisted Processing*

The wide range of deposition and etching processes involving ionized species falls under the umbrella of plasma-assisted thin film processing (Rossnagel, 1991). Plasma discharges and their applications in materials processing have been detailed in the monograph edited by Einspruch and Brown (1984) and in the textbook by Lieberman and Lichtenberg (1994).

In plasma-enhanced CVD (PECVD), the reactants are often similar to those used in CVD. Much bond breaking occurs in the gas phase and heterogeneous chemistry also occurs in forming the deposited film. As in

1.7 Thin Film Processes and Their Diagnostics Needs 43

"ordinary" CVD, heterogeneous processes are often promoted by heating the substrate. Diagnostics needs include those listed in Table 1.9 to probe the film during CVD and the gas during plasma etching. Plasma-enhanced deposition has been discussed by Reif and Kern (1991), Lucovsky *et al.* (1991), Konuma (1992), and Hess and Graves (1993). Cathodic arc plasma deposition of thin films has been reviewed by Johnson (1989).

Etching in a discharge where the electrodes are directly powered, also known generically as plasma etching or sometimes dry etching, is often characterized as being either reactive ion etching (RIE) or "plasma etching." In RIE, the wafer is placed on the cathode that is capacitively powered by a radio frequency source (usually at ~13.56 MHz), while the counterelectrode and chamber walls are grounded to form the larger anode. The gas pressure is ~5–150 mTorr and ion energies are ~200–500 eV. Ions accelerate across the sheath, normal to the surface, often leading to anisotropic etching. In plasma etching, the electrode holding the wafer is grounded, along with the chamber walls, while the counterelectrode is driven by the RF and is the smaller electrode. The selectivities are better and surface damage is less in plasma etching than in RIE because of the lower ion bombardment energies (~50–200 eV), but the etch patterns tend to deviate more from ideal anisotropic profiles. Because pressures are higher in plasma etching (~100–1000 mTorr) than in RIE, contamination problems are worse. Plasma electron and ion densities of ~10^9–3×10^{10}/cc and etch rates of ~500–5000 Å/min are typical for both types of dry etching. In remote plasma processing, the plasma is produced outside the processing chamber and then flows to the wafer (Lucovsky *et al.*, 1991; Rudder *et al.*, 1993). Plasma-enhanced etching has been extensively reviewed in the volume edited by Manos and Flamm (1989) and has also been detailed by Smith (1984) (plasma etching), Lehmann (1991), van Roosmalen *et al.* (1991), and Donnelly (1994).

Fast, selective etching of a material relative to that of the mask is central to photolithography. These dry etching processes have replaced wet etch methods in many photolithographic steps. Oxygen plasmas are also used to strip resists after masked etching (ashing). Plasma etching methods are also used for surface cleaning, as in removing monolayers of oxides and hydrocarbons on a surface before epitaxial deposition. Since the wafer is at a relatively low temperature during plasma processing, dopants and impurities are not redistributed during the process. Also, plasma processes can be implemented in dry cluster tools.

New plasma sources have been developed for etching and deposition that are capable of ionizing a relatively large fraction of the neutral species, sometimes with better control of species energy. When used for etching, they improve etching selectivity, produce less damage near the surface, and etch faster *vis-à-vis* these other methods; and, when the source can be operated at lower pressures, they leave fewer residues on the surface. In magnetically enhanced reactive ion etching (MERIE) (or RF magnetron etching), magnets are added to an RIE reactor to confine the plasma above the wafer. The plasma density is $\sim 10^{10}$–10^{11}/cc, the etch rate is ~ 1000–$10{,}000$ Å/min, and the selectivity is improved *vis-à-vis* RIE. In electron cyclotron resonance (ECR) sources, microwaves (~ 2.45 GHz) excite a plasma that is confined by large magnets. Several other sources can be termed inductively coupled plasmas. In helical resonator (HR) plasmas, RF power is applied to a coil that surrounds the insulating tube through which the gas flows, and this power is inductively (and sometimes capacitively) coupled to the discharge. Magnetic confinement increases the plasma density near the wafer. The helicon plasma is similar, except that the RF power is coupled into a helicon wave in the plasma. In transformer-coupled plasma (TCP) (or inductively coupled plasma, ICP) reactors, RF powers a flat spiral coil near a dielectric window on the plasma chamber. In ECR, HR, helicon, and TCP sources the operating pressure is relatively low (~ 0.5–10 mTorr), the plasma density is relatively high ($\sim 10^{11}$–10^{12}/cc), the etch rates are fast (~ 1000–6000 Å/min), the ion energies are low (~ 10–100 eV), and the selectivity is very good. These high-density plasma sources have been reviewed by Moslehi *et al.* (1992), Lieberman and Gottscho (1994), Popov (1994), Donnelly (1994), and Burggraaf (1994).

Plasma etching involves etching by ions and radicals. Although the fluxes of ions and radicals in plasma etching depend on the process parameters, these fluxes are not controlled independently. Dry etching with separate control of the ion and chemical fluxes is more precisely called ion-beam etching. Etching can occur by separate beams of ions, such as Ar ions, and reactive molecules, such as Cl_2, that are incident on the surface. An ECR plasma source can be used to dissociate the reactive gas so that a beam of reactive radicals or atoms, such as Cl for Cl_2, hits the surface. In chemical-beam etching (CBE), only a reactive gas beam hits the surface. In ion-beam etching (IBE), only an ion beam impinges on the surface. In chemically assisted ion-beam etching (CAIBE), both ion and reactive beams are incident on the surface. In radical-beam etching (RBE), a reactive gas beam passes through a plasma region, so radicals hit the surface. In radical-beam/ion-beam etching (RBIBE), the reactive gas beam passes through a plasma region, so ion and radical beams etch the surface. Ion-assisted etching has been reviewed by Puckett *et al.* (1991). While the various plasma etching techniques are widely used in manufacturing, ion-beam beam etching is currently used mostly in research.

Plasma characterization is usually performed by using optical methods such as OES (Chapter 6) and LIF (Chapter 7), as well as nonoptical methods, such as current–voltage characterization and Langmuir probes. Table 1.9 lists the process parameters that can be probed by optical diagnostics during plasma etching, and shows where these probes are described in this volume.

1.7 Thin Film Processes and Their Diagnostics Needs

Although many of these optical probes are similar to those employed for CVD, the use of these spectroscopies for *in situ* optical monitoring during plasma processing can be complicated by the plasma glow. This optical emission consists of sharp and broad spectral features, as is described in Chapter 6. For example, this can affect photoluminescence (PL) monitoring of a surface during plasma etching (Chapter 14), because the PL spectrum cannot be spectrally resolved from the plasma light, and Raman scattering from the gas and wafer, because the Raman peaks are very weak (Chapter 12). This problem can be avoided by using pulsed lasers for excitation and gated detection, instead of continuous-wave (cw) lasers and steady-state detection. This optical background in plasmas does not affect some real-time measurements, such as absorption probing (Chapter 8) and reflection interferometry (Section 9.5).

The detection and control of particulates are particularly important during plasma processing. More generally, the appearance of particulates during many types of processing (including plasma processing and CVD) on the wafer, in process liquids, or in gases (both upstream and downstream from the tool) is a serious contamination issue (Tolliver, 1988; Donovan, 1990; Steinbrüchel, 1994). Particle control is essential during *in situ* processing in cluster tools. Laser light scattering is used extensively for process-related studies and for real-time control, as is detailed in Section 11.1. That section covers only briefly the very important and broad area of wafer inspection. In some instances LIF can also be used to examine particles. Particle filters are also employed for post-processing analysis (Breiland and Ho, 1993).

One particularly important diagnostic goal during plasma etching is determining precisely when the etching of a layer is completed; this is one example of endpoint detection. More generally, endpoint detection and control are important in many thin film processes, such as in deposition whose endpoint is reached when the desired film thickness has been deposited. Endpoint detection is essential in dry etching when the thickness and surface composition of the underlying film are critical. Table 1.2 lists several optical and nonoptical methods for endpoint control during plasma chemical etching (Marcoux and Foo, 1981; Roland *et al.*, 1985; Auciello and Flamm, 1989a,b) and other processes. Each of the techniques commonly used to monitor etching, with the exception of reflection, measures a parameter that is averaged over the wafer; reflection monitors only the region impinged by the laser. Optical emission spectroscopy (OES) and mass spectrometry (MS) are the only methods that directly detect etch products. In some instances they have roughly the same sensitivity, which increases with the etch rate and the area being etched, and both have advantages over the other methods when a masked film is being etched. For strong optical emitters, OES is more sensitive than mass spectrometry.

Optical emission spectroscopy (Section 6.3.1.2) is used extensively as an endpoint detector in plasma reactors, and OES instrumentation is often supplied by the manufacturer integrated into the tool. Reflection can serve as a probe of thickness and temperature, not just as an endpoint probe (Chapter 9). The magnitude of the reflection signal is insensitive to the etch rate and the area being etched. Different reflection probes can be used. Interference patterns change as the film thickness changes, which affects reflection (interferometric metrology, Section 9.5.1) as well as other optical probes such as OES (Section 6.3.1.3). Ellipsometric methods, which are more complex, can also be employed to obtain monolayer control (Section 9.11). In evaluating resist processing, scatterometry (latent image evaluation, Section 11.2.2.2) and transmission (Section 8.3) are useful for endpoint detection. Scatterometry can also be important in controlling the critical dimensions (CD) during dry etching (Section 11.2.2.1). Photoemission optogalvanic spectroscopy (Section 17.3) is sensitive to the surface through changes in the work function. Impedance and Langmuir probe monitoring depend on changes in the discharge that occur at the endpoint, when more reactive species remain in the discharge and fewer etch products are produced.

Related to endpoint detection is the need to monitor etch uniformity during plasma etching, which can be accomplished by using optical diagnostics. Another application of optical probes is detecting plasma tool malfunctioning, such as that caused by vacuum leaks.

1.7.3 *Rapid Thermal Processing*

In rapid thermal processing (RTP), lamps are used to heat a substrate quickly to high temperatures to promote a rapid reaction in the cold-wall reactor, such as rapid deposition (RTP-CVD), annealing (RTA), oxidation (RTO), reflow, diffusion, and film reaction (Peyton *et al.*, 1990; Gelpey *et al.*, 1993; Fair, 1993; Moslehi *et al.*, 1994a,b). Figure 13.4 shows several typical designs for RTP reactors; Roozeboom (1993) has detailed design features of these reactors. The rapid thermal switch turns the process on and off so rapidly that in RTP-CVD thin epitaxial layers can be grown with abrupt interfaces. The short-time, high-temperature steps in RTP minimize the diffusion of dopants. Rapid thermal processing is flexible enough that several different types of processing steps can be performed in the same chamber; it also has a very low thermal budget. Moreover, all thermal fabrication steps required in CMOS technologies can be performed with RTP (Moslehi *et al.*, 1994a,b). Real-time diagnostics of RTP must be fast, in the millisecond-to-second range. Of great importance in rapid thermal processing is measuring temperature fast and assessing temperature uniformity across the wafer (Table 1.5).

1.7 Thin Film Processes and Their Diagnostics Needs

1.7.4 *Photon-Assisted Growth and Processing*

Photons can be used to stimulate growth in three ways. In pyrolytic growth the laser heats the substrate and/or gas; this form of CVD is sometimes called laser CVD. In photolytic growth, photons photodissociate species with little or no concomitant heating. In pulsed-laser deposition (PLD), a pulsed laser ejects material from a target by thermal (evaporation) and/or nonthermal (electronic) mechanisms, both of which are called laser ablation, which then travels to and deposits on the substrate. Pulsed-laser deposition is actually one type of evaporation. In addition to these pyrolytic, photolytic, and laser ablation mechanisms, the photoproduction of electron/hole pairs (in semiconductors) can be important during laser-assisted etching. Thermal and plasma-enhanced deposition and etching techniques are much more widely used in production than are these photo-CVD methods.

The volume edited by Ehrlich and Tsao (1989) surveys laser-assisted etching (including that by laser ablation), growth, doping, and oxidation. Herman (1989) and Eden (1991) have written critical reviews of the methods and mechanisms of laser-assisted growth, which also discuss the use of optical diagnostics for process analysis. The book edited by Hubler and Chrisey (1994) surveys pulsed-laser deposition, and Saenger (1993) and Geohegan (1994) have detailed the use of diagnostics to study the plume formed in PLD. Optical techniques, including OES (Section 6.3.4), LIF (Section 7.2.2), absorption (Section 8.2.2.1.2), pyrometry (thermal emission) (Section 13.5), laser interferometry and photography (Chapter 10), and photodeflection (Section 15.2) have been used to study PLD laser plasmas, along with several nonoptical spectroscopies, including mass spectrometry and ion probing. Probes of the laser ablation plume need to have nanosecond time resolution and spatial imaging capability.

Optical probes have also been used to monitor laser etching. For example, the products that desorb from the surface during laser etching have been observed by LIF (Sections 7.2.2 and 15.1).

1.7.5 *Other Techniques*

Other thin film processes have been reviewed by Vossen and Kern (1978, 1991), Sze (1983), Marcus (1983), Wolf and Tauber (1986), McGuire (1988), Moreau (1988), Elliott (1989), Rao (1993), and Middleman and Hochberg (1993).

1.7.5.1 Etching

Dry etching by plasmas has been discussed in Section 1.7.2. Other dry processes are also important, such as low-damage neutral-beam etching, including atomic-layer etching, and ion-beam etching. Wet chemical etching is still very important in many phases of processing, including cleaning. It has been reviewed by Kern and Deckert (1978). Several of the diagnostics described for plasma etching (Table 1.9) can also be applied to each of these other etching processes. Furthermore, the monitoring of surface properties (Table 1.3) can be useful when etching with processes that have monolayer control.

1.7.5.2 Surface Cleaning and Passivation

The measurement and control of impurity levels on wafers are very important in integrated circuit processing (Kern, 1993; Granneman, 1994). Common molecular contaminants include O_2, H_2O, and hydrocarbons. Cleaning issues in single-wafer processing are detailed by Moslehi et al. (1992). These contaminants can be measured by using surface optical spectroscopies (Table 1.3), such as attenuated total internal reflection spectroscopy (Section 9.7.2). Particulates on the wafer are also a major problem because they lead to defects. Once, air-borne particles in the clean room were the main source of particles. This is no longer so; instead, a majority of particles are now formed in the equipment itself (O'Hanlon and Parks, 1992). Particles can be detected by elastic scattering, as described in Section 11.1.

Linked to the surface contamination issue is the passivation of surfaces (Kern, 1993; Granneman, 1994). In surface passivation the surface is treated to remove contaminants, be they atoms/radicals/molecules or defects, presumably in a manner preventing recontamination. Surface passivation can be accomplished by wet or plasma processing. For example, HF passivation of Si(111) can leave a monolayer of atomic hydrogen (see Section 9.7.2). This passivation layer decreases the sticking coefficient for many impurities (Granneman, 1994) and greatly decelerates the rate of surface oxidation.

1.7.5.3 Other Large-Area Thin Film Processes

Several other large-area thin film processes are important in semiconductor processing and other applications; these are discussed in the above-cited volumes on thin film and integrated circuit processing and by Moslehi et al. (1992) and Barna et al. (1994). Several of these processes are listed in Table 1.1, along with some of their diagnostics needs.

Oxides, such as silicon dioxide, can be produced by the oxidation of the substrate in a wet or dry furnace or by rapid thermal oxidation (Katz, 1983), and by the deposition of the oxide by thermal, plasma (Adams, 1983), and rapid thermal CVD. Nitridation is similar. Ellipsometry (Section 9.11) has found widespread use for *ex situ* determination of the thickness and dielectric constant of these insulators, and has also been used for real-time *in situ* monitoring during insulator formation.

1.7 Thin Film Processes and Their Diagnostics Needs

Other processes include annealing (including rapid thermal annealing, RTA), doping, and silicide formation (Moslehi *et al.*, 1992).

1.7.5.4 Photoresist Patterning

Process control of the patterning steps in photolithography (Moreau, 1988; Einspruch and Watts, 1987) is essential when tight control of the linewidth or critical dimension (CD) is desired. If each step in the photolithographic process is not under satisfactory open-loop control, then feedback from diagnostics is needed so that closed-loop control can be implemented. This is clearly significant because there are more photolithographic patterning steps in microelectronics fabrication than any other type of step (Luckock, 1987). As depicted in Figure 1.8, the photoresist processing steps are as follows: (1) cleaning, (2) adding an adhesion promoter, (3) depositing and spin coating the photoresist (coating), (4) prebaking to remove the casting solvent and improve adhesion (softbaking), (5) exposing the photoresist with ultraviolet radiation through the mask, (6) developing, (7) postbaking [not shown], (8) etching the underlying film (or depositing a film for the lift-off process), and (9) stripping the patterned resist. These resists

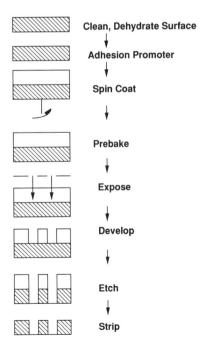

Figure 1.8 Typical steps of photoresist processing in photolithography (from Moreau, 1988).

are generally organic film-forming polymers that are photosensitized with molecular additives.

The specific *in situ* diagnostics needs of resist processing have been detailed by Barna *et al.* (1994); some important requirements are summarized in Table 1.1. Optical diagnostics have also been used to monitor and investigate many of these steps, and can be used for closed-loop control of each step and endpoint control (Table 1.2). For example, reflection interferometry (Section 9.5.1) has been used for real-time monitoring of photoresist coating and development, and for process development to measure solvent uptake and dissolution in organic polymer films and polymer shrinking during drying and curing. Sometimes measuring absorption by the photoresist film can be used as its own process monitor (Sections 8.3 and 9.7.2.3). Scatterometry (laser light scattering) can be used for real-time control of the CD during photoresist exposure (the "latent" image, Section 11.2.2.2). Resist stripping in plasmas (ashing) has been monitored by optical emission spectroscopy (Section 6.3.1). Resist dissolution has also been followed by using ellipsometry (Section 9.11.4). These probes can be used for endpoint detection during manufacturing.

References

A. C. Adams, in *VLSI Technology* (S. M. Sze, ed.), Chapter 3, p. 93. McGraw-Hill, New York, 1983.

R. J. Adrian, *Annu. Rev. Fluid Mech.* **23,** 261 (1991).

D. E. Aspnes, *Thin Solid Films* **233,** 1 (1993).

D. E. Aspnes, *Surf. Sci.* **307–309,** 1017 (1994a).

D. E. Aspnes, *Mater. Res. Soc. Symp. Proc.* **324,** 3 (1994b).

O. Auciello and D. L. Flamm, *Plasma Diagnostics,* Vol. 1. Academic Press, Boston, 1989a.

O. Auciello and D. L. Flamm, *Plasma Diagnostics,* Vol. 2. Academic Press, Boston, 1989b.

G. G. Barna, M. M. Moslehi, and Y. J. Lee, *Solid State Technol.* **37**(4), 57 (1994).

A. S. Bergendahl, D. V. Horak, P. E. Bakeman, and D. J. Miller, *Semicond. Int.* **13**(10), 94 (1990).

W. G. Breiland, and P. Ho, in *Chemical Vapor Deposition* (M. L. Hitchman and K. F. Jensen, eds.), Chapter 3, p. 91. Academic Press, Boston, 1993.

C. R. Brundle, C. A. Evans, Jr., and S. Wilson, eds. *Encyclopedia of Materials Characterization: Surfaces, Interfaces, Thin Films* (Butterworth, Stoneham, MA/ Heinemann, London, 1992).

P. Burggraaf, *Semicond. Int.* **17**(5), 56 (1994).

S. W. Butler, *J. Vac. Sci. Technol. B* **13,** 1917 (1995).

F. J. Cadieu, *Phys. Thin Films* **16,** 145 (1992).

F. G. Celii, Y. C. Kao, H.-Y. Liu, L. A. Files-Sesler, and E. A. Beam, III, *J. Vac. Sci. Technol. B* **11,** 1014 (1993a).

References

F. G. Celii, Y. C. Kao, E. A. Beam, III, W. M. Duncan, and T. S. Moise, *J. Vac. Sci. Technol. B* **11,** 1018 (1993b).

R. W. Cheek, J. A. Kelber, J. G. Fleming, R. S. Blewer, and R. D. Lujan, *J. Electrochem. Soc.* **140,** 3588 (1993).

J. T. Cheung and H. Sankur, *CRC Crit. Rev. Solid State Mater. Sci.* **15,** 63 (1988).

P. P. Chow, in *Thin Film Processes II* (J. L. Vossen and W. Kern, eds.), Chapter II-3, p. 133. Academic Press, Boston, 1991.

J. R. Creighton, *Appl. Surf. Sci.* **82/83,** 171 (1994).

J. R. Creighton and J. E. Parmeter, *CRC Crit. Rev. Solid State Mater. Sci.* **18,** 175 (1993).

C. J. Davis, I. P. Herman, and T. R. Turner, *Process Module Metrology, Control, and Clustering,* Vol. 1594. SPIE, Bellingham, WA, 1992.

M. K. Debe, *Prog. Surf. Sci.* **24,** 1 (1987).

F. L. Degertekin, J. Pei, B. T. Khuri-Yakub, and K. C. Saraswat, *Appl. Phys. Lett.* **64,** 1338 (1994).

C. V. Deshpandey and R. F. Bunshah, in *Thin Film Processes II* (J. L. Vossen and W. Kern, eds.), Chapter II-2, p. 79. Academic Press, Boston, 1991.

N. G. Dhere, in *Phys. Thin Films* **16,** 1 (1992).

V. M. Donnelly, in *Plasma Diagnostics* (O. Auciello and D. L. Flamm, eds.), Vol. 1, Chapter 1, p. 1. Academic Press, Boston, 1989.

V. M. Donnelly, in *The Encyclopedia of Advanced Materials* (D. Bloor, R. J. Brook, M. C. Flemings, and S. Mahajan, eds.), p. 1156. Pergamon, Oxford, 1994.

R. P. Donovan, ed., *Particle Control for Semiconductor Manufacturing.* Dekker, New York, 1990.

R. W. Dreyfus, J. M. Jasinski, R. E. Walkup, and G. S. Selwyn, *Pure Appl. Chem.* **57,** 1265 (1985).

A. C. Eckbreth, *Laser Diagnostics for Combustion Temperature and Species,* 2nd ed. Gordon & Breach, Luxembourg, 1996.

J. G. Eden, in *Thin Film Processes II* (J. L. Vossen and W. Kern, eds.), Chapter III-3, p. 443. Academic Press, Boston, 1991.

D. J. Ehrlich and J. Y. Tsao, eds., *Laser Microfabrication: Thin Film Processing and Lithography.* Academic Press, New York, 1989.

N. G. Einspruch and D. M. Brown, eds., *VLSI Electronics Microstructure Science,* Vol. 8. Academic Press, Orlando, FL, 1984.

N. G. Einspruch and R. K. Watts, eds., *VLSI Electronics Microstructure Science,* Vol. 16, Academic Press, Orlando, FL, 1987.

D. J. Elliott, *Integrated Circuit Fabrication Technology,* 2nd ed. McGraw-Hill, New York, 1989.

R. B. Fair, ed., *Rapid Thermal Processing Science and Technology.* Academic Press, Boston, 1993.

L. C. Feldman and J. W. Mayer, *Fundamentals of Surface and Thin Film Analysis.* North-Holland Publ., New York, 1986.

J. E. Fouquet and J. L. Merz, eds., *IEEE J. Sel. Top. Quantum Electron.* **1**(4) (1995).

M. H. Francombe, in *Phys. Thin Films* **17,** 225 (1993).

M. H. Francombe and J. L. Vossen, eds., *Physics of Thin Films: Advances in Research and Development,* Vols. 1–17. Academic Press, Boston, 1963–1993.

P. Freymuth, *Rev. Sci. Instrum.* **64,** 1 (1993).

J. C. Gelpey, J. K. Elliott, J. J. Wortman, and A. Ajmera, eds., *Mater. Res. Soc. Symp. Proc.* **303** (1993).

D. B. Geohegan, in *Pulsed Laser Deposition of Thin Films* (G. Hubler and D. B. Chrisey, eds.), Chapter 5, p. 115. Wiley (Interscience), New York, 1994.

J. George, *Preparation of Thin Films.* Dekker, New York, 1992.

R. Glang, in *Handbook of Thin Film Technology* (L. I. Maissel and R. Glang, eds.), Chap. 1, McGraw–Hill, New York, 1983.

R. A. Gottscho and T. A. Miller, *Pure Appl. Chem.* **56,** 189 (1984).

E. H. A. Granneman, *J. Vac. Sci. Technol. B* **12,** 2741 (1994).

I. P. Herman, *Chem. Rev.* **89,** 1323 (1989).

M. A. Herman and H. Sitter, *Molecular Beam Epitaxy: Fundamentals and Current Status.* Springer-Verlag, Berlin, 1989.

N. Hershkowitz, in *Plasma Diagnostics* (O. Auciello and D. L. Flamm, eds.), Vol. 1, Chapter 3, p. 113. Academic Press, Boston, 1989.

D. W. Hess and D. B. Graves, in *Chemical Vapor Deposition* (M. L. Hitchman and K. F. Jensen, eds.), Chapter 7, p. 385, Academic Press, Boston, 1993.

L. Hesselink, *Annu. Rev. Fluid Mech.* **20,** 421 (1988).

M. L. Hitchman and K. F. Jensen, eds., *Chemical Vapor Deposition: Principles and Applications.* Academic Press, Boston, 1993.

Y.-M. Houng, *CRC Crit. Rev. Solid State Mater. Sci.* **17,** 277 (1992).

G. Hubler and D. B. Chrisey, eds., *Pulsed Laser Deposition of Thin Films.* Wiley (Interscience), New York, 1994.

K. F. Jensen and W. Kern, in *Thin Film Processes II* (J. L. Vossen and W. Kern, eds.), Chapter III-1, p. 283. Academic Press, Boston, 1991.

P. C. Johnson, *Phys. Thin Films* **14,** 130 (1989).

L. E. Katz, in *VLSI Technology* (S. M. Sze, ed.), Chapter 4, p. 131. McGraw Hill, New York, 1983.

W. Kern, ed., *Handbook of Semiconductor Wafer Cleaning Technology.* Noyes, Park Ridge, NJ, 1993.

W. Kern and V. S. Ban, in *Thin Film Processes* (J. L. Vossen and W. Kern, eds.), Chapter III-2, p. 258. Academic Press, New York, 1978.

W. Kern and C. A. Deckert, in *Thin Film Processes* (J. L. Vossen and W. Kern, eds.), Chapter V-1, p. 401. Academic Press, New York, 1978.

G. S. Kino and S. S. C. Chim, *Appl. Optics* **29,** 3775 (1990).

G. S. Kino and T. R. Corle, *IEEE Circuits and Devices* **CD 6**(2), 28 (1990).

M. Konuma, *Film Deposition by Plasma Techniques.* Springer-Verlag, Berlin, 1992.

S. V. Krishnaswamy, B. R. McAvoy, and M. H. Francombe, *Phys. Thin Films* **17,** 145 (1993).

B. S. Krusor and G. A. N. Connell, *Phys. Thin Films* **15,** 143 (1991).

T. F. Kuech and K. F. Jensen, in *Thin Film Processes II* (J. L. Vossen and W. Kern, eds.), Chapter III-2, p. 369. Academic Press, Boston, 1991.

W. Lauterborn and A. Vogel, *Annu. Rev. Fluid Mech.* **16,** 223 (1984).

H. W. Lehmann, in *Thin Film Processes II* (J. L. Vossen and W. Kern, eds.), Chapter V-1, p. 673. Academic Press, Boston, 1991.

M. A. Lieberman and R. A. Gottscho, *Phys. Thin Films* **18,** 1 (1994).

M. A. Lieberman and A. J. Lichtenberg, *Principles of Plasma Discharges and Materials Processing.* Wiley, New York, 1994.

M. Liehr and G. W. Rubloff, *J. Vac. Sci. Technol. B* **12,** 2727 (1994).

References

L. Luckock, *SPIE* **775,** 289 (1987).
G. Lucovsky, D. V. Tsu, R. A. Rudder, and R. J. Markunas, in *Thin Film Processes II* (J. L. Vossen and W. Kern, eds.), Chapter IV-2, p. 565. Academic Press, Boston, 1991.
H. A. Macleod, *Thin-Film Optical Filters,* 2nd ed. McGraw-Hill, New York, 1989.
L. I. Maissel and R. Glang, *Handbook of Thin Film Technology.* McGraw-Hill, New York, 1983.
D. M. Manos and D. L. Flamm, *Plasma Etching: An Introduction.* Academic Press, Boston, 1989.
P. J. Marcoux and P. D. Foo, *Solid State Technol.* **24**(4), 115 (1981).
R. B. Marcus, in *VLSI Technology* (S. M. Sze, ed.), Chapter 12, p. 507. McGraw-Hill, New York, 1983.
N. A. Masnari, *J. Vac. Sci. Technol. B* **12,** 2749 (1994).
J. F. McGilp, *J. Phys.: Condens. Matter* **2,** 7985 (1990).
J. F. McGilp, *Prog. Surf. Sci.* **49,** 1 (1995).
G. E. McGuire, ed., *Semiconductor Materials and Process Technology Handbook.* Noyes, Park Ridge, NJ, 1988.
T. K. McNab, *Semicond. Int.* **13**(9), 58 (1990a).
T. K. McNab, *Semicond. Int.* **13**(11), 86 (1990b).
W. Merzkirch, *Flow Visualization,* 2nd ed. Academic Press, Orlando, 1987.
S. Middleman and A. K. Hochberg, *Process Engineering Analysis in Semiconductor Device Fabrication.* McGraw-Hill, New York, 1993.
L. M. Miller and J. J. Coleman, *CRC Crit. Rev. Solid State Mater. Sci.* **15,** 1 (1988).
R. L. Moon and Y.-M. Huong, in *Chemical Vapor Deposition* (M. L. Hitchman and K. F. Jensen, eds.), Chapter 6, p. 245. Academic Press, Boston, 1993.
W. M. Moreau, *Semiconductor Lithography: Principles, Practices, and Materials.* Plenum, New York, 1988.
M. M. Moslehi, R. A. Chapman, M. Wong, A. Paranjpe, H. N. Najm, J. Kuehne, R. L. Yeakley, and C. J. Davis, *IEEE Trans. Electron Devices* **ED-39,** 4 (1992).
M. M. Moslehi, C. J. Davis, A. Paranjpe, L. A. Velo, H. N. Najm, C. Schaper, T. Breedijk, Y. J. Lee, and D. Anderson, *Solid State Technol.* **37**(1), 35 (1994a).
M. M. Moslehi, A. Paranjpe, L. A. Velo, and J. Kuehne, *Solid State Technol.* **37**(5), 37 (1994b).
J. Musil, J. Vyskocil, and S. Kadleč, *Phys. Thin Films* **17,** 79 (1993).
H. N. Norton, *Handbook of Transducers.* Prentice-Hall, Englewood Cliffs, NJ, 1989.
J. F. O'Hanlon and H. G. Parks, *J. Vac. Sci. Technol. A* **10,** 1863 (1992).
M. A. Olmstead, *Surf. Sci. Rep.* **6,** 159 (1987).
M. B. Panish and H. Temkin, *Gas Source Molecular Beam Epitaxy.* Springer-Verlag, Berlin, 1993.
R. Parsons, in *Thin Film Processes II* (J. L. Vossen and W. Kern, eds.), Chapter II-4, p. 177. Academic Press, Boston, 1991.
J. Pei, F. L. Degertekin, B. T. Khuri-Yakub, and K. C. Saraswat, *Appl. Phys. Lett.* **66,** 2177 (1995).
D. Peyton, H. Kinoshita, G. Q. Lo, and D. L. Kwong, *SPIE* **1393,** 295 (1990).
C. Pickering, in *The Handbook of Crystal Growth* (D. T. J. Hurle, ed.), Vol. 3, p. 817. North-Holland Publ., Amsterdam, 1994.
O. A. Popov, *Phys. Thin Films* **18,** 121 (1994).

P. R. Puckett, S. L. Michel, and W. E. Hughes, in *Thin Film Processes II* (J. L. Vossen and W. Kern, eds.), Chapter V-2, p. 749. Academic Press, Boston, 1991.

G. K. Rao, *Multilevel Interconnect Technology.* McGraw-Hill, New York, 1993.

R. R. Reeber, *J. Am. Ceram. Soc.* **76**(2), 261 (1993).

R. Reif and W. Kern, in *Thin Film Processes II* (J. L. Vossen and W. Kern, eds.), Chapter IV-1, p. 525. Academic Press, Boston, 1991.

W. Richter, *Philos. Trans. R. Soc. London, Ser. A* **344,** 453 (1993).

E. Ritter, *Phys. Thin Films* **8,** 1 (1975).

T. A. Roberts and K. E. Gray, *MRS Bulletin* **20**(5), 43 (1995).

J. P. Roland, P. J. Marcoux, G. W. Ray, and G. H. Rankin, *J. Vac. Sci. Technol. A* **3,** 631 (1985).

F. Roozeboom, in *Rapid Thermal Processing Science and Technology* (R. B. Fair, ed.), Chapter 9, p. 349. Academic Press, Boston, 1993.

S. M. Rossnagel, in *Thin Film Processes II* (J. L. Vossen and W. Kern, eds.), Chapter II-1, p. 11. Academic Press, Boston, 1991.

R. A. Rudder, R. E. Thomas, and R. J. Nemanich, in *Handbook of Semiconductor Wafer Cleaning Technology* (W. Kern, ed.), Chapter 8, p. 340. Noyes, Park Ridge, NJ, 1993.

K. L. Saenger, *Process. Adv. Mater.* **3**(2), 63 (1993).

K. C. Saraswat, P. P. Apte, L. Booth, Y. Chen, P. C. P. Dankoski, F. L. Degertekin, G. F. Franklin, B. T. Khuri-Yakub, M. M. Moslehi, C. Schaper, P. J. Gyugyi, Y. J. Lee, J. Pei, and S. C. Wood, *IEEE Trans. Semicond. Manuf.* **SM-7,** 159 (1994).

C. Schneider, L. Pfitzner, and H. Ryssel, *Top. Conf. Manuf. Sci. Natl. Symp. Am. Vac. Soc. 40th,* Orlando, FL, *1993, Presentation TC1-TuM10,* unpublished.

D. K. Schroder, *Semiconductor Material and Device Characterization.* Wiley, New York, 1990.

G. S. Selwyn, *Optical Diagnostic Techniques for Plasma Processing.* AVS Press, New York, 1993.

Semiconductor Industry Association, *The National Technology Roadmap for Semiconductors.* SIA, San Jose, CA, 1994.

R. Shadmehr, D. Angell, P. B. Chou, G. S. Oehrlein, and R. S. Jaffe, *J. Electrochem. Soc.* **139,** 907 (1992).

A. Sherman, *Chemical Vapor Deposition for Microelectronics.* Noyes, Park Ridge, NJ, 1987.

D. L. Smith, in *VLSI Electronics Microstructure Science* (N. G. Einspruch and D. M. Brown, eds.), p. 253. Academic Press, Orlando, FL, 1984.

J. P. Stagg, J. Christer, E. J. Thrush, and J. Crawley, *J. Cryst. Growth* **120,** 98 (1992).

C. Steinbrüchel, *Phys. Thin Films* **18,** 289 (1994).

G. B. Stringfellow, *Organometallic Vapor-Phase Epitaxy: Theory and Practice.* Academic Press, Boston, 1989.

S. M. Sze, ed., *VLSI Technology.* McGraw-Hill, New York, 1983.

H. R. Thorsheim and J. E. Butler, in *Synthetic Diamond: Emerging CVD Science and Technology* (K. E. Spear and J. P. Dismukes, eds.), Chapter 7, p. 193. Wiley, New York, 1994.

D. L. Tolliver, ed., *Handbook of Contamination Control in Microelectronics.* Noyes, Park Ridge, NJ, 1988.

References

J. Y. Tsao, *Materials Fundamentals of Molecular Beam Epitaxy.* Academic Press, Boston, 1993.

J. Y. Tsao, T. M. Brennan, J. F. Klem, and B. E. Hammons, *Appl. Phys. Lett.* **55,** 777 (1989).

A. J. van Roosmalen, J. A. G. Baggerman, and S. J. H. Brader, *Dry Etching for VLSI.* Plenum, New York, 1991.

J. L. Vossen and W. Kern, eds., *Thin Film Processes.* Academic Press, Boston, 1978.

J. L. Vossen and W. Kern, eds., *Thin Film Processes II.* Academic Press, Boston, 1991.

W. D. Westwood, *Phys. Thin Films* **14,** 1 (1989).

D. J. Whitehouse, *Handbook of Surface Metrology.* Institute of Physics Publishing, Bristol, 1994.

S. Wolf, *Silicon Processing for the VLSI Era,* Vol. 2. Lattice Press, Sunset Beach, CA, 1990.

S. Wolf and R. N. Tauber, *Silicon Processing for the VLSI Era,* Vol. 1. Lattice Press, Sunset Beach, CA, 1986.

D. P. Woodruff and T. A. Delchar, *Modern Techniques of Surface Science.* Cambridge Univ. Press, Cambridge, UK, 1994.

W.-J. Yang, ed., *Handbook of Flow Visualization.* Hemisphere, New York, 1989.

CHAPTER 2

The Properties of Light

Electromagnetic radiation can be described as electromagnetic waves, with electric and magnetic fields oscillating at a radial frequency ω, or as quantized particles, photons, each with energy $\hbar\omega$. The wave picture is adequate in describing most of the properties of light needed for optical diagnostics. In particular, the interaction of light with matter is usually handled well in the semiclassical approximation in which matter is treated quantum mechanically but light is treated classically as waves. The full quantum picture, in which both matter and light are quantized, is helpful in some cases, such as in laser theory, nonlinear optics, and Raman scattering. Except for referring to photons in scattering, emission, and detection processes, only the wave picture of light will be used here.

This chapter surveys those aspects of optics relevant to diagnostics. Comprehensive descriptions of optics are given by Born and Wolf (1970), Klein and Furtak (1986), Hecht (1987), Smith (1990), Saleh and Teich (1991), and Bass (1995). There are many slight variations in notation and convention between different areas in optics (Holm, 1991), and even variations within given disciplines, so there is no "standard" notation. For example, many of the "Nebraska optical conventions" in ellipsometry (Azzam and Bashara, 1977; Hauge *et al.,* 1980) differ from those now used in ellipsometry and in much of the rest of optics. Gaussian CGS units will be used in this volume. (Consult the appendix to Chapter 4 for equivalent expressions in rationalized MKSA (SI) units.)

The propagation of light beams, which is determined by Maxwell's equations, can be described at varying levels of approximation. The lowest order is geometric optics, in which beams are composed of rays that propagate in a straight line forever (except upon reflection or refraction) and whose shapes never change; i.e., they never "diffract." This approximation is best when $\lambda \ll D$, where λ is the light wavelength and D is the lateral dimension of the optical element (such as a lens or aperture). The paraxial approximation is common where all rays make, at most, small angles to a given axis—the optic axis. A plane wave is represented by an infinite set of parallel rays. The propagation, imaging, and collection of light in optical diagnostics are often treated well enough by the geometric optics approximation.

Still, several properties of light that are ignored in geometric optics are important in optical diagnostics. The wave properties of light lead to interference and diffraction. The polarization of light leads to important phenomena in reflection and scattering studies. Each of these properties of light is described by various manipulations of Maxwell's equations. For example, diffraction can be treated by either scalar or vector forms of the Kirchhoff integral (Jackson, 1975).

For any given propagation direction of light, the oscillating electric field **E** has two degrees of freedom. That is, for propagation in the z direction, the electric field polarization can point anywhere in the xy plane. Two basis vectors are needed to describe **E**, which can be the unit vectors in the x and y direction, **x** and **y**, respectively:

$$\mathbf{E} = E_x \mathbf{x} + E_y \mathbf{y} \tag{2.1}$$

For a plane wave these components are

$$E_x(z,t) = \text{Re}\{E_{0x} e^{-i(\omega t - kz + \delta_x)}\} \tag{2.2}$$

$$E_y(z,t) = \text{Re}\{E_{0y} e^{-i(\omega t - kz + \delta_y)}\} \tag{2.3}$$

where E_{0x}, E_{0y}, δ_x, and δ_y are real. (The x, y, and z axes are defined here relative to the propagation direction of the beam. In some problems they are defined relative to the medium, as in the reflection probes described in Figure 9.36.) The magnitude of the electric field vector is E. $k = 2\pi\tilde{n}/\lambda$, where $\tilde{n} = n + ik$ is the complex index of refraction. Unless otherwise specified, λ is the wavelength in vacuum. The wavelength in a medium with refractive index n is $\lambda_{\text{medium}} = \lambda/n$. In gases, $n \approx 1$ and so $\lambda_{\text{air}} \approx \lambda$ (which is a good approximation in many diagnostics applications). Absorption of the beam occurs when $k > 0$, as is discussed in Section 2.1. Over a path of length l, the optical path length is nl.

The wavelength and frequency ν are related by $\lambda_{\text{medium}} \nu = c/n$. The radial frequency ω (units of rad/sec) is related to the frequency ν (in Hz) by $\omega = 2\pi\nu$. Common units of wavelength are angstroms (Å), nanometers (nm), and microns [micrometers] (μm), which are interrelated by 10,000 Å = 1000 nm = 1 μm.

The energy of a photon is $\mathcal{E}_{\text{photon}} = h\nu = \hbar\omega$, where h is Planck's constant (6.626×10^{-27} erg sec) and $\hbar = h/2\pi$. One common energy unit is the electron volt (1 eV = 1.602×10^{-12} ergs); 1 eV/atom corresponds to 11,600 K (using the Boltzmann constant $k_B = 1.3807 \times 10^{-16}$ erg/K) and to 23.06 kcal/mole. The relation $\lambda(\text{in } \mu\text{m}) \mathcal{E}_{\text{photon}}(\text{in eV}) = 1.124$ follows from $\lambda\nu = c$. Following common spectroscopic practice, wavenumber units (cm^{-1}), corresponding to $1/\lambda = \mathcal{E}_{photon}/hc$, are often used as "energy" units. A 1-μm wavelength photon has an energy of 1.24 eV = 10^4 cm^{-1}. Also, 1 cm^{-1} = 30 GHz = 3×10^{10} Hz.

2.1 Propagation

Both E_{0x} and E_{0y} are independent of transverse position in the plane wave approximation. This assumes that the beam intensity is uniform everywhere transverse to the propagation direction and has constant phase in that plane. These assumptions can never be totally accurate, but in many cases they are "close enough" to experimental conditions to be assumed with little error. Spherical waves that emanate from a point, with $E \propto \exp(ikr)/r$, are useful in describing scattering.

Two different conventions are used in describing the "strength" of optical beams. The terminology that is defined first is common among those working with lasers in optics, physics, and chemistry (Yariv, 1989; Steinfeld, 1985), and will be used in most of this volume. The alternative convention, that of radiometry, is described afterwards.

The *intensity* of an optical beam, I, is the magnitude of the Poynting vector averaged over an optical cycle,

$$I = \frac{c}{8\pi} \frac{n}{\mu} E_0^2 = \frac{c}{8\pi} \left(\frac{\varepsilon}{\mu}\right)^{1/2} E_0^2, \qquad (2.4)$$

where n is the refractive index = $\sqrt{\mu\varepsilon}$, μ is the magnetic permeability, and ε is the dielectric constant. In most optical diagnostics applications $\mu = 1$ and $n = \sqrt{\varepsilon}$. The units of intensity are erg/sec cm^2 in CGS units; the mixed MKS/CGS units W/cm^2 are more often used in practice. (See the appendix to Chapter 4.) Closely related to intensity is the flux of photons, which is $I/h\nu$, with units photons/cm^2 sec. *Power* P (W) is the intensity integrated over an area normal to the propagation direction **k**. For pulsed lasers, such quantities can be integrated over the length of a *pulse length* t_p. The *fluence*, F_{pulse}, is the intensity integrated over the pulse (J/cm^2), while the *pulse energy*, U_{pulse}, is the power integrated over the pulse (J). The average power of a pulsed laser is the pulse energy times the pulse repetition rate.

The terminology of radiometry (Levi, 1968) is analogous, but it is better suited to describing incoherent sources. It is often employed in describing thermal (blackbody) radiation, which will be discussed in Chapter 13, and will be used in this book only in that chapter. The radiation power crossing a surface is the *radiant flux* (Φ_e, units W). Related to this is the radiant power crossing or impinging per unit area, which is the *irradiation* or *irradiance* (E_e, W/m^2), and that emitted per unit area, which is the *radiant emittance* (M_e, W/m^2). The *radiant intensity* (I_e, W/sr) is the radiant flux per unit solid angle, while the *radiance* (L_e, W/m^2 sr) is the radiant intensity emitted per unit area. The adjective *spectral* is placed before each of these parameters to denote the parameter per unit frequency or wavelength. Units of photometry, in which the adjective *radiant* used in radiometry is replaced by *luminous,* describe light in terms of the sensitivity of the human eye, and are not used in optical diagnostics.

Absorption decreases the power of a propagating beam. Using Equations 2.2–2.4, both the intensity I (plane waves) and power $\mathcal{P}$ decrease exponentially as

$$I(z) = I(0) \exp(-\alpha z) \tag{2.5}$$

where the absorption coefficient $\alpha = 4\pi k/\lambda$. This is commonly known as Beer's law (or the Beer–Lambert law).

Most continuous-wave (cw) lasers are made to lase in a single transverse mode, the TEM$_{00}$ mode, which permits the tightest possible focus (Yariv, 1989; Verdeyen, 1989; Saleh and Teich, 1991). The transverse profiles of the electric field and intensity of these beams are gaussian, and these modes are often called gaussian beams. A TEM$_{00}$ gaussian beam propagating in the z direction has an electric field $E_0 \exp(-i\omega t)\exp(-r^2/w_0^2)$ at $z = 0$, where it has a minimum spot size; the beam waist ("radius") is w_0 and the beam curvature $\mathcal{R} = \infty$. This could be at the output of a laser (with a flat output mirror) or at the focus of a lens. The electric field of this beam propagating in a transparent medium with $n = 1$ is

$$E(r,z) = E_0 \frac{w_0}{w(z)} \exp(-i\omega t) \exp\left[+i(kz - \zeta(z)) + i\frac{kr^2}{2q(z)}\right] \tag{2.6}$$

where

$$w^2(z) = w_0^2\left(1 + \frac{z^2}{z_0^2}\right) \tag{2.7}$$

$$\mathcal{R}(z) = z\left(1 + \frac{z_0^2}{z^2}\right) \tag{2.8}$$

$$\frac{1}{q(z)} = \frac{1}{\mathcal{R}(z)} - i\frac{\lambda}{\pi w^2(z)} \tag{2.9}$$

$$\zeta(z) = \tan^{-1}(z/z_0) \tag{2.10}$$

and

$$z_0 = \frac{\pi w_0^2}{\lambda} \tag{2.11}$$

The beam intensity varies as

$$I(r,z) = \frac{2\mathcal{P}}{\pi w^2(z)} \exp\left[\frac{-2r^2}{w^2(z)}\right] \exp(-\alpha z) \tag{2.12}$$

so w is the radius where the intensity falls to $1/e^2$ of the peak. Unlike Equation 2.6, this expression includes the possibility of absorption. Sometimes Equation 2.12 is expressed instead as $I \propto \exp[-r^2/w_I^2(z)]$, where this "intensity" spot size w_I is $\sqrt{2}$ times smaller than the "electric field" spot size in Equations 2.6–2.12. This beam remains fairly collimated for a distance $\sim z_0$, the Rayleigh range, and then diffracts with a diffraction (half) angle $\theta_D = \lambda/\pi w$, ($n = 1$, as in a gas), so $w \approx \theta_D z$ for $z \gg z_0$. The TEM$_{00}$ mode is the lowest-order gaussian mode. Higher-order gaussian modes are described by the same w and $\mathcal{R}$.

Pulsed lasers often operate with multiple transverse modes; when needed, they can be made to lase in a single mode (TEM$_{00}$ mode). Propagation of such a multimode laser is characterized by the divergence angle, which is usually much larger than the divergence of the TEM$_{00}$ mode. The transverse profile of multimode lasers can sometimes be approximated by a flat-top beam.

2.2 Imaging

In geometric optics approximation, each ray is imaged by a lens according to

$$\frac{1}{d_1} + \frac{1}{d_2} = \frac{1}{f} \tag{2.13}$$

where f is the focal length of the lens, and d_1 and d_2 are the respective distances from the source and image to the lens. With a convex lens, the image is magnified by d_2/d_1 and it is inverted.

In the geometric optics approximation, "perfect" lenses image from a point to a point. In reality, however, point-to-point imaging does not occur for two reasons. Even within the geometric optics framework, blurring occurs owing to spherical aberration, astigmatism, coma, chromatic aberration, and the like. Moreover, the geometric optics assumption of $\lambda/D_{lens} \to 0$, where D_{lens} is the diameter of the lens, is not valid; therefore, wave optics must be used.

Using wave optics, a plane wave focused by a perfect lens has an intensity profile

$$I(r) = I(r = 0) \left(\frac{2J_1(\pi D_{lens} r/\lambda f)}{\pi D_{lens} r/\lambda f} \right)^2 \tag{2.14}$$

where J_1 is the Bessel function of the first kind of order 1. Therefore the diameter of the focused spot is

$$d_{focus} = 2.44 \frac{\lambda f}{D_{lens}} = 2.44 \lambda f_\# \tag{2.15}$$

where $f_\#$ is the f-number of the lens which is equal to f/D_{lens}. This is the diameter of the Airy central disk in the far-field diffraction pattern, which contains ~84% of the total energy. If the diameter of the light source $D_{light} < D_{lens}$, then D_{light} replaces D_{lens} in this equation.

Section 5.2.1.1 describes the collection and imaging of incoherent light generated in optical emission (OES), fluorescence (laser-induced fluorescence LIF and photoluminescence PL), and incoherent scattering (elastic and Raman) diagnostics by using Equations 2.13–2.15. Sometimes these signals are produced by focused lasers. The focusing of lasers can be analyzed for two extreme regimes. For both, aberrations due to imperfections in the imaging optics will be neglected.

A collimated gaussian beam (TEM$_{00}$; Equations 2.6–2.12) with spot size w_0 passing through a perfect lens with focal length f and diameter D_{lens} (with diameter $D_{lens} \gg 2w_0$) comes to focus a distance f after the lens if $f \ll z_0$ (Equation 2.11). The "diameter" of the beam at the focus, i.e., twice the spot size w_f, is

$$d_{focus} = 2w_f = \frac{2\lambda f}{\pi w_0} = \frac{4}{\pi} \lambda f_\# \qquad (2.16)$$

where $f_\#$ is the effective f-number which is equal to $f/2w_0$ (Yariv, 1989; Verdeyen, 1989; Saleh and Teich, 1991). The cited references also describe other cases, for example, that in which the beam is not collimated at the lens and that in which $f > z_0$.

Light from extended sources, such as lamps and lasers that are not diffraction-limited (e.g., many multimode solid-state and excimer lasers), cannot be as tightly focused as these lowest-order gaussian beams. If the full divergence angle of the source is σ_{div}, the spot diameter at the focus is

$$d_{focus} \sim 2f \tan(\sigma_{div}/2) \sim \sigma_{div} f \qquad (2.17)$$

where the last relation holds for the usual case of small divergence angles. The typical beam divergence of an excimer laser is ~2–4 mrad. The spot size at the focus can be one to two orders of magnitude larger than the prediction of Equation 2.16.

2.3 Polarization Properties of Light

The polarization properties of light are discussed in many of the above-cited volumes, including that by Azzam and Bashara (1977) and the book by Kliger *et al.* (1990). The electric field, say for a plane wave, can be expressed in terms of a Jones vector:

$$\mathbf{E} = \begin{bmatrix} E_{0x} e^{-i\delta_x} \\ E_{0y} e^{-i\delta_y} \end{bmatrix} = \begin{bmatrix} E_x \\ E_y \end{bmatrix} \qquad (2.18)$$

2.3 Polarization Properties of Light

This formalism makes it easy to treat beam propagation through different media by using matrix methods. Every optical event, such as transmission or reflection at an interface or propagation in a nonpolarizing or polarizing medium, and every optical element, which represents the cumulative effect of several of these events, can be represented by a 2 × 2 matrix. The electric field after the process is related to that before by

$$\begin{bmatrix} E_x \\ E_y \end{bmatrix}_{\text{after}} = \begin{bmatrix} M_{xx} & M_{xy} \\ M_{yx} & M_{yy} \end{bmatrix} \begin{bmatrix} E_x \\ E_y \end{bmatrix}_{\text{before}} \quad (2.19)$$

A cascade of optical elements is represented by a product of matrices.

When $\delta_x = \delta_y$, **E** vibrates in one plane and the light is linearly polarized. Figure 9.1 shows linearly polarized light impinging on a solid at an angle of incidence θ. The electric field can be linearly polarized perpendicular to the plane of incidence and parallel to the interface, as in Figure 9.1 (left). From the ultraviolet to the infrared this is called *s*- or σ-polarized light; in the microwave region it is more commonly known as a transverse electric (TE) wave. When the electric field is parallel to the plane of incidence, as in Figure 9.1 (right), the field is *p*- or π-polarized, which is also described as a transverse magnetic (TM) wave.

When E_x and E_y have different phases, the light is generally elliptically polarized and the electric field vector does not stay in the same line. The sense of the polarization is right-handed if, at a given *z*, **E** moves clockwise versus time with the beam propagating toward the observer. Instantaneously, **E** forms a right-handed helix in space. This corresponds to $\sin(\delta_y - \delta_x) > 0$. It is left-handed if the field moves counterclockwise and forms a left-handed helix, so that $\sin(\delta_y - \delta_x) < 0$. This definition of handedness is common among optical physicists and chemists, and will be adopted here. The opposite convention is used by some, including astronomers and engineers (Bohren and Huffman, 1983; Holm, 1991).

As seen in Figure 2.1, for elliptically polarized radiation the electric field vector traverses an elliptical trajectory with a semimajor axis a_E and a semiminor axis b_E. The amplitude $A_E = (a_E^2 + b_E^2)^{1/2}$, and $A_E^2 \propto I$. The ellipticity e_E is defined as b_E/a_E and ranges from −1 to 1. It is positive for right-handed light and negative for left-handed light. Ellipticity is also described by the ellipticity angle $\eta_E = \tan^{-1}(b_E/a_E)$, with $0 < \eta_E \leq 45°$ for right-handed polarization and $-45° < \eta_E < 0$ for left-handed polarization. For linearly polarized light, $e_E = \eta_E = 0$. The azimuth γ_E is the angle that the major axis of the ellipse makes with a reference orientation (*x* axis) in a counterclockwise sense, for an observer looking into the beam. γ_E can range from −90° to 90°.

For circularly polarized light, $E_{0y} = E_{0x}$, and the trajectory in Figure 2.1 is a circle. The light is right-handed if $\delta_y - \delta_x = 90°$ ($e_E = 1$) and left-handed if $\delta_y - \delta_x = -90°$ ($e_E = -1$).

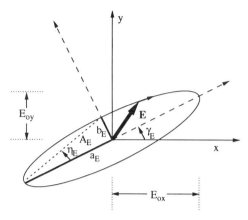

Figure 2.1 The trajectory of the electric field vector for right-handed elliptically polarized radiation with a semimajor axis a_E and a semiminor axis b_E.

The state of polarization can be represented by a point on the surface of the Poincaré sphere with a longitude at twice the azimuth, $2\gamma_E$, and latitude at twice the ellipticity angle, $2\eta_E$. Linear polarizations are along the equator, and the north and south poles represent right- and left-handed circularly polarized light, respectively.

A beam can be completely described either by E_{0x}, E_{0y}, δ_x and δ_y or by following the four Stokes parameters:

$$S_0 = \langle E_x E_x^* \rangle + \langle E_y E_y^* \rangle = \langle E_{0x}^2 \rangle + \langle E_{0y}^2 \rangle \tag{2.20}$$

$$S_1 = \langle E_x E_x^* \rangle - \langle E_y E_y^* \rangle = \langle E_{0x}^2 \rangle - \langle E_{0y}^2 \rangle \tag{2.21}$$

$$S_2 = 2 \operatorname{Re} \langle E_x^* E_y \rangle = 2 \langle E_{0x} E_{0y} \cos(\delta_y - \delta_x) \rangle \tag{2.22}$$

$$S_3 = 2 \operatorname{Im} \langle E_x^* E_y \rangle = 2 \langle E_{0x} E_{0y} \sin(\delta_y - \delta_x) \rangle \tag{2.23}$$

where S_0 is the intensity of the beam (I_0), S_1 is the difference in the intensities measured after linear polarizers that are alternately aligned along the x and y directions ($I_x - I_y$), S_2 is the difference in intensities measured after linear polarizers that are alternately aligned $+45°$ and $-45°$ relative to the x direction ($I_{+45°} - I_{-45°}$), and S_3 is the difference in intensities measured after polarizers that alternately transmit right- and left-handed circularly polarized light ($I_r - I_l$). (More precisely, these Stokes parameters as defined are really proportional to these intensities.) For right-handed light $S_3 > 0$, and for left-handed light $S_3 < 0$. Three normalized Stokes parameters can be defined with S_1, S_2, and S_3 normalized by S_0. This definition of the Stokes parameters is common in ellipsometry (Section 9.11). In other fields they may be labeled and/or defined differently. For example, in scattering theory (Section 11.1.1), S_0, S_1, S_2, and S_3 are called I, Q, U, and V, respectively (Bohren and Huffman, 1983).

For elliptically polarized light

$$S_0 = A_E^2 \tag{2.24}$$

$$S_1 = A_E^2 \cos 2\eta_E \cos 2\gamma_E \tag{2.25}$$

$$S_2 = A_E^2 \cos 2\eta_E \sin 2\gamma_E \tag{2.26}$$

$$S_3 = A_E^2 \sin 2\eta_E \tag{2.27}$$

In general $S_0^2 \geq S_1^2 + S_2^2 + S_3^2$. The equality holds for totally polarized light, and the inequality holds for partially or unpolarized light. The degree of polarization is defined as $P = (S_1^2 + S_2^2 + S_3^2)^{1/2}/S_0$.

In analogy with the Jones vector notation, beams can be expressed as a four-element column vector, the Stokes vector, with the four (real) Stokes parameters as elements. Optical processes can be represented by 4×4 Mueller matrices. This Mueller notation can handle polarized and partially polarized light, whereas Jones matrices can describe only completely polarized light. Partially polarized light can also be described by the 2×2 coherency matrix $\bar{J}$, which is defined as $\bar{J}_{xx} = \langle E_x E_x^* \rangle$, $\bar{J}_{xy} = \langle E_x E_y^* \rangle$, $\bar{J}_{yx} = \langle E_y E_x^* \rangle$, and $\bar{J}_{yy} = \langle E_y E_y^* \rangle$; this matrix is closely related to the Stokes parameters.

References

R. M. A. Azzam and N. M. Bashara, *Ellipsometry and Polarized Light.* North-Holland Publ., Amsterdam, 1977.
M. Bass, ed., *Handbook of Optics,* 2nd ed., Vol. 1. McGraw-Hill, New York, 1995.
C. F. Bohren and D. R. Huffman, *Absorption and Scattering of Light by Small Particles.* Wiley, New York, 1983.
M. Born and E. Wolf, *Principles of Optics,* 4th ed. Pergamon, Oxford, 1970.
P. S. Hauge, R. H. Muller, and C. G. Smith, *Surf. Sci.* **96,** 81 (1980).
E. Hecht, *Optics,* 2nd ed. Addison-Wesley, Reading, MA, 1987.
R. T. Holm, in *Handbook of Optical Constants of Solids II* (E. D. Palik, ed.), Chapter 2, p. 21. Academic Press, Boston, 1991.
J. D. Jackson, *Classical Electrodynamics,* 2nd ed. Wiley, New York, 1975.
M. V. Klein and T. E. Furtak, *Optics,* 2nd ed. Wiley, New York, 1986.
D. S. Kliger, J. W. Lewis, and C. E. Randall, *Polarized Light in Optics and Spectroscopy.* Academic Press, Boston, 1990.
L. Levi, *Applied Optics: A Guide to Optical System Design,* Vol. 1. Wiley, New York, 1968.
B. E. A. Saleh and M. C. Teich, *Fundamentals of Photonics.* Wiley, New York, 1991.
W. J. Smith, *Modern Optical Engineering: The Design of Optical Systems,* 2nd ed. McGraw-Hill, New York, 1990.
J. I. Steinfeld, *Molecules and Radiation: An Introduction to Modern Molecular Spectroscopy,* 2nd ed. MIT Press, Cambridge, MA, 1985.
J. T. Verdeyen, *Laser Electronics,* 2nd ed. Prentice-Hall, Englewood Cliffs, NJ, 1989.
A. Yariv, *Quantum Electronics,* 3rd ed. Wiley, New York, 1989.

CHAPTER 3

The Structure of Matter

The electronic and nuclear structure of matter is surveyed in this chapter as each type of structure relates to the optical spectroscopies used for diagnostics. This discussion forms the basis of the discussion of spectroscopy in Chapter 4 and lays the groundwork for the applications of optical diagnostics described in subsequent chapters.

Chapters 3 and 4 are mere overviews of the basics of a very large topic—the spectroscopies of gases and solids. This treatment is not intended to be comprehensive, as only the simplest spectroscopic systems are presented. In both of these chapters it is assumed that the reader has at least a little background in this area. Readers requiring either an introduction to spectroscopy or a detailed treatment on spectroscopy beyond that provided here should consult the many excellent textbooks and monographs that are available in this area. General references are White (1934), Condon and Shortley (1959), Kuhn (1962), Herzberg (1937, 1945, 1950, 1966), Levine (1975), Huber and Herzberg (1979), Wilson *et al.* (1980), Eisberg and Resnick (1985), and Steinfeld (1985) for the structure of atoms and molecules; and Ashcroft and Mermin (1976), Harrison (1980), Burns (1985), Eisberg and Resnick (1985), and Kittel (1986) for the structure of solids. Böer (1990) surveys the physics of semiconductors.

3.1 Separation of Electronic and Nuclear Motion

Electrons are assumed to respond instantaneously to any change in nuclear motion because they are 10^3–10^4 times lighter than atomic nuclei. In this Born–Oppenheimer approximation, the electronic energy levels are determined with the nuclear positions fixed and parameterized. The nuclear motion and structure are then determined by using these calculated electronic energy curves. The total energy of any state is the sum of the electronic and nuclear motion energy, and transitions between these states can be considered as separate transitions between electronic and nuclear motion states. Coupling between electronic and nuclear motion cannot be totally ignored, as is seen by some transition selection rules and resonance effects in spectroscopy and by electron–phonon coupling in solids; these effects are not very important in diagnostics.

3.1.1 *Nuclear Motion*

If a species has $\mathcal{N}$ atoms, it has $3\mathcal{N}$ degrees of freedom associated with nuclear motion. Three degrees of freedom are associated with translation. In nonlinear molecules and solids, three degrees of freedom describe rotation, while linear molecules have only two rotational degrees of freedom. This leaves $3\mathcal{N} - 6$ degrees of freedom for vibrations in nonlinear molecules and solids and $3\mathcal{N} - 5$ in linear molecules. Although translation is not important in describing structure, it is significant because of Doppler shifts and broadening. Rotational motion is important in molecules, but of no significance in the diagnostics of solids. Vibrations are important in both molecules and solids.

To first-order, vibrational motion can be decoupled into separate, uncoupled normal modes, each with its own harmonic frequency. Each mode can be quantized as a harmonic oscillator, and the vibrational excitation of that mode is given by the occupation quantum number for that mode. The atomic motion of a mode is sometimes classified by its symmetry. While the frequency and symmetry are well known for each mode of a molecule for species with relatively small $\mathcal{N}$ (Nakamoto, 1986; Herzberg, 1945, 1950, 1966; Huber and Herzberg, 1979, Wilson *et al.,* 1980), this is not so in solids because $\mathcal{N} \sim 10^{22}$. Instead, in solids with translational symmetry (crystals) this motion is described in terms of collective lattice vibrations, which are labeled by a mode branch and a propagation wavevector **q** (Kittel, 1986). Crystals have a small number of mode branches, $3n$ where n is a small integer. q ranges between two finite limits, 0 and π/a, in some simple situations where a is the lattice constant. Since there are $\sim 10^7$ evenly spaced values of q, the wavevector can be considered a continuous parameter. Quantized lattice vibrations in solids are called phonons.

The effects of anharmonicity, i.e., deviations from harmonic behavior, are evident in molecules and solids. In molecules, the spacing between vibrational states in a given mode decreases and overtone transitions ($|\Delta v| > 1$; see below), which are forbidden in the harmonic approximation, become weakly allowed. At very high energies the characterization of vibrational states in terms of harmonic modes breaks down. In solids, anharmonicity leads to several important effects, including thermal expansion, phonon energies that change with temperature, and the decay of phonons into lower-energy phonons.

3.2 Energy Levels in Atoms and Molecules

The total energy of a state can be expressed as

$$E_{total} = E_{elect} + E_{vib} + E_{rot} \tag{3.1}$$

where the terms on the right-hand side represent the energy associated with the electronic, vibrational, and rotational degrees of freedom, respectively. With the characteristic differences between adjacent rotational, vibrational, and electronic states denoted as ΔE_{rot}, ΔE_{vib}, and ΔE_{elect}, respectively, $\Delta E_{rot} \ll \Delta E_{vib} \ll \Delta E_{elect}$ in most molecules. Consequently, the energy level structure due to molecular rotation can be associated with each vibrational state, and the structure due to vibration can be associated with each electronic state. This is illustrated in Figure 3.1. A vibrational level in an electronic state can be called a vibronic state; a given rotational level is then a rovibronic level.

Light can induce three types of absorption and emission transitions between molecular states: (1) pure rotational transitions, in which no change occurs in the vibrational or electronic state; (2) vibrational–rotational transitions, in which the vibrational state of the molecule changes, usually with some change in rotational state, but the electronic state does not change; and (3) electronic transitions, which occur between two different electronic states, usually with some change in the vibrational and rotational state.

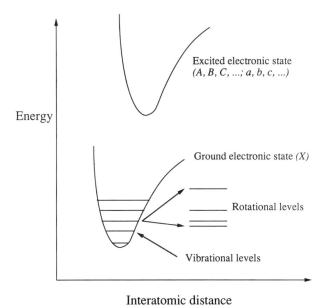

Figure 3.1 Schematic depicting the relative energies of electronic, vibrational, and rotational states.

Raman transitions can also occur, with analogs of these three cases. Each of these transitions is detailed in Chapter 4.

Often energy is expressed in wavenumbers (cm^{-1}), which actually represent the energy divided by hc.

3.2.1 Rotational States

The energy of a rotational state in a linear molecule is

$$\mathcal{E}_{\text{rot,linear}} = B_v J(J+1) - D_v[J(J+1)]^2 + H_v[J(J+1)]^3 + \cdots \quad (3.2)$$

where J is the rotational quantum number 0, 1, 2, ... When B_v, D_v, etc. are expressed in wavenumbers (cm^{-1}), the right-hand side of this equation must be multiplied by hc to obtain the units of energy. Each state has a degeneracy g_J of $2J + 1$; i.e., there are $2J + 1$ states with the same energy. The rigid rotor approximation gives only the first term with $B_v = h/(8\pi^2 cI)$ (units of cm^{-1}), where I is the moment of inertia. Levels J and $J + 1$ are separated by $2B_v(J + 1)$ in this approximation. The second term is the first higher-order correction for centrifugal distortion. B_v and D_v depend on the vibrational level and the electronic state. For B_v this dependence is approximately

$$B_v = B_e - \alpha_e(v + \tfrac{1}{2}) + \gamma_e(v + \tfrac{1}{2})^2 + \delta_e(v + \tfrac{1}{2})^3 + \cdots \quad (3.3)$$

where B_e is the rotational constant for the particular electronic state and v is the vibrational quantum number. $\alpha_e = 6\sqrt{\omega_e x_e B_e^3}/\omega_e - 6B_e^2/\omega_e$, which is usually <0. ($\omega_e$ and x_e are defined below.) For example, for H_2, $H^{35}Cl$, and $^{35}Cl_2$ the rotational constant B_e is 60.8, 10.6, and 0.244 cm^{-1}, respectively, for the ground electronic state.

In nonlinear molecules, the moments of inertia about the three principal axes, I_A, I_B, and I_C, can be different. A rotational constant, A_v, B_v, and C_v, is associated with each, with $A_v = h/(8\pi^2 cI_A)$, etc. Since the axes are labeled so that $I_A \leq I_B \leq I_C$, it is clear that $A_v \geq B_v \geq C_v$. In asymmetric tops, A_v, B_v, and C_v are unequal. In symmetric tops, two moments of inertia are equal with $I_A \neq I_B = I_C$ for prolate symmetric tops and $I_A = I_B \neq I_C$ for oblate symmetric tops.

The energy of a prolate symmetric top is

$$\mathcal{E}_{\text{rot,symmetric top}} = B_v J(J+1) + (A_v - B_v)K^2 - D_{vJ}J^2(J+1)^2$$
$$- D_{vJK}J(J+1)K^2 - D_{vK}K^4 + \cdots \quad (3.4)$$

where K is the quantum number associated with the projection of rotation on the symmetry axis, $K = -J, -J+1, \ldots, J$. For oblate symmetric tops, A_v is replaced by C_v. The first two terms assume the rigid rotor approximation, while the next three terms account for centrifugal distortion. As with linear molecules, A_v, B_v, and C_v depend on the specific vibrational and electronic state. The $K = 0$ state is nondegenerate, while the other levels are twofold degenerate.

3.2 Energy Levels in Atoms and Molecules

For spherical tops, $I_A = I_B = I_C$, and only the J quantum number is important. The rotational energy of each state is given by Equation 3.2 for linear molecules.

3.2.2 Vibrational States

The vibrational energy of a diatomic molecule is

$$E_{vib} = (v + \tfrac{1}{2})\omega_e - (v + \tfrac{1}{2})^2 \omega_e x_e + (v + \tfrac{1}{2})^3 \omega_e y_e + \cdots \qquad (3.5)$$

where v is the vibrational quantum number, which can be 0, 1, 2, ... Again, the right-hand side must be multiplied by hc when ω_e is expressed in terms of cm^{-1}. For example, for H_2, $H^{35}Cl$, and $^{35}Cl_2$ the vibrational constant ω_e is 4395, 2990, and 565 cm^{-1}, respectively, for the ground electronic state.

The harmonic oscillator approximation gives the first term in Equation 3.5, where $\omega_e = \sqrt{k_f/\mu_m}$, k_f is the effective force constant, and μ_m is the reduced mass of the molecule. The levels are equally spaced in this approximation. When the effect of anharmonicity is included, additional terms appear, such as the last two terms in Equation 3.5, and the levels become closer together with increasing energy. The fundamental transition has energy ω_v:

$$\omega_v = E_{vib}(v=1) - E_{vib}(v=0) = \omega_e - 2\omega_e x_e - \frac{13}{4}\omega_e y_e + \cdots \qquad (3.5a)$$

In a molecule with n vibrational modes, each mode is labeled by its quantum number $v_i = 0, 1, 2, \ldots$. The vibrational energy of a state with quantum numbers $(v_1, v_2, v_3, \ldots)$ is

$$E_{vib} = \sum_{i=1}^{n} (v_i + \tfrac{1}{2})\omega_{i,e} - \sum_{i=1}^{n} (v_i + \tfrac{1}{2})^2 \omega_{i,e} x_{i,e} \\ - \sum_{i \neq j} (v_i + \tfrac{1}{2})(v_j + \tfrac{1}{2}) q_{ij} + \cdots \qquad (3.6)$$

The first term is the harmonic approximation, and the subsequent terms are due to anharmonicity.

Characteristic vibrational frequencies of molecules are given by Herzberg (1945, 1950, 1966), Huber and Herzberg (1979), and Nakamoto (1986).

3.2.3 Electronic States

3.2.3.1 Atoms

Electronic states in atoms are designated by the electron configuration and the spectroscopic term symbol. The electron configuration gives the number of electrons in each shell, with principal quantum number $n = 1$, 2, 3, ..., and subshell, with orbital angular momentum quantum number $l = 0, 1, 2, \ldots, n - 1$, which are represented by $s, p, d, \ldots$, respectively. For example, the electron configuration for the ground state of Cl is $1s^2 2s^2 2p^6 3s^2 3p^5$, where the numbers preceding the l designation are the quantum numbers of the shell n and the superscripts represent the number of electrons in each nl subshell (Figure 6.33a).

Each electron configuration corresponds to the one or more states that are formed by the coupling of the individual electron spins (s_i) and orbital angular momenta (l_i) to form the total spin S and the total orbital angular momentum L, and by the spin–orbit coupling of L and S to form the total angular momentum J. This type of coupling, L–S or Russel–Saunders coupling, is dominant in lower-Z atoms. In higher-Z atoms j–j coupling begins to dominate in which individual electron spins and orbital angular momenta couple first to form j_i, which then couple to form J. For L–S coupling the term symbol designation is $^{2S+1}L_J$, where $L = 0, 1, 2, \ldots$ is represented by $S, P, D, \ldots$, and $2S + 1$ is the multiplicity of the state. The relative energies of the different spectroscopic states for a given electron configuration depend on the details of the interactions, including electron–electron repulsion; these are summarized by Hund's rules. The ground state of Cl has term symbol $^2P_{1/2}$, so the ground-state designation is $3p^5$ $^2P_{1/2}$, where the closed subshells have been excluded in the electron configuration, as is customary. The first excited state of Cl has the same electron configuration, but has a different term symbol $^2P_{3/2}$.

In atoms with two valence electrons having orbital angular momenta l_1 and l_2 coupled by L–S coupling, L ranges from $|l_1 - l_2|, |l_1 - l_2| + 1, \ldots, l_1 + l_2$. Terms for which the albegraic sum $l_1 + l_2$ is odd are called odd terms, and those for which it is even are called even terms. The odd terms are distinguished from even terms by a superscript 0 after the term symbol (Figure 6.33a). Therefore, two (nonequivalent) p electrons can couple to form a state with $L = 1$, $S = 0$, and $J = 1$, with term symbol $^1P_1$ (among other possible states), while an s and a p electron can couple to form a state having the same L, S, and J, but which is instead designated as $^1P_1^0$ (White, 1934).

3.2.3.2 Molecules

Electronic states in molecules are labeled in a similar way. Electron occupation in molecular orbitals can be written as the electron configuration in atomic orbitals. The specific electronic state can also be designated by a state designation analogous to the spectroscopic term symbol. For linear molecules, the spin S, orbital angular momentum L, and molecular rotational angular momentum R (which is normal to the molecular axis) can couple in various ways to form the total angular momentum J. In Hund's case (a) the spin S and orbital angular momentum L each couple to the molecular axis to give projections Σ and Λ, respectively, and $\Omega = \Sigma + \Lambda$;

3.2 Energy Levels in Atoms and Molecules

here the state designation is $^{2\Sigma+1}\Lambda_\Omega$, where $\Lambda = 0, 1, 2, \ldots$ is represented by $\Sigma, \Pi, \Delta, \ldots$. When the linear molecule is symmetric with respect to a plane perpendicular to the axis, as in homonuclear diatomic molecules, the inversion symmetry can be either even, designated by g, or odd, designated by u. Also, Σ states can be symmetric $(+)$ or antisymmetric $(-)$ with respect to reflection. Electronic states in linear molecules have designations of the form $^1\Sigma^+$, $^1\Sigma_g^+$, $^3\Pi$, and $^3\Pi_g$. In this type of coupling, R couples with Ω to form the total angular momentum J.

In Hund's case (c), which is important for molecules with higher-Z atoms, L and S couple off-axis to give a constant projection Ω along the axis, which is then a good quantum number. Hund's case (b) always applies to Σ states ($\Lambda = 0$) and often to $\Lambda \neq 0$ in linear molecules of low atomic number where $S \neq 0$. In this case, Λ and R couple to form N, and N and S couple to form J, which can have all integer values from $|N - S|$ to $N + S$. Other types of coupling are described in previously cited spectroscopy volumes.

Preceding the electronic symbol of a molecule is a label that usually gives the relative energy of the state, with the ground state being X. Excited states with the same spin multiplicity as the ground state are labeled A, B, C, etc, in order of increasing energy. States of different multiplicity are labeled $a, b, c, \ldots$, in order of increasing energy. For example, the ground electronic state of CO is $X\,^1\Sigma^+$ and the first excited state is $A\,^1\Pi$.

For nonlinear molecules, the state is designated by a group-theoretic symmetry element preceded by a $2S + 1$ superscript. For example, the ground electronic state of CF_2 is $\tilde{X}\,^1A_1$ and the first excited state is $\tilde{A}\,^1B_1$.

The vibrational level in the electronic state is often denoted by appending the vibrational quantum number after the electronic state, as with $v = 1$.

Figure 3.2 shows the energy potential diagram of N_2, N_2^+ and N_2^-. This depicts the electronic energy states of each, along with their corresponding vibrational structure.

3.2.4 Population of a State

Optical spectroscopy can be used to probe the population density of each of these states. In thermal equilibrium at temperature T the density of molecules in a particular state i with energy $\mathcal{E}_i$ and degeneracy g_i is given by the Boltzmann expression

$$N_i = N \frac{g_i \exp(-\mathcal{E}_i/k_B T)}{Z_t} \tag{3.7}$$

where N is the density of all molecules and $i = e, v, J$, where e, v, and J designate the electronic, vibrational, and rotational state quantum numbers, respectively. Consequently, the population of a given rotational and vibrational level in a given electronic state N_i can also be denoted by N_{evJ}. When

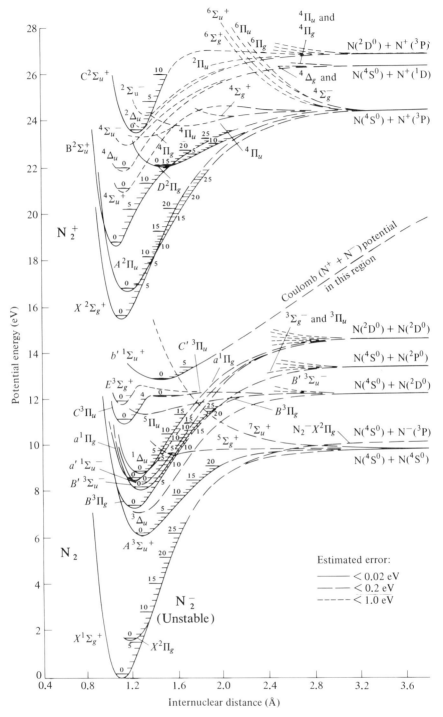

Figure 3.2 Potential energy level diagram for N_2, N_2^+, and N_2^-, showing the electronic states of each and their vibrational energy structure. (From Gilmore, 1966, and Steinfeld, 1985). Reprinted by permission.

3.2 Energy Levels in Atoms and Molecules

the electronic state is assumed to be the ground state, this can be written as N_{vJ}. Z_t is the total partition function:

$$Z_t = \sum_j g_j \exp(-\mathcal{E}_j/k_B T) \tag{3.8}$$

which is a sum over all molecular states with quantum numbers j.

Under many conditions Z_t can be approximated as the product of the separate partition functions corresponding to the electronic, vibrational, and rotational energies, where the sum is over only the respective energy states. Then $Z_t = Z_{elect} Z_{vib} Z_{rot}$. Usually $Z_{elect} \approx 1$. For a (nondegenerate) harmonic oscillator, Equation 3.8 can be used to obtain the vibrational partition function. Using Equation 3.5 (in the harmonic limit) and $g_v = 1$, it is

$$Z_{vib} = \sum_v \exp(-v\hbar\omega_v/k_B T) = \frac{1}{1 - \exp(-\hbar\omega_v/k_B T)} \tag{3.9}$$

For a diatomic molecule (in a $\Lambda = 0$ state), Equation 3.8 can be used to obtain the rotational partition function. Using Equation 3.2 and the degeneracy $g_J = 2J + 1$, in the rigid rotor approximation it is

$$Z_{rot} = \sum_J g_I (2J + 1)\exp(-hcB_v J(J + 1)/k_B T) \tag{3.10a}$$

$$= \frac{k_B T}{hcB_v} \qquad \text{when } B_v \ll k_B T \text{ and } g_I = 1 \tag{3.10b}$$

The population of a vibrational band is therefore

$$N_{ev} = N_e \frac{\exp(-v\hbar\omega_v/k_B T)}{Z_{vib}} \tag{3.11}$$

where N_e is the population density in the particular electronic state. ω_v in Equations 3.9 and 3.11 refers to the fundamental vibrational transition energy in the electronic state of interest. This is usually the ground electronic state for absorption, laser-induced fluorescence (LIF), and Raman scattering, for which N_e is usually $\sim N$, and an excited electronic state for optical emission.

The population of a rotational state for a diatomic molecule can be written as

$$N_{evJ} = \frac{N}{Z_{rot} Z_{vib} Z_{elect}} \frac{g_I (2J + 1)\exp(-B_v J(J + 1)/k_B T)}{Z_{rot}} \tag{3.12a}$$

$$= N_{ev} \frac{g_I (2J + 1)\exp(-B_v J(J + 1)/k_B T)}{Z_{rot}} \tag{3.12b}$$

In Equations 3.10 and 3.12 B_v refers to the rotational constant in the particular vibrational level that is being probed.

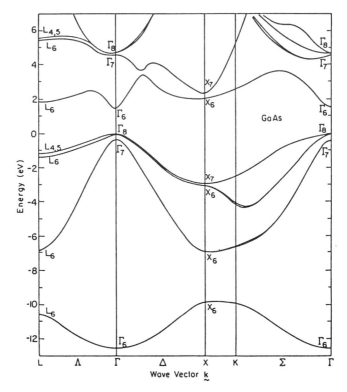

Figure 3.3 The electronic band structure of GaAs. (From M. L. Cohen and J. R. Chelikowsky, *Electronic Structure and Optical Properties of Semiconductors,* p. 103, Fig. 8.21, 1988. Copyright © Springer-Verlag.)

In molecules with inversion symmetry, such as homonuclear diatomic molecules, and symmetric linear molecules, such as $C^{16}O_2$, quantum mechanical restraints on the molecular wave function lead to an additional factor, g_I, in Equations 3.10 and 3.12, where I is the nuclear spin (Eisberg and Resnick, 1985). For example, the ratio $g_I(J_{odd}):g_I(J_{even})$ is $3:1$ in H_2 and $1:2$ in N_2, while for the ground electronic state of $^{16}O_2$ $g_I(J_{even}) = 0$, so only odd J states exist. In such cases, absorption and Raman spectra can have rotational lines with alternating intensities, as in H_2 and N_2 (as seen in Figures 12.6, 12.7, and 17.9), or they may have missing rotational lines, as in $^{16}O_2$.

3.3 Energy Levels in Solids

3.3.1 *Electronic Structure*

Electronic band theory and lattice dynamics in semiconductors are surveyed in the book edited by Paul (1982), and in textbooks by Ashcroft and

3.3 Energy Levels in Solids

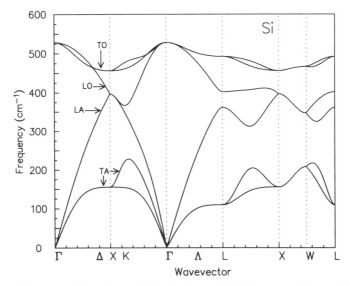

Figure 3.4 Phonon dispersion in Si (R. Eryigit and I. P. Herman, 1995, unpublished).

Mermin (1976) and Kittel (1986). Cohen and Chelikowsky (1988) discuss the electronic structure of semiconductors.

A typical electronic band structure is depicted in Figure 3.3, which shows the structure of GaAs. The density of states for a transition is large at the critical points (CPs), and consequently the dielectric function ε has distinctive features at these points. The CPs are often labeled $\mathcal{E}_0$, $\mathcal{E}_1$, $\mathcal{E}_2$, etc., and often have spin–orbit split features $\mathcal{E}_0 + \Delta_0$, $\mathcal{E}_1 + \Delta_1$, etc. The lowest critical-point energy corresponds to the fundamental absorption edge (band gap, $\mathcal{E}_{bg}$). For example, for GaAs in Figure 3.3, $\mathcal{E}_0$ (~1.52 eV at 300 K) is due to the fundamental gap at $\Gamma_8^v - \Gamma_6^c$ (Γ_8 in the valence band $-\Gamma_6$ in the conduction band) and the split-off feature (1.86 eV) $\mathcal{E}_0 + \Delta_0$ is due to $\Gamma_7^v - \Gamma_6^c$. $\mathcal{E}_1$ (3.04 eV) and $\mathcal{E}_1 + \Delta_1$ (3.25 eV) are due to $\Gamma_{4;5}^v - \Gamma_6^c$ and $\Gamma_6^v - \Gamma_6^c$, respectively. Higher-energy CPs $\mathcal{E}_0'$, $\mathcal{E}_2$, and $\mathcal{E}_1'$ for GaAs and CPs for other semiconductors are described in Cohen and Chelikowsky (1988).

This description is appropriate for bulk samples (substrates) and most thin films. In very thin films, such as quantum wells, additional features appear due to the effects of confinement.

3.3.2 Vibrational Structure (Phonons)

When there are n atoms per unit cell, there are $3n$ vibrational branches. There are three acoustic modes, where the atoms in the unit cell move in the same direction as $\mathbf{q} \rightarrow \mathbf{0}$, and $3n - 3$ optic modes, where the atoms in the unit cell move in different directions as $\mathbf{q} \rightarrow \mathbf{0}$. In sufficiently symmetric crystals, each triplet of modes corresponds to one longitudinal mode, where the atomic motion is along the direction of phonon propagation, and two transverse modes, where this motion is normal to phonon propagation.

The phonon dispersion in silicon is shown in Figure 3.4. The triply degenerate structure at the Γ point ($\mathbf{q} = \mathbf{0}$) near 521 cm^{-1} denotes the phonons observed in Raman scattering. This corresponds to two transverse optical (TO) phonons and one longitudinal optical (LO) phonon. In polar semiconductors, such as GaAs, this triply degenerate peak splits into a doubly degenerate TO mode (nonpolar modes) and a singly degenerate LO mode (polar mode) at slightly higher energy. Table 4.4 gives the Γ-point phonon peaks for several semiconductors, which correspond to the observed Raman shifts (Mitra and Massa, 1982).

References

N. W. Ashcroft and N. D. Mermin, *Solid State Physics*. Holt, Rinehart, & Winston, New York, 1976.

K. W. Böer, *Survey of Semiconductor Physics*. Van Nostrand-Reinhold, New York, 1990.

G. Burns, *Solid State Physics*. Academic Press, Orlando, FL, 1985.

M. L. Cohen and J. R. Chelikowsky, *Electronic Structure and Optical Properties of Semiconductors*. Springer-Verlag, Berlin, 1988.

E. U. Condon and G. H. Shortley, *The Theory of Atomic Spectra*. Cambridge Univ. Press, Cambridge, UK, 1959.

R. Eisberg and R. Resnick, *Quantum Physics of Atoms, Molecules, Solids, Nuclei, and Particles*, 2nd ed. Wiley, New York, 1985.

F. R. Gilmore, *Potential Energy Curves for N_2, NO, O_2 and Corresponding Ions*, RAND Corporation Memorandum RM-4034-1-PR (April 1966).

W. A. Harrison, *Electronic Structure and the Properties of Solids*. Freeman, San Francisco, 1980.

G. Herzberg, *Atomic Spectra and Atomic Structure*. Dover, New York, 1937.

G. Herzberg, *Molecular Spectra and Molecular Structure. II. Infrared and Raman Spectra of Polyatomic Molecules*. Van Nostrand-Reinhold, New York, 1945.

G. Herzberg, *Molecular Spectra and Molecular Structure. I. Spectra of Diatomic Molecules*. Van Nostrand-Reinhold, New York, 1950.

G. Herzberg, *Molecular Spectra and Molecular Structure. III. Electronic Spectra and Electronic Structure of Polyatomic Molecules*. Van Nostrand-Reinhold, New York, 1966.

K. P. Huber and G. Herzberg, *Molecular Spectra and Molecular Structure. IV. Constants of Diatomic Molecules*. Van Nostrand-Reinhold, New York, 1979.

C. Kittel, *Introduction to Solid State Physics*, 6th ed. Wiley, New York, 1986.

H. G. Kuhn, *Atomic Spectra*, Academic Press, New York, 1962.

I. N. Levine, *Molecular Spectroscopy*. Wiley (Interscience), New York, 1975.

S. S. Mitra and N. E. Massa, in *Handbook of Semiconductors* (W. Paul, ed.), Vol. 1, Chapter 3, p. 81. North-Holland Publ., Amsterdam, 1982.

References

K. Nakamoto, *Infrared and Raman Spectra of Inorganic and Coordination Compounds,* 4th ed. Wiley, New York, 1986.

W. Paul, ed., *Handbook of Semiconductors,* Vol. 1. North-Holland Publ., Amsterdam, 1982.

J. I. Steinfeld, *Molecules and Radiation: An Introduction to Modern Molecular Spectroscopy,* 2nd ed. MIT Press, Cambridge, MA, 1985.

H. E. White, *Introduction to Atomic Spectra.* McGraw-Hill, New York, 1934.

E. B. Wilson, Jr., J. C. Decius, and P. C. Cross, *Molecular Vibrations: The Theory of Infrared and Raman Vibrational Spectra.* Dover, New York, 1980.

CHAPTER 4

Interactions of Light with Matter for Spectroscopy

This chapter presents the fundamental interactions between light and the probed material that underlie most of the optical diagnostics discussed in subsequent chapters. Here the focus is on the quantum mechanical aspects of these interactions whereas the following chapters make little reference to quantum mechanics. The structure of matter, as surveyed in Chapter 3, is used to describe the transitions that can be induced by light in atoms, molecules, and solids, i.e., the spectroscopy of these materials. Emphasis is placed on linear processes, such as absorption, spontaneous emission, and spontaneous Raman scattering. Nonlinear processes are summarized at the end of this chapter.

As stated in the introduction to Chapter 3, this presentation is meant to serve as a reference to assist those working in optical diagnostics who have some background in spectroscopy; it is not intended to be comprehensive. Those needing either a more pedagogical introduction or a more detailed treatment should consult the books cited in this chapter and the previous chapter.

In this chapter Gaussian CGS units will be used. Consult the appendix to this chapter for expressions in rationalized MKSA (SI) units.

4.1 Dipole Moments and Polarization

Much of the interaction of light with gases and solids can be described by considering the electric dipole moment μ or the electric polarization **P**, which is the electric dipole moment per unit volume. In linear optics, which dominates when the electric field strengths are much smaller than atomic fields, **P** is linearly related to the externally applied electric field **E** by

$$\mathbf{P} = \chi\mathbf{E} \tag{4.1}$$

Similarly, the electric dipole moment $\boldsymbol{\mu}$ of a given species (atom, molecule, etc.) is linearly related to the local electric field $\mathbf{E}^l$ by

$$\boldsymbol{\mu} = \hat{\alpha}\mathbf{E}^l \tag{4.2}$$

With N dipoles per unit volume, $\mathbf{P} = N\hat{\alpha}\mathbf{E}^l$. As is seen below, $\mathbf{E}^l \approx \mathbf{E}$ for gases but not for solids.

In Equations 4.1 and 4.2 χ is the electric susceptibility, which is a macroscopic parameter, and $\hat{\alpha}$ is the electric polarizability, which is a microscopic parameter. (While the latter term is often called α, the designation $\hat{\alpha}$ is used here to distinguish it from the absorption coefficient.) In isotropic media these parameters are scalars, while in anisotropic materials and media in applied magnetic fields, they are second-rank tensors $\overleftrightarrow{\chi}$ and $\overleftrightarrow{\hat{\alpha}}$. Isotropic media are assumed here (with no magnetic fields) unless otherwise noted.

The electric displacement vector $\mathbf{D}$ is related to $\mathbf{E}$ by $\mathbf{D} = \mathbf{E} + 4\pi\mathbf{P} = \varepsilon\mathbf{E}$, where ε is the dielectric constant or function—which, again, may be a scalar or tensor. This constitutive relation gives $\varepsilon = 1 + 4\pi\chi$, or $\overleftrightarrow{\varepsilon} = \overleftrightarrow{\mathbf{I}} + 4\pi\overleftrightarrow{\chi}$. A tilde will be placed over ε ($\tilde{\varepsilon}$) and the complex index of refraction n ($\tilde{n}$) as a reminder that they are complex.

The local electric field usually differs from the applied electric field (from the light source) because of the field produced by the local polarization. For isotropic media, these fields are related by $\mathbf{E}^l = \mathbf{E} + 4\pi\mathbf{P}/3$ (Jackson, 1975; Debe, 1987), the last term being the Lorenz field produced by local dipoles. This local field correction leads to

$$\chi = \frac{N\hat{\alpha}}{1 - \tfrac{4}{3}N\pi\hat{\alpha}} \tag{4.3}$$

which becomes $\chi = N\hat{\alpha}$ for low densities, as in gases. The microscopic polarizability is related to the dielectric constant by the Lorentz–Lorenz formula

$$\hat{\alpha} = \frac{3}{4\pi N}\left(\frac{\tilde{\varepsilon} - 1}{\tilde{\varepsilon} + 2}\right) \tag{4.4}$$

In linear optics, the dipole moment of a species, such as a molecule, is more generally expressed as

$$\boldsymbol{\mu} = \boldsymbol{\mu}_0 + (\hat{\alpha} + \hat{\alpha}_R)\mathbf{E}^l \tag{4.5}$$

for an isotropic medium. Here $\boldsymbol{\mu}_0$ is the permanent electric dipole moment, which is responsible for pure rotational infrared absorption. The molecular polarizability $\hat{\alpha}$ is the polarizability of a molecule with its atoms in their equilibrium positions. It has contributions from induced dipoles from electronic bonds $\hat{\alpha}_e$ and from vibrations of the nuclei $\hat{\alpha}_n$, so $\hat{\alpha} = \hat{\alpha}_e + \hat{\alpha}_n$. $\hat{\alpha}_e$ is responsible for reflection, refraction, and absorption in the visible and

4.1 Dipole Moments and Polarization

ultraviolet, while $\hat{\alpha}_n$ is similarly responsible for these phenomena in the infrared. Elementary excitations of materials, such as material deformations or vibrations with normal coordinates Q_i, also perturb the polarization, to give the dynamic Raman contribution

$$\hat{\alpha}_R \sim \sum_i \frac{\partial \hat{\alpha}_e}{\partial Q_i} Q_i \tag{4.6}$$

This leads to spontaneous Raman scattering.

4.1.1 Material Parameters for Linear Optical Spectroscopies

The linear electromagnetic properties of a material are described by its complex dielectric function:

$$\tilde{\varepsilon} = \varepsilon' + i\varepsilon'' \tag{4.7}$$

which is a function of frequency ω (or wavelength λ) and temperature T. Since in some communities the convention is $\tilde{\varepsilon} = \varepsilon' - i\varepsilon''$, the adopted convention must be known before using a theoretical result, as is seen in Chapter 9. The real and imaginary parts of $\tilde{\varepsilon}$ are often called ε_1 and ε_2, respectively. However, since this can be confusing when discussing multimedium systems, they are called ε' and ε'' here. (The ε_1, ε_2 notation has not been changed in the figures reprinted from the literature.) Figure 9.6 shows the optical dielectric functions for several common semiconductors.

Closely related is the complex index of refraction $\tilde{n} = \sqrt{\mu_m \tilde{\varepsilon}}$, where μ_m is the magnetic permeability. (The magnetic permeability is usually called μ; however, this symbol is reserved here for the dipole moment.) Since most materials treated here are not magnetic, $\mu_m = 1$ and $\tilde{n} = \sqrt{\tilde{\varepsilon}}$. The complex index of refraction can be expressed in terms of its real and imaginary parts as

$$\tilde{n} = n + ik \tag{4.8}$$

As with the dielectric function, theoretical results must be used with caution because of the alternative conventions for $\tilde{n}$, such as $\tilde{n} = n - ik$ or $n(1 \pm ik)$; also, κ is sometimes used instead of k. The choice of the positive sign in Equation 4.8 is dictated by the form for the electric field in Equations 2.2 and 2.3, and by the desire to make $k > 0$ for absorption. These various conventions are discussed by Holm (1991).

Using $\tilde{\varepsilon} = \tilde{n}^2$, the real and imaginary parts of the dielectric constant and index of refraction are related by

$$\varepsilon' = n^2 - k^2 \tag{4.9a}$$

$$\varepsilon'' = 2nk \tag{4.9b}$$

The inverse relations can be obtained from

$$n^2 = \frac{\sqrt{\varepsilon'^2 + \varepsilon''^2} + \varepsilon'}{2} \quad (4.10a)$$

$$k^2 = \frac{\sqrt{\varepsilon'^2 + \varepsilon''^2} - \varepsilon'}{2} \quad (4.10b)$$

Using Equations 2.2 and 2.3, the electric field decays as $\exp(-2\pi kz/\lambda)$, where k is positive and λ is the free-space wavelength, and the intensity decays as $\exp(-4\pi kz/\lambda)$. This is Beer's law for absorption (Equation 2.5), with the absorption coefficient

$$\alpha = 4\pi k/\lambda \quad (4.11a)$$

$$= 8\pi^2 \chi''/\lambda \quad (4.11b)$$

with $n = 1$. Note that k is often called the extinction coefficient by members of the optics community. The term "extinction coefficient" is also used by some members of the chemical community to mean the absorption coefficient referenced to base 10 (Section 8.1.1). Thus the exact definition of this term must be known before using a result. The absorption coefficient can also be expressed as $\alpha = N\sigma$, where N is the density of absorbing species (1/cm^3) and σ is the absorption cross-section, which has units of area (cm^2).

As with χ and $\hat{\alpha}$, the scalar forms for $\tilde{\varepsilon}$ and $\tilde{n}$ assume isotropic materials, such as gases and cubic solids in the absence of a magnetic field. This is satisfactory for most optical diagnostics, although the full tensor form is sometimes needed to describe $\tilde{\varepsilon}$ in reflection studies [e.g., reflection difference spectroscopy (RDS), Section 9.10] and is usually needed in nonlinear optics (Chapter 16). The linear optical properties of any material are totally specified by knowing any of the interrelated complex parameters $\tilde{\varepsilon}$, $\tilde{n}$, or χ (or $\hat{\alpha}$). The real parts of each [ε', n, or χ' (or $\hat{\alpha}'$)] describe the dispersive (or refractive) properties of the medium and their imaginary parts [ε'', k, or χ'' (or $\hat{\alpha}''$)] describe absorption (and gain in laser amplifying media).

Sometimes information is available about ε' or ε'' only. The other parameter can be determined by using the Kramers–Kronig relations (Stern, 1963), which relate the real and imaginary parts of "well-behaved" analytic functions. In physical terms, the complex function described by these relations must linearly relate a response to an excitation in a causal manner. Such a function must be analytic in the upper half of the complex plane. The dielectric constant relates **D** and **E** in such a manner. ε' and ε'' of the dielectric function are related by using the Kramers–Kronig relations

$$\varepsilon'(\omega) = \varepsilon(\infty) + \frac{2}{\pi} P \int_0^\infty \frac{\omega' \varepsilon''(\omega')}{\omega'^2 - \omega^2} d\omega' \quad (4.12)$$

$$\varepsilon''(\omega) = -\frac{2\omega}{\pi} P \int_0^\infty \frac{\varepsilon'(\omega')}{\omega'^2 - \omega^2} d\omega' \quad (4.13)$$

4.2 Quantum Mechanics of the Interaction of Light with Matter

where P is the Cauchy principal value of the integral taken at $\omega' = \omega$. These integrals must extend over a suitably wide frequency range to be accurate.

The analog for the complex index of refraction is

$$n(\omega) = 1 + \frac{2}{\pi} P \int_0^\infty \frac{\omega' k(\omega')}{\omega'^2 - \omega^2} d\omega' \qquad (4.14)$$

Because $\tilde{n}$ does not relate physical parameters but is derived from $\tilde{\varepsilon}$, the inverse relation does not automatically follow, but it can follow under certain conditions (Stern, 1963; Toll, 1956).

Ohta and Ishida (1988) have compared several numerical integration methods for performing Kramers–Kronig transformations.

4.2 Quantum Mechanics of the Interaction of Light with Matter

The interaction of light with matter is often treated as a time-dependent perturbation induced by the Hamiltonian $\mathcal{H}' = -\boldsymbol{\mu} \cdot \mathbf{E}$, where $\boldsymbol{\mu}$ is the electric dipole moment operator and $\mathbf{E}$ is the oscillating electric field. $\boldsymbol{\mu} = e\mathbf{r}$, where e is the charge of an electron and $\mathbf{r}$ is the position operator. Such electric dipole coupling is usually the strongest interaction. This formulation is suitable for atoms and molecules because they have sharp levels. It is more convenient to use the closely related form $\mathcal{H}' = -(e/mc)\,\mathbf{p} \cdot \mathbf{A}$ to describe transitions between bands in solids, where $\mathbf{p}$ is the momentum operator and $\mathbf{A}$ is the vector potential.

Figure 4.1 illustrates several optical processes. Figure 4.1a shows absorption from a lower state i to an excited state j. The wavefunction of each state can be represented by ϕ_i or in Dirac notation as $|i\rangle$. Spontaneous emission (or radiative decay) from j to state k is shown in Figure 4.1b. Laser-induced fluorescence (LIF) and photoluminescence (PL) involve ab-

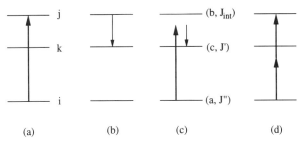

Figure 4.1 Energy-level diagrams of several optical processes: (a) Absorption from a lower state i to an excited state j; (b) spontaneous emission (or radiative decay) from j to state k; (c) spontaneous Raman scattering from i (or a or J'') to k (c, J') through intermediate state j (b, J_{int}); (d) two-photon absorption (TPA).

sorption from $j \leftarrow i$ and then spontaneous emission from $j \rightarrow k$. In spectroscopic descriptions of a transition, the symbol for the higher-energy state is written on the left and the lower state on the right, separated by an arrow whose direction denotes absorption or emission. Figure 4.1c depicts spontaneous Raman scattering from i to k, while Figure 4.1d shows two-photon absorption (TPA).

A light beam is absorbed by a medium if its frequency ω is resonant with a transition in the medium and if electromagnetic radiation can couple the two levels. Quantum mechanically, this latter criterion means that the matrix element of the interaction Hamiltonian $\mathcal{H}'$ that couples the two states does not vanish—that is, $\mathcal{H}'_{ij} = \langle i|\mathcal{H}'|j\rangle \neq 0$—and is the basis of transition selection rules.

Some expressions will be presented in frequency units (ν), which are often used in the chemistry and quantum-electronics communities, while others will be presented in radial frequency units (ω), which are often used in the solid-state physics community.

4.2.1 *Absorption and Spontaneous Emission*

This transition rate between lower level 1 and upper level 2 is given by Fermi's Golden Rule (Yariv, 1989; Liboff, 1992):

$$R_{12} = \frac{2\pi}{\hbar}|\mathcal{H}_{12}'|^2 \, g(\mathcal{E}) \tag{4.15}$$

where $g(\mathcal{E})$ is the density of states—or, equivalently, the lineshape—expressed in terms of energy $\mathcal{E} = \hbar\omega$ near the resonance. This is the rate both of absorption and of stimulated emission.

The energy per unit volume absorbed in the medium per unit time is $N_{net}\hbar\omega R_{12}$, where N_{net} is the net density of absorbers; it is also equal to αI, where α is the absorption coefficient at frequency ω, as in Beer's law (Equation 2.5), and I is the light intensity at frequency ω. Using Equation 4.15, this gives

$$\alpha(\omega) = \frac{\lambda^2}{4n^2} A_{21} \left(\frac{g_2}{g_1} N_1 - N_2 \right) g(\omega) \tag{4.16a}$$

$$\alpha(\nu) = \frac{\lambda^2}{8\pi n^2} A_{21} \left(\frac{g_2}{g_1} N_1 - N_2 \right) g(\nu) \tag{4.16b}$$

where N_1 and g_1 are the population density and degeneracy of the lower state, respectively, and N_2 and g_2 are those for the excited state. Since N_2 is usually ~ 0 under processing conditions, this stimulated emission term can be ignored. λ is the vacuum wavelength and

$$A_{21} = \frac{8\pi\omega^3|\boldsymbol{\mu}_{21}|^2 n}{3hc^3} = \frac{64\pi^4\nu^3|\boldsymbol{\mu}_{21}|^2 n}{3hc^3} \tag{4.17}$$

where A_{21} is the Einstein "A" coefficient and μ_{21} is the dipole moment transition matrix element. The factor $|\mu_{21}|^2 = (\Sigma |\mu_{2i,1j}|^2)/g_2$, where the sum is over all possible transitions from the states of the degenerate upper level to those of the lower level. Sometimes Equation 4.17 is written without the factor of 3 in the denominator because the dipole moment along one axis is actually being used, with $|\mu_{21,x}|^2 = |\mu_{21}|^2/3$. This assumption is often valid, as for transitions in molecules involving rotational level changes (Kovács, 1969). $A_{21} = 1/\tau_{rad}$, where τ_{rad} is the radiative decay (or spontaneous emission) lifetime for transitions from state 2 to state 1. (Also, $\gamma_{rad} = 1/\tau_{rad}$.) $g(\omega)$ is the spectral lineshape, which is normalized so $\int g(\omega)d\omega = 1$. $g(\nu) = 2\pi g(\omega)$ is similarly normalized.

For such a transition between two sharp levels, the susceptibility χ can be expressed in a form obtained by using the classical simple free electron (or Lorentz oscillator) model,

$$\chi = \frac{Ne^2/m}{\omega_0^2 - \omega^2 - i\Gamma^{(\omega)}\omega} = \frac{Ne^2/4\pi^2 m}{\nu_0^2 - \nu^2 - i\Gamma^{(\nu)}\nu} \tag{4.18}$$

where N is the density of the species, e is the charge of the electron, m is its mass, $\omega_0(\nu_0)$ is the resonant frequency, and Γ is the damping factor in the corresponding frequency units. This equation assumes the low-density limit where $\chi = N\hat{a}$.

With $\chi = \chi' + i\chi''$, which ensures a positive absorption coefficient in Equation 4.11b, separation into real and imaginary parts and simplification assuming near-resonance conditions ($\omega \approx \omega_0$) gives

$$\chi' = \frac{Ne^2}{2m\omega_0} \frac{\omega_0 - \omega}{(\omega - \omega_0)^2 + (\Gamma^{(\omega)}/2)^2} \tag{4.19a}$$

$$= \frac{Ne^2}{8\pi^2 m\nu_0} \frac{\nu_0 - \nu}{(\nu - \nu_0)^2 + (\Gamma^{(\nu)}/2)^2} \tag{4.19b}$$

and

$$\chi'' = \frac{N\pi e^2}{2m\omega_0} \frac{\Gamma^{(\omega)}/2\pi}{(\omega - \omega_0)^2 + (\Gamma^{(\omega)}/2)^2} \tag{4.20a}$$

$$= \frac{Ne^2}{8\pi m\nu_0} \frac{\Gamma^{(\nu)}/2\pi}{(\nu - \nu_0)^2 + (\Gamma^{(\nu)}/2)^2} \tag{4.20b}$$

where the last factor in χ'' is seen to be a normalized lorentzian profile $g(\omega)$ or $g(\nu)$ [Equation 4.20a or 4.20b], which also appears in Equation 4.16.

Valid quantum mechanical expressions can be obtained by substituting $e^2/2m\omega_0 = e^2/4\pi m\nu_0$ by $|\mu_{21,x}|^2/\hbar$ or, equivalently, by multiplying the right-hand sides of Equations 4.19 and 4.20 by the oscillator strength $f_{12} = 4\pi m\omega_0 |\mu_{21}|^2/3\hbar e^2 = 8\pi^2 m\nu_0|\mu_{21}|^2/3\hbar e^2$. Also, N is replaced by $(g_2/g_1)N_1 - N_2$. Equation 4.20 is then seen to be consistent with Equations 4.11b and 4.16.

Another way to make the classical model quantum mechanically valid is by replacing e and m by the effective charge (e^*) and mass (m^*, m_c, m_h) of the particle or quasiparticle (electron, hole, etc.).

4.2.1.1 Spectral Lineshapes

In Equation 4.16 $g(\nu)$ is the spectral lineshape, which is usually normalized. General discussions about spectral lineshapes in atoms and molecules can be found in Mitchell and Zemansky (1971), Corney (1977), Demtröder (1982), and Verdeyen (1989). When all species have the same resonant frequency ν_0, $g(\nu)$ is a lorentzian:

$$g(\nu) = \frac{\Gamma_L^{(\nu)}/2\pi}{(\nu - \nu_0)^2 + (\Gamma_L^{(\nu)}/2)^2} \qquad (4.21)$$

where the FWHM (full width at half maximum) $\Gamma_L^{(\nu)} = 1/\pi T_2$ is the sum of the rates of relaxation processes due to spontaneous emission, collisions, etc. [The analogous expression for $g(\omega)$ uses $\Gamma_L^{(\omega)} = 2\pi\Gamma_L^{(\nu)}$ as the broadening factor.] This is known as homogeneous broadening. In a gas, $\Gamma_L^{(\nu)} = \gamma_{rad}^{(\nu)} + \Sigma_n k_n^{(\nu)} p_n$, where the first term describes radiative decay and the second term describes collisional decay due to each gas, with partial pressure p_n. $\gamma_{rad}^{(\nu)} \approx 10^6 – 10^9$ Hz for UV transitions and is much smaller for IR transitions. Typically, $k_n^{(\nu)} \sim 10^6$ Hz/Torr.

When the resonant frequency ν_0 is different for different species inhomogeneous broadening occurs. This can be the result of molecules with different velocities absorbing at slightly different frequencies, which is the basis of Doppler broadening in gases, or spatially varying electric, magnetic, or strain fields. In gases in thermal equilibrium, the Maxwell–Boltzmann distribution function for speeds leads to the gaussian lineshape for Doppler broadening:

$$g(\nu) = \sqrt{\frac{4 \ln 2}{\pi}} \; \frac{1}{\Gamma_D^{(\nu)}} \; \exp\left(-4\ln 2 \left(\frac{\nu - \nu_0}{\Gamma_D^{(\nu)}}\right)^2\right) \qquad (4.22)$$

where the Doppler width $\Gamma_D^{(\nu)} = \sqrt{(8 \ln 2 \, k_B T)/mc^2} \, \nu_0$ has been expressed as the FWHM and ν_0 is the transition frequency for the molecule at rest. If the average velocity of the gas $\langle \mathbf{v} \rangle$ is not zero, as in a plume during pulsed laser deposition or in an electron cyclotron resonance (ECR) plasma, then the center frequency ν_0 is replaced by the Doppler-shifted frequency $\nu_0(1 + \langle v_z \rangle/c)$, for measurements made by a detector placed in the $+z$ axis.

Neutral gases are homogeneously broadened (Equation 4.21) when $\Gamma_L^{(\nu)} \gg \Gamma_D^{(\nu)}$ ($p \gg$ a few Torr—in the IR) and inhomogeneously broadened (Equation 4.22) when $\Gamma_L^{(\nu)} \ll \Gamma_D^{(\nu)}$ ($p \ll$ a few Torr). When $\Gamma_L^{(\nu)} \approx \Gamma_D^{(\nu)}$, the convolution of the lorentzian and the gaussian gives an intermediate lineshape, a Voigt profile.

In plasmas, lineshapes are affected by Doppler broadening, collisional (resonance) broadening, and Stark broadening due to shifting of the resonance frequency by collisions with electrons and ions (Griem, 1964, 1974; Hughes, 1975; Bekefi, 1976).

In optically thick gas media, corrections for self-absorption (or radiation trapping), i.e., absorption of emitted light, may be needed when analyzing emission spectra (Holstein, 1947). Self-absorption can lead to unusual lineshapes in gases where the temperature, electric fields (plasmas), etc. vary with position along the probed region (Griem, 1964, 1974; see also Figure 8.27).

Multiple resonances may be treated as sums of Equation 4.21 with different resonant frequencies. This picture is valid for the sharp resonances in gases and for isolated features in solids, such as dopant and excitonic resonances, and vibrations. Band-to-band transitions in solids require integration of the lineshape over the density of states; this leads to expressions such as Equation 4.40.

4.2.1.2 Validity of Beer's Law

Transmission through a medium is described by Beer's law, $I(z) = I(0)\exp(-\alpha z)$ (Equation 2.5), where the absorption coefficient α is given by Equation 4.11. Beer's law assumes several conditions:

(1) The absorption coefficient is the same for all spectral components of the optical beam. This is not true when spectrally narrow lines are probed by broad optical sources. Corrections to Beer's law for spectrally broad sources are discussed in Section 8.1.1.

(2) The laser pulse lengths are assumed to be so long that transient effects can be ignored. This is a good assumption for the continuous-wave (cw) lasers and pulsed lasers with pulse widths >1 nsec that are usually used in diagnostics.

(3) The laser intensities are assumed to be so low that saturation effects are not important. Saturation occurs at high intensities when the absorption rate is comparable to relaxation rates of the states in the transition. This tends to equalize the population of the two states and also broadens the spectral line (power broadening) (Yariv, 1989). This occurs when $\mu E/h \gtrsim \Gamma_L^{(\nu)}$ or, equivalently, when $I \gtrsim I_{sat}$, where $I_{sat} = (h\Gamma_L^{(\nu)}/\mu)^2$. At linecenter ($\nu = \nu_0$), saturation decreases the absorption coefficient from its unsaturated value α to $\alpha_{sat} = \alpha/(1 + I/I_{sat})$ for homogeneously broadened lines and to $\alpha/\sqrt{1 + I/I_{sat}}$ for Doppler-broadened (inhomogeneously broadened) lines (Yariv, 1989). For a very strong transition, $\mu = 1$ D (1 D (Debye) = 10^{-18} esu · cm), saturation effects become significant for $I \gtrsim I_{sat} \approx 4 \times 10^3$ W/cm² for a 1-GHz linewidth, which is typical in the visible, and for $\gtrsim I_{sat} \approx 10$ W/cm² for a 50-MHz linewidth, which is represen-

tative of the infrared. I_{sat} is 100× larger for transitions with $\mu = 0.1$ D, which are still quite strong.

Saturation effects can be important in cw laser spectroscopy of gases when spectrally narrow lasers are tightly focused and pressures are low. They are not significant in absorption (transmission) probing by infrared diode lasers, atomic resonance lamps, or spectrally broad, incoherent sources (Section 8.2). However, saturation can be significant when focused pulsed lasers are used, such as in pulsed-dye laser-induced fluorescence (LIF) experiments (Chapter 7). At first glance, extreme saturation ($I \gg I_{sat}$) appears to be desirable because the LIF signal ($\propto \alpha I$) would then be independent of I for homogeneously broadened transitions, thereby simplifying calibration. However, such extreme saturation does not occur across the whole beam; thus it is usually wiser to avoid saturation in LIF and instead account for the linear dependence of the LIF signal on I. Note that when $I \gtrsim I_{sat}$, different transitions have different degrees of saturation when their transition moments differ. Under typical conditions, saturation is not significant in the *in situ* diagnostics of films and wafers.

(4) The laser intensities are assumed to be so low that nonlinear optical effects are not important. At high laser intensities, nonlinear phenomena such as multiphoton absorption can occur (Section 4.4 and Chapter 16).

These considerations are particularly relevant when using LIF (Chapter 7) and absorption (Chapter 8) diagnostics.

4.2.2 *Spontaneous Raman Scattering*

Scattering by a material can be described in terms of the induced dipole moment in molecules (Equation 4.2) or the induced electric polarization in solids (Equation 4.1) by using Equations 4.5 and 4.6. The total scattering cross-section for scattering from incident frequency ω_i to ω_s is

$$\sigma = \frac{8\pi}{3} \frac{\omega_i^4 \hat{\alpha}^2}{c^4} \tag{4.23}$$

for a scalar $\hat{\alpha}$. The differential scattering cross-sections $d\sigma/d\omega_s$, $d\sigma/d\Omega$, and $d^2\sigma/d\Omega d\omega_s$ describe scattering per unit frequency of the scattered light ($d\omega_s$), per solid angle ($d\Omega$), and per $d\omega_s d\Omega$, respectively. Furthermore, $d^2\sigma/d\Omega d\omega_s = (d\sigma/d\Omega)g(\omega_s)$, where $g(\omega_s)$ is the lineshape of the scattered light. Details about Raman scattering can be found in Long (1977), Hayes and Loudon (1978), Cardona (1982), Ferraro and Nakamoto (1994), and Herman (1996) and other citations in Chapter 12.

The static part of the polarizability $\hat{\alpha}$ is responsible for Rayleigh (elastic) scattering and pure rotational Raman scattering in molecules. The polarizability and susceptibility (χ) can also depend on vibration through the

4.2 Quantum Mechanics of the Interaction of Light with Matter

characteristic displacement Q that oscillates with frequency ω_v to give the motion $u(Q)$, as in Equation 4.6. To first order, χ (in scalar form) can be expanded as

$$\chi(Q) = \chi(Q=0) + (d\chi/dQ)Q \qquad (4.24)$$

The analogous expression for the polarizability is Equation 4.6. Classically, vibrational Raman scattering in molecules and solids occurs because the dipole moment μ and the polarization P oscillate at and therefore radiate at $\omega_i \pm \omega_v$. The second term in Equation 4.24 leads to the derived polarizability tensor. A vibrational transition in a molecule is Raman-allowed if the polarizability varies with the normal coordinate of the vibrational mode (Q).

In the macroscopic description of Raman scattering, the fluctuating components of the polarization lead to scattering. The spectral differential cross-section for scattering photons from frequency ω_i to ω_s is (Hayes and Loudon, 1978)

$$\frac{d^2\sigma}{d\Omega d\omega_s} = \frac{1}{16\pi^2} \frac{\omega_i \omega_s^3}{c^4} vV \frac{n_s}{n_i} | \mathbf{e}_s \cdot \overleftrightarrow{d\chi}/d\mathbf{Q} \cdot \mathbf{e}_i |^2 \langle u(\mathbf{Q})u^*(\mathbf{Q}) \rangle_{\omega_s} \qquad (4.25)$$

where v is the scattering volume, V is the sample volume, n is the index of refraction, and $\mathbf{e}_s$ and $\mathbf{e}_i$ are the polarization vectors of the scattered and incident beams, respectively. The last term is the power spectrum of the fluctuations of the vibration, such as a vibration in a molecule or a phonon in a solid. The tensor properties of this cross-section are sometimes expressed in terms of the Raman tensor $\overleftrightarrow{\mathbf{R}}$, and so $d^2\sigma/d\Omega d\omega_s$ is proportional to $|\mathbf{e}_s \cdot \overleftrightarrow{\mathbf{R}} \cdot \mathbf{e}_i|^2$. The angular dependence of Equation 4.25 is discussed by Louisell (1990). Since $\omega_s \approx \omega_i$, the frequency dependence for Raman scattering is similar to the $\propto \omega_i^4$ dependence for elastic scattering far below resonance.

4.2.2.1 Molecules

Spontaneous Raman scattering from molecules in a gas is a coherent two-step process, as from level a to level c in Figure 4.1c. In vibrational Raman scattering in gases, these two levels are different vibrational levels in the same electronic state. Because the rotational quantum numbers of a and c can also differ, this is also called vibrational–rotational Raman scattering. When these two states are different rotational states in the same vibrational level, it is called pure rotational Raman scattering. In vibrational and rotational scattering, the intermediate state b is an excited electronic state. Levels a and c can be different electronic states, which is always true for Raman scattering in atoms. Such electronic Raman scattering is relatively seldom used in diagnostics.

Quantum mechanically, the term in the electric polarizability describing scattering from rotational–vibrational (rovibronic) states a to c is

$$\hat{\alpha}_{ij} = \frac{e^2}{\hbar}\sum_b \left(\frac{\langle c|r_j|b\rangle\langle b|r_i|a\rangle}{(\omega_{ab} - \omega_i - i\Gamma_b^{(\omega)})} + \frac{\langle b|r_j|a\rangle\langle c|r_i|b\rangle}{(\omega_{bc} + \omega_i - i\Gamma_b^{(\omega)})}\right) \quad (4.26)$$

where r_i is the Cartesian component of the electron position operator, $\Gamma_b^{(\omega)}$ is a damping term that describes the width of level b, and ω_{ab} is the frequency of the a–b transition (> 0) (Rousseau et al., 1979). Given the Born–Oppenheimer approximation and the separation of the rotational and vibrational degrees of freedom, the wavefunction can be factored into three terms. For state a, this gives a vibrational term ($a_v(\mathbf{R})$) and a rotational term $a_r(\mathbf{R})$, both of which are functions of the internuclear coordinate $\mathbf{R}$, as well as an electronic term $a_e(\mathbf{r},\mathbf{R})$, which is a function of the electron coordinate $\mathbf{r}$ and is parameterized by $\mathbf{R}$.

Integrating the rotational wavefunction terms in the matrix elements over molecular orientation gives the dipole selection rules for the rotational quantum number. Integrating the electronic terms (a_e and c_e in the ground electronic state, and b_e in an excited electronic state) over the electronic coordinate gives an $\mathbf{R}$-dependent electronic dipole moment $\mathbf{M}$ that can be expanded about the equilibrium configuration ($\mathbf{R}_0$), as $\mathbf{R} = \mathbf{R}_0 + \mathbf{u}$, to give $M(R) = M(R_0) + [\partial M/\partial R|_{R_0}]u + \cdots$. (For simplicity, M and R are written as scalars here.) Ignoring the nonresonant second term in Equation 4.26 gives

$$\hat{\alpha}_{ij} = |M(R_0)|^2 \frac{1}{\hbar}\sum_b \frac{\langle c_v|b_v\rangle\langle b_v|a_v\rangle}{(\omega_{ab} - \omega_i - i\Gamma_b^{(\omega)})}$$
$$+ M(R_0)\,\partial M/\partial R|_{R_0}\frac{1}{\hbar}\sum_b \frac{\langle c_v|R|b_v\rangle\langle b_v|a_v\rangle + \langle c_v|b_v\rangle\langle b_v|R|a_v\rangle}{(\omega_{ab} - \omega_i - i\Gamma_b^{(\omega)})} + \cdots \quad (4.27)$$

where the remaining integration is over vibrational coordinates.

Far from resonance ($|\omega_{ab} - \omega_i| \gg 0$), the numerator of the first term sums to $\langle c_v|a_v\rangle$, which leads to no change in vibrational state, i.e., Rayleigh and pure rotational Raman scattering (static polarizability tensor). Far from resonance, the numerator of the second term sums to $\langle c_v|R|a_v\rangle$, which means that the vibrational quantum number v changes by ± 1 (fundamental scattering) (derived polarizability tensor). Overtone scattering ($|\Delta v| > 1$) is weak, but stronger near resonance (Rousseau et al., 1979).

Classically, pure rotational Raman scattering is due to the anisotropy in the static polarizability tensor, as given by $2\hat{\alpha}_{zz} - \hat{\alpha}_{xx} - \hat{\alpha}_{yy}$. This is analogous to the requirement of a nonvanishing permanent electric dipole moment for allowed infrared transitions between rotational levels. Vibrational Raman scattering can occur (Raman-active mode) if at least one of the derivatives of the polarizability tensor with respect to the normal coordinate, $\partial\hat{\alpha}/\partial Q$, does not vanish when evaluated at the equilibrium position. Infrared vibrational transitions are allowed (IR or infrared-active mode)

4.2 Quantum Mechanics of the Interaction of Light with Matter

when at least one derivative of the dipole moment (transition dipole moment) does not vanish at the equilibrium position.

4.2.2.2 Solids

For Stokes scattering of phonons in solids, the substitution $\langle uu^* \rangle = (\hbar/2\mathcal{N}\omega_p)[n(\omega_p) + 1]g(\omega_s)$ is made in Equation 4.25, where $n(\omega_p)$ is the phonon occupation number $1/[\exp(\hbar\omega_p/k_B T) - 1]$ in thermal equilibrium, ω_p is the phonon frequency, $g(\omega_s)$ is the normalized lineshape whose linewidth is ~ 1/phonon lifetime, and $\mathcal{N}$ is the number of scatterers in the volume. The expression for anti-Stokes scattering is similar, with the corresponding anti-Stokes values used for ω_s, $n(\omega_p)$, and χ in Equation 4.25 and the bracket in the expression for $\langle uu^* \rangle$ replaced by $n(\omega_p)$.

Quantum mechanical time-dependent perturbation theory provides the framework for the microscopic description of vibrational Raman scattering in solids (Hayes and Loudon, 1978; Cardona, 1982). Photon interactions are mediated by the $-(e/mc)\mathbf{p} \cdot \mathbf{A}$ term in the Hamiltonian, where $\mathbf{p}$ is the electron momentum operator and $\mathbf{A}$ is the vector potential. Quantum mechanically, first-order vibrational Raman scattering in solids is a three-step process. First the incoming photon ω_i creates a virtual electron–hole pair through an intermediate level. Then the electron or hole interacts with the lattice, leading to the creation or annihilation of a phonon. Finally, the electron and hole recombine to form the scattered photon ω_s, leaving the material in the initial electronic state. The most important term from third-order perturbation theory gives

$$\mathbf{e}_s \cdot d\chi/dQ \cdot \mathbf{e}_i \propto \frac{\langle v|\mathbf{p} \cdot \mathbf{e}_s|c_2\rangle \langle c_2|\mathcal{H}_{e\text{-}p}|c_1\rangle \langle c_1|\mathbf{p} \cdot \mathbf{e}_i|v\rangle}{(\omega_i \pm \omega_p - \omega_0)(\omega_i - \omega_0)} \quad (4.28)$$

where v is the initial valence band, c_1 and c_2 are the intermediate conduction band states, $\mathcal{H}_{e\text{-}p}$ is the electron–phonon interaction Hamiltonian, $\hbar\omega_0$ is the critical-point energy (perhaps the band gap), and $+/-$ correspond to anti-Stokes/Stokes transitions. The $\mathbf{p} \cdot \mathbf{A}$ term induces the first and third steps, which involve interband transitions.

One effect that contributes to the electron–phonon coupling for both nonpolar and polar vibrations is the change in the electronic potential energy with the lateral displacement of atoms in a unit cell. This is characterized by the deformation potential, and is short-range in nature. Lattice vibrations modulate the electronic energy structure, which modulates the electric susceptibility to give terms like $d\chi/dQ$ due to interband transitions. A second effect contributes only for polar vibrations; this effect is caused by the modulation of χ by the quasistatic, macroscopic electric field, and is longer-range in nature. This interaction is described by the Frohlich interaction, and has two components: a $\mathbf{q}$-independent three-band interband term, which is allowed and can be characterized by the electro-optic tensor, and a **q**-dependent two-band intraband term, which is dipole-forbidden away from resonance.

The interband terms in Equation 4.28 have ingoing and outgoing resonances when either ω_i or the scattered photon frequency $\omega_s = \omega_i \pm \omega_p$ is near a critical-point energy. While these peaks can be resolved in some cases, there is often one strong resonant Raman peak near $\omega_i \pm \omega_p/2$.

4.3 Spectroscopy

The basics of absorption, emission, and spontaneous Raman scattering spectroscopies are presented in this section. Details about these and other optical spectroscopies are given in Chapters 6–17. Demtröder (1982), Steinfeld (1985), and Svanberg (1991) are excellent general references on laser spectroscopies.

4.3.1 *Absorption and Emission Spectroscopies*

4.3.1.1 Atoms and Molecules

The transition matrix elements in Equations 4.15–4.17 determine the selection rules for molecular transitions, as well as the strengths of these transitions. This is important for transmission, optical emission, and laser-induced fluorescence probes. For simplicity, this discussion will focus on diatomic molecules having no electronic angular momentum along the axis. Detailed accounts of molecular spectroscopy can be found in Herzberg (1945, 1950, 1966), Levine (1975), Huber and Herzberg (1979), Steinfeld (1985), and Bernath (1995). Atomic spectroscopy is described in White (1934), Herzberg (1937), Condon and Shortley (1959), Kuhn (1962), Corney (1977), and Bernath (1995). Additional sources of spectroscopic data are cited in Chapters 6, 7, and 8.

For electric-dipole-allowed transitions, which are the strongest transitions, the matrix element of the dipole moment operator (Section 4.2) between the initial and final states must be nonzero for absorption or emission to occur. Transitions that are "electric-dipole forbidden" may occur, though quite weakly, by electric quadrupole or magnetic dipole interactions. Most transitions discussed in this volume are electric-dipole allowed. One notable exception is the $^2P_{1/2} \leftarrow {}^2P_{3/2}$ transition in Cl and F atoms, which is magnetic-dipole allowed.

The quantum numbers of the state with the lower energy are labeled by a double prime (″) and those of the higher energy state are labeled by a single prime (′), whether the transition involves absorption or emission.

4.3.1.1.1 *Transition Energies*

Vibrational–rotational transitions are electric-dipole allowed if the vibration induces a change in the electric dipole moment, which means that the

4.3 Spectroscopy

matrix element of the dipole moment between the two vibrational states is nonzero. In the strongest transitions, the v_i quantum number changes for only a single mode (which is the only mode for diatomics), and it can change only by $\Delta v_i = \pm 1$ in the harmonic approximation. For electric-dipole transitions the concomitant change in rotational quantum number is $\Delta J = +1$ (the R branch) and $\Delta J = -1$ (the P branch), where $\Delta J = J' - J''$, and $\Delta M = 0, \pm 1$, where M is the z component of J. $\Delta J = 0$ (the Q branch) can occur in diatomic molecules only when the electronic angular momentum Σ is not zero. Higher-order transitions can occur due to anharmonicity, such as overtones where a single v_i changes, with $\Delta v_i = \pm 2, 3, \ldots$, or combination bands, where more than a single v_i changes.

Using Equation 3.2, the transition energies for P-, Q-, and R-branch transitions in linear molecules are

$$\nu_P(J) = \nu_0 - (B_{v'} + B_{v''})J + (B_{v'} - B_{v''})J^2 \qquad (4.29)$$

$$\nu_Q(J) = \nu_0 + (B_{v'} - B_{v''})J + (B_{v'} - B_{v''})J^2 \qquad (4.30)$$

$$\nu_R(J) = \nu_0 + 2B_{v'} + (3B_{v'} - B_{v''})J + (B_{v'} - B_{v''})J^2 \qquad (4.31)$$

where J is the rotational quantum number of the lower state (which is equal to J'') and ν_0 is the vibrational energy change between vibrational levels v' and v'' ($= v' - 1$ for fundamental transitions) which can be obtained from Equations 3.5 and 3.6. Generally, $B_{v'}$ and $B_{v''}$, which correspond to different v, are unequal because of rotational–vibrational interactions (Equation 3.3). Although the higher-order terms in J in Equation 3.2 have not been included in these expressions, they can be quite important. Equation 3.4 can be used to obtain similar expressions for symmetric tops, using the appropriate selection rules.

In the rigid rotor approximation (with $B_{v'} = B_{v''} = B$), the spectrum consists of a picket fence of R- and P-branch lines equally spaced by $2B$, respectively increasing and decreasing with increasing J, with a gap at ν_0 if there is no Q-branch. When higher-order terms are included, the pattern changes. If $B_{v'} > B_{v''}$, for high enough J, $\nu_P(J)$ starts to increase with J, while for $B_{v'} < B_{v''}$, $\nu_R(J)$ starts to decrease with J.

In pure rotational transitions J (and K) change, but all v_i remain unchanged. They are electric-dipole allowed if the molecule has a permanent electric dipole moment. Pure rotational transitions are forbidden in homonuclear diatomic molecules, such as N_2, and other molecules with reflection symmetry, such as CO_2. For pure rotational transitions, $\Delta J = 1$, and the transition energies are given by Equation 4.31 with $\nu_0 = 0$ and $B_{v'} = B_{v''}$.

Electronic transitions are electric-dipole allowed if the transition electric dipole moment matrix element between the initial and final electronic states is nonzero. The concomitant selection rules can be determined from the symmetry of the states. For atoms with L–S coupling, $\Delta L = 0, \pm 1$ ($\Delta l = \pm 1$ for individual electrons), $\Delta S = 0$, and $\Delta J = 0, \pm 1$ (but not $0 \leftrightarrow 0$). [These transitions couple odd and even terms in atoms with two valence electrons, since $\Delta l = \pm 1$ for individual electrons and usually only one electron makes a transition (while $\Delta l = 0$ for the others).] For j–j coupling, $\Delta J = 0, \pm 1$, $\Delta j_i = 0, \pm 1$.

For diatomic molecules described by Hund's cases (a) and (b), $\Delta \Lambda = 0$ (parallel transitions) or ± 1 (perpendicular transitions) is allowed. For Σ–Σ transitions, $+ \leftrightarrow +$ and $- \leftrightarrow -$ are allowed and $+ \leftrightarrow -$ is forbidden, and in homonuclear diatomic molecules $u \leftrightarrow g$ is allowed, while $u \leftrightarrow u$ and $g \leftrightarrow g$ are forbidden. Changes in spin are forbidden, so $\Delta \Sigma = 0$, but they are observed as weak transitions in molecules with higher Z. Also, $\Delta \Omega = 0$, ± 1 for Hund's cases (a) and (c), and $\Delta N = 0, \pm 1$ for Hund's case (b).

In molecular electronic transitions, changes in rotational quantum numbers are $\Delta J = \pm 1$ for diatomic molecules (Σ states). Also $\Delta J = 0$ is allowed if Λ is not zero in either of the two states, i.e., if at least one of the states is not a Σ state. However, there is no strict selection rule for the change in vibrational states. As discussed below, the transition probability is governed by the Franck–Condon factor (Equation 4.35). The transition energy is given by Equations 4.29–4.31, where ν_0 is the band origin which depends on the difference of electronic and vibrational energies. Figure 6.2 shows the vibrational progression in optical emission from N_2. The rotational structure is often characterized by overlapping structure with a sharp onset on one side, which is called the bandhead, and shading on the other side. The bandhead is often given for spectral identification. From Equations 4.29–4.31, it is seen that bands shade to the red (bandhead to the blue) when $B_{v'} < B_{v''}$ (excited electronic state has larger internuclear distance than the lower state) and to the blue when $B_{v'} > B_{v''}$ (excited electronic state has smaller internuclear distance than the lower state). When $B_{v'} \approx B_{v''}$ there is no clear shading or bandhead. Blue shading in a N_2 emission band can be seen in Figure 6.32.

Many examples of emission and absorption spectra can be found in later chapters. For example, Figures 6.6 (O, F), and 6.21 and 7.6 (Si) illustrate atomic emission lines. Figure 8.18 shows the rotational structure in the vibrational-rotational absorption spectrum of arsine. Figures 6.2 (N_2), 7.5 (Si_2), 7.9 (As_2), 7.12 (SiN), 7.27 (AlO), and 17.9 (N_2^+) illustrate rotational and vibrational structure in electronic transitions; these transitions can be called rovibronic and vibronic bands. Figure 8.23 shows the fairly unstructured electronic absorption spectrum in several polyatomic molecules.

4.3.1.1.2 *Transition Strengths*

The absorption coefficient of a molecule is given by Equation 4.16, using the appropriate number of molecules in the initial state (Equations 3.7–3.12) and the appropriate transition matrix elements. Often, these transition elements can be factored into terms describing the changes in elec-

4.3 Spectroscopy

tronic, vibrational, and rotational states. The rate of spontaneous emission (Equation 4.17) also depends on these same matrix elements.

For a vibrational–rotational transition, the matrix element term can be factored as:

$$\mu^2(v',J'; v'',J'') = \mu_{vib}^2(v', v'') S(J',J'') \tag{4.32}$$

For a harmonic oscillator, $\mu_{vib}^2(v',v'') = (v'' + 1) \mu_{vib}^2(1, 0)$, describes absorption from v'' to $v' = v'' + 1$. $S(J',J'')$, often called the Hönl–London factor or the line strength, equals $\Sigma |\mu_{rot}(J',J'')|^2$, where μ_{rot} is that part of the dipole moment matrix element that has been integrated over the angle orientations of the nuclei. The sum is over M' (i.e., the z-components of the upper state J') for emission and over M'' for absorption. The Hönl–London factor depends on J', J'', Λ', and Λ'', and is sometimes written as $S(J)$, with $J = J''$. At times $S(J)$ is replaced by $(2J + 1)h(J)$, where the $2J + 1$ term is the rotational state degeneracy.

The dependence of J in the absorption coefficient α, aside from the rotational-energy-dependent term in the exponent, is $S(J)$. As is seen below, this factor is important for all optical processes in molecules and must be included in any quantitative analysis, such as absolute determinations of density and in thermometry. The sum rules for $S(J)$ demonstrate that strengths of vibrational transitions are independent of molecular rotation, when integrated over all rotational structure. For transitions between Σ states in diatomic molecules ($\Lambda' = \Lambda'' = 0$), $S(J'') = J'' + 1$ for the R branch and J'' for the P branch. The Hönl–London factors are more complex when $\Lambda \neq 0$ and for nonlinear molecules. Kovács (1969) lists these factors for general transitions in diatomic molecules. Using the expression for the population of a rotational state (Equation 3.12), the absorption coefficient for a vibrational-rotational transition at temperature T is

$$\alpha(\nu) = G\mu_{vib}^2(v'') [N_{ev} S(J'')\exp(-\mathcal{E}_{J'}/k_BT)/Z_{rot}]g(\nu) \tag{4.33}$$

where $G = 8\pi^3\nu/3hc$. (See the appendix to this chapter for the rationalized MKSA expression.)

For electronic transitions, the matrix element term is

$$\mu^2(e',v',J'; e'',v'',J'') = \mu_e^2 f(v',v'')S(J',J'') \tag{4.34}$$

where μ_e is the electronic matrix element. $f(v',v'')$ is the Franck–Condon factor, which is the square of the overlap integral of the vibrational wavefunctions in the upper (e') and lower states (e''):

$$f(v',v'') = \left|\int \phi_{e'v''}^*(R)\phi_{e'v'}(R)R^2 dR\right|^2 \tag{4.35}$$

Consequently, there is no strict selection rule governing which electronic–vibrational transitions are allowed. The expression for the absorption coefficient is therefore the same as in Equation 4.33 with $\mu_{vib}^2(v',v'')$ replaced by $\mu_e^2 f(v',v'')$:

$$\alpha(\nu) = GN_{ev}\, \mu_e^2\, f(v',v'')S(J'')[\exp(-\mathcal{E}_{J'}/k_BT)/Z_{\text{rot}}]\, g(\nu) \qquad (4.36)$$

Equations 4.17 and 4.34 can be used to determine the power emitted per unit volume p_{OES} due to spontaneous emission from an excited electronic state $e'(v',J')$ (in rotational thermal equilibrium) to a lower state $e''(v'',J'')$ integrated over all solid angle (OES, Chapter 6):

$$p_{\text{OES}} = G'N_{ev}\, \mu_e^2\, f(v',v'')\, S(J'')[\exp(-\mathcal{E}_{J'}/k_BT)/Z_{\text{rot}}]g(\nu) \qquad (4.37)$$

where G' is $64\pi^4\nu^4/3c^3$.

As in absorption and emission, rotational state changes in LIF, spontaneous pure rotational and vibrational–rotational Raman scattering, and coherent anti-Stokes Raman scattering (CARS) affect the transition rates. These latter processes involve two sequential changes in (rotational) states, while the former processes, absorption and emission, involve only one. Using the notation in Figure 7.1, the overall rotational factor for LIF is thus $(2J'' + 1)h(J'')\, h(J''') = S(J'')\, S(J''')/(2J''' + 1)$, which involves a product of two Hönl–London factors. (The rotational state factor for Raman scattering is discussed in Section 4.3.2.1.1.)

The steady-state flux from LIF (Section 7.1.1) is obtained by combining the expressions for absorption (Equation 4.36) and emission (Equation 4.37), which, with using the notation in Figure 7.1, is

$$p_{\text{LIF}} = G''\, \mu_{e21}^2 f_{21}(v',v'')\, \mu_{e23}^2 f_{23}(v',v''')\, \frac{N_{ev}\, S(J'')\, S(J''')}{(2J''' + 1)} \\ \times [\exp(-\mathcal{E}_{J'}/k_BT)/Z_{\text{rot}}]g(\nu_{21})\, g(\nu_{23}) \qquad (4.38)$$

where $G'' = GG'(I_{\text{inc}}/\gamma_t h\nu_{23})$. G is evaluated at the incident frequency ν_{23} and G' at the emitted frequency, I_{inc} is the incident intensity, and γ_t is the rate of relaxing the intermediate level 2 (see Equation 7.1 and the text thereafter). This expression is valid for LIF when the excitation is monochromatic and when the emission is spectrally resolved. If either condition is relaxed, the observed LIF signal is described by this equation, appropriately summed over J' and/or J''.

Consequently, the signal S corresponding to an absorption, emission, LIF, spontaneous Raman scattering, or CARS process has a rotational dependence that can be expressed as $S(J) = G'''f(J)\exp(-\mathcal{E}(J)/k_BT)$, where G''' is a factor independent of J. (For absorption and emission $f(J)$ is the Hönl–London factor.) If the system is in thermal equilibrium, the temperature can be obtained by plotting this equation as

$$\ln[S(J)/f(J)] = -\mathcal{E}(J)/k_BT + \text{constant} \qquad (4.39)$$

This is a straight line with slope $-1/k_BT$, from which T can be determined. Note that the temperature dependence of G''', $\propto 1/T$ from $1/Z_{\text{rot}}$, does not affect this procedure. $f(J)$ includes all J-dependent matrix element factors (including corrections to the rigid rotor model), J-dependent population degeneracy factors, and factors due to nuclear spin statistics. For spontaneous Raman scattering it also includes a factor of $v_{\text{scattered}}^4$, which is also a function of J through the change in energy. In limited thermal equilibrium this temperature is called the rotational temperature T_{rot}, which often is close to the translational temperature T_{trans}, as discussed in Section 18.1. The "vibrational" temperature T_{vib} and the "electronic" temperature T_{elect} can be determined in a similar manner.

4.3.1.2 Solids

For electronic band-to-band transitions in solids, the sums over the densities of states in Equation 4.18 lead to nonlorentzian lineshapes for optical parameters (Pankove, 1971; Aspnes, 1980; Cardona, 1982; Lynch, 1985). The features in the dielectric function $\tilde{\varepsilon}(\omega)$ affect refraction and absorption, and are due to interband transitions that are often analyzed in terms of the lineshape for that particular critical point (CP). In the parabolic band approximation

$$\tilde{\varepsilon}(\omega) = C - Ae^{i\phi_{\text{ex}}}(\hbar\omega - \mathcal{E}_i + i\Gamma_L^{(E)})^{n'} \quad (4.40)$$

where constant C, amplitude A, threshold energy $\mathcal{E}_i$, broadening $\Gamma_L^{(E)}$ (in energy units), and excitonic phase angle ϕ_{ex} are for that particular critical point. C is the nonresonant contribution from other CPs. The exponent n' is $-\frac{1}{2}$ for a one-dimensional (1D) CP, 0 for a 2D CP [giving $\ln(\hbar\omega - \mathcal{E}_i + i\Gamma_L^{(E)})$], and $\frac{1}{2}$ for a 3D CP; for discrete excitons $n' = -1$. Optical properties of semiconductors have been reviewed by Pankove (1971), Balkanski (1980), Cohen and Chelikowsky (1988), and Böer (1990). Available data on the dielectric functions of semiconductors are described in Section 9.3 and Table 9.1. The dependence of the critical-point energy on temperature is discussed in Section 18.4.2.

For allowed direct transitions in semiconductors, as in GaAs and ZnSe, the absorption coefficient is

$$\alpha_{\text{direct}}(\hbar\omega) = A^*(\hbar\omega - \mathcal{E}_{\text{bg}})^{1/2} \quad (\hbar\omega \geq \mathcal{E}_{\text{bg}}) \quad (4.41a)$$

with

$$A^* = e^2 \left(2 \frac{m_h m_e}{m_h + m_e}\right)^{3/2} /nch^2 m_e \quad (4.41b)$$

where m_e and m_h are the effective masses of electrons and holes, respectively, e is the charge, and n is the index of refraction.

For indirect transitions between indirect valleys, as in Si and Ge, the absorption coefficient is

$$\alpha_{\text{indirect}}(\hbar\omega) = \alpha_a(\hbar\omega) + \alpha_e(\hbar\omega) \quad (4.42)$$

The first term involves the absorption of a phonon (of energy $\mathcal{E}_p$)

$$\alpha_a(\hbar\omega) = \frac{\alpha(\mathcal{E}_a)}{\exp(\mathcal{E}_p/k_B T) - 1} \qquad (\mathcal{E}_a \geq 0) \qquad (4.43)$$

where $\mathcal{E}_a = \hbar\omega - \mathcal{E}_{bg} + \mathcal{E}_p$, and the second term involves the emission of a phonon

$$\alpha_e(\hbar\omega) = \frac{\alpha(\mathcal{E}_e)}{1 - \exp(-\mathcal{E}_p/k_B T)} \qquad (\mathcal{E}_e \geq 0) \qquad (4.44)$$

where $\mathcal{E}_e = \hbar\omega - \mathcal{E}_{bg} - \mathcal{E}_p$. Only the former term is important for below-gap absorption, whose use in probing temperature is described in Section 8.4.

Theoretical calculations give $\alpha(\mathcal{E}) = A\mathcal{E}^2$ for $\mathcal{E} > 0$ in these two equations (Pankove, 1971). For Si, a better fit is given by $\alpha(\mathcal{E}) = A(\mathcal{E} - \mathcal{E}_x)^2 + B\mathcal{E}^{1/2}$ for $\mathcal{E} > \mathcal{E}_x$, $= B\mathcal{E}^{1/2}$ for $\mathcal{E}_x > \mathcal{E} > 0$, and $= 0$ for $\mathcal{E} < 0$ (Macfarlane et al., 1958; Jellison and Lowndes, 1982). More realistically, the right-hand side of Equation 4.42 must be summed for each type of phonon with energy $\mathcal{E}_p$, which assists absorption [each with its own absorption parameters A, B and $\mathcal{E}_x$ (and $\mathcal{E}_p$)]. Two phonons are important in silicon absorption: transverse optical and acoustic phonons (Jellison and Lowndes, 1982).

The absorption coefficient α due to absorption by free (intrinsic and extrinsic) carriers can be obtained by using Equations 4.11b and 4.18:

$$\alpha_{\text{free carrier}} = \frac{Ne^2\lambda^2}{8\pi^2 m_e nc^3 \tau} \qquad (4.45)$$

where N is the density of free carriers with effective mass m_e and τ is the relaxation time. Since this scattering time depends on λ, $\alpha_{\text{free carrier}}$ depends on λ as λ^p, where p varies from ~2–3.5 for free electrons in different semiconductors.

The linear optical properties of surfaces have been detailed by Feibelman (1982) and Olmstead (1987).

4.3.2 Spontaneous Raman Scattering

4.3.2.1 Molecules

Pure rotational Raman scattering occurs if the static polarizability of the molecule, in the reference frame fixed to the molecule, is anisotropic. For example, it is allowed in homonuclear diatomic molecules, but not in methane (CH_4). Vibrational–rotational Raman scattering occurs when the polarizability has a component that can oscillate at the mode frequency; this component is called the dynamic polarizability.

The polarizability (and susceptibility) described in Section 4.2.2 was referenced to a fixed species axis. For randomly oriented species, such as

4.3 Spectroscopy

rotating molecules in a gas, the grains in polycrystalline material, and molecules in most liquids, the polarization dependence must be averaged over angle when the overall medium is analyzed in reference to fixed laboratory axes. This can be done in terms of two parameters—the mean polarizability a and its anisotropy γ (Long, 1977; Eckbreth, 1996):

$$a = \frac{1}{3}(\hat{\alpha}_{xx} + \hat{\alpha}_{yy} + \hat{\alpha}_{zz}) \qquad (4.46)$$

$$\gamma^2 = \frac{1}{2}[(\hat{\alpha}_{xx} - \hat{\alpha}_{yy})^2 + (\hat{\alpha}_{yy} - \hat{\alpha}_{zz})^2 + (\hat{\alpha}_{zz} - \hat{\alpha}_{xx})^2 + 6(\hat{\alpha}_{xy}^2 + \hat{\alpha}_{yz}^2 + \hat{\alpha}_{zx}^2)] \qquad (4.47)$$

These orientation-averaged polarizabilities are then

$$\overline{(\hat{\alpha}_{xx})^2} = \overline{(\hat{\alpha}_{yy})^2} = \overline{(\hat{\alpha}_{zz})^2} = \frac{1}{45}(45a^2 + 4\gamma^2) \qquad (4.48a)$$

$$\overline{(\hat{\alpha}_{xy})^2} = \overline{(\hat{\alpha}_{yz})^2} = \overline{(\hat{\alpha}_{zx})^2} = \frac{1}{15}\gamma^2 \qquad (4.48b)$$

Therefore cross-sections are proportional to $(45a^2 + 4\gamma^2)/45$ for polarized scattering, such as the scattering of x-polarized light to x-polarized light, and $\gamma^2/15$ for depolarized scattering, such as from x-polarized light to y- or z-polarized light. (See Chapter 12.) Consult Hirschfeld (1973) and Long (1977) for these factors for general scattering geometries.

In scattering from molecules, a_0 and γ_0 from the equilibrium (static) polarizability contribute to Rayleigh (elastic) scattering and pure rotational Raman scattering, respectively. a' and γ' from the derived or dynamic polarizability contribute to vibrational Raman scattering.

Using the original quantum mechanical expression for the polarizability, Equation 4.26, and Equation 4.23 (together with what follows), the differential cross-section can be expressed as

$$\left.\frac{d\sigma}{d\Omega}\right|_{Raman} = \frac{\omega_i \omega_s^3}{c^4} \left|\sum_b \left(\frac{\mu_{cb}\mu_{ba}}{\hbar(\omega_{ab} - \omega_i - i\Gamma_b^{(\omega)})} + \frac{\mu_{cb}\mu_{ba}}{\hbar(\omega_{bc} + \omega_i - i\Gamma_b^{(\omega)})}\right)\right|^2 \qquad (4.49)$$

4.3.2.1.1 Pure Rotational Scattering

Rotational Raman scattering from diatomic molecules such as H_2 and N_2, which are common dominant species in thin film processing, is presented as an example. The ground electronic state is assumed to be $^1\Sigma$, so the orbital angular momentum about the axis is zero, which simplifies the discussion. Long (1977) details the more general case.

In pure rotational scattering of such diatomics, the selection rule is $\Delta J = +2$ ($\Delta v = 0$). This gives rise to two S-branches with anti-Stokes lines at

$$\nu_{\text{anti-Stokes}} = \nu_i + (4J + 6)B \qquad [J \geq 0] \qquad (4.50)$$

which are upshifted in frequency by $6B, 10B, 14B, \ldots$, relative to the laser line at ν_i, and Stokes lines at

$$\nu_{\text{Stokes}} = \nu_i - (4J-2)B \quad [J \geq 2] \quad (4.51)$$

which are downshifted by $6B, 10B, 14B, \ldots$. (This is seen in Figure 12.6 for H_2 and Figure 12.7 for N_2.) These lines are equally spaced by $4B$, except for large J, where centrifugal deviations from the rigid rotor model (Equation 3.2) alter this spacing.

For pure rotational Raman scattering from a diatomic molecule, the differential cross-section can be expressed as (Schrötter and Klöckner, 1979)

$$\left.\frac{d\sigma}{d\Omega}\right|_{\text{Raman,rot}} = \frac{16\pi^4}{c^4} \nu_s^4 \, b_{J',J''} \, F_{00}(J) \frac{4\gamma_0^2}{45} \quad (4.52)$$

for polarized scattering (z polarization, incident along the x axis and detected along the y axis), where ν_s is the scattered frequency and $b_{J',J''}$ are the Placzek–Teller coefficients for a transition between rotational states J'' and J'. Sometimes, they are expressed as $b_{J',J''} = L(J)/(2J+1)$, where $L(J)$ is the line strength in the rigid rotor approximation. $F_{00}(J)$ accounts for corrections for vibrational–rotational coupling for $v=0 \leftrightarrow v=0$ transitions. For a diatomic molecule with its axis aligned along the z direction, $\gamma_0 = (\hat{\alpha}_{zz} - \hat{\alpha}_{xx})_0$, where the subscript zero denotes static polarizability. Tabulations of rotational cross-sections may be found in Eckbreth (1996). At 488 nm, $d\sigma/d\Omega|_{\text{Raman,rot}}$ is 5.4 and 1.0×10^{-30} cm^2/sr for N_2 ($J'' = 6 \to J' = 8$) and H_2 ($1 \to 3$), respectively.

For a diatomic molecule in a $^1\Sigma$ (ground) electronic state, the Placzek–Teller coefficients are

$$b_{J+2,J} = \frac{3(J+1)(J+2)}{2(2J+1)(2J+3)} \quad (4.53a)$$

$$b_{J-2,J} = \frac{3J(J-1)}{2(2J+1)(2J-1)} \quad (4.53b)$$

$$b_{J,J} = \frac{J(J+1)}{(2J-1)(2J+3)} \quad (4.53c)$$

The last coefficient describes the Q-branch, which can be important in vibrational–rotational Raman scattering in diatomics.

The scattered intensity is $\propto N_{evJ} \, (d\sigma/d\Omega)_{\text{Raman,rot}}$ (Equation 12.1), where N_{evJ} includes a factor of $2J+1$ (Equation 3.12). Alternating line intensities are sometimes seen because of nuclear spin statistics (Section 3.2.4). As detailed in Section 4.3.1.1.2, the rotational Raman spectrum can be used to obtain the rotational temperature.

The identification of species can be difficult in rotational Raman scattering of a multicomponent mixture because the Raman shifts are small and

4.3 Spectroscopy

overlapping. It can be best used when there is one dominant species. Its sole advantage over vibrational scattering is that the cross-section for rotational Raman scattering for a single transition is usually an order of magnitude larger than that for an entire vibrational branch (Eckbreth, 1996).

The special case of $\Delta v = 0$, $\Delta J = 0$ is actually Rayleigh scattering. The scattering cross-section for Rayleigh scattering from a molecule with polarizability $\hat{\alpha}_0$ is (Drake, 1988)

$$\left.\frac{d\sigma}{d\Omega}\right|_{\text{Rayleigh}} = \frac{16\pi^4}{c^4} v_s^4 a_0^2 \qquad (4.54)$$

for polarized scattering as in Equation 4.52.

4.3.2.1.2 Vibrational–Rotational Scattering

Symmetry arguments show that vibrational Raman scattering of a given mode is allowed if an $\hat{\alpha}_{ij}$ exists whose symmetry species is the same as that of the mode; that is, it transforms the same way under symmetry operations (Long, 1977), which also means that the second term in Equation 4.27 is nonzero. Since the dipole moment μ is a vector, with components along x, y, and z, and $\overleftrightarrow{\alpha}$ is a second-rank tensor, with components that transform as x^2, y^2, z^2, xy, yz, and zx, in molecules with a center of inversion a given mode can be only infrared-active, Raman-active, or both infrared- and Raman-inactive. Modes in molecules without a center of symmetry may be both infrared- and Raman-active, although one of the two is often decidedly stronger.

In vibrational Raman scattering from a given mode with quantum number v, the Stokes spectrum corresponds to $\Delta v = 1$, while the anti-Stokes spectrum corresponds to $\Delta v = -1$. There is further structure in the spectrum due to changes in the rotational quantum numbers, giving rise to a vibrational–rotational Raman spectrum that is qualitatively similar to that in pure rotational scattering (Long, 1977). In linear molecules, such as N_2 and CO_2, where the rotational quantum number is J, the selection rules for vibrational–rotational scattering (and pure rotational scattering) from totally symmetric vibrational species are $\Delta J = 0, \pm 2$; for degenerate vibrational species, $\Delta J = \pm 1$ is also allowed. The $\Delta J = -2, -1, 0, 1,$ and 2 branches are also called the O, P, Q, R, and S branches, respectively. Long (1977) gives the selection rules for pure rotational scattering and vibrational–rotational scattering from symmetric-top and spherical-top molecules.

In vibrational–rotational Stokes scattering, the $\Delta J = \pm 2$ selection rule gives rise to a series of O-branch and S-branch lines shifted down in frequency from the laser line v_i, and at

$$v_O = v_i - [v_0 + 2B_{v'} - (3B_{v'} + B_{v''})J + (B_{v'} - B_{v''})J^2] \quad \text{for } J \geq 2 \quad (4.55)$$

$$v_S = v_i - [v_0 + 6B_{v'} + (5B_{v'} - B_{v''})J + (B_{v'} - B_{v''})J^2] \quad \text{for } J \geq 0 \quad (4.56)$$

where v_0 is the separation of the v' and v'' vibrational levels and $v' = v'' + 1$ for fundamental transitions. These bands look qualitatively like those in pure rotational Raman scattering except that since usually $B_{v'} < B_{v''}$, the spacing of the O-branch lines increases with J, while the spacing of S-branch lines decreases with J. For large enough J, the shifts of S-branch lines no longer increase with J, but reverse sign to form a bandhead. The $\Delta J = 0$ selection rule gives Q-branch lines at

$$v_Q = v_i - [v_0 + (B_{v'} - B_{v''})J + (B_{v'} - B_{v''})J^2] \tag{4.57}$$

which are closely spaced, sometimes overlapping to form a big peak, and are near but usually less than v_0.

The polarizability for a Raman-allowed vibrational transition can be expressed in terms of Placzek polarizability theory as (Eckbreth, 1996)

$$\hat{\alpha}_{v''v'} = \left(\frac{\partial \hat{\alpha}}{\partial Q}\right)_0 (v + 1)^{1/2} \left(\frac{\hbar}{2m\omega_e}\right)^{1/2} \tag{4.58}$$

for the Stokes $\Delta v = +1$ [v ($=v''$) to $v + 1$ ($=v'$)] transition, where ω_e is the harmonic frequency expressed in rad/sec. For the anti-Stokes $\Delta v = -1$ transition, $v + 1$ is replaced by v.

For polarized scattering, the cross-section for vibrational scattering from v to $v + 1$, integrated over all rotational transitions, is $\propto \hat{\alpha}^2 \propto (v + 1)(a'^2 + \frac{4}{45}\gamma'^2)$, where a' and γ' are the parameters for the derived polarizability. Because of anharmonicity, each of these transitions occurs at a slightly different frequency (Equation 3.5). Stokes transitions that scatter from $v = 0$ to 1 dominate, but at higher temperatures Stokes scattering from higher v can occur, which are weaker by $(v + 1)\exp(-vhc\omega_v/k_B T)$. Summed over all v, the differential Raman cross-section for vibrational scattering is

$$\left.\frac{d\sigma}{d\Omega}\right|_{\text{Raman,vib};zz} = \frac{h(v_i - v_0)^4 \left(a'^2 + \frac{4}{45}\gamma'^2\right)}{8mc^4 v_0[1 - \exp(-hc\omega_v/k_B T)]} \tag{4.59}$$

where v_0 is the vibrational frequency (in Hz) and ω_v is the fundamental frequency (in cm^{-1}, Equation 3.5a). Tabulations of vibrational cross-sections may be found in Schrötter and Klöckner (1979), Eckbreth (1996), and Yariv (1989). At 488 nm, $d\sigma/d\Omega|_{\text{Raman,vib}}$ is 0.68 and 1.32×10^{-30} cm^2/sr for N_2 ($v_0 = 2330.7$ cm^{-1}) and H_2 (4160.2 cm^{-1}), respectively. These cross-sections increase rapidly with laser frequency.

For polarized scattering, the cross-sections for the individual vibrational–rotational lines in a diatomic molecule for $\Delta v = +1$ are

$$\left.\frac{d\sigma}{d\Omega}\right|_{\text{Raman,vib};Q} = \frac{16\pi^4}{c^4} v_s^4 (v + 1) \left(a'^2 + \frac{4}{45} b_{J,J} \gamma'^2\right) \quad (\Delta J = 0) \tag{4.60}$$

$$\left.\frac{d\sigma}{d\Omega}\right|_{\text{Raman,vib};S,O} = \frac{16\pi^4}{c^4} v_s^4 (v + 1) \frac{4}{45} b_{J\pm 2,J} \gamma'^2 \quad (\Delta J = \pm 2) \tag{4.61}$$

so the Q branch is stronger than the O and S branch transitions, because it includes effects of the mean polarizability in addition to the anisotropic term. The Placzek–Teller coefficients ($b_{J',J''}$) for rigid rotor diatomics are given in Equation 4.53. Vibrational–rotational coupling corrections ($F_{01}(J)$) can also be included, as in Equation 4.52 (Drake, 1988).

The relative strength of the rotational lines depends on the product of the thermal population of the rotational state and the rotational matrix element. The scattered intensity is $\propto N_{evJ}(d\sigma/d\Omega)_{\text{Raman,vib}}$, where N_{evJ} is given by Equation 3.12 and both terms depend on J (Long, 1977; Weber, 1979). As with pure rotational scattering, alternating line intensities are sometimes seen because of nuclear spin statistics (Section 3.2.4), and the spectrum can be used to obtain rotational temperatures (Equation 4.39).

4.3.2.2 Solids

Symmetry determines which optical phonon modes are Raman-active (i.e., have a nonzero cross-section for first-order scattering), infrared-active (i.e., have a nonzero infrared absorption cross-section), Raman- and IR-active, or both Raman- and IR-inactive (silent). Raman transitions from state 1, with irreducible representation Γ_1 of the point group of the crystal or molecule, to state 2 with representation Γ_2, are symmetry-allowed if $\Gamma_2^* \times \Gamma_\chi \times \Gamma_1$ contains the totally symmetric representation. Γ_χ is the representation generated by the components of the second-rank susceptibility tensor $\overleftrightarrow{\chi}$. In centrosymmetric crystals, modes can be only Raman-active, IR-active, or silent. This is similar to vibrational Raman scattering in molecules. Polarization selection rules are given in Tables 4.1–4.3 for backscattering configuration from diamond and zincblende crystals.

Table 4.4 gives representative Raman frequencies (Mitra and Massa, 1982). In diamond structure materials, there is a single optic mode which is threefold degenerate at zone center (LO, 2TO); it is Raman-active. In backscattering, longitudinal optical (LO) scattering is allowed from (100) and (111) surfaces, while transverse optical (TO) scattering is allowed from (110) and (111) surfaces (Tables 4.1–4.3). In diamond, Si and Ge the LO/TO mode produces first-order Raman peaks at 1332, 521, and 301 cm^{-1}, respectively. In zincblende materials, such as GaAs and ZnSe, the optical phonon has the same polarization selection rules as those for diamond. In these polar materials, the TO and LO modes split due to electrostatic interactions (Table 4.4). TO phonon scattering is induced by the deformation potential, while LO scattering is usually stronger because it is induced by deformation potential and electrostatic effects. Many metals have only one atom per primitive cell, and therefore have no optic modes that could be observed by Raman scattering. Even when allowed, Raman spectra of metals is weak because of the large absorption coefficient.

Table 4.1 Raman Polarization Selection Rules for Backscattering from a (001) Surface for Given Atomic Displacements[a]

Polarization configuration[b]	Phonon modes		
	LO(z_1)	TO(x_1)	TO(y_1)
$z_1(x_1,x_1)\bar{z}_1$	0	0	0
$z_1(x_1,y_1)\bar{z}_1$	d^2	0	0
$z_1(y_1,y_1)\bar{z}_1$	0	0	0
	LO(z_1)	TO(x_1')	TO(y_1')
$z_1(x_1',x_1')\bar{z}_1$	d^2	0	0
$z_1(x_1',y_1')\bar{z}_1$	0	0	0
$z_1(y_1',y_1')\bar{z}_1$	d^2	0	0

[a] From F. Pollak, in *Analytical Raman Spectroscopy* (J. G. Grasselli and B. J. Bulkin, eds.), Chapter 6. Copyright © 1991 John Wiley & Sons, Inc. Reprinted by permission of John Wiley & Sons, Inc.
[b] $x_1 = <100>$, $y_1 = <010>$, $z_1 = <001>$; $x_1' = <110>$, $y_1' = <1\bar{1}0>$.

In crystals, sharp peaks are seen due to $\mathbf{q} = 0$ phonons. Deviations from this observation can indicate many things about this material (Pollak, 1991). For example, in microcrystalline and amorphous semiconductors, the spectrum is often downshifted and broadened asymmetrically due to the relaxation of $\mathbf{q}$ conservation.

The effects of strain (ε_{ij}) can be seen as a shift and/or splitting of the $\mathbf{q} = 0$ phonon peak. In zincblende semiconductors, these changes can be characterized by the deformation potentials p, q, and r (Anastassakis *et al.*, 1970, Cerdeira *et al.*, 1972). The shift in the frequency of the phonon in backscattering along the [001] direction (i.e. the LO phonon singlet, Table 4.1) due to biaxial strain in the (001) plane is

$$\Delta\omega = \frac{1}{2\omega_0}[p\varepsilon_{zz} + q(\varepsilon_{xx} + \varepsilon_{yy})] \quad (4.62a)$$

where ω_0 is the unperturbed phonon frequency, the strains $\varepsilon_{xx} = \varepsilon_{yy}$ and $\varepsilon_{zz} = -(2C_{12}/C_{11})\varepsilon_{xx}$, and C_{ij} are the elastic constants of the material. For strained films on a substrate, ε_{xx} is the in-plane lattice mismatch strain. The shift in the frequency of the phonon in backscattering along the [111] direction (i.e., the TO phonon doublet, Table 4.3) due to biaxial strain in the (111) plane is

$$\Delta\omega = \frac{1}{2\omega_0}[(p + 2q)\varepsilon^d - r\varepsilon^0] \quad (4.62b)$$

where $\varepsilon^d = \varepsilon_{xx} = \varepsilon_{yy} = \varepsilon_{zz} = [2C_{44}/(2C_{44} + C_{11} + 2C_{12})]\varepsilon_\parallel$ and $\varepsilon^0 = \varepsilon_{xy} =$

4.3 Spectroscopy

Table 4.2 Raman Polarization Selection Rules for Backscattering from a (110) Surface for Given Atomic Displacements[a]

Polarization configuration[b]	Phonon modes		
	LO(z_2)	TO(x_2)	TO(y_2)
$z_2(x_2,x_2)\bar{z}_2$	0^c	0	0
$z_2(x_2,y_2)\bar{z}_2$	0	0	d^2
$z_2(y_2,y_2)\bar{z}_2$	0^c	d^2	0^c
	LO(z_2)	TO(x_2')	TO(y_2')
$z_2(x_2',x_2')\bar{z}_2$	0	$\frac{2}{3}d^2$	$\frac{1}{3}d^2$
$z_2(x_2',y_2')\bar{z}_2$	0	$\frac{1}{3}d^2$	0
$z_2(y_2',y_2')\bar{z}_2$	0	0	$\frac{4}{3}d^2$

[a] From F. Pollak, in *Analytical Raman Spectroscopy* (J. G. Grasselli and B. J. Bulkin, eds.), Chapter 6. Copyright © 1991 John Wiley & Sons, Inc. Reprinted by permission of John Wiley & Sons, Inc.
[b] $x_2 = <001>$, $y_2 = <1\bar{1}0>$, $z_2 = <110>$; $x_2' = <\bar{1}12>$, $y_2' = <1\bar{1}1>$.
[c] Becomes nonzero with surface electric fields or for $q \neq 0$ phonons.

$\varepsilon_{yz} = \varepsilon_{xz} = -[(C_{11} + 2C_{12})/(2C_{44} + C_{11} + 2C_{12})]\varepsilon_\parallel$. For strained films on a substrate, $\varepsilon_\parallel$ is the in-plane lattice mismatch strain.

In resonance vibrational Raman scattering, higher-order terms can also be important; these can induce strong peaks that are not seen in the nonreso-

Table 4.3 Raman Polarization Selection Rules for Backscattering from a (111) Surface for Given Atomic Displacement[a]

Polarization configuration[b]	Phonon modes		
	LO(z_3)	TO(x_3)	TO(y_3)
$z_3(x_3,x_3)\bar{z}_3$	$\frac{1}{3}d^2$	0	$\frac{2}{3}d^2$
$z_3(x_3,y_3)\bar{z}_3$	0	$\frac{2}{3}d^2$	0
$z_3(y_3,y_3)\bar{z}_3$	$\frac{1}{3}d^2$	0	$\frac{2}{3}d^2$

[a] From F. Pollak, in *Analytical Raman Spectroscopy* (J. G. Grasselli and B. J. Bulkin, eds.), Chapter 6. Copyright © 1991 John Wiley & Sons, Inc. Reprinted by permission of John Wiley & Sons, Inc.
[b] $x_3 = <1\bar{1}0>$, $y_3 = <11\bar{2}>$, $z_3 = <111>$.

Table 4.4 Characteristic Raman Shifts in Crystalline Diamond and Zincblende Semiconductors Due to the Scattering of $q = 0$ (Zone Center) Transverse Optical (TO) and Longitudinal Optical (LO) Phonons (cm^{-1}) at Ambient Temperature

Semiconductor	TO	LO
Si	521	521
Ge	301	301
AlAs	361	404
GaP	367	403
GaAs	269	292
GaSb	231	240
InP	304	345
InAs	219	243
InSb	185	197
ZnS	274	350
ZnSe	207	253
ZnTe	179	206
CdTe	141	168

nant Raman spectrum (Cardona, 1982). For example, a breakdown can occur in the symmetry properties for LO phonon scattering, so that LO phonons may be seen in a Raman spectrum for a given polarization configuration when $\hbar\omega_i$ is near a resonance, while it is forbidden off-resonance. Also, in some cases multiple LO phonon lines can be seen at $\omega_s = \omega_i \pm n\omega_{LO}$ with $n = 2, 3, 4, \ldots$. Since the intraband Frolich interaction depends on **q**, it does not contribute to nonresonant first-order scattering; but at resonance it can make "forbidden" first-order and higher-order scattering quite strong.

Second-order vibrational Raman scattering is due to the scattering of two phonons that shift ω_i by $\omega_{p1}(\mathbf{q}_1) \pm \omega_{p2}(\mathbf{q}_2)$. It is usually very weak compared to first-order scattering, and is not important for diagnostics.

4.4 Nonlinear Optical Interactions

Nonlinear optical effects are important at high laser intensities, where the electric field strengths start becoming comparable to the fields inside the material. Including these effects, the ith Cartesian component of the dipole moment can be expressed as

4.4 Nonlinear Optical Interactions

$$\mu_i = \mu_{0,i} + \sum_j \hat{\alpha}_{ij} E_j^l + \sum_{jk} \hat{\beta}_{ijk} E_j^l E_k^l + \sum_{jkl} \hat{\gamma}_{ijkl} E_j^l E_k^l E_l^l + \cdots \quad (4.63)$$

The first term is the permanent dipole moment, while in the second, linear, term, the polarizability $\hat{\alpha}$ now includes all linear terms from Equations 4.2, 4.5, and 4.6, including Raman scattering. The second-order term is characterized by $\hat{\beta}$, the hyperpolarizability, while the third-order term is characterized by $\hat{\gamma}$, the second hyperpolarizability. The electric polarization can be similarly characterized:

$$P_i = P_{0,i} + \sum_j \chi_{ij}^{(1)} E_j(\omega_1) + \sum_{jk} \chi_{ijk}^{(2)} E_j(\omega_1) E_k(\omega_2) \\ + \sum_{jkl} \chi_{ijkl}^{(3)} E_j(\omega_1) E_k(\omega_2) E_l(\omega_3) + \cdots \quad (4.64)$$

where $P_{0,i}$ is the macroscopic electric polarization. (See the appendix to this chapter for expressions in rationalized MKSA units.) The electric field terms in Equations 4.63 and 4.64 represent fields that can oscillate either at the same frequency or at different frequencies, as has been written explicitly in the latter equation.

$\chi^{(1)}$ is the linear susceptibility, which in previous sections was simply called χ, as in Equation 4.1. $\chi^{(2)}$ and $\chi^{(3)}$ are the second- and third-order susceptibilities, respectively, and are related to $\hat{\beta}$ and $\hat{\gamma}$ (Debe, 1987). $\chi^{(2)}$ and $\hat{\beta}$ describe processes in which two fields ($\omega = \omega_1, \omega_2$) couple to form a third field at $\omega_1 \pm \omega_2$, such as frequency summation ($\omega_1 + \omega_2$), second-harmonic generation (SHG, $2\omega_1$) (Figure 4.2a), frequency subtraction (or difference generation) ($\omega_1 - \omega_2$), optical rectification ($\omega_1 - \omega_1 = 0$), and the electro-optic effect ($\omega_1 + \omega_2$, with $\omega_2 = 0$). They also describe spontaneous

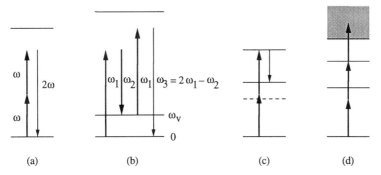

Figure 4.2 Energy-level diagrams of nonlinear optical processes: (a) Second harmonic generation (SHG); (b) coherent anti-Stokes Raman scattering (CARS); (c) two-photon LIF, which is resonant two-photon absorption via an intermediate level (dashed line) (also see TPA in Figure 4.1d), followed by the fluorescence of a single photon; (d) resonant-enhanced MPI (REMPI) for a 2 + 1 process, corresponding to two-photon absorption, followed by photoionization with the absorption of another photon.

hyper-Raman scattering through terms like $(d\chi^{(2)}/dQ)Q$ and $(d\hat{\beta}/dQ)Q$; this leads to scattering at $2\omega \pm \omega_p$ that is too weak for diagnostics. $\chi^{(3)}$ and $\hat{\gamma}$ describe processes in which three fields ($\omega = \omega_1, \omega_2, \omega_3$) couple to form a fourth at $\omega_1 \pm \omega_2 \pm \omega_3$, such as a third harmonic generation (THG, $3\omega_1$), degenerate four-wave mixing ($\omega_1 - \omega_1 + \omega_1$), coherent anti-Stokes Raman scattering (CARS, $2\omega_1 - \omega_2$, with $\omega_1 - \omega_2 = \omega_v$—a vibrational frequency) (Figure 4.2b), coherent Stokes Raman scattering (CSRS, $2\omega_2 - \omega_1$, with $\omega_1 - \omega_2 = \omega_v$), and stimulated Raman scattering ($\omega_1 - \omega_1 + \omega_2$, with $\omega_1 - \omega_2 = \omega_v$).

An alternative way of expressing the electric susceptibility has been given by Geurts and Richter (1987) and McGilp (1990) which clearly describes how χ depends on the oscillating electric field $E(\omega)$, any static field $E(0)$, and material deformations Q. With considerable simplifications assumed in the notation, and assuming no static polarization, the electric polarization is

$$P = \chi E(\omega) \qquad (4.65)$$

where

$$\chi = \chi^{(1)} + \frac{\partial \chi^{(1)}}{\partial Q} Q + \frac{\partial^2 \chi^{(1)}}{\partial Q\, \partial E(0)} QE(0) + \chi^{(2)} E(\omega) + \cdots \qquad (4.66)$$

In this last equation, the first term on the right-hand side describes absorption, reflection, and refraction; the second and third terms describe Raman scattering; and the fourth term describes the interaction of two waves with a third, i.e., second harmonic generation, etc.

These nonlinear terms are further discussed in Tolles and Harvey (1981), Shen (1984), Levenson and Kano (1988), and Boyd (1992), and in the references cited in Chapter 16. Of these, the probes most often used in film diagnostics are second harmonic generation and CARS, which are discussed in Chapter 16.

The $\chi^{(3)}$ terms leading to CARS can be separated into a term describing the resonance of interest χ_{CARS}^{res} and a nonresonant term χ^{NR}, which is due to the tails of other resonances:

$$\chi_{CARS}^{(3)} = \chi_{CARS}^{res} + \chi^{NR} \qquad (4.67)$$

Usually, only the real part of χ^{NR} is significant.

Classically, the resonant term can be related to $\partial \hat{\alpha}/\partial Q|_0$, as in Equations 4.24 and 4.58, by (Eckbreth, 1996)

$$\chi_{CARS}^{res}(2\omega_1 - \omega_2) = \frac{N}{2m} \left(\frac{\partial \hat{\alpha}}{\partial Q}\right)_0^2 \frac{1}{\omega_v^2 - (\omega_1 - \omega_2)^2 - i\Gamma(\omega_1 - \omega_2)} \qquad (4.68)$$

where N is the density of oscillators with frequency ω_v and Γ is the linewidth of the transition. This can be related to the differential cross-section for

4.4 Nonlinear Optical Interactions

spontaneous vibrational Raman scattering from $v = 0$ to 1 by using Equation 4.58

$$\left.\frac{d\sigma}{d\Omega}\right|_{\text{Raman}} = \frac{\hbar}{2m\omega_v}\left(\frac{\partial\hat{\alpha}}{\partial Q}\right)_0^2 \frac{\omega_2^4}{c^4} \quad (4.69)$$

to give the classical expression

$$\chi_{\text{CARS}}^{\text{res}}(2\omega_1 - \omega_2) = \frac{Nc^4(v+1)}{\hbar\omega_2^4}\left(\left.\frac{d\sigma}{d\Omega}\right|_{\text{Raman}}\right) \quad (4.70a)$$

$$\times \frac{\omega_v}{\omega_v^2 - (\omega_1 - \omega_2)^2 - i\Gamma(\omega_1 - \omega_2)}$$

The quantum mechanical expression is (Druet and Taran, 1981)

$$\chi_{\text{CARS}}^{\text{res}}(2\omega_1 - \omega_2) = \frac{(N_i - N_j)c^4}{\hbar\omega_1\omega_2^3}\left(\left.\frac{d\sigma}{d\Omega}\right|_{\text{Raman}}\right) \quad (4.70b)$$

$$\times \frac{1}{\omega_v - (\omega_1 - \omega_2) - i\Gamma}$$

where N_i and N_j are the respective densities in the lower and upper states, and the differential cross-section is now the resonant term in Equation 4.49. (Note that because of the use of different definitions for the forms of the electric field and polarization in the literature, expressions like Equations 4.68 and 4.70 often have different coefficients in different sources. Also see the appendix to this chapter for conversions to Rationalized MKSA units.)

Another type of nonlinear interaction is multiphoton absorption. Figure 4.1d illustrates resonant two-photon absorption (TPA), which can be detected by the fluorescence of a single photon. This technique is known as two-photon LIF (Figure 4.2c) and is discussed in Section 7.2.1.2. The initial and final levels in the TPA are not connected by an electric-dipole transition ($\Delta l = 0, \pm 2$), but both the initial and final levels are connected by electric-dipole transitions ($\Delta l = \pm 1$) to an intermediate "virtual" level (dashed line in the figure). Single-photon fluorescence ($\Delta l = \pm 1$) can occur from the pumped state, as in ordinary LIF. When this intermediate level lies in between the initial and final state (as shown), TPA can be resonantly enhanced by using two different photon frequencies, so that this intermediate level is nearly resonant with one of the photons. Unless this enhancement possibility exists and is absolutely necessary for detection, the use of one laser is preferred, and is common in diagnostics. (For the example depicted in Figure 4.2c, there is some enhancement by the intermediate state even with one photon frequency.)

The rate of two-photon absorption with a single laser (with frequency v) is given by the corresponding rate for one-photon absorption, Equation

4.15, but with $\mathcal{H}'_{12} = \mu_{12}E$ replaced by $\Sigma\, 2(\mu_{2i}E)(\mu_{i1}E)/(\mathcal{E}_2 - \mathcal{E}_i - h\nu)$, which is summed over all possible intermediate states i. The lineshape factor is now resonant for $\mathcal{E}_2 - \mathcal{E}_1 = 2h\nu$. Therefore the rate of TPA varies as I^2, where I is the laser intensity. Unless other processes affect the excited state, the rate of two-photon LIF is also $\propto I^2$, whereas the rate of single-photon LIF is $\propto I$.

Multiphoton absorption leading to photoionization is known as multiphoton ionization (MPI), and is discussed in Section 17.1. When one of the absorption steps is a resonant multiphoton absorption process, it is known as resonant-enhanced MPI or REMPI. This is illustrated in Figure 4.2d for a 2 + 1 process, corresponding to two-photon absorption, followed by photoionization with the absorption of another photon. Multiphoton absorption can be stimulated using a single laser wavelength or several wavelengths. Laser saturation (Section 4.2.1.2) is also a nonlinear process.

4.5 Heating by the Probing Laser

Absorption of the optical beam can heat the gas or solid sample. Chapter 15 describes several diagnostics for which this is desired. In laser desorption, a pulsed laser heats the surface to desorb adlayers. In thermal wave or photothermal spectroscopy the optical source heats one region, and thermal/acoustic waves are generated and detected elsewhere. However, in most cases, such heating is a perturbation that must be avoided. Some processes, such as chemical vapor deposition (CVD), are very sensitive to changes of even a few degrees. Moreover, when an optical probe is used to measure temperature, it must do so without changing it.

Detailed calculations of the temperature profile in an absorbing gas or solid during laser heating are straightforward, and some are described in Section 15.1. Still, simple estimates are sufficient to ensure that laser heating is not a problem. They involve thermophysical parameters, like the specific heat C and thermal conduction coefficient K, that depend on the material and vary with temperature T. Optical parameters, including the absorption coefficient α and reflectance R (for solids), usually depend on wavelength and the local temperature. Moreover, the fraction of light absorbed can vary during the thin film processes, because changing film thickness can change the fraction of light transmitted into the sample.

The temperature rise due to heating by a cw laser is determined by the steady-state balance between light absorption in the volume and thermal conduction away from this region. Consider a laser beam with power $\mathcal{P}$ and gaussian TEM$_{00}$ profile with width w ($1/e$ intensity decay) impinging on a semi-infinite medium with absorption coefficient α (cm^{-1}). The maximum temperature rise $\Delta T_{\max}$ is

$$\Delta T_{max} = \frac{(1-R)\mathcal{P}}{2\sqrt{\pi}\,Kw} N(W) \tag{4.71}$$

where R is the reflectance at the interface and $W = \alpha w$ (Lax, 1977, 1978). This expression assumes that the thermophysical and optical parameters do not vary with temperature. When the absorption depth $1/\alpha \ll w$ ($W \to \infty$), $N \to 1$. As the absorption depths gets larger, the temperature rise decreases; and when $1/\alpha \gg w$ ($W \to 0$), $N \to (W/\sqrt{\pi})[\ln(2/W) - \gamma_e/2]$, where $\gamma_e = 0.57721\ldots$ is Euler's constant. Equation 4.71 also applies when the thickness of the absorber is finite but much greater than $1/\alpha$. This expression can also be used for cw laser heating of gases.

The maximum temperature rise by pulsed-laser heating depends on thermal diffusion during the pulse. Consider a laser of intensity I, pulse width t_p, and therefore fluence $F_{pulse} = It_p$. It has a flat-top profile with radius w, so the pulse energy is $\pi w^2 F_{pulse}$. As above, it impinges on a semi-infinite medium. The thermal diffusion constant $D_t = K/\rho_d C$, where ρ_d is the density, so $\rho_d C$ is the specific heat per unit volume. Heat flows a characteristic distance $L_d = (2D_t t_p)^{1/2}$ during the laser pulse, which is assumed to be much smaller than w. So

$$\Delta T_{max} \approx \frac{(1-R)F_{pulse}}{L_{larger}\,\rho_d C} \tag{4.72}$$

where L_{larger} is the larger of L_d and the absorption depth $1/\alpha$.

When thermal diffusion can be neglected ($1/\alpha \gg L_d$), the temperature rise is

$$\Delta T_{max} \approx \frac{\alpha\,(1-R)F_{pulse}\,e^{-\alpha z}}{\rho_d C} \tag{4.73}$$

This form is usually valid for laser heating of gases. The maximum temperature rise is that at the surface $z = 0$.

When thermal diffusion is dominant ($L_d \gg 1/\alpha$), the maximum average temperature rise in the heated layer is

$$\Delta T_{max} \approx \frac{(1-R)F_{pulse}}{\rho_d C\sqrt{2D_t}}\frac{1}{\sqrt{t_p}} = \frac{(1-R)\,I}{\rho_d C\,\sqrt{2D_t}}\sqrt{t_p} \tag{4.74}$$

For more exact expressions, see Ready (1971) and the discussion in Section 15.1.

APPENDIX
Converting Units and Equations from the Gaussian CGS to the Rationalized MKSA System

Equations in this volume are expressed in gaussian CGS units. Equivalent forms for several expressions in Chapters 2 and 4 are presented here in rationalized MKSA

Table 4A.1 Converting Expressions from Gaussian CGS Units to Rationalized MKSA Units[a]

	Replace	By
Speed of light	c	$(\mu_0\varepsilon_0)^{-1/2}$
Electric field	E	$\sqrt{4\pi\varepsilon_0}\, E$
Electric displacement vector	D	$\sqrt{4\pi/\varepsilon_0}\, D$
Linear electric susceptibility or polarizability ($x = \chi^{(1)}$ or $\hat{\alpha}$)	x	$x/4\pi$
Charge, polarization, or dipole moment ($y = q$ (or e), P, or μ)	y	$y/\sqrt{4\pi\varepsilon_0}$
Dielectric constant	ε	$\varepsilon/\varepsilon_0$
Permeability	μ	μ/μ_0

[a] From J. D. Jackson, *Classical Electrodynamics*, 2nd ed. Copyright © 1975 John Wiley & Sons, Inc. Reprinted by permission of John Wiley & Sons, Inc.

(SI) units. The question of units rarely arises after the introductory chapters, Chapters 2–4. Consult the appendices in Boyd (1992) and Jackson (1975) for detailed prescriptions for converting between these two sets of units. Table 4A.1 shows how to convert an expression from CGS to MKSA units. Table 4A.2 shows useful unit conversions between these two systems. Table 4A.3 gives values for constants used in the rationalized MKSA system.

In the CGS system, the units of P and E are statvolt/cm ($=$esu^2/cm), while in the MKSA system, the units of P are C/m^2 and of E are V/m. The electric field in MKSA units is 3×10^4 times that expressed in CGS units.

Following the discussion of (scalar) linear dielectric theory in Section 4.1, and using Table 4A.1, $\mathbf{D} = \varepsilon_0\mathbf{E} + \mathbf{P} = \varepsilon_0\mathbf{E}(1 + \chi^{(1)})$, $\mathbf{P} = \varepsilon_0\chi^{(1)}\mathbf{E}$, the constitutive relation is $\varepsilon = \varepsilon_0(1 + \chi^{(1)})$ and the complex index of refraction $\tilde{n} = \sqrt{\mu_m\tilde{\varepsilon}/\mu_0\tilde{\varepsilon}_0}$. In both units

Table 4A.2 Units Equivalence (MKSA = CGS)

Distance	1 m = 100 cm
Mass	1 kg = 1000 g
Time	1 sec = 1 sec
Force	1 newton (N) = 10^5 dyne
Energy	1 joule (J) = 10^7 ergs ($=1/4.184$ calorie (cal))
Charge	1 coulomb (C) = 2.998×10^9 statcoulomb (1 statcoulomb = 1 esu)
Voltage	1 volt (V) = 1/299.8 statvolt
Resistance	1 ohm (Ω) = 1.139×10^{-12} s/cm
Capacitance	1 farad (F) = 0.899×10^{12} cm
Inductance	1 henry (H) = 1.113×10^{-12} s^2/cm

Appendix

$\chi^{(1)}$ is dimensionless. The value of $\chi^{(1)}$ expressed in MKSA units is 4π times that expressed in CGS units.

Using the conversions for $\chi^{(1)}$ and the charge e in Table 4A.1, the expressions given in Chapter 4 for the linear susceptibility need to be divided by ε_0. So $\chi\ (= \chi^{(1)})$ for the classical simple free electron model (Equation 4.18, Section 4.2.1) becomes

$$\chi = \frac{Ne^2/m\varepsilon_0}{\omega_0^2 - \omega^2 - i\gamma\omega} \tag{A.1}$$

with analogous changes for Equations 4.19 and 4.20.

Using the transformation for the dipole moment μ, the Einstein "A" coefficient in Equation 4.17 (Section 4.2.1) becomes

$$A_{21} = \frac{2\omega^3|\mu_{21}|^2 n}{3hc^3\varepsilon_0} \tag{A.2}$$

This transformation can also be used to convert the factor G in the absorption coefficient in Equations 4.33 and 4.36 to $2\pi^2\nu/3hc\varepsilon_0$ and G' in Equation 4.37 to $16\pi^3\nu^4/3c^3\varepsilon_0$.

In Section 4.2.2, the total scattering cross-section (Equation 4.23) becomes

$$\sigma = \frac{\omega_i^4 \hat{\alpha}^2}{12\pi c^4 \varepsilon_0^2} \tag{A.3}$$

when the conversion of the polarizability $\hat{\alpha}$ is used. The factor in brackets also appears in corresponding expressions for the differential scattering cross sections $d\sigma/d\omega_s$, $d\sigma/d\Omega$, and $d^2\sigma/d\Omega d\omega_s$.

The time-averaged intensity of an optical beam in Equation 2.4 (Section 2.1) becomes

$$I = \frac{n}{2}\left(\frac{\varepsilon_0}{\mu_0}\right)^{1/2} E_0^2 = \frac{n}{2Z_0} E_0^2 \tag{A.4}$$

The units of intensity are W/m^2 in MKSA units and erg/cm^2-sec in CGS units. When dealing with practical matters such as intensity and power, those working with expressions in CGS units sometimes lapse into using more practical units, such as the watt (W), the MKSA unit for power, and W/cm^2, a hybrid MKSA/CGS unit, for intensity. It is useful to remember that the electric field envelope (E_0) expressed in CGS units is given by $0.091\sqrt{I(W/cm^2)}$ (for $n = 1$). (This assumes that the form for the electric field is given by Equations 2.2 and 2.3. If instead, the form $E = E_0 \exp(-i\omega t)$ + c.c. is used, then the right-hand-sides of Equations 2.4 and A.4 should

Table 4A.3 Constants Used in the Rationalized MKSA System

Permittivity of free space	$\varepsilon_0 = 8.85 \times 10^{12}$ F/m
Permeability of free space	$\mu_0 = 4\pi \times 10^{-7}$ H/m
Impedance of free space	$Z_0 = (\mu_0/\varepsilon_0)^{1/2} = 377\ \Omega$

be multiplied by four and, for a given intensity, E_0 is half that given by this expression.)

The field-induced terms in the electric polarization, from Equation 4.64, are represented by either of two different MKSA conventions:

$$P_i = \varepsilon_0 \left[\sum_j \chi_{ij}^{(1)} E_j + \sum_{jk} \chi_{ijk}^{(2)} E_j E_k + \sum_{jkl} \chi_{ijkl}^{(3)} E_j E_k E_l + \cdots \right] \quad (A.5)$$

or

$$P_i = \sum_j \varepsilon_0 \chi_{ij}^{(1)} E_j + \sum_{jk} \chi_{ijk}^{(2)} E_j E_k + \sum_{jkl} \chi_{ijkl}^{(3)} E_j E_k E_l + \cdots \quad (A.6)$$

The linear polarization term, the analog of Equation 4.1, is the same in both conventions, while the nonlinear terms are different. Using Equation A.5, $\chi^{(2)}$ is replaced by $\varepsilon_0 \chi^{(2)}/(4\pi\varepsilon_0)^{3/2}$ and $\chi^{(3)}$ by $\varepsilon_0 \chi^{(3)}/(4\pi\varepsilon_0)^2$, while with Equation A.6, $\chi^{(2)}$ is replaced by $\chi^{(2)}/(4\pi\varepsilon_0)^{3/2}$ and $\chi^{(3)}$ by $\chi^{(3)}/(4\pi\varepsilon_0)^2$. The conversions are similar for the higher order polarizability terms in Equation 4.63. As an example, using this replacement for $\chi^{(3)}$ in Equation A.5 and for the polarizability, the right hand side of Equation 4.68 for the classical CARS susceptibility needs to be multiplied by ε_0. (Other changes may also be necessary when the conventions for the electric field and polarization are different from those adopted here.) Boyd (1992) should be consulted for more details.

References

E. Anastassakis, A. Pinczuk, E. Burstein, F. H. Pollak, and M. Cardona, *Solid State Commun.* **8,** 133 (1970).

D. E. Aspnes, in *Handbook on Semiconductors* (M. Balkanski, ed.), Vol. 2, Chapter 4A, p. 109. North-Holland Publ., Amsterdam, 1980.

M. Balkanski, ed., *Handbook of Semiconductors,* Vol. 2. North-Holland Publ., Amsterdam, 1980.

G. Bekefi, ed. *Principles of Laser Plasmas.* Wiley, New York, 1976.

P. F. Bernath, *Spectra of Atoms and Molecules.* Oxford Univ. Press (Clarendon), Oxford, 1995.

K. W. Böer, *Survey of Semiconductor Physics.* Van Nostrand-Reinhold, New York, 1990.

R. W. Boyd, *Nonlinear Optics.* Academic Press, Boston, 1992.

M. Cardona, in *Light Scattering in Solids II.* (M. Cardona and G. Güntherodt, eds.), Chapter 2. Springer-Verlag, Berlin, 1982.

F. Cerdeira, C. J. Buchenauer, F. H. Pollak, and M. Cardona, *Phys. Rev. B* **5,** 580 (1972).

M. L. Cohen and J. R. Chelikowsky, *Electronic Structure and Optical Properties of Semiconductors.* Springer-Verlag, Berlin, 1988.

E. U. Condon and G. H. Shortley, *The Theory of Atomic Spectra.* Cambridge Univ. Press, Cambridge, UK, 1959.

A. Corney, *Atomic and Laser Spectroscopy.* Oxford Univ. Press (Clarendon), Oxford, 1977.

M. K. Debe, *Prog. Surf. Sci.* **24,** 1 (1987).

References

W. Demtröder, *Laser Spectroscopy: Basic Concepts and Instrumentation.* Springer-Verlag, Berlin, 1982.

M. C. Drake, *Mater. Res. Soc. Symp. Proc.* **117**, 203 (1988).

S. A. J. Druet and J. P. Taran, *Prog. Quantum Electron.* **7**, 1 (1981).

A. C. Eckbreth, *Laser Diagnostics for Combustion Temperature and Species,* 2nd ed. Gordon & Breach, Luxembourg, 1996.

P. J. Feibelman, *Prog. Surf. Sci.* **12**, 287 (1982).

J. R. Ferraro and K. Nakamoto, *Introductory Raman Spectroscopy.* Academic Press, Boston, 1994.

J. Geurts and W. Richter, *Springer Proc. Phys.* **22**, 328 (1987).

H. R. Griem, *Plasma Spectroscopy.* McGraw-Hill, New York, 1964.

H. R. Griem, *Spectral Line Broadening by Plasmas.* Academic Press, New York, 1974.

W. Hayes and R. Loudon, *Scattering of Light by Crystals.* Wiley, New York, 1978.

I. P. Herman, "Raman Scattering", *The Encyclopedia of Applied Physics,* Vol. 15, (G. L. Trigg, ed.) p. 587. VCH, New York, 1996.

G. Herzberg, *Atomic Spectra and Atomic Structure.* Dover, New York, 1937.

G. Herzberg, *Molecular Spectra and Molecular Structure. II. Infrared and Raman Spectra of Polyatomic Molecules.* Van Nostrand-Reinhold, New York, 1945.

G. Herzberg, *Molecular Spectra and Molecular Structure. I. Spectra of Diatomic Molecules.* Van Nostrand-Reinhold, New York, 1950.

G. Herzberg, *Molecular Spectra and Molecular Structure. III. Electronic Spectra and Electronic Structure of Polyatomic Molecules.* Van Nostrand-Reinhold, New York, 1966.

T. Hirschfeld, *Appl. Spectrosc.* **27**, 389 (1973).

R. T. Holm, in *Handbook of Optical Constants of Solids II* (E. D. Palik, ed.), Chapter 2, p. 21. Academic Press, Boston, 1991.

T. Holstein, *Phys. Rev.* **72**, 1212 (1947).

K. P. Huber and G. Herzberg, *Molecular Spectra and Molecular Structure. IV. Constants of Diatomic Molecules.* Van Nostrand-Reinhold, New York, 1979.

T. P. Hughes, *Plasmas and Laser Light.* Wiley, New York, 1975.

J. D. Jackson, *Classical Electrodynamics,* 2nd ed. Wiley, New York, 1975.

G. E. Jellison and D. H. Lowndes, *Appl. Phys. Lett.* **41**, 594 (1982).

I. Kovács, *Rotational Structure in the Spectra of Diatomic Molecules.* American Elsevier, New York, 1969.

H. G Kuhn, *Atomic Spectra.* Academic Press, New York, 1962.

M. Lax, *J. Appl. Phys.* **48**, 3919 (1977).

M. Lax, *Appl. Phys. Lett.* **33**, 786 (1978).

M. D. Levenson and S. S. Kano, *Introduction to Nonlinear Laser Spectroscopy.* Academic Press, Boston, 1988.

I. N. Levine, *Molecular Spectroscopy.* Wiley (Interscience), New York, 1975.

R. L. Liboff, *Introductory Quantum Mechanics,* 2nd ed. Addison-Wesley, Reading, MA, 1992.

D. A. Long, *Raman Spectroscopy.* McGraw Hill, New York, 1977.

W. H. Louisell, *Quantum Statistical Properties of Radiation.* Wiley, New York, 1990.

D. W. Lynch, in *Handbook of Optical Constants of Solids* (E. D. Palik, ed.), Chapter 10, p. 189. Academic Press, Boston, 1985.

G. G. Macfarlane, T. P. McLean, J. E. Quarrington, and V. Roberts, *Phys. Rev.* **111,** 1245 (1958).

J. F. McGilp, *J. Phys. Condens. Matter* **2,** 7985 (1990).

A. C. G. Mitchell and M. W. Zemansky, *Resonance Radiation and Excited Atoms.* Cambridge Univ. Press, Cambridge, UK, 1971.

S. S. Mitra and N. E. Massa, in *Handbook of Semiconductors,* (W. Paul, ed.), Vol. 1, Chapter 3, p. 81. North-Holland, Amsterdam, 1982.

K. Ohta and H. Ishida, *Applied Spectroscopy* **42,** 952 (1988).

M. A. Olmstead, *Surf. Sci. Rep.* **6,** 159 (1987).

J. I. Pankove, *Optical Processes in Semiconductors.* Dover, New York, 1971.

F. H. Pollak, in *Analytical Raman Spectroscopy* (J. G. Grasselli and B. J. Bulkin, eds.), Chapter 6. Wiley, New York, 1991.

J. F. Ready, *Effects of High-Power Laser Radiation.* Academic Press, New York, 1971.

D. L. Rousseau, J. M. Friedman, and P. F. Williams, in *Raman Spectroscopy of Gases and Liquids* (A. Weber, ed.), Chapter 6. Springer-Verlag, Berlin, 1979.

H. W. Schrötter and H. W. Klöckner, *Raman Spectroscopy of Gases and Liquids* (A. Weber, ed.), Chapter 4. Springer-Verlag, Berlin, 1979.

Y. R. Shen, *The Principles of Nonlinear Optics.* Wiley, New York, 1984.

J. I. Steinfeld, *Molecules and Radiation: An Introduction to Modern Molecular Spectroscopy,* 2nd ed. MIT Press, Cambridge, MA, 1985.

F. Stern, *Solid State Phys.* **15,** 299 (1963).

S. Svanberg, *Atomic and Molecular Spectroscopy: Basic Aspects and Practical Applications.* Springer-Verlag, Berlin, 1991.

J. S. Toll, *Phys. Rev.* **104,** 1760 (1956).

W. M. Tolles and A. B. Harvey, in *Chemical Applications of Nonlinear Raman Spectroscopy* (A. B. Harvey, ed.), Chapter 1, p. 1. Academic Press, New York, 1981.

J. T. Verdeyen, *Laser Electronics,* 2nd ed. Prentice-Hall, Englewood Cliffs, NJ, 1989.

A. Weber, in *Raman Spectroscopy of Gases and Liquids* (A. Weber, ed.), Chapter 3. Springer-Verlag, Berlin, 1979.

H. E. White, *Introduction to Atomic Spectra.* McGraw-Hill, New York, 1934.

A. Yariv, *Quantum Electronics,* 3rd ed. Wiley, New York, 1989.

CHAPTER 5

Diagnostics Equipment and Methods

An overview of the experimental techniques used in optical diagnostics is presented in this chapter; these techniques are detailed further in the discussion of each probe in the succeeding chapters. Section 5.1 presents the components of these systems, while Section 5.2 discusses techniques for light collection, imaging, and signal analysis.

5.1 Optical Components

The components presented here are commonly used in fundamental research and in development studies of thin film processing. Many are also used to make the compact, integrated sensor packages needed for real-time monitoring. Figure 5.1 is a schematic of the building blocks of optical detection. An optical source is used to interact with the medium for all diagnostics other than optical emission spectroscopy (OES) and pyrometry. The optical signal is spectrally analyzed, either before or after this interaction, and then is detected to give an electronic signal that is interpreted. Demtröder (1982) and Bass (1995) provide overviews of the optical sources, detectors, and other instrumentation used in optical spectroscopy.

5.1.1 *Optical Sources*

Both incoherent and coherent light sources are used in optical diagnostics. The specific requirements of spatial and temporal coherence of the "coherent" laser sources depend on the application.

Broadband incoherent light is available from discharge lamps and glowbars (thermal radiation); these sources are listed in Table 5.1. In many applications, these sources are spectrally narrowed by using a monochroma-

119

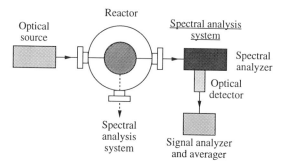

Figure 5.1 Schematic of the building blocks for optical detection.

tor or transmission filter (dielectric, colored glass, holographic, etc.). The wavelength of the radiation can be tuned by sending it through a monochromator to disperse it (so that a narrow spectral band can be selected) and by scanning it. "Spectroscopic" forms of diagnostics examine the interaction as a function of wavelength, often by using spectrally tuned radiation from these incoherent sources; spectroscopic analysis with these sources is common in transmission/absorption (Chapter 8) and ellipsometry, reflectometry, and modulated reflectance (Chapter 9) spectroscopies. Glowbars are commonly used in Fourier transform infrared (FTIR) spectroscopy (Sec-

Table 5.1 Optical Sources Often Used in Diagnostics

Incoherent sources	
White light sources	Tungsten lamps (visible), deuterium lamps (UV, ~160–350 nm), Xe lamps, glowbars (IR)
Atomic resonance lamps	Hollow cathode lamps
Coherent sources	
cw lasers	
He–Ne	632.8, 594.1, 543.5 nm (~1 mW)
Semiconductor diode	Fixed wavelength: AlGaAs/GaAs heterostructure (~670 nm) Tunable: lead-salt (3–34 μm) (Section 8.2.1.1)
He–Cd	325 nm (UV), 441.6 nm (blue) (~10 mW)
Argon-ion	488.0, 514.5 nm, 335.8–358.8 nm (~1–10 W)
Krypton-ion	647.1 nm (many other visible and UV lines) (~1 W)
Tunable	Dye (Figure 5.2), Ti–sapphire, optical parametric oscillators
Pulsed lasers	
Excimer	193 nm (ArF), 248 nm (KrF), 308 nm (XeCl), 351 nm (XeF)
Nd:YAG	1.06 μm [530 nm (2×), 355 nm (3×), 266 nm (4×)]
N_2	337.1 nm
Dye laser	(Figure 5.2.)
Nonlinear mixing	Frequency doubling, summation and difference generation in nonlinear crystals; frequency tripling in gases
Stimulated Raman shifting	

5.1 Optical Components

tions 8.2.1.2 and 9.7). Atomic resonance lamps (often called atomic absorption, or AA, lamps) are inexpensive sources of resonant light that are well suited for absorption (Sections 8.2.2.1.3 and 8.2.2.1.4) and fluorescence (Chapter 7) spectroscopies. They provide radiation that is resonant with atomic transitions, with a lineshape that is usually somewhat broadened because of the nature of the plasma emission source. Resonance lamp radiation is often spectrally filtered because it contains spontaneous emission from many transitions.

Lasers can be classified and characterized by their (1) wavelength λ, (2) wavelength tunability, (3) spectral linewidth $\delta\lambda$ (which is related to temporal and longitudinal spatial coherence) and mode structure, (4) beam quality (lateral profile, beam divergence, lateral spatial coherence), (5) temporal properties [continuous-wave lasers (cw) versus pulsed lasers with pulse width t_p], and by (6) practical issues such as cost, size (footprint), lifetime, repair and maintenance requirements, and overall complexity (single component versus a chain of interacting lasers involving frequency conversion).

Table 5.1 lists cw lasers commonly used in thin film diagnostics. Low-power, fixed-frequency cw lasers are needed in interferometry (Section 9.5) and elastic laser light scattering (Chapter 11). Helium–neon lasers (632.8 nm) have often been used in these applications, but are being increasingly replaced by fixed-wavelength semiconductor heterostructure diode lasers, which are often cheaper, smaller, and more robust; tunable semiconductor diode lasers are discussed below. Raman scattering (Chapter 12) often uses higher-power fixed-wavelength cw lasers, such as argon-ion and krypton-ion lasers. These lasers also lase on several lines in the UV (see Table 5.1); the visible lines can also provide UV light by frequency doubling, either external to the cavity (which provides lower power) or internally (providing higher power at higher cost) by using nonlinear mixing crystals. Helium–cadmium lasers are convenient, relatively inexpensive, low-power sources of fixed-line, cw UV radiation. Continuous-wave GaAs/AlGaAs diode lasers can also pump Nd:YAG (1.06 μm), which can be frequency-doubled or -tripled.

High-powered cw lasers can also be used to pump dye lasers (Figure 5.2) or titanium–sapphire lasers to provide tunable cw radiation for use in LIF (Chapter 7) and absorption (Chapter 8) studies; these tunable lasers can also be frequency-doubled to extend the spectral range into the ultraviolet. The fine frequency control possible with cw tunable lasers, which is necessary for very high-resolution spectroscopy, has been needed in only a few investigations of thin film diagnostics, as in some LIF investigations described in Section 7.2.1.1; consequently, pulsed tunable lasers, which are simpler to operate, are usually preferred. Tunable infrared semiconductor diode lasers have been used extensively to probe infrared absorption fea-

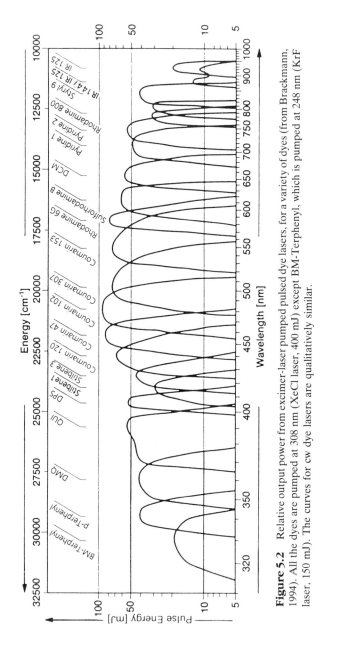

Figure 5.2 Relative output power from excimer-laser pumped pulsed dye lasers, for a variety of dyes (from Brackmann, 1994). All the dyes are pumped at 308 nm (XeCl laser, 400 mJ) except BM-Terphenyl, which is pumped at 248 nm (KrF laser, 150 mJ). The curves for cw dye lasers are qualitatively similar.

5.1 Optical Components

tures in gases with high spectral resolution, and are detailed in Section 8.2.1.1.

Table 5.1 also lists common pulsed lasers. Pulsed lasers are preferred to cw lasers when higher intensities are needed to induce a process, such as for laser-induced thermal desorption (Section 15.1) and nonlinear optical processes (Chapter 16), and when gated detection is needed to remove any steady-state background light (LIF, Chapter 7). Also, the operation of pulsed lasers is usually simpler, as with dye laser operation, because the laser gain is higher. Excimer lasers (pulse width ~ 20 nsec) and Nd:YAG lasers (ω; $\lambda = 1.06$ μm)—sometimes frequency-doubled (2ω; 530 nm), frequency-tripled by frequency summation ($3\omega = 2\omega + \omega$; 355 nm), or frequency-quadrupled ($4\omega = 2 \times 2\omega$; 265 nm) in nonlinear mixing crystals—are used for laser-induced thermal desorption (Section 15.1) and for LIF (Chapter 7) when there is a fortuitous resonance. The Nd:YAG laser lines and harmonics are employed in nonlinear optical applications, for example, as the fixed wavelength in coherent anti-Stokes Raman scattering (CARS) (Section 16.1) and for surface second-harmonic generation (Section 16.2). The Nd:YAG harmonics and the XeCl excimer laser (308 nm) are also used to pump dye lasers (Figure 5.2) to obtain tunable radiation for LIF (Chapter 7), absorption (Section 8.2.2), (the tunable laser for) CARS (Section 16.1), and resonant-enhanced multiphoton ionization (REMPI) (Section 17.1). If lower peak intensities are acceptable, nitrogen lasers (337.1 nm) can also be used to pump the dye.

When tunable ultraviolet radiation is needed, such as for LIF or absorption spectroscopy, pulsed dye lasers can be frequency-doubled or frequency-summed with a fixed-wavelength, pulsed laser. Common frequency mixing crystals are KDP (potassium dihydrogen phosphate) and BBO (β-barium borate). They must be phase-matched for efficient conversion, which is often accomplished by angle tuning. To obtain even shorter wavelengths (shorter than ~250–300 nm), these tunable, shorter wavelengths can be summed again with fixed-wavelength lasers. The fundamental limitation in obtaining short-wavelength radiation ($\leq$200 nm) this way is absorption in the mixing crystals. Third harmonic generation in gases can provide even shorter wavelengths. For example, 118 nm can be obtained by tripling 355 nm in Xe/Ar mixtures (Section 17.1). Other sources of tunable radiation are higher-order stimulated Raman scattering, which occurs in high pressures of gases such as H_2 and leads to a series of Stokes ($S_{1,2,3,...}$)- and anti-Stokes ($AS_{1,2,3,...}$)-shifted lines, and optical parametric oscillation. These nonlinear optical shifting techniques are discussed by Yariv (1989).

Incoherent sources are usually cheaper than lasers, and should be used instead of lasers for real-time monitoring during manufacturing. One notable exception is for applications that require fixed-frequency (nonresonance) operation, such as elastic light scattering; low-power, cw semiconductor diode lasers and He–Ne lasers are usually cheaper than and otherwise superior to incoherent sources for such applications. Moreover, diode lasers are preferred to He–Ne lasers because they are smaller. Stabilization of the source intensity is necessary in absorption monitoring (Chapter 8). This is true for both incoherent and coherent sources, although lasers tend to be noisier than lamps.

5.1.2 *Optical Detectors*

Sensors of electromagnetic radiation from the infrared to the ultraviolet measure either optical power (the magnitude of the Poynting vector, Equation 2.4), and are called thermal detectors, or the flux of photons, and are called quantum detectors. Thermal detectors are often used to measure the power of lasers and lamp sources, while quantum detectors are often employed to measure weak optical signals when high quantum efficiency, low noise, and fast response are needed.

Thermal detectors measure the temperature rise due to the absorption of light. Unlike quantum detectors, thermal detectors have a very broad wavelength response and are fairly insensitive to wavelength, making them valuable in the mid- and far-IR; they are usually much slower than quantum detectors. In pyroelectric detectors, the change of the electric polarization of a ferroelectric is measured as temperature changes. (Deuterated) triglycine sulfate [(D)TGS] pyroelectric detectors are commonly used in Fourier transform infrared spectrometers (Section 5.1.3.2). Measurements of laser power are often made with a thermocouple-based detector.

Quantum detectors are routinely used for detection in optical diagnostics. In most cases, the absorption of an optical photon in a material leads to the production of an electron, which is then detected. Quantum detectors are available as single-element detectors with lateral dimensions ~1–25 mm, and as one- and two-dimensional arrays of small single-element detectors, each with lateral dimension ~25 μm. These multichannel detectors, sometimes generically called OMAs (optical multichannel analyzers), permit parallel detection for rapid data collection and analysis of spectroscopic and imaging data.

In photoconductive quantum detectors, a free carrier is produced when the energy of the incident photon is above the band gap of the semiconductor element, thereby changing the resistance of the material. In semiconductor photodiodes, electron–hole pairs are formed in the depletion region of a p–n junction as a result of absorption. They are separated by the electric fields in the depletion region and then are detected. The noise in photodiode detection is discussed in Section 5.2.3.1. In the unbiased, photovoltaic mode, the voltage induced by these charges is measured directly, while in the reverse-biased photoconductive mode, the reverse-bias current is measured. The spectral response of such semiconductor-based detectors is shown in Figure 13.3c in terms of the detectivity D^*, which gives the reciprocal of

5.1 Optical Components

the minimum detectable power normalized to detector dimensions and frequency bandwidth. Silicon and germanium detectors are sensitive to photon energies larger than their indirect gaps, corresponding to wavelengths shorter than ~1.1 and 1.7 μm, respectively. PbS, PbSe, InAs, and HgCdTe detectors are sensitive in the near-infrared and are often used for detection in infrared absorption spectroscopy (Sections 8.2.1 and 8.4) and pyrometry (Chapter 13).

In photomultipliers (PMT), photoemission from the cathode is internally amplified by ~10^6 and then detected as a current or as individual electron bunches, which is the photon-counting mode. Both modes provide high-sensitivity, low-noise detection in the visible and ultraviolet. The noise in PMT detection is discussed in Section 5.2.3.1. The current mode is sufficient for many optical diagnostics applications, such as OES (Chapter 6) and LIF (Chapter 7). For very low light levels, the improved signal/noise ratio provided by photon counting is needed, as is sometimes required for spontaneous Raman scattering (Chapter 12). The spectral response of PMTs range from ~200–1100 nm, as shown in Figure 5.3.

When noise levels are suitably low, the signal from the single-element detector can be used without any signal averaging; otherwise some type of averaging is necessary. Signal averaging methods for transient signals in-

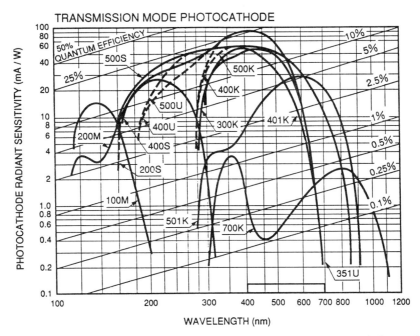

Figure 5.3 The spectral response for typical photomultipliers with transmission mode photocathodes (from Hamamatsu Corp., 1995).

clude boxcar integration and transient digitizing. Lock-in detection is used to detect modulated "steady-state" signals. Photon counting for photomultiplier signals is yet another method. See Section 5.2.3.1 for more details about signal averaging and noise.

The simplest multielement detectors are integrated units of two adjacent semiconductor photodiodes or four diodes in a quadrant arrangement. These are useful for determining properties of adjacent regions, as used in reflection interference thermometry (Section 9.5.2, Figure 9.30), and to detect lateral movement in an optical beam, as in photothermal beam deflection spectroscopy (Section 15.2, Figure 15.4; Section 15.2.1, Figure 15.8a).

The development of 1D- and 2D-array detectors with ~256–1024 detectors (pixels) per row or column has greatly advanced parallel detection techniques. One early multichannel detector was the silicon-intensified target (SIT) vidicon; it is rarely used at present. More recent array detectors include photodiode arrays (PDAs), position-sensitive resistive anodes or Mepsicrons, and charge-coupled-device (CCD) array detectors. The performance characteristics of these array detectors have been compared by Chang and Long (1982), Acker *et al.* (1988), and Tsang (1989).

Linear (1D) photodiode arrays consist of 512–1024 silicon photodiodes (~25 μm wide, ~2 mm high). Each element serves as a photodiode with an effective exit slit width of a single pixel (25 μm). Photodiode arrays are still used to detect dispersed OES radiation (~200–1000 nm; Chapter 6) and for rapid analysis in spectroscopic ellipsometry (Section 9.11).

Since PDAs have high readout noise, intensified versions have been developed to detect weak signals, as are common in spontaneous Raman scattering. The light imaged on a semitransparent photocathode produces electrons that are amplified (gain $\sim 5 \times 10^6$) by a multichannel plate (MCP). These electrons are converted into photons on a phosphor screen, and then this light image is detected by the diode array. Use of the MCP greatly improves the signal-to-noise ratio and also permits gated detection down to ~10 nsec by applying a fast-voltage pulse to remove temporarily the reverse-bias on the MCP. However, the effective "exit slit width" of this intensified detector is ~3–4 pixels (~75–100 μm) *vis-à-vis* the ~1 pixel resolution of the unintensified version. Also, this device is susceptible to damage at high light levels, as is the PMT. Intensified PDAs are rarely used except for gating applications, because they have been supplanted by CCD array detectors. As is seen below, CCD arrays have better sensitivity, better exit slit resolution, are much less sensitive to damage from high light levels (when used without an MCP), and are less expensive.

Position-sensitive resistive anodes or Mepsicrons are essentially photomultipliers that are subdivided into a 2D array. As such, they have signal/noise ratios as good as those for CCD arrays and operate in the same wavelength region. They are extremely linear; in contrast, CCD arrays

5.1 Optical Components

saturate when the electron well is full. However, they are less suitable for diagnostics than are CCD arrays because they are more costly and are subject to damage at high light levels.

Charge-coupled-device arrays are 2D arrays of MOS capacitors, each of which is a square element with lateral dimensions typically in the range $\sim$22–28 μm. When photons of energy greater than the indirect band gap of Si ($\lambda < 1080$ nm) are incident, electron–hole pairs are formed. The minority carriers (electrons) are trapped in the potential wells of the depletion regions, which are formed with appropriate bias, and these electrons are stored in the wells until readout. For spectroscopic applications, these arrays are often rectangular and $\sim\frac{1}{2}$ or 1 inch long (Figure 5.4), while for imaging they are typically square. CCD arrays are cooled for low-noise operation. Details about CCD operation and applications in spectroscopy and imaging can be found in Bilhorn *et al.* (1987), Permberton *et al.* (1990), and Esser and Theuwissen (1993).

In front-illuminated CCD arrays, the light must pass through >50 μm of silicon. Figure 5.5 (Princeton Instruments, Inc., 1994) shows the quantum efficiency (QE) for a CCD camera with an EEV model CCD02 chip. The peak quantum efficiency of the usual front-illuminated chip is already a very high 45%. But because photons with energies much greater than the band gap are strongly absorbed near the front surface and recombination can occur before the electrons are trapped in the well, the QE of these cameras in the UV is relatively low. Various types of coatings can be applied to increase the QE. In the example shown in this figure, the QE is quite high in the ultraviolet (to 200 nm) when the surface is coated with a lumogen

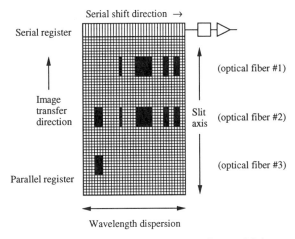

Figure 5.4 Schematic of a CCD array, with simulated dispersed light patterns collected from three regions in a reactor by using three optical fibers. Fiber #2 brings in emissions from species that are also seen by fibers #1 and #3.

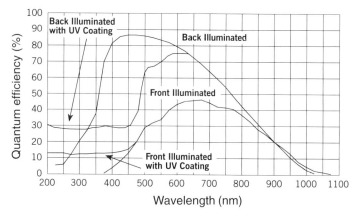

Figure 5.5 Quantum efficiencies of a CCD array camera with either back- or front-illuminated EEV model CCD02 chips, with different coatings (from Princeton Instruments, Inc., 1994).

organic UV coating that converts UV light to visible radiation. Another way to lessen the absorption problem is by backthinning the substrate of the array to less than 15 μm and then illuminating the back. With various types of antireflection coatings applied, the peak QE can increase to ~85% in this back-thinned, back-illuminated chip, and ultraviolet response improves, as is seen in Figure 5.5.

Very low-light level spectroscopic signals are typically detected with CCD arrays because of the advantages of these arrays and the inherent advantage of parallel detection (the Fellgett advantage, Section 5.1.3). In spectroscopic applications the effective dimension of the exit slit is the pixel dimension (~22–28 μm). To minimize noise and contributions from cosmic rays, the vertical binning of pixels (parallel to the exit slit) can be limited to those pixels actually receiving light. The 2D nature of CCD arrays combined with the ability to bin vertically also allows simultaneous acquisition and analysis of several signals, which can be multiple spectra from different regions in one reactor or spectra from different reactors. This capability permits multiplexing of many reactors in a fabrication plant to a single CCD array camera by using fiber optics.

Readout times for large CCD arrays can sometimes be longer than 1 sec. The shifting mechanism used in the readout can lead to a residual image ("lag") after the signal has supposedly been read. These two features can limit the use of CCD arrays for fast switching applications. However, performance can be improved if the CCD array is gated with an MCP, as in the intensified photodiode array. Although the intensified CCD array (ICCD) camera has only the same signal/noise level as an ordinary CCD, while being significantly more expensive, it is much faster. The ICCD

5.1 Optical Components

camera can be gated on for periods as short as ~5 nsec, and has been used to examine thermal emission from particulates in pulsed-laser desorption (Section 13.5).

These array detectors are limited to the wavelength region of Si absorption (~200–1100 nm). Other multielement detectors enable parallel detection in the infrared, and have been made using different materials, such as Ge, PtSi, and HgCdTe.

5.1.3 *Components for Spectroscopic Analysis*

Spectroscopic analysis in optical diagnostics can be performed in four ways:

(1) Light can be propagated through a filter, such as an interference or a colored-glass filter, which preferentially transmits in a given wavelength band. This is suitable for relatively low-resolution, fixed-wavelength applications, such as selecting a given atomic emission line from a resonance lamp for absorption studies (Section 8.2.2.1) or the wavelength band for detection of thermal radiation (pyrometry, Chapter 13). A bandpass filter can transmit 25–70% of the band of interest, which usually exceeds the transmission factor of dispersive spectrometers, and it is far less expensive. Interference filters should be used only with collimated light.

(2) Light can be analyzed by a dispersive spectrometer with a diffraction grating or a prism as the dispersive element. The radiation is dispersed at the exit focal plane. Wavelength-selected radiation is measured either by a single-element detector placed behind the exit slit (in the exit focal plane) after which the spectrometer is scanned to other wavelengths (serial detection), or by a multielement detector placed in the exit focal plane that detects many wavelengths simultaneously (parallel detection). This is detailed in Section 5.1.3.1. Use of grating spectrometers is common in many diagnostics applications, including OES (Chapter 6), LIF (Chapter 7), spectroscopic reflection studies (Chapter 9), and Raman scattering (Chapter 12); prism-based spectrometers are also used in spectroscopic reflection studies.

(3) Light can be analyzed with a Fabry–Perot interferometer. This is commonly used for very high-resolution spectroscopy, and sometimes requires spectral prefiltering, perhaps by a grating spectrometer. Although the high resolution provided by a Fabry–Perot is rarely needed in optical diagnostics, there are some important applications in OES (Chapter 6) and LIF spectroscopy (Chapter 7). See also Section 10.1.

(4) Light can be analyzed by Fourier transform (FT) analysis by using a Michelson interferometer. This is detailed in Sections 5.1.3.2, 8.2.1, 9.7, 10.1, and 13.4.

Satisfactory spectral analysis and detection are often limited by how long a signal at a given wavelength needs to be analyzed, which is determined, in part, by what fraction of the light signal entering the spectrometer is actually detected. This leads to two considerations that are important in assessing spectral analysis and signal detection.

The first is the improvement in the signal/noise ratio (S/N) with parallel detection, i.e., the simultaneous detection of many wavelength elements. This is known as the Fellgett advantage (Johnston, 1991). If a signal is integrated over a time t, the signal $S \propto t$ and the random noise $N \propto t^{1/2}$, so the $S/N \propto t^{1/2}$. If M spectral elements are each analyzed for a time t in a serial arrangement, the spectrum is measured in time $T = Mt$. If each spectral element is measured for time T in a parallel (or multiplexing) detection arrangement, S/N improves by $M^{1/2}$. This means that S/N is greatly improved when parallel detection is used with a dispersive spectrometer, by ~ 30 when an array detector with $M \sim 1000$ elements is used, vis-à-vis serial detection. Another advantage of array detection vis-à-vis mechanical scanning of the wavelength is improved wavelength stability, which makes background subtraction and ratioing much more accurate. Since Fourier transform spectroscopy analyzes all wavelengths at once, it also enjoys the Fellgett advantage, with similar S/N improvement.

The slits in any dispersive spectrometer set the spectral resolution, but they also limit the throughput (*étendue*) of signal into the spectrometer. However, all collected light is spectrally analyzed in Fourier spectroscopy, which is known as the Jacquinot advantage (Johnston, 1991).

5.1.3.1 Dispersive Spectrometers

In a dispersive spectrometer, sometimes called a monochromator, the signal is focused onto the entrance slit and the detector is placed after the exit slit (in serial detection). The Czerny–Turner arrangement shown in Figure 5.6 is one common design for grating spectrometers (Davis, 1970; James and Sternberg, 1969). A curved mirror, which is usually spherical, images the source from the entrance slit into a parallel beam that is directed to the diffraction grating; a second curved mirror focuses the diffracted

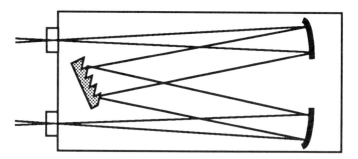

Figure 5.6 Grating spectrometer with a Czerny–Turner design.

5.1 Optical Components

beam onto the exit slit. The spectrometer focal length f is the distance from the exit plane to the last focusing mirror, which is also the distance from the entrance plane to the first focusing mirror.

In a spectrograph, there is no exit slit. The exit focal plane is flat (flat field) and a wide spectral region can be imaged by a multichannel parallel detector, such as a CCD array, photodiode array, or (historically) a film plate. In some instruments there is a mirror near the exit plane that can be rotated to direct the spectrally dispersed beam either to an exit slit for use as a spectrometer (or monochromator) or to a port where the focal plane is flat for use as a spectrograph.

The diffraction grating is commonly a square, of length L, with v grooves per millimeter that run normal to the plane of incidence of the light. The groove density determines the spectral dispersion, which can be obtained from the grating equation $m\lambda = d(\sin \theta_m - \sin \theta_i)$ (Equation 11.21), where m is the diffraction order (an integer), d is the groove separation equal to $1/v$, and θ_i and θ_m are the angles of incidence and diffraction, defined so that $\theta_m = \theta_i$ for reflection ($m = 0$) (Figure 5.7a). (In some conventions the diffraction angle θ_m is defined as the negative of the angle defined here.) The grooves are blazed to maximize first-order diffraction ($m = 1$) at a specified wavelength (blaze wavelength λ_b for the given blaze angle). Diffraction efficiency, shown in Figure 5.7b for a typical grating, is quite dependent on polarization. The grating is more efficient for light polarized along the grooves when $\lambda < \lambda_b$ and for light linearly polarized perpendicular to the grooves when $\lambda > \lambda_b$. A polarization scrambler can be placed before the spectrometer if it is necessary to avoid the polarization sensitivity of the instrument. At $2\lambda_b/3$ and $2\lambda_b$ the grating efficiency typically falls to half that at λ_b; this determines the useful range for the grating. A grating can usually be used in first order near λ_b or in mth order near λ_b/m where

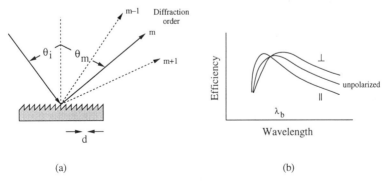

Figure 5.7 (a) Angles of incidence and diffraction for a diffraction grating; (b) diffraction efficiency for a ruled diffraction grating for light polarized parallel and perpendicular to the grooves and for unpolarized light, as a function of wavelength (λ_b is the blaze wavelength).

$m = 2, 3, \ldots$. Gratings can be ruled mechanically or made holographically. Holographic gratings have superior rejection of scattered light and freedom from ghosts due to periodic mechanical errors, which is important in Raman scattering, and can have diffraction efficiencies as high as those for ruled gratings. In many instruments only one grating is mounted, while in others any one of the two or three gratings mounted on a turret can be selected. A more thorough discussion about diffraction gratings is given by Zissis (1995).

The grating is usually the limiting factor in determining the throughput of light in a spectrometer. The $f_\#$ of a spectrometer is given by $f/1.1284L$. [The 1.1284 factor $(2/\pi^{1/2})$ converts the area of the square grating into an equivalent circle with diameter d, and $f_\# = f/d$.] The half-angle subtended by the grating at the slit is $\tan^{-1}(1/2f_\#)$. ($1/2f_\#$ is also known as the numerical aperture NA.) As the $f_\#$ of a spectrometer decreases, more light can be collected and dispersed. If the $f_\#$ of the lens that focuses (parallel) light into the entrance slit is smaller than that of the spectrometer, then light spills over the focusing mirrors and grating (overfilling) and is lost as signal; in fact, it adds to the scattered light background. If it is larger, the light signal is imaged onto only part of the grating (underfilling), and the spectral resolution is less than optimum (see Equation 5.2).

However, such $f_\#$ matching of the lens and spectrometer maximizes the spectrometer throughput only if the slit width is larger than the lateral dimension of the image at the slit. More generally, to maximize the fraction of incident light that can be coupled through the spectrometer, η_c, the beam should be imaged onto the entrance slit to maximize the product of three factors: (1) the transmission losses in the imaging system × (2) the fraction of imaged area that is transmitted through the slit × (3) the fraction of light transmitted through the slit that impinges on the grating (Webb, 1989). For overfilling conditions this last factor is $1/[2f_\# \tan(\sin^{-1}(NA))]$, where $f_\#$ is that of the spectrometer and $\sin^{-1}(NA)$ is the half-angle of the light imaged onto the slit (NA = numerical aperture); for $f_\#$ matching and underfilling conditions, this factor is 1. The coupling of optical fibers to dispersive spectrometers is described in Section 5.1.4.

One way to assess the light collection and throughput experimentally is by back-illuminating the spectrometer at the exit slit and monitoring the beam as it traverses the collection optics, making sure that no external optical element causes any loss of light (other than by reflection). This procedure also facilitates alignment of a focused laser beam in the thin film processing chamber with the image of the entrance slit.

The linear dispersion at the exit focus is $d\lambda/dx = D = \cos\theta_m/mvf$. Assuming equal entrance and exit slit widths w (as is common in practice), the spectral resolution due to dispersion is

5.1 Optical Components

$$\Delta\lambda = \frac{w}{mvf}\cos\theta \qquad (5.1)$$

Spectral resolution improves as the slit width decreases until the limit set by diffraction is reached:

$$\Delta\lambda = \lambda/mN \qquad (5.2)$$

where N is the number of grooves covered by the light. $\lambda/\Delta\lambda$ is the resolving power of the instrument. With proper $f_\#$ matching, the entire grating is covered and $N = vL$.

Light-gathering ability is improved by using smaller $f_\#$ optics and spectrometers (as $\propto 1/(f_\#)^2$), which means that f should be minimized and grating size L maximized. Since spectral dispersion and (usually) spectral resolution are $\propto f$, there is a tradeoff between improving throughput and spectral resolution. The throughput also depends on the reflectance factor for each mirror (~92% for aluminized mirrors in the visible range) and the diffraction efficiency for each grating for the given polarization (Figure 5.7b) in the particular diffraction order.

For a given $f_\#$, there is also a tradeoff between the throughput in a grating spectrometer used for serial detection and spectral resolution that depends on the relative size of the image on the entrance and exit slits and the slit widths. If the lateral dimension of this image on the entrance slit q_1 is $>w$, then decreasing w linearly decreases the optical signal entering the spectrometer. (The entrance slit does not affect throughput when $q_1 < w$.) If, for a peak with spectral width $\Delta\lambda_{signal}$, dispersion leads to a lateral spread $q_2 = \Delta\lambda_{signal}/D$ on the exit slit that is $>w$, then decreasing w linearly decreases exit slit transmission. (Decreasing w does not affect this transmission when $q_2 < w$.) When w is increased, overall throughput increases, however with concomitant instrumental broadening of sharp spectral features.

Consider the spectral analysis of optical emission in a plasma. In such analyses it is common for q_1 to be $>w$, unless the collected light has been spatially filtered. If the widths of the entrance and exit slits are halved, the spectral resolution is improved by a factor of 2 (from Equation 5.1, assuming the diffraction limit of Equation 5.2 has not been reached). However, this occurs with decreased signal throughput. If $q_2 > w$ (broad spectral features or a continuum), then the throughput decreases by a factor of 4 (a factor of 2 at each slit) when w is halved. If $q_2 \ll w$ (sharp spectral peaks), the throughput only decreases by a factor of 2 when w is halved (because of the entrance slit).

The slit width is chosen as a compromise between spectral resolution and signal throughput. When scanning the spectrometer for serial detection, this compromise depends on the distribution of narrow and broad spectral features; both are common in plasma-induced emission (PIE) because there are sharp atomic lines and broader molecular bands.

When a parallel detector is used, the effective exit slit width is the pixel size p_{pix} (or effective pixel width in intensified detectors). This determines the best possible spectral resolution, which is Dp_{pix}. Because all of the light is collected by some element, the throughput is not affected by this effective exit slit width. However, throughput and resolution are still affected by the entrance slit width $w_{entrance}$. If $w_{entrance} > p_{pix}$, the throughput linearly increases with increasing $w_{entrance}$ (as long as $q_1 > w_{entrance}$), but the spectral resolution degrades as $\sim Dw_{entrance}$. If signal throughput is not a limiting factor, it is best to set $w_{entrance} = p_{pix}$. There is no advantage in setting $w_{entrance} < p_{pix}$. Parallel detectors must be placed in a flat focal plane, and therefore, spectrographs must be used (or spectrometers with suitably flat focal planes, i.e., those with very large f).

Typically, single-stage spectrometers with $f \approx 0.3\text{–}0.5$ m and $f_\# \approx 4\text{–}6$ are used for diagnostics applications such as plasma-induced emission, laser-induced fluorescence, and white light source dispersion for spectroscopic reflection studies. When important, interference from lines at different grating orders must be eliminated, and can be prevented by placing absorbing filters before the spectrometer. For instance, light near 400 nm, observed in second order, cannot be distinguished from 800-nm light in first order unless such filters are used. (In spectroscopic reflection studies covering a wide range of wavelengths, prism spectrometers, which have poorer resolution, are often used to avoid this problem of overlapping orders in grating-based instruments.) In Raman scattering, the dispersion provided by such grating spectrometers is sufficient for Raman shifts greater than ~ 300 cm^{-1}. For smaller shifts, longer focal lengths may be needed. Of greater importance in Raman scattering is removing stray light due to the incident laser. This is accomplished either by placing a prefilter that reflects or absorbs the laser line (such as a holographic notch filter) in front of a single-stage spectrometer, by increasing the focal length of the single-stage spectrometer, or by using a triple spectrometer (a subtractive double spectrometer that acts as a filter, followed by a single-stage spectrometer) or double (two-stage) spectrometer. Use of multiple-stage spectrometers can be very expensive. Also, such spectrometers usually have relatively low throughput because they contain many optical elements.

The focal length of the curved mirrors in the spectrometer is that in the plane of Figure 5.6, and the focal plane is determined by this focal length. Since the light beams are not normal to these spherical mirrors, the focal length out of the plane is different and astigmatism (vertical blurring) occurs along the exit slit. In imaging spectrometers or spectrographs, a circular "point" source at the entrance slit more closely images to a circular point at the exit focal plane, because this aberration is lessened *vis-à-vis* conventional spectrometers either by adding a correcting optics stage after the spectrometer or by replacing the spherical mirrors by toroidal mirrors for better off-axis imaging. Imaging (or astigmatic) spectrographs are particu-

5.1 Optical Components

larly useful when multiplexing many spectra onto a two-dimensional CCD array. The pixels are binned perpendicular to the dispersion axis to integrate the signal at each wavelength region. With less optical aberration, the signal is spread over fewer pixels and so fewer pixels need to be binned; this leads to lower noise and fewer unwanted background events due to cosmic rays. Also, this means that more spectra can be acquired at once, and perhaps light from more parts of a reactor or from more reactors can be monitored simultaneously by using fiber optics.

5.1.3.2 Fourier Transform Spectrometers

Fourier transform (FT) spectrometers employ a form of a Michelson interferometer, as illustrated in Figure 5.8 (Johnston, 1991; Back, 1991). When used for infrared absorption spectroscopy, which is the dominant application, the application of these spectrometers is called FTIR (Fourier transform infrared) spectroscopy (or FTIRS). In FTIRS, light with intensity I_i and electric field $E(t)$, which has contributions over the frequency range of interest, impinges on the interferometer. I_i is averaged over several optical cycles, while $E(t)$ includes the oscillation at optical frequencies, as in Equation 2.2. This input beam is split by a beamsplitter into two beams with equal intensities (Figure 5.8). The beams in the two legs are reflected back to the beamsplitter by two mirrors, where they are recombined and then detected, often with a DTGS pyroelectric detector, to give I_m. The position of one of these reflecting mirrors is fixed a distance d from the beamsplitter, while the distance of the other mirror from the beamsplitter is $x = d + \delta$, which is scanned with time. For the scanning mirror, $x = d$ is the zero point distance (ZPD). The intensity I'_m is then measured as

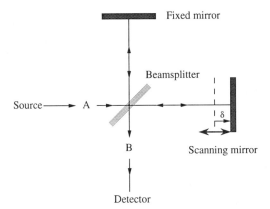

Figure 5.8 Schematic of a Fourier transform spectrometer, detailing the operation of the Michelson interferometer portion.

δ of this second mirror is scanned (across $x = d$); this trace can be expressed in terms of δ or τ, the transit time difference between the two paths $\tau = 2\delta/c$. The input signal I_i must be constant during this scan.

For a single wavelength λ, tracking the beam gives

$$I'_m(\delta) = 0.5\ I_i(\bar{\nu})[1 + \cos(2\pi\delta\bar{\nu})] \tag{5.3}$$

where $\bar{\nu} = 1/\lambda$. Only the modulated component $I_m = 0.5\ I_i(\bar{\nu})\cos(2\pi\delta\bar{\nu})$, called the interferogram, is important. Including instrumental factors, such as the output of the source at each $\bar{\nu}$ [$I_i(\bar{\nu})$] and the detector efficiency, and sample absorption, this expression can be generalized to give

$$I_m(\delta) = B(\bar{\nu})\cos(2\pi\delta\bar{\nu}) \tag{5.4a}$$

For a continuum source, this becomes

$$I_m(\delta) = \int_0^\infty B(\bar{\nu})\cos(2\pi\delta\bar{\nu})d\bar{\nu} \tag{5.4b}$$

The Fourier transform of I_m gives $B(\bar{\nu})$, which contains all the information about the absorption:

$$B(\bar{\nu}) = \int_{-\infty}^{\infty} I_m(\delta)\cos(2\pi\delta\bar{\nu})d\delta \tag{5.5}$$

$B(\bar{\nu})$ is multiplied by an apodization function to correct for finite mirror movement. To correct for the wavelength dependence of the passive and active optical components, the spectrum is often ratioed with one taken with the sample removed from the beam path.

Data analysis of the FT interferogram is performed rapidly by using the fast Fourier transform (FFT) procedure. A spectrum with n spectral points requires n^2 operations using ordinary Fourier transformation and only $n \log_2 n$ steps with the FFT algorithm; this analysis time is comparable to that required for spectral analysis with dispersive spectrometers with multiplexing. Fast Fourier transform analysis is conducted with 2^m points (m integral). An ideal interferometer produces an interferogram that is the same whether the scanning mirror moves forward or backward from its zero path position (ZPD). The interferogram can then be decomposed into cosine waves, and the amplitude of each wave corresponds to one frequency component of the spectrum. Because of imperfections, the interferogram can be asymmetric with respect to the ZPD, and contain sine functions as well. There are standard techniques that correct this phase error (Johnston, 1991).

A broad-band light source, such as that from a glowbar is usually used for FTIR spectroscopy. The sample can be placed either in position A in Figure 5.8, between the source and the interferometer stage, or in position B, after the interferometer and before the detector. Position B is preferred in diagnostics so that no emission from the sample is analyzed.

5.1 Optical Components

Fourier transform methods are also used for analysis of thermal radiation (Section 13.4) and spontaneous Raman scattering. In FT-Raman, a narrow frequency laser, often a low-noise semiconductor diode laser or a Nd:YAG laser (1.06 μm), irradiates a sample. The collected light is directed to the spectrometer for analysis.

Fourier transform spectroscopy enjoys the Fellgett and Jacquinot advantages (Johnston, 1991). In diagnostics applications, where identification of a feature and measuring its intensity are often more important than the high-resolution spectroscopic measurement of feature lineshapes, the Jacquinot advantage is not always significant. However, it has clear advantages in many cases, and FT spectroscopy is often used in infrared absorption experiments (Section 8.2.1) and for surface infrared reflectometry (Section 9.7). It has also been used to examine the plume during pulsed-laser ablation (Section 10.1).

Fourier transform methods are also being applied to the visible (FTVIS) and ultraviolet (FTUV).

5.1.4 *Other Optical Components*

Smith (1990) describes the components and the design of imaging systems. The imaging systems described in Section 5.2.1 assume aberration-free thin lenses. Lenses (and imaging arrangements) must be designed that are sufficiently free of chromatic aberration and other optical aberrations (such as spherical aberration, coma, astigmatism, and field curvature and distortion). All transmitting optical components (lenses, beamsplitters, etc.) must transmit in the wavelength range of interest.

Optical fibers are often useful in delivering light to a reactor and bringing light collected from the reactor to the detection system, as in Figure 9.14. This is particularly true when long distances are involved or when reactor geometry makes it difficult to direct the beams with mirrors. Fibers also enable multiplexing a single laser source and detection system with many reactors. Divided fiber bundle systems (bifurcated, trifurcated, . . .) can be employed for multiplexing and for other applications, as is illustrated in Figure 5.9.

Plastic and compound glass optical fibers are only useful in the visible. Silica fibers, with a useful transmission range of 0.2–2.3 μm, must be used for the ultraviolet. Fluoride and chalcogenide fibers can be used in the infrared. Chalcogenide fibers are transparent in most of the region from 900 to 3300 cm^{-1}, but with limited transmission near 2200 cm^{-1}, while fluoride optical fibers transmit from 2200 to 20,000 cm^{-1} (see Section 8.2.1). Multimode step fibers are usually preferred to multimode graded index and single mode fibers for optical diagnostics applications (Webb, 1989).

Imaging optics can couple into the fiber, as diagrammed in Figure 5.10. To optimize coupling into a fiber, the imaging system must maximize the

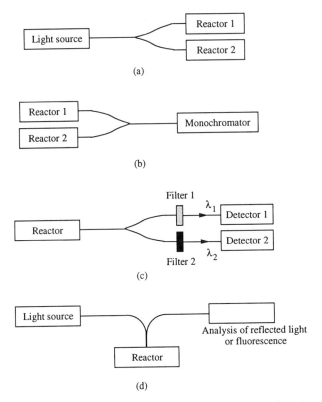

Figure 5.9 Applications of bifurcated optical fiber cables in optical diagnostics.

product of (1) the imaged area that is within the fiber core for a single fiber or the overall diameter for a fiber bundle × (2) the fraction of incident light that is within the solid angle of the acceptance cone of the fiber. As seen in the figure, the (linear) acceptance half angle is $\theta_a = \sin^{-1}(NA)$ of

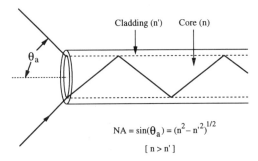

Figure 5.10 Imaging optics for coupling into and from a multimode step index optical fiber.

the fiber, where NA is the numerical aperture of the fiber. Light is easily coupled into fibers from extended sources, such as plasmas, without lenses; however, with the concomitant loss of spatial resolution. This is often done in endpoint detection.

The light leaving the optical fiber has an area defined by the core and the emittance angle, which equals the acceptance angle. Coupling of a fiber, or any other light source, into a grating spectrometer is discussed in Section 5.1.3.1 (also see Webb, 1989). Since multimode step index fibers typically have cores that are wider than typical spectrometer slit widths and $f_\#$'s that are smaller than those for most spectrometers, three coupling strategies should be evaluated to maximize transmission through the spectrometer: (1) imaging with a lensing system to an even smaller spot on the slit to more nearly match the aperture (slit) size, (2) imaging to a larger spot to more nearly match the spectrometer $f_\#$, or (3) butt coupling by putting the fiber at the slit. Since single fibers are circular and the slits are rectangular, there is usually a very large loss at the slit. Fiber optic devices called linear arrays, ribbon arrays, or line arrays transform a circular bundle of fibers to a rectangular array and can improve this coupling factor greatly. One common way to use these arrays is by butt coupling.

To ensure good coupling, the relative position of the lens used to couple into or from an optical fiber must be carefully controlled; this is usually accomplished with xyz translation and rotation stages.

The use of fiber optics for *in situ* monitoring is illustrated by the work of Moslehi *et al.* (1994), who have described the use of chalcogenide fibers to determine the emissivity of a silicon wafer at 5.4 μm in real time and the development of an *in situ* fiberoptic scatter sensor to determine surface roughness, layer thickness, and the spectral emissivity of silicon wafers. The use of optical fibers in optical-based thermometry has been reviewed by Grattan and Zhang (1995).

Polarizing optics are important in reflectance difference spectroscopy (Section 9.10) and ellipsometry (Section 9.11). Table 5.2 is a brief listing of common optical elements that can be used to monitor polarization and to polarize light. Bennett (1995) gives a detailed account of polarization optics.

5.2 Signal Collection and Analysis

The optical apparatus needed to deliver light to the medium to be probed and to collect light from the medium for analysis depends on the nature of the spectroscopy. It also depends on the stated requirements of spatial and temporal resolution and the need for spatial profiling. For example, usually a collimated optical source is delivered to the medium in transmission and specular reflection spectroscopies, and a collimated light source is collected. In LIF, PL, Raman scattering, and nonlinear optical probing,

Table 5.2   Common Polarizing Optics[a]

Type	Properties
Polarizers—to separate or analyze polarizations	
Beamsplitter cubes	Splits into an undeviated beam and one reflected at right angles. Two contacted right-angle prisms, with dielectric coatings at contacted interface; coatings designed for wavelength bands and can be totally or partially polarizing
Glan–Thompson polarizer	One linear polarization is transmitted undeviated, the other is totally internally reflected; cemented calcite prisms, for 230–5000 nm
Glan–Foucault or Glan–air polarizer	Higher-power version of Glan–Thompson, with air between prisms
Wollaston prism	Both linear polarizations somewhat deviated; contacted calcite or crystal quartz prisms
Polaroid	Transmits one linear polarization, absorbs other
Reflection at Brewster's angle	Only s (σ) light is reflected
Retarders—to modify polarization	
Half-wave plate	Rotates linear polarization by 2θ when rotated by θ; birefringent plate, designed for specific wavelength(s)
Quarter-wave plate	Converts from linear to circular polarization and vice versa; birefringent plate, designed for specific wavelength(s)
Fresnel rhomb	Converts from linear to circular polarization and vice versa, two internal reflections, achromatic; a double (half-wave) Fresnel rhomb is an achromatic, in-line rotator of linear polarization
Babinet compensator	Variable retardance across element, can be used as half- or quarter-wave retarder at specific points; two birefringent wedges
Soleil compensator	Uniform retardance across element that can be varied, can be used as half- or quarter-wave retarder at specific points, no deviation; two wedges and a slab, each birefringent

[a] See Bennett (1995) for details.

the optical beam is usually focused onto the medium, as is discussed in Section 2.2, and light emitted into a wide range of angles must be collected. In elastic scattering, commonly, a collimated laser hits the medium and, depending on the topography, either collimated beams leave the material (periodically patterned surfaces) or widely dispersed beams leave it (particles or randomly roughened surfaces). Section 4.5 discusses the possibility of sample heating by the probing laser, which is a concern when focused cw and pulsed lasers are used. No optical source is needed in OES and pyrometry, but the collection of signals emitted over a wide range of solid angles is still an issue in these diagnostics.

5.2.1 *Light Collection, Imaging, and Spatial Mapping*

Collecting light emitted over a wide solid angle is central to many optical probes, such as OES, LIF, and Raman scattering. When spatially resolved measurements are desired with a specified spatial resolution, the details of the imaging system become very important (Smith, 1990). Imaging is treated here in the geometric optics approximation, assuming aberration-free thin lenses. With the size of the imaged volume well defined, the optical probe can spatially map the gas or surface in a reactor. Mapping can provide valuable information about the operation of the reactor, such as the density and temperature profiles in the gas phase, and can be accomplished in several ways depending on the optical spectroscopy.

Three mapping methods are surveyed here and their applications are detailed in succeeding chapters: (1) imaging a confined 3D volume (often used for OES, LIF, Raman scattering), (2) direct 2D imaging (LIF), and (3) tomographic reconstruction from measurements integrated over a line (1D) (OES, absorption). These approaches can entail translating lasers and the collection optics (or, for small laboratory systems, moving the chamber) and the use of array detectors.

Absorption (transmission) spectroscopy integrates density along a line. The laser (or reactor) can be translated to obtain spatially resolved information, such as at various distances between the electrodes of a plasma reactor or at different heights above the wafer in a chemical vapor deposition (CVD) chamber. These data can be used to obtain a 3D profile of density by employing tomographic reconstruction.

While the emphasis of this section in on imaging and mapping gas-phase regions, many of these considerations also apply to solid materials. A 2D map of a surface, film, or wafer can be obtained by transmission, reflection, elastic scattering, Raman scattering, thermal wave, and other solid-state optical spectroscopies by (1) scanning the optical beam across the surface in two dimensions, by reflection from a rotating or translating mirror or by acousto-optic or electro-optic deflection, (2) translating the surface in two dimensions, (3) scanning an optical line (formed by a cylindrical lens) relative to the surface and imaging the signal onto a 1D array detector, or (4) exposing the material to an unfocused beam and imaging the two-dimensional response with a 2D array detector. When these probes are strongly absorbed by the medium and the absorption coefficient depends on wavelength, 2D maps obtained at different wavelengths can sometimes be used to obtain a 3D picture of the material.

As is discussed further in the next chapter, Figure 6.5 illustrates several ways to image a 3D volume with lenses (a, b) and optical fibers (c, d), with spectral analysis of the collected light. The dispersed light is detected either with a single- or multi-element detector. In Figure 6.5a the reactor can be translated to map the volume in the reactor, while in Figures 6.5b and 6.5c, the optics can be translated for mapping. One of the advantages of using fiber optics for light collection during spatial profiling is that the fiber can be translated to examine different regions without concern about alignment with the spectrometer/detector assembly (Figure 6.5c). Several optical fibers can be used to collect light from various regions in one reactor [Figure 6.5d, as in Splichal and Anderson (1991)] or from different reactors, and this light can be spectrally analyzed simultaneously by an astigmatic monochromator and a 2D array detector (CCD array). Figure 5.11 shows an apparatus that permits collection of backscattering light, as well as real-time viewing with a microscope.

The intensity of the electronic signal recorded by the detector depends on the efficiency of the transmission of light through the optical elements and the efficiency of detection (Tsang, 1989). The product of the transmittance of the incident laser and the detected light through all lenses, filters, and beamsplitters (and reflectance from mirrors) η_l is typically 0.1–0.8. In spectral analysis, a fraction η_m of the light is transmitted through the monochromator due to imperfect mirror reflectance and efficiency of grating diffraction into the desired order; this ranges from ~0.1 for multistage monochromators to ~0.6 for optimized single-stage monochromators. The fraction of light coupled into the spectrometer and transmitted to the exit slit is η_c, which was discussed for a grating spectrometer in Section 5.1.3.1. The fraction of this light that is transmitted through the exit slit is η_e. The

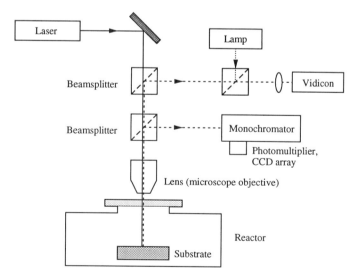

Figure 5.11 Experimental apparatus that permits laser illumination of a sample, the collection of backscattered or back-emitted light, and real-time observation with a microscope.

5.2 Signal Collection and Analysis

product $\eta_c \eta_e$ is at most ~0.7, and is usually much smaller. The overall optical efficiency is $\eta_o = \eta_l \eta_m \eta_c \eta_e$. Typical quantum efficiencies of detection are $\eta_d \sim 0.1–0.5$. The total optical collection and detection efficiency factor is $\eta_t = \eta_o \eta_d$.

5.2.1.1 Imaging Confined 3D Regions

Incoherent light is usually imaged from the reactor to the spectrometer and detector by using a series of lenses; sometimes fiber optics and reflective objects are also used. Several configurations are depicted in Figure 6.5. Webb (1989) and Selwyn (1993) have discussed different ways of coupling light into monochromators with fibers. Selwyn (1993) has also discussed a periscope design for spatially resolved measurements in a plasma. Several other clever schemes are presented in other chapters. For example, Figure 6.35 depicts an assembly that can collect and spectrally analyze optical emission from a laser ablation plume as a function of distance from the surface (which is similar to Figure 6.5b).

The imaging lenses are chosen to define and to image the observation region. Sometimes apertures are placed in the beam path to define the observation region, as is shown in Figure 5.12. Other criteria are also important in imaging. The lenses and all optical materials must transmit the light, which necessitates the use of special materials for collection in the UV and IR. For example, quartz optical fibers must be used to acquire

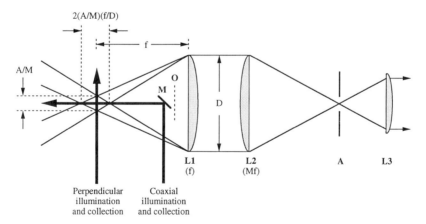

Figure 5.12 Imaging a confined 3D volume with either coaxial illumination and light collection (using a mirror or beamsplitter **M**) or perpendicular illumination and light collection. In scattering experiments these are called backscattering and right-angle scattering configurations, respectively. The collection lenses **L1** and **L2**, depicted here as planoconvex lenses, have focal lengths f and Mf. This diagram depicts a magnification $M = 1$. The imaging volume is defined by the aperture **A**. Sometimes an obscuration disk **O** is placed before the collection lens **L1** to define the imaged volume better (from Eckbreth and Davis, 1977; Eckbreth, 1996).

UV OES spectra in the interesting 200–300 nm regime; glass fibers must not be used. The lenses must be chosen and the overall optical system must be designed so that aberrations, including spherical aberration, astigmatism, coma, and chromatic aberration, do not limit performance. Using doublet lenses corrected for spherical and chromatic aberrations is often helpful. Sometimes commercial camera lenses and microscope objectives provide good performance at low cost.

A typical collection system is shown in Figure 5.12 (Eckbreth, 1996). Light collected a distance d_1 from a thin convex lens, with focal length F, is imaged a distance d_2 away given by

$$\frac{1}{d_1} + \frac{1}{d_2} = \frac{1}{F} \qquad (5.6)$$

The image is inverted and magnified by $M = d_2/d_1$. This single lens should be optimized for this conjugation ratio. Alternatively, this lens can be replaced by two field lenses, each designed for infinite conjugation ratio, with focal lengths $f = d_1$ and $Mf = d_2$, where $f = (1 + 1/M)F$. This latter arrangement is shown in Figure 5.12, and is quite common because it is quite versatile and because large lenses with infinite conjugation ratio (planoconvex lenses) are readily obtainable from vendors. The distance between the lenses can be changed because the first lens makes the beam parallel. With the diameter of both lenses equal to D, the $f_\#$ is defined by the first lens $f_\# = f/D$. Sometimes the numerical aperture of an optical element is defined as $NA = 1/(2\ f_\#)$.

The second lens images the light into an aperture, with diameter A, which defines the focal volume. This aperture may be the detector element, the slit of the spectrometer, a fiber optic, or an inserted aperture. In the last case, another lens is placed after the aperture to image the light onto the detector or into the monochromator.

When this infinite conjugation pair is made with two planoconvex lenses, spherical aberration is minimized by orienting the lenses so that the flat sides face the source and detector apertures and the curved sides face the parallel beam propagating between the lenses. When a grating spectrometer is used, the distance between these two lenses can be adjusted so that an image of the grating falls on the source collection lens, as verified by back-illumination of the spectrometer. An aperture can then be placed at this image to prevent overfilling of the grating, which will minimize the scattered light background in the spectrometer. For optimum illumination of the spectrometer slits, achromatic doublets optimized for infinite conjugation ratio are far superior to simple planoconvex lenses for use in this two-lens collection system because they are simultaneously corrected for spherical and chromatic aberrations.

The solid angle of light collected from the on-axis point $d_1 = f$ from the lens is

5.2 Signal Collection and Analysis

$$\Delta\Omega = \frac{\pi}{4}\left(\frac{D}{d_1}\right)^2 = \frac{\pi}{4}\frac{1}{f_\#^2} \tag{5.7}$$

If light emission is uniform over 4π steradians, then the lens collects only $1/16f_\#^2$ of all the light emitted. Only 0.7% of the emitted photons are collected with $f/3$ optics ($f_\# = 3$) common in OES studies. Lenses with $f_\# \approx 1$ are often used in Raman scattering because the scattered signals are so weak. [In Raman scattering from semiconductors, the effective solid angle of collection is much smaller than that given by Equation 5.7 because of refraction (Section 12.2).] The *étendue,* or throughput, is the product of the solid angle and the area of the collected object (or the image). The *étendue* does not change as an incoherent beam propagates in a lossless optical system.

The imaged volume (Figure 5.12) is defined by the diameter of the source imaged through the aperture, A/M, and the depth of focus, $2f_\#A/M$. This region is approximately cylindrical, so the imaged volume is

$$V_{im} = \frac{\pi}{2}f_\#\left(\frac{A}{M}\right)^3 \tag{5.8}$$

Although light outside this region does not form an image at the aperture, many rays from this outer region pass through the system and contribute to the signal. This is significant only if light emanates from these outer regions, as in plasma emission and in coaxial collection of LIF and Raman scattering; then these outer regions can account for 50% of the signal. In coaxial geometry the laser is directed along the axis with mirror **M** or by a beamsplitter placed between the two lenses **L1** and **L2**. Because such rays often pass through the center of the collecting lens, placing an obscuration disk in the center of this lens preferentially blocks these rays and provides a cleaner image, as is detailed below (Eckbreth and Davis, 1977). Light outside the imaged volume is not significant in right-angle collection during LIF and Raman scattering (Figure 5.12).

In OES from a plasma, the magnitude of the signal is determined by the collection optics. If the power emitted per unit volume into 4π steradians is

$$p_{OES} = N^*h\nu_{em}\gamma_{rad} \tag{5.9}$$

where N^* is the density of excited species, ν_{em} is the frequency of the emission, and γ_{rad} is the rate of radiative decay (Equations 4.17 and 6.7), then the total collected power is

$$\mathcal{P}_{OES} = p_{OES}V_{im}\frac{\Delta\Omega}{4\pi} = \frac{\pi}{32}p_{OES}\frac{1}{f_\#}\left(\frac{A}{M}\right)^3 \tag{5.10}$$

The detected electronic signal is

$$S_{OES} = \eta_t(\mathcal{P}_{OES}/h\nu_{em}) \tag{5.11}$$

where η_t is a product of the optical efficiency through the spectrometer and detector quantum efficiency.

When a laser with power $\mathcal{P}_{inc}$ and intensity I_{inc} induces emission or scattering, as in LIF and spontaneous Raman scattering, the collected signal depends on the contribution within the imaged volume. In LIF, the fluorescence power emitted per unit volume into 4π steradians is

$$p_{LIF} = \alpha I_{inc} \frac{\nu_{em}}{\nu_{inc}} \frac{\gamma_{rad}}{\gamma_t} \tag{5.12}$$

in steady state, where α is the absorption coefficient and γ_t is the total relaxation rate from the excited state, including that due to spontaneous emission ($\gamma_{rad} = A_{21}$). This assumes an idealized three-level system. In spontaneous Raman scattering, the scattered power density per unit volume over all solid angles and per solid angle are, respectively,

$$p_{Raman} = N\sigma_{Raman} I_{inc} \frac{\nu_s}{\nu_i} \tag{5.13a}$$

$$\tilde{p}_{Raman} = N \frac{d\sigma_{Raman}}{d\Omega} I_{inc} \frac{\nu_s}{\nu_i} \tag{5.13b}$$

where ν_i and ν_s are the incident and scattered frequencies, and σ_{Raman} and $d\sigma_{Raman}/d\Omega$ are the total and differential Raman scattering cross-sections.

If the exciting laser is so weakly focused that its waist w is larger than the imaged object height and depth of focus, i.e., $w > A/M$ and $2f_\# A/M$, then the total collected power $\mathcal{P}_i$ and the detected electronic signal S_i are obtained by using Equations 5.10 and 5.11 with p_i (i = LIF, Raman) replacing p_{OES}.

When the laser focus is tighter than the object height and depth of focus, the probed volume depends on the details of the geometry (Figure 5.12). If the laser enters coaxially with the axis of the collection system, as in backscattering, then this volume is determined by the laser cross-sectional area ($\sim \pi w^2$) and the depth of focus $2f_\# A/M = l_\parallel$. If the laser enters normal to the collection axis, in right-angle or 90° scattering configuration, then the volume is determined by the laser area and height of the imaged source region $A/M = l_\perp$. Equation 5.10 can still be used with the appropriate p_i (Equation 5.12 or 5.13), but with the $I_{inc}V_{im}$ product replaced by $\mathcal{P}_{inc}l_{\parallel,\perp}$. For example, the collected power in an LIF experiment is

$$\mathcal{P}_{LIF} = \frac{\alpha \mathcal{P}_{inc}}{8} \frac{\nu_{em}}{\nu_{inc}} \frac{\gamma_{rad}}{\gamma_t} \frac{1}{f_\#} \frac{A}{M} \tag{5.14a}$$

for backscattering/coaxial collection and

$$\mathcal{P}_{LIF} = \frac{\alpha \mathcal{P}_{inc}}{16} \frac{\nu_{em}}{\nu_{inc}} \frac{\gamma_{rad}}{\gamma_t} \frac{1}{f_\#^2} \frac{A}{M} \tag{5.14b}$$

5.2 Signal Collection and Analysis

for right-angle collection. Again, the detected signal is given by Equation 5.11, $S_i = \eta_t (\mathcal{P}_i/h\nu_i)$. Since the ratio of this signal to any background in the imaged volume varies as $\sim 1/l^2$, improving spatial resolution also improves this ratio. Diffusion of the species excited in LIF (or of excited carriers in PL) before they radiatively decay, can broaden the emission region.

Right-angle scattering has two strong advantages. Spurious scattering of the incident laser is less serious than with coaxial collection. Also, the spatial resolution is better and more clearly defined with right-angle scattering. One reason is that $l_\parallel > l_\perp$. Another reason is that in the coaxial geometry light is collected not only from the imaged region spanned by $f \pm f_\# A$ before the lens (for $M = 1$) but also along the optic axis from $f/2$ to ∞ before the lens ($M = 1$). Even though the collected solid angle is smaller outside the imaged region, this still worsens the spatial resolution. This extended contribution does not occur in right-angle scattering.

Coaxial scattering does have some advantages. Light from the imaged volume from other sources, such as from optical emission, constitutes a background for LIF and Raman scattering experiments. Eckbreth (1996) has shown that in comparing different systems, one with an $l_\parallel$ equal to the $l_\perp$ of the other, the collected background light is $8f_\#^3$ ($\gg 1$) larger for right-angle scattering than for backscattering. Another advantage of coaxial collection is that it requires only one port on the chamber, while right-angle scattering requires two. This can be another important advantage for *in situ* probing in a reactor by LIF or light scattering.

Eckbreth and Davis (1977) have shown that placing an obscuration disk of diameter O in front of the collection lens (Figure 5.12) better defines the observed volume in coaxial collection, because the disk limits a greater fraction of the collected solid angle for the regions outside the "imaged" volume than inside it. With the disk in place, collected LIF and scattering comes only between $f \pm f_\# A(D/O)$ for $M = 1$. The fraction of light from the imaged region bounded by $f \pm f_\# A$ increases to $0.5(1 + O/D)$, while the total amount of light from the imaged region decreases only by a factor $1 - O^2/D^2$. Obscuration also helps better define the region for OES measurements.

So far, spatial resolution has been addressed only for linear optical processes involving one laser or no lasers. In crossed-beam CARS (BOX-CARS), and any other process defined by two intersecting lasers (two-color MPI or LIF, etc.), only the medium in the intersection volume gives a signal, so the imaged region is well defined. In CARS, the signal is a coherent beam, all of which can be delivered to the detector.

In each example of imaging described in this section, the profile to be measured—density, temperature, etc.—can be mapped by translating the lasers and the collection objects, or for small laboratory reactors, the reactor itself.

5.2.1.2 Direct 2D Mapping

A wide span of wavelengths can be simultaneously detected by using array detectors, such as 1D photodiode arrays and 2D CCD arrays. Similarly, large spatial regions can be imaged at once by using these arrays. Sometimes they are used to image both wavelength and physical spaces at once.

A 2D image can be projected onto a CCD array, much as it can be on a photographic plate, although with real-time analysis. A narrow-pass filter placed before the detector limits analysis to the selected spectral feature. For example, Geohegan (1992) has used a gated-CCD (ICCD) array to obtain 2D images of the plume formed in pulsed-laser deposition with excellent time resolution, as is seen in Figures 6.29 and 6.30 (in Section 6.3.4). A CCD camera can also be used to obtain a wavelength-dispersed spectrum across the source region that is imaged along the spectrometer slit. Also, light from several, not necessarily adjacent, regions in the reactor can be monitored at once by using a CCD array. Light from each region is delivered to a different position along the entrance slit by optical fibers. The exit plane contains a series of spectra from each region. These last two applications often require an imaging spectrometer to minimize aberrations.

The linear photodiode array can be used to map emission across a line at a given wavelength in two ways: (1) a narrow-pass filter can be placed in front of the PDA to select the wavelength range, or (2) the array can be mounted at the exit slit of a spectrometer, however rotated 90° from its usual orientation.

Hanson and Seitzman (1989) have reviewed the use of fluorescence imaging in two and three dimensions (the light-sheet technique) in flames. In 2D imaging of flames, the laser is loosely focused with a long focal length spherical lens to $\sim$200–500 μm. A cylindrical lens telescope then expands the beam to form a sheet that cuts across the flame. Laser-induced fluorescence is collected at right angles from this cylindrical sheet of light, and is filtered to remove scattered laser light and other emission. It is then imaged on an array detector, such as a CCD array, to give a 2D profile. This planar LIF method is discussed in Section 7.1.2.1 and illustrated in Figure 7.4. Scanning the excitation optics or the flame provides a 3D image. These methods can also be used for thin film diagnostics.

5.2.1.3 Tomographic Reconstruction

In tomographic imaging or computer-assisted tomography (CAT), a 3D or 2D picture of a system is obtained by probing the object along lines, by tracking absorption or emission (1D integration) (Figure 5.13). Deans (1983, 1985) and Kak and Slaney (1988) have reviewed tomographic principles. Burns (1994) has reviewed optical tomography for spectroscopy.

5.2 Signal Collection and Analysis

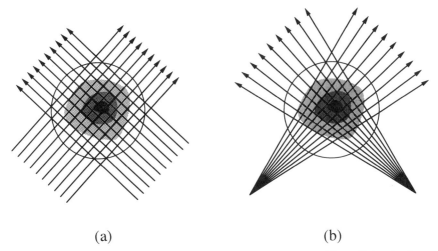

Figure 5.13 Scanning patterns for tomographic reconstruction by (a) parallel projections and (b) fan beam projections.

The Radon transform (Radon, 1917; Deans, 1983, 1985) is useful in analyzing data integrated over a line because it reduces a function in the xy plane into a line integral. The Radon transform of a function $f(x,y)$ is

$$\hat{f}(p,\phi) = \int_{-\infty}^{\infty}\int_{-\infty}^{\infty} f(x,y)\,\delta(p - x\cos\phi - y\sin\phi)dx\,dy \quad (5.15)$$

This integral is along the line defined by $p = x\cos\phi + y\sin\phi$, which is defined by the perpendicular vector from the origin to the line; this vector has length p and makes an angle ϕ with the x axis.

This Radon transform can represent the results in a linear absorption experiment. For a beam propagating along the q axis, Beer's law (Equation 2.5) is $I_{out}/I_{in} = \exp(-\int\alpha(q)dq)$, where the absorption coefficient can vary with q because of variations in the density of the absorbing species, usually in the electronic ground state. So $\ln(I_{in}/I_{out}) = \int\alpha(q)dq$ represents the line integral $\hat{f}(p,\phi)$ along this q axis. The Radon transform also describes optical emission integrated over a line when the collected light is apertured so that rays parallel to the line are collected. Moreover, the solid angle collected for each emitting point along the line must be approximately the same or these changes must be included in the analysis (see Section 6.3.6). This OES mapping gives the spatial profile of the excited-state density.

When measurements are performed for a range of p and ϕ, the data can be converted, or reconstructed, into $f(x,y)$ by using an inverse Radon transform. In optical diagnostics, this function can then be converted into a density profile. Figure 5.13 shows that data can be acquired by parallel projection (a) along a series of M equally spaced parallel lines (different

p) at one angle ϕ. This is then repeated for N different ϕ spanning 360°. In fan beam projections (b) the angular orientation from a point is scanned, and then this is repeated from different points. In either case, this planar analysis is then repeated for a different plane, i.e., for different z. Such extensive collection of data is needed for asymmetric 2D and 3D profiles.

Eckbreth (1996) has surveyed several of the methods used for reconstructing asymmetric profiles in optical analysis of flames, while Hesselink (1989) detailed optical tomography and reconstruction as applied to flow visualization. In the Fourier transform method, the measured Radon transform is Fourier-transformed from p-space to its conjugate ω-space, giving

$$\hat{f}_{FT}(\omega,\phi) = \int_{-\infty}^{\infty} \hat{f}(p,\phi) \, e^{-i\omega p} \, dp \tag{5.16}$$

Then, this is inverted to give

$$f(x,y) = \frac{1}{4\pi^2} \int_0^{\pi} d\phi \int_{-\infty}^{\infty} \hat{f}_{FT}(\omega,\phi) e^{i\omega(x \cos \phi + y \sin \phi)} |\omega| \, d\omega \tag{5.17}$$

Fast Fourier transforms or the even faster convolution methods are used to evaluate this integral. Another reconstruction algorithm is based on linear superposition. $f(x,y)$ is estimated, perhaps by assuming that the projected line integral is equally distributed along the line, and then $\hat{f}(p,\phi)$ is evaluated and compared to the data set. Then a better $f(x,y)$ is determined, and the cycle is repeated until it converges. Such algorithms are also known as the back-projection and filtered back-projection approaches. In the algebraic method, the xy plane is divided into $M \times N$ elements and $f(x,y)$ is converted into an array; $f(x,y)$ and $\hat{f}(p,\phi)$ are then related by a series of linear equations that are solved.

For circularly symmetric profiles, as is common in many thin film reactors (cylindrical plasma chambers), data need be taken only for M different p at one ϕ (for each z) (Figure 5.13). The Radon transform then becomes identical to the Abel transform f_A (Deans, 1983, 1985; Vest, 1979). Choosing $\phi = 0$ and defining $r^2 = p^2 + y^2$, Equation 5.15 becomes

$$\hat{f}(p) = f_A(p) = 2 \int_{|p|}^{\infty} \frac{f(r) r \, dr}{(r^2 - p^2)^{1/2}} \tag{5.18}$$

Inversion of the Abel transform gives $f(r)$. Deans (1985) points out that one way to do this is by using a Hankel transform (Barrett, 1984):

$$f(r) = -\frac{1}{\pi} \int_r^{\infty} \frac{f'_A(p) dp}{(p^2 - r^2)^{1/2}} \tag{5.19}$$

where $f'_A(p) = df_A(p)/dp$. More generally, the Abel transform can be inverted by Fourier methods (Bracewell, 1978), Laplace transform methods (Sneddon, 1972), and by Radon transform methods (Deans, 1983). Hughey and Santavicca (1982) have compared the errors resulting by reconstruction by "onion peeling," Abel transformation, and Fourier convolution techniques. "Onion peeling" is an algebraic method in which the axisymmetric region is divided into M concentric rings, for M lateral measurements.

This type of tomographic reconstruction has been used to map OES profiles in plasma reactors (Section 6.3.6). Although little such work has been conducted in thin film reactors by using absorption diagnostics, which is analogous to x-ray-computed tomography, such an application is certainly feasible and interesting.

5.2.2 *Temporal Resolution*

The temporal response requirements in optical diagnostics depends on the application. In real-time control it is determined by the maximum time the process can wait for feedback control to correct an error. This can range from many milliseconds in rapid thermal processing to several seconds in molecular beam epitaxy (MBE) and metalorganic chemical vapor deposition (MOCVD) to minutes for even slower processes. The limiting factor in the feedback may be the time required to acquire the desired data, and what the desired data set is, or the time needed for data analysis and interpretation. Given how long the process can continue astray and still be corrected, this leads to a tradeoff that determines the minimum amount of data that needs to be collected for a sufficiently rapid response. For example, while spectroscopic analysis at many wavelengths may be required for detailed analysis of the process, the much more rapid data acquisition and analysis at only a few wavelengths may suffice for control applications.

The temporal resolution needed for some fundamental-research and process-development studies may be much faster than this characteristic process time. In pulsed-laser deposition (PLD) and laser-induced thermal desorption (LITD), the desorbed/ablated species need to be probed with a resolution in the range of ~ 1 nsec–1 μsec. In PLD the characteristic process times are much slower, ~ 1 sec, because the deposition rate is ~ 1 Å/sec.

Temporal resolution can be determined either by the detection electronics or by the appearance and duration of the laser pulse. Fast electronic gates and image converters can be used. When they are also employed to obtain images, these devices are called high-speed cameras, such as those used in intensified-CCD array (ICCD) photography, high-speed framing photography, and streak photography. In some cases, an entire 2D image must be acquired with a multielement detector (camera) within the set time limits. Alternatively, pulsed lasers can be used, one to initiate the event and a delayed pulse to probe it. Since the pulse length of the laser determines the timing, a slower detector can be used. In imaging, slow cameras can be employed with the laser pulse as a flashbulb, as in interference imaging by Schlieren photography or (scattering imaging by) shadow photography (Chapter 10). These methods are discussed more comprehensively by Geohegan (1994) and described in the discussions of PLD diagnostics in Chapters 6, 7, 8, 10, and 13.

5.2.3 *Signal Processing*

5.2.3.1 Noise and Signal Averaging

The optical beam, spectroscopic interaction, and detection process all contribute to noise in the signal (Yariv, 1985; Saleh and Teich, 1991). The maximum acceptable level of noise depends on the process. For example, in absorption monitoring, noise and drifts in the amplitude stability of the optical source can be very important. The amplitude stability of coherent and incoherent light sources used in absorption measurements is discussed in Chapter 8. Frequency stability, which is of great concern in high-resolution spectroscopy, is usually less significant in optical diagnostics.

The probabilistic nature of quantum detection leads to shot noise. If there are S counts of signal per second, then there are St counts in t sec and the shot noise is $\sqrt{St}$. The measured signal is therefore usually within $St \pm \sqrt{St}$, and the signal-to-noise ratio (S/N) is $\sqrt{St}$, which increases with collection time. Dark current is the current measured even in the absence of a light signal. Although the signal due to dark current can be subtracted from the measured signal count rate to give the signal due to the optical beam, it still contributes to shot noise. Thermal (or Johnson or Nyquist) noise contributes a signal $\sim k_B T \Delta\nu$, where $\Delta\nu$ is the frequency bandwidth of detection $\sim 1/t$. Cooling the detector lessens these latter two sources of noise.

For a photomultiplier, the signal-to-noise (voltage) ratio is (Yariv, 1985)

$$\left(\frac{S}{N}\right)^2 = \frac{2(\mathcal{P}e\eta_d/h\nu_s)^2 G^2}{2G^2 e(i_c + i_d)\Delta\nu + 4k_B T \Delta\nu/R} \tag{5.20}$$

where $\mathcal{P}$ is the incident power, η_d is the detection efficiency, ν_s is the photon frequency, G is the PMT gain ($\sim 10^6$), i_c is the signal current equal to $\mathcal{P}e\eta_d/h\nu_s = Se$, i_d is the dark current, and R is the output load of the PMT. When dark and thermal noise can be neglected, this ratio is $\sqrt{St}$.

For a photodiode, the signal-to-noise (voltage) ratio is (Yariv, 1985)

$$\left(\frac{S}{N}\right)^2 = \frac{2(\mathcal{P}e\eta_d/h\nu_s)^2}{3e^2\mathcal{P}\eta_d\Delta\nu/h\nu_s + 2ei_d \Delta\nu + 4k_B T \Delta\nu/R_L} \tag{5.21}$$

where R_L is the load resistance.

The minimum detectable power is obtained by setting $S/N = 1$. This is inversely related to the detectivity for semiconductor photodiode detection, which is plotted in Figure 13.3c.

5.2 Signal Collection and Analysis

Signal averaging is often needed to lessen the impact of noise on the signal and the detector. There are several common ways to reduce noise levels when acquiring steady-state and transient signals. For steady-state signals, averaging the signal over a long time is often sufficient, as in CCD array collection (see the discussion of parallel detection in Section 5.1.3) and photon counting in PMT detection, or integrating the (constant) signal with an RC circuit.

Lock-in detection is frequently used to reduce noise in steady-state signals. The light intensity is amplitude-modulated at frequency (ν) by passing it through a mechanical chopper (which is the most common method), acousto-optic deflector, or electro-optic modulator. This usually results in a square-wave light intensity modulation, whose major sinusoidal frequency component is at ν. For linear interactions, the signal is modulated mostly at ν. The signal is analyzed by a lock-in detector, which senses the component of the signal modulated at ν, which may be partially out of phase with the amplitude modulation. When analyzed with a time constant τ, the signal noise is then reduced to that between $\nu \pm 1/\tau$. ν should not be chosen at or near AC power line frequencies, 50 or 60 Hz, or near their harmonics or subharmonics, and must be much less than the slowest characteristic rate of the light–matter interaction, such as the slowest relevant relaxation rate.

Pulsed signals can be analyzed by collecting data in time bins and averaging the results of many pulses in those bins. In such "boxcar averaging," one or more fixed time bins are set relative to the laser pulse during the light generation process, and sometimes immediately before and much after the laser pulse, to get a background level. One time bin can also be slowly scanned in time. Pulsed signals can also be digitized and then averaged (transient digitizer).

Winefordner and Rutledge (1985) have compared the detection limits of several optical spectroscopies by examining the signal-to-noise ratios for each.

5.2.3.2 Signal Calibration

Signal calibration may entail (1) determining the absolute magnitude of the signal intensity, (2) correcting for wavelength-dependent sensitivities, or (3) calibrating the wavelength scale.

Absolute calibration of signal intensity is difficult. It is very important in pyrometry (Chapter 13); however, it is often unnecessary for other applications, as in OES and LIF, where comparison to a reference signal is sometimes helpful. Of more importance for OES and LIF is the wavelength dependence of the system response (the transmittance, wavelength dependence of the imaging system, etc.), which can distort the relative signal strengths at different wavelengths. This can be particularly important in thermometry where the relative heights of peaks are used to determine temperature (Chapter 18). The emission spectra may need to be corrected for the wavelength dependence of the detector efficiency and the spectrometer throughput. The latter is determined by the efficiency of diffraction gratings in spectrometers, which depends both on wavelength and polarization (Figure 5.7b) The effect of polarization can be removed by placing a polarization scrambler before the entrance slit. When the linear polarization of the signal is analyzed before dispersion by the monochromator, as is common in spontaneous Raman scattering, another method is also common. Any linear polarization can be analyzed by transmission through a (rotatable) half-wave plate, which is followed by a fixed polarizer that transmits only one, fixed linear polarization into the monochromator. This arrangement removes the grating dependence on polarization.

Wavelength calibration of grating spectrometers is often performed by reference to atomic emission lines.

References

W. P. Acker, B. Yip, D. H. Leach, and R. K. Chang, *J. Appl. Phys.* **64,** 2263 (1988).

D. M. Back, *Phys. Thin Films* **15,** 265 (1991).

H. H. Barrett, *Prog. Opt.* **21,** 219 (1984).

M. Bass, ed., *Handbook of Optics,* 2nd ed., Vol. 2. McGraw-Hill, New York, 1995.

J. M. Bennett, in *Handbook of Optics,* 2nd ed., Vol. 2, Chapter 3, p. 3.1. McGraw-Hill, New York, 1995.

R. B. Bilhorn, J. V. Sweedler, P. M. Epperson, and M. B. Denton, *Appl. Spectrosc.* **41**(7), 114 (1987).

R. N. Bracewell, *The Fourier Transform and Its Applications,* 2nd ed. McGraw-Hill, New York, 1978.

U. Brackmann, *Lambdachrome® Laser Dyes,* 2nd ed. Lambda Physik GmbH, Göttingen, Germany, 1994.

D. H. Burns, *Appl. Spectrosc.* **48,** 12A (1994).

R. K. Chang and M. B. Long, in *Light Scattering in Solids II* (M. Cardona and G. Güntherodt, eds.), Chapter 3. Springer-Verlag, Berlin, 1982.

S. P. Davis, *Diffraction Grating Spectrographs.* Holt, Rinehart & Winston, New York, 1970.

S. R. Deans, *The Radon Transform and Some of Its Applications.* Wiley (Interscience), New York, 1983.

S. R. Deans, in *Mathematical Analysis of Physical Systems* (R. E. Mickens, ed.), Chapter 3, p. 81. Van Nostrand-Reinhold, New York, 1985.

W. Demtröder, *Laser Spectroscopy: Basic Concepts and Instrumentation.* Springer-Verlag, Berlin, 1982.

A. C. Eckbreth, *Laser Diagnostics for Combustion Temperature and Species,* 2nd ed. Gordon & Breach, Luxembourg, 1996.

A. C. Eckbreth and J. W. Davis, *Appl. Opt.* **16,** 804 (1977).

References

L. J. M. Esser and A. J. P. Theuwissen, in *Handbook on Semiconductors, Completely Revised Edition* (C. Hilsum, ed.), Chapter 4, p. 389. Elsevier, Amsterdam, 1993.

D. B. Geohegan, *Appl. Phys. Lett.* **60,** 2732 (1992).

D. B. Geohegan, in *Pulsed Laser Deposition of Thin Films* (G. Hubler and D. B. Chrisey, eds.), Chapter 5, p. 115. Wiley (Interscience), New York, 1994.

K. T. V. Grattan and Z. Y. Zhang, *Fiber Optic Fluorescence Thermometry.* Chapman & Hall, London, 1995.

Hamamatsu Corp., product information, 1995.

R. K. Hanson and J. M. Seitzman, in *Handbook of Flow Visualization,* (W.-J. Yang, ed.), Chapter 15, p. 219. Hemisphere, New York, 1989.

L. Hesselink, in *Handbook of Flow Visualization* (W.-J. Yang, ed.), Chapter 20, p. 307. Hemisphere, New York, 1989.

B. J. Hughey and D. A. Santavicca, *Combust. Sci. Technol.* **29,** 167 (1982).

J. F. James and R. S. Sternberg, *The Design of Optical Spectrometers.* Chapman & Hall, London, 1969.

S. Johnston, *Fourier Transform Infrared: A Constantly Evolving Technology.* Ellis Horwood, New York, 1991.

A. C. Kak and M. Slaney, *Principles of Computerized Tomographic Imaging.* IEEE Press, New York, 1988.

M. M. Moslehi, C. J. Davis, A. Paranjpe, L. A. Velo, H. N. Najm, C. Schaper, T. Breedijk, Y. J. Lee, and D. Anderson, *Solid State Technol.* **37**(1), 35 (1994).

J. E. Pemberton, R. L. Sobocinski, M. A. Bryant, and D. A. Carter, *Spectroscopy* **5**(2), 26 (1990).

Princeton Instruments, Inc., *Catalog of High Performance Cameras,* p. 30. Princeton, Instruments, Princeton, NJ, 1994.

J. Radon, *Gese. Wisse. Leipzig Mathe. Phys. Kl.* **69,** 262 (1917).

B. E. A. Saleh and M. C. Teich, *Fundamentals of Photonics.* Wiley, New York, 1991.

G. S. Selwyn, *Optical Diagnostic Techniques for Plasma Processing.* AVS Press, New York, 1993.

W. J. Smith, *Modern Optical Engineering: The Design of Optical Systems,* 2nd ed. McGraw-Hill, New York, 1990.

I. N. Sneddon, *The Use of Integral Transforms.* McGraw-Hill, New York, 1972.

M. P. Splichal and H. M. Anderson, *SPIE* **1594,** 189 (1991).

J. C. Tsang, in *Light Scattering in Solids V* (M. Cardona and G. Güntherodt, eds.), Chapter 6, p. 233. Springer-Verlag, Berlin, 1989.

C. M. Vest, *Holographic Interferometry.* Wiley, New York, 1979.

M. J. Webb, *Spectroscopy* **4,** 26 (1989).

J. D. Winefordner and M. Rutledge, *Appl. Spectrosc.* **39**(3), 377 (1985).

A. Yariv, *Optical Electronics,* 3rd ed. Holt, Rinehart & Winston, New York, 1985.

A. Yariv, *Quantum Electronics,* 3rd ed. Wiley, New York, 1989.

G. J. Zissis, in *Handbook of Optics* (M. Bass, ed.), 2nd ed., Vol. 2, Chapter 5, p. 5.1. McGraw-Hill, New York, 1995.

CHAPTER **6**

Optical Emission Spectroscopy

Optical emission spectroscopy (OES) is the analysis of light that is emitted from a medium in the absence of external optical excitation. Significant levels of such optical emission (OE) are present only in special circumstances. Optical emission is important during plasma-assisted processing, such as plasma-assisted etching and deposition, and sputter deposition processes, where gas-phase species are promoted to excited electronic states by collisions with energetic electrons. These excited states relax to lower states, which may be the ground electronic state, by spontaneous emission. This emitted radiation is then spectrally dispersed and detected. Optical emission from a plasma is also called plasma-induced emission (PIE). Although OE between vibrational levels in the same electronic state also occurs, it is weaker (Equation 4.17) and of less interest.

More generally, all materials in thermal equilibrium at a given temperature T emit light that is called blackbody radiation or thermal radiation. Monitoring the intensity and spectral distribution of this optical emission is a valuable probe of the temperature of the surface; this is known as optical or infrared pyrometry, which is discussed in Chapter 13.

The emission processes in OES are depicted in Figure 6.1. Because only excited species in plasmas are detected by OES, the observed spectrum gives information about the excited-state density and does not directly reflect the ground-state population profile. The densities of excited-state species are typically $<10^{-4}$ of the ground-state density. Furthermore, while emission from specific intermediates and products may dominate OE, emission from the chemically dominant species and interesting highly reactive species may not be observable at all. Still, when such emission is present, OES is a powerful, yet simple diagnostic that is useful for practical real-time monitoring. The absence of an external excitation source makes OES an inexpensive and robust candidate for real-time control, but it limits the

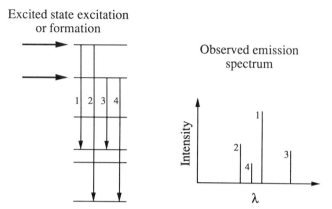

Figure 6.1 Energy levels involved in typical OES experiments.

versatility of this method. Nonetheless, OES is probably the most widely used optical probe of plasma-assisted processing, and is often available in OEM (original-equipment manufacturer) processing equipment for manufacturing.

While OES has proved to be very useful in obtaining a fundamental and process-oriented understanding of gas-phase and surface events in thin film processing in glow discharges, it is having an even greater impact on real-time monitoring and control of plasma-assisted processing. This is due, in part, to its relative simplicity and low cost. Perhaps the most important example is endpoint detection, which indicates the completion of film etching. Optical emission spectroscopy is routinely used to monitor gas-phase population during process development and for endpoint detection with real-time control sensors during manufacturing. Controlling film composition by using OE data during reactive sputtering is also important. Optical emission spectroscopy is more easily adapted for real-time monitoring and control than other optical diagnostics because it requires no optical source. Very good spatial and temporal resolution is possible.

Optical emission during plasma processing can come from neutral or ionized atoms, radicals, or molecules that have been electronically excited. This type of emission has been critically reviewed by Gottscho and Miller (1984), Dreyfus et al. (1985), and Donnelly (1989). Selwyn (1993) has described the practical application of OES for process monitoring during plasma processing, and has detailed the required instrumentation. Reviews by Greene (1978) and Greene and Sequeda-Osorio (1973) have detailed earlier OES work on glow discharge etching and sputter deposition.

Optical emission from other types of film processes is less frequent. For example, emission from thermal reactions involved in film processing is rare, although some processes are chemiluminescent. Chemiluminescence from O_2, OH, and possibly SiO during thermal chemical vapor deposition (CVD) of silicon dioxide from SiH_4/O_2 mixtures has been reported by Van de Weijer et al. (1988). Emission has also been seen during photosensitized reactions, such as from $HgNH_3$ excimers during the 254-nm Hg-photosensitized reaction of SiH_4–NH_3 to form silicon nitride films (Fuyuki et al., 1990). Fluorescence can also be monitored during ultraviolet-laser–assisted deposition. During KrF laser (248 nm) excitation of GeH_4 to grow Ge films, Osmundsen et al. (1985) found that the intensities of optical emission from Ge at 303.9 nm ($5s\ ^1P_1 \to 4p^2\ ^1D_2$) and 265.2 nm ($5s\ ^3P_1 \to 4p^2\ ^3P_0$) and from GeH ($A\ ^2\Delta \to X\ ^2\Pi$) varied quadratically with laser intensity. This suggested that Ge and GeH have a common precursor that is produced by the absorption of two KrF laser photons.

Most PIE features have been assigned and are readily identifiable. Emission from atoms results in sharp lines, while that from molecules is broader and sometimes structured, as is seen in the N_2 emission band in Figure 6.2. Bandheads of characteristic molecular emission lines have been tabulated by Pearse and Gaydon (1976). Spectroscopic information can also be obtained from Herzberg (1937, 1950, 1966), Barrow (1973, 1975), Suchard (1975), Suchard and Melzer (1976), Huber and Herzberg (1979), and Jacox (1988, 1990, 1994). Emission lines from atoms and ions of the elements, along with their intensities, have been tabulated by Moore (1949, 1952, 1958), Corliss and Bozman (1962), Moore and Merrill (1968), Striganov and Sventitskii (1968), Harrison (1969), Zaidel' et al. (1970), Parsons and McElfresh (1971), Bashkin and Stoner (1975, 1978, 1981, 1982), and Phelps (1982).

Quantitative analysis of the intensity of OES signals must be made with caution as the intensity depends on the product of the density of species, the efficiency of excitation, and the rate of spontaneous emission; this is

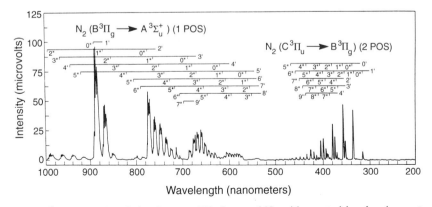

Figure 6.2 Optical emission from an RF plasma of N_2, with spectral band assignments (from Harshbarger et al., 1977). See Figure 3.2 for the electronic energy levels of N_2.

detailed below. Calibration to obtain absolute densities is relatively difficult with OES and relatively easy using absorption spectroscopy; however, OES signals are much larger (Greene, 1978). Although laser-induced fluorescence (LIF) signals can be even larger than those for OES, it is still difficult to obtain absolute densities with LIF. The competitive advantage of OES over these other techniques is in its use as a monitor of relative, not absolute, conditions for process monitoring, without an external optical source. For example, endpoint detection relies on changes in emission intensities with time. During steady-state processing, the most reliable correlations with process parameters come by ratioing the emission intensities of different spectral features.

6.1 Mechanisms for Optical Emission

Plasma-induced emission can result from several different excitation processes (Donnelly, 1989). For example, it can come from electron (e^-) impact excitation,

$$A + e^- \rightarrow A^* + e^- \tag{6.1}$$

electron impact dissociation,

$$AB + e^- \rightarrow A^* + B + e^- \tag{6.2}$$

ion impact,

$$A^+ + e^- (+ M) \rightarrow A^* (+ M) \tag{6.3}$$

or chemiluminescent recombination,

$$A + BC \rightarrow AB^* + C \tag{6.4}$$

where A and B are atoms, radicals, or molecules, AB and BC are radicals or molecules, the asterisk (*) indicates the excited species that emits light, and e^- (+ M) may be either a neutral species, a negative ion, an electron plus a third body (M), or a surface. Reaction 6.3 includes species that are excited by collisions with fast neutrals or ions, as described by Phelps (1990, 1991, 1992).

Each mechanism has been identified as being important in different plasmas (Donnelly, 1989). Optical emission from excited F atoms in CF_4/O_2 and excited Cl in Cl_2 discharges used for etching are excited by electron impact as in Reaction 6.1, while electron impact dissociation (Reaction 6.2) contributes to Cl emission in the momentary cathode sheath in a Cl_2 discharge (Gottscho and Donnelly, 1984). Hydrogen atoms are excited in part due to Reaction 6.3 in a DC H_2 discharge (Cappelli et al., 1985). Recombination (Reaction 6.4) leads to emission in Cl_2 discharges, which

6.1 Mechanisms for Optical Emission

is very weak, and SiF$_3^*$ emission in F-atom-containing discharges during Si etching by

$$\text{SiF}_2 + \text{F (or F}_2) \rightarrow \text{SiF}_3^* \ (+\text{F}) \tag{6.5}$$

(Donnelly and Flamm, 1980). SiF$_3^*$ emission is useful in probing etching processes and as an endpoint detector.

Only a very small fraction of the species in the plasma are excited in steady-state conditions. The emission intensity I_{A^*} from A* depends on the radiative lifetime τ_{rad} and the excitation mechanism. For Reaction 6.1

$$N_{A^*} = \left(\frac{N_A}{\gamma_t}\right) \int_{\mathcal{E}_0}^{\infty} N(\mathcal{E})\sigma_A(\mathcal{E})v(\mathcal{E})d\mathcal{E} \tag{6.6}$$

where N_{A^*} is the density of excited species A*, N_A is the density of ground-state species A, γ_t is the total rate of decay of A*, $\sigma_A(\mathcal{E})$ is the electron impact excitation cross-section that depends on electron energy $\mathcal{E}$, $N(\mathcal{E})$ is the electron energy distribution function (density of electrons with energies between $\mathcal{E}$ and $\mathcal{E} + d\mathcal{E}$), $\mathcal{E}_0$ is the threshold energy for exciting A from the ground state to the emitting state, and $v(\mathcal{E})$ is the electron speed $(2\mathcal{E}/m_e)^{1/2}$. The power per unit volume emitted by this excited species is (Equation 5.9)

$$p_{\text{OES}} = N_{A^*} h\nu_{\text{em}} \gamma_{\text{rad}} \tag{6.7a}$$

where γ_{rad} is the rate of radiative decay (spontaneous emission; Equation 4.17) from the excited state and $h\nu_{\text{em}}$ is the energy of the emitted photon. This is related to the detected signal in the next subsection. The effect of radiation trapping (self absorption) has been detailed by Holstein (1947) [see Figure 8.27].

In a discharge, N_A depends on heterogeneous events in the film process itself (etching or deposition), as well as gas-phase events. For example, in plasma etching of Si by Cl$_2$, optical emission from SiCl depends on the electron impact excitation of desorbed SiCl, as well as the electron-induced dissociation and excitation of SiCl$_x$ ($x = 2 - 4$)—which could occur in several steps or in a single step. In sputter deposition of metals, N_{metal} is determined by the sputtering process.

Equations 6.6 and 6.7a describe only excitation, and hence emission, from states directly excited by electrons. Collisions of excited molecules with heavier species redistribute part of this population to other levels in the excited electronic state, which also fluoresce and are detected. The emission profile from a given excited electronic state in a molecule to a lower electronic state consists of a series of transitions between different vibronic levels in the two states. For a given vibronic level in the excited state, the intensity of each emission line allowed by the rotational selection rules is determined by the Franck–Condon factor that describes vibrational level overlap and by the rotational matrix element; this is given by Equation

4.37. Since many states are excited by electron collisions, in contrast to the more selective laser absorption in LIF, the resulting spectrum is more complex in comparison.

Doppler broadening dominates the measured emission profile of excited atoms in steady-state discharges if there is small instrumental broadening, slow radiative decay, little Stark broadening, and low enough pressure to make collisional broadening insignificant (Section 4.2.1.1). If the temperature determined from the Doppler linewidth is the gas temperature, it is likely that electron impact (Reaction 6.1) has excited the atom because there is little translational energy given to an atom during electron impact, while much more is imparted in Reactions 6.2–6.4. The narrow emission peaks in Figure 6.3 suggest that F* and Ar* are produced by electron impact in a $CF_4/O_2/Ar$ discharge (Gottscho and Donnelly, 1984). Time-resolved measurements of Cl* emission in a Cl_2/Ar discharge (Figure 6.4) show a narrow profile at the anode at the maximum applied voltage during the cycle, suggesting Reaction 6.1, and a narrow profile superimposed upon a broader feature at the cathode at the minimum applied voltage, suggesting both Reactions 6.1 and 6.2 (Gottscho and Donnelly, 1984).

6.2 Instrumentation

Optical emission spectroscopy involves the collection, spectral dispersion, and detection of light. Because OES from plasmas is often very strong, the light collection and detection efficiencies need not always be optimized.

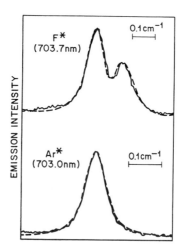

Figure 6.3 Time-averaged F* (upper, $CF_4/O_2/Ar$ discharge) and Ar* (lower, pure Ar discharge) emission lineshapes in the sheath near the electrode recorded by a scanning Fabry–Perot interferometer (from Gottscho and Donnelly, 1984).

6.2 Instrumentation

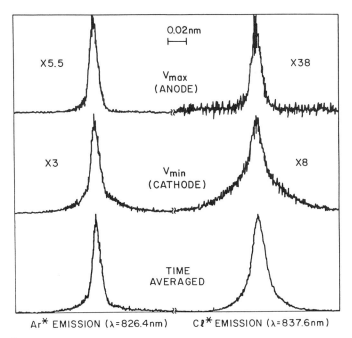

Figure 6.4 Ar* (left) and Cl* (right) emission lineshapes from the sheath near the (momentary) electrode, recorded by a scanning Fabry–Perot interferometer as a function of time during the RF cycle in an RF Cl$_2$/Ar discharge (from Gottscho and Donnelly, 1984).

However, efficient detection is needed when rapid analysis is important and only a small fraction of the surface is being processed, as is true for endpoint detection during plasma etching of patterned films. Typical experimental setups for OES are shown in Figure 6.5. Emission from a specific volume in the plasma chamber is imaged onto the entrance slit of a spectrometer by a series of lenses. The required spatial resolution can be achieved by using the techniques described in Section 5.2.1. Since emission from 200–900 nm is often collected, UV-grade fused-silica lenses should be used. The location of the imaged volume can be laterally scanned by moving the collection lenses or, in research reactors only, by translating the plasma reactor itself. The collected light can be imaged directly onto the spectrometer slits or it can be delivered to it by optical fibers.

As discussed in Chapter 5, if the power emitted per unit volume is p_{OES}, the total collected power in the OE signal is (Equation 5.10)

$$\mathcal{P}_{OES} = p_{OES} \, V_{im} \frac{\Delta\Omega}{4\pi} = \frac{\pi}{32} p_{OES} \frac{1}{f_\#} \left(\frac{A}{M}\right)^3 \qquad (6.7b)$$

where V_{im} is the volume imaged into the monochromator, $\Delta\Omega$ is the collected solid angle, and $f_\#$ is the f-number of the collection system with

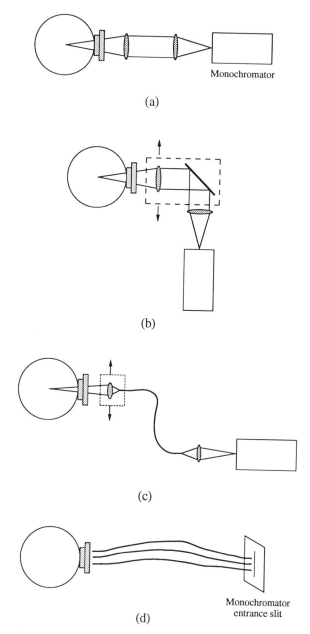

Figure 6.5 Typical experimental apparatus used for OES, depicting light collection by (a) a pair of lenses, as in Figure 5.12; (b) a pair of lenses, with the capability of mapping by translating the lens/mirror combination; (c) an optical fiber, which can be translated for mapping; and (d) a series of optical fibers, which provides individual spectra for each when an astigmatic monochromator is used with a 2D array detector (CCD array). The lens system for (a) and (b) is more accurately depicted, such as with planoconvex lenses, in Figure 5.12. Lenses are needed for coupling to and from the fibers in (d). In each case there is a detector after the monochromator.

6.2 Instrumentation

magnification M and aperture diameter A, as discussed in Section 5.2.1.1. The detected signal intensity is $S_{OES} = \eta_t(\mathcal{P}_{OES}/h\nu_{em})$, where η_t is a product of the optical efficiency through the spectrometer and the detector quantum efficiency.

Light dispersed by the monochromator is detected by a photomultiplier as the spectrometer is scanned. A 0.25-m focal length spectrometer gives ~0.05 nm resolution (1200 groove/mm grating, ~15 μm slits) (Equation 5.1), and this usually gives sufficient resolution for detailed spectroscopy. When resolution is not a concern, slit widths can be opened to increase light collection or the monochromator can be replaced by a narrow-band transmission (interference) filter (~5–15 nm bandpass) centered on the emission wavelength of interest; in this latter case the filter must transmit the desired line without interference from other emission features. GaAs and S-20 response photomultipliers (Section 5.1.2) have good quantum efficiency from 200–900 nm. Detection of shorter wavelengths (<190 nm) requires the removal of oxygen throughout the collection path and inside the spectrometer to avoid absorption by O_2, and is rarely done. Longer wavelength transitions (>900 nm) from electronic transitions are relatively weak and rare, while those from vibrational transitions are much weaker than electronic transitions. Analysis of such IR emission involves the use of necessarily noisier detectors and IR signals must compete with the background from blackbody radiation.

One way to maintain good spectral resolution over a wide spectral range with rapid data analysis is to use a spectrometer with a fast turning or spinning grating. Instruments are available that can scan the visible in ~1 sec, which is much faster than that obtainable in spectrometers geared for research.

Alternatively, the dispersed light can be detected by a parallel detector such as a diode array, an intensity-enhanced diode array, or a charge-coupled device (CCD) array detector. Parallel detection enhances the rate of signal collection of an entire spectrum *vis-à-vis* that possible by scanning (or spinning) the grating in the monochromator, enabling rapid analysis. The wavelength accuracy is also superior when the grating is fixed. When CCD cameras are not available, the lower-sensitivity diode arrays are usually satisfactory because of the strong OE signals. One potential disadvantage in using parallel detectors is poor spectral resolution when a broad spectral region is being monitored at one time, because these detectors often have (only) 1000 or 500 pixels, each of which is ~25 μm wide. For example, if the whole visible range (~420–720 nm) is monitored with a 1000-channel detector, the resolution is only 0.3 nm, which may not be satisfactory. In intensity-enhanced diode arrays, the ultimate resolution is only 3–4 pixels, so the resolution would be ~1 nm. The spectral resolution can be improved by examining a smaller wavelength range with the same detector or by using a CCD array with a smaller pixel size (Section 5.1.3.1).

Sometimes it is difficult to analyze rapidly a complex spectrum acquired by a multichannel detector. Wangmaneerat *et al.* (1992), who have discussed the use of multivariate analysis (Section 19.1) to analyze OE emission spectra during plasma etching of silicon nitride acquired with a photodiode array, have shown that such analysis can be fast enough for *in situ* monitoring of several plasma parameters. Splichal and Anderson (1991) and Anderson and Splichal (1994) have used the multivariate statistical algorithms and statistical design methods of chemometrics to correlate optical emission spectra during plasma etching, acquired with a CCD array, with discharge electrical parameters and etch rate.

The monochromator/parallel detector combination is the most powerful, albeit the most costly, system; the monochromator/photomultiplier system is less expensive, though it is still very versatile (Singer, 1988). Use of grating spectrometers and elaborate (parallel) detectors is necessary in research and development, but can add considerable expense and complications for practical monitoring and control. Once the spectral emission pattern of a plasma process has been well characterized, OES can sometimes be monitored by using a narrow-band interference filter that transmits only the spectral feature of interest. This mode of operation is cheaper and more reliable; when it is feasible, it is preferred for real-time monitoring and control, such as endpoint control. Monitored optical emission lines must be chosen that do not interfere with each other and that correlate with the process parameter of interest: etching or deposition rate, or (for deposition) film composition.

6.3 Applications in Processing

Optical emission spectroscopy can be used to measure plasma etching or deposition rates if the emission feature can be correlated with the species responsible for the process or, for etching, a product. Table 6.1 gives a representative set of species often detected in OES.

Table 6.1 Representative Species Detected by OES[a]

Type	Species detected
Atoms	Al, Ar, As, Ba, Br, C, Cl, Cr, Cu, F, Fe, Ga, Ge, H, In, Nb, Ni, O, Si, Ti, Y, Zr
Diatomics	AlCl, AlF, BCl, BaO, C_2, CCl, CH, CN, CO, Cl_2, CuF, CuO, GaCl, H_2, NH, N_2, NO, OH, S_2, SiBr, SiCl, SiH, YO
Larger neutrals	BCl_2, CCl_2, CH_2, CF_2, CO_2, SiF_3
Ions	Al^+, Ar^+, Ba^+, Cl^+, Cl_2^+, CO^+, CO_2^+, Cu^+, F^+, Ge^+, N_2^+, O^+, Y^+

[a] Specific transitions and wavelengths are given in the text and in the references cited in this chapter.

6.3 Applications in Processing

Sometimes OES is used to characterize the plasma itself so that the thin film process can be understood better, as is described in Section 6.1. For example, Harshbarger (1982) correlated the relative intensities of N_2 and N_2^+ emission bands in nitrogen plasmas operated at 13.5 and 0.5 MHz, which could be used for etching, with their different electron temperatures. Vaudo *et al.* (1994) performed a high-resolution study of a N_2 electron cyclotron resonance (ECR) plasma to examine emission from atomic N. This was done to assess the production of N atoms from this MBE-compatible source for use in the molecular-beam epitaxy (MBE) of III–V nitride semiconductors, such as GaN. McKillop *et al.* (1989a,b) used high-resolution OES (using a grating spectrometer) to measure the Doppler profiles of Ar and Ar^+ emission in a divergent magnetic field ECR source, so that the translational energies of these species could be determined. O'Neill *et al.* (1993) used a Fabry–Perot interferometer/grating spectrometer tandem for high-resolution OES to examine the radial motion of Ar ions in a radio-frequency-induction plasma reactor, and found that the Doppler width, and hence the Ar^+ energy, decreases with increasing pressure. This system was also used to measure the Doppler profiles of Ar^+ and Cl^+ in a divergent field ECR reactor (O'Neill *et al.*, 1990). The high-resolution methods used in these latter studies are not needed in the vast majority of OE applications, because in most cases the emission intensity, and not the spectral linewidth, is monitored. Heinrich *et al.* (1989b) examined Doppler shifts in the OE from fast and slow ion beams to determine ion velocities during reactive ion beam etching (RIBE).

In most studies of plasmas and plasma-assisted processing, visible and UV emission from electronic transitions is monitored. Sometimes, monitoring infrared emission is useful. For example, O'Neill (1991) used an infrared imaging camera to map 4.7-μm emission from a N_2O discharge to investigate vibrational excitation processes in discharges.

The subsequent discussion is subdivided into the application of OES for different types of film processes. The use of optical emission for temperature measurements and tomographic reconstruction of OE for spatial profiling are both deferred to separate sections at the end of this chapter.

6.3.1 *Plasma Etching*

Optical emission spectroscopy has been used widely to probe the mechanism of plasma etching, and to monitor and control the progress of etching. In one early example, Harshbarger *et al.* (1977) used OES both to probe and to control the etching of silicon by $CF_4 + O_2$ plasmas. Figure 6.6 shows optical emission from a $CF_4 + O_2$ plasma without a Si wafer present. The spectrum shows sharp atomic peaks from F* and O*, and other structure from electronically excited CO. The F* and O* emission decreases greatly when a silicon wafer is introduced because these atoms are consumed

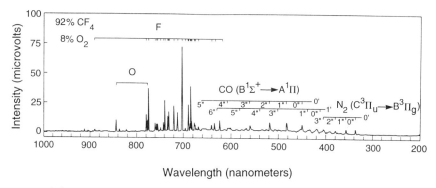

Figure 6.6 Optical emission from an RF plasma of 92% CF$_4$/8% O$_2$ without etching (no substrate present) with species identifications and spectral band assignments (from Harshbarger et al., 1977).

during etching; the CO bands are unaffected. (This is also seen for Si$_3$N$_4$ etching in Figure 6.7.) It is the return of the F and O emissions after the silicon layer has been etched that is the basis of one endpoint detection scheme (see below). Figure 6.8 shows how the emissions from F* (704 nm), O* (843 nm), and CO* (482 nm) and the Si etching rate vary with the fraction of O$_2$ in the CF$_4$/O$_2$ mixture. For a very low O$_2$ fraction, the F and CO emissions increase rapidly with O$_2$ fraction, as does the Si etch rate, while the O emission increases slowly. This occurs by CF$_3$ + O → COF$_2$ + F, CF$_2$ + O → COF + F, and CF$_2$ + O → CO + 2F, as has been shown by Plumb and Ryan (1986). For higher O$_2$ fractions, these first three signals decrease, while O emission increases rapidly. These data suggest that for low O$_2$ percentages, most of the O atoms produced are used to remove

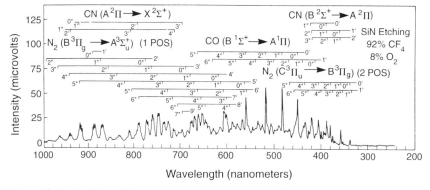

Figure 6.7 Optical emission from an RF plasma of 92% CF$_4$/8% O$_2$ while etching plasma deposited silicon nitride, with species identifications and spectral band assignments (from Harshbarger et al., 1977).

6.3 Applications in Processing

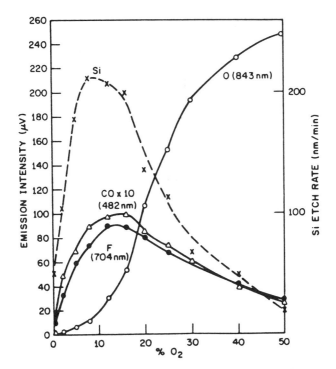

Figure 6.8 Emission intensities of several excited species and silicon etch rate as a function O_2 fraction in CF_4/O_2 plasma etching of silicon (from Harshbarger et al., 1977).

carbon (and form CO), and this frees up F atoms needed for etching. At higher $O_2/(CF_4 + O_2)$ fractions there is simply less CF_4 to stimulate etching.

Many studies have been conducted on similar etching species. In one such study, d'Agostino et al. (1982) examined OE from RF discharges of a series of fluorinated compounds. Field et al. (1984) used OES to identify F, F^+, O, O^+, C, CO, CO^+, CO_2^+, and CF_2 in a CF_4/O_2 discharge. d'Agostino et al. (1981) have shown how adding O_2 to a CF_4 plasma greatly changes the identity of emitting, and presumably also ground-state, species. Mutsukura et al. (1994) used spatially resolved OES to determine sheath thicknesses in CF_4, CH_4, and O_2 plasmas; and in CF_4 plasmas they used this sheath thickness and the OE intensity to control the etch depth during etching of Si.

Other representative OES studies of plasma etching include the following: Si etching in CCl_4 plasmas, where Si, Cl, Cl_2, CCl, SiCl, Cl^+, and Cl_2^+ OE were monitored (Clarke et al., 1990); $CF_4/O_2/Ar$ discharges with a Cu electrode, where Cu and CuF were seen (Nieman et al., 1990); tungsten etching in CF_4 and SF_6 discharges, where F was monitored (Tang and Hess,

1984); and fluorocarbon plasmas with an RF-driven Al cathode, where CF_2, AlF, and C were observed (Selwyn and Kay, 1985).

Pearton *et al.* (1993) surveyed optical emission during ECR etching of several gas mixtures used for etching III–V compound semiconductors, dielectrics, and metals. BCl_3/Ar plasmas were dominated by the broad molecular continuum of BCl_3 from 350 to 690 nm, sharp features near 270 nm due to BCl and BCl_2 emissions, and line emission from Cl (725.7–858.6 nm) and Ar. These features were much stronger in the ECR source than downstream in the chamber. The GaAs etch rate was seen to scale with the emission intensity of the 837.5-nm Cl line. The Cl_2^+ emission band from ~340 to 480 nm, which is a series of vibronic bands, was also found in BCl_3/Cl_2 discharges. (Continuum bands near 258 and 305 nm due to Cl_2^* emission can also be seen in chlorine discharges.) In CH_4/H_2/Ar discharges, intense H (434.0, 486.1, and 656.3 nm), CH (431.4 nm), and Ar emission were observed. In CCl_2F_2/Ar, strong continuum emission from CF_2 and CCl_2 was seen, along with F, Cl, and Ar atomic lines. During GaAs etching, a GaCl line (338.3 nm) and two Ga lines (403.3 and 417.2 nm) were easily discernible. In SF_6/Ar discharges used for etching SiN_x, SiO_2 and W, there were atomic F lines from ~620 to 840 nm and a molecular continuum between ~340 and 480 nm.

Many other studies utilizing optical emission from other glow discharge reactors are discussed in the following subsections in the context of specific applications. For example, Gottscho *et al.* (1982) used optical emission from In and InCl (Ga) products to monitor the rate of etching of InP (GaAs) in CCl_4 plasmas. The use of OES to determine temperature profiles in glow discharges is described in Section 6.3.5. The OE spectral "fingerprint" can be used to characterize process input parameters and results. One example is described in Sections 19.1 and 19.2.3 where Shadmehr *et al.* (1992) used a neural network to determine the effect of process parameters on RIE etching of Si by CHF_3/O_2 by using *in situ* measurements with OES and residual gas analysis (Figure 19.5). In most optical emission studies, measurements are made *in situ* to analyze intermediates and products in the plasma during the thin film process. Using OE for a slightly different purpose, Danner *et al.* (1983) sent the products from a process, atomic fluorine etching of silicon, into a separate discharge downstream where they analyzed optical emission to monitor the products from the reactor.

Ordinarily, OE is collected and analyzed with time resolution from many milliseconds to a few seconds, which is sufficient for real-time monitoring. Faster response is useful in some instances. Flamm and Donnelly (1986) used time-resolved OES to monitor species near the grounded electrode sheath of a Cl_2/Ar discharge excited at 220 kHz and 13.2 MHz. Spatially and spectrally resolved emission was detected by a photomultiplier, whose output was averaged by a boxcar integrator. This integrator was triggered synchronously with the waveform at a given phase, and this phase was slowly

6.3 Applications in Processing

scanned across one RF cycle (Figure 7.14a). Figure 6.9 shows emission from Cl, Ar, Cl⁺, and Cl_2^+ during one RF cycle at 220 kHz near the center of the discharge. The observed OE peaks occur at slightly different times in the cycle for these four species, and the times also depend on the observation region. The time lags near the sheaths depend on the time required to accelerate positive ions across the sheath to the grounded electrode. The ions cannot respond at all during a cycle with 13.2-MHz RF. In an earlier study, Donnelly et al. (1985) investigated the effect of excitation frequency (10 kHz–25 MHz) on plasma etching by monitoring the time-resolved OE from the Cl_2/Ar/He plasmas. Monitored emissions included the following: Cl_2^+ (408, 430, 455 nm; $A\,^2\Pi_u \to X\,^2\Pi_g$), Cl⁺ (384.4, 385.1, 386.1, 413.3 nm), Cl (725.7, 741.4, 754.7, 808.7 nm), Ar (750.4, 751.4, and 763.5 nm), and He (587.6 nm). Fujiwara et al. (1995) used an intensified photodiode array to time resolve OE from pulse-modulated Cl_2 and HCl plasmas.

In another application of time-resolved OES, Herman et al. (1994) used boxcar integration to observe transient changes in optical emission after laser-induced thermal desorption of the surface layer during plasma etching of Si by Cl_2 in a helical resonator. Figure 6.10 shows optical emission from Si, SiCl, and Cl_2 (called plasma-induced emission, PIE). As is seen in this figure (LD-PIE, laser desorption-PIE), optical emission from Si and SiCl increased in the ~10 μsec after desorption due to direct electron excitation of (these) desorbed species and other collisional events involving other desorbed species. Laser-induced fluorescence from the desorbed SiCl is also seen in the figure. Laser-induced thermal desorption is discussed in more detail in Section 15.1.

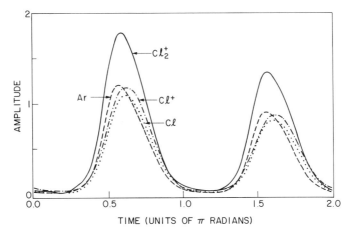

Figure 6.9 Time-resolved optical emission from several species near the center of a 220-kHz RF chlorine discharge (from Flamm and Donnelly, 1986).

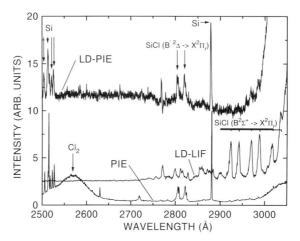

Figure 6.10 Spectra during etching of Si(100) by Cl_2 in a helical resonator, showing steady-state plasma-induced emission (PIE, bottom trace) from Si*, Cl_2^*, and SiCl*; the transient increase in PIE due to laser desorption (LD, 248-nm XeCl laser) of adsorbates from the surface (LD-PIE, top trace) from Si* and SiCl*; and laser-induced fluorescence from laser-desorbed SiCl (LD-LIF, middle trace) (from Herman et al., 1994).

Optical emission can also be used to characterize the plasma and the film process. Splichal and Anderson (1991) examined OE in the standardized GEC Reference Cell plasma etch reactor at various positions using the imaging arrangement depicted in Figure 6.5d. They found that the emission fingerprint, as determined by multivariate statistics, correlates well with the results of electrical probe measurements and with the silicon dioxide etch rate. Figure 6.11a shows the spectral variance in OE near the anode during 27 experiments of CF_4/CHF_3 plasma etching of silicon dioxide under different conditions. The univariate correlation for the etch rate, i.e., the correlation with each wavelength, seen in Figure 6.11b, is best near 370 nm where the etchant CF_x emission (strong from 220 to 320 nm) is weak. It is even better (~0.8 from 325 to 405 nm) for OE near the cathode (on which the wafer rests). Multivariate analysis, which uses the entire spectrum, is superior to univariate analysis and can be used to predict the etch rate, as is demonstrated in Figure 6.11c.

In addition to species identification, density determination, and endpoint control, OE can be used to detect tool and process malfunction in plasmas used for etching and other processes by monitoring emission from discharge-excited impurities. Selwyn (1993) has given several such examples: (a) N_2 emission at 337.1 nm can indicate an air leak; (b) OH emission at 306.4 nm can indicate dissociation of water vapor; (c) CO emission at 483.5 nm can indicate residual photoresist on the surface, (d) Al emission at 396 nm can indicate excessive sputtering of an aluminum electrode, and

6.3 Applications in Processing

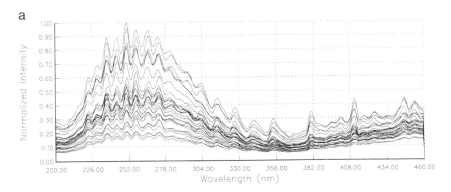

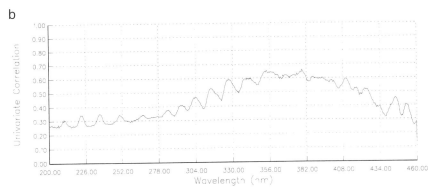

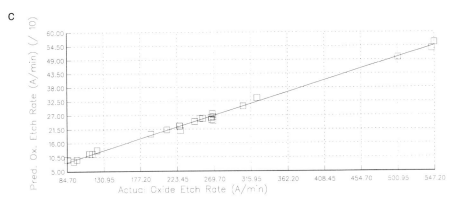

Figure 6.11 (a) Normalized spectral variance of optical emission during CF_4/CHF_3 plasma etching of silicon dioxide under different experimental conditions, with OE from the anode fiber channel. (b) Univariate correlation of oxide etch rate versus wavelength from the data in (a). (c) Chemometric prediction of oxide etch rate, from multivariate analysis, versus measured rate using OE from the cathode fiber channel. [From M. P. Splichal and H. M. Anderson, Application of chemometrics to optical emission spectroscopy for plasma processing, in *Process Module Metrology, Control, and Clustering, SPIE* **1594,** 189 (1991).]

(e) Cu emission at 324.8 and 327.4 nm can indicate sputtering of exposed copper or brass.

Optical emission has also been used to probe the plasma during the formation of particles, as is illustrated in Figure 11.8.

6.3.1.1 Actinometry and Density Calibration

Optical emission intensities reflect excited-state densities, which are often a small and variable fraction of the ground-state density. While this uncertainty does not affect the use of OES as a real-time monitor, it does affect the usefulness of OES data for understanding process steps. Several studies have assessed how OES data can be related to species density.

Actinometry is a calibrated procedure that has been used to obtain ground-state species densities from OES data. In actinometry, a small amount of a nonperturbing species X (the actinometer), a known standard, is added to the discharge to help probe quantitatively features of species A in the original discharge. If X* and A* are both produced by the same route (say, Reaction 6.1) and decay radiatively, and if they are formed by processes with similar excitation cross-section dependences and threshold energies, one would expect that

$$N_A = cN_X \left(\frac{I_A}{I_X}\right) \tag{6.8}$$

where c is a constant and I_y is the OE intensity from the excited species. To ensure that N_X is spatially uniform, as well as accurately known, it is best to use an atomic species, i.e., an inert gas, as the actinometer. Although actinometry is not absolutely necessary if other quantitative optical spectroscopies, such as absorption and LIF, are available, it is still much simpler and cheaper than these other methods. Since absorption features are sometimes very weak and not all species can be probed by LIF, it is also more general than many other methods. Actinometry is applicable to all types of plasma processing, not just etching.

Coburn and Chen (1980, 1981) demonstrated the use of actinometry by adding Ar to a CF_4/O_2 etching discharge. The emission intensity ratio F* (703.7 nm)/Ar* (750.4 nm) was found to monitor N_F well, while emission from F* alone was shown to be an unreliable indicator of the F atom concentration. This calibration procedure appears to be reasonable because the energies of the emitting states are close, ~14.5 eV for the atomic F transition and ~13.5 eV for that in argon, so emissions should track each other for different discharge conditions if both emitting states are excited by energetic electrons. Hanish et al. (1995) used this actinometric scheme to provide real-time feedback control of CF_4/Ar etching plasmas.

Equation 6.8 has been verified for Ar actinometry in several plasmas. Donnelly et al. (1984a,b) showed it determines N_F well in $CF_4/O_2/Ar$ and

6.3 Applications in Processing

NF_3/Ar etching discharges by measuring N_F independently by downstream chemical titration (Figure 6.12). Again using Ar as the actinometer, Ibbotson et al. (1983) have shown that the Br 700.5-nm line, normalized by the Ar emission intensity at 750.4 nm, is a good measure of N_{Br} in Br_2/Ar plasmas; N_{Br} was determined by absorption measurements of Br_2. In Cl_2 plasmas with 1% Xe, Donnelly (1996) monitored Cl_2^* (305.0 nm)/Xe* (823.2 nm) to deduce the fractional dissociation of Cl_2.

In another test of this method, Gottscho and Donnelly (1984) used a scanning Fabry–Perot interferometer in tandem with a grating spectrometer to conduct high-resolution emission profile measurements, as mentioned earlier. The narrow emission peaks in Figure 6.3 suggest that both F* (703.7 nm) and Ar* (703.0 nm) are produced by electron impact in a $CF_4/O_2/Ar$ discharge, suggesting that N_F can be measured by actinometry using these lines. Their time-resolved measurements of Cl* (837.6 nm) emission in a Cl_2/Ar discharge (Figure 6.4; see Figure 7.14a for the timing strategy) show a narrow profile at the momentary anode at the maximum applied voltage during the cycle, suggesting Reaction 6.1, and a narrow profile superimposed upon a broader feature at the momentary cathode at the minimum applied voltage, suggesting both Reactions 6.1 and 6.2. Ar* (826.4 nm) always shows narrow lines (Reaction 6.1). Therefore, actinometry is a valid method in Cl_2/Ar plasmas only under special observation conditions. These authors also cite other assessments of actinometry.

In some discharges actinometry either cannot be used or must be used with caution. In CF_4/O_2 discharges, Walkup et al. (1986) showed that the

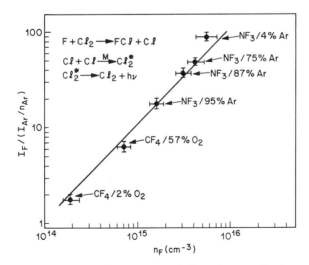

Figure 6.12 Calibration of fluorine-to-argon emission ratio versus fluorine atom concentration, determined by downstream titration, for different RF discharge mixtures (from Donnelly et al., 1984b).

ratio O* (844.6 nm, $3p\,^3P \to 3s\,^3S$)/Ar* (750.4 nm) generally has the same dependence on CF_4 fraction as does the ground-state O density determined by two-photon LIF, except for very small CF_4 fractions. However, in general the ratio O* (777.4 nm, $3p\,^5P \to 3s\,^5S$)/Ar* (750.4 nm) does not. Emission lineshapes measured with a Fabry–Perot interferometer showed broadening beyond that expected from instrumental and Doppler contributions for O* (777.4 nm), based on measurements of Ar* emission; in some cases this was also true for O* (844.6 nm) emission. These data suggest that the primary source of O* (844.6 nm) and Ar* (750.4 nm) is electron impact (Reaction 6.1), while the primary source of O* (777.4 nm) is dissociative excitation of O_2 (Reaction 6.2). However, dissociative excitation also contributes some to O* (844.6 nm) production. Overall, this study suggests that actinometry using the ratio O* (844.6 nm)/Ar* (750.4 nm) is still generally valid. In contrast, Booth et al. (1991) have found that this ratio correlates well with the O_2 concentration, but not well with the O atom concentration, measured by resonance absorption at 130 nm, in ECR plasmas of O_2/Ar (plus added SF_6, N_2, or Kr). This would suggest that dissociative excitation of O_2 is the most important mechanism for producing the 844.6-nm emission in the plasma studied.

There is uncertainty as to the use of CF_2 OE as a probe of ground-state density in CF_4/O_2 plasmas, even with actinometric normalization. Buchmann et al. (1990) observed that the ground-state CF_2 density, measured by LIF, and the excited-state $CF_2(A)$ density decrease in a similar way as the O_2 fraction is increased, which suggests that OE does represent the N_{CF_2} well for this system. However, Hancock et al. (1990) found that CF_2 OE measurements normalized by actinometry generally do not give the ground-state population.

Actinometry can also be employed to determine $\sigma_A(\mathcal{E})$ and $\mathcal{E}_0$ for a given species. If the added species has a smaller ionization potential than the gases in the initial discharge, the electron temperature will get colder and optical emission requiring hot electrons for excitation will be quenched (Donnelly, 1989). To examine how the electron distribution in a CF_4/O_2 discharge changes with O_2, d'Agostino et al. (1981) examined emission from F, O, CO, and CO_2, whose emitting levels range from 8 to 20 eV, and from added N_2, whose $C\,^3\Pi_u$ level has energy 11.28 eV, which fluoresces to $B\,^3\Pi_g$.

In some circumstances, optical emission may be useful for determining relative ground-state densities, even without calibration. For example, Greenberg and Hargis (1989) have shown that unnormalized OE and LIF of S_2 scale with O_2 the same way in SF_6/O_2 etching plasmas, so OE from the excited sulfur dimer is a good measure of ground-state population.

6.3.1.2 Endpoint Detection and Monitoring

Often of more practical significance during plasma etching is the sudden appearance or disappearance of characteristic emission lines when the film

6.3 Applications in Processing

is etched away. Consider the etching of a uniform film on a substrate. During etching, emission lines related to etch products, which will be called X*, appear (or get stronger), while those related to reactive etchants (Y*) get weaker because the etchants are consumed during etching. When etching is complete, X* emission gets weaker, that from Y* gets stronger, and new emission lines (Z*) due to (the hopefully slow) etching of the substrate appear.

Selwyn (1993) has pointed out that etching of films is often monitored by product emission (X*), for example, oxide and polymer films by CO* emission, nitride films by N_2^* or CN* (with carbon-containing etchants), polysilicon films by SiF*, and Al films by AlCl*. Polysilicon etching by fluorine-containing etchants can also be monitored by following emission by F*, a reactant (Y*). Riley et al. (1990) have monitored the endpoint of etching of W on TiN by looking for a sharp rise in N_2^* (Z*) emission near 380 nm from the underlying TiN. Representative emission lines used in

Table 6.2 Representative Emission Lines Used in Endpoint Detection of Plasma Etching[a]

Monitored species	Wavelength (nm)
Al	308.2, 309.3, 396.1
AlCl	261.4
As	235.0
CF_2	251.9
Cl	725.6, 741.4
CN	289.8, 304.2, 387.0
CO	292.5, 302.8, 313.8, 325.3, 483.5, 519.8
F	685.4, 703.7, 712.8
Ga	417.2
H	486.1, 656.5
In	325.6
N	674.0
N_2	315.9, 337.1
NO	247.9, 288.5, 289.3, 303.5, 304.3, 319.8, 320.7, 337.7, 338.6
O	777.2, 844.7
OH	281.1, 306.4, 308.9
S	469.5
Si	288.2
SiCl	287.1
SiF	777.0

[a] From Singer (1988), Selwyn (1993), and other sources cited in the text.

endpoint detection are listed in Table 6.2. Marcoux and Foo (1981), Roland *et al.* (1985), and Singer (1988) have surveyed optical and nonoptical methods of endpoint detection during plasma etching, including that by optical emission (see Table 1.2). Selwyn (1993) has discussed several of the potential operational problems that can occur in using OES for endpoint detection in a manufacturing line.

Several issues are important for automated endpoint control by real-time OE monitoring. Characterization of a precise endpoint must be determined with an algorithm that minimizes under- and overetching. This is often done by monitoring the derivative of the OE signal and compensating for signal noise (H. Litvak, private communication, Luxtron Corp., 1994) or by integrating the OE signal. Endpoint detection is complicated by nonuniformities in etch rate across the film. Furthermore, emission intensities—and their changes—are much smaller when etching a film overlaid by a patterned mask, such as a photoresist, rather than an unpatterned film. Even though the etching process is usually designed to etch the film selectively over the mask, the mask is still slowly etched and its gas-phase products also emit light, which could "mask" the monitoring of the film and the underlying film. Etching of reactor materials also can contribute to the background. Still, the endpoint of etching when only a small fraction, <1%, of the area is exposed can be determined with suitable OE signal collection and analysis. (Other endpoint monitors, such as discharge voltage, can also be sensitive to etching small areas.)

Figure 6.13a shows the OE during plasma etching of SiO_2 by a CHF_3/CF_4/Ar mixture (Luxtron Corp., 1994). In this sample, contact holes in the overlying resist mask exposed <1% of the SiO_2 surface. A bifurcated multifiber optical cable collects OE and then splits into two bundles, each directed to separate monochromators for analysis of the OE from the etch product and a reference OE signal (dual sensor). The "endpoint" signal is the 483.5-nm line from the CO etch product. The 2-nm-wide bandpass on the monochromator for the "background" signal transmitted the 488-nm Ar line, as well as the 486-nm line from H atoms that is sometimes present due to water vapor impurities. Since much of the noise in the CO* signal is also present in the Ar* signal, the endpoint signal can be corrected by normalizing the CO* emission intensity by that from Ar*. This is seen in Figure 6.13b. (Note the different time scales in Figures 6.13a and 6.13b.) Qualitatively similar traces are obtained for larger exposed areas, without normalization (single sensor).

In a similar application, Pearton *et al.* (1994) used optical emission to determine the endpoint of the ECR etching of 30-μm-diameter via holes in GaAs and InP substrates. Etching of vias in GaAs in a BCl_3/Cl_2 discharge was monitored by Ga* (417.2 nm) emission. Via hole etching of InP in a Cl_2/CH_4/H_2/Ar discharge was monitored by the In* 325.6 nm atomic line,

6.3 Applications in Processing

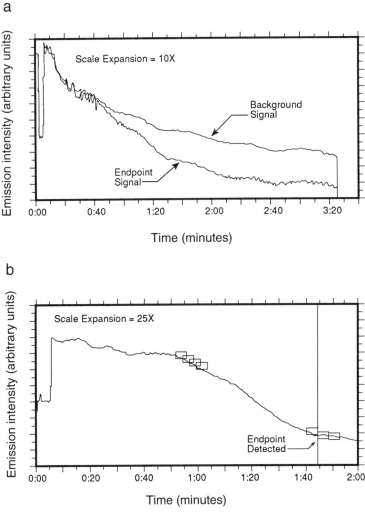

Figure 6.13 Emission traces versus time during plasma etching ($CHF_3/CF_4/Ar$ mixture) of SiO_2 with low exposed area (<1%) through contact holes in the resist mask: (a) emission from dual sensors detecting the "endpoint" signal from the CO* product and the plasma "background" from Ar* and H*; (b) the endpoint detection trace after correction using the "background" emission (from Luxtron Corp., 1994).

as is seen in Figure 6.14. Monitoring the derivative of the emitted intensity gives an accurate indication of the endpoint.

In an early application of endpoint detection during microelectronics processing, Degenkolb *et al.* (1976) studied the use of OES during oxygen plasma stripping ("ashing") of photoresists. Figure 6.15 (upper) shows

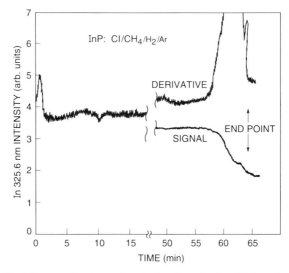

Figure 6.14 Real-time monitoring and endpoint detection by OES of the In* 325.6-nm line during via-hole etching of InP in a Cl$_2$/CH$_4$/H$_2$/Ar discharge (from Pearton *et al.*, 1994).

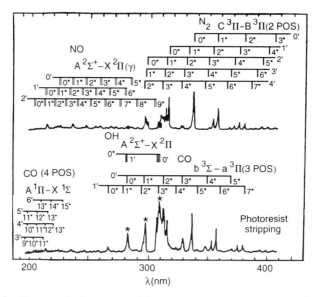

Figure 6.15 Plasma emission spectra of (top) an uncoated silicon wafer and (bottom) of a silicon wafer coated with photoresist, during the stripping of the photoresist, in an O$_2$ RF plasma (from Degenkolb *et al.*, 1976). The N$_2$ and NO emission bands are due to impurities in the gas.

6.3 Applications in Processing

strong emission from NO* and N_2^* molecular bands, due to nitrogen impurities, during etching of bare Si wafers. During etching of Si wafers coated with photoresist, additional emission bands from CO* and OH* were found (Figure 6.15 lower). CO and OH are among the volatile byproducts that are formed in the plasma oxidation and degradation of the polymer backbone. Decreases in combined OH* and CO* bands at 283.0 nm and the CO* emission at 297.7 nm were found to signify the endpoint of resist removal, and the time-integrated intensity of the CO* band at 297.7 nm was seen to be nearly linearly proportional to the amount of photoresist removed. In another application of OE endpoint detection of plasma etching of photoresist, Roland *et al.* (1985) used the sharp rise of reactant F* emission at 704 nm to signal the end of plasma planarization of photoresist on oxynitride films.

Curtis and Brunner (1978) and Curtis (1980) monitored the endpoint for etching Al films in chlorine plasmas by a decrease in AlCl* emission at 261.4 nm. Korman (1982) used the respective decrease of Al* emission at 309.8 nm and SiCl* at 287.1 nm to monitor Al and Si etching by CCl_4 in a batch-loaded planar plasma etcher. He also monitored the increase in emission from reactive species at the endpoint, at 199.1 nm, presumably from HCl, and 308.9 nm, from Cl_2 or CCl. In monitoring plasma etching of Si films in a fluorine environment, there is an increase in F* emission (704 nm) when a Si film has been etched away totally, since fluorine atoms now appear and are no longer consumed by the etching process (Harshbarger *et al.*, 1977). Multiple species can be probed. During plasma etching of patterned photoresist on aluminum, the etch rate of Al can be monitored by AlCl emission ($A\,^1\Pi \rightarrow X\,^1\Sigma^+$, 261.4 nm), and that of the resist by CCl emission ($A\,^2\Delta \rightarrow X\,^2\Pi$, 278, 279 nm) (Krogh *et al.*, 1987).

Collot *et al.* (1991) demonstrated how OES and laser interferometry can be used to monitor selective and nonselective reactive ion etching (RIE) of III–V heterostructures. By monitoring with either method, it is easily seen that RIE with CCl_2F_2/He mixtures is selective for GaAs over AlGaAs, i.e., in a GaAs/AlGaAs heterostructure, etching stops, and optical emission from Ga* ceases, after the GaAs layer has been totally etched. Since RIE etching in the environmentally more acceptable $SiCl_4$/He mixture is nonselective, real-time monitoring must be used to monitor, and consequently control, the etching of such structures, such as stopping it at the GaAs/AlGaAs interface. (For environmentally acceptable selective etches, this is not necessary.) Figure 6.16 shows the emission spectrum during the etching of AlGaAs, while Figure 6.17 shows that after the GaAs overlayer has been removed and the $Al_{0.40}Ga_{0.60}As$ layer is then being etched, the Ga* emission (417.2 nm) decreases, the AlCl* emission (261.4 nm) signal increases, and the As* emission (235.0 nm) is unchanged. Derivative signals are seen to determine the interface more clearly than the signal itself. The time needed to etch through the AlGaAs layer as measured by OES is seen

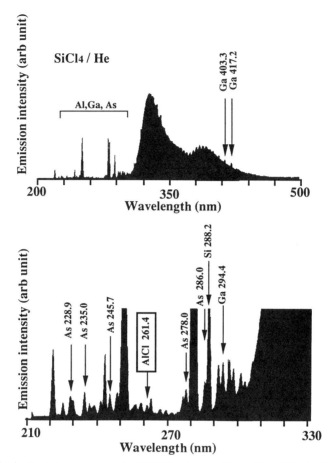

Figure 6.16 Emission spectrum of a SiCl$_4$/He plasma while etching AlGaAs (from Collot et al., 1991).

to exceed that measured by reflectometry because of etching nonuniformity across the wafer. AlGaAs layers as thin as 5 nm can be detected by OES. This method was also used to monitor the etching of the emitter of a GaAs/AlGaAs heterobipolar junction transistor (HBT). Similarly, by monitoring In* emission at 325.6 nm during the CH$_4$/H$_2$ RIE etching, they were able to observe In$_{0.53}$Ga$_{0.47}$As layers as thin as 3 nm sandwiched between InP layers, as is seen in Figure 6.18. In related studies, Thomas et al. (1995), monitored InP and GaAs etching in Cl$_2$/Ar plasmas by OES.

Richter et al. (1994) monitored the 265.1-nm Ge* emission line to determine the endpoint of etching of the SiGe layer during RIE etching (Cl$_2$/SiCl$_4$/N$_2$) of Si/Si$_{0.78}$Ge$_{0.22}$/Si heterostructures. Again, OE endpoint detection is important because this etch is not selective for SiGe vis-à-vis Si. Orloff

6.3 Applications in Processing

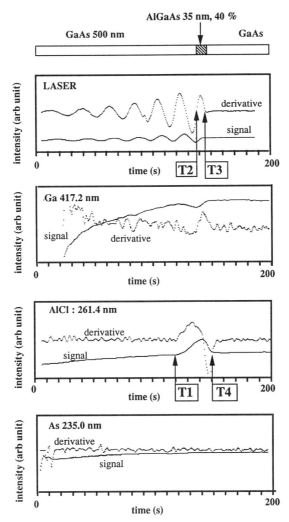

Figure 6.17 Real-time monitoring of the nonselective etching of the depicted GaAs/ $Al_{0.4}Ga_{0.6}As$ heterostructure by $SiCl_4$/He RIE. Optical emission and laser interferometry signals are shown, along with their derivatives and the endpoint determination times Ti (from Collot *et al.*, 1991).

(1992) monitored Zn* emission at 472 nm to determine the endpoint of the RIE of unpatterned and patterned ZnS films in a hydrogen plasma.

Wangmaneerat *et al.* (1992) used partial least-squares analysis of the emission spectrum acquired by a photodiode array to find the endpoint of plasma etching of silicon nitride. This multivariate analysis showed that the signal-to-noise ratio could be improved by a factor of 2 if the 655-nm feature due to N* is used, rather than the often-monitored F* peak at 703.8 nm.

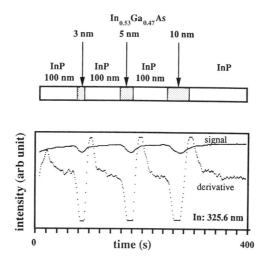

Figure 6.18 Monitoring the etching of thin $In_{0.53}Ga_{0.47}As$ layers by In* optical emission during CH_4/H_2 RIE (from Collot et al., 1991).

In a related endpoint detection scheme, Danner and Hess (1986) monitored the endpoint of aluminum etching during plasma etching by remotely plasma-generated Cl atoms by monitoring chemiluminescence from Cl atom recombination.

Optical emission spectroscopy can also be used to monitor the endpoint of tool cleaning. For example, blue CO* emission is seen during the cleaning of tools with an oxygen plasma while oil or polymers still coat the walls. It disappears when the walls are clean of carbon-containing films, and is replaced by the straw-yellow emission of O_2^+* (Selwyn, 1993).

6.3.1.3 Optical Emission Interferometry

Optical emission from the plasma that is reflected from the wafer exhibits an interference pattern that depends on the composition and thickness of the film(s) on the wafer. This is analogous to optical interferometry in reflection using a laser (or other optical) source (Section 9.5) and pyrometric interferometry (Section 13.4). Monitoring this reflected emission is a potentially important optical probe that combines the strengths of laser interferometry and optical emission for endpoint detection. Heinrich et al. (1989a,b) developed this method for use during sputtering of an oxide layer on a Si wafer by an argon-ion beam. Figure 6.19 shows the time-dependent emission of several Ar* and Ar+* [Ar(I)* and Ar(II)*] emission lines. The slow oscillations are due to multipass interference, which can give the film thickness; the use of shorter wavelengths gives finer thickness control. The decreasing modulation depth with time is due to inhomogeneity of the etch

6.3 Applications in Processing

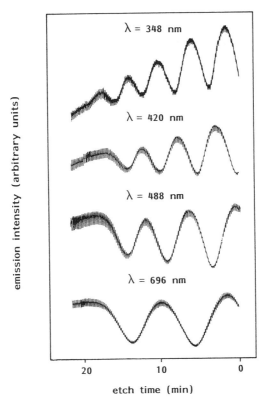

Figure 6.19 Emission at several wavelengths from Ar* and Ar^{+*} detected after reflection from the substrate during the sputtering of silicon oxide on a Si wafer by an argon-ion beam. The film thickness (and the endpoint) can be determined from the slow oscillations in these optical emission interferometry traces. The rapid oscillations are due to sample rotation (from Heinrich *et al.*, 1989a).

depth across the wafer, and the rapid oscillations, caused by wafer rotation, can be removed. This technique was used as an etch-rate monitor and endpoint detector.

Angell and Oerhrlein (1991) developed this endpoint monitor further for etching SiO_2 on Si by examining OES reflected at a grazing angle. Usually etching is continued past the observed endpoint to ensure that all of the oxide has been removed, because of uncertainties in the signal and etch nonuniformities across the wafer. However, with such overetching, ions can damage the silicon substrate. This can be avoided by trying to detect a "flexible endpoint" ~300 Å before the endpoint is reached. At this time, the etching recipe can be changed to lessen this damage and to determine the etch stop more accurately. They monitored the 297.7-nm CO* line in a CHF_3/O_2 RIE plasma with optical fibers placed at normal

($\theta = 0°$) and grazing (75°) angles. The model and observations (Figure 6.20) show that at grazing angle the last interference maximum is observed 300 nm before the interface is reached, which is nearer the interface than for normal-angle collection. The extra modulation in the 75° observation, which is responsible for this effect, is attributed to the phase reversal of the reflection coefficient for *p*-polarized light r_p when the angle of incidence passes through Brewster's angle.

6.3.2 Plasma-Enhanced CVD

While OES is important both for process analysis and for real-time monitoring during plasma etching, it is primarily useful for investigating process mechanisms for PECVD studies.

Several groups have monitored optical emission from a variety of species during the deposition of diamond and diamond-like films, including H (Balmer series: Hα, 656.2 nm; Hβ, 486.1 nm; Hγ, 434.0 nm), C (248 nm),

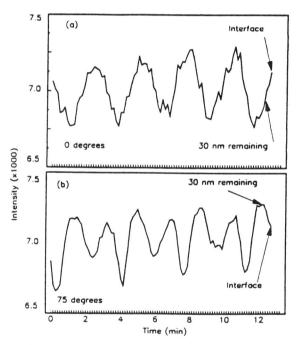

Figure 6.20 Optical emission from the 297.7-nm CO* line in a CHF$_3$/O$_2$ RIE plasma, which etches SiO$_2$ atop Si, collected with a optical fibers placed at (a) normal ($\theta = 0°$) and (b) grazing (75°) angles of incidence. Part of the emission is reflected from the wafer and therefore monitors the wafer thickness by interferometry. A "flexible endpoint" can be detected by using the grazing incidence geometry in (b), allowing a change in discharge recipe before the real endpoint at the SiO$_2$/Si interface (from Angell and Oehrlein, 1991).

6.3 Applications in Processing

CH ($C\,^2\Sigma^+ \to X\,^2\Pi$, ~314.3 nm; $B\,^2\Sigma \to X\,^2\Pi$, 390 nm; $A\,^2\Delta \to X\,^2\Pi$, 431.5 nm), CH$^+$ ($A\,^1\Pi \to X\,^1\Sigma$, ~400 nm; $B\,^1\Delta \to A\,^1\Pi$, $b\,^3\Sigma \to a\,^3\Pi$, ~360 nm), C$_2$ (Swan bands, $d\,^3\Pi_g \to a\,^3\Pi_u$, 563.6, 516.5, and 473.7 nm), H$_2$ (463.4, 581 nm), and CH$_2$, due to the main deposition species, as well as CO, OH, Si, CN, and NH, due to additives to the discharge, such as O$_2$, and to impurities. Such studies have been conducted in glow and arc discharges, microwave plasma jets, and flames. Balestrino *et al.* (1993) found a correlation of diamond-film quality with the appearance of emission at 431 nm (CH) and the absence of an emission band at 505–517 nm (C$_2$) for CH$_4$/H$_2$, CH$_4$/CO$_2$, and C$_2$H$_2$/CO$_2$ plasmas. T. D. Mantei (private communication, 1995) has observed strong increases in the OES signal from OH, CO$^+$, and CHO as O$_2$ was added to a CH$_4$/H$_2$ discharge, which demonstrate that oxygen helps to tie up the carbon in the discharge. Other representative work has been performed by Matsumoto *et al.* (1985), Mucha *et al.* (1989), Inspektor *et al.* (1989), Mitsuda *et al.* (1990), Zhang *et al.* (1990), Yalamanchi and Harshavardhan (1990), Gomez-Aleixandre *et al.* (1993), and Reeve and Weimer (1995). [Also see the review on vapor phase diagnostics of diamond CVD by Thorsheim and Butler (1994).]

Optical emission has frequently been used to investigate the PECVD of silicon and silicon-containing films. Dautremont-Smith and Lopata (1983) showed that spectrally unresolved emission in Ar/SiH$_4$/N$_2$O plasmas used to deposit silicon dioxide, linearly tracked the deposition rate at higher RF powers. This emission was due mostly to the atomic Ar lines from ~700 to 850 nm. In a silane RF discharge used to grow a-Si:H, Mataras *et al.* (1989) observed different spatial profiles for SiH optical emission from $A\,^2\Delta \to X\,^2\Pi$ and SiH LIF observed with excitation and emission between the same two bands. Tochikubo *et al.* (1990) examined Si, SiH, H, and H$_2$ emission in a space- and time-resolved manner in silane RF glow discharges excited at 100 kHz and 13.56 MHz. Optical emission spectroscopy has also been used during PECVD in discharges with silane and silane/hydrogen for Si deposition (Kampas and Griffith, 1981; Matsuda and Tanaka, 1982; Vanier *et al.*, 1984); SiCl$_4$/H$_2$ (Bruno *et al.*, 1986); silane and germane (Hata *et al.*, 1987); and silane/ammonia for silicon nitride deposition (Ho *et al.*, 1989), in which Si, H, SiH, NH, and N$_2$ were identified, as seen in Figure 6.21.

6.3.3 *Sputter Deposition*

Although, sputtering, and in particular reactive sputtering, has many features in common with plasma etching and deposition, it is often treated as a distinct process, and will be treated as such in this discussion. Optical emission spectroscopy is a practical technique for controlling film composition during reactive sputtering in critical applications by real-time monitoring of the gas phase by OES. (Alternatively, the surface can be monitored by reflectometry or ellipsometry). Greene (1978) has reviewed the early

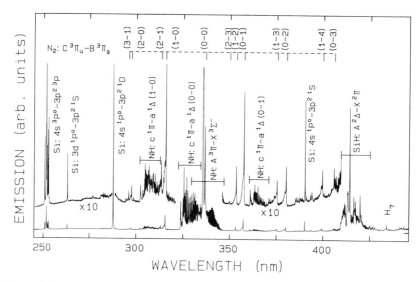

Figure 6.21 Optical emission from a silane/ammonia plasma for the deposition of silicon nitride, with spectral band assignments (from Ho *et al.*, 1989).

studies of optical emission monitoring during sputtering, and Schiller *et al.* (1987) has discussed the stabilization of DC sputter deposition by using OES.

Optical emission spectroscopy has been used to monitor the deposition of metals in glow-discharge sputtering systems in a series of studies by Greene and co-workers (Greene and Sequeda-Osorio, 1973; Greene *et al.*, 1975; Greene, 1978). The emission spectrum from Monel K-500 alloy (65.33% Ni, 29.31% Cu, 1.01% Fe by weight) DC-sputtered in Ar is shown in Figure 6.22 (Greene *et al.*, 1975). The unidentified features are

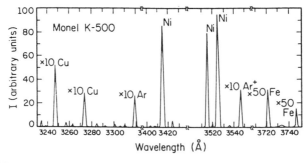

Figure 6.22 Optical emission from a glow discharge with a Monel K-500 target (from Greene *et al.*, 1975).

6.3 Applications in Processing

due to excited argon. Figure 6.23 shows that the Cu* emission intensity (3247.5 Å) is linearly proportional to the sputtering rate of the Cu target in diode and triode systems, as the target voltage is varied at fixed argon pressure. Optical emission at the In* line at 4104 Å also tracks the sputtering rate of the In target in a DC diode glow as the argon pressure is varied at constant voltage. Similar trends are seen for the In emission line in both the cathode fall and negative glow regions.

The composition of films sputtered from compound and alloy targets can vary with time owing to preferential sputtering (Greene et al., 1978; Greene, 1978). Although electron spectroscopies, such as Auger analysis, can profile these sputtered films only after deposition, OES can be used to monitor such processes in real time. Greene et al. (1978) used OES to investigate the preferential sputtering of an Inconel 718 alloy target (53.25% Ni, 18.14% Fe, 18.01% Cr). Figure 6.24 shows Ni*, Fe*, and Cr* emission lines monitored as a function of time. The relative intensities changed in the first 12 min, during which time ~1.5 μm of the target was removed. Afterward, the emission intensities no longer varied in time, indicating a steady-state sputtering condition in which the composition of the sputtered material equals that of the bulk. [Greene and co-workers also compared these results with those from absorption measurements. After this work on sputtering metal films, this group performed early OES studies of RIE etching of GaAs (Klinger and Greene, 1981).]

Similarly, OES can be used to test the impurity of sputter targets. Selwyn (1993) has shown that 4% Cu in Al is easily seen in an neon sputter plasma

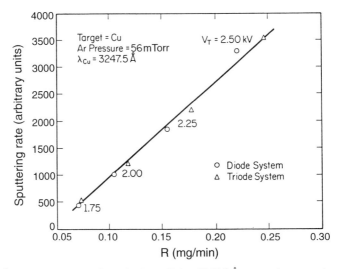

Figure 6.23 The emission intensity from Cu* at 3247.5 Å versus the sputtering rate from a Cu target in DC diode and triode glow discharges (from Greene, 1978).

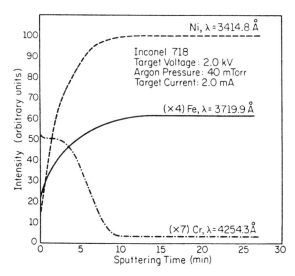

Figure 6.24 Emission intensity from Ni*, Fe*, and Cr* during the initial transient sputtering period of an Inconel 718 target (from Greene et al., 1978).

by the 521.82-nm Cu* emission, which is surrounded by aluminum and neon emission lines.

Optical emission spectroscopy can also be used to monitor and control the deposition of metal compound films by reactive sputtering and related processes. Such control has been analyzed by Westwood (1989) in terms of the parameter $f/\mathcal{P}$, where f is the reactive gas flow and $\mathcal{P}$ is the sputtering power. Affinito and Parsons (1984) detailed control-loop sputtering of Al in Ar/N$_2$ mixtures using Al* emission at 396.1 nm.

Pang et al. (1994) have used OES to monitor the RF sputter deposition of TiN thin films in the presence of N$_2$/Ar gas mixtures. Since the effectiveness of TiN films as a diffusion barrier in multilayer ohmic contact structures depends on their stoichiometry, it is important to control growth conditions to achieve the optimal composition. Figure 6.25 shows that for fixed N$_2$/Ar ratios, the ratio of N to Ti in the TiN films is linearly proportional to the ratio of the intensity of the N$_2^+$ line at 391.4 nm to that of the Ti line at 364.2 nm. Consequently, OES can be used as an *in situ* process monitor of TiN composition. This Ti line is well resolved from all nitrogen (and argon)-related features. At lower pressures, the TiN deposition rate is also found to be linearly proportional to this N$_2^+$*/Ti* emission ratio. This N$_2^+$* emission line correlated with film properties better than the N$_2^*$ emission features. (See Section 9.11 for ellipsometric monitoring of the TiN composition.)

Optical emission spectroscopy can be used to control the triode ion plating of hard coatings such as TiN, HfN, and ZrN. Salmenoja et al. (1985)

6.3 Applications in Processing

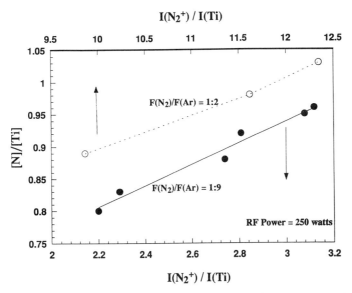

Figure 6.25 The ratio of the atomic concentrations [N]/[Ti] during sputter deposition of TiN films versus the ratio of the optical emission intensities of N_2^+* (391.4 nm) and Ti* (364.2 nm) at different nitrogen-to-argon flow ratios, for *in situ* process monitoring of TiN composition (from Pang *et al.*, 1994).

found the N_2^* (357.7 nm)/Ti* (364.3 nm) ratio to be a good monitor of the composition in TiN coatings for N atomic fractions of ~40–50%. Similarly, Salmenoja and Korhonen (1986) found the N_2^{+*} (391.4 nm)/Zr* (389.0, 389.1 nm) ratio could monitor the composition of ZrN coatings for N atomic fractions of ~20–55%.

Reactive DC magnetron sputtering of transparent and conductive films of indium tin oxide (ITO) has several advantages over other techniques, such as its high sputtering rate and the ability to use large targets. However, the rate of ITO deposition changes with time, even when the sputtering parameters (pressure, gas composition, discharge current) are kept constant, because the oxide coverage on the target changes with time. Enjouji *et al.* (1983) showed that the deposition rate of ITO using an In–Sn metallic target in a magnetron with an argon/oxygen mixture can be controlled by using OES. Figure 6.26 shows optical emission from such a discharge with an 80% Ar/20% O_2 mixture. They found that for a given target current, the ITO deposition rate varies linearly with the intensity of the In* (451.1 nm) emission. Different lines were obtained for each current. The implicitly changing variable was the O_2 fraction in the mixture, which varied from 5 to 19%. (The ITO films were more highly oxygenated with high oxygen fractions.) Moreover, as Figure 6.27 shows, these data can be plotted on a universal curve with the ITO deposition rate linearly propor-

192   Chapter 6. Optical Emission Spectroscopy

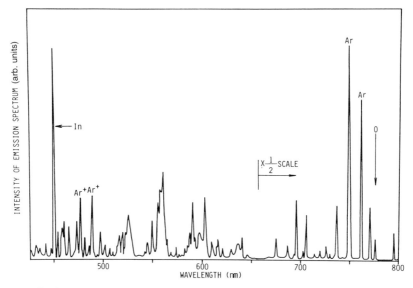

Figure 6.26 Optical emission from an 80% Ar/20% O$_2$ discharge in a magnetron with an In–Sn target, for depositing indium tin oxide (ITO) (from Enjouji *et al.*, 1983).

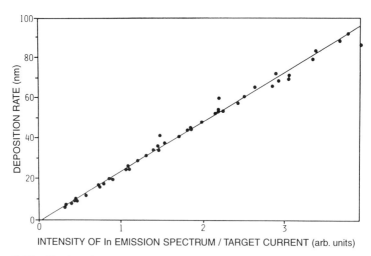

Figure 6.27 The deposition rate of indium tin oxide (ITO) by magnetron sputtering versus the intensity of In* (451.1 nm) emission (Figure 6.26) normalized by the target current. This linear relationship can be used for controlling the thickness of the ITO layer (from Enjouji *et al.*, 1983).

6.3 Applications in Processing

tional to the In* emission intensity normalized by the target current. The implicit variables are the O_2 fraction and the target current (0.5–1.1 A). Bhushan (1987) observed that the intensity of the Nb emission lines integrated from 528.1 to 536.5 nm gave a measure of the rate of magnetron sputter deposition of NbN films.

Tsuji and Hirokawa (1991) found that the sputtering rates of silicon and carbon in a glow discharge with an Ar–H_2 gas mixture correlated with optical emission intensities. The Si* (288.2 nm)/Ar* (750.4 nm) ratio correlates, though nonlinearly, with the silicon sputtering rate with a silicon cathode. All data with varying operating current and H_2 fraction fit on one universal curve. Similarly, for a carbon cathode the CH* (431.4 nm)/Ar* (750.4 nm) ratio correlates with the carbon sputtering rate.

Aita and Marhić (1981, 1983) used optical emission to probe reactive species during the reactive RF diode sputtering of zinc oxide targets in argon/oxygen and neon/oxygen mixtures.

6.3.4 Pulsed-Laser Deposition and Other Laser Processing

Optical emission spectroscopy is a widely used probe of the plume formed in pulsed-laser deposition (Saenger, 1993; Geohegan, 1994a). Optical emission is usually collected from the side and imaged on the entrance port of the monochromator so that a slice at a constant distance from the surface is sampled. Space- and time-resolved measurements are often made to determine the density and time of flight for a particular identified species. Rapid 2D imaging (fast photography) of optical emission is discussed below and in Geohegan (1992a). The peak strength and lineshape of the optical emission signal can be followed as a function of time after the pulse and distance from the target surface.

During the initial stages of plume expansion, in approximately the first millimeter of expansion, there is very bright continuum emission due to Bremsstrahlung emission (free–free electron transitions) and broadened emission lines from atoms and ions (Geohegan, 1994a). After expansion to a few millimeters from the surface, the OES signal is dominated by atomic, ionic, and molecular lines. These emission lines are seen many microseconds after the plume formation. Since the radiative lifetimes of the emitting species are typically ~10 nsec, this longer emission denotes continued collisional excitation. When the line spectra of excited species are no longer observable by OES, the ground-stage species can still be measured by absorption. Weak broad-banded emission seen at much longer times, ~10 μsec–1 msec, has been attributed to blackbody radiation from particulates in the plume, and is discussed in Section 13.5.

Optical emission spectroscopy has been used to probe the ablation plume during pulsed-laser deposition from many targets, with much work concentrating on depositing high-T_c superconductors. Figure 6.28 shows a high-resolution optical emission spectrum of the plume during KrF laser (248 nm) ablation of $YBa_2Cu_3O_{7-x}$ (Girault et al., 1989). This shows emission from several atomic (Ba, Cu, Y), ionic (Ba^+, Cu^+, Y^+) and diatomic (BaO, CuO, YO) species. Figure 8.26 shows that absorption by ground-state species is still strong long after the OES signal has disappeared in the plume formed in laser ablation from $YBa_2Cu_3O_{7-x}$ (Geohegan, 1992a). Fried et al. (1991) used a gated diode array to follow emission during ablation from targets made of $YBa_2Cu_3O_{7-x}$ and related materials (CuO, BaO_2, Y_2O_3), and found that metal oxides formed in the plume when oxygen was present—which could explain the improved film quality with an oxygen background gas.

Hastie et al. (1994) have conducted similar OE diagnostics during the PLD of oxide dielectrics, such as $BaTiO_3$ and $PbZr_{0.53}Ti_{0.47}O_3$ (PZT). They also demonstrated that the integrated spectral emission intensity approximately tracks the deposition rate during the PLD of Fe_3O_4 and Ag, which can be useful for real-time monitoring. Hermann et al. (1995) used time- and space-resolved optical emission measurements to probe the plume during excimer laser irradiation of titanium targets in a low-pressure nitrogen atmosphere, which leads to the PLD of TiN films. Other OES studies of the PLD plume include those on targets of high-T_c superconductors (Saenger, 1989; Zheng et al., 1989; Geyer and Weimer, 1989; Yoo et al., 1989; Wu et al., 1989; Scott et al., 1990; Dyer et al., 1990; Geohegan, 1992b; Sakeek et al., 1994), polymers (Dyer and Sidhu, 1988), graphite (Rohlfing, 1988; Schenck et al., 1990), Al (Knudtson et al., 1987), Cu/CuO (Saenger,

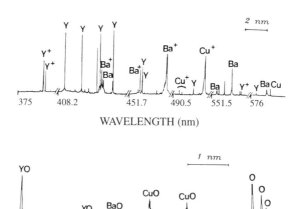

Figure 6.28 High-resolution spectrum of emission from the plasma plume after KrF-excimer laser ablation of a YBaCuO target (from Girault et al., 1989).

6.3 Applications in Processing

1989), Ge (for germanium oxide deposition; Vega et al., 1993), and ceramics such as Al_2O_3 and ZrO_2 (Voss et al., 1993). The use of OES to determine temperature profiles in the plume is described in Section 6.3.5.

Several different fast photography methods have been used by several groups to monitor plume emission as a function of position and time. In very early work, performed long before pulsed-laser deposition was developed, Ready (1963) studied the plume formed in laser ablation of carbon into air by using high-speed framing photography with 10-nsec resolution. This method has the advantage that a series of 2D images can be taken of the same plume. More recently, Scott et al. (1990) used this method during laser ablation from YBaCuO to show that plasma expansion depends on interplasma interactions and collisions with the background gas, and at higher pressures the latter results in the formation of a shock wave and instabilities. Kasuya and Nishina (1990) observed fast- and slow-velocity components in the plume of laser ablated graphite using a streak camera. Dyer and Sidhu (1988) used framing and streak photography to follow the plume in laser ablation of polyimide, while Yoo et al. (1989) noted fast (~25 psec) and slow (~1 nsec) emission in laser ablation of $YBa_2Cu_3O_{7-x}$ using a streak camera. Geohegan (1992b) used a gated, intensified CCD array (ICCD) to image plasma emission from KrF-laser-ablated YBaCuO surfaces as a function of time. Figure 6.29a shows these 2D photographs at different times, while Figure 6.29b plots the profile at one time. In vacuum, the plume has stationary and expanding components, while in oxygen a shockwave-like third component develops at 1.0 μsec. Figure 6.30 shows what occurs in the deposition phase of PLD, when the ablated material hits the substrate. The beam partially reflects when it hits the substrate, and the reflected portion of the leading edge of the plume interacts with the slower part of the plume still heading toward the surface (Geohegan, 1994b). Emission observed at much later times has been attributed to blackbody radiation (Section 13.5). Hastie et al. (1994) have used a gated CCD camera to take photographic images of $BaTiO_3$ PLD plumes. Fast images of the PLD plume have also been taken using LIF, absorption, etc.

Monitoring optical emission can also be important during laser CVD. Heszler et al. (1993) monitored OE during the ArF-excimer laser-assisted CVD of W from WF_6; they observed short-lived emission attributed to W atoms and ions as well as spectrally broad, longer-lived emission attributed to W clusters. During the ArF-excimer CVD of SiC from Si_2H_6/C_2H_2 mixtures, Mizunami et al. (1993) monitored emission from CH (431.5 nm), C_2 (516 nm), SiH (414 nm), and Si (252, 391 nm).

6.3.5 Thermometry

Optical emission samples the excited-stage energy distributions, and the temperatures obtained by using OES characterize these excited states.

a

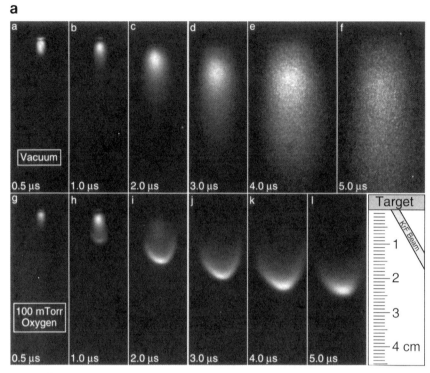

Figure 6.29 (a) Gated, intensified CCD array (ICCD) two-dimensional imaging of visible plasma emission from KrF-laser-ablated YBaCuO surfaces as a function of time (20-nsec exposure times), either in vacuum or in oxygen; (b) Intensity of the emission spectra acquired 1.0 μsec after laser ablation of YBaCuO plotted against the two-dimensional position (from Geohegan, 1992b).

When rotational structure is resolvable, Equations 4.37 and 4.39 are used to obtain the rotational temperature T_{rot} by plotting $\ln(S_{OES}(J'')/S(J''))$ versus $\mathcal{E}_{J'}$, where S_{OES} is the OES signal and S is the Hönl–London factor. For larger molecules (triatomics and bigger, and heavier diatomics), individual rotational lines can be resolved only with very high-resolution spectrometers. In these cases, T_{rot} can be determined instead by simulating the overall emission profile or by adding small amounts of other gases whose rotational bands are more easily resolved. Note that the Hönl–London factor $S(J)$ in Equation 4.37 is often more complicated than that given in Section 4.3.1.1.2, as for linear molecules with $\Lambda \neq 0$ (Kovács, 1969) and for nonlinear molecules. Optical emission spectroscopy has been used to measure temperatures in glow-discharge plasmas and in the plasma plume in PLD. Such temperature diagnostics in thermal plasmas have been reviewed by Fauchais *et al.* (1989) and are not covered here.

6.3 Applications in Processing

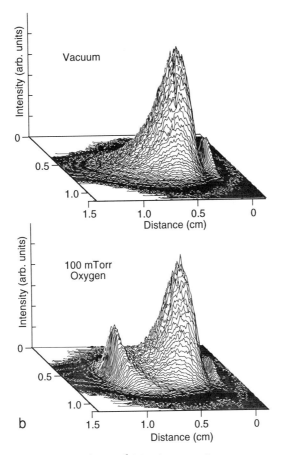

Figure 6.29 (*Continued*)

Using OES measurements, d'Agostino *et al.* (1981) found T_{vib} to be ~4200 K in a CF_4/O_2 discharge, and T_{rot} for CO and (added) N_2 to be ~500 K. Davis and Gottscho (1983) used OES to determine T_{rot} in a N_2 glow-discharge plasma, as used in silicon nitride deposition, from the N_2 second positive ($C\,^3\Pi_u \to B\,^3\Pi_u$) emission and the N_2^+ first negative ($B\,^2\Sigma_u^+ \to X\,^2\Sigma_g^+$) emission (see Figure 3.2). With 25 W powering the discharge, both methods gave T_{rot} ~500 K, in agreement with LIF probing of the $N_2^+\ X\,^2\Sigma_g^+$ ground state. Figure 6.31 shows the fit of the normalized N_2^+ first negative emission lines used to obtain T_{rot}; the linear dependence indicates a thermal distribution. As part of this study, Davis and Gottscho also studied a CCl_4 plasma, with 2% N_2 added to help obtain T_{rot}. T_{rot} can be determined without added N_2 from the CCl created in the discharge only with high-resolution excitation and analysis LIF (Chapter 7) or by

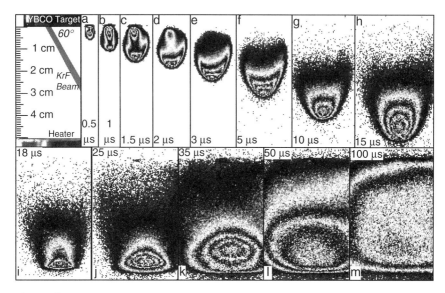

Figure 6.30 Contour plot of emission intensity during PLD, as in Fig. 6.29, but including later times and when the plume impinges on a metal target at room temperature (from Geohegan, 1994b). Reprinted by permission of Kluwer Academic Publishers.

OES with a high-resolution spectrometer. The addition of this N_2 actinometer permits OES analysis with a moderate-resolution spectrometer. T_{rot} values obtained from N_2 second positive emission were found to be near, but somewhat higher than that obtained from CCl LIF; the difference was attributed to possible sputtering of chlorocarbon deposits to form cold CCl. For both discharges, the collection optics were scanned to map the temperature profiles.

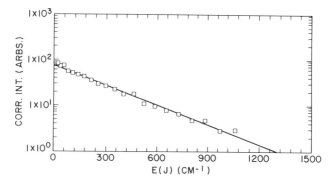

Figure 6.31 Plot of corrected N_2^{+*} first negative emission intensities as a function of rotational energy for the ionic ground state. Using this plot, the rotational temperature was determined to be 430 ± 10 K (from Davis and Gottscho, 1983).

6.3 Applications in Processing

Gottscho and Donnelly (1984) measured the translational temperature in glow discharges by measuring the Doppler width from OE by using high-resolution spectroscopy. Tsu *et al.* (1995) used this method to determine T_{trans} for argon neutrals and ions in an ECR plasma.

OE thermometry has also been used during the plasma deposition of thin films. For example, Reeve and Weimer (1995) found that for both the C_2 and CH emission spectra T_{rot} and T_{vib} differed widely at all locations in a DC arcjet diamond deposition reactor, except near the substrate. Moreover, T_{rot} for C_2 and T_{vib} for CH appeared to track the plasma gas temperature, while T_{vib} for C_2 and T_{rot} for CH were anomalously high.

In earlier work on optical emission thermometry, Phillips (1975) determined T_{rot} in a nitrogen discharge by analyzing unresolved bands in the second positive system emission spectrum. The rotational lineshape was assumed to have the form

$$g(\Delta\lambda) = \frac{a - (2\Delta\lambda/W)^2}{a + (a-2)(2\Delta\lambda/W)^2} \tag{6.9}$$

and was summed over the emission at the different wavelengths in the rotational band. This lineshape has a maximum of $g(0) = 1$, a width at half-maximum W, and wings that extend to $\pm \frac{1}{2} W a^{1/2}$. Porter and Harshbarger (1979) used this method to determine the temperature in an RF glow discharge similar to that used in semiconductor processing, finding $T_{rot} = 595 \pm 30$ K for 1000-W RF power (13.56 MHz) and 1.0 Torr N_2. The $v' = 0 \to v'' = 3$ (Figure 6.32) and $1 \to 4$ transitions in the second positive system were analyzed because Phillips had found them less sensitive than other bands to self absorption. Oshima (1978) analyzed the unresolved $B\,^1\Sigma$–$A\,^1\Pi$ CO emission near 5590 Å in C_2F_6/O_2 discharges.

Visible/ultraviolet optical emission from transitions between electronic bands is monitored in the overwhelming majority of OE studies. Using a different approach, Knights *et al.* (1982) used a modified Connes-type interferometer with ultimate resolution of 0.0009 cm^{-1} to resolve IR emission (and absorption) near 2200 cm^{-1}, due to vibrational transitions, for temperature measurements in a SiH_4 discharge. From the rotational lines in SiH vibrational emission $1 \to 0$ and $2 \to 1$, $T_{rot} = 485 \pm 30$ K was determined for SiH, which is much less than $T_{vib} = 2000 \pm 200$ K for the SiH vibrational mode. T_{vib} was estimated to be ~850 K for the SiH_4 mode near 2120 cm^{-1} from hot-band absorption.

Optical emission spectroscopy has also been used to determine the temperature of the atoms and molecules in the plume during PLD. The temperature of particulates measured by blackbody-radiation detection is described in Section 13.5.

Knudtson *et al.* (1987) determined the electronic temperature of the plume during PLD of Al from the relative emission intensity of Al(II) states. Figure 6.33a shows the emission lines used and Figure 6.33b plots

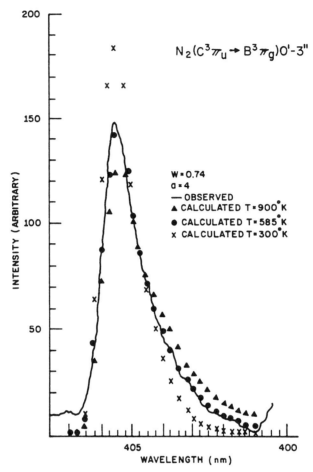

Figure 6.32 Comparison of the observed emission intensity for the $v' = 0 \to v'' = 3$ band in the second positive system of N_2 in a nitrogen RF glow discharge with simulated band profiles assuming different rotational temperatures (from Porter and Harshbarger, 1979).

the relative emission signals between states m and n, S_{mn}. These emission intensities are corrected for the frequency v_{mn} and radiative decay rate A_{mn} by the left-hand side of

$$\ln\left(\frac{S_{mn}}{g_m A_{mn} v_{mn}}\right) = \ln\frac{N}{Z} - \frac{\mathcal{E}_m}{k_B T_{\text{elect}}} \quad (6.10)$$

where $\mathcal{E}_m$ and g_m are the energy and degeneracy of the excited state, respectively. Equation 6.10 is a modified version of Equation 4.39. The straight line fit in Figure 6.33b indicates that there is limited thermal equilibrium, and the slope gives an electronic temperature T_{elect} of 8130 ± 270 K in the measured volume.

6.3 Applications in Processing

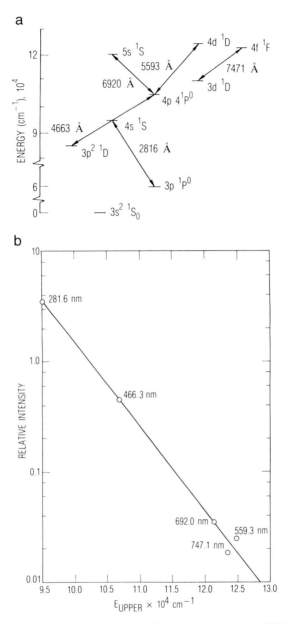

Figure 6.33 (a) Energy-level diagram showing the transitions of Al(II)* emission lines that are seen in the plume during pulsed laser deposition (PLD) of Al. (b) Normalized emission intensities of the Al(II)* lines from (a) versus the energy of the upper level of the respective transition. The slope of this line gives an electronic temperature of 8130 ± 270 K (from Knudtson et al., 1987).

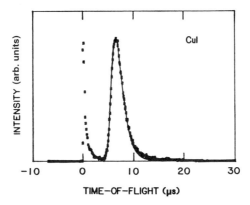

Figure 6.34 Time of flight (TOF) spectrum of optical emission from Cu atoms in the plume after laser ablation of a YBaCuO target 7.2 cm from the target (from Zheng *et al.*, 1989).

Zheng *et al.* (1989) measured the time of flight (TOF) of excited Y, Ba, and Cu atoms and Ba ions ablated from a YBaCuO target at 248 nm (KrF laser). As is seen in Figure 6.34 for Cu, the speed distribution of the detected species of mass m was fit to the Maxwell–Boltzmann speed (v) distribution, $f(v) \propto v^3 \exp[-m(v - v_0)^2/2k_B T_{trans}]$ with stream (average) speed normal to the surface v_0 and an effective center-of-mass translational temperature T_{trans}. This is expected in isentropic supersonic expansion. Optical emission

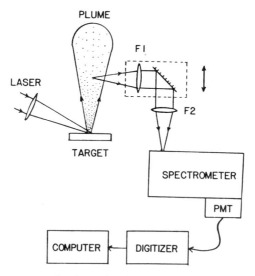

Figure 6.35 Apparatus used to determine the TOF spectrum of products of laser ablation. The assembly of the F1 lens and beam-turning mirror can be translated to examine the laser plume as a function of distance from the target (from Zheng *et al.*, 1989).

was detected using a collimation lens/right-angle-turning mirror assembly that delivered the light to a lens that focused it into the spectrometer (Figure 6.35). By translating the assembly along the axis of plume streaming, optical emission could be examined as a function of distance from the surface.

Rohlfing (1988) observed emission from several C_2 bands during laser ablation of graphite. Analysis of the $d\ ^3\Pi_g$–$a\ ^3\Pi_u$ Swan band near 560 nm gave $T_{rot} = 1060$ K for the $\Delta v = -1$ sequence and $T_{vib} \approx 10,400$ K.

6.3.6 2D and 3D Profiles by Imaging and Tomographic Reconstruction

Optical emission spectroscopy images of a plasma can be obtained by scanning the imaged region by translating the collection optics or the reactor. The spatial resolution possible with this method has been discussed in Section 5.2.1.1 (see Equation 5.8). Two-dimensional imaging of plasma plumes in PLD has been described above.

Another approach to mapping is computer-assisted tomography (CAT), which can be used to obtain a three-dimensional view of the gas-phase region in a reactor from various slices obtained by optical diagnostics, as is described in Section 5.2.1.3. Tomographic reconstruction of optical emission intensities has been used to examine the uniformity of plasma density.

Radon transform analysis is needed for general tomographic reconstruction. With circular symmetry, Radon transforms reduce to Abel transforms, which are generally used for analysis. For parallel-plate reactors, with the plates parallel to the xy plane, the emitted light can be collected by using a pinhole assembly to ensure sampling a cylindrical volume. If this sampled volume is a line parallel to the y axis, passing through x and z_0, then the intensity at the detector is

$$I(x,z_0) = 2 \int_x^R \frac{r\ \Omega(x,r)}{\sqrt{r^2 - x^2}} i(r,z_0) dr \qquad (6.11)$$

where z_0 is the height above the substrate, R is the radius of the plasma region, and $i(r,z_0)$ is intensity at the detector due to emission between r and $r + dr$ at height z_0 ($r = \sqrt{x^2 + y^2}$). This intensity i is determined from the power emitted per unit volume (p_{OES}) and geometric collection factors. Measurements are made as a function of x for a 2D analysis in the parallel-projections method depicted in Figure 5.13a (for one angle); z_0 can then be varied to examine emission from different planes. The Ω factor accounts for possible reabsorption of the emitted light along the collection path, with absorption coefficient α:

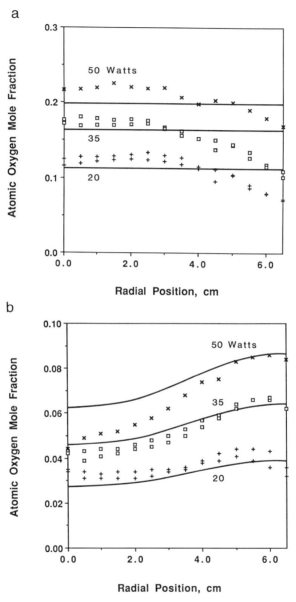

Figure 6.36 Atomic oxygen mole fraction versus radial position in an O_2 plasma reactor determined from tomographic reconstruction of OES data (844.6 nm from O^*, using actinometric calibration with the 750.4-nm Ar^* line) for (a) an empty reactor and (b) a loaded reactor (electrode coated with Ag_2O), for different RF powers (from Economou *et al.*, 1989).

6.3 Applications in Processing

$$\Omega(x,r) = \frac{\exp\{-\alpha(\sqrt{R^2 - x^2} - \sqrt{r^2 - x^2})\} + \exp\{-\alpha(\sqrt{R^2 - x^2} + \sqrt{r^2 - x^2})\}}{2} \quad (6.12)$$

This factor can also be used to account for differences in the collected solid angle along the path.

When reabsorption can be neglected ($\Omega = 1$), Equation 6.11 reduces to

$$I(x,z_0) = 2 \int_x^R \frac{r\, i(r,z_0)}{\sqrt{r^2 - x^2}} dr \quad (6.13)$$

which is similar to Equation 5.18. This Abel transform can be inverted to give

$$i(r,z_0) = -\frac{1}{\pi} \int_r^R \frac{dI(x,z_0)/dx}{\sqrt{x^2 - r^2}} dx \quad (6.14)$$

as in Equation 5.19.

In a series of studies Economou and co-workers determined the radial profile of reactive species during plasma etching by using this Abel transform analysis. They used argon actinometry to help calibrate the densities of the emitting reactive species. As reviewed by Economou *et al.* (1991), they mapped densities during plasma etching of polysilicon by SF_6 (with the 704-nm F* line), photoresist etching by O_2 (844.6-nm O*), and polysilicon etching by Cl_2 (837.6-nm Cl*). In the first two studies, the 750.4-nm Ar*

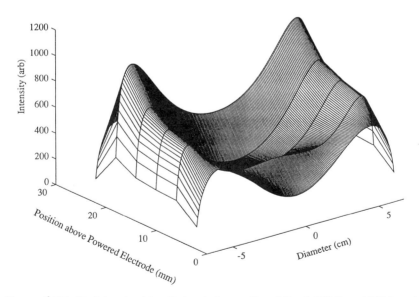

Figure 6.37 Radial and axial optical emission profiles of the Ar(I)* line at 750.4 nm in a GEC reference cell operated as an Ar RF discharge, obtained by Abel inversion (from Pender *et al.*, 1993).

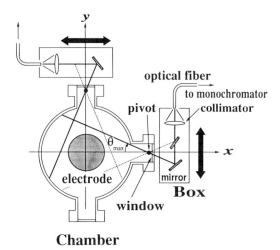

Figure 6.38 Schematic illustration of a discharge chamber with optical emission collection optics for tomographic analysis (from Miyake *et al.*, 1992).

line was monitored and used for calibration, while in the last study the 811.5-nm Ar* line was used. In each case, 20 line-integrated OES measurements were made across the 100-mm-diameter wafer. Using tomographic reconstruction of OES data, Figure 6.36 (Economou *et al.*, 1989) shows that the atomic oxygen concentration decreases greatly when the polymer substrate is added (loaded reactor), and the oxygen density is lowest near the center. Analogous studies have demonstrated how the atomic chlorine radial profile changes during plasma etching of polysilicon when conditions change so that a plasmoid is formed (Aydil and Economou, 1991). Concomitant etch-rate measurements (see Section 9.5.1, Figures 9.20 and 9.21) show that the etch rate is greatest in the center when there is a plasmoid.

Pender *et al.* (1993) used the Abel inversion to determine the radial optical emission profile in the standard parallel plate GEC reference cell, which was operated as an Ar RF glow discharge. The optical system imaged and telescoped down the parallel light collected across the chamber, at a given height between the electrodes, into a spectrometer set to transmit either the Ar(I)* line at 750 nm or the Ar(II)* line at 428 nm. A photodiode array was mounted at the exit port rotated by 90° to image this horizontal slice transmitted through the slit, and the signal was analyzed with Abel inversion. This was repeated for different heights above the lower electrode to give the pattern shown in Figure 6.37. The emission profile was observed to have an annular pattern, which mimicked the etch nonuniformity observed when using this type of discharge. Beale *et al.* (1994) used tomographically resolved Ar emission to characterize a planar RF inductively coupled Ar plasma.

6.3 Applications in Processing

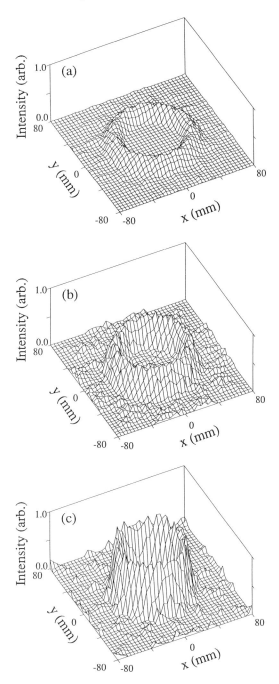

Figure 6.39 Reconstructed two-dimensional slices of Ar(I)* emission of three axial locations in a ring-shaped DC magnetron discharge, at (a) 11 mm, (b) 7 mm, and (c) 3 mm (from Miyake *et al.,* 1992).

Miyake et al. (1992) have applied the more general Radon tomographic methods to reconstruct the optical emission profile of several species in a ring-shaped DC magnetron discharge with aluminum electrodes. They have labeled this technique emission-selected computer tomography (ESCT). Figure 6.38 shows a cross-sectional view of the chamber and tomographic system. Optical emission was collected from two ports at 90° to each other. In each case, the direction of the slice of collected emission was pivoted about a point and controlled by the position of a mirror (along x or y) to obtain information about the radial profile. This is the fan-beam projection method depicted in Figure 5.13b. The mirror systems were also translated normal to the electrodes (and the page), which defines the z axis. Three species were monitored: Ar* ($3p_5 \rightarrow 1s_4$, 419.8 nm), Ar$^+$* ($4p\ ^4D_{7/2} \rightarrow 4s\ ^4P_{5/2}$, 434.8 nm), and Al* ($4s\ ^2S \rightarrow 3p\ ^2P$, 396.2 nm). A series of 100 scans in the xy plane were reconstructed with an inverse discrete Radon transform (Radon, 1917) using the algebraic reconstruction technique (Kouris et al., 1982). Figure 6.39 displays reconstructed two-dimensional slices of the Ar* emission. The emission is a maximum at the radius where the magnetic field is parallel with the electrode and at the radius of the ring-like pattern that erodes into the electrode.

References

J. Affinito and R. R. Parsons, *J. Vac. Sci. Technol. A* **2,** 1275 (1984).
C. R. Aita and M. E. Marhić, *J. Appl. Phys.* **52,** 6584 (1981).
C. R. Aita and M. E. Marhić, *J. Vac. Sci. Technol. A* **1,** 69 (1983).
H. M. Anderson and M. P. Splichal, *SPIE* **2091,** 333 (1994).
D. Angell and G. S. Oehrlein, *Appl. Phys. Lett.* **58,** 240 (1991).
E. S. Aydil and D. J. Economou, *J. Appl. Phys.* **69,** 109 (1991).
G. Balestrino, M. Marinelli, E. Milani, A. Paoletti, I. Pinter, A. Tebano, and P. Paroli, *Appl. Phys. Lett.* **62,** 879 (1993).
R. F. Barrow, ed., *Molécules Diatomiques: Bibliographie Critique de Données Spectroscopiques* [Diatomic Molecules: A Critical Bibliography of Spectroscopic Data], Vol. I. Éditions du Centre National de la Recherche Scientifique, Paris, 1973.
R. F. Barrow, ed., *Molécules Diatomiques: Bibliographie Critique de Données Spectroscopiques* [Diatomic Molecules: A Critical Bibliography of Spectroscopic Data], Vol. II. Éditions du Centre National de la Recherche Scientifique, Paris, 1975.
S. Bashkin and J. O. Stoner, Jr., *Atomic Energy Levels and Grotrian Diagrams:* Vol. I (1975), Vol. II (1978), Vol. III (1981), Vol. IV (1982). North-Holland, Amsterdam.
D. F. Beale, A. E. Wendt, and L. J. Mahoney, *J. Vac. Sci. Technol. A* **12,** 2775 (1994).
M. Bhushan, *J. Vac. Sci. Technol. A* **5,** 2829 (1987).
J. P. Booth, O. Joubert, J. Pelletier, and N. Sadeghi, *J. Appl. Phys.* **69,** 618 (1991).

References

G. Bruno, P. Capezzuto, G. Cicala, and F. Cramarossa, *Plasma Chem. Plasma Process.* **6,** 109 (1986).
L.-M. Buchmann, F. Heinrich, P. Hoffmann, and J. Janes, *J. Appl. Phys.* **67,** 3635 (1990).
A. L. Cappelli, R. A. Gottscho, and T. A. Miller, *Plasma Chem. Plasma Process.* **5,** 317 (1985).
P. E. Clarke, D. Field, and D. F. Klemperer, *J. Appl. Phys.* **67,** 1525 (1990).
J. W. Coburn and M. Chen, *J. Appl. Phys.* **51,** 3134 (1980).
J. W. Coburn and M. Chen, *J. Vac. Sci. Technol.* **18,** 353 (1981).
P. Collot, T. Diallo, and J. Canteloup, *J. Vac. Sci. Technol. B* **9,** 2497 (1991).
C. H. Corliss and W. R. Bozman, *Experimental Transition Probabilities for Spectral Lines of Seventy Elements,* NBS Monograph 53. U.S. Government Printing Office, Washington, DC, 1962.
B. J. Curtis, *Sold State Technol.* **23**(4), 129 (1980).
B. J. Curtis and H. J. Brunner, *J. Electrochem. Soc.* **125,** 829 (1978).
R. d'Agostino, F. Cramarossa, S. De Benedictis, and G. Ferraro, *J. Appl. Phys.* **52,** 1259 (1981).
R. d'Agostino, F. Cramarossa, and S. De Benedictis, *Plasma Chem. Plasma Process.* **2,** 213 (1982).
D. A. Danner and D. W. Hess, *J. Appl. Phys.* **59,** 940 (1986).
D. A. Danner, D. L. Flamm, and J. A. Mucha, *J. Electrochem. Soc.* **130,** 905 (1983).
W. C. Dautremont-Smith and J. Lopata, *J. Vac. Sci. Technol. B* **1,** 943 (1983).
G. P. Davis and R. A. Gottscho, *J. Appl. Phys.* **54,** 3080 (1983).
E. O. Degenkolb, C. J. Mogab, M. R. Goldrick, and J. R. Griffiths, *Appl. Spectrosc.* **30,** 520 (1976).
V. M. Donnelly, in *Plasma Diagnostics* (O. Auciello and D. L. Flamm, eds.), Vol. I, Chapter 1, p. 1. Academic Press, Boston, 1989.
V. M. Donnelly, *J. Vac. Sci. Technol. A* **14,** 1076 (1996).
V. M. Donnelly and D. L. Flamm, *J. Appl. Phys.* **51,** 5273 (1980).
V. M. Donnelly, D. E. Ibbotson, and D. L. Flamm, in *Ion Beam Modification of Surfaces: Fundamentals and Applications* (O. Auciello and R. Kelly, eds.), Chapter 8. Elsevier, Amsterdam, 1984a.
V. M. Donnelly, D. L. Flamm, W. C. Dautremont-Smith, and D. J. Werder, *J. Appl. Phys.* **55,** 242 (1984b).
V. M. Donnelly, D. L. Flamm, and R. H. Bruce, *J. Appl. Phys.* **58,** 2135 (1985).
R. W. Dreyfus, J. M. Jasinski, R. E. Walkup, and G. S. Selwyn, *Pure Appl. Chem.* **57,** 1265 (1985).
P. E. Dyer and J. Sidhu, *J. Appl. Phys.* **64,** 4657 (1988).
P. E. Dyer, A. Issa, and P. H. Key, *Appl. Phys. Lett.* **57,** 186 (1990).
D. J. Economou, S.-K. Park, and G. D. Williams, *J. Electrochem. Soc.* **136,** 188 (1989).
D. J. Economou, E. S. Aydil, and G. Barna, *Solid State Technol.* **34**(4), 107 (1991).
K. Enjouji, K. Murata, and S. Nishikawa, *Thin Solid Films* **108,** 1 (1983).
P. Fauchais, J. F. Coudert, and M. Vardelle, in *Plasma Diagnostics* (O. Auciello and D. L. Flamm, eds.), Vol. I, Chapter 7, p. 349. Academic Press, Boston, 1989.
D. Field, A. J. Hydes, and D. F. Klemperer, *Vacuum* **34,** 563 (1984).
D. L. Flamm and V. M. Donnelly, *J. Appl. Phys.* **59,** 1052 (1986).
D. Fried, G. P. Reck, T. Kushida, and E. W. Rothe, *J. Appl. Phys.* **70,** 2337 (1991).

N. Fujiwara, T. Maruyama, and M. Yoneda, *Symp. Dry Process., 17th,* Tokyo, *1995,* p. 51 (1995).
T. Fuyuki, B. Allain, and J. Perrin, *J. Appl. Phys.* **68,** 3322 (1990).
D. B. Geohegan, *Thin Solid Films* **220,** 138 (1992a).
D. B. Geohegan, *Appl. Phys. Lett.* **60,** 2732 (1992b).
D. B. Geohegan, in *Pulsed Laser Deposition of Thin Films* (G. Hubler and D. B. Chrisey, eds.), Chapter 5, p. 115. Wiley (Interscience), New York, 1994a.
D. B. Geohegan, in *Excimer Lasers* (L. D. Laude, ed.), Series E: Applied Sciences, Vol. 265, NATO Series. Kluwer Academic, Dordrecht, Norrell, MA, 1994b.
T. J. Geyer and W. A. Weimer, *Appl. Phys. Lett.* **54,** 469 (1989).
C. Girault, D. Damiani, J. Aubreton, and A. Catherinot, *Appl. Phys. Lett.* **55,** 182 (1989).
C. Gomez-Aleixandre, O. Sanchez, A. Castro, and J. M. Albella, *J. Appl. Phys.* **74,** 3752 (1993).
R. A. Gottscho and V. M. Donnelly, *J. Appl. Phys.* **56,** 245 (1984).
R. A. Gottscho and T. A. Miller, *Pure Appl. Chem.* **56,** 189 (1984).
R. A. Gottscho, G. Smolinky, and R. H. Burton, *J. Appl. Phys.* **53,** 5908 (1982).
K. E. Greenberg and P. J. Hargis, Jr., *Appl. Phys. Lett.* **54,** 1374 (1989).
J. E. Greene, *J. Vac. Sci. Technol.* **15,** 1718 (1978).
J. E. Greene and F. Sequeda-Osorio, *J. Vac. Sci. Technol.* **10,** 1144 (1973).
J. E. Greene, F. Sequeda-Osorio, and B. R. Natarajan, *J. Appl. Phys.* **46,** 2701 (1975).
J. E. Greene, B. R. Natarajan, and F. Sequeda-Osorio, *J. Appl. Phys.* **49,** 417 (1978).
G. Hancock, J. P. Sucksmith, and M. J. Toogood, *J. Phys. Chem.* **94,** 3269 (1990).
P. D. Hanish, J. W. Grizzle, M. D. Giles, and F. L. Terry, Jr., *J. Vac. Sci. Technol. A* **13,** 1802 (1995).
G. R. Harrison, *M. I. T. Wavelength Tables.* MIT Press, Cambridge, MA, 1969.
W. R. Harshbarger, *Solid State Technol.* **25**(4), 126 (1982).
W. R. Harshbarger, R. A. Porter, T. A. Miller, and P. Norton, *Appl. Spectrosc.* **31,** 201 (1977).
J. W. Hastie, D. W. Bonnell, A. J. Paul, and P. K. Schenck, *Mater. Res. Soc. Symp. Proc.* **334,** 305 (1994).
N. Hata, A. Matsuda, and K. Tanaka, *J. Appl. Phys.* **61,** 3055 (1987).
F. Heinrich, H.-P. Stoll, and H.-C. Scheer, *Appl. Phys. Lett.* **55,** 1474 (1989a).
F. Heinrich, H.-P. Stoll, H.-C. Scheer, and P. Hoffmann, *SPIE* **1188,** 185 (1989b).
I. P. Herman, V. M. Donnelly, K. V. Guinn, and C. C. Cheng, *Phys. Rev. Lett.* **72,** 2801 (1994).
J. Hermann, A. L. Thomann, C. Boulmer-Leborgne, B. Dubreuil, M. L. De Giorgi, A. Perrone, A. Luches, and I. N. Mihailescu, *J. Appl. Phys.* **77,** 2928 (1995).
G. Herzberg, *Atomic Spectra and Atomic Structure.* Dover, New York, 1937.
G. Herzberg, *Molecular Spectra and Molecular Structure. I. Spectra of Diatomic Molecules.* Van Nostrand-Reinhold, New York, 1950.
G. Herzberg, *Molecular Spectra and Molecular Structure. III. Electronic Spectra and Electronic Structure of Polyatomic Molecules.* Van Nostrand-Reinhold, New York, 1966.
P. Heszler, P. Mogyorósi, and J. O. Carlsson, *Appl. Surf. Sci.* **69,** 376 (1993).
P. Ho, R. J. Buss, and R. E. Loehman, *J. Mater. Res.* **4,** 873 (1989).
T. Holstein, *Phys. Rev.* **72,** 1212 (1947).

References

K. P. Huber and G. Herzberg, *Molecular Spectra and Molecular Structure. IV. Constants of Diatomic Molecules.* Van Nostrand-Reinhold, New York, 1979.

D. E. Ibbotson, D. L. Flamm, and V. M. Donnelly, *J. Appl. Phys.* **54,** 5974 (1983).

A. Inspektor, Y. Liou, T. McKenna, and R. Messier, *Surf. Coatings Technol.* **39/40,** 211 (1989).

M. E. Jacox, *J. Phys. Chem. Ref. Data* **17,** 269 (1988).

M. E. Jacox, *J. Phys. Chem. Ref. Data* **19,** 1387 (1990).

M. E. Jacox, *Vibrational and Electronic Energy Levels of Polyatomic Transient Molecules,* Monograph No. 3, *J. Phys. Chem. Ref. Data* (1994).

F. J. Kampas and R. W. Griffith, *J. Appl. Phys.* **52,** 1285 (1981).

A. Kasuya and Y. Nishina, *High Temp. Sci.* **27,** 473 (1990).

R. E. Klinger and J. E. Greene, *Appl. Phys. Lett.* **38,** 620 (1981).

J. C. Knights, J. P. M. Schmitt, J. Perrin, and G. Guelachvili, *J. Chem. Phys.* **76,** 3414 (1982).

J. T. Knudtson, W. B. Green, and D. G. Sutton, *J. Appl. Phys.* **61,** 4771 (1987).

C. S. Korman, *Solid State Technol.* **25**(4), 115 (1982).

K. Kouris, H. Tuy, A. Lent, G. T. Herman, and R. M. Lewitt, *IEEE Trans. Med. Imaging* **MI-1,** 161 (1982).

I. Kovács, *Rotational Structure in the Spectra of Diatomic Molecules.* American Elsevier, New York, 1969.

O. Krogh, H. Slomowitz, Y. Melaku, and H. O. Blom, *J. Electrochem. Soc.* **134,** 2045 (1987).

Luxtron Corp., Application Notes, (1994). See also U.S. Patent 5,208,644 (1993).

P. J. Marcoux and P. D. Foo, *Solid State Technol.* **24**(4), 115 (1981).

D. Mataras, S. Cavadias, and D. Rapakoulias, *J. Appl. Phys.* **66,** 119 (1989).

A. Matsuda and K. Tanaka, *Thin Solid Films* **92,** 171 (1982).

O. Matsumoto, H. Toshima, and Y. Kanzaki, *Thin Solid Films* **128,** 341 (1985).

J. S. McKillop, J. C. Forster, and W. M. Holber, *J. Vac. Sci. Technol. A* **7,** 908 (1989a).

J. S. McKillop, J. C. Forster, and W. M. Holber, *Appl. Phys. Lett.* **55,** 30 (1989b).

Y. Mitsuda, K. Tanaka, and T. Yoshida, *J. Appl. Phys.* **67,** 3604 (1990).

S. Miyake, N. Shimura, T. Makabe, and A. Itoh, *J. Vac. Sci. Technol. A* **10,** 1135 (1992).

T. Mizunami, N. Toyama, and T. Uemura, *J. Appl. Phys.* **73,** 2024 (1993).

C. E. Moore, *Atomic Energy Levels,* Vol. I (1949); Vol. II (1952); Vol. III (1958). NBS Circular 467.

C. E. Moore and P. W. Merrill, *Partial Grotrian Diagrams of Astrophysical Interest,* National Standard Reference Data Series, NSRDS-NBS 23. 1968.

J. A. Mucha, D. L. Flamm, and D. E. Ibbotson, *J. Appl. Phys.* **65,** 3448 (1989).

N. Mutsukura, Y. Fukasawa, Y. Machi, and T. Kubota, *J. Vac. Sci. Technol. A* **12,** 3126 (1994).

G. C. Nieman, S. D. Colson, S. G. Hansen, and G. Luckman, *J. Appl. Phys.* **67,** 6728 (1990).

J. A. O'Neill, *J. Vac. Sci. Technol. A* **9,** 669 (1991).

J. A. O'Neill, W. M. Holber, and J. B. O. Caughman, *SPIE* **1392,** 516 (1990).

J. A. O'Neill, M. S. Barnes, and J. H. Keller, *J. Appl. Phys.* **73,** 1621 (1993).

G. J. Orloff, *J. Vac. Sci. Technol. A* **10,** 3065 (1992); erratum: *ibid.* **12,** 3248 (1994).

M. Oshima, *Jpn. J. Appl. Phys.* **17,** 1157 (1978).

J. F. Osmundsen, C. C. Abele, and J. G. Eden, *J. Appl. Phys.* **57,** 2921 (1985).

Z. Pang, M. Boumerzoug, R. V. Kruzelecky, P. Mascher, J. G. Simmons, and D. A. Thompson, *J. Vac. Sci. Technol. A* **12,** 83 (1994).
M. L. Parsons and P. M. McElfresh, *Flame Spectroscopy: Atlas of Spectral Lines.* IFI/Plenum, New York, 1971.
R. W. B. Pearse and A. G. Gaydon, *The Identification of Molecular Spectra,* 4th ed. Chapman & Hall, London, 1976.
S. J. Pearton, T. A. Keel, A. Katz, and F. Ren, *Semicond. Sci. Technol.* **8,** 1889 (1993).
S. J. Pearton, F. Ren, C. R. Abernathy, and C. Constantine, *Mater. Sci. Eng. B* **23,** 36 (1994).
J. Pender, M. Buie, T. Vincent, J. Holloway, M. Elta, and M. L. Brake, *J. Appl. Phys.* **74,** 3590 (1993).
F. M. Phelps, III, *M. I. T. Wavelength Tables,* Vol. 2. MIT Press, Cambridge, MA, 1982.
A. V. Phelps, *J. Phys. Chem. Ref. Data* **19,** 653 (1990).
A. V. Phelps, *J. Phys. Chem. Ref. Data* **20,** 557 (1991).
A. V. Phelps, *J. Phys. Chem. Ref. Data* **21,** 883 (1992).
D. M. Phillips, *J. Phys. D* **8,** 507 (1975).
I. C. Plumb and K. R. Ryan, *Plasma Chem. Plasma Process.* **6,** 205 (1986).
R. A. Porter and W. R. Harshbarger, *J. Electrochem. Soc.* **126,** 460 (1979).
J. Radon, *Ber. Verh. Saechs. Akad. Wiss. Leipzig, Math.-Naturwiss. Kl.* **69,** 262 (1917).
J. F. Ready, *Appl. Phys. Lett.* **3,** 11 (1963).
S. W. Reeve and W. A. Weimer, *J. Vac. Sci. Technol. A* **13,** 359 (1995).
H. H. Richter, A. Wolff, B. Tillack, and T. Skaloud, *Mater. Sci. Eng. B* **27,** 39 (1994).
P. E. Riley, T. E. Clark, E. F. Gleason, and M. M. Garver, *IEEE Trans. Semicond. Manuf.* **SM-3,** 150 (1990).
E. A. Rohlfing, *J. Chem. Phys.* **89,** 6103 (1988).
J. P. Roland, P. J. Marcoux, G. W. Ray, and G. H. Rankin, *J. Vac. Sci. Technol. A* **3,** 631 (1985).
K. L. Saenger, *J. Appl. Phys.* **66,** 4435 (1989).
K. L. Saenger, *Process. Adv. Mater.* **3**(2), 63 (1993).
H. F. Sakeek, T. Morrow, W. G. Graham, and D. G. Walmsley, *J. Appl. Phys.* **75,** 1138 (1994).
K. Salmenoja and A. S. Korhonen, *Vacuum* **36,** 33 (1986).
K. Salmenoja, A. S. Korhonen, and M. S. Sulonen, *J. Vac. Sci. Technol. A* **3,** 2364 (1985).
P. K. Schenck, D. W. Bonnell, and J. W. Hastie, *High Temp. Sci.* **27,** 483 (1990).
S. Schiller, U. Heisig, C. Korndörfer, G. Beister, J. Reschke, K. Steinfelder, and J. Strümpfel, *Surf. Coatings Technol.* **33,** 405 (1987).
K. Scott, J. M. Huntley, W. A. Phillips, J. Clarke, and J. E. Field, *Appl. Phys. Lett.* **57,** 922 (1990).
G. S. Selwyn, *Optical Diagnostic Techniques for Plasma Processing.* AVS Press, New York, 1993.
G. S. Selwyn and E. Kay, *Plasma Chem. Plasma Process.* **5,** 183 (1985).
R. Shadmehr, D. Angell, P. B. Chou, G. S. Oehrlein, and R. S. Jaffe, *J. Electrochem. Soc.* **139,** 907 (1992).
P. H. Singer, *Semicond. Int.,* August, p. 66 (1988).
M. P. Splichal and H. M. Anderson, *SPIE* **1594,** 189 (1991).

References

A. R. Striganov and N. S. Sventitskii, *Tables of Spectral Lines of Neutral and Ionized Atoms*. IFI/Plenum, New York, 1968.

S. N. Suchard, ed., *Spectroscopic Data*, Vol. 1, *Heteronuclear Diatomic Molecules*, Parts A and B. IFI/Plenum, New York, 1975.

S. N. Suchard and J. E. Melzer, eds., *Spectroscopic Data*, Vol. 2. *Homonuclear Diatomic Molecules*. IFI/Plenum, New York, 1976.

C. C. Tang and D. W. Hess, *J. Electrochem. Soc.* **131,** 115 (1984).

S. Thomas III, K. K. Ko, and S. W. Pang, *J. Vac. Sci. Technol. A* **13,** 894 (1995).

H. R. Thorsheim and J. E. Butler, in *Synthetic Diamond: Emerging CVD Science and Technology* (K. E. Spear and J. P. Dismukes, eds.), Chapter 7, p. 193. Wiley, New York, 1994.

F. Tochikubo, A. Suzuki, S. Kakuta, Y. Terazono, and T. Makabe, *J. Appl. Phys.* **68,** 5532 (1990).

D. V. Tsu, R. T. Young, S. R. Ovshinsky, C. C. Klepper, and L. A. Berry, *J. Vac. Sci. Technol. A* **13,** 935 (1995).

K. Tsuji and K. Hirokawa, *Thin Solid Films* **205,** 6 (1991).

P. Van de Weijer, B. H. Zwerver, and J. L. G. Suijker, *Chem. Phys. Lett.* **153,** 33 (1988).

P. E. Vanier, F. J. Kampas, R. R. Corderman, and G. Rajeswaran, *J. Appl. Phys.* **56,** 1812 (1984).

R. P. Vaudo, J. W. Cook, Jr., and J. F. Schetzina, *J. Vac. Sci. Technol. B* **12,** 1232 (1994).

F. Vega, C. N. Afonso, and J. Solis, *J. Appl. Phys.* **73,** 2472 (1993).

A. Voss, E. W. Kreutz, J. Funken, M. Alunović, and H. Sung, *Appl. Surf. Sci.* **69,** 174 (1993).

R. E. Walkup, K. L. Saenger, and G. S. Selwyn, *J. Chem. Phys.* **84,** 2668 (1986).

B. Wangmaneerat, T. M. Niemczyk, G. Barna, and D. M. Haaland, in *Plasma Processing* (G. S. Mathad and D. V. Hess, eds.), p. 115. Electrochem. Soc., Pennington, NJ, 1992.

W. D. Westwood, *Phys. Thin Films* **14,** 1 (1989).

X. D. Wu, B. Dutta, M. S. Hegde, A. Inam, T. Venkatesan, E. W. Chase, C. C. Chang, and R. Howard, *Appl. Phys. Lett.* **54,** 179 (1989).

R. S. Yalamanchi and K. S. Harshavardan, *J. Appl. Phys.* **68,** 5941 (1990).

K. M. Yoo, R. R. Alfano, X. Guo, M. P. Sarachik, and L. L. Isaacs, *Appl. Phys. Lett.* **54,** 1278 (1989).

A. N. Zaidel', V. K. Prokof'ev, S. M. Raiskii, V. A. Slavnyi, and E. Ya. Shreider, *Tables of Spectral Lines*. IFI/Plenum, New York, 1970.

F. Zhang, Y. Zhang, Y. Yang, G. Chen, and X. Jiang, *Appl. Phys. Lett.* **57,** 1467 (1990).

J. P. Zheng, Z. Q. Huang, D. T. Shaw, and H. S. Kwok, *Appl. Phys. Lett.* **54,** 280 (1989).

CHAPTER 7

Laser-Induced Fluorescence

In laser-induced fluorescence (LIF) a gas-phase molecule or atom is excited by a laser, and the spontaneous emission to a lower state, i.e., fluorescence, is spectrally analyzed and detected (Figure 7.1). This is sometimes also called laser-induced fluorescence spectroscopy (LIFS) or laser-excited fluorescence (LEF). The species is detected either by a selective identification in absorption or fluorescence, and can be identified unambiguously.

In dispersed fluorescence experiments, the molecule is excited at a fixed, resonant transition and the fluorescence spectrum is dispersed by a monochromator. The monochromator is scanned to obtain a spectrum that looks like an emission spectrum. In excitation LIF, either broadband fluorescence or the fluorescence at a specific wavelength is monitored as the excitation wavelength is scanned with a tunable laser; this gives a spectrum very similar to the absorption spectrum. Figure 7.2 shows both the dispersed fluorescence and the fluorescence excitation spectrum of HSiCl taken by Ho and Breiland (1983) during Si chemical vapor deposition (CVD) from dichlorosilane. When spectral interference with other species is minimal, a species can be monitored by LIF more simply by exciting it at a single wavelength and monitoring emission from it in a single-wavelength region. In such cases, a fixed-wavelength laser can be used for excitation if it is resonant. A typical experimental configuration used to perform laser-induced fluorescence is depicted in Figure 7.3 in Section 7.1.1.

The term *laser-induced fluorescence,* or *LIF,* will be used in this chapter to denote all processes in which the gas-phase species absorbs optical radiation and then radiatively decays. Although this term is appropriate in the vast majority of cases, it is technically incorrect in some instances. For example, when incoherent light sources are used for excitation, "laser"-induced fluorescence is incorrect and the term *resonance fluorescence* is sometimes preferable. Moreover, the "fluorescence" part of LIF is not always appropriate. The term *luminescence* applies more generally to any type of relaxation by light emission, while the terms *fluorescence* and *phos-*

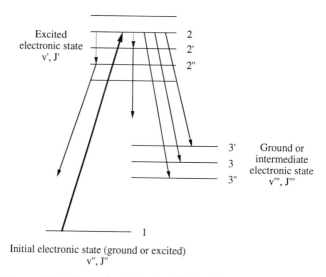

Figure 7.1 Energy levels involved in laser-induced fluorescence.

phorescence denote luminescence due to transitions that are spin-allowed and -forbidden, respectively, and are therefore relatively fast and slow.

Species identification is usually unambiguous when LIF is used to probe atomic, diatomic, and some triatomic species because of the highly resolved and unique structure in the absorption and emission spectra (Figure 7.2). For many atoms, including several group III, IV, and V atoms and metal atoms, LIF excitation occurs in the visible or near ultraviolet; but for other atoms, such as H, O, Cl, and N, the excitation wavelengths are so short that it is more convenient to excite the atoms by using two-photon absorption, which is followed by the emission of a single photon (two-photon LIF; see Section 7.2.1.2). Selective excitation and well-resolved fluorescence is possible with diatomics and sometimes with triatomics. Spectral resolution gets worse with larger molecules, i.e., those with more atoms or heavier atoms, because the densities of states and transitions become larger; a continuum develops for sufficiently large molecules (heavy triatomics and larger molecules). Concomitantly, the ability to identify a species becomes more difficult or impossible. Also, nonradiative decay becomes an increasingly important pathway for relaxation in molecules as the number of their atoms increases, and the fluorescence quantum efficiency becomes unacceptably low for large molecules.

The sensitivity of LIF can be very high. Wormhoudt *et al.* (1983) have calculated the minimum density detectable by LIF for many of the atoms and radicals that are important in thin film processing; this detection limit is $\sim 10^6$–10^8/cm^3 for many of these species. Detection levels of 10^8–10^{10} molecules/cm^3 ($\sim 10^{-8}$–10^{-6} Torr) are routine. Laser-

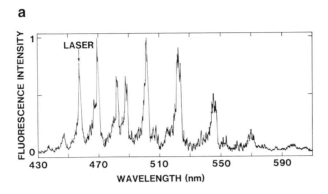

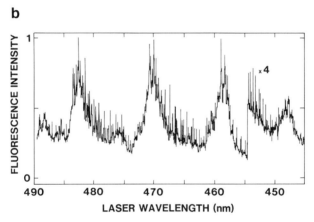

Figure 7.2 Laser-induced fluorescence spectra of HSiCl during Si CVD using a dichlorosilane reactant: (a) Dispersed fluorescence excited at 457.5 nm, (b) fluorescence excitation spectrum, excited between 445 and 490 nm (from Ho and Breiland, 1983).

induced fluorescence is much more sensitive than absorption because absorption spectroscopies measure the small differences between incident and transmitted beams.

There is usually no background in LIF measurements, except in plasmas where LIF must be distinguished from any background optical emission (OE). Although optical emission spectroscopy (OES) is a simpler probe of plasmas than is LIF, it detects only the excited-state species, which usually constitute a small fraction of the total population. Laser-induced fluorescence can probe population in any state, including the ground state, which is usually the most highly populated state.

The strength of LIF is that it can be used to identify species and to measure their concentration nonperturbatively, in the ground state or any other state. Good spatial resolution can be obtained, and concentration profiles can be mapped by spatially scanning a focused laser in the reactor or by scanning the reactor. Although LIF is very good in making relative density measurements, the absolute densities are uncertain because of collisional relaxation and uncertain absorption cross-sections (Breiland and Ho, 1993). Absolute measurements of concentration by LIF often require reference to a calibration measurement, such as absorption. To determine temperature by using LIF, however, only the measurement of the relative populations of different states is needed.

Although LIF can be used to probe many smaller species, it may not be capable of probing all species present in a thin film reactor or even the majority species, for the above-cited reasons. Even when several species can be probed by LIF, it is difficult to probe each one because they have different excitation wavelengths; usually, the same laser system cannot be used for identifying more than one or two species in the reactor.

Laser-induced fluorescence has been used extensively as a probe during CVD and plasma chemical processing. In studying thin film processes that are limited by heterogeneous reactions, LIF can be used to detect species that have desorbed from the surface or products of gas-phase reactions involving these desorbed species. The use of LIF during thin film processing has been reviewed by Gottscho and Miller (1984), Dreyfus *et al.* (1985), Singh and Hopman (1987), Donnelly (1989), and Breiland and Ho (1993). Laser-induced fluorescence is primarily a tool for probing species densities to obtain a complete description of the process and for conducting fundamental studies. Laser-induced fluorescence can provide valuable information about the process mechanism even when it can probe only one or a few minority species. Although applications in manufacturing are certainly feasible, LIF is less useful for real-time monitoring and control than are some other optical diagnostics because it requires relatively expensive laser systems and usually requires calibration for absolute density measurements.

Resonant-enhanced multiphoton ionization (REMPI) is a method that evolved from LIF investigations. It is more sensitive, but less versatile, than LIF and is discussed in Section 17.1. Another related technique is laser-induced photofragment emission (LIPE), in which the exciting laser predissociates the parent molecule, and one of the fragments is electronically excited and fluoresces. The predissociation rate is slow enough that the absorption spectrum still has sharp, resolvable features. Both REMPI and LIPE are more destructive than LIF because the probe generates new species that can alter the process chemistry.

7.1 Experimental Considerations

Normally, LIF is induced by exciting a species from its ground electronic state to an excited electronic state. In molecules, fluorescence is usually

7.1 Experimental Considerations

monitored back down to the ground state, but to a vibrational–rotational level other than the initial level. Excitation and fluorescence wavelengths are chosen to maximize the Franck–Condon factors for both transitions while maintaining a large enough difference in wavelength so that the background due to scatter of the laser is minimized in the monochromator. Monitoring the fluorescence to another excited electronic state is sometimes favorable. The spectroscopy needed to perform LIF is surveyed in Chapters 3 and 4, and presented in Herzberg (1937, 1945, 1950, 1966), Moore (1949, 1952, 1958), Moore and Merrill (1968), Barrow (1973, 1975), Bashkin and Stoner (1975, 1978, 1981, 1982), Suchard (1975), Suchard and Melzer (1976), Okabe (1978), Huber and Herzberg (1979), Wormhoudt *et al.* (1983), and Jacox (1988, 1990, 1994), and references cited therein.

7.1.1 *Signal Analysis*

A typical experimental arrangement used for LIF measurements is shown in Figure 7.3. As detailed in Section 5.2.1.1, fluorescence is usually collected at right angles to the direction of laser excitation. Sometimes, excitation and fluorescence collection are collinear (Figure 5.12), with the laser directed into the axis of LIF collection by a small mirror or a beamsplitter

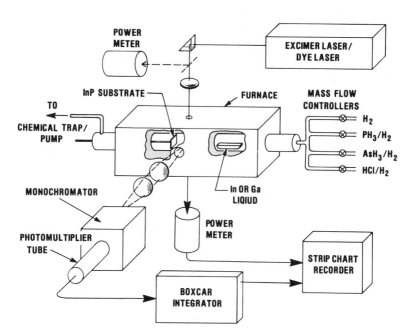

Figure 7.3 Schematic apparatus used for LIF measurements in a CVD reactor (from Donnelly and Karlicek, 1982).

with special spectral or polarization properties. Right-angle collection gives a better defined collection volume, i.e., better spatial resolution, and superior rejection of scatter from the incident laser, whereas coaxial LIF requires only one port on the reactor (not the two needed in right-angle collection) and collects less background light, such as OE in plasma processing. Light collected outside the "imaged" volume in coaxial LIF degrades spatial resolution, and this contribution can be decreased by placing an obscuration disk in front of the collection lens (Section 5.2.1.1). Diffusion of excited molecules out of the excitation volume can decrease LIF light collection, although this is usually unimportant for the typically fast radiative decay times $\tau_{\text{rad}} \sim 10$ nsec.

The collected LIF signal depends on the contribution within the imaged volume. If a species is excited from state 1 to a specific state 2, the fluorescence power emitted per unit volume from that state 2 to final state 3 (Figure 7.1) is (Donnelly, 1989)

$$p_{\text{LIF}}(\nu_{23}) = \alpha I_{\text{inc}} \frac{\nu_{23}}{\nu_{21}} \frac{\gamma_{\text{rad}(2 \to 3)}}{\gamma_{\text{rad}(2 \to \text{all})} + \sum_n k_n p_n} \quad (7.1)$$

in steady state. The incident laser has intensity I_{inc} (power $\mathcal{P}_{\text{inc}}$) and frequency ν_{21}. The absorption coefficient $\alpha = N_1 \sigma_{1 \to 2}$, where N_1 is the density of the species in the initial state; thus α is proportional to the density. The frequency of the emitted photon is ν_{23}. In the last factor, $\gamma_{\text{rad}(2 \to 3)}$ is the rate of radiative decay (or spontaneous emission) ($= A_{23}$) from the directly excited state 2 to the final state 3. The denominator includes the rates of all processes that relax level 2 and is divided into two terms. The first term is the rate of radiative decay from state 2 to state 3 and to other states (3′, 3″, . . .), which could be different electronic, vibrational, or rotational states that can be observed with the monochromator set for different ν. The second term describes collisional relaxation from state 2 to other states 2′, 2″, . . . ; these other states can fluoresce to give LIF at a different ν. In this collisional term, k_n is the relaxation rate constant for each species n (including the fluorescing species) and p_n is the partial pressure for n. The sum of all of these rates in this denominator is the total relaxation rate γ_t.

The strength of the LIF signal depends on the specific quantum numbers of the states involved and the details of the spectral analysis and detection. Equations 4.34 and 4.38 show that the LIF signal in a (diatomic) molecule excited from state 1 (with vibrational and rotational quantum numbers v'', J'') to 2 (v', J') and fluorescing to state 3 (v''', J''') is proportional to the population density of state 1 and to the product of the squares of the transition matrix elements for the absorption step (μ_{21}^2) and the emission step (μ_{23}^2). Each of the transition element terms is a product of the electronic matrix element term μ_e^2, the Franck–Condon factor describing the overlap

7.1 Experimental Considerations

of vibrational states $f(v',v'')$, and the Hönl–London factor $S(J',J'')$ describing the dependence on rotation. Excitation and analysis transitions should be chosen to maximize the Franck–Condon and Hönl–London factors. In the absence of collisional relaxation, this expression describes the LIF spectrum.

In emission LIF, if the laser is spectrally narrow and the line density of the molecule is low, a single state 1 can be excited to a single state 2 and emission to all possible states 3 is determined by the Franck–Condon distribution and the rotational selection rules; this emission is resolved by spectral analysis of the collected LIF. Equation 4.38 gives the LIF emission intensity per unit volume for transitions to each final state v''', J'''. If, instead, the laser is spectrally broad and the species has a relatively high line density, many states 1 are excited to many states 2 in LIF excitation. Then quantitative analysis of the LIF signal to obtain absolute densities becomes even more difficult and external calibration is definitely needed.

In excitation LIF performed with a spectrally narrow laser, transitions between different combinations of states 1 and 2 are resonantly excited as the laser is tuned. The fluorescence over a specific wavelength range is monitored, which may be chosen to include all possible rotational transitions for a given vibronic band ($v' \rightarrow v'''$). Then the observed LIF signal is due to the sum of the contributions of transitions to all final states (v''', J''') in that wavelength band. Since this procedure integrates over rotational structure in emission, the LIF signal does not change much when significant rotational relaxation occurs within the excited electronic state. With this removal of the dependence on J''', the excitation LIF spectrum is similar to the absorption spectrum, as given by Equation 4.36.

The rotational temperature T_{rot} can be obtained by exciting populations from different J'' in a given vibronic band (Equation 4.39), while the vibrational temperature (for a given vibrational mode) T_{vib} can be obtained by exciting populations from different v'' (Chapter 18). The use of LIF to measure temperature in combustion environments and in flames is detailed by Eckbreth (1996) and Rensberger et al. (1989); LIF thermometry during thin film processing is discussed later in this chapter.

Quantitative analysis of LIF signals to obtain ground-state densities is complicated by several factors. The discussion so far, which includes Equation 7.1, has assumed that the excitation transition is unsaturated, and so the LIF signal is proportional to the incident intensity. The effect of saturation is addressed below. Collisions can redistribute population from the excited state to other states, which can also fluoresce. In excitation LIF, fluorescence from states populated by collisions can still be collected. If only that from the directly excited state is collected, Equation 7.1 shows that collisions decrease the LIF signal. Because of the uncertainties in laser intensities and collection efficiency, Equations 4.38, 5.14, and 7.1 are rarely useful for the absolute determination of density. Sometimes, absorption measurements or LIF measurements performed on known densities are used as an absolute calibration standard.

Since the matrix elements in Equation 4.38 are highly sensitive to polarization, quantitative analysis of densities from LIF measurements would be expected to be sensitive to the choice of the polarizations of the incident and analyzed emitted light. However, Kinsey (1977) has noted that, at least for diatomics, the LIF intensity is independent of the polarization of the incident light if the detector senses all polarization states with equal efficiency.

A certain fraction of the emitted fluorescence (power per unit volume) p_{LIF}, given by Equations 4.38 and 5.12, is collected and sent for analysis. This collected power $\mathcal{P}_{LIF}$ depends on the collected solid angle $\Delta\Omega$ and the imaged volume V_{im}, and is given by Equation 5.14 for coaxial (backscattering) and right-angle collection. Diffusion of excited molecules out of the field of view is not included in this analysis, and is expected to be important only in special cases. The detected signal is $S_{LIF} = \eta_t(\mathcal{P}_{LIF}/h\nu_{32})$, where η_t includes the factors for transmission of the collected light through the optics (spectrometer, etc.) and the detector efficiency.

The effect of beam absorption before the point of observation must be included if quantitative analysis of concentration is desired (Donnelly and Karlicek, 1982; Donnelly, 1989). If absorption is linear, the effective excitation intensity is $I_{eff} = TI_{inc}\exp(-N\sigma z)$ (Equation 2.5), where σ and N are the absorption cross-section and density of the absorbing species, respectively, and z is the distance from the input window to the observation region. The transmittance T of the input window depends on reflection by the window and absorption due to any film buildup during processing. Attenuation due to beam absorption by other gas-phase species should also be included. If LIF is measured halfway through the cell, this exponential dependence implies that $I_{eff} = (I_{inc}I_{out})^{1/2}$, where I_{out} is the intensity leaving the reactor. This expression includes window losses if the entrance and exit windows are symmetric.

If S_{LIF} depends linearly on the incident laser intensity I_{inc}, saturation and other nonlinear effects are probably not important. A sublinear dependence on intensity suggests that the absorption transition is saturated. Sometimes this can make calibrated determination of densities easier because the signal is less strongly dependent on local intensity variations. A superlinear dependence implies that more than one photon is absorbed in the process leading to emission. Two or more photons can be absorbed by the same species, or photons can be absorbed both by the initial molecule, which can dissociate, and by the photodissociation product. Since one or more transitions in a multiphoton process can be saturated and other processes can also be involved, determining the number of absorbed photons can be tricky. The fact that $S_{LIF} \propto (I_{inc})^n$ indicates that n photons are involved only if each transition is not saturated and no other processes are important.

7.1 Experimental Considerations

Use of laser intensities high enough to saturate or partially saturate the excited transition (Section 4.2.1.2) can affect LIF measurements in two ways. For homogeneously broadened lines, saturation adds an additional term BI_{inc} to the (γ_t) denominator in Equation 7.1. As I_{inc} gets very large, the denominator becomes proportional to I_{inc} and the LIF signal becomes independent of the laser intensity. This cancellation suggests that careful analysis of laser powers and focal spot sizes is no longer necessary. The second reason why saturation is important is that as I_{inc} gets large, $BI_{inc} + \gamma_{rad(2\rightarrow 3)}$ becomes larger than the collisional term. Consequently, severe saturation decreases the effect of collisions on the LIF signal, which simplifies analysis (Pfefferle *et al.*, 1988).

Saturated excitation in LIF is widely used in combustion studies because of these benefits, as has been detailed by Eckbreth (1996). However, there are several problems in using saturated LIF for absolute density measurements, and many avoid saturation conditions because of these problems (Breiland and Ho, 1993). Because lasers have nonuniform transverse profiles, signals are never truly independent of the laser intensity (Salmon and Laurendeau, 1985). This problem could be avoided by expanding the laser to overfill the sample so there is a uniform intensity; however, this is usually not possible in thin film processing reactors. Furthermore, if pulsed lasers are used, boxcar sampling must be employed to measure the signal only at the peak of the laser pulse. These complications arise from regions that are only partially saturated, either from the weak intensity tails of the beam profile or the weak intensity during the rise and fall of the pulse. Also, since different transitions in molecules have different Hönl–London factors, they would display different degrees of saturation.

The consequence of saturation, called *power broadening* (Section 4.2.1.2), must either be avoided or else included in lineshape analysis. For example, Goeckner and Goree (1989) have detailed how power broadening increases the spectral linewidth beyond the Doppler width (Equation 4.22), and how this affects LIF measurements of ion temperatures in plasmas (T_{trans}).

7.1.2 *Instrumentation*

Steady-state processes, such as CVD and plasma chemical processing, can be probed with LIF by using continuous-wave (cw) or pulsed lasers (Section 5.1.1). Signal averaging techniques are used to minimize noise and the contribution of background light, which can be very strong during plasma processing. When cw lasers are used, the laser is chopped and the LIF signal is analyzed with lock-in detection. When pulsed lasers are used, the LIF signal is analyzed by using boxcar integration or a transient digitizer. The large steady-state background during plasma processing is very effectively removed by using pulsed-laser sources that have small duty cycles, such as excimer-laser-based dye lasers. Another advantage of pulsed lasers is that they frequency-double with higher efficiency than do cw lasers, since the conversion efficiency is proportional to the laser intensity. This is significant because ultraviolet radiation is often needed for LIF excitation. Pulsed lasers are also preferred when non–steady-state processes are examined, such as excimer-laser-induced deposition and etching and rapid thermal processing, and for synchronous LIF detection of plasma processing (Gottscho et al., 1984).

Tunable lasers in the ultraviolet and visible are usually required for LIF excitation. However, fixed-frequency lasers are sometimes fortuitously resonant, and dispersed fluorescence LIF can be performed. Potential lines from pulsed lasers include those from excimer lasers, at 351 nm (XeF laser), 308 nm (XeCl), 248 nm (KrF), and 193 nm (ArF); the harmonics of Nd:YAG at 532 nm (2ω), 355 nm (3ω), and 266 nm (4ω); and lines from cw lasers, including those from the argon-ion laser, at 488 and 514.5 nm.

Tunable-wavelength radiation is needed to obtain an excitation LIF spectrum, as is needed for temperature measurements, and more generally when resonant excitation with fixed-frequency lasers is not possible. Usually a pulsed laser, such as an excimer laser or a Nd:YAG harmonic, is used to pump a dye laser to obtain tunable light. Pulsed dye lasers are simpler to operate than are cw dye lasers, and they can be frequency-mixed with much higher efficiency. Dye lasers are usually operated down to ~350–400 nm. Although operation at somewhat shorter wavelengths is possible, it is often better to obtain shorter-wavelength radiation by frequency mixing because of the shorter lifetimes of UV dyes. The dye laser can be frequency-doubled (second harmonic generation) in crystals such as KDP (potassium dihydrogen phosphate, ~270–350 nm) or BBO (β-barium borate, ~220–295 nm). The original or frequency-doubled dye laser can also be summed in such nonlinear crystals with a fixed line source, such as the Nd:YAG laser or one of its harmonics, to obtain tunable radiation in the ultraviolet. Multiorder Stokes or anti-Stokes Raman scattering of pulsed lasers in high-pressure gases, such as H_2 and D_2, is another convenient source of high-power laser radiation.

Optical parametric oscillators (OPOs) may well replace dye lasers in many tunable-wavelength applications. Integrated with solid-state lasers, such as frequency-doubled Nd:YAG lasers, the OPO system is an all–solid-state laser system that would require lower maintenance and could be smaller than dye laser-based systems.

In LIF dispersion experiments the laser frequency is fixed, and the emission is spectrally dispersed and detected by using a scanning monochromator and a photomultiplier or a nonscanning monochromator and a multichannel detector. In LIF excitation measurements, the laser frequency is varied and the spectrometer setting is fixed, often with the spectrometer slits set wide enough to detect an entire vibronic band. This emission can

7.2 Applications

also be detected without a spectrometer, using either a broadband filter or, if there is no other fluorescence or source of light, with the bare detector.

Laser-induced fluorescence emission can be detected at the excitation wavelength only when pulsed lasers are used, and then only when the radiative decay time of the excited state is much longer than the laser pulse width and fluorescence detection is gated to begin after the laser pulse. This is usually not possible with the pulsed lasers cited in this section because their pulse widths (~20 nsec) are on the order of characteristic radiative decay rates; picosecond-scale lasers can be used for this application. This type of operation is relatively rare, even when the radiative lifetime is long.

7.1.2.1 Imaging

Experimental methods for LIF imaging and mapping in two and three dimensions are described in Section 5.2.1. Good spatial resolution can be achieved by focusing the laser to a spot and imaging that spot into the spectrometer. The spatial resolution obtainable in collinear and right-angle collection configurations is described in Section 5.2.1.1. This confined three-dimensional geometry can be mapped by scanning the laser and collection optics.

Hanson (1988), Hanson and Seitzman (1989), Eckbreth (1996), and Cappelli and Paul (1990) have detailed how LIF imaging has been used to study flames (Figure 7.4). In 2D imaging of flames, the laser is loosely focused with a long-focal-length spherical lens to ~200–500 μm. A cylindrical lens telescope then expands the beam to form a sheet that cuts across the flame. Laser-induced fluorescence is collected at right angles from this cylindrical sheet of light, and is filtered to remove scattered laser light and other emission. It is imaged on an array detector, such as a charge-coupled device (CCD) array, to give a 2D profile. Scanning the excitation optics or the flame provides a 3D image. This planar laser-induced fluorescence (PLIF) has been used to image radical concentrations in flames (Dyer and Crosley, 1982; Pfefferle *et al.*, 1988; Hanson and Seitzman, 1989), and, for example, to image Ar metastables in an RF GEC plasma reactor (McMillin and Zachariah, 1995) in thin film studies.

7.2 Applications

7.2.1 *Probing Gas-Phase Species*

7.2.1.1 Single-Photon LIF

Laser-induced fluorescence has been used extensively in studying deposition and etching. Table 7.1 lists representative neutral atoms, diatomics, triatomics, and ions that have been detected by LIF during thin film process-

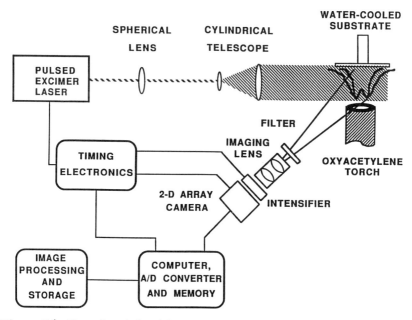

Figure 7.4 Planar laser-induced fluorescence imaging (from Cappelli and Paul, 1990).

ing. Since several sets of excitation and emission wavelengths have been used in these studies, representative transitions are listed in the text along with the cited application and not in this table. Singh and Hopman (1987) have discussed the potential use of LIF to detect the presence of molecules and ions that have high-lying first excited states and are hard to excite with a single photon, such as F_2 and F_2^+, and larger (tetratomic and pentatomic) species, such as CF_3, CF_3^+, CF_4, CF_4^+, SiF_3^+, SiF_4, and SiF_4^+.

Table 7.1 Representative Species Detected by LIF[a]

One-photon LIF
Atoms Al, As, Ba, Cu, Fe, Ga, Ge, In, Si, Te, Y
Diatomics AlO, As_2, BCl, BaO, C_2, CCl, CF, CN, Cu_2, CuO, GaCl, InCl, NH, OH, P_2, SO, Si_2, SiBr, SiCl, SiF, SiH, SiN, SiO, YO
Triatomics C_2H, CF_2, HSiCl, SO_2, SiF_2
Ions Ar^+, Cl^+, Cl_2^+, Cu^+, N_2^+

Two-photon LIF
Atoms Cl, H, N, O, Zn

[a] Specific transitions and wavelengths are given in the text and in the references cited in this chapter.

7.2 Applications

Sedgwick *et al.* (1975) and Smith and Sedgwick (1977) conducted early studies of LIF detection and spatial mapping of species during Si CVD. The species that fluoresced strongly in chlorosilane/H_2 gas mixtures in this work was assumed to be $SiCl_2$, but was later identified to be HSiCl (Ho and Breiland, 1983).

Ho, Breiland and co-workers have made extensive use of LIF to measure and profile the densities of the transient species that are important in the CVD of silicon films, and they used these data to compare with the predictions of detailed models of Si CVD. Specifically, they detected Si_2 during silane CVD (Ho and Breiland, 1984), HSiCl during dichlorosilane CVD (Ho and Breiland, 1983), and Si during silane and dichlorosilane CVD (Breiland *et al.*, 1986b; Ho and Breiland, 1988; Ho *et al.*, 1994). Breiland *et al.* (1986a) used LIF to map the density of Si_2 during the CVD of Si from $SiH_4/H_2/He$ mixtures, and they compared these profiles and those of other species obtained by Raman scattering with CVD models. In this study, they pumped the $H\,^3\Sigma_u^-\,(v' = 5)$–$X\,^3\Sigma_g^-\,(v'' = 0)$ transition at 391 nm with a XeCl laser-pumped dye laser, and detected emission from the $5 \rightarrow 2$ band to obtain the Si_2 density, as is seen in Figure 7.5. Figure 7.2 shows excitation and dispersed LIF spectra of HSiCl obtained by Ho and Breiland (1983). Breiland *et al.* (1986b) measured the Si atom density as a function of height over the susceptor during Si CVD from silane, by using a frequency-doubled dye laser to excite $(3p)4s\,^3P_2^0 \leftarrow 3p^2\,^3P_1$ at 250.7 nm and monitoring

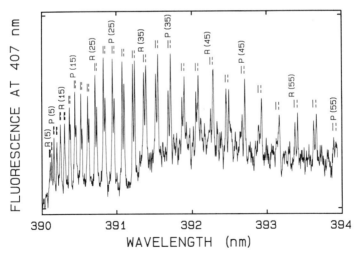

Figure 7.5 Fluorescence excitation spectrum of Si_2 obtained by monitoring the $H\,^3\Sigma_u^-(v' = 5) \rightarrow X\,^3\Sigma_g^-(v'' = 2)$ emission band at 407 nm while scanning the dye laser over the $5 \leftarrow 0$ band. The labels mark the positions for the $5 \leftarrow 0$ rovibronic absorption spectrum, with dashed lines for *P*-branch and solid lines for *R*-branch transitions (from Ho and Breiland, 1984).

fluorescence from $4s\ ^3P_2^0 \rightarrow 3p^2\ ^3P_2$ at 251.6 nm. The dispersed fluorescence spectrum of Si atoms is seen in Figure 7.6. Figure 7.7 plots the Si atom density as a function of height above the susceptor as detected by LIF and as predicted by their models. Ho and Breiland (1988) measured much smaller Si atom densities during Si CVD when SiH_2Cl_2 was used as the reactant than they had measured in their earlier work with silane as the reactant. Ho *et al.* (1994) used LIF in an intensive investigation of the Si atom profile in a rotating-disk CVD reactor, with either silane or disilane reactants.

Roth *et al.* (1984) measured the spatial distribution of Si atoms in a silane glow discharge that can be used for the deposition of hydrogenated amorphous silicon. By exciting $4s\ ^3P_1^0 \leftarrow 3p^2\ ^3P_0$ at 251.43 nm and monitoring fluorescence from $4s\ ^3P_1^0 \rightarrow 3p^2\ ^3P_2$ at 252.85 nm, they showed that the Si atom profile has sharp boundaries at the sheath regions. Hata *et al.* (1987) used LIF to map the profile of Si and Ge in silane and germane plasmas, used for plasma-assisted CVD.

Donnelly and Karlicek (1982) have investigated how LIF and laser-induced photofragment emission (LIPE) can be used to probe species, such

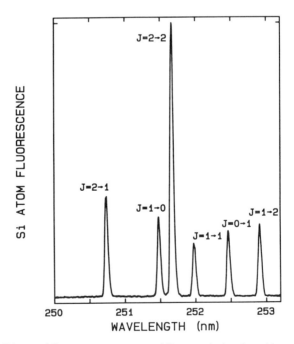

Figure 7.6 Dispersed fluorescence spectrum of Si atoms during deposition, with excitation of the $4s\ ^3P^0(J=2) \leftarrow 3p^2\ ^3P(J=1)$ transition at 250.7 nm and emission at the corresponding transitions (from Breiland *et al.*, 1986b).

7.2 Applications

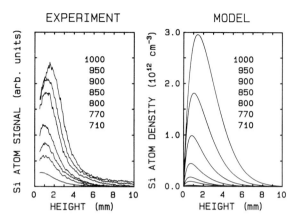

Figure 7.7 Silicon-atom density versus height above the susceptor as determined by LIF experiments and by a CVD model, during the CVD of silicon from silane for a series of susceptor temperatures (highest curve corresponds to the top-listed temperature, etc.) (from Breiland et al., 1986b).

as PH_3, P_2, AsH_3, As_2, InCl, and GaCl, that are important in the CVD of InP and InGaAsP epitaxial layers (Figure 7.3). Excitation was achieved with an excimer laser operating at 193 nm (ArF laser) or 248 nm (KrF), or with a tunable dye laser pumped at 308 nm (XeCl). Typical species concentrations were significantly higher than the detection limits. The species identified in this study can also be probed in the CVD of GaAs and the etching of InP, GaAs, and InGaAsP.

In CVD, phosphorus is often introduced as PH_3 and arsenic as AsH_3, both in H_2 backgrounds. At high temperatures (~700°C), P_2 and P_4 can exist in addition to PH_3, due to the equilibria: $PH_3 \leftrightarrow \frac{1}{2}P_2 + \frac{3}{2}H_2$ and $2P_2 \leftrightarrow P_4$. Donnelly and Karlicek (1982) detected PH_3 by emission from PH_2 and PH that was formed by LIPE at 193 nm. Photofragmentation by the absorption of one photon led to a broadbanded feature from 450 to 600 nm due to PH_2 $\tilde{A}\,^2A_1 \to \tilde{X}\,^2B_1$, while LIPE with two photons gave sharper features near 342 nm due to PH $A\,^3\Pi \to X\,^3\Sigma^-$. Although this PH* emission is stronger, absolute calibration is less certain because it is a two-photon process. Diatomic phosphorus (P_2) excited at 193 nm $C\,^1\Sigma_u^+(v' = 11) \leftarrow X\,^1\Sigma_g^+(v'' = 0)$ emitted an extensive band from 195 to 370 nm due to $C\,^1\Sigma_u^+(v' = 11,10,9,8,7) \to X\,^1\Sigma_g^+(v'' = 0\text{–}33)$ laser-induced fluorescence. They detected AsH_3 by one-photon LIPE which produced AsH_2 $\tilde{A}\,^2A_1 \to \tilde{X}\,^2B_1$ emission, and As_2 by the absorption at 248 nm $A\,^1\Sigma_u^+ \leftarrow X\,^1\Sigma_g^+$, which gave $A \to X$ LIF emission in structured bands from 250 to 340 nm. By using 193-nm excitation, they also saw As emission lines that they attributed to either AsH_3 or As_2/As_4.

Indium can be transported for CVD by the reaction of HCl gas with the liquid metal at 700°C to form InCl. Similarly, GaCl can be made for CVD. Higher chlorides such as In_2Cl_4 and $InCl_3$ are not formed at high concentrations at this higher temperature, but may be present as products in (lower-temperature) etching processes. Donnelly and Karlicek (1982) detected InCl by LIF by using a dye laser to excite either $A\ ^3\Pi_{0^+} \leftarrow X\ ^1\Sigma^+$ or $B\ ^3\Pi_1 \leftarrow X\ ^1\Sigma^+$. The B state probably collisionally relaxes to the A state. They saw strong $A \rightarrow X$ emission with 340–360 nm excitation. Although both the excitation and emission transitions are formally forbidden because they involve a change in spin between triplet and singlet states, they are relatively strong for InCl because of the large atomic number (Z) of In. Spin–orbit coupling, which relaxes spin conservation, increases rapidly with Z and leads to a large effect. Analogous LIF for the GaCl A–X transition was found to be $\sim 10^3$ weaker than that in InCl because the transition is "more forbidden" for the lower-Z Ga. They also found that a two-photon absorption process in InCl at 193 nm leads to strong In atomic emission at 451 nm. A similar process occurs in GaCl to produce Ga emission at 417 nm.

In addition to using LIF for *in situ* monitoring of reactants during steady-state operation, Donnelly and Karlicek (1982) employed these methods to monitor "turn-on" and "turn-off" delays in a CVD reactor. Laser-induced fluorescence probing of InCl showed that it took ~ 30 sec for InCl to appear and ~ 55 sec until it reached its steady-state value in the reactor because of the transit time for gas flow and the time for the HCl/liquid metal reaction to attain steady state. Also, they used LIF of As_2 to demonstrate that a significant amount of arsenic remains in the reactor for several minutes after the flow of AsH_3 is stopped, as is seen in Figure 7.8. This LIF continues because arsenic on the walls, formed during CVD, continues to leave the walls after CVD.

Laser-induced fluorescence has also been used to probe species during the CVD of diamond films. Cappelli and Paul (1990) observed large densities of C_2H during diamond-film deposition in an oxyacetylene flame by using planar laser-induced fluorescence, which images LIF in two dimensions (see Figure 7.4). They also imaged optical emission from C_2. Meier *et al.* (1991) measured the OH distribution near the filament in a diamond growth environment ($H_2/CH_4/O_2$ mixtures) by exciting and observing emission in the $A\ ^2\Sigma^+ - X\ ^2\Pi_i$ transition. Monitoring H atoms during diamond deposition is described later in the discussion of two-photon LIF.

Van de Weijer and Zwerver (1989) probed several potential transient species by LIF as part of their study of chemiluminescence during CVD of SiO_2 by silane/oxygen mixtures. They successfully detected OH [$A\ ^2\Sigma(v = 1) \leftarrow X\ ^2\Pi(v = 0)$, 281 nm; $A(v = 1) \rightarrow X(v = 1)$, 312 nm] and SiO [$A\ ^1\Pi(v = 0$–$6) \leftarrow X\ ^1\Sigma(v = 0$–$2)$, 220–237 nm; $A(v = 0$–$6) \rightarrow X(v = 0$–$16)$, 214–350 nm] by LIF, but could not detect O, Si, SiH, and SiH_2.

7.2 Applications

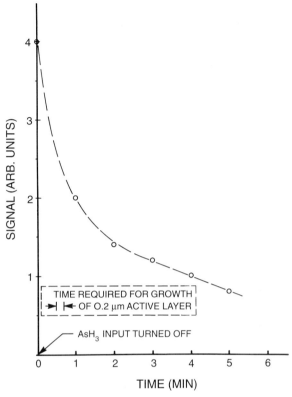

Figure 7.8 Laser-induced fluorescence of As_2 excited at 248 nm and monitored at 289 nm after turning off the AsH_3 flow into a CVD reactor, while maintaining the flow of H_2 (from Donnelly and Karlicek, 1982).

For the SiO molecules, they used LIF to measure $T_{rot} = 700 \pm 50$ K and $T_{vib} = 1100 \pm 50$ K during CVD.

Laser-induced fluorescence has also been used to probe gas-phase species present in laser-assisted deposition. Marinero and Jones (1985) used a pulsed dye laser to probe transiently the Cu atoms formed during laser photolysis of $Cu(hfac)_2$ in the photochemical laser deposition of copper. Brewer (1987) and Jensen *et al.* (1988) used LIF to detect Te atoms produced during excimer-laser photodissociation of dialkyltellurides, such as diethyltellurium (DETe), for the photolytic deposition of CdTe and HgTe epilayers. Using a tunable dye laser for LIF excitation, they showed that 248-nm photolysis of DETe produces only the $^3P_2$ ground state of the Te atom (and not the $^3P_{0,1}$ spin–orbit excited states), and this process occurs with a probability linearly proportional to laser fluence up to saturation, indicating a single-photon dissociation. Suzuki *et al.* (1988) used LIF to probe Ga atoms spatially and temporally during KrF-laser-assisted deposition from trimethylgallium. Use of LIF to study species ejected from surfaces during the surface ablation process in pulsed-laser deposition (PLD) is described in Section 7.2.2.

Resonance fluorescence can be used to measure the atomic beam flux in molecular-beam epitaxy (MBE). McClintock and Wilson (1987) excited the Ga beam with radiation from a Ga hollow cathode lamp, and transmitted the collected fluorescence through a bandpass filter and detected it with a photomultiplier. Potential transitions for excitation and analysis in Ga are $4s^2 5s\ ^2S_{1/2} - 4s^2 4p\ ^2P_{1/2}$ at 403 nm, which involves the ground state, and $4s^2 5s\ ^2S_{1/2} - 4s^2 4p\ ^2P_{3/2}$ at 417 nm. McClintock and Wilson excited with both wavelengths and detected at 417 nm. Background scatter at 417 nm was determined with the Ga shutter closed, and subtracted. As the authors noted, it is wiser to excite at 403 nm and detect at 417 nm; this would then have the very positive feature of "LIF" of having no background signal with no absorber. The flux of In atoms was also monitored by exciting with the 410-nm line from an In hollow cathode lamp and detecting at 451 nm. The logarithms of the Ga and In LIF signals were proportional to $1/T_{oven}$, as expected. Preliminary measurements were also made for Al beams (394, 396 nm). Hebner and Killeen (1990) used resonance fluorescence to monitor atomic In produced by the pyrolysis of trimethylindium (TMIn) during metalorganic chemical vapor deposition (MOCVD). Ground-state atoms $5p\ ^2P^0_{1/2}$ were excited to $6s\ ^2S_{1/2}$ by 410.2-nm radiation from an indium hollow cathode lamp that passed through a bandpass filter centered at 400 nm. Fluorescence to $5p\ ^2P^0_{3/2}$ (at 451.1 nm) was transmitted through a bandpass filter centered at 450 nm and detected by a photomultiplier. The fluorescence signal was linear with TMIn flow rate.

Absorption probes for atomic flux are described in Section 8.2.2.1.4. They are more sensitive to source and detector noise than is LIF, but are easier to calibrate because LIF signals depend on the solid angle collected. Both LIF and absorption monitoring are affected by changes in window transmission and by thermal and photoinduced changes in the relative populations of the lowest-lying states, such as $^2P_{1/2}$ and $^2P_{3/2}$ in Ga.

Laser-induced fluorescence has also been used to measure the flux of molecules during MBE. Smilgys and Leone (1990) used excitation LIF to resolve the states of As_2 emanating from an As_4 oven cracker source in an MBE reactor. $A\ ^1\Sigma_u^+(v' = 5) \leftarrow X\ ^1\Sigma_g^+(v'' = 0 - 4)$ in As_2 was excited by using an excimer-laser (308 nm) -pumped dye laser (Coumarin 480) that was frequency-doubled in a β-barium borate crystal to give tunable radiation from 230 to 250 nm. Unresolved $A\ ^1\Sigma_u^+ (v' = 5) \rightarrow X\ ^1\Sigma_g^+ (v'')$ fluorescence was detected after transmission through a broadband interference filter, and was averaged with a boxcar integrator. Figure 7.9 shows As_2 LIF from a heated reference cell, together with a simulated spectrum. Both P

7.2 Applications

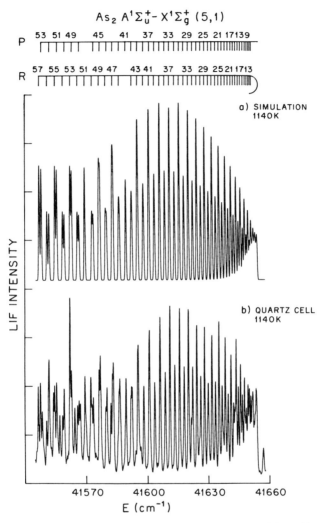

Figure 7.9 Laser-induced fluorescence of the (5,1) band of the $A\ ^1\Sigma_u^+ - X\ ^1\Sigma_g^+$ transition in As$_2$ at 1140 K: (a) Simulation; (b) experimental spectrum of As$_2$ formed during the thermal dissociation of As$_4$ (from Smilgys and Leone, 1990).

and R branches are seen (there is no Q branch for $^1\Sigma - ^1\Sigma$ transitions), with a 5:3 intensity alternation due to nuclear spin statistics. Attempts to obtain a rotational temperature (Equations 4.38 and 4.39) were complicated by the perturbation of many A levels by nearby electronic states and the accidental overlap of P- and R-branch lines. Still, the rotational band was consistent with thermalization with the cracker source at 1050 K. The vibrational band intensities for $v'' = 0-3$, integrated over only part of the rotational band to avoid problems due to perturbations, fit a Boltzmann distribution with $T_{vib} = 1020 \pm 100$ K, which was consistent with the source temperature.

The use of LIF diagnostics for process development of sputter deposition is discussed here; more fundamental studies of LIF with sputtering are described in Section 7.2.2. Jellum and Graves (1990) used LIF to probe sputtered Al atoms in parallel-plate reactors with aluminum electrodes. They considered exciting the $J = 1/2$ state of the $3p\,^2P^0$ ground level to $4s\,^2S$ (394.4 nm), $3d\,^2D$ (308.2 nm), or $4d\,^2D$ (256.8 nm) and monitoring emission to the $J = 3/2$ state of the ground level. When they wanted to map the Al atom density quantitatively, they excited the $4d\,^2D$ state because its emission line (257.5 nm) has the smallest absorption cross-section of the three potential emission lines and is therefore the least sensitive to radiation trapping. They calibrated LIF signals by absorption measurements at 394.4 nm to determine atom densities absolutely.

Jellum and Graves (1990) found that when the cathode voltage was decreased below a certain value, the sputter rate decreased and the LIF profile changed dramatically. Figure 7.10 shows that LIF emission (at 396.15 nm, with excitation at 394.4 nm) becomes much stronger and (in the DC glow) the peak LIF signal moves nearer the anode (to the right). They interpreted this as evidence of laser ablation of Al atoms from Al particulates, which was followed by LIF of Al atoms; they also saw that elastic light scattering from particles peaked where the Al LIF was enhanced the most. Spears *et al.* (1986) made similar observations about the LIF from Si atoms in RF discharges of silane.

Laser-induced fluorescence has also been employed extensively to study glow discharges used for etching. Hargis and Kushner (1982) and Pang and

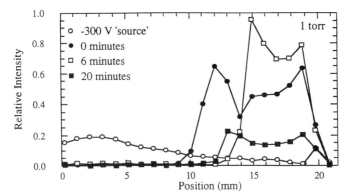

Figure 7.10 Spatially resolved LIF profile of aluminum atoms under laser-particle ablation conditions in a DC argon discharge with parallel-plate aluminum electrodes (from Jellum and Graves, 1990).

7.2 Applications

Brueck (1983) detected CF_2 in CF_4/O_2 discharges by LIF by exciting and then monitoring the $\tilde{A}\ ^1B_1-\tilde{X}\ ^1A_1$ transition. For excitation, these researchers used fortuitous resonances of fixed-line lasers—the former using a KrF laser at 248 nm that coincides with the (0,6,0) ← (0,0,0) vibronic band, and the latter a frequency-quadrupled Nd:YAG laser at 266 nm that is resonant with (0,2,0) ← (0,1,0). Pang and Brueck (1983) found that the CF_2 density decreases with O_2 fraction, while the Si and SiO_2 etch rates increase, which suggests that the density of F, the probable etching species, also increases. Hargis et al. (1984) showed that the SiO_2/Si etch ratio is a unique function of the CF_2 density determined from LIF in CHF_3/O_2 discharges. Ninomiya et al. (1986) used fixed frequency lasers to detect CF ($A\ ^2\Sigma^+-X\ ^2\Pi_r$), as well as CF_2, in CF_4/O_2 etching discharges. Hansen et al. (1988) used LIF to monitor the formation and decay of these species in a time-resolved manner in a pulsed discharge; they also monitored OES from F*. In addition, they showed that CF can be monitored by exciting $B(v' = 2)-X(v'' = 0)$ by using an ArF laser (193 nm). In other studies in this etching plasma, Buchmann et al. (1990) found no simple correlation between the densities of CF_2 measured by LIF and CF_2^+ measured by quadrupole mass spectrometry, while Matsumi et al. (1986) used LIF to measure both SiF_2 and CF_2 during silicon etching.

In contrast to these studies of dispersed LIF emission, Booth et al. (1987) used a tunable laser to obtain laser excitation spectra of CF and CF_2. For CF, the ($v' = 0$, $v'' = 0$) $A-X$ band near 233 nm was excited and the (0,3) vibronic band near 255.5 nm was detected. The rotational profiles of CF indicated Boltzmann distributions with temperatures that increased with input RF power from 324 ± 15 K for 50 W to 443 ± 30 K for 200 W. For CF_2, the $A(0,5,0) \leftarrow X(0,0,0)$ transition near 252 nm was excited and the $A(0,5,0) \rightarrow X(0,3,0)$ band near 265 nm was detected.

In a related etching system, Kitamura et al. (1989) studied the competition between SiO_2 etching and polymer formation in C_2F_6 RF plasmas by measuring the spatial profile of CF_2 by using LIF. They found that the etch rate of SiO_2 was proportional to the CF_2 density, while the deposition rate of a fluorocarbon polymer on SiO_2 was proportional to the nth power of the CF_2 density, where n = 2.5–3.0.

Davis and Gottscho (1983) have detailed how to determine rotational temperatures in RF discharges from LIF measurements, and have compared the use of LIF and OES in determining T_{rot} from different electronic states (of either the same or different species) in a given discharge. They measured T_{rot} of ground-state CCl radicals present in CCl_4/N_2 discharges by LIF of CCl using the $A\ ^2\Delta(v' = 0)-X\ ^2\Pi(v'' = 0)$ band by exciting the $^0P_{12}$ branch with a narrow-linewidth dye laser and detecting fluorescence in the Q_1 branch head, as is seen in Figure 7.11. T_{rot} was ≈550 K and slowly increased with RF power. T_{rot} for CCl was comparable to that determined by N_2 OES measurements, and was ≈400 K colder than estimates of T_{vib}

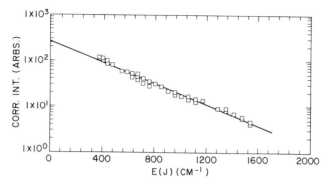

Figure 7.11 Determination of the rotational temperature of CCl from LIF measurements in a CCl$_4$ discharge (for one specific set of discharge parameters), using the natural logarithm of the normalized LIF signal (Equation 4.39). The temperature determined from the slope was 540 ± 10 K; the lower electrode was heated to 573 ± 3 K (from Davis and Gottscho, 1983).

in CCl. They also verified that in N$_2$ glow-discharge plasmas, T_{rot} obtained from LIF of N$_2^+$ $B\ ^2\Sigma_u^+ - X\ ^2\Sigma_g^+$ (see Figure 3.2), with dye laser excitation of the R branch of the (0,0) band and emission detected by simultaneously scanning the R branch of the (0,1) band, agreed with measurements from N$_2$ and N$_2^+$ OES (Section 6.3.5). In these studies, Davis and Gottscho (1983) also scanned the collection optics to map the temperature profile. In related studies, Gottscho et al. (1983a,b) scanned the collection optics to map the CCl density profile.

Donnelly et al. (1982) used LIF to determine the relative density of Cl$_2^+$ in a Cl$_2$ RF etching plasma as a function of discharge parameters. Cl$_2^+$ OE from $A\ ^2\Pi_i \to X\ ^2\Pi_i$ is responsible for the blue color of the chlorine discharge. When they tuned a ~20-nsec-long dye laser to 386.4 nm to excite Cl$_2^+$ $A(v' = 10) \leftarrow X(v'' = 0)$, emission at this same wavelength transiently increased by approximately two orders of magnitude for about 1 μsec. Despite this long radiative decay, it proved simpler to monitor $v' = 10 \to v'' = 1$ emission at 396 nm. The Cl$_2^+$ densities measured for different RF frequencies from 0.1 to 13 MHz exhibited a minimum near 1 MHz. Within experimental error, the rotational temperatures of Cl$_2^+$ determined from excitation LIF were independent of frequency and position in the discharge.

Greenberg and Hargis (1990) used LIF to show that both SO and SO$_2$ exist in SF$_6$/O$_2$ plasma etching discharges and comprise 2–3% of the total sulfur-bearing species. SO$_2$ was detected by exciting the $\tilde{B}\ ^1B_1 - \tilde{X}\ ^1A_1$ and $\tilde{C}\ ^1B_2 - \tilde{X}\ ^1A_1$ transitions using a KrF laser, tunable from 248 to 249 nm, and a frequency-doubled dye laser tuned to 235 nm. Densities were determined by LIF measurements on known densities of SO$_2$. The KrF laser was used to excite the SO $B\,^3\Sigma - X\,^3\Sigma$ transition. Absolute SO densities were estimated by reference to spontaneous Stokes–vibrational Raman measurements of N_2. T_{vib} for SO was found to be 880 ± 80 K.

Walkup et al. (1984) used LIF to detect several diatomic gas-phase products of the ion bombardment of surfaces in plasmas. They measured the excitation LIF spectrum of SiN ($B\,^2\Sigma^+ \leftarrow X\,^2\Sigma^+$) for N_2 plasmas when a Si substrate was placed on the cathode, which is seen in Figure 7.12, and determined that the SiN concentration was $\sim 10^8/\text{cm}^3$. They detected SiO during O_2/Ar plasma etching of Si and Ar sputtering of solid SiO_2 by scanning across $A\,^1\Pi(v' = 3) \leftarrow X\,^1\Sigma^+(v'' = 0)$ and detecting $A(v' = 3) \rightarrow X(v'' = 2)$ emission. Under some conditions, they observed SiF ($A\,^2\Sigma^+ - X\,^2\Pi$) during CF_4/O_2 etching of Si. In each case, they found a Boltzmann rotational distribution with T_{rot} in the ~ 400–500 K range. Vibrational distributions were approximately Boltzmann. For SiN, $T_{vib} = 740 \pm 40$ K, while for SiO, $T_{vib} = 4700 \pm 500$ K.

Selwyn (1987) detected As atoms that sputter from GaAs wafers and arsenic-doped n-type Si wafers in argon plasmas by using either one of two LIF schemes. In one, an ArF laser excited the $5s\,^4P_{3/2} \leftarrow 4p^3\,^4S_{3/2}$ transition at 193.76 nm and $5s\,^4P_{3/2} \rightarrow 4p^3\,^2D_{5/2}$ emission at 245 nm was detected. In the second scheme, he mixed frequency-doubled dye laser radiation with 1.06 μm from a Nd:YAG laser to give 228.8 nm. This light was absorbed by a metastable As level $5s\,^2P_{3/2} \leftarrow 4p^3\,^2D_{5/2}$, and led to emission at 278 and 275 nm, corresponding to $5s\,^2P_{3/2} \rightarrow 4p^3\,^2P_{3/2,1/2}$. Selwyn (1988) sug-

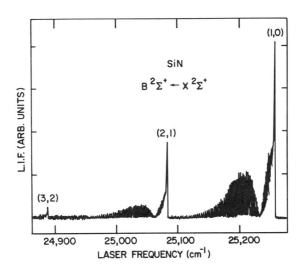

Figure 7.12 Laser-induced fluorescence spectrum of SiN in a DC N_2 plasma with a Si cathode, with the bandheads for the vibrational bands indicated (from Walkup et al., 1984).

gested that this LIF technique could be used for endpoint control during the etching of Si layers with different types of dopants.

Doppler-shifted laser-induced fluorescence spectroscopy (DSLIF) is a high-resolution spectroscopy that has been used to probe the velocity distribution of ions in electron cyclotron resonance (ECR) plasmas. Monochromatic radiation of frequency ν excites all molecules that are Doppler-shifted into the resonance frequency ν_0. Therefore, all molecules with velocity components v_z along the direction of laser propagation are excited if v_z/c is within $(\nu - \nu_0)/\nu_0 \pm \Gamma_L^{(\nu)}/\nu_0$, where $\Gamma_L^{(\nu)}$ is the homogeneous linewidth. For inhomogeneously broadened transitions, the frequency of a laser can be tuned so it selectively excites a given velocity group, as long as the laser linewidth is much less than the Doppler width. Excited-state fluorescence comes only from that excited-velocity group. As the laser is tuned across the absorption profile, this excitation-LIF signal directly gives the distribution function for v_z. Such measurements give the translational temperature (from the width) and the average speed $\langle v_z \rangle$ (from LIF profiles that are not centered about $\nu = \nu_0$ or are asymmetric). The velocity components along different directions can be obtained by directing the laser into the reactor at different angles.

Den Hartog et al. (1990) and Woods et al. (1991) employed this technique to probe the Doppler profile of a N_2^+ transition in a N_2 electron cyclotron resonance plasma, using it to determine the temperature perpendicular to the magnetic field along the symmetry axis. They used a nitrogen-laser pumped dye laser that was frequency narrowed to 0.025 cm^{-1} with an external etalon. In a series of investigations, Trevor et al. (1990), Nakano et al. (1991), and Sadeghi et al. (1991) probed the velocity distribution of Ar^+ in Ar/He ECR plasmas by DSLIF. They used narrowed pulsed dye lasers and a 1-MHz linewidth cw argon-ion-laser-pumped dye laser that operated on a single mode to make extensive measurements both parallel and perpendicular to the magnetic field and both on and off the symmetry axis, all as a function of ECR plasma parameters. The experimental apparatus shown in Figure 7.13a permits probing the parallel component of velocity (along the z direction) or the radial component (along r) in a diverging ECR reactor. Figure 7.13b shows the velocity profile of argon ions obtained with the cw laser. Giapis et al. (1993) measured ion densities in this ECR plasma with (ordinary) LIF and absorption spectroscopies.

Gottscho et al. (1991) and Nakano et al. (1992) then extended this work to measuring the velocity distribution of Cl^+ in Cl_2/He ECR plasmas. While the DSLIF profile of Ar^+ directly gives the ion velocity distribution function, it does not do so for Cl^+ because of complications from isotope shifts and Zeeman splitting. These researchers utilized laser saturation spectroscopy (Letokhov and Chebotayev, 1977) to resolve features that overlap because the Cl isotope shifts are smaller than the spectral linewidth; then they analyzed the DSLIF profile by using several fitting procedures to obtain

7.2 Applications

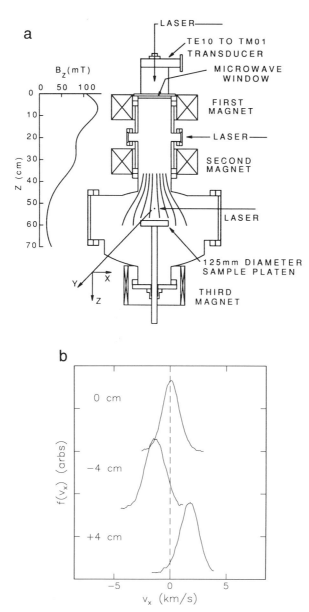

Figure 7.13 (a) Introduction of lasers into the ECR system; (b) LIF line profiles of Ar⁺ obtained by using cw laser excitation, expressed in terms of the distribution of speeds along the *x* axis in the ECR reactor (from Sadeghi *et al.,* 1991).

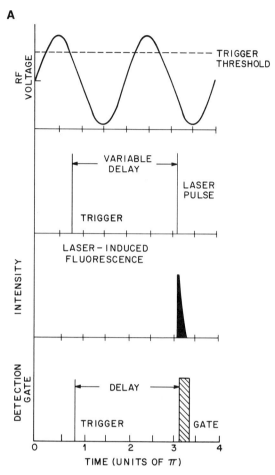

Figure 7.14 (A) Depiction of timing for synchronous detection during RF excitation of a discharge by triggering the laser pulse for LIF and the detection gate for plasma-induced emission; (B) absolute densities of N_2^+ in the ground state from time-resolved LIF in (b) N_2 and (d) N_2/Cl_2 discharges and in the excited state from time-resolved plasma-induced emission (c) (from Gottscho et al., 1984).

the velocity profile of Cl^+. Use of DSLIF to probe sputtering is discussed below, where it has been called Doppler-shift laser fluorescence spectroscopy (Husinsky, 1985).

Usually, pulsed lasers are employed for LIF diagnostics because pulsed dye lasers are more convenient to use than are cw dye lasers. However, the temporal resolution made possible by using these pulsed sources is not often utilized. In a series of studies, Gottscho and co-workers combined the temporal resolution possible with dye lasers with the spatial resolution capability of LIF and developed time-resolved LIF (TRLIF) to monitor

7.2 Applications

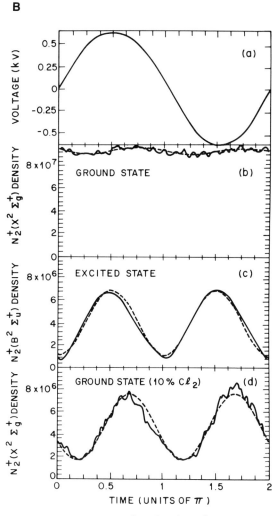

Figure 7.14 (*Continued*)

ion dynamics in RF glow discharges. Gottscho et al. (1984) pulsed a dye laser synchronously with 20–90 kHz N_2/Cl_2 plasmas, as is shown in Figure 7.14A, to probe ground-state ion densities. Ground-state $N_2^+ \, X\,^2\Sigma_g^+$ was measured by LIF and excited-state $N_2^+ \, B\,^2\Sigma_u^+$ was measured by gated OES, as is illustrated in Figure 7.14B (also see Figure 3.2). Analogous measurements were made on Cl_2^+. Spatially resolved measurements showed that near the electrodes ion dynamics can be explained by a periodically expanding and contracting sheath.

Gottscho and Mandich (1985) used this technique to probe local electric fields in a 13.56-MHz BCl_3 discharge by level mixing induced by the electric field. The $Q(5)$ transition in BCl is forbidden in the absence of an electric field, while the $P(6)$ and $R(4)$ transitions are allowed. By timing the LIF laser with the waveform maximum or minimum, they used the laser to probe BCl near the momentary anode or cathode, respectively. As is shown in Figure 7.15, the relatively small and large $Q(5)$ fluorescence at these two times indicates relatively small and large electric field amplitudes, respectively, in the probed region.

Laser-induced fluorescence can also be used to monitor products that evolve from the surface during processes other than deposition and etching. Walkup and Raider (1988) used LIF to monitor SiO evolution during the dry oxidation of silicon by O_2. At relatively low O_2 flux to the surface, silicon etching occurred and SiO was released into the gas by the reaction $O_2(g) + 2Si(s) \rightarrow 2SiO(g)$, which is known as "active" oxidation. At higher O_2 fluxes, a protective oxide (SiO_2) formed, and LIF detected no significant release of SiO to the gas phase during this "passive" oxidation. By exciting the A–X $3 \leftarrow 0$ vibronic band and monitoring the A–X $3 \rightarrow 4$ band, they found that a sensitivity down to $\sim 3 \times 10^6$ SiO molecules/cm^3 was possible.

Laser-induced fluorescence can also be used to probe extremely small volumes by using very tightly focused lasers (microLIF). In one such application, Magnotta (1987, also personal communication, 1987) and Herman

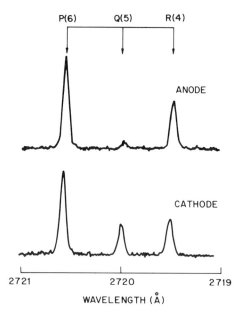

Figure 7.15 Spectrally resolved LIF from parity-mixed rotational levels in BCl due to the sheath electric field in an RF discharge of BCl_3 (from Gottscho and Mandich, 1985).

7.2 Applications

et al. (1991) focused a cw laser to ~1 µm and scanned it across the surface to heat the region near the surface and induce a local thermal reaction (direct laser writing); this laser (488 nm, argon-ion laser) also resonantly excited LIF in gas-phase species that evolved from the microreaction on the surface. Magnotta (1987, also personal communication, 1987) detected HSiCl LIF during local laser CVD of Si from $SiCl_4/H_2$ mixtures and during local laser etching of Si by HCl. Herman *et al.* (1991) examined microLIF from HSiCl during laser etching of Si by HCl, and showed that the microLIF intensity nearly tracked the volumetric etching rate; they also examined microLIF from AlO during laser thermal decomposition of the oxide layer on an Al film (Figure 7.16, $B\,^2\Sigma^+ - X\,^2\Sigma^+$). Even though the laser was focused to ~1 µm, the region probed in these studies was probably $\geqslant 100$ µm^3 because LIF was collected in backscattering configuration (Figures 5.11 and 5.12). The spatial resolution would improve, and could approach 1–10 µm^3, if appropriate apertures were used (confocal microscopy) or if a right-angle configuration were used. MicroLIF can also be used to probe the gas-phase region immediately above the surface with very high spatial resolution during conventional (nonlaser, large-area) etching of patterned surfaces, as is illustrated in Figure 7.17. In such process development studies, the different local chemistries that occur on the top and sidewalls of the patterning mask and on the film surface (in the mask hole) could be probed, along with "die-scale" flows.

Flows in reactors can be visualized in several ways (Section 1.7.1.2; Table 1.10). Flow visualization by "LIF," though by no means the most common method, has some strong features and does not have the disadvantages of other common techniques. For example, particle lag is a problem when particles are injected for elastic light scattering probing, and insensitivity

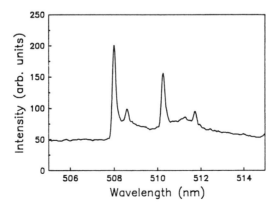

Figure 7.16 MicroLIF of AlO desorbed from the aluminum oxide layer on an Al film in an argon ambient, using 488 nm (from Herman *et al.*, 1991).

at low pressure and lack of quantitative results are problems with methods that probe density variations, such as interference holography and Schlieren photography (Itoh *et al.*, 1985).

Itoh *et al.* (1985) seeded the flow into a low-pressure reactor (pressures down to 0.5 Torr) with biacetyl and excited this molecule at 435 nm by using a pulsed dye laser. The flow field was visualized by monitoring phosphorescence from this molecule near 512 nm by using a 2D intensified photodiode array camera. They noted that it is better to use a seed molecule that phosphoresces, i.e., emits light as a result of a forbidden transition between different spin states and has a relatively long radiative lifetime ($>10^{-4}$ sec), than one that fluoresces, i.e., radiatively relaxes via an allowed transition and therefore has a shorter radiative lifetime ($\sim 10^{-6}$–10^{-8} sec), because it travels a longer distance before radiative decay. (Consequently, this is actually laser-induced phosphorescence.) Biacetyl (radiative lifetime ~ 1.5–2 msec) radiates long enough to trace flows with speeds of 1–10 m/sec, which are typical of low-pressure CVD. Visser *et al.* (1990) used this method to show that memory effects during gas-switching and the associated diffusion, increase residence times by several seconds in MOCVD reactors. They also noted potential problems in using biacetyl at temperatures above 250°C, such as thermal decomposition and a decreased luminescence intensity.

7.2.1.2 Two-Photon LIF and Other Higher-Order Processes

The one-photon LIF process described in the previous section requires excitation and detection in the vacuum ultraviolet (VUV) for atoms such as Cl, O, N, and H. This is often inconvenient because of the lack of VUV optical sources and VUV absorption by gases in the reactor and window materials. An alternative LIF scheme is two-photon LIF, in which two-photon absorption (TPA) is followed by the emission of a single photon, which is detected (Figure 7.18). This process is described in Section 4.4.

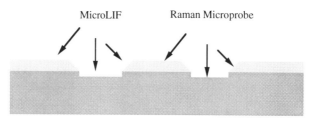

Figure 7.17 Schematic of microLIF and Raman microprobe scattering during etching of a patterned film.

7.2 Applications

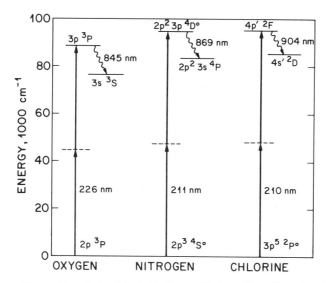

Figure 7.18 Two-photon LIF in O, N, and Cl atoms (from Donnelly, 1989).

Heaven et al. (1982) detected Cl atoms downstream from a Cl_2 discharge by two-photon LIF using 210-nm photons and the excitation scheme shown in Figure 7.18. Selwyn et al. (1987) measured the concentration of Cl atoms with three-dimensional spatial resolution in RF etching plasmas of $CClF_3$ and CCl_2F_2 by using two-photon LIF. However, they excited the "spin-forbidden" transition $3p^44p\ ^4S^0 \leftarrow 3p^5\ ^2P^0$ with two 233.3-nm photons and observed emission in the $3p^44p\ ^4S^0 \rightarrow 3p^44s\ ^4P$ multiplet (726–775 nm), as is seen in Figure 7.19, instead of the spin-allowed transition used by Heaven et al., because the generation of the 233.3-nm light is simpler. Still, the TPA of this "forbidden" transition permitted sufficient excitation. Selwyn et al. claimed that this method also detects chlorine-containing negative ions because of laser-induced photodetachment, followed by TPA of Cl:

$$Cl^- + h\nu \rightarrow Cl + e^- \quad (7.2)$$

or

$$XCl^- + h\nu \rightarrow X + Cl + e^- \quad (7.3)$$

$$Cl + 2h\nu \rightarrow Cl^* \quad (7.4)$$

This additional path also explains why they observed that the LIF intensity varied more nearly as I_{inc} than I_{inc}^2.

Using this same spin-forbidden, two-photon LIF transition, Ono et al. (1992) measured Cl atom concentrations in RF, RF magnetron, and ECR Cl_2 discharges. Figure 7.20 shows that as Cl_2 pressure decreases, the Cl

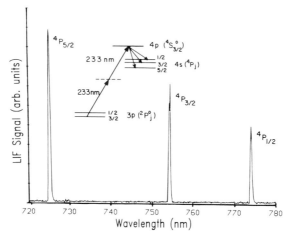

Figure 7.19 Emission spectrum from two-photon LIF of ground-state Cl in a CCl_2F_2 RF plasma, along with the energy-level diagram (from Selwyn et al., 1987).

atom density in each reactor decreases; however, the fraction of chlorine that exists as atoms increases. This, along with electrical measurements, showed that the ratio of neutral Cl flux to ion flux toward the substrate decreases almost linearly with pressure in each of the three discharges. In detecting Cl atoms via this TPA process in a Cl_2/Ar RF etching plasma, Sappey and Jeffries (1989) noticed amplification of the spontaneous emission near 750-nm emission due to stimulated emission.

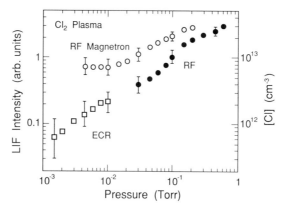

Figure 7.20 Two-photon LIF signals of Cl at 725.6 nm measured in various plasma reactors as a function of Cl_2 pressure (from Ono et al., 1992).

7.2 Applications

Bischel et al. (1981) demonstrated the detection of O and N atoms by using the excitation schemes shown in Figure 7.18. Di Mauro et al. (1984) used two-photon LIF to detect O atoms in an O_2 plasma, and estimated a detection limit $<10^{13}$ atoms/cm^3. Walkup et al. (1986) measured the ground-state density of O atoms in CF_4/O_2 RF discharges by using this method. Under usual conditions the LIF signal was more nearly proportional to I_{inc} than to I_{inc}^2 because the 226-nm laser used for TPA also photoionizes O ($3p\ ^3P$), thereby depleting the population of the $3p\ ^3P$ state. Consequently, they needed to calibrate the LIF signal to determine the absolute O density. They did this by using the 226-nm laser to produce O atoms from O_2 with the discharge off and to excite two-photon LIF in O; they then calibrated the LIF signal by using the well-known photodissociation cross-section. Fine structure due to two-photon absorption into the $2p\ ^3P_{0,1,2}$ fine structure states was also seen. The LIF data showed that ground-state O atoms account for ~2–15% of the plasma composition, depending on discharge conditions. They used these measurements to confirm that optical emission actinometry of O* (844.6 nm)/Ar* (750.4 nm) generally reflects ground-state O density, while that using O* (777.4 nm)/Ar* (750.4 nm) generally does not. (See Section 6.3.1.1.)

Selwyn (1986, 1988) obtained spatially resolved profiles of ground-state O atoms in O_2/Ar RIE plasmas by using two-photon LIF. Loading the electrode was observed to affect the uniformity of O atoms in the gas very strongly, as is seen in Figure 7.21, which plots the O-atom density across an aluminum electrode that was half covered by graphite. Two millimeters above the electrode, the density is lower above the graphite than above the aluminum, probably because the graphite consumes oxygen atoms. These differences are seen to be much smaller 25 mm above the substrate. In related experiments, Selwyn measured the vertical profile of O atoms and found it to be strongly affected by Kapton which partially covered the electrode. Tserepi and Miller (1995) used two-photon LIF to probe the spatial distribution and temporal evolution of oxygen atoms in a parallel plate RF discharge.

One-photon absorption and LIF measurements of H atoms must be made in the VUV, at 121.6 nm and shorter wavelengths. Several higher-order LIF schemes for detecting H atoms with longer-wavelength radiation are illustrated in Figure 7.22 (Goldsmith and Laurendeau, 1990). The most commonly used process is two-photon excitation to $n = 3$ with 205 nm, which is followed by fluorescence of the 656-nm Balmer α line (Lucht et al., 1983). Park et al. (1992) generated 205-nm radiation by stimulated Raman scattering of 193-nm radiation from an ArF laser in D_2 (the first Stokes line, S_1), and they used it to detect H atoms in silane plasmas. The small tunability of the ArF laser, over 1 nm, permits resonant excitation of the two-photon transition by this method. Czarnetzki et al. (1994) also used this two-photon LIF method to perform spatially resolved, absolute

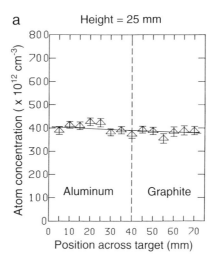
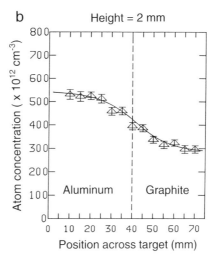

Figure 7.21 Spatial profile of O-atom density measured by two-photon LIF across an aluminum electrode with the right half covered by graphite, for two different heights above the electrode (from Selwyn, 1986).

measurements of H atoms in a silane discharge. The scale for absolute calibration was determined by performing two-photon LIF on H atoms in a flow tube reactor, where the hydrogen atom density was measured by titration.

Hydrogen atoms have also been detected during the CVD of diamond films. Meier *et al.* (1990) detected and mapped H atoms by using this two-photon LIF method during flame-assisted CVD in methane/hydrogen mixtures. They used 205-nm radiation generated by stimulated Raman

7.2 Applications

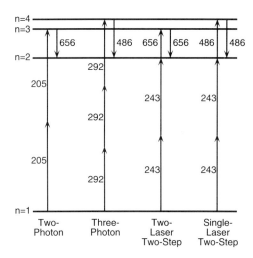

Figure 7.22 Energy-level diagrams for multiphoton-excited fluorescence detection of H atoms [wavelengths in nm] (from Goldsmith and Laurendeau, 1990; figure provided courtesy of Sandia National Laboratories and the Optical Society of America).

scattering in H_2 (third anti-Stokes line) of frequency-doubled 551-nm dye laser radiation. Tserepi *et al.* (1992) measured H-atom concentrations in a 10-MHz H_2 discharge by two-photon LIF, and calibrated this signal by titration. The 205-nm photons were made by summing 614 nm from a dye laser and 307 nm, the frequency-doubled dye laser, in a BBO crystal. Hydrogen-atom concentrations were measured to be $\sim 10^{14}/cm^3$ in the center of an unloaded reactor, and the concentration near the electrode was found to increase in the presence of a Si or GaAs substrate. Schäfer *et al.* (1991) used this method to map the atomic hydrogen concentration near hot filaments.

The other H-atom detection schemes depicted in Figure 7.22 employ lasers that do not require a Raman shifting step, and have been used in combustion studies. Three-photon excitation at 292 nm leads to 486-nm fluorescence; however, very high intensities are required in this nonresonant three-photon process (Aldén *et al.*, 1984). Although the two-laser, two-step process (TLATS) requires two lasers—one at 243 nm (doubling a 630-nm dye laser, and then mixing 315 nm with 1064 nm from a Nd:YAG laser; or the scheme given below) and one that saturates $3p$–$2s$ at 656 nm—Goldsmith (1985) has suggested that it may provide more reliable results than the two- and three-photon processes that use a single laser wavelength. Goldsmith and Laurendeau (1990) modified this process to a single-laser, two-step process (SLATS) that needs only one laser system. Two-photon excitation of $n = 2$ uses high-intensity radiation at 243 nm, generated by frequency doubling a 486-nm dye laser in BBO. This 486-nm radiation is then used to excite $n = 4$.

Two-photon LIF of Zn atoms is described in the Section 7.2.2 (Arlinghaus *et al.*, 1989a,b).

7.2.2 *Probing Processes at Surfaces*

Although LIF probes species in the gas phase, it can been used to probe processes that occur on surfaces and has proved to be a very effective surface diagnostic in numerous fundamental surface science and process development investigations. For example, LIF has been used in many investigations of state-specific scattering of particles from surfaces whose goal was the fundamental interactions of particles with surfaces. Also, the interaction of OH, NO, NH_3, and H_2 with transition-metal surfaces has been studied (Lin and Ertl, 1986; Zacharias, 1988), in part to understand catalytic processes. Laser-induced fluorescence can be used to monitor species that are scattered from a surface or to probe species that have been ejected from the surface by resistive or laser heating, laser ablation, or ion bombardment. While the discussion of Section 7.2.1 often involved LIF detection of species that directly desorbed from the surface during the course of a thin film process or those that were formed from these desorbed species, in this section the aim is to use LIF to probe the specific interaction of these species with a surface to improve the understanding thin film processes.

Ho *et al.* (1989) developed a spatially resolved LIF technique for the fundamental study of the reactivity of transient intermediate species on surfaces, and they used this IRIS (imaging of radicals interacting with surfaces) technique to study surface reactions important in CVD and plasma surface processing. As seen in Figure 7.23, a beam of radicals from a plasma source is directed to the surface of interest at an angle, say 30° from the surface normal. A dye laser is tuned to excite the radical of interest and excites both incident and scattered species. For dispersed fluorescence experiments, the collected radiation is dispersed by a spectrometer and detected by an intensified diode array. For the scattering experiments, light along the laser path is collected and imaged along the long axis of the diode array, and an optical filter is placed before the array to block scattered laser light. They conducted experiments with the laser positioned at one of several specified distances from the surface, and determined the surface reactivity by comparing the LIF profiles with and without the surface present. Figure 7.24 shows this for the scattering of NH.

By using this method, Ho *et al.* (1989) found that SiH from a silane glow discharge reacts with a 0.94 probability with the depositing amorphous hydrogenated silicon surface, and that desorbing SiH appears to leave the surface with a cosine angular distribution. Ho *et al.* (1991) and Buss *et al.* (1993) observed that SiO radicals from a $SiCl_4/O_2$ plasma react very slowly, if at all (probability $\ll 0.25$), with the depositing silicon oxide film. Similarly,

7.2 Applications

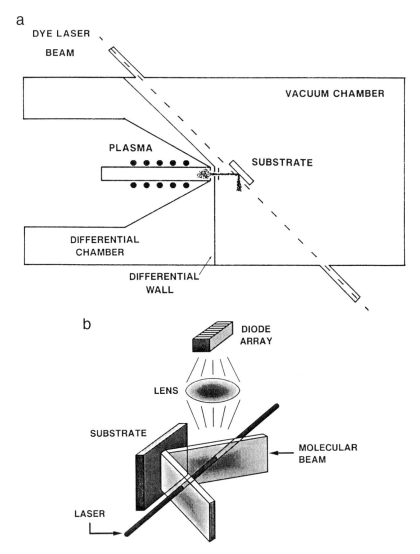

Figure 7.23 Experimental apparatus for IRIS LIF detection to determine the reactivity of species with surfaces. The top view is shown in (a), with specular scattering of the molecular beam depicted, and details of the interaction region and the collection optics are illustrated in (b). [Reprinted with permission from Fisher *et al.* (1992). Copyright 1992 American Chemical Society.]

Fisher *et al.* (1992) found that NH essentially does not react at all with silicon nitride surfaces (Figure 7.24). In contrast to these IRIS studies that showed either unity or zero reaction probability, Fisher *et al.* (1993) found that ground-state OH reacts with both oxidized Si_3N_4 and PMMA (poly-

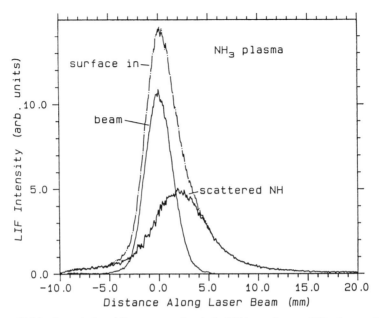

Figure 7.24 Laser-induced fluorescence signal of a NH beam from an NH_3 plasma without a surface (solid line) and with a surface present 2 mm from the laser (dash-dot line). The small solid curve is the difference of these signals in this IRIS detection scheme. [Reprinted with permission from Fisher *et al.* (1992). Copyright 1992 American Chemical Society.]

methylmethacrylate) surfaces with a 0.6 ± 0.05 probability at room temperature; this probability decreases to 0 for temperatures above 500 K. These latter IRIS findings have been summarized by Buss *et al.* (1994).

Laser-induced fluorescence has been used to probe species desorbing from a surface that is heated conventionally, as in temperature-programmed desorption (TPD), or by laser heating, which is called laser-induced thermal desorption (LITD) or laser-desorption spectroscopy (LDS) (Section 15.1). Laser excitation of these same species on the surface produces no fluorescence because of rapid relaxation by the surface and film. Other detection methods, including REMPI and mass spectrometry, can also be used to probe desorbed species.

Herman *et al.* (1994) used LIF to probe SiCl that is desorbed from the surface during plasma etching of Si by Cl_2 in a helical resonator reactor by heating the surface with a XeCl laser (308 nm). This XeCl laser also excited the SiCl $B\ ^2\Sigma^+ - X\ ^2\Pi_{3/2,1/2}$ transition ($v' = 0 \leftarrow v'' = 3,4$), and LIF from the same electronic transition was observed, as is seen in Figures 6.10 and 7.25. Since the same ~20-nsec-long laser pulse both heated the surface layer and excited the SiCl LIF, the observed fluores-

7.2 Applications

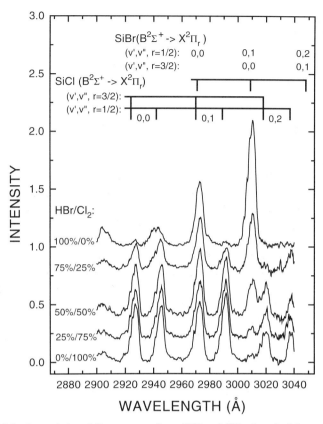

Figure 7.25 Laser-induced fluorescence from SiCl and SiBr desorbed from the surface layer during the etching of Si in a helical resonator plasma reactor with various Cl_2/HBr mixtures, using a XeCl laser (308 nm) for laser desorption and LIF excitation (from Cheng et al., 1995).

cence can come only from species that are directly desorbed from the surface. Cheng et al. (1994) demonstrated that this LD-LIF signal (laser desorption-LIF) is a real-time, *in situ* probe of the thickness of the surface adlayer during etching that is very useful for process development, as is detailed in Section 15.1 (Figures 15.1 and 15.2). Cheng et al. (1995) used LD-LIF to probe the relative chlorination and bromination of the steady-state surface layer during Si etching by Cl_2/HBr mixtures, by monitoring the relative intensities of SiCl and SiBr LIF as shown in Figure 7.25. The XeCl laser also excites $B \leftarrow X$ in SiBr. More details can be found in Section 15.1.

The activation energies for thermal desorption of group III atoms during MBE growth of III–V compound semiconductors and simulated MBE growth have been measured by using temperature-programmed desorption and LIF. Carleton and Leone (1987) probed the desorption of Ga from Si(100), Strupp *et al.* (1992) probed the laser desorption of In and Ga from Si(100), and Alstrin *et al.* (1992) probed the desorption of As_2 from Si(100) and Si(111) dosed with As_4. Kuo *et al.* (1993) examined In desorption from InP layers during gas-source MBE.

Thoman *et al.* (1986) have developed a model for the concentration of reactive gas-phase species adjacent to a surface, which they tested by investigating the interaction of CF_2, measured by using LIF, with Si surfaces.

Laser-induced fluorescence has been used to probe the plume emitted by the ablated surface during pulsed laser deposition (PLD) (Saenger, 1993; Geohegan, 1994). To achieve good spatial resolution, the laser is usually directed normal to the plume and the fluorescence is measured in the direction normal to both the plume and the laser.

In a series of studies, Dreyfus and co-workers used LIF to examine the plumes from a series of different ablated targets. Dreyfus *et al.* (1986) investigated 248-nm KrF laser ablation of Al_2O_3 by LIF of Al atoms, with $4s\ ^2S_{1/2} \leftarrow 3p\ ^2P_{1/2}$ excitation at 394.4 nm and $4s\ ^2S_{1/2} \rightarrow 3p\ ^2P_{3/2}$ emission at 396.2 nm, and AlO molecules, through the $B\ ^2\Sigma^+ - X\ ^2\Sigma^+$ transition near 445 nm. They conducted time-of-flight (TOF) measurements by varying the time between the excimer laser, used to ablate the surface, and the dye

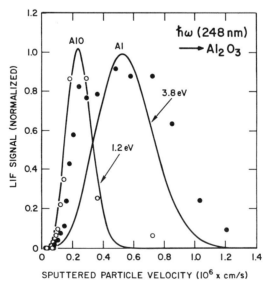

Figure 7.26 Time-of-flight velocity distributions for Al atoms and AlO molecules during excimer laser ablation of Al_2O_3, with Maxwell–Boltzmann distributions shown for comparison (from Dreyfus *et al.*, 1986).

7.2 Applications

laser, used to excited LIF at a fixed position in the plume. The sputtered-particle kinetic energy normal to the target peaked at ~4 eV for Al and ~1 eV for AlO, as is seen in Figure 7.26. Although the velocity distribution was not strictly Maxwellian, the velocity spread suggested a translational "temperature" in AlO that was much higher than the corresponding T_{rot} and T_{vib}, which were measured to be ~600 K. This suggested that PLD particle ejection was not due to thermal evaporation but to electronic ablation. Figure 7.27 shows LIF from ablated AlO, which, in this case, is very similar to the spectrum expected from a Boltzmann distribution with $T_{rot} = 500$ K.

Dreyfus (1990, 1991) probed Cu atoms, Cu$^+$, and Cu$_2$ during 193- and 351-nm laser ablation of Cu. Ground-state Cu atoms were probed at the $^4D^0_{1/2} - {}^2S_{1/2}$ transition at 222.6 nm, which has a significantly large oscillator strength, even though it is "spin forbidden." Metastable Cu$^+$ ions (2.7 eV) were probed at the $^3P^0_2 - {}^3D_3$ transition at 224.7 nm because vacuum UV light would have been needed to probe ground-state Cu$^+$ by one-photon LIF. Diatomic copper (Cu$_2$) was probed in the $B\,^1\Sigma_u^+ - X\,^1\Sigma_g^+$ transition with $2 \leftarrow 0$ vibrational band excitation at 449.8 nm and $2 \rightarrow 2$ emission at 460.7 nm. Following these three species enabled essentially simultaneous analysis of thermal vaporization of Cu, multiphoton ionization of Cu vapor to form Cu$^+$, and electron–atom collision-induced ionization and dissociation (of Cu$_2$). Sappey and Gamble (1991) also used LIF to probe Cu and Cu$_2$ in the plume during laser ablation of Cu, and subsequently Sappey and Gamble (1992) used planar LIF (see Figure 7.4) to monitor these species spatially in the plume.

Pappas *et al.* (1992) studied the PLD of diamond-like films by 248-nm KrF laser ablation of graphite. The most probable kinetic energy of the observed C$_2$ was found to be 12 eV, by using TOF LIF on the Swan band of C$_2$ molecules with $d\,^3\Pi_g(v' = 0) \leftarrow a\,^3\Pi_u(v'' = 0)$ excitation and $d(v' =$

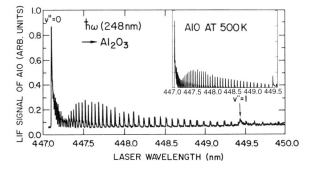

Figure 7.27 Laser-induced fluorescence of AlO during excimer-laser ablation of Al$_2$O$_3$, with simulated LIF spectrum at 500 K in inset (from Dreyfus *et al.*, 1986).

0) → $a(v'' = 1)$ fluorescence detection. Srinivasan *et al.* (1987) probed C_2 [(2,0), 438.2 nm Swan band excitation; (2,1), 471.5 nm detection], as well as CN [$B\,^2\Sigma - X\,^2\Sigma$; (0,0), 399.3 nm excitation; (0,1), 419.5 nm detection], from laser-ablated polyimide targets.

Otis and Dreyfus (1991) used these methods to examine the ground-state populations of cationic and neutral Y, Ba, and Cu and neutral YO, BaO, and CuO during excimer laser ablation of $YBa_2Cu_3O_{7-x}$ (high-temperature superconductor) targets. Lynds *et al.* (1989) also studied the laser ablation of these targets by LIF. Cappelli *et al.* (1990) used planar laser-induced fluorescence (Figure 7.4) to produce the 2D images of neutral ground-state Ba atoms during laser ablation of Ba in Figure 7.28. Barium was excited by 307 nm ($7p\,^1P^0 \leftarrow 6s^2\,^1S$), and fluorescence at 472.8 nm ($7p\,^1P^0 \to 5d\,^1D$) was imaged at right angles to the incident laser sheet.

Wang *et al.* (1991) probed excimer-laser ablation of Al by exciting Al atoms at $4s\,^2S \leftarrow 3p\,^2P_{1/2}$ with 394.4 nm and monitoring $4s\,^2S \to 3p\,^2P_{3/2}$ fluorescence at 396.2 nm. Measurements by LIF TOF indicated a non-Maxwellian distribution, while measurements of the Doppler width, by scanning the laser wavelength, indicated a velocity profile consistent with the TOF profile. AlO was observed by LIF in low-pressure backgrounds of O_2. Hermann *et al.* (1995) used time- and space-resolved LIF measurements to probe the plume during excimer laser irradiation of titanium

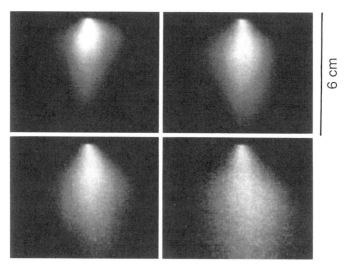

Figure 7.28 Single-shot planar laser-induced fluorescence images of Ba atoms taken at various times after the laser ablation of Ba in 10 mTorr He, delayed by 1.5 μs (top left); 2.5 μs (top right); 4.0 μs (bottom left); 6.0 μs (bottom right) (from Cappelli *et al.*, 1990).

targets in a low-pressure nitrogen atmosphere, which leads to the PLD of TiN films.

By using TOF and Doppler-shifted two-photon LIF, Arlinghaus et al. (1989a,b) found that the Zn atoms ejected during 308-nm ablation of ZnS targets fit a Maxwellian profile. The Zn transition $4d\,^1D_2 \leftarrow 4s^2\,^1S_0$ was excited by the absorption of two 320-nm photons, and then $4d\,^1D_2 \to 4p\,^1P_1^0$ emission at 636 nm was detected. The exciting laser was a narrowband cw dye laser oscillator amplified by a three-stage high-power pulsed dye laser amplifier chain.

Husinsky (1985) has reviewed the use of Doppler-shift laser fluorescence spectroscopy (DSLFS) to detect particles sputtered from surface by ions, atoms, or electrons and to measure their velocity distribution. This diagnostic is particularly useful in learning about the fundamental processes in ion sputtering for film deposition. Arlinghaus et al. (1989b) used TOF and Doppler-shifted two-photon LIF to detect Zn atoms sputtered from Zn surfaces by argon-ion bombardment. In earlier work, Young et al. (1984) used TOF and Doppler-shifted one-photon LIF to detect sputtered Fe atoms by $y\,^5D_4^0 \leftarrow a\,^5D_4$ excitation at 302.065 nm and $y\,^5D_4^0 \to a\,^5F_5$ detection at 382.043 nm.

References

M. Aldén, A. L. Schawlow, S. Svanberg, W. Wendt, and P.-L. Zhang, *Opt. Lett.* **9,** 211 (1984).
A. L. Alstrin, R. V. Smilgys, P. G. Strupp, and S. R. Leone, *J. Chem. Phys.* **97,** 6864 (1992).
H. F. Arlinghaus, W. F. Calaway, C. E. Young, M. J. Pellin, D. M. Gruen, and L. L. Chase, *J. Appl. Phys.* **65,** 281 (1989a).
H. F. Arlinghaus, W. F. Calaway, C. E. Young, M. J. Pellin, D. M. Gruen, and L. L. Chase, *J. Vac. Sci. Technol. A* **7,** 1766 (1989b).
R. F. Barrow, ed., *Molécules Diatomiques: Bibliographie Critique de Données Spectroscopiques* [Diatomic Molecules: A Critical Bibliography of Spectroscopic Data], Vol. I. Éditions du Centre National de la Recherche Scientifique, Paris, 1973.
R. F. Barrow, ed., *Molécules Diatomiques: Bibliographie Critique de Données Spectroscopiques* [Diatomic Molecules: A Critical Bibliography of Spectroscopic Data], Vol. II. Éditions du Centre National de la Recherche Scientifique, Paris, 1975.
S. Bashkin and J. O. Stoner, Jr., *Atomic Energy Levels and Grotrian Diagrams,* Vol. I (1975), Vol. II (1978), Vol. III (1981), Vol. IV (1982). North-Holland, Amsterdam.
W. K. Bischel, B. E. Perry, and D. R. Crosley, *Chem. Phys. Lett.* **82,** 85 (1981).
J. P. Booth, G. Hancock, and N. D. Perry, *Appl. Phys. Lett.* **50,** 318 (1987).
W. G. Breiland and P. Ho, in *Chemical Vapor Deposition* (M. L. Hitchman and K. F. Jensen, eds.), Chapter 3, p. 91. Academic Press, Boston, 1993.

W. G. Breiland, M. E. Coltrin, and P. Ho, *J. Appl. Phys.* **59,** 3267 (1986a).
W. G. Breiland, P. Ho, and M. E. Coltrin, *J. Appl. Phys.* **60,** 1505 (1986b).
P. D. Brewer, *Chem. Phys. Lett.* **141,** 301 (1987).
L.-M. Buchmann, F. Heinrich, P. Hoffmann, and J. Janes, *J. Appl. Phys.* **67,** 3635 (1990).
R. J. Buss, P. Ho, and M. E. Weber, *Plasma Chem. Plasma Process.* **13,** 61 (1993).
R. J. Buss, P. Ho, E. R. Fisher, and W. G. Breiland, *Mater. Res. Soc. Symp. Proc.* **334,** 51 (1994).
K. L. Carleton and S. R. Leone, *J. Vac. Sci. Technol. B* **5,** 1141 (1987).
M. A. Cappelli and P. H. Paul, *J. Appl. Phys.* **67,** 2596 (1990).
M. A. Cappelli, P. H. Paul, and R. K. Hanson, *Appl. Phys. Lett.* **56,** 1715 (1990).
C. C. Cheng, K. V. Guinn, V. M. Donnelly, and I. P. Herman, *J. Vac. Sci. Technol. A* **12,** 2630 (1994).
C. C. Cheng, K. V. Guinn, I. P. Herman, and V. M. Donnelly, *J. Vac. Sci. Technol. A* **13,** 1970 (1995).
U. Czarnetzki, K. Miyazaki, T. Kajiwara, K. Muraoka, T. Okada, M. Maeda, A. Suzuki, and A. Matsuda, *J. Vac. Sci. Technol. A* **12,** 831 (1994).
G. P. Davis and R. A. Gottscho. *J. Appl. Phys.* **54,** 3080 (1983).
E. A. Den Hartog, H. Persing, and R. C. Woods, *Appl. Phys. Lett.* **57,** 661 (1990).
L. F. DiMauro, R. A. Gottscho, and T. A. Miller, *J. Appl. Phys.* **56,** 2007 (1984).
V. M. Donnelly, in *Plasma Diagnostics* (O. Auciello and D. L. Flamm, eds.), Vol. I, Chapter 1, p. 1. Academic Press, Boston, 1989.
V. M. Donnelly and R. F. Karlicek, *J. Appl. Phys.* **53,** 6399 (1982).
V. M. Donnelly, D. L. Flamm, and G. Collins, *J. Vac. Sci. Technol.* **21,** 817 (1982).
R. W. Dreyfus, *High Temp. Sci.* **27,** 503 (1990).
R. W. Dreyfus, *J. Appl. Phys.* **69,** 1721 (1991).
R. W. Dreyfus, J. M. Jasinski, R. E. Walkup, and G. S. Selwyn, *Pure Appl. Chem.* **57,** 1265 (1985).
R. W. Dreyfus, R. Kelly, and R. E. Walkup, *Appl. Phys. Lett.* **49,** 1478 (1986).
M. J. Dyer and D. R. Crosley, *Opt. Lett.* **7,** 382 (1982).
A. C. Eckbreth, *Laser Diagnostics for Combustion Temperature and Species,* 2nd ed. Gordon & Breach, Luxembourg, 1996.
E. R. Fisher, P. Ho, W. G. Breiland, and R. J. Buss, *J. Phys. Chem.* **96,** 9855 (1992).
E. R. Fisher, P. Ho, W. G. Breiland, and R. J. Buss, *J. Phys. Chem.* **97,** 10287 (1993).
D. B. Geohegan, in *Pulsed Laser Deposition of Thin Films* (G. Hubler and D. B. Chrisey, eds.), Chapter 5, p. 115. Wiley (Interscience), New York, 1994.
K. P. Giapis, N. Sadeghi, J. Margot, R. A. Gottscho, and T. C. J. Lee, *J. Appl. Phys.* **73,** 7188 (1993).
M. J. Goeckner and J. Goree, *J. Vac. Sci. Technol. A* **7,** 977 (1989).
J. E. M. Goldsmith, *Opt. Lett.* **10,** 116 (1985).
J. E. M. Goldsmith and N. M. Laurendeau, *Opt. Lett.* **15,** 576 (1990).
R. A. Gottscho and M. L. Mandich, *J. Vac. Sci. Technol. A* **3,** 617 (1985).
R. A. Gottscho and T. A. Miller, *Pure Appl. Chem.* **56,** 189 (1984).
R. A. Gottscho, G. P. Davis, and R. H. Burton, *Plasma Chem. Plasma Process.* **3,** 193 (1983a).
R. A. Gottscho, G. P. Davis, and R. H. Burton, *J. Vac. Sci. Technol. A* **1,** 622 (1983b).
R. A. Gottscho, R. H. Burton, D. L. Flamm, V. M. Donnelly, and G. P. Davis *J. Appl. Phys.* **55,** 2707 (1984).

References

R. A. Gottscho, T. Nakano, N. Sadeghi, D. J. Trevor, and R. W. Boswell, *SPIE* **1594,** 376 (1991).

K. E. Greenberg and P. J. Hargis, Jr., *J. Appl. Phys.* **68,** 505 (1990).

S. G. Hansen, G. Luckman, and S. D. Colson, *Appl. Phys. Lett.* **53,** 1588 (1988).

R. K. Hanson, *J. Quant. Spectrosc. Radiat. Transfer* **40,** 343 (1988).

R. K. Hanson and J. M. Seitzman, in *Handbook of Flow Visualization* (W.-J. Yang, ed.), Chapter 15, p. 219. Hemisphere, New York, 1989.

P. J. Hargis, Jr. and M. J. Kushner, *Appl. Phys. Lett.* **40,** 779 (1982).

P. J. Hargis, Jr., R. W. Light, and J. M. Gee, *Springer Ser. Chem. Phys.* **39,** 526 (1984).

N. Hata, A. Matsuda, and K. Tanaka, *J. Appl. Phys.* **61,** 3055 (1987).

M. Heaven, T. A. Miller, R. R. Freeman, J. C. White, and J. Bokar, *Chem. Phys. Lett.* **86,** 458 (1982).

G. A. Hebner and K. P. Killeen, *J. Appl. Phys.* **67,** 1598 (1990).

I. P. Herman, H. Tang, and P. P. Leong, *Mater. Res. Soc. Symp. Proc.* **201,** 563 (1991).

I. P. Herman, V. M. Donnelly, K. V. Guinn, and C. C. Cheng, *Phys. Rev. Lett.* **72,** 2801 (1994).

J. Hermann, A. L. Thomann, C. Boulmer-Leborgne, B. Dubreuil, M. L. De Giorgi, A. Perrone, A. Luches, and I. N. Mihailescu, *J. Appl. Phys.* **77,** 2928 (1995).

G. Herzberg, *Atomic Spectra and Atomic Structure.* Dover, New York, 1937.

G. Herzberg, *Molecular Spectra and Molecular Structure. II. Infrared and Raman Spectra of Polyatomic Molecules.* Van Nostrand-Reinhold, New York, 1945.

G. Herzberg, *Molecular Spectra and Molecular Structure. I. Spectra of Diatomic Molecules.* Van Nostrand-Reinhold, New York, 1950.

G. Herzberg, *Molecular Spectra and Molecular Structure. III. Electronic Spectra and Electronic Structure of Polyatomic Molecules.* Van Nostrand-Reinhold, New York, 1966.

P. Ho and W. G. Breiland, *Appl. Phys. Lett.* **43,** 125 (1983).

P. Ho and W. G. Breiland, *Appl. Phys. Lett.* **44,** 51 (1984).

P. Ho and W. G. Breiland, *J. Appl. Phys.* **63,** 5184 (1988).

P. Ho, W. G. Breiland, and R. J. Buss, *J. Chem. Phys.* **91,** 2627 (1989).

P. Ho, R. J. Buss, and M. E. Weber, *Mater. Res. Soc. Symp. Proc.* **204,** 483 (1991).

P. Ho, M. E. Coltrin, and W. G. Breiland, *J. Phys. Chem.* **98,** 10138 (1994).

K. P. Huber and G. Herzberg, *Molecular Spectra and Molecular Structure. IV. Constants of Diatomic Molecules.* Van Nostrand-Reinhold, New York, 1979.

W. Husinsky, *J. Vac. Sci. Technol. B* **3,** 1546 (1985).

F. Itoh, G. Kychakoff, and R. K. Hanson, *J. Vac. Sci. Technol. B* **3,** 1600 (1985).

M. E. Jacox, *J. Phys. Chem. Ref. Data* **17,** 269 (1988).

M. E. Jacox, *J. Phys. Chem. Ref. Data* **19,** 1387 (1990).

M. E. Jacox, Vibrational and Electronic Energy Levels of Polyatomic Transient Molecules, *J. Phys. Chem. Ref. Data Monograph #3* (1994).

G. M. Jellum and D. B. Graves, *J. Appl. Phys.* **67,** 6490 (1990).

J. E. Jensen, P. D. Brewer, G. L. Olson, L. W. Tutt, and J. J. Zinck, *J. Vac. Sci. Technol. A* **6,** 2808 (1988).

J. L. Kinsey, *Annu. Rev. Phys. Chem.* **28,** 349 (1977).

M. Kitamura, H. Akiya, and T. Urisu, *J. Vac. Sci. Technol. B* **7,** 14 (1989).

C.-H. Kuo, C. Choi, G. N. Maracas, and T. C. Steimle, *J. Vac. Sci. Technol. B* **11,** 833 (1993).

V. S. Letokhov and V. P. Chebotayev, *Nonlinear Laser Spectroscopy.* Springer, New York, 1977.

M. C. Lin and G. Ertl, *Annu. Rev. Phys. Chem.* **37,** 587 (1986).

R. P. Lucht, J. T. Salmon, G. B. King, D. W. Sweeney, and N. M. Laurendeau, *Opt. Lett.* **8,** 365 (1983).

L. Lynds, B. R. Weinberger, D. M. Potrepka, G. G. Peterson, and M. P. Lindsay, *Physica C (Amsterdam)* **159,** 61 (1989).

F. Magnotta, in *Microbeam Analysis—1987* (R. H. Geiss, ed.), p. 153. San Francisco Press, San Francisco, 1987.

E. E. Marinero and C. R. Jones, *J. Chem. Phys.* **82,** 1608 (1985).

Y. Matsumi, S. Toyoda, T. Hayashi, M. Miyamura, H. Yoshikawa, and S. Komiya, *J. Appl. Phys.* **60,** 4102 (1986).

J. A. McClintock and R. A. Wilson, *J. Cryst. Growth* **81,** 177 (1987).

B. K. McMillin and M. R. Zachariah, *J. Appl. Phys.* **77,** 5538 (1995).

U. E. Meier, K. Kohse-Hoinghaus, L. Schafer, and C.-P. Klages, *Appl. Opt.* **29,** 4993 (1990).

U. E. Meier, L. E. Hunziker, D. R. Crosley, and J. B. Jeffries, in *Proceedings of the Second International Symposium on Diamond Materials,* p. 202. Electrochemical Society, Pennington, NJ, 1991.

C. E. Moore, *Atomic Energy Levels,* NBS Circular 467, Vol. I (1949); Vol. II (1952); Vol. III (1958).

C. E. Moore and P. W. Merrill, *Partial Grotrian Diagrams of Astrophysical Interest,* National Standard Reference Data Series, NSRDS-NBS 23, 1968.

T. Nakano, N. Sadeghi, and R. A. Gottscho, *Appl. Phys. Lett.* **58,** 458 (1991).

T. Nakano, N. Sadeghi, D. J. Trevor, R. A. Gottscho, and R. W. Boswell, *J. Appl. Phys.* **72,** 3384 (1992).

K. Ninomiya, K. Suzuki, S. Nishimatsu, and O. Okada, *J. Vac. Sci. Technol. A* **4,** 1791 (1986).

H. Okabe, *Photochemistry of Small Molecules.* Wiley, New York, 1978.

K. Ono, T. Oomori, M. Tuda, and K. Namba, *J. Vac. Sci. Technol. A* **10,** 1071 (1992).

C. E. Otis and R. W. Dreyfus, *Phys. Rev. Lett.* **67,** 2102 (1991).

S. Pang and S. R. J. Brueck, *Mater. Res. Soc. Symp. Proc.* **17,** 161 (1983).

D. L. Pappas, K. L. Saenger, J. J. Cuomo, and R. W. Dreyfus, *J. Appl. Phys.* **72,** 3966 (1992).

W. Z. Park, M. Tanigawa, T. Kajiwara, K. Muraoka, M. Masuda, T. Okada, M. Maeda, A. Suzuki, and A. Matsuda, *Jpn. J. Appl. Phys.* **31,** 2917 (1992).

L. D. Pfefferle, T. A. Griffin, and M. Winter, *Appl. Opt.* **27,** 3197 (1988).

K. J. Rensberger, J. B. Jeffries, R. A. Copeland, K. Kohse-Höinghaus, M. L. Wise, and D. R. Crosley, *Appl. Opt.* **28,** 3556 (1989).

R. M. Roth, K. G. Spears, and G. Wong, *Appl. Phys. Lett.* **45,** 28 (1984).

N. Sadeghi, T. Nakano, D. J. Trevor, and R. A. Gottscho, *J. Appl. Phys.* **70,** 2552 (1991); erratum: *ibid.* **71,** 3648 (1992).

K. L. Saenger, *Process. Adv. Mater.* **3**(2), 63 (1993).

J. T. Salmon and N. M. Laurendeau, *Appl. Opt.* **24,** 1313 (1985).

A. D. Sappey and T. K. Gamble, *Appl. Phys. B* **53,** 353 (1991).

A. D. Sappey and T. K. Gamble, *J. Appl. Phys.* **72,** 5095 (1992).

A. D. Sappey and J. B. Jeffries, *Appl. Phys. Lett.* **55,** 1182 (1989).

L. Schäfer, C.-P. Klages, U. Meier, and K. Kohse-Höinghaus, *Appl. Phys. Lett.* **58**, 571 (1991).
T. O. Sedgwick, J. E. Smith, Jr., R. Ghez, and M. E. Cowher, *J. Cryst. Growth* **31**, 264 (1975).
G. S. Selwyn, *J. Appl. Phys.* **60**, 2771 (1986).
G. S. Selwyn, *Appl. Phys. Lett.* **51**, 167 (1987).
G. S. Selwyn, *J. Vac. Sci. Technol. A* **6**, 2041 (1988).
G. S. Selwyn, L. D. Baston, and H. H. Sawin, *Appl. Phys. Lett.* **51**, 898 (1987).
O. N. Singh and H. J. Hopman, *Microelectron. J.* **18**, 39 (1987).
R. V. Smilgys and S. R. Leone, *J. Vac. Sci. Technol. B* **8**, 416 (1990).
J. E. Smith, Jr. and T. O. Sedgwick, *Thin Solid Films* **40**, 1 (1977).
K. G. Spears, T. J. Robinson, and R. M. Roth, *IEEE Trans. Plasma Sci.* **PS-14**, 179 (1986).
R. Srinivasan, B. Braren, and R. W. Dreyfus, *J. Appl. Phys.* **61**, 372 (1987).
P. G. Strupp, A. L. Alstrin, B. J. Korte, and S. R. Leone, *J. Vac. Sci. Technol. A* **10**, 508 (1992).
S. N. Suchard, ed., *Spectroscopic Data*, Vol. 1, *Heteronuclear Diatomic Molecules*, Parts A and B. IFI/Plenum, New York, 1975.
S. N. Suchard and J. E. Melzer, eds., *Spectroscopic Data*, Vol. 2. *Homonuclear Diatomic Molecules*. IFI/Plenum, New York, 1976.
H. Suzuki, K. Mori, M. Kawasaki, and H. Sato, *J. Appl. Phys.* **64**, 371 (1988).
J. W. Thoman, Jr., K. Suzuki, S. H. Kable, and J. I. Steinfeld, *J. Appl. Phys.* **60**, 2775 (1986).
D. J. Trevor, N. Sadeghi, T. Nakano, J. Derouard, R. A. Gottscho, P. D. Foo, and J. M. Cook, *Appl. Phys. Lett.* **57**, 1188 (1990).
A. D. Tserepi and T. A. Miller, *J. Appl. Phys.* **77**, 505 (1995).
A. D. Tserepi, J. R. Dunlop, B. L. Preppernau, and T. A. Miller, *J. Vac. Sci. Technol. A* **10**, 1188 (1992).
P. Van de Weijer and B. H. Zwerver, *Chem. Phys. Lett.* **163**, 48 (1989).
E. P. Visser, C. A. M. Govers, and L. J. Giling, *J. Cryst. Growth* **102**, 529 (1990).
R. E. Walkup and S. I. Raider, *Appl. Phys. Lett.* **53**, 888 (1988).
R. Walkup, P. Avouris, R. W. Dreyfus, J. M. Jasinski, and G. S. Selwyn, *Appl. Phys. Lett.* **45**, 372 (1984).
R. E. Walkup, K. L. Saenger, and G. S. Selwyn, *J. Chem. Phys.* **84**, 2668 (1986).
H. Wang, A. P. Salzberg, and B. R. Weiner, *Appl. Phys. Lett.* **59**, 935 (1991).
R. C. Woods, R. L. McClain, L. J. Mahoney, E. A. Den Hartog, H. Persing, and J. S. Hamers, *SPIE* **1594**, 366 (1991).
J. Wormhoudt, A. C. Stanton, and J. Silver, *SPIE* **452**, 88 (1983).
C. E. Young, W. F. Calaway, M. J. Pellin, and D. M. Gruen, *J. Vac. Sci. Technol. A* **2**, 693 (1984).
H. Zacharias, *Appl. Phys. A* **47**, 37 (1988).

CHAPTER **8**

Transmission (Absorption)

Transmission through, and reflection from, a medium both depend on the refractive (n) and absorptive (k) properties of the medium. Consequently, it is not always straightforward to classify interactions into the separate categories of transmission, reflection, and absorption spectroscopies. Transmission clearly monitors absorption in gases (except for the reflection from the chamber windows). In probing solids, however, transmission depends on the reflections at the ambient/material interfaces as well as on absorption in the material. This chapter covers spectroscopies that probe absorption in a medium by transmission, while Chapter 9 discusses most reflection spectroscopies, whether they depend primarily on refractive or absorptive effects. For example, infrared reflection-absorption spectroscopy (IRRAS) and attenuated total internal reflection (ATR) spectroscopy, which probe the region near the surface, are treated in Section 9.7. However, probes that involve a roundtrip transmission through a wafer, which is aided by a single reflection at a wafer surface, are closely tied to "single-pass" transmission probes and thus are included in this chapter (Section 8.4).

This chapter covers only those "absorption" spectroscopies in which resonant light enters the medium and is partially absorbed, after which the transmitted beam is detected. Spectroscopies in which the absorption of light is monitored by other means are treated in other chapters. For example, in laser-induced fluorescence (LIF, Chapter 7) and photoluminescence (PL, Chapter 14), absorption induces emission of light at a different wavelength, which is then detected. In some cases the absorption of light leads to the production of electrons, as in photoionization of atoms and molecules, including resonant multiphoton ionization and optogalvanic spectroscopy of plasmas, and visible/near-ultraviolet photoelectron spectroscopy of solids (photoemission). These are described in Chapter 17. Absorbed light can produce thermal and acoustic waves. Light absorption monitored by these optoacoustic/thermal wave spectroscopies is covered in Chapter 15.

Absorption and excitation laser-induced fluorescence in gases and excitation photoluminescence in solids probe the same properties of a system.

Each has its strengths and weaknesses. When nonradiative relaxation processes are very fast relative to radiative decay, fluorescence is very weak and LIF (and PL) cannot be used; rapid relaxation has only a minor effect on absorption. This is particularly significant in the infrared because spontaneous emission rates are very slow in the IR ($\sim v^3$). Because absorption measures the difference in the incident and transmitted beams, laser- and detector-related noise must be small in absorption measurements when the absorbed fraction is small; LIF is usually measured relative to a zero background and is therefore more sensitive. The absolute measurement of density is easier with absorption, because LIF analysis of density critically depends on relaxation processes, the solid angle of light collected, and quantum efficiency. Absorption integrates density along the propagation path and requires tomographic reconstruction to achieve 3D resolution, while LIF has better spatial resolution because the exciting laser can be tightly focused to probe, and then to map, small volumes.

Absorption primarily serves two diagnostics needs: (1) determining gas-phase species identity and concentration for process studies and (2) controlling the wafer temperature and the flow of gas-phase reactants for real-time monitoring. Other applications include the real-time analysis of thin films. This chapter is unique among those detailing linear optical diagnostics in that it addresses the application of one type of optical probe to investigate both the gas above the wafer (process-state parameters) and the wafer itself (wafer-state parameters).

8.1 Experimental Considerations

8.1.1 *Signal Analysis*

The fraction of the beam remaining after propagation through a distance z in a medium with absorption coefficient α is given by Beer's law:

$$I(z) = I(0) \exp(-\alpha z) \qquad (8.1)$$

which follows from Maxwell's equations (Equation 2.5). This exponential decay applies to the light intensity I, as written here, and to the power in the beam $\mathcal{P}$. The range of validity of Beer's law is discussed in Section 4.2.1.2, which set limits on the laser intensity and pulse length. As defined here, α has units of 1/cm (or 1/m) and is a function of frequency v. The effective absorption length (or penetration depth) in the medium is $\sim 1/\alpha$, assuming that $1/\alpha < l$, the sample length. $I(z)/I(0)$ is the transmittance, while $1 - I(z)/I(0)$ is the absorbance (or absorptance or absorbed fraction).

Since α is proportional to the density of absorbing species N (1/cm^3), it can be expressed as $\alpha = N\sigma$, and the effective optical thickness (or depth) [or absorption path length] per pass of length l in Equation 8.1 is $\alpha l = N\sigma l$, which is 2.303 × the optical density. (This so-defined "optical depth" is the same as the absorbance when the sample is optically thin, $\alpha l \ll 1$.) σ is the absorption cross-section, which has units of area (cm^2). Since the partial pressure of a species in a gas is proportional to N, the absorbed fraction can also be written as $\alpha' pl$, where $\alpha' = \alpha/p$ is the absorption coefficient α per unit pressure. In some communities α', or the closely related parameter σ, is called the absorption coefficient instead of α. There should not be any confusion in terminology if the units are stated clearly.

Another confusion can arise because the chemical community sometimes expresses absorption in terms of the extinction (or molar absorption) coefficient $\bar{\varepsilon}$, which is a parameter very closely related to σ. (Although the extinction coefficient is usually denoted as ε, it is indicated by $\bar{\varepsilon}$ here to avoid confusion with the dielectric constant.) With $\bar{\varepsilon}$, the exponential decay due to absorption is expressed in base 10, and Beer's law becomes $I(z) = I(z = 0) 10^{-\bar{\varepsilon} c l}$, where c is the concentration, i.e., density, expressed in moles/liter, so $\bar{\varepsilon}$ has units of liters/mole cm and is equal to $2.61 \times 10^{20} \sigma$ (expressed in cm^2). In this case, knowing the units is not enough to avoid confusion. However, when units of liters/mole are given, the correct base is most likely 10. Note that k, the complex part of the refractive index (in Equation 4.8), is sometimes also called the extinction coefficient; however, k is proportional, though not equal, to $\bar{\varepsilon}$.

Gas-phase infrared spectra and normal-mode analyses of many compounds are given in Herzberg (1945), Wilson et al. (1980), and Nakamoto (1986). Infrared absorption data may also be found in Wormhoudt et al. (1983) and Jacox (1984), and in references cited therein. Some representative vibrational IR absorption bands are listed in Table 8.1. Compilations of electronic states (and vibrational and rotational constants for each molecular state), lines, and spectra can be found in Herzberg (1937, 1950, 1966), Moore (1949, 1952, 1958), Moore and Merrill (1968), Barrow (1973, 1975), Suchard (1975), Suchard and Melzer (1976), Bashkin and Stoner (1975, 1978, 1981, 1982), Pearse and Gaydon (1976), Okabe (1978), Huber and Herzberg (1979), Jacox (1988, 1990, 1994), and Rothschild (1989).

Quantitative analysis of the measured absorption spectrum to determine density in gases and composition in solids depends on the spectral absorption profile, i.e., the lineshapes of the individual lines and their density, and instrumental resolution. Equations 4.16, 4.33, and 4.36 give the absorption coefficient for a single transition. Several factors affect this absorption profile: (1) the resonance of the transition energy and the photon energy; (2) dynamic factors, such as the transition dipole moment, which also lead to the selection rules; (3) the density of absorbers in the resonant state; and (4) the spectral lineshape of the transition. These relations can be used to determine density only if the absorber is probed by an optical source with a spectral width that is

Table 8.1 Typical Bond Vibration Energies for Absorption and ATR (Attenuated Total Internal Reflection) Spectroscopies[a]

Vibration [on]	Approximate energy[b] (cm^{-1})
As—H stretch [on GaAs]	2100
C—H stretch [on diamond]	2900
C—H bend [on diamond]	1250
Ga—H stretch (symmetric) [on GaAs]	1850
Ga—H—Ga stretch (asymmetric, bridging) [on GaAs]	1200–1750
Ge—H stretch [in Ge]	1800
H—Ge—H stretch [in Ge]	2000
N—H stretch [in SiN$_x$]	3350
N—H bend [in SiN$_x$]	1060
O—H stretch [in SiOH]	3650
O—H stretch of H$_2$O	3350
O—H bend in H$_2$O	1620
Si—H stretch [in Si]	2000
H—Si—H stretch [in Si]	2094
H—Si—H bend [in Si]	845–890
H—Si—H wag [in Si]	650
Ga—Br (GaBr)	263
Ga—Cl (GaCl)	365
Ge—Br (GeBr)	297
Ge—Cl (GeCl)	408
Ge—F (GeF)	665
Si—Br (SiBr)	425
Si—Cl (SiCl)	535
Si—F (SiF)	857
Si—F$_x$ stretches [in Si]	~830–1015
Si—N stretch [in SiN$_x$]	880
B—O stretch [in BSG, BPSG]	1300–1400
C—O stretch	900–1200
C=O stretch	1600–1800
P=O stretch [in PSG, BPSG]	1330
Si—O$_{1,2}$ stretches [in SiO$_x$]	1040–1150
Si—O$_{1,2}$ bends [in SiO$_x$]	800–876
O—Si—O rock, wag [in SiO$_x$]	450–475

[a] Compiled from several references cited in Chapters 8 and 9; consult Nakamoto (1986) for more details.

[b] The specific energy depends on the details of the bonding, e.g., SiH$_2$ vs SiH, etc. Approximate energies come either from the denoted [solid] or (gas-phase molecule).

8.1 Experimental Considerations

much narrower than the transition linewidth. Otherwise, they need to be modified, as is discussed below. Typically, spectral lines in gases are homogeneously (collisionally) broadened for total pressures $p \gg 5$ Torr at 10 μm and $\gg 200$ Torr at 500 nm (Equation 4.21), while they are Doppler-broadened at lower pressures (Equation 4.22). Since $\gamma_{\text{rad}}^{(v)} \propto v^3$ (Equation 4.17), radiative decay often dominates broadening in the ultraviolet and rarely affects the profile in the infrared. Stark broadening can be important in plasmas. The relative merits of infrared and ultraviolet absorption in gas-phase diagnostics are presented in Section 8.2.

Accurate determination of density in a gas depends on the species to be probed, the reactor conditions, and the properties of the optical source used for probing (Mitchell and Zemansky, 1971; Corney, 1977; Blass and Halsey, 1981; Baltayan et al., 1992). In particular, it depends on the relative size of the optical source linewidth Δv_L and the spectral linewidth of the transition $\Gamma^{(v)}$. Four cases can be identified:

(i) $\Delta v_L \ll \Gamma^{(v)}$, low line density (resolvable spectral lines)

In this extreme, Equations 4.16, 4.33, and 4.36 can be used directly to obtain the density. This is common in probing by tunable infrared diode laser absorption. The effect of spectral linewidth on the measurement is important whether density is determined from the peak absorption strength or the integrated strength, or by fitting the absorption profile.

(ii) $\Delta v_L \sim$ or $\gg \Gamma^{(v)}$, low line density (resolvable spectral lines)

When the frequency width of the optical source is comparable to or larger than the linewidth, the absorbed fraction is less and the measured linewidth is larger than in case (i). For very broad sources, $\Delta v_L \gg \Gamma^{(v)}$, the absorbed fraction is smaller by a factor of $\sim \Gamma^{(v)}/\Delta v_L$, and the measured linewidth is $\sim \Delta v_L$. Detection of low densities is clearly aided by using narrow optical sources. Still, such less-than-optimal conditions are not uncommon. The linewidths of atomic resonance lamps can be the same or even somewhat larger than the atomic widths of the probed species (because of higher temperatures and discharge-related broadening); this also can occur in the self-absorption of plasma-induced emission (PIE). This condition is also common when dye lasers are used, although dye lasers can be frequency-narrowed by using etalons so that condition (i) applies. This condition also generally applies to white-light lamp sources that are spectrally narrowed by transmission through a grating monochromator; condition (i) can be achieved by using a Fabry–Perot interferometer in tandem with the monochromator.

The effect of a broad source can be included by integrating the optical source profile $I(v)$ over the spectral profile. Several such cases have been examined in Mitchell and Zemansky (1971). In one common situation, the laser or emission line has a gaussian linewidth, Δv_L, that is centered at the same frequency as a Doppler-broadened absorption feature, which is also a gaussian, with a $\Gamma_D^{(\nu)}$ linewidth. With the ratio $\Delta\nu_L/\Gamma_D^{(\nu)} = x$, the absorbed fraction (absorbance) A is

$$A = \frac{I(l=0) - I(l)}{I(l=0)} = \sum_{n=1}^{\infty} (-1)^{n+1} \frac{(\alpha_0 l)^n}{n!\sqrt{1+nx^2}} \quad (8.2)$$

where l is the absorption length and α_0 is the spectral absorption coefficient at linecenter, given by Equations 4.16 and 4.22:

$$\alpha_0 = \frac{1}{\Gamma_D^{(\nu)}} \sqrt{\frac{\ln 2}{\pi}} \frac{\lambda^2 g_2}{4\pi g_1} A_{21} N_1 \quad (8.3)$$

where λ is the wavelength at linecenter, g_1 and g_2 are the degeneracies of the lower and upper levels, respectively, and A_{21} is the Einstein coefficient.

For narrow optical linewidths $x \to 0$, Equation 8.2 reduces to Beer's law (Equation 8.1). For very broad optical linewidths $x \to \infty$, the absorbed fraction becomes $\alpha_0 l/x = [(\alpha_0 \Gamma_D^{(\nu)})l]/\Delta\nu_L$, as expected. Deviations from Beer's law are significant for $x > 1$ since only a fraction of the light is resonant, $\Gamma_D^{(\nu)}/\Delta\nu_L = 1/x$. The transmitted beam can be thought of as the sum of two beams: one, corresponding to a fraction $\sim 1/x$ of the incident beam, is absorbed according to Beer's law, and the other, a $\sim 1 - 1/x$ fraction, is unattenuated. For "optically thick" samples, $\alpha l \gg 1$, the transmitted fraction is $\sim 1 - 1/x$, which differs from that in case (i), where the transmitted fraction approaches 0.

(iii) $\Delta\nu_L \gg$ line separation $> \Gamma^{(\nu)}$, low line density

Individual lines are not resolved, although they could be with a narrower source. The apparent linewidth is $\Delta\nu_L$ and the absorbed fraction at a given frequency ν is $\sim [\Gamma_D^{(\nu)}/\Delta\nu_L] l \Sigma \alpha_{0i}$, which is summed over all absorption resonances within $\nu \pm \Delta\nu_L$. There are deviations from Beer's law, as in case (ii).

(iv) $\Gamma^{(\nu)} >$ line separation, overlapping lines

This condition starts becoming significant as the number of atoms in a molecule increases and as the atoms become heavier, and occurs in the ultraviolet absorption spectrum before it does in the infrared. Often, $\Delta\nu_L$ is narrower than any remaining structure in the profile, and the density can be determined from the known absorption coefficient at that frequency without further calibration. Resolving different species and determining their densities in a multicomponent mixture can be difficult. This is particularly true in the ultraviolet because the absorption features of each species are broad and can overlap.

8.1.2 *Instrumentation*

A typical experimental apparatus used in transmission/absorption diagnostics is shown in Figure 8.1. Both broadband incoherent sources and lasers are commonly used to probe absorption. In the Fourier-transform infrared

8.1 Experimental Considerations

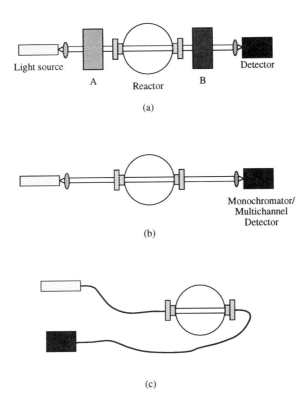

Figure 8.1 Schematic of different experimental approaches for transmission experiments: with (a) a transmission filter, dispersive spectrometer, or Fourier transform spectrometer placed either before (A) or after (B) the reactor, which are often not needed with laser sources, (b) multichannel detection of the transmitted dispersed "white" light, and (c) fiber optic coupling, with lens couplers to and from the fibers and wavelength analysis components not shown.

(FTIR) analysis of vibrational modes, the continuum source is often a glowbar. Electronic transitions in the visible are probed with a tungsten lamp and dispersed with a grating monochromator before transmission through the sample, while in the ultraviolet ($\sim$ 160–350 nm) a deuterium lamp can be used; an alternative source is a Xe arc lamp. These visible/UV continuum sources are often employed to probe diatomic and polyatomic species. Other incoherent sources provide resonant emission lines. Atomic resonance lamps are inexpensive sources that can be used to probe atoms. Hollow cathode lamps of many metal atoms and group III, IV, and V atoms emit resonance light in the visible and near-UV that can excite their respective ground-state atoms (Price, 1972; Setser, 1979). Also, the absorption of light emitted by species due to plasma emission (OE, PIE) can be quantitatively monitored; this is called self-absorption. Tunable lasers are more likely to be used in absorption studies than are fixed-wavelength lasers because of the rarity of fortuitous resonances. Tunable diode lasers are frequently used for infrared absorption, while dye lasers are used for visible and ultraviolet studies. Difference frequency mixing of a fixed-frequency laser with a tunable laser is another source of tunable infrared radiation. Some atomic species of interest, such as halogen atoms, O, and N, absorb in the vacuum ultraviolet (110–160 nm), which requires vacuum enclosures of the optical paths.

Absorption measurements integrate density along a line, so 3D resolution is lost. Still, the probe beam can be translated relative to the substrate for valuable 1D mapping. Tomographic reconstruction can be used to retrieve 3D mapping from various absorption trajectories, and is discussed in Section 5.2.1.3. Although absorption tomography is widely used in other areas, it has rarely been used in thin film processing.

One significant problem in using absorption as a probe of thin film processes is transmission loss at the reactor windows. This is particularly serious when the absorbed fraction is small, as is common in probing gas-phase species, and in chemical vapor deposition (CVD) applications, when thin films often deposit on the windows as well as on the substrate. This problem can be mitigated by preventing reactive species from reaching the window by flowing a nonreactive gas, such as argon, across it.

Absorption can be monitored directly for large absorbed fractions without any need to reduce noise. For small absorbed fractions, the signal-to-noise ratio (S/N) becomes poor and must be improved. In standard dispersive and transform spectrometers that use low-brightness, incoherent lamp sources, S/N approaches 1 for absorbed fractions near 10^{-4}. This noise is due to fluctuations in the source, Johnson noise, and shot noise; it becomes less important with increased integration time (Haller and Hobbs, 1991). Haller and Hobbs (1991) have detailed how S/N can be made shot-noise-limited by using coherent, high-brightness lasers.

One way to improve S/N is to modulate the amplitude of the optical source at a frequency f, e.g., with a mechanical chopper, and to detect the transmitted beam by phase-sensitive detection at this same frequency with a lock-in amplifier. This phase-sensitive detection reduces the noise bandwidth and removes background light from the signal, which is particularly important in suppressing the optical background during plasma processing; this method is widely used. While this improves S/N greatly, such synchronous detection rarely attains the shot-noise limit because the excess noise spectrum of lasers usually extends beyond the 1–10-kHz frequency attainable by chopping and because the modulation shifts low-frequency noise into the passband (Haller and Hobbs, 1991). When probing species in a discharge, the discharge voltage can be modulated at f (instead of the light source). In this case, noise can be reduced even further if the laser is chopped at f_1, the discharge at f_2, and the signal is detected at $f_1 \pm f_2$.

In double-beam differential absorption, which is often used in absorption spectrometers, a source is divided into two beams—one that is transmitted through the sample and one that serves as a reference beam. The two beams are detected individually and the resulting photocurrents are either subtracted or divided. While both procedures improve S/N, neither technique usually attains shot-noise-limited performance. Haller and Hobbs (1991) have designed and operated an inexpensive electronic circuit noise canceller that analyzes these two signals to achieve shot-noise-limited performance (also see Hobbs, 1990, 1991). This noise cancellation operates on the principle that, for linear detectors, when the DC photocurrents of the two arms cancel, the noise at all frequencies will then cancel. This technique permits the detection of 10^{-6} absorbed fractions even with noisy diode lasers.

Very small absorptions in the infrared are often detected by using frequency-modulation spectroscopy; this is described in the next section.

8.1.2.1 Frequency-Modulation Spectroscopy

The sensitivity in measuring small absorbed fractions $\Delta I/I$ can be improved dramatically by frequency modulation of the laser frequency and homodyne detection of the signal photocurrent at the modulation frequency ω_m or a harmonic. Laser source noise becomes very small with detection at these high modulation frequencies. Also, frequency modulation (FM) removes the baseline slopes seen in direct absorption. Frequency modulation is achieved in infrared diode laser absorption spectroscopy by adding a frequency-modulation component to the DC injection current, which is itself varied in time for wavelength scanning. This FM technique is widely used with diode lasers and can also be applied to other laser systems, such as dye lasers, by external electro-optic modulation of the laser frequency. When ω_m is less than the frequency half-width of the absorption line $\Gamma_{1/2}$, this method is commonly called wavelength-modulation spectroscopy (WMS), harmonic detection, or derivative spectroscopy. When $\omega_m > \Gamma_{1/2}$, it is called frequency-modulation spectroscopy (FMS). The relative sensitivity of the various FM techniques has been addressed by Silver (1992) and Bomse *et al.* (1992).

In wavelength-modulation spectroscopy the instantaneous frequency is

$$\omega(t) = \omega_0 + [m\Gamma_{1/2}]\sin(\omega_m t) \qquad (8.4)$$

where m is the modulation index. The signal is $\propto \alpha(\omega(t))$ and can be decomposed into a sum of terms at the harmonics of ω_m. For low m, the signal profile of the nth harmonic is the nth derivative of the α lineshape. The modulation index m can be chosen to optimize the amplitude of this signal. Second harmonic ($n = 2$) detection is more common than first harmonic analysis because the signal peaks at the absorption linecenter frequency and the baseline slope is removed with $n = 2$, and not with $n = 1$ (Silver, 1992). With FM in the kHz range, WMS routinely achieves sensitivities (absorbed fractions) of 10^{-4}–10^{-5} and can achieve 5×10^{-6}; and when ω_m is increased to approach $\Gamma_{1/2}$, the sensitivity can be improved to the level of FMS, ~ low to mid-10^{-7} range with a 1-Hz detection bandwidth (Bomse et al., 1992). It is usually easier to extract lineshapes from WMS than from FMS. [Note that while researchers in this area often work with the radial frequency ω (rad/sec), they also often cite frequencies in Hz, corresponding to $\omega/2\pi$.]

The modulation frequencies in FMS are so high, typically greater than several hundred megahertz, that the laser noise term is very small and the detection sensitivity can be limited only by detector quantum noise (Silver, 1992). In one-tone FMS, there is modulation at ω_m; and when the laser is scanned across the absorption profile, there are positive and negative peaks in the homodyne signal. In two-tone FMS, there is modulation at two closely spaced frequencies ω_{m1} and ω_{m2}, and the detected frequency is $|\omega_{m1} - \omega_{m2}| \ll \omega_{m1}, \omega_{m2}$. Since this corresponds to frequencies of ~1–10 MHz, less expensive detectors are needed than the >150-MHz-speed detectors needed for one-tone FMS. Although theoretical sensitivities of ~10^{-8} are possible with one-tone FMS, in practice wide-bandwidth WMS can achieve comparable ~10^{-6}–10^{-7} sensitivities in some cases (Bomse et al., 1992).

One way to confirm densities and lineshapes measured with frequency modulation is to compare such measurements with those made on the same transition by amplitude modulating (chopping) the beam. This calibration has to be performed on relatively strong transitions to achieve good signal/noise with amplitude modulation, and then can be applied to weaker transitions.

8.2 Gas-Phase Absorption

This section details the use of absorption spectroscopy in diagnostics of molecules, radicals, and atoms. Although both infrared and visible/ultraviolet absorption are important, the former often has strong advantages.

Infrared radiation usually probes molecular vibrations, while visible/ultraviolet light probes transitions between different electronic states. Analysis of infrared transitions more easily identifies molecules and radicals because infrared absorption spectra tend to have sharp structure, even for "medium"-sized species (say, composed of 5 atoms) [Figures 8.13, 8.15, and 8.16], due to the relatively low density of spectral lines and their narrow linewidths, while the features in the electronic absorption spectra begin to overlap for molecules with ≥3 atoms (Figure 8.23) and therefore become

8.2 Gas-Phase Absorption

harder to distinguish. There are several reasons for this: (1) Spectral linewidths tend to be larger at shorter wavelengths because of the increased importance of Doppler broadening ($\sim \nu$) and radiative decay ($\sim \nu^3$). (2) There are relatively strict vibrational selection rules for pure vibrational (–rotational) transitions, while the Franck–Condon principle (Section 4.3.1.1.2) leads to many potential vibrational transitions and consequently much higher line densities in electronic spectra. (3) Electronic transitions can be to predissociative or dissociative states, leading to a mild or extensive broadening of the absorption lineshape. Moreover, excitation to a level that dissociates strongly perturbs the process.

Consequently, IR absorption is generally a better diagnostic of molecules and radicals than is absorption and LIF in the visible/ultraviolet; this is particularly true in larger species. Even when IR and UV absorption profiles are both broad and unresolved, it is usually easier to identify mixtures of gases from the distinctive peaks in their IR spectrum. Still, diatomic species are readily probed in the visible/ultraviolet with tunable dye lasers, while atoms are often probed with dye lasers and with radiation from atomic resonance lamps.

Since excitation LIF probes the absorption spectrum and dispersive LIF probes a spectrum closely related to it, the relative merits of absorption and LIF spectroscopies for diagnostics must be addressed. Because radiative rates are slow in the infrared (Equation 4.17) and IR detectors are relatively noisy, it is better to probe infrared transitions by absorption than by LIF for diagnostics, although the latter has been demonstrated and used in many spectroscopic and kinetics studies. Laser-induced fluorescence is often preferred in probing electronic transitions because it is very sensitive due to efficient detection and the rapid rate of radiative decay at shorter wavelengths. The background levels are essentially zero in LIF measurements, while absorption measures the difference between transmission with and without the sample, which is often very small. Furthermore, better spatial resolution is attainable with LIF than with absorption. One advantage of absorption is that it is easier calibrate to obtain absolute densities. Sometimes visible/electronic LIF signals are calibrated with data from a single absorption calibration run. Virtually all molecules can be detected by infrared absorption, exceptions being homonuclear diatomic molecules like N_2. Only smaller molecules tend to be observable by LIF or OES.

8.2.1 *Infrared (Vibrational) Absorption*

High-resolution infrared (IR) absorption spectroscopy is now a well-developed probe of vibrational–rotational transitions of molecules, radicals, and ions during thin film processing. Even at low density, species can be identified noninvasively and nonperturbatively, and their concentrations can be determined. Even minor components of complex mixtures can be identified because the rovibronic spectra have sharp, patterned features that are unique to the species. It is also used extensively for more fundamental studies of gas-phase reaction dynamics and collisional energy transfer. Infrared spectroscopy is also sensitive to several atomic transitions of interest, including atomic Cl and F.

Although dispersive (grating and prism) infrared monochromators are very useful for many infrared transmission experiments, they have neither the resolution and sensitivity nor the speed needed for many diagnostics applications. They have been supplanted by infrared diode laser absorption spectroscopy (IR-DLAS) and FTIR spectroscopy (FTIRS). Figure 8.2 shows the experimental setup for IR-DLAS, in which a diode laser is transmitted through the gas, often in a multipass arrangement. The intensity of the transmitted light is measured as the wavelength of the diode laser is tuned and carefully tracked. In FTIRS (Section 5.1.3.2) light from a broadband infrared source, such as a glowbar, is transmitted through the

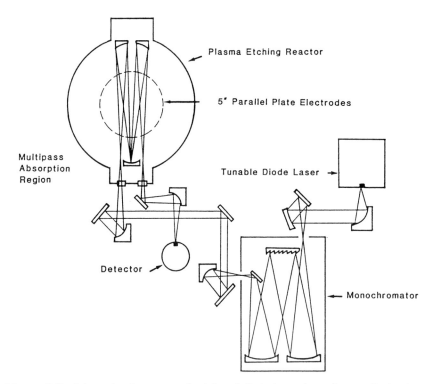

Figure 8.2 Schematic of apparatus for infrared diode laser absorption monitoring in a plasma etching reactor with multipass absorption arrangement. Parts of the laser are sometimes split off from the main beam and are separately transmitted through a reference gas cell and an etalon for wavelength calibration; this is not shown here (from Wormhoudt et al., 1987).

8.2 Gas-Phase Absorption

sample and then analyzed by the monochromator (Figure 8.1). Sensing from remote reactors and simultaneous probing of many reactors is possible with either technique by using IR-transmitting optical fibers. Chalcogenide fibers are transparent in most of the region from 900 to 3300 cm^{-1}, but with limited transmission near 2200 cm^{-1}, while fluoride optical fibers transmit from 2200 to 20,000 cm^{-1} (Salim et al., 1994, 1995).

The high spectral resolution capability of both methods improves absorption measurements in two ways: (1) A spectral line with linewidth $\Gamma^{(\nu)}$ cannot be analyzed well unless the measurement resolution $\delta\nu < \Gamma^{(\nu)}$; this condition can be achieved by using these spectroscopies, particularly IR-DLAS. (2) The detection of small absorbed fractions improves, as is seen for case (i) described earlier.

Both IR-DLAS and FTIRS are powerful methods, and each is very useful for probing the gas above the surface in plasma processing and CVD. Each has distinctive strengths: (1) Longer path lengths (l) are possible with IR-DLAS, because of the coherence of diode lasers and the possibility of multipassing in a White-cell arrangement, so it is sensitive to smaller concentrations than FTIRS. (2) The ultimate spectral resolution is at least ~10× better in IR-DLAS (typically 0.0003–0.0005 cm^{-1}) than in FTIRS (typically 0.05 cm^{-1}, at best ~0.005 cm^{-1}), which is important in resolving Doppler profiles and measuring small concentrations. Infrared diode laser absorption also has better temporal resolution (Butler *et al.*, 1986). (3) Each diode laser can cover only a small spectral region, say 200 cm^{-1}, and therefore may be useful for only one species. FTIR spectroscopy is more versatile because it covers the vibrational absorption features of all species. (4) Both spectroscopies integrate density along a line, but the lateral dimensions of the diode laser (~1 mm) gives better lateral resolution. This advantage is lost if the laser is multipassed in the reactor.

In either case, IR spectroscopy can determine several parameters of interest:

(1) The species (atom, molecule, radical, ion) can be identified by its characteristic absorption pattern.

(2) The absorption strength gives the average density along the laser path, by using Equation 4.16 if the population distribution is thermal. The decision to use either the peak absorption strength or the integrated line intensity depends on the frequency width of the source. In IR-DLAS, the laser linewidth is typically smaller than the spectral width, while in FTIR spectroscopy it can be broader.

(3) Measuring the absorbed fraction across a single resonance gives the absorption lineshape. At low pressures, the profile is a gaussian which gives the translational temperature T_{trans} from the Doppler width $\propto \sqrt{T_{\text{trans}}}$ (after Equation 4.22), as has been demonstrated by Flynn and co-

workers (Hershberger *et al.*, 1988; Park *et al.*, 1991). (Sub-Doppler spectroscopy is also possible.) At higher pressures, pressure broadening determines the profile (lorentzian). Lineshape analysis is easier with diode lasers because of their narrow linewidths.

(4) Tuning across absorptions from several different rotational states determines the rotational temperature T_{rot} for a given vibrational state from the known Boltzmann distribution (and dependence of the matrix elements on rotational quantum number) (Equation 4.36). T_{rot} approximately equals T_{trans} in the gas phase in many thin film systems (Section 18.1).

(5) Tuning across lines from different vibrational levels determines the vibrational temperature T_{vib}.

While dispersive, FTIR, and diode laser IR diagnostics are useful for process development, they can be too expensive, slow, and unreliable for real-time monitoring during manufacturing. In applications where the IR absorption feature of the species to be monitored is well-resolved from interfering absorptions, these systems can be replaced by a glowbar source, which is spectrally narrowed by (relatively cheap) IR transmission filters and chopped (for lock-in detection) [see below].

8.2.1.1 IR Diode Laser Absorption Spectroscopy

Infrared diode laser absorption spectroscopy (IR-DLAS) allows virtually unambiguous identification of species, even in a complex gas mixture. The techniques and applications in tunable diode laser spectroscopy have been reviewed by Eng *et al.* (1980), Kuritsyn (1986), and Mantz (1994, 1995). Agrawal and Dutta (1986) have described these tunable, long-wavelength semiconductor diode lasers in detail. IR-DLAS has been used for many applications, including high-resolution molecular spectroscopy, chemical dynamics, detection of trace impurities (including pollutants), and combustion diagnostics, as well as diagnostics of the gas phase in thin film processing (Hirota, 1985; Hirota, and Kawaguchi, 1985). Using cw diode lasers, temporal resolutions ranging from 10^{-7} to 1 sec are attainable. Even faster responses are possible with frequency upconversion (Moore *et al.*, 1987).

IR-DLAS typically uses lead salt diode lasers. Such lasers made from the ternaries $PbS_{1-x}Se_x$, $Pb_{1-x}Sn_xTe$, $Pb_{1-x}Sn_xSe$, the quaternaries $Pb_{1-x}Sn_x$-Se_yTe_{1-y}, and related compounds lase from ~ 2.5 to 46.2 μm. The first two ternaries themselves provide a tuning range of 3 to ~ 34 μm. $Pb_{1-x}Cd_xS$ extends the tuning range to shorter wavelengths. The band gaps of these ternaries at 77 K are shown in Figure 8.3 as a function of composition. Once a composition is chosen so the band gap is near the range of interest, these lasers are usually scanned by changing temperature to tune the band gap. Coarse tuning is performed by scanning the temperature of the laser mount, as is seen in Figure 8.4. The modes are finely tuned by varying the

8.2 Gas-Phase Absorption

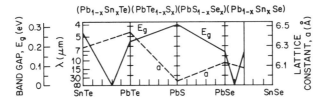

Figure 8.3 Band gap (solid line) and lattice constant (dashed line) of indicated ternary semiconductors that are often used in lead-salt semiconductor infrared diode lasers, at 77 K as a function of composition x (from Agrawal and Dutta, 1986).

current through the diode, which actually fine-tunes the diode temperature, as is illustrated in Figure 8.5. For a given composition, the overall (coarse) tuning range can be ~100 to ~250 cm^{-1}, while the fine-tuning range of a given mode is ~0.2–1.5 cm^{-1} between mode hops (Figure 8.5). For example,

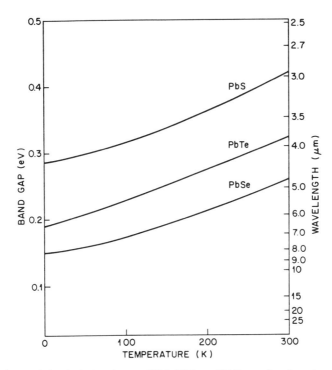

Figure 8.4 Variation in the band gaps of PbS, PbTe, and PbSe as a function of temperature. This shows the capability of tuning the wavelength of infrared diode lasers by changing temperature (from Agrawal and Dutta, 1986).

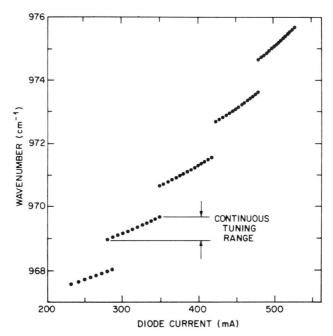

Figure 8.5 Wavelength tuning of the modes of $Pb_{1-x}Sn_xTe$ diode lasers as a function of injection current. This shows regions of continuous wavelength tuning and discrete wavelength jumps, corresponding to longitudinal-mode hopping (from Agrawal and Dutta, 1986).

the diode laser used by Sun *et al.* (1994) has a tuning range of 843–1060 cm^{-1} when the temperature is scanned from 20 to 107 K. A single longitudinal mode is selected by passing the diode output through a monochromator. The linewidth of each mode is typically ~10 MHz (0.00033 cm^{-1}) (or smaller), which is usually narrower than the absorption linewidth. The laser can be frequency-modulated to ~100 kHz by modulating the bias current (WMS). The total output over all modes is typically a few milliwatts for 3–8 μm and a few hundred microwatts for longer wavelengths.

Two portions of the beam are split from the laser before it passes through the reaction chamber and are sent through an etalon (or Fabry–Perot interferometer) and a reference gas cell. (This is not shown in Figure 8.2.) The reference cell is filled with gases with absorption transitions near those of the gas to be probed. Several wavelength standards are used, including bands in N_2O, CO_2, HCN, OCS, NO, H_2O, CO, NH_3, SO_2, and H_2CO (Eng *et al.*, 1980). Sometimes, a discharge is operated in this cell so excited-state absorption can be used for wavelength calibration. Precise interpolation between these absorption features is made by using the interference fringes in the transmission through the etalon.

8.2 Gas-Phase Absorption

The beams transmitted through the reaction chamber, reference cell, and the etalon are detected by infrared photodetectors, such as HgCdTe photoconductive detectors, and processed with lock-in amplifiers (when the laser is chopped). The laser can propagate through the chamber once or, alternatively, several times (typically ~ 25 times) in a White-cell-type multipass arrangement (Figure 8.2) to increase the absorption length, with a concomitant loss in spatial resolution. One common arrangement is that by Herriott et al. (1964) where two spherical mirrors are separated by less than twice their (equal) focal lengths. The beam enters through a hole in one mirror, the reflected beams trace an elliptical or circular pattern on both mirrors, and the laser exits the cell through the entrance hole, though at a slightly different angle. By adjusting the distance between the two mirrors, the number of passes can be set to 6, 10, 14, ... The monochromator is often placed after the sample to eliminate angular motion caused by the grating. Either reactants or products can be probed by appropriately choosing and tuning the laser diode.

In IR-DLAS, the laser frequency is slowly tuned across the resonances of interest, and the transmitted signal is monitored. In direct absorption spectroscopy, the laser power is chopped and the signal is detected with a lock-in amplifier. The signal-to-noise ratio can be improved further and the background more efficiently removed by using frequency-modulated absorption spectroscopy (Section 8.1.2.1), which has become the standard IR-DLAS technique. Sensitivities of absorbed fractions (αl) of 10^{-7} and lower have been reported by using modulation methods with low-noise lasers, but 10^{-6} is perhaps a more realistic goal in a diagnostics environment.

The absorbed fraction in the reaction chamber is $\alpha'pl$, where α' is the absorption coefficient (cm^{-1} Torr^{-1}), p is the partial pressure of the absorbing gas (Torr), and l is the path length (cm). This comes from Beer's law (Equation 8.1), assuming an optically thin sample. Assuming a conservative estimate of $\Delta I/I = \alpha'pl = 10^{-5}$, the minimum detectable pressures for $l = 10$ cm in a single pass are: 4.5×10^{-8} Torr CO_2, 5.0×10^{-7} Torr CO, 4.1×10^{-7} Torr H^{35}Cl, 7.6×10^{-7} Torr H_2O, 1.6×10^{-7} Torr N_2O, $< 1.0 \times 10^{-4}$ Torr CH_3, and 3×10^{-5} Torr Cl (G. W. Flynn, private communication, 1992). Wormhoudt et al. (1983) have also determined this detection limit for many of the atoms, molecules, and radicals that are important in thin film processing.

8.2.1.1.1 Applications

Butler and co-workers used diode laser absorption as a diagnostic in a series of studies of the organometallic vapor-phase epitaxy (OMVPE) of III–V semiconductors. In particular, Butler et al. (1986) used a White-cell-type multipass arrangement with 20 to 26 passes to probe the decomposition of AsH_3 and of trimethylgallium (TMGa), trimethylaluminum (TMAl), and trimethylindium (TMIn) in an OMVPE reactor. In addition to probing these reactants they monitored the intermediate product CH_3, along with the stable products CH_4 and C_2H_6. At the wavelengths probed, the lighter species AsH_3 (895 cm^{-1}), CH_3 (607.03 cm^{-1}), CH_4 (1295 cm^{-1}), and C_2H_6 (830 cm^{-1}) had well resolved vibrational–rotational features with linewidths ~ 0.003 cm^{-1}, and thus improved detectivity vis-à-vis FTIRS, while the heavier species TMAl (695 cm^{-1}), TMGa (588 cm^{-1}), and TMIn (730 cm^{-1}) had only broad featureless absorption profiles in the wavelength range examined. Since the rotational lines are much more closely spaced in the heavier species, the relative advantages of the diode laser technique are lost. Figure 8.6 compares the CH_3 Q(6,6) absorption feature, with the absorption in the N_2O calibration gas and the transmission through the etalon.

This study showed that CH_3 radicals are formed in the OMVPE reactor and react with AsH_3. Gaskill et al. (1988) studied this reaction in more detail by using time-resolved diode laser absorption. CH_3 radicals were produced by KrF excimer laser (248 nm, 30-nsec pulse length) photolysis of CH_3I, and probed at 606.12032 cm^{-1}. The reaction of CH_3 with AsH_3 is seen by the relatively fast consumption of CH_3 in Figure 8.7 when AsH_3 is added. Accurate determination of the rate constants can be obtained with this method.

Sharp interfacial abruptness during growth by OMVPE can be achieved only with rapid switching of reactant gases. Sillmon et al. (1986) were able to devise and optimize such a switching system by examining the time-resolved absorption of either a reactant, such as TMGa at 588 cm^{-1} or TMAl at 695 cm^{-1}, or a rapidly produced product, such as CH_3 at 607.02405 cm^{-1}, that is produced by the decomposition of TMIn over a hot GaAs substrate (Figure 8.8).

Celii et al. (1988) probed CH_3 (606 cm^{-1}), C_2H_2 (716 cm^{-1}), and C_2H_4 (950 cm^{-1}) during filament-assisted deposition of diamond films from 0.5% methane in hydrogen (Figure 8.9). With the sensitivity of this setup, neither the CH_2 radical nor any of the stable hydrocarbons C_2H_6 (ethane), C_3H_4 (allene and propyne), C_3H_6 (propene and cyclopropane), and C_3H_8 (propane) were seen in the 600–1000 cm^{-1} region.

Itabashi et al. (1988) measured the SiH_3 radical density in a pulsed SiH_4/H_2 discharge, which can be used to deposit a-Si films, by passing the resonant diode laser 40 times through the discharge. Loh and Jasinski (1991) used tunable infrared diode laser spectroscopy to probe the kinetics of reactions of potential importance in Si CVD, including the reactions SiH_3 + SiH_3, H, SiD_4, and Si_2H_6. The silyl radical (SiH_3) was monitored near 726.9 cm^{-1} in the ν_2 band. They found that it is a long-lived species under typical CVD conditions and is therefore potentially important during plasma and photochemical deposition of silicon. SiH radicals have been detected by Davies et al. (1985) by using IR-DLAS.

8.2 Gas-Phase Absorption

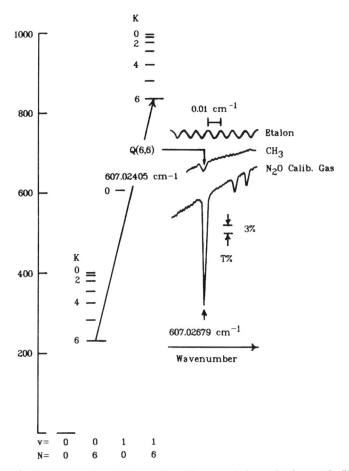

Figure 8.6 Monitoring the $Q(6,6)$ rotational feature of the ν_2 fundamental vibrational mode of the CH_3 radical in an OMVPE reactor, which is formed by the thermal decomposition of TMGa in the reactor. The CH_3 transmission spectrum is shown along with calibration transmission spectra through an etalon and a reference N_2O cell. The rotational energy level structure of CH_3 is also shown (in cm^{-1}) (from Butler et al., 1986).

Accurate measurements of concentrations require accurate values of absorption cross-sections. Using such data, Davies and Martineau (1990) made absolute density measurements of CH_4 (1300 cm^{-1}), C_2H_4 (~950 cm^{-1}), and CH_3 (608.301 cm^{-1}) in a CH_4 deposition reactor as a function of reactor conditions, as is seen in Figure 8.10. Wormhoudt (1990) performed similar measurements of CF_2 and C_2F_6 in CF_4/Ar RF plasmas and of CH_3 in a CH_4/Ar plasma. Woods et al. (1991) made absolute measurements of CF_2 and relative measurements of CF_3 in electron cyclotron reso-

282   Chapter 8. Transmission (Absorption)

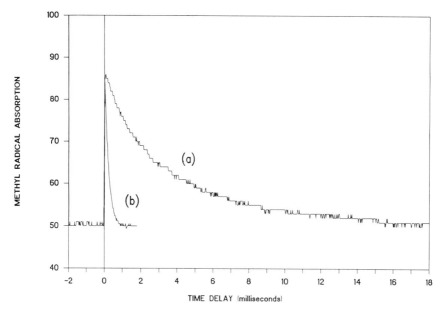

Figure 8.7 Monitoring CH$_3$ in a flow cell by IR-DLAS for (a) CH$_3$I and Ar and (b) with AsH$_3$ added (from Gaskill *et al.*, 1988).

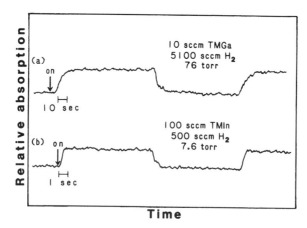

Figure 8.8 The on/off switching characteristics of an OMVPE reactor as measured by IR-DLAS for (a) a conventional and (b) an improved injection gas delivery system (from Sillmon *et al.*, 1986).

8.2 Gas-Phase Absorption

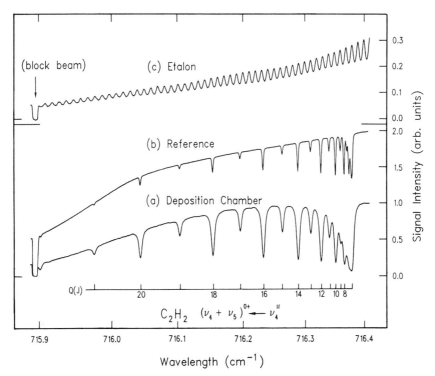

Figure 8.9 IR-DLAS scan of (a) C_2H_2 in a reactor for filament-assisted growth of diamond films (20 passes, effective length ~0.5 m) compared to the transmission through (b) C_2H_2 in a reference cell and (c) an etalon (from Celii et al., 1988).

nance (ECR) plasmas with $CF_4/CHF_3/O_2$ mixtures. Haverlag et al. (1994) used IR-DLAS with first-derivative detection to probe CF_2, and found that the partial pressure of CF_2 is around 1–5% of the total pressure in parallel-plate RF discharges operating with either CF_4, CHF_3, C_2F_6, or CF_2Cl_2. Davies et al. (1981) conducted early measurements of ν_1 absorption in CF_2 by using diode lasers.

Oh et al. (1995) used IR-DLAS to probe CF_2 intermediates during the CF_3H/CF_4 plasma etching of silicon and silicon dioxide in a gaseous electronics conference (GEC) reference cell. They also measured the etch product CF_2O, which is a potential endpoint detector of the etching of SiO_2 over poly-Si.

Diode laser absorption can also be used to be probe atoms *in situ*. Wormhoudt et al. (1987) probed Cl atoms in a Cl_2 glow discharge by the magnetic-dipole-allowed transition $^2P_{1/2} \leftarrow {}^2P_{3/2}$ at 882.36 cm^{-1} (11.33 μm) between spin–orbit split levels of the ground electronic level. Although

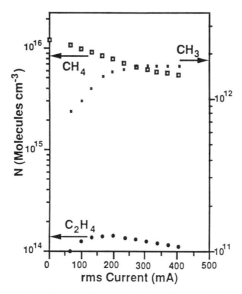

Figure 8.10 Current dependence of the number densities of CH_4, C_2H_4, and CH_3 in a CH_4 plasma (350 mTorr methane) measured by IR-DLAS (from Davies and Martineau, 1990).

the transition matrix element of this magnetic-dipole-allowed transition is smaller than that of electric-dipole molecular vibrational transitions, this transition is detectable because all the population participates in this transition, while it is spread over many rotational states in vibrational mode measurements in molecules. They used a White-cell arrangement (20 passes, ~ 250 cm). Measurements were made by direct absorption using a mechanically chopped laser and lock-in detection (Figure 8.11a), and the signal-to-noise ratio was further improved by modulating the laser frequency with triangular wave current modulation, with the lock-in amplifier set for twice the frequency of the modulation. Figures 8.11b and 8.11c respectively show this signal before and after subtraction of the superimposed etalon signal.

Wormhoudt *et al.* (1987) measured atomic-chlorine concentrations of $1.8-6.6 \times 10^{14}/cm^3$ in this discharge, representing atomic-chlorine fractions of 3-8%. Measurement of the Doppler width gave a translational temperature of 770 ± 100 K. This was in agreement with T_{rot} determined from OES of the nitrogen second-positive band in the discharge seeded with N_2. Related work by Richards *et al.* (1987) and Richards and Sawin (1987) measured the Cl-atom density for a wide range of Cl_2 and CF_3Cl plasma conditions, and showed IR-DLAS to be a more accurate probe of Cl-atom concentration than optical emission actinometry. Stanton and Kolb

8.2 Gas-Phase Absorption

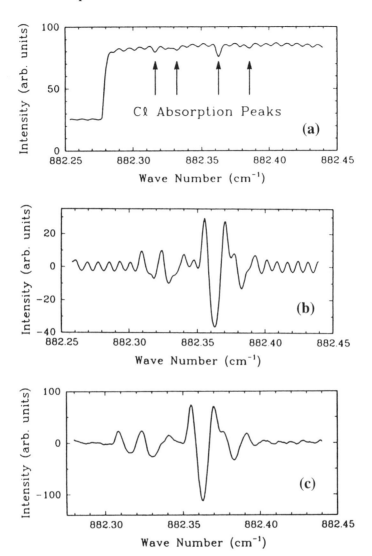

Figure 8.11 IR-DLAS spectra of Cl atoms in an RF Cl_2 glow discharge for (a) a direct absorption scan, (b) a corresponding second-derivative scan, and (c) the scan in (b) after subtraction of the superimposed etalon background scan (from Wormhoudt et al., 1987). [Note the slightly different abscissa scale in (c).]

(1980) and Loge et al. (1994) have made similar measurements of $^2P_{1/2} \leftarrow {}^2P_{3/2}$ at 404 cm^{-1} (25 μm) in atomic F.

Sun, Whittaker, and co-workers have combined the two modulation schemes for IR-DLAS, and developed a method they called combined wavelength and frequency modulation spectroscopy (CWFMS) (Sun and

Whittaker, 1992; Sun et al., 1993). By using this modulation method, they found the limit for the detectable absorbed fraction was 5.3×10^{-8} with an AlGaAs diode laser and 1.9×10^{-7} with a lead-salt diode laser, and applied this diagnostic to several thin film processing systems. Sun and Whittaker (1993) detected SF_6 concentrations as low as 1.3 μTorr (normalized to a 1-m path, 1-Hz bandwidth) in real time in a plasma etching chamber. Sun et al. (1994) showed that the endpoint of etching silicon films on silicon dioxide and etching silicon dioxide films on silicon can be detected in a SF_6 etching chamber by tuning and then locking the diode laser to detect the Si etching product SiF_4 (1023 cm^{-1}). As seen in Figure 8.12, this probe

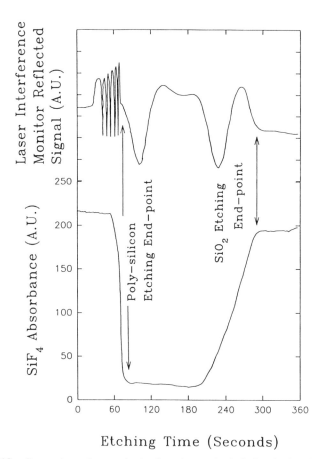

Figure 8.12 Comparison of two endpoint detection methods during the SF_6 plasma etching of a 5000-Å polysilicon/5000-Å SiO_2/Si substrate system: He–Ne laser interferometry (632.8 nm) and tunable infrared diode laser absorption spectroscopy of SiF_4 (from Sun et al., 1994).

senses the large increase in SiF$_4$ product during the Si etching cycle. Exposed areas as small as 33 mm^2 can be detected with this probe. Sun *et al.* (1993) made *in situ* measurements of CF$_3$ and CF$_2$ intermediates in several configurations of low-pressure C$_2$F$_6$ discharge reactors that are used for plasma etching of silicon dioxide. By using CWFM diode laser absorption, they also probed changes in reactor absorption during CVD of methylsilazane.

Although IR-DLAS can also be used to probe ionic species in plasmas, there has been little study of the ions that are present in thin-film-processing plasma reactors by this method. Multipass diode laser absorption has, however, been used in plasmas. Foster and McKellar (1984) studied the IR absorption spectrum near 5 μm of HN$_2^+$, DN$_2^+$, and DCO$^+$ in a discharge. To improve signal/noise, the discharge was pulsed on and off at 9 kHz and the transmitted beam was analyzed with a lock-in amplifier to sense changes at this frequency. Davies and Rothwell (1985) analyzed the ν_2 mode in HCS$^+$ in a CO/H$_2$S/He discharge.

Other sources of narrow-band, tunable infrared radiation can be used in absorption studies. Difference-frequency mixing between an argon-ion laser and a tunable dye laser in a LiNbO$_3$ crystal can generate ~ 10 μW from 2250 to 4400 cm^{-1}. While this is a broader range than possible with a single diode laser, difference-frequency mixing is more expensive and complicated, and is less easily modulated for improved detection sensitivity.

Using a very different type of IR modulation spectroscopy, Gudeman *et al.* (1983) utilized a color center laser to perform velocity-modulated infrared laser spectroscopy on molecular ions (HCO$^+$) in a discharge. The polarity of the glow discharge was reversed at a frequency of several kilohertz to alternate Doppler red- and blue-frequency shifts in the absorption frequency. Lock-in detection permitted the analysis of the infrared absorption spectrum of the ions with little background from the much more abundant neutral species.

8.2.1.2 FTIR and Dispersive Infrared Spectroscopy

Most *in situ* studies of vibrational absorption are now conducted by using FTIR spectroscopy because of its greatly improved performance, including better resolution, absorption detection sensitivity, and speed, *vis-à-vis* dispersive infrared spectroscopy, which usually employs prisms for dispersion. Still, these more conventional dispersive spectrometers can also be used when αl is large.

Several earlier infrared studies used dispersive monochromators to probe processes in real time. Nishizawa and coworkers used *in situ* IR absorption as a monitor in a series of investigations of plasma etching and the deposition of Si and GaAs. Nishizawa and Hayasaka (1982) monitored the IR absorption peaks of stable molecules (reactants and products, including SiF$_4$) as a function of time during the reactive ion etching (RIE) of Si by CF$_4$, C$_2$F$_6$

or C_3F_8. In real-time IR absorption during epitaxial growth of silicon by $SiCl_4/H_2$ CVD, Nishizawa and Nihira (1978), Nishizawa and Saito (1981), and Nishizawa (1982) observed $SiHCl_3$, SiH_2Cl_2, HCl, $SiCl_2$, and $SiCl_3$ products. The latter two are intermediates, noted by the arrows at the emission peaks at 500 and 570 cm^{-1}, respectively, in Figure 8.13. They were not seen in IR measurements on gas sampled through a quartz capillary, which were conducted to probe the reaction spatially. In later studies, Nishizawa et al. (1990) used infrared absorption to probe the gas phase during silicon molecular layer epitaxy by SiH_2Cl_2 and H_2.

Nishizawa and Kurabayashi (1983) used the sampling method shown in Figure 8.14 to study of GaAs MOCVD. Figure 8.15 shows the infrared absorption peak at 2080 cm^{-1} seen in heated $TMGa/AsH_3/H_2$ mixtures that is not present in either $TMGa/H_2$ or AsH_3/H_2, and which is apparently due to a product related to deposition. Nishizawa et al. (1986, 1987a,b, 1988) studied a variety of thermal and photoenhanced GaAs epitaxy processes by using mixtures such as $GaCl_3/AsH_3/H_2/N_2$. In one of these, a study of photolytic deposition of GaAs, infrared absorption showed that KrF and ArF lasers photolyze gas-phase TMGa, while only ArF laser irradiation decomposes AsH_3 (Nishizawa et al., 1987a).

In early work, Poll et al. (1982) studied the use of infrared spectroscopy to evaluate the partial pressures of potential reactants and products in plasma

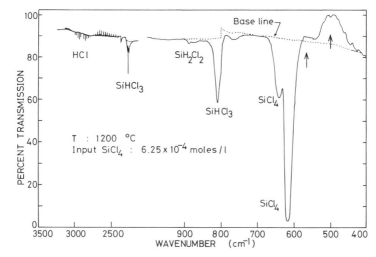

Figure 8.13 *In situ* IR absorption spectrum of flowing reactants during the epitaxial growth of Si from $SiCl_4$ and hydrogen. The peaks at 500 and 570 cm^{-1} (indicated by the arrows) are due to the intermediate products $SiCl_2$ and $SiCl_3$, respectively; these peaks are above the baseline, showing that these two products are emitting in the infrared (from Nishizawa and Nihira, 1978).

8.2 Gas-Phase Absorption

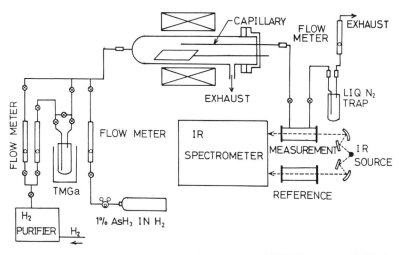

Figure 8.14 Schematic diagram of atmospheric-pressure MOCVD system with IR absorption analysis of gases sampled through a quartz capillary (from Nishizawa and Kurabayashi, 1983).

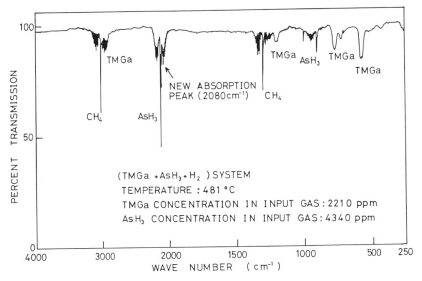

Figure 8.15 The infrared absorption spectrum of reactants sampled (as in Figure 8.14) from a GaAs MOCVD reactor operating with TMGa/AsH$_3$/H$_2$ mixtures. The absorption peak at 2080 cm^{-1} is due to a product that is not formed with either TMGa/H$_2$ or AsH$_3$/H$_2$ mixtures (from Nishizawa and Kurabayashi, 1983).

etching with fluorinated gases, such as CF_4, C_2F_4, C_2F_6, n-C_3F_6, C_3F_8, SiF_4, and COF_2, and the conversions in $CF_4 + O_2$ and C_2F_4 glow discharges. Two probing geometries were considered: (1) probing the reaction zone and the pumping line in flow systems and (2) probing in closed systems.

The technique of FTIR spectroscopy (FTIRS) [Section 5.1.3.2] is versatile because a wide wavelength range is probed and leads to excellent signal-to-noise because the entire spectrum is monitored simultaneously by the probe beam, which is not apertured (Back, 1991). FTIRS is usually performed by diverting the IR beam with external mirrors into the reactor and then back into the spectrometer. Xi et al. (1994) used IR transmitting fibers to couple the FTIR spectrometer to an MOCVD reactor. The use of FTIRS to probe adsorbates on surfaces is covered in Section 9.7.

Cleland and Hess (1988) have shown how to correct for the distortion in FTIR spectra due to finite resolution. They found this to be important in quantitative analysis of rotational lines in a rovibronic band to obtain T_{rot}, and used it to determine T_{rot} for a N_2O discharge. N_2O is used as a reactant in the plasma-assisted CVD of SiO_x, SiO_xN_y, and phosphosilicate glass.

FTIRS has been used as a real-time diagnostic in other plasma systems as well. Cleland and Hess (1989) monitored gas-phase species during Cl_2 plasma etching. Al_2Cl_6, $AlCl_3$, and perhaps $AlCl$ were observed during etching of Al, while $SiCl_4$ was the only infrared-absorbing product seen during the etching of heavily n-type doped polycrystalline Si. O'Neill et al. (1990) monitored absorption during plasma etching of Si by either CF_4, CF_3Cl, CF_2Cl_2, $CFCl_3$, or CCl_4. Typical spectra are shown in Figure 8.16, along with peak assignments. A higher degree of fluorination than chlorination was seen in the products when both F and Cl were present in the reactant. No absorption features due to transient species, such as CF_2 and CF_3, were seen, within the $\alpha'pl = 0.005$ FTIR detection limit. Goeckner et al. (1994) used FTIRS to show that CF_4 and CF_3 each account for about 20% of the total density in a CF_4 ECR plasma. The only gas-phase product Nishikawa et al. (1994) saw with FTIRS during the etching of Si in an ECR Cl_2 plasma was $SiCl_4$.

This method has also been used to probe many types of CVD processes. Using FTIRS to probe the gas above the substrate, Kobayashi et al. (1991) showed that the major silicon-bearing product in the selective deposition of tungsten from $WF_6 + SiH_4$ is $SiHF_3$, while SiF_4 is a minor product. In a later study, Kobayashi et al. (1993) used FTIR spectroscopy to probe the absorption by surface adsorbates by infrared reflection–absorption spectroscopy (Section 9.7.1) during this deposition process, as well as to probe the gas above it.

Mazzarese et al. (1989) used *in situ* FTIRS to probe in a heated cell the gas phase reaction of trimethylgallium (TMGa) (peaks at 580 and 770 cm^{-1}) and NH_3 (931, 968, and 1627 cm^{-1} for NH_3; 749 and 1191 cm^{-1} for ND_3),

8.2 Gas-Phase Absorption

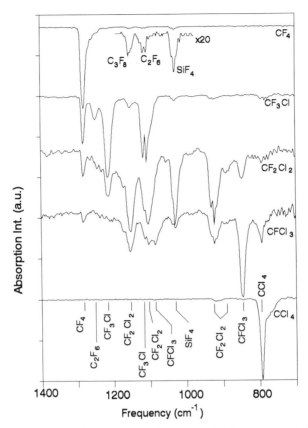

Figure 8.16 FTIR absorption spectra of CF_4, CF_3Cl, CF_2Cl_2, $CFCl_3$, and CCl_4 plasmas, indicating products. There are different vertical scales for each spectrum (from O'Neill *et al.*, 1990).

which is important in GaN MOCVD. TMGa decomposition was the same with either H_2 or N_2 carrier gases. With NH_3 (+TMGa) the primary observed gas-phase product was CH_4 (peak at 3020 cm^{-1}), while with ND_3 it was CH_3D (2200 cm^{-1}). Therefore, hydrogen from the ammonia is incorporated in this product. Such isotopic labelling studies are easily monitored by infrared absorption. They are best performed in static systems because such labelled molecules are expensive.

Xi *et al.* (1994) and Salim *et al.* (1995) investigated the thermal decomposition of three arsenic precursors for GaAs OMVPE and MOMBE (metalorganic molecular-beam epitaxy), *tris*(dimethylamino)phosphine (DMAP), *tris*(dimethylamino)arsine (DMAAs), *tris*(dimethylamino)stibine (DMASb), by using optical-fiber-based FTIRS. Chalcogenide and fluoride optical fibers were used to deliver the IR radiation to the reactor and back to the instrument. (Conventional silica-based fibers absorb in this IR region.) The use of fibers permits remote sensing, while the use of external mirrors to direct the beams is limited to probing only nearby reactors. Also, fiber coupling enables the simultaneous monitoring of many different reactors by a single FTIR spectrometer. These researchers identified homolysis of the E–N (E = P, As, Sb) bond to form dimethylaminyl radicals as the key gas-phase path, as is seen in Figure 8.17 where the transformation of DMAAs into products, determined from their characteristic IR peaks, is plotted against temperature.

Armstrong *et al.* (1992) studied the pyrolysis of precursors for GaAs MOCVD, such as arsine, trimethylgallium (TMGa), and triethylgallium (TEGa), by using *in situ* and *ex situ* FTIRS. For example, the *in situ* spectra in Figure 8.18 illustrate the decomposition of arsine at elevated temperatures. FTIRS enabled the identification of gas-phase reaction products methylarsine, $(CH_3)AsH_2$ and $(CH_3)_2AsH$, with TMGa and ethylarsine, $(C_2H_5)AsH_2$, with TEGa. Fan *et al.* (1992) sampled gas-phase reactants and products under the growth conditions for MOVPE of InP by using *ex situ* FTIRS.

In the investigations cited so far, the IR diagnostics were used to elucidate the process chemistry in the thin film reactor. A quite different application of IR spectroscopy involves monitoring the delivery of reactants to the

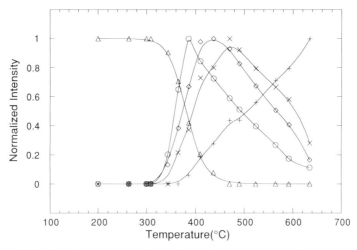

Figure 8.17 Gas-phase decomposition of arsenic precursor *tris*(dimethylamino)arsine (DMAAs) in H_2 monitored *in situ* by optical-fiber-based FTIRS [(△) parent molecule, (○) dimethylamine, (+) methane, (×) methyleneimine, and (◇) methylmethyleneimine]. [Reprinted with permission from Salim *et al.* (1995). Copyright 1995 American Chemical Society.]

8.2 Gas-Phase Absorption

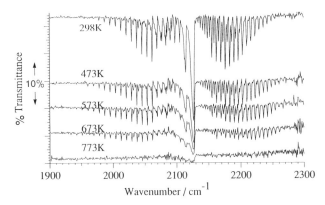

Figure 8.18 *In situ* FTIRS measurements of the As—H stretch region of arsine for arsine in a GaAs MOCVD reactor, at various substrate temperatures (from Armstrong *et al.*, 1992).

reactor. The delivery of stable, volatile gas-phase reactants to CVD reactors is regulated quite well by using conventional flowmeters. However, the delivery of condensible gases, derived by bubbling the flow gas through a liquid source or by evaporating (low vapor pressure) solid state precursors, and highly reactive gases, such as those generated *in situ*, is relatively unregulated and this can lead to large uncertainties in CVD growth. O'Neill *et al.* (1994) have demonstrated how FTIRS can be used to monitor and, consequently, control these flows on line.

They monitored the two reactive gases in the atmospheric-pressure CVD growth of silicon dioxide films: TEOS (tetraethoxysilane or tetraethylorthosilicate) and ozone (O_3). TEOS, delivered by bubbling He through the liquid TEOS, was monitored by integrating the CH stretch absorption at 3000 cm^{-1}. The thickness of the grown silicon dioxide film was shown to increase linearly with this TEOS absorption signal. The ozone signal, monitored by the integrated absorption strength of the ν_1 band at 1054 cm^{-1}, depends on the stability of the ozonator and the "passivation" of the feed line, as is illustrated in Figure 8.19. O'Neill *et al.* also demonstrated how the percentage of phosphorus in the doped oxide can be predicted from IR measurements of TMP (trimethylphosphine) that is added to the gas mixture; both TMP and TMB (trimethylborane), the boron dopant precursor, are also delivered by the bubbler system.

The deposition of many other films with conformal coverage at low temperature, such as titanium nitride, aluminum, aluminum oxide, and copper, often relies on precursors derived from liquid sources. With the real-time control provided by such IR monitoring, diagnostics of the film itself may not be necessary in some cases. A simpler and less expensive version of this diagnostic has been commercialized, in which infrared radia-

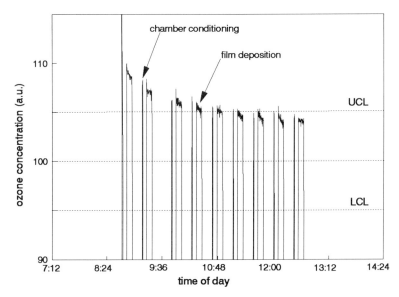

Figure 8.19 Trend chart of the concentration of ozone, determined by FTIRS, during the deposition of CVD oxide films on nine wafers, using TEOS/ozone mixtures. Chamber conditioning and film deposition are shown for each run. UCL and LCL are the established upper and lower concentration limits, respectively. The drift in the ozone concentration during early runs was due to unstable operation of the ozonator (from O'Neill et al., 1994).

tion from a glowbar is filtered about the desired absorption band and modulated for synchronous (lock-in) detection (MKS Instruments, 1995).

The decomposition of TEOS/O_3 mixtures has also been investigated by other groups. By using *in situ* FTIRS in a closed system, Kawahara et al. (1992) has suggested that CH_3CHO may be the first gas-phase product, which then gets transformed into CO_2 and H_2O, through an HCHO intermediate. Mucha and Washington (1994) used FTIRS analysis of the products of TEOS/O_3 reactions to show that each TEOS molecule could be eliminating one acetaldehyde (CH_3CHO) molecule and one acetic acid (CH_3COOH) molecule. (See Section 8.3 for another IR study of TEOS decomposition.)

Salim et al. (1994, 1996) also used FTIRS to monitor the concentrations of several organometallic precursors *in situ* in an OMVPE reactor; light was delivered to and from the reactor by using infrared optical fibers. Salim et al. (1994) determined the transient changes in precursor pressure in the reactor during turn-on and turn-off delivery, along with the temporal variations in precursor pressure under steady-state delivery conditions. Salim et al. (1996) improved the detection limit by an order of magnitude by replacing the HgCdTe (MCT) detector used by Salim et al. (1994) with

8.2 Gas-Phase Absorption

an InSb detector. The detection limit in a 1 sec scan improved to 0.006 Torr for TTBAl (tritertiarybutylaluminum), 0.02 Torr for TEGa (triethylgallium) and TESb (triethylantimony), and to 0.05 Torr for TMGa and TMIn. Salim *et al.* (1996) found good agreement between concentration measurements made by FTIRS and ultrasonic techniques (see Section 1.6).

8.2.1.3 IR Absorption Thermometry

The rotational temperature T_{rot} of a gas can be determined by absorption through the dependence of α on J, as seen in Equations 4.33, 4.36, and 4.39. This can be performed by using a scanning laser or an FTIR spectrometer. As cited above, Cleland and Hess (1988) determined T_{rot} by using FTIRS to probe absorption by individual vibrational/rotational lines of the $2\nu_1$ harmonic and $\nu_1 + \nu_3$ combination bands of N_2O in an N_2O discharge (Figure 8.20). Also, Knights *et al.* (1982) used a modified Connes-type interferometer with a very high ultimate resolution of 0.0009 cm^{-1} to resolve IR absorption and emission for temperature measurements in a SiH_4 discharge near 2200 cm^{-1}; these are detailed in the OES discussion in Section 6.3.5. Farrow (1985) probed the absorption by the ν_3 vibrational mode of BCl_3 in a BCl_3 RF discharge by using a line-tunable CO_2 laser (near 10.6 μm). T_{rot} was determined by simulating the band contour at different temperatures. Haverlag *et al.* (1994) found that the rotational temperatures of CF_4, CF_2, and HF are all close to room temperature in RF plasmas of several fluorocarbons, by comparing their FTIR absorption spectra with rotational distribution simulations. Using the related methods of millimeter

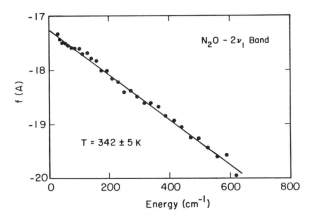

Figure 8.20 Normalized line absorption strength ($f(A) \propto$ left-hand side of Equation 4.39) vs rotational energy from the *in situ* FTIR absorption spectrum of N_2O in a N_2O RF glow discharge, which is plotted this way to determine the rotational temperature (from Cleland and Hess, 1988).

and submillimeter wave absorption, Clark and De Lucia (1981) found that $T_{rot} \approx T_{trans}$ for OCS in an OCS discharge. Their work is described further in the discussion of limited thermal equilibrium in Section 18.1.

8.2.2 Ultraviolet/Visible (Electronic) Absorption

Molecules can also be monitored by absorption between electronic transitions. Although it is sometimes simpler to monitor these transitions by LIF because it has higher sensitivity or instead to monitor these molecules by IR absorption, there are still some important diagnostics applications of ultraviolet/visible absorption.

Typical UV absorption cross-sections are $\sigma \sim 10^{-19}$–10^{-16} cm^2 (Rothschild, 1989). For $\sigma = 10^{-17}$ cm^2, this means that the absorbed fraction ($N\sigma l$) is 1% for a density of 1×10^{14}/cm^3 (3 mTorr) over a 10-cm path length. By using very stable white-light sources, lock-in detection, and longer path lengths, the detection sensitivity can be improved to $\sim 10^{-3}$–10^{-4} (Donnelly, 1989). The detection sensitivity of ultraviolet/visible absorption is poorer than that for LIF and IR absorption, and often only major species can be monitored. Molecules with strong bands at >200 nm can be detected by using standard UV quartz optics, while those with strong bands that are only <200 nm require vacuum UV techniques. Overlap of bands from different species can hinder the analysis of multicomponent mixtures.

8.2.2.1 Applications

8.2.2.1.1 CVD and Plasma Processing

Ibbotson *et al.* (1983) probed the absorption by Br_2 in a Br_2 discharge used for GaAs etching. They propagated chopped radiation from a tungsten filament lamp through the discharge and analyzed it with a scanning monochromator. The absorption spectrum was integrated over the entire band to account for temperature effects, and this integrated line strength was then used to obtain N_{Br_2} and to infer N_{Br}. This study showed N_{Br_2} is approximately halved and N_{Br} doubled as the RF frequency applied to the discharge increases from 0.1 to 15 MHz, and that $N_{Br} \approx N_{Br_2}$ near 1.5 MHz, as is seen in Figure 8.21.

In a series of studies, Karlicek and co-workers investigated the use of ultraviolet absorption spectroscopy to probe processes during MOCVD (MOVPE) of InGaAsP. Karlicek *et al.* (1984) monitored the thermal decomposition of Me_3In (trimethylindium) and its adducts $Me_3In \cdot PMe_3$ and $Me_3In \cdot PEt_3$, which can be employed in the MOCVD of InP. These gases have absorption profiles that peak between 195 and 211 nm, and absorption was measured by using a commercial UV/visible spectrometer. Karlicek *et al.* (1986) presented the UV absorption spectra of several of the gas-phase reactants and intermediates involved in the growth of InGaAsP alloys by

8.2 Gas-Phase Absorption

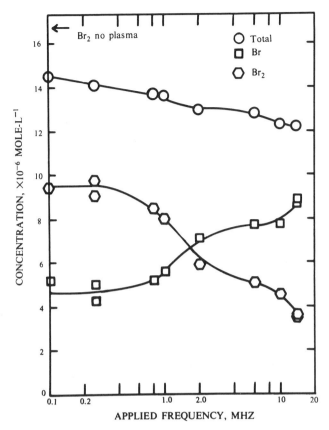

Figure 8.21 Absolute concentrations of Br and Br_2 in a plasma as a function of RF frequency, as determined by absorption spectroscopy of Br_2 ($\lambda \approx 415$ nm) (from Ibbotson *et al.*, 1983).

hydride VPE, including the room-temperature spectra of PH_3, AsH_3, and HCl, and the high-temperature (700°C) spectra of InCl, GaCl, PH_3, P_2, P_4, As_2, and As_4. Measurements were made with a D_2 lamp, a grating monochromator for dispersion, and a diode array for detection. These data were then used to study PH_3 pyrolysis and monitor the In/Ga ratio for metal transport from an alloy source. They found that the components in individual subsystems often can be analyzed, such as $PH_3/P_2/P_4$ or $AsH_3/As_2/As_4$, but the analysis of multicomponent systems can be difficult, as for $InCl/PH_3/P_2/P_4$, because of overlapping features and the often dominating features of a particular absorber.

In situ sensing is sometimes complicated by the design of a reactor. Film growers are reluctant to modify these designs to incorporate diagnostics for fear, often unwarranted, that the film properties will change. Karlicek and Bloemeke (1985) developed the remote optical monitoring scheme depicted in Figure 8.22 which allows optical access in unmodified reactors operating inside a furnace. A quartz light pipe delivers light from a flashlamp to the hydride transport InGaAsP VPE reactor, and a second one, 2.3 inches away, collects the transmitted beam and delivers it to the monochromator and photomultiplier. They monitored absorption by InCl, GaCl, and P_4 this way, and, in particular, studied the transient behavior in the reactor as the gas flow was changed. P_4 absorption was monitored as the PH_3 flow rate was changed, and InCl absorption was monitored after the HCl flow by the In source was stopped. This latter absorption decayed very slowly because residual InCl that was dissolved in the In continued to supply InCl to the reactor.

In similar studies, Hebner et al. (1989a,b), used *in situ* UV absorption to measure partial pressures of reactants in several types of MOCVD reactors, with reactants such as $Ga(CH_3)_3$, $Al(CH_3)_3$, AsH_3, $In(CH_3)_3$, $Sb(CH_3)_3$, and $Zn(C_2H_5)_2$, as is seen in Figure 8.23. A deuterium lamp was transmitted through the growth cell, dispersed in a monochromator, and detected with a diode array; for improved stability, the light was chopped and detected by a photomultiplier and lock-in analysis. Even though the first three reactants are spectrally similar (the last three have distinct spectra that peak at different wavelengths), they claimed that it is possible to identify the contribution each makes to the net absorption in the reaction chamber. They also used UV absorption to monitor the transient response of flows for each reactant, as is illustrated in Figure 8.24. Haigh and O'Brien (1984b) studied MOCVD of InP in a flow tube by monitoring UV absorption near 215 nm. Fujita et al. (1989) used UV absorption between 190 and 400 nm to probe the absorption of Cd, Te, Zn, S, and Se alkyls and to monitor several of these *in situ* during photolysis in an MOCVD reactor.

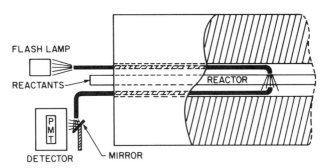

Figure 8.22 Schematic of apparatus used for remote optical absorption monitoring in a VPE reactor using quartz light pipes (from Karlicek and Bloemeke, 1985).

8.2 Gas-Phase Absorption

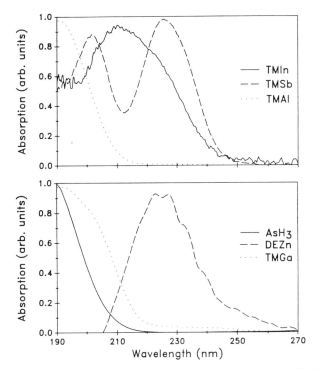

Figure 8.23 Ultraviolet absorption spectra of potential precursors for MOCVD of III–V semiconductors: trimethylindium (TMIn), trimethylantimony (TMSb), trimethylaluminum (TMAl), arsine (AsH_3), diethylzinc (DEZn), and trimethylgallium (TMGa); peak absorbances are normalized to 1 (from Hebner *et al.*, 1989a).

O'Neill and Singh (1994) have used UV absorption spectroscopy to detect CF_2 radicals ($A \leftarrow X$) in inductively coupled high-density plasmas of $C_2F_4H_2$ and CF_4, which selectively etch SiO_2 *vis-à-vis* Si. With 300 W applied power and a pressure of 10 mTorr, they found that CF_2 constitutes at least 17% of the gas in the reactor. Optical emission signals from CF and CF_2 were found to correlate more strongly with the electron density than with the density of CF_2. The etch selectivity was tracked by the concentration of CF_2 determined from absorption, rather than by the OE signals. O'Neill and Singh (1995) found that in plasmas containing C_2F_6 and CF_4, the concentration of CF_2 determined by UV absorption spectroscopy depends strongly on the applied power and pressure, as well as the amount of polymer on the walls. They observed that changes in the state of conditioning of the reactor, and consequently the amount of polymer on the walls, greatly affect the CF_2 density and therefore the etching rates of the plasma.

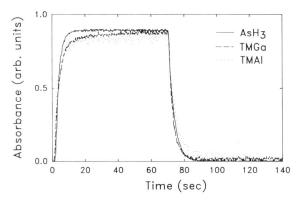

Figure 8.24 Time response of reactants in a horizontal MOCVD reactor monitored in real time by UV absorption. The reactant gas is turned on at 0 sec and off at 70 sec (from Hebner *et al.*, 1989a).

Wamsley *et al.* (1990), Lawler *et al.* (1991), and Giapis *et al.* (1993) used ultraviolet/visible absorption to probe the concentrations of several species in plasmas. In particular, Giapis *et al.* (1993) monitored self-absorption of plasma-induced emission (PIE, OE) in a helicon-wave excited plasma by reflecting that radiation back into the reactor.

Jasinski *et al.* (1984) used frequency-modulation absorption spectroscopy to detect SiH_2 radicals in silane and disilane discharges used to deposit a-Si. In this form of FM absorption spectroscopy (Bjorklund, 1980; Gerhtz *et al.*, 1985) [Section 8.1.2.1], a laser—in this case a single-frequency dye laser (~5730 Å, frequency ω_L)—is focused into a $LiTaO_3$ electro-optic modulator, driven here by RF at ~800 MHz (corresponding to ω_{RF}). Leaving the modulator is the laser, with frequency ω_L, and two sidebands, at $\omega_L \pm \omega_{RF}$. There is no amplitude modulation of the laser when the sidebands are equally strong, as they are after the crystal and after the absorption cell when ω_L is tuned either exactly to the linecenter of an absorption or very far from all resonances. When the laser is tuned near resonance, there is differential absorption by the sidebands. This results in amplitude modulation of the transmitted beam at ω_{RF}, which is demodulated by a double-balance mixer. This gives the derivative of the absorption lineshape for the SiH_2 transition, as is seen in Figure 8.25. The detection sensitivity is improved, to $\sim 10^{-6}$, by chopping the laser at ω_c (corresponding to 2 kHz) and the discharge voltage at ω_p (1.1 kHz), and sending the output from the double-balance mixer to a lock-in amplifier tuned to $\omega_c + \omega_p$. This latter modulation is not possible in thermal CVD reactors. Relatively little work has proceeded with this method because modulated spectroscopy with infrared diode lasers is simpler and more powerful.

8.2 Gas-Phase Absorption

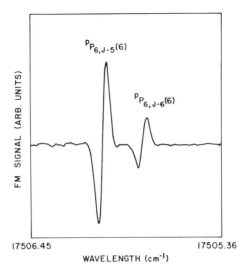

Figure 8.25 Frequency-modulation absorption signals from two single rotational lines of the 2–0 vibronic band of SiH$_2$ in a SiH$_4$ glow discharge (from Jasinski *et al.*, 1984).

Another very sensitive form of absorption spectroscopy is intracavity laser spectroscopy, in which the reactor is placed within the laser cavity so that the effect of absorption resonances is amplified. Small absorptions per pass, due either to very low densities or very weak transitions, lead to proportionately larger decreases in the output power of a low-gain, cw laser. This is monitored in the output of a tunable or broad-bandwidth laser. By monitoring absorption resonances in the ~10-cm^{-1}-wide spectral bandwidth of a dye laser this way, O'Brien and Atkinson (1986) detected SiH$_2$ in a silane discharge and Miller *et al.* (1989) observed B and BH$_2$ in a B$_2$H$_6$ discharge. Since intracavity intensities are much larger than the output intensities of low-gain cw lasers, intracavity absorption produces very large LIF and optoacoustic signals. Although intracavity absorption spectroscopy is very sensitive, it is not a very convenient diagnostic.

8.2.2.1.2 *Pulsed-Laser Deposition*

Time-resolved absorption can be used to detect the atoms and molecules formed in the plume during pulsed-laser deposition (PLD). Either single-wavelength sources (fixed-wavelength lasers, tunable lasers, hollow cathode atomic lamps) or broadbanded sources can be used. For example, a pulsed xenon flashlamp, which emits a ~1-μsec pulse that covers the visible and UV, can be fired at an arbitrary time after the laser pulse. Absorption at a single wavelength can be monitored by fixing the monochromator wavelength and averaging the photomultiplier (PMT) signal over many pulses; or instead, many wavelengths can be monitored at once by using multichannel detection. Geohegan (1992) has described an experimental apparatus that can be used for absorption (and other probes) in pulsed-laser deposition. The peak strength and lineshape of the absorption signal can be followed as a function of time after the pulse and the distance from the target surface.

Figure 8.26b shows the absorption spectrum of a plume from laser ablation of $YBa_2Cu_3O_{7-x}$ (Geohegan, 1994). Features due to ground-state Ba, Y, and YO are seen that last for several microseconds. Similar features are observed in the OES spectrum in Figure 8.26a; however, these emissions from the excited states diminish much faster. The lineshapes of atomic lines in absorption and emission are generally much broader than the linewidth due to radiative decay because of line broadening due to the plasma. Spectral broadening and self-absorption is illustrated in Figure 8.27, where the 455.4-nm Ba^+ ($6p\ ^2P_{3/2} \rightarrow 6s\ ^2S_{1/2}$) emission line from the hotter central region in the plume is reabsorbed by the cooler outer areas [with less line broadening] (Geohegan, 1994). (See Section 6.1.) Other studies of transient absorption during PLD of this material have been performed by Geohegan and Mashburn (1989), Cheung *et al.* (1991), and Sakeek *et al.* (1991); they followed many species, including Y, Ba, Cu, Ba^+, and YO.

Gilgenbach and Ventzek (1991) and Ventzek *et al.* (1992) used dye laser resonance absorption photography to follow the plume during ablation from polymer and aluminum targets. Mitzner *et al.* (1993) probed the absorption by Ca, Ca^+, and CaF after the KrF laser ablation of CaF_2, by using a broadband pulsed dye laser and a diode array. Kunz *et al.* (1990) used this technique to follow carbon plumes after ablation.

8.2.2.1.3 *Atomic Absorption for Density Measurements*

Absorption has been used to monitor atoms in many types of deposition processes, including sputtering, plasma-assisted deposition, laser-assisted deposition, MOCVD, (ordinary) evaporation, and molecular beam epitaxy (MBE). They are described here, with the exception of applications in the last two processes, which are presented in Section 8.2.2.1.4. Resonance hollow cathode lamps are very convenient sources of light for these absorption experiments (Stirling and Westwood, 1970; Price, 1972; Setser, 1979); other sources have also been used to monitor atoms. One early application of atomic absorption probing was by Stirling and Westwood (1970), who used it to investigate the sputtering of aluminum.

Greene (1978) has discussed the use of atomic absorption in monitoring sputter deposition and also used it to calibrate OES measurements which were also performed during sputtering. He measured the Cu atom density in DC diode and triode glow discharge sputtering systems with the 324.754 nm Cu resonance line from a lamp, and found that the density was linear with the Cu target sputter rate; he also mapped the Cu density as a

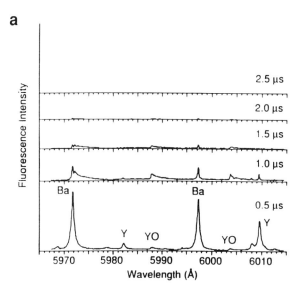

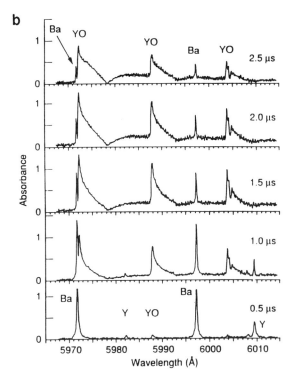

Figure 8.26 Comparison of (a) optical emission and (b) absorption spectra of the plasma plume formed after laser ablation of a YBa$_2$Cu$_3$O$_{7-x}$ pellet. The emission spectra show that the excited-state species disappear by 2.5 μsec, while the absorption spectra demonstrate that the ground-state species last much longer (from Geohegan, 1992).

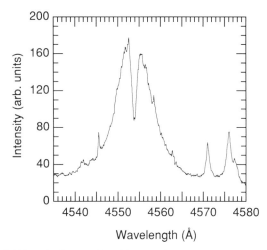

Figure 8.27 Self-broadening and self-absorption observed in the 4554-Å Ba⁺ ($6p\,^2P_{3/2} \to 6s\,^2S_{1/2}$) emission line measured 1 mm above a YBa$_2$Cu$_3$O$_{7-x}$ pellet after laser ablation. The central part of the spectral line, the dip, is absorbed by the cooler region surrounding the hotter central region of the plume (from Geohegan, 1992).

function of distance from the target surface. Jellum and Graves (1990) monitored the absorption of 394.4-nm radiation by Al atoms in a sputtering system to calibrate Al densities for concurrent LIF measurements. The excimer-laser pumped dye laser used in this experiment was suitably attenuated to avoid saturation. They needed to use Equation 8.2 in their analysis because the laser linewidth of 0.2 cm^{-1} was ~3.1× larger than the Doppler width of the Al atoms.

Osmundsen *et al.* (1985) used a pulsed xenon lamp to probe the Ge atoms that were formed during the photolytic growth of Ge films by KrF-laser (248 nm) excitation of GeH$_4$. They determined the temporal and spatial dependence of Ge atoms by analyzing the $4p5s\,^1P_1 \leftarrow 4p^2\,^1S_0$ absorption at 422.6 nm. Although the detection system resolution (monochromator/optical multichannel analyzer) was ~30–60× larger than the spectral linewidth, absorbances as large as 1% were consistently measured.

Tachibana *et al.* (1982) measured the Si atom density in a pulsed silane discharge by using a Si hollow cathode lamp. The populations of the five sublevels of the ground configuration $3p^2$ ($^3P_0$, $^3P_1$, $^3P_2$, $^1D_2$, and $^1S_0$) were measured by using different resonances lines, and they indicated a thermal distribution with $T \approx 300$ K.

Haigh and O'Brien (1984a) concluded that, within the detection limit, no atoms are produced due to gas-phase processes during the pyrolysis of TEGa and TMIn that occur in MOCVD. They used radiation from In

8.2 Gas-Phase Absorption

(303.9 nm) and Ga (287.4 nm) hollow cathode lamps in this study. It should be noted that In atoms have been observed under conditions similar to those in this study by using resonance fluorescence, as is described in Section 7.2.1.1.

8.2.2.1.4 *Atomic Flux Monitoring*

Careful control of the temperature of the atomic effusion cell often does not provide sufficient control of the atomic flux during MBE because the beam flux can change slowly over time with source-material depletion or physical redistribution in the cell. One way to monitor and control the deposition rate during MBE or any evaporation process is by monitoring the atomic flux to the surface. The film thickness can then be regulated by controlling the source shutter. Although flux monitoring with a calibrated ionization gauge is very good, the calibration sometimes changes with conditions. Atomic absorption spectroscopy has been shown to be an accurate and relatively inexpensive real-time monitor of atomic flux during physical vapor deposition, and has attracted strong interest.

In one common arrangement (Figure 8.28), radiation from a hollow cathode lamp for the specific element of interest is chopped and propagated above the substrate. The light transmitted through the atomic beam is

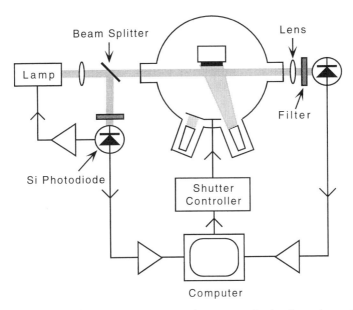

Figure 8.28 Schematic of the apparatus to monitor reactant flux by absorption, and consequently the MBE growth rate, and to control the MBE shutter (from Chalmers and Killeen, 1993).

passed through an interference filter, detected by a photodiode or photomultiplier, and analyzed by lock-in detection. If the absorption path length is ~1%, as is typical, and the film thickness and hence the flux must be controlled with ~1%, then the noise in the light source and detector must be less than 0.01%. Sometimes this can be achieved by analyzing the signal by lock-in detection and normalizing the transmitted power by the incident power. However, this may not be sufficient if the lamp is very noisy or drifts much and if the absorption path lengths are small. The optical flux from the lamp must then be feedback-stabilized by actively controlling the lamp voltage, as is illustrated in Figure 8.28 for MBE growth (Chalmers and Killeen, 1993). A deposition accuracy of 1% can be achieved during a several-hour growth run. Figure 8.29 shows the rate of growth of GaAs by MBE determined by flux monitoring, with stabilization.

Flux monitoring is a good probe of film thickness only if the impinging flux is not reevaporated. Reevaporated atoms do not form a deposit and are, in fact, counted twice by the probe—before impinging the surface and after desorption. The deposited thickness is best calibrated relative to standard runs rather than to calculated absorption path lengths, because this latter method is complicated by the Doppler shifts seen by the atomic beam and by the different linewidths of the emitting species in the lamp and the atomic beam. Errors can appear if the calibration between the absorbed fraction and film thickness is linearly extrapolated for different conditions, such as different oven temperatures T (Klausmeier-Brown et al., 1992). This can occur because the absorbed fraction varies linearly with density $N(T)$, while the deposition rate varies with the flux to the surface $N(T)v(T)$ (v is the component of velocity normal to the surface). Therefore, the absorbed fraction is not strictly proportional to the deposition rate. The different linewidths of the light source and absorber can also lead to nonlinearities [Condition (ii) in Section 8.1.1].

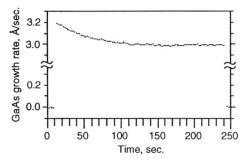

Figure 8.29 GaAs growth rate determined by monitoring absorption by Ga by resonant light in an MBE chamber, showing the flux transient due to effusion cell cooling after the shutter was opened at $t \sim 10$ sec (from Chalmers and Killeen, 1993).

8.2 Gas-Phase Absorption

Klausmeier-Brown *et al.* (1992) measured the atomic flux of Bi, Sr, Ca, and Cu by absorption during the atomic layer-by-layer MBE growth of high T_c superconducting films and heterostructures. Critical control of film thickness is necessary in this growth because the two-dimensional layer growth is not self-limiting, as it is with the epitaxial growth of some semiconductors such as GaAs. Control of the Ca and Sr fluxes proved to be easy because of the large absorption path for these elements, while measuring Bi and Cu was more difficult because the absorptions for each were <1%. Benerofe *et al.* (1994) developed a double-beam atomic absorption system to probe Sr and Ru during the electron beam evaporation growth of epitaxial films of the ferromagnetic perovskite $SrRuO_3$ (Figure 8.30). The reference arm path compensated for lamp fluctuations, and the normalized absorption signal was fed back to the electron beam evaporator. The method is stable (<1% drift/h), fast (~200 msec), and sensitive ($S/N \approx 10$–200) for 1 Å/sec growth rate. Absolute calibration of atomic density is complicated by deviations from Beer's law caused by the broader Doppler width of the lamp emission profile (~2 GHz) *vis-à-vis* that of the evaporated atoms (~0.5 GHz) [Condition (ii), Section 8.1.1].

Chalmers and Killeen (1993) monitored the Al (~394 nm) and Ga (~405 nm) fluxes during MBE growth of AlAs/GaAs distributed Bragg

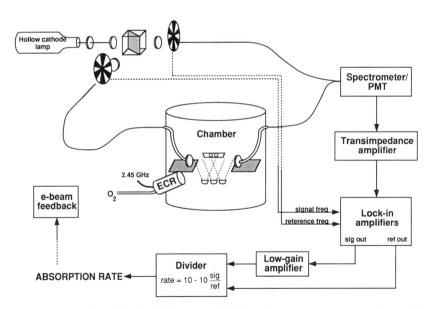

Figure 8.30 Schematic of a dual-beam atomic absorption spectroscopy apparatus used to control the electron beam sources in an evaporator deposition chamber (from Benerofe *et al.*, 1994).

reflectors (DBRs). Reevaporation is not a problem for AlGaAs growth below ~650°C. The lamp output varied by as much as 10% without regulation, but only <0.1% with regulation. Use of the two-beam referencing with an unregulated lamp decreased noise to ~0.2%; since this was not sufficient, source regulation was also needed. By carefully controlling the layer thicknesses, the center wavelengths of the DBRs were controlled to 0.3% of the target wavelength, which was found to be much better than relying on the effusion cell temperature for growth rates.

Lu and Guan (1995) have described a different dual beam scheme that improves the base-line stability in probing flux by atomic absorption. Changes due to viewport coating and minor optical alignment shifts are corrected by transmitting radiation from both the hollow cathode lamp and a (reference) xenon lamp (filtered by a monochromator) through the same reactor and reference paths; the two lamps are individually controlled by a timing circuit.

8.2.3 Laser Magnetic Resonance

Many reactive species, including atoms, radicals, and some molecules common in plasma processing, have a nonzero magnetic moment μ_m because of uncoupled spins or angular momentum. For example, in an atom where there is Russel–Saunders coupling, the total orbital angular momentum **L** and total spin **S** couple to form the total angular momentum **J**. Then $\mu_m = -g_J\mu_B \mathbf{J}$, where g_J is the Landé g factor and μ_B is the Bohr magneton. A magnetic field H, say along the z direction, splits this level in $2J + 1$ sublevels with energies $g_J\mu_B H m_J$, where $m_J = -J, -J + 1, \ldots, J$, which is known as the Zeeman effect.

The applied magnetic field can be varied until the tuned states come into resonance with a fixed-frequency electromagnetic source. Absorption is then monitored as a function of magnetic field. This paramagnetic resonance is a signature of the species. Transitions between split m_J levels in the same electronic state are in the microwave region, such as $\Delta m_J = \pm 1$ transitions in the $^3P_2$ ground state of the O atom. This magnetic resonance method is called electron paramagnetic resonance (EPR). Transitions between magnetic tuned levels in different states can sometimes be tuned into resonance with fixed-frequency infrared and far-infrared lasers. Examples of this laser magnetic resonance (LMR) include transitions between the Zeeman-tuned $m_J = 1$, $^3P_1$ level and $^3P_0$ in O atoms, and transitions between rotational levels in OH $^2\Pi_{3/2}$. To improve signal/noise a modulated magnetic field is superimposed upon the tuned (static) field, and derivative signals are analyzed, as in Section 8.1.2.

Cook and Miller (1989) have reviewed the use of LMR as a plasma discharge diagnostic. Atoms detected by LMR include C, O, Si, and Cl, and radicals include CH, SiH, GeH, OH, NH, AsH, CF, CH_2, AsH_2, CH_2F,

CH_3, SiH_3, and GeH_3. Other discussions of LMR may be found in Evenson et al. (1980), Evenson (1981), McKellar (1981), Davies (1981), Thrush (1981), and Hills (1984). Although LMR is very sensitive, it is not a widely applied diagnostic. Application of the magnetic field is rarely convenient, and it can perturb the thin film process, particularly in plasma processing. In most cases, it would be better to use infrared diode laser spectroscopy.

8.3 Transmission through Adsorbates or Thin Films

The previous section examined absorption by gas-phase species, which is important in assessing process-state parameters. This section and the next section address the use of absorption to probe wafer-state parameters.

The identity, density, and orientation of adsorbates on transparent substrates can be monitored by their characteristic IR or UV/visible absorption features determined in transmission experiments. However, in most cases such analysis is performed more conveniently in reflection (Chapter 9). For example, surface infrared reflectometry spectroscopies are very sensitive to absorption by the vibrational modes of adlayers (Section 9.7). Reflection ellipsometry (Section 9.11) and surface photoabsorption (Section 9.9) are very sensitive to the electronic structure and absorption of adlayers. Similarly, even though thin films can be monitored by absorption, reflection measurements often have advantages. Sometimes, for example, film absorption cannot be monitored in transmission because of absorption by the substrate. Nevertheless, transmission probes are sometimes useful.

Osgood and co-workers monitored chemisorbed and physisorbed adlayers of dimethylcadmium (DMCd) (Chen and Osgood, 1983a,b) and diethylzinc (DEZn) (Krchnavek et al., 1987) on fused silica by ultraviolet absorption. This was done both before and after excimer laser irradiation to determine the bonding and the nature of the photolytically produced film. Because of the weak absorption per adlayer, transmission was monitored through several fused-silica substrates, with adsorbates on both surfaces of each.

The growth or etching of a film can be monitored by transmitting a light beam that is absorbed by the film, but not by the substrate. This method is simple and inexpensive to implement, but it is not versatile. Selectivity in absorption between the processed film and substrate, which may be overlaid by other films, is not always possible. Moreover, transmission is sensitive for thicknesses up to ~3 absorption lengths, $3/\alpha$. Monitoring the transmission of several wavelengths, with different α, could overcome this shortcoming. (Reflection interferometry does not have this problem because the film is transparent to the probe wavelength chosen.) Transmission can be used to monitor only very thin films of metals, a few hundred angstroms thick, and cannot be used in semiconductor processing when the film has a larger bandgap than the wafer. Although it is easy to implement transmission probes of films for fundamental and process development studies, implementation for real-time monitoring during manufacturing can be difficult.

Transmission probes have been used to monitor film deposition in real time during laser-assisted film growth. For example, Allen *et al.* (1985) monitored the transmission of a He–Ne laser (632.8 nm) during the laser-assisted deposition of W and Ni films promoted by cw CO_2 laser heating of the quartz substrate. King *et al.* (1987) and Osmundsen *et al.* (1985) measured the transmission at 632.8 nm to monitor the growth of Ge films on a SiO_2 substrate during the excimer-laser-assisted deposition from GeH_4.

Moreau (1988) and Watts *et al.* (1988) have described how absorption can be used to monitor the processing of photoresist films. Since the resin and solvent in the resist are transparent from 310 to 450 nm and the photoactive compound (PAC) absorbs in this region (Figure 8.31), monitoring absorption in this range determines the amount of PAC. Moreover, the absorbance of the PAC changes upon UV exposure, as is seen in Figure 8.31. In most positive resists, the PAC is naphthoquinone diazide which is transformed into indenecarboxylic acid upon exposure to UV light. As described by Watts *et al.* (1988), resist absorbance is usually determined by measuring the fraction of light reflected from the resist/wafer interface (giving two passes in the film), because transmission cannot be used with these opaque Si wafers. Monitoring of the step of resist exposure to UV is

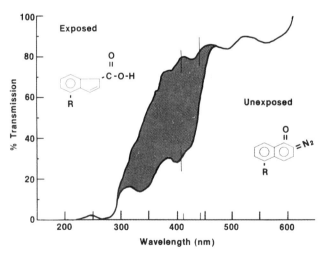

Figure 8.31 UV spectrum of a typical optical positive resist, showing the decrease in absorption from 310 to 450 nm on exposure (from Watts *et al.*, 1988).

8.3 Transmission through Adsorbates or Thin Films

shown in Figure 8.32. Watts *et al.* (1988) also used absorption measurements made before and after the coating step, which determines the amount of PAC, to help monitor the changes of thickness due to solvent evaporation during softbaking. (The resist thickness is determined by the amounts of PAC, solvent, and resin, and the amount of PAC is constant during softbaking.) These thickness measurements were performed by reflection interferometry (Section 9.5.1).

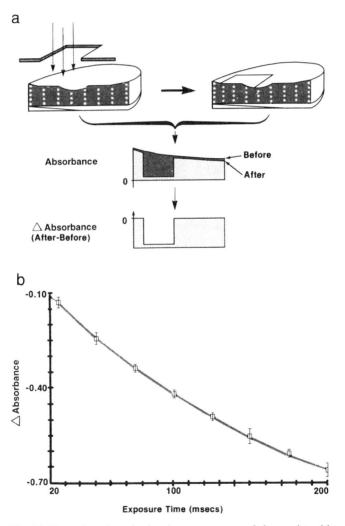

Figure 8.32 (a) Illustration of monitoring the exposure step of photoresist, with patterned exposure, by absorption. (b) Absorbance change in photoresist vs. exposure time for a 436-nm line source (from Watts *et al.*, 1988).

Howes et al. (1994) used real-time FTIR absorption spectroscopy to monitor the kinetics of chemically amplified resists. During the bake step, a peak at 982 cm^{-1} grew, which they associated with the formation of an ether linkage during crosslinking. FTIRS was also used to monitor the loss of solvent during the preexposure bake. See Section 9.7.2.3 for work on monitoring photoresist processing by using attenuated total internal reflection (ATR) spectroscopy.

Franke et al. (1994) have reviewed the use of infrared absorption spectroscopy to characterize dielectric thin films on silicon wafers; they also covered other infrared probes, such as infrared reflection and emission spectroscopies. Examples of the application of IR absorption include the work of Adhihetty et al. (1991) on the use of multivariate calibration methods to determine precisely the composition of borophosphosilicate glass (BPSG) on silicon, and that by Niemczyk et al. (1994) on the use of these techniques for at-line characterization of arsenosilicate glass films.

The fundamental steps during deposition are sometimes investigated by using infrared absorption to investigate the deposition process on porous films or substrates; use of porous media increases the surface coverage and therefore the absorbance for a given depth. For example, Tedder et al. (1991) studied the dissociative adsorption of TEOS (tetraethoxysilane) in a porous SiO_2 layer (on a Si wafer) by using FTIR transmission spectroscopy. Dillon et al. (1991) used FTIRS to monitor the decomposition of H_2O (D_2O) and NH_3 (ND_3) on porous silicon. Ott et al. (1995) probed infrared absorption by surface adsorbates after the alternating steps of exposure to trimethylaluminum and to water vapor in the atomic-layer-controlled deposition of Al_2O_3 films on porous alumina.

8.4 Transmission through Substrates for Thermometry

Although pyrometry and thermocouples have been used quite often for wafer thermometry during MBE and rapid thermal processing (RTP), these methods have serious shortcomings that can lead to large errors in determining the wafer temperature (Chapters 13 and 18). The measurement of wafer temperature by optical transmission at a wavelength near the band gap has many attractive features, and is sometimes superior to these two more conventional diagnostics. Since it is useful only if the wafer transmits the chosen wavelength (at least partially), use of the transmission probe is constrained by the properties of the substrate and the overlaying films.

The temperature of a semiconductor wafer can be monitored by transmission because absorption can be sensitive to (1) the dependence of the band gap E_{bg} of the substrate material on temperature, (2) the thermal population of phonons (in phonon-assisted absorption, which is very important in indirect gap semiconductors), and (3) the thermal population of

8.4 Transmission through Substrates for Thermometry

intrinsic free carriers (which is important in analyzing free-carrier absorption).

Section 4.3.1.2 presents expressions for the absorption coefficient due to the three most important processes: (1) direct transitions (Equation 4.41), which is significant in semiconductors such as GaAs and InP; (2) indirect transitions between indirect valleys [phonon-assisted transitions] (Equations 4.42–4.44), which are important in Si and Ge; and (3) free-carrier absorption (Equation 4.45), which is important for $\mathcal{E} < \mathcal{E}_{bg}$.

Section 18.4.2 describes the physical origin of the temperature dependences of these processes in more detail. The first two processes are sensitive to the temperature dependence of the band gap energy $\mathcal{E}_{bg}$ in semiconductors. One way to characterize this is by the Varshni expression (Varshni, 1967), which is the empirical fit (Equation 18.7):

$$\mathcal{E}_{bg}(T) = \mathcal{E}_{bg}(0) - \frac{\alpha_v T^2}{T + \beta_v} \tag{8.5}$$

where α_v and β_v are the Varshni coefficients. Table 18.2 gives these coefficients for several semiconductors.

Measuring the wavelength of the onset of strong absorption is a useful temperature probe in direct gap semiconductors. Kyuma et al. (1982) developed a fiber optic temperature sensor based on the increase of the absorption edge wavelength with increasing T. This was a self-contained unit with the output from an LED transmitted through a fiber to the wafer sensor (a small piece of a semiconductor), and back through another fiber to the detector. A second LED, operating at a different λ, was used as a reference for more precise measurements.

Wafer transmission is fast becoming a very important real-time probe during film processing. In early work, Hellman and Harris (1987) monitored the GaAs absorption edge for wafer thermometry, using the MBE substrate heater as the light source. The band gap of GaAs decreases by ~50 meV per 100°C in the temperature range used in MBE growth. They cited an accuracy of better than 10°C and a precision of ±2°C. More detailed studies of substrate absorption were conducted in continuing work by this group (Lee et al., 1991). Lee et al. (1991) also proposed another temperature probe, in which the peaks in the transmission from (and reflection through) a quarter-wave filter of alternating GaAs/AlAs layers are monitored. This method senses the temperature dependence of the refractive index, and is analogous to reflection interferometry described in Section 9.5.2.

Real-time thermometry becomes necessary when the wafer temperature changes during growth. This can occur in MBE when the amount of radiation the wafer absorbs from the heating lamps changes during deposition, which is particularly important when depositing semiconductors with smaller band gaps on substrates with larger band gaps. Without proper monitoring and control, this can lead to large variations in the quality of the material grown. Shanabrook *et al.* (1993) monitored the transmission of infrared light from the substrate heater through the GaAs substrate to control temperature during MBE, as is illustrated in Figure 8.33. IR light imaged from a portion of the wafer was transmitted by an optical fiber to a monochromator, where the dispersed light was detected with either a Ge or Si photodiode. The temperature was determined by determining the point of inflection $\mathcal{E}_P(T)$ in the transmitted spectrum, or equivalently, the peak in the first derivative of the transmitted light, where the transmitted light is normalized by the infrared light monitored with the wafer absent. This inflection point is a measure of the onset of absorption above the band gap for direct band-to-band absorption.

Kirillov and Powell (1992) demonstrated that for a GaAs wafer of thickness h (in cm), $\mathcal{E}_P$ varies with temperature as

$$\mathcal{E}_P(T,h) = \mathcal{E}_P(T=0,h) - \frac{k_1 T^2}{T + k_2} \quad (8.6)$$

where $k_1 = 5.85 \times 10^{-4}$ eV/K and $k_2 = 204$ K, and the $T = 0$ inflection point depends on wafer thickness as

$$\mathcal{E}_P(T=0,h) \cong \mathcal{E}_{bg} + \frac{1}{B} \ln\left(\frac{1}{hA}\right) \quad (8.7)$$

where $A = 5000$ cm^{-1}, $B = 110$ eV^{-1}, and $\mathcal{E}_{bg} = 1.525$ eV. Equations 8.6

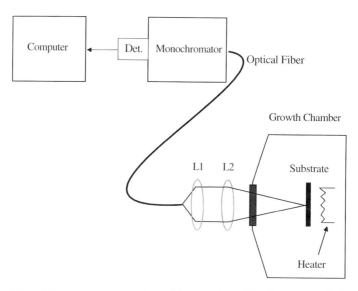

Figure 8.33 Schematic representation of transmission of IR from the radiative heater through a substrate in an MBE growth chamber to measure the substrate temperature (from Shanabrook *et al.*, 1993).

8.4 Transmission through Substrates for Thermometry

and 8.7 can be obtained by using the Varshni expression (Equation 8.5), the absorption coefficient for direct transitions (Equation 4.41), and Beer's law (Equation 8.1).

Using this method, Shanabrook and co-workers saw the wafer temperature increase by as much as ~150°C during the MBE deposition of films with band gaps smaller than the GaAs substrate, such as GaSb or InAs, because of increased wafer heating by film absorption (Figure 8.34). The thermocouple attached to the substrate did not indicate these temperature changes (Shanabrook *et al.*, 1993), while optical pyrometry agreed well with transmission experiments, within ±7°C (Katzer and Shanabrook, 1993) only when corrected for the usual sources of error in pyrometric measurements. These errors include (1) stray light from sources of light, such as the MBE effusion cells and substrate heater, and even control lights on electronics displays, (2) MBE coatings on the optical viewport, and (3) changes in emissivity during growth. D. S. Katzer (Naval Research Laboratory, private communication, 1995) was able to improve the mobility of HEMT (high electron mobility transistor) structures when he monitored the substrate temperature by infrared transmission during the MBE growth of the GaAs/AlGaAs heterojunctions used to make these devices.

These studies were conducted with wafers that were polished on both sides. When this technique is used with wafers that are roughened on one

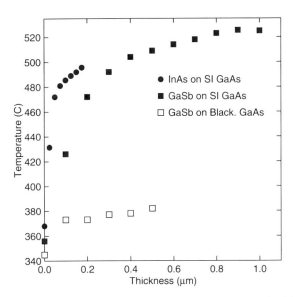

Figure 8.34 Change in substrate temperature during the MBE growth of GaSb and InAs on GaAs substrates as a function of film thickness, measured by optical transmission (from Shanabrook *et al.*, 1993).

side, there can be changes in the absorption paths and transmission losses from scattering, both due to roughness. Therefore, calibration errors can occur when data for wafers polished on both sides are used in experiments where the wafers are polished only on one side, and even when the calibration wafers have different degrees of backside roughness.

The strength of this thermometric technique is its excellent Run-to-Run precision. Although good accuracy is possible, it is not always necessary for real-time monitoring and control. A temperature diagnostic based on wafer transmission has been marketed commercially.

A closely related technique has been reported by Weilmeier *et al.* (1991) and Johnson *et al.* (1993) to monitor temperature during MBE growth on wafers that are polished on one side and textured on the other. It combines features of transmission and laser light scattering (Section 11.2.1). White light from a tungsten–halogen lamp is focused on the polished side of the wafer. Near-band-edge and below-gap radiation is transmitted through the GaAs wafer, diffusely reflected from the backside, and transmitted through the wafer again. The light reflecting from the wafer is then collected at angle removed from that of specularly reflected light. This diffusely reflected light is then spectrally analyzed as in the "direct" transmission method. The temperature sensitivity of this technique, and that of the closely related methods described earlier, is seen by the shift in the band edge with temperature in Figure 8.35. Diffuse reflection of the white-light source from the front surface is a potential background signal. For example, the background

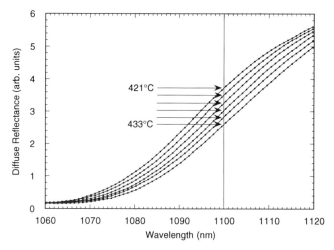

Figure 8.35 Successive diffuse reflection spectra of a GaAs substrate that is slowly increasing in temperature. These spectra monitor transmission through a wafer, diffuse reflection at the back surface, and transmission through the wafer again (from Johnson *et al.*, 1993).

level was observed to increase after oxide desorption, which roughened the front surface. The relative sensitivity of this method is reported to be better than 1°C. Johnson et al. (1994) used this method to monitor the temperature distribution of GaAs substrates in an MBE system for different indium-free wafer mounting techniques. The lateral spatial resolution was 3 mm and the temperature resolution was 0.4°C.

This thermometry technique is not limited to direct gap semiconductors. Analogous measurements have been made in indirect gap semiconductors by Sturm et al. (1990), who monitored the temperature of silicon wafers in a quartz-walled RTP reactor by measuring the transmission of semiconductor lasers operating at 1.3 and 1.55 μm (Figure 8.36). This is a potential real-time sensor of wafer temperature during RTP. The diode lasers were modulated and the transmitted signals were detected by lock-in detection to avoid interference from the RTP lamps. The transmission signal was normalized by its value at room temperature to remove the dependence on laser power, detector efficiency, reflection at the interfaces, and back-surface roughness, and is shown in Figure 8.37. Multiple reflections in the wafer were not significant. This method is sensitive to the changes in (below-band gap) absorption, and is weakly affected by changes in wafer thickness due to thermal expansion. A change in 1°C was easily detected.

Sturm and Reaves (1992) showed that in lightly doped Si wafers from 500 to 800°C absorption at 1.30 μm is predominantly by phonon-assisted band-to-band absorption (Equations 4.42–4.44), while at 1.55 μm it is domi-

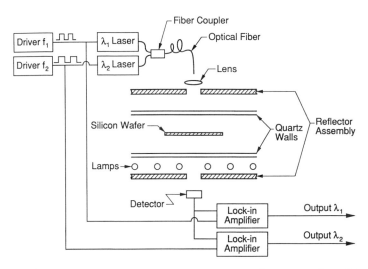

Figure 8.36 Schematic diagram of a rapid thermal processing system adapted for wafer thermometry by infrared transmission measurements at two wavelengths (from Sturm and Reaves, 1992). (© 1992 IEEE.)

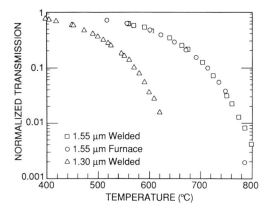

Figure 8.37 Optical transmission of a silicon wafer at elevated temperature at 1.3 and 1.55 μm, normalized by the transmission at room temperature (from Sturm et al., 1990).

nated by free-carrier absorption by electrons and holes (Equation 4.45) and depends on the doping in the wafer. The relative importance of these two effects at these wavelengths is seen in Figure 8.38 for n-type doped Si. Neither wavelength is useful for measuring the temperature of Si wafers below 500°C because of weak absorption, but use of shorter wavelengths, ~1.05–1.15 μm, should enable measurements down to 200°C and lower. Wavelengths longer than 1.55 μm are needed to measure temperatures above ~800°C in silicon to increase the transmitted signals.

Sturm et al. (1991) showed that this technique could be used to determine temperature during RTP growth of GeSi layers on Si. They examined the transmitted fraction for bare Si wafers and with GeSi overlayers; in regimes of interest the deposit layers have little effect on the diagnostic.

Adel et al. (1992) and Duek et al. (1993) have developed a probe that monitors light that is incident on the backside of a semiconductor wafer and is then reflected. This technique senses near-band-gap absorption in a semiconductor wafer by internal reflection of incident light that is transmitted through the wafer, and it has been used to monitor the backside temperature of Si wafers during plasma etching. At photon energies above the band gap, only light reflected from the back surface is detected, while below the band gap light transmitted through the wafer, reflected from the front surface, and transmitted through the wafer again is also observed. The method developed by Johnson et al. (1993), and cited above, is very similar, except that it senses from the front of the wafer, rather than from the back. Roth et al. (1994) extended this band-edge reflection method to GaAs, InP, and $Cd_{0.96}Zn_{0.04}Te$ substrates, and used it to probe temperature during the MBE growth of ZnTe on Si.

8.4 Transmission through Substrates for Thermometry

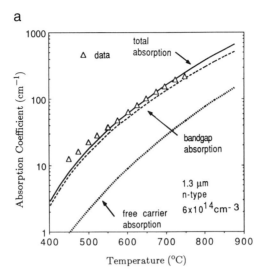

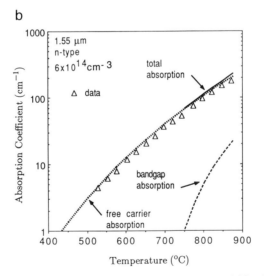

Figure 8.38 Absorption coefficient of *n*-type Si at (a) 1.30 μm and (b) 1.55 μm vs. temperature. Experimental data are compared to the model for free-carrier, band-to-band (band gap), and total absorption (from Sturm and Reaves, 1992). (© 1992 IEEE.)

In another thermometric application of transmission, Friberg and Kajanto (1989) have proposed a temperature sensor that is based on near-band-gap absorption in single-crystal silicon.

References

M. E. Adel, Y. Ish-Shalom, and H. Gilboa, *SPIE* **1803,** 290 (1992).
I. S. Adhihetty, J. A. McGuire, B. Wangmaneerat, T. M. Niemczyk, and D. M. Haaland, *Anal. Chem.* **63,** 2329 (1991).
G. P. Agrawal and N. K. Dutta, *Long-Wavelength Semiconductor Lasers.* Van Nostrand-Reinhold, New York, 1986.
S. D. Allen, R. Y. Jan, S. M. Mazuk, and S. D. Vernon, *J. Appl. Phys.* **58,** 327 (1985).
S. R. Armstrong, R. D. Hoare, M. E. Pemble, I. M. Povey, A. Stafford, A. G. Taylor, and J. O. Williams, *J. Cryst. Growth* **124,** 10 (1992).
D. M. Back, *Phys. Thin Films* **15,** 265 (1991).
P. Baltayan, F. Hartmann, I. Hikmet, and N. Sadeghi, *J. Chem. Phys.* **97,** 5417 (1992).
R. F. Barrow, ed., *Molécules Diatomiques: Bibliographie Critique de Données Spectroscopiques* [Diatomic Molecules: A Critical Bibliography of Spectroscopic Data], Vol. I. Éditions du Centre National de la Recherche Scientifique, Paris, 1973.
R. F. Barrow, ed., *Molécules Diatomiques: Bibliographie Critique de Données Spectroscopiques* [Diatomic Molecules: A Critical Bibliography of Spectroscopic Data], Vol. II. Éditions du Centre National de la Recherche Scientifique, Paris, 1975.
S. Bashkin and J. O. Stoner, Jr., *Atomic Energy Levels and Grotrian Diagrams,* Vol. I (1975), Vol. II (1978), Vol. III (1981), Vol. IV (1982). North-Holland, Amsterdam.
S. J. Benerofe, C. H. Ahn, M. M. Wang, K. E. Kihlström, K. B. Do, S. B. Arnason, M. M. Fejer, T. H. Geballe, M. R. Beasley, and R. H. Hammond, *J. Vac. Sci. Technol. B* **12,** 1217 (1994).
G. C. Bjorklund, *Opt. Lett.* **5,** 15 (1980).
W. E. Blass and G. W. Halsey, *Deconvolution of Absorption Spectra.* Academic Press, New York, 1981.
D. M. Bomse, A. C. Stanton, and J. A. Silver, *Appl. Opt.* **31,** 718 (1992).
J. E. Butler, N. Bottka, R. S. Sillmon, and D. K. Gaskill, *J. Cryst. Growth* **77,** 163 (1986).
F. G. Celii, P. E. Pehrsson, H.-T. Wang, and J. E. Butler, *Appl. Phys. Lett.* **52,** 2043 (1988).
S. A. Chalmers and K. P. Killeen, *Appl. Phys. Lett.* **63,** 3131 (1993).
C. J. Chen and R. M. Osgood, Jr., *Appl. Phys. A* **31,** 171 (1983a).
C. J. Chen and R. M. Osgood, Jr., *Chem. Phys. Lett.* **98,** 363 (1983b).
N. H. Cheung, Q. Y. Ying, J. P. Zheng, and H. S. Kwok, *J. Appl. Phys.* **69,** 6349 (1991).
W. W. Clark, III and F. C. De Lucia, *J. Chem. Phys.* **74,** 3139 (1981).
T. A. Cleland and D. W. Hess, *J. Appl. Phys.* **64,** 1068 (1988).
T. A. Cleland and D. W. Hess, *J. Vac. Sci. Technol. B* **7,** 35 (1989).
J. M. Cook and T. A. Miller, in *Plasma Diagnostics* (O. Auciello and D. L. Flamm, eds.), Vol. 1, Chapter 6, p. 313. Academic Press, Boston, 1989.
A. Corney, *Atomic and Laser Spectroscopy.* Oxford Univ. Press (Clarendon), Oxford, 1977.
P. B. Davies, *J. Phys. Chem.* **85,** 2599 (1981).
P. B. Davies and P. M. Martineau, *Appl. Phys. Lett.* **57,** 237 (1990).

P. B. Davies and W. J. Rothwell, *J. Chem. Phys.* **83,** 1496 (1985).
P. B. Davies, W. Lewis-Bevan, and D. K. Russell, *J. Chem. Phys.* **75,** 5602 (1981).
P. B. Davies, N. A. Isaacs, S. A. Johnson, and D. K. Russell, *J. Chem. Phys.* **83,** 2060 (1985).
A. C. Dillon, P. Gupta, M. B. Robinson, A. S. Bracker, and S. M. George, *Mater. Res. Soc. Symp. Proc.* **204,** 339 (1991).
V. M. Donnelly, in *Plasma Diagnostics* (O. Auciello and D. L. Flamm, eds.), Vol. 1, Chapter 1, p. 1. Academic Press, Boston, 1989.
R. Duek, N. Vofsi, S. Mangan, and M. Adel, *Semicond. Int.,* July, p. 208 (1993).
R. S. Eng, J. F. Butler, and K. J. Linden, *Opt. Eng.* **19,** 945 (1980).
K. M. Evenson, *Faraday Discuss. Chem. Soc.* **71,** 7 (1981).
K. M. Evenson, R. J. Saykally, D. A. Jennings, R. F. Curl, and J. M. Brown, *Chem. Biochem. Appl. Lasers* **5,** 95 (1980).
G. H. Fan, R. D. Hoare, M. E. Pemble, I. M. Povey, A. G. Taylor, and J. O. Williams, *J. Cryst. Growth* **124,** 49 (1992).
L. A. Farrow, *J. Chem. Phys.* **82,** 3625 (1985).
S. C. Foster and A. R. W. McKellar, *J. Chem. Phys.* **81,** 3424 (1984).
J. E. Franke, T. M. Niemczyk, and D. M. Haaland, *Spectrochim. Acta Part A* **50A,** 1687 (1994).
A. T. Friberg and I. Kajanto, *Rev. Sci. Instrum.* **60,** 2764 (1989).
Y. Fujita, S. Fujii, and T. Iuchi, *J. Vac. Sci. Technol. A* **7,** 276 (1989).
D. K. Gaskill, V. Kolubayev, N. Bottka, R. S. Sillmon, and J. E. Butler, *J. Cryst. Growth* **93,** 127 (1988).
D. B. Geohegan, in *Laser Ablation of Electronic Materials: Basic Mechanisms and Applications* (E. Fogarassy and S. Lazare, eds.), p. 73. North-Holland Publ., Amsterdam, 1992.
D. B. Geohegan, in *Pulsed Laser Deposition of Thin Films* (G. Hubler and D. B. Chrisey, eds.), Chapter 5, p. 115. Wiley (Interscience), New York, 1994.
D. B. Geohegan and D. N. Mashburn, *Appl. Phys. Lett.* **55,** 2345 (1989).
M. Gerhtz, G. C. Bjorklund, and E. A. Whittaker, *J. Opt. Soc. Am. B* **2,** 1510 (1985).
K. P. Giapis, N. Sadeghi, J. Margot, R. A. Gottscho, and T. C. J. Lee, *J. Appl. Phys.* **73,** 7188 (1993).
R. M. Gilgenbach and P. L. G. Ventzek, *Appl. Phys. Lett.* **58,** 1597 (1991).
M. J. Goeckner, M. A. Henderson, J. A. Meyer, and R. A. Breun, *J. Vac. Sci. Technol. A* **12,** 3120 (1994).
J. E. Greene, *J. Vac. Sci. Technol.* **15,** 1718 (1978).
C. S. Gudeman, M. H. Begemann, J. Pfaff, and R. J. Saykally, *Phys. Rev. Lett.* **50,** 727 (1983).
J. Haigh and S. O'Brien, *J. Cryst. Growth* **67,** 75 (1984a).
J. Haigh and S. O'Brien, *J. Cryst. Growth* **68,** 550 (1984b).
K. L. Haller and P. C. D. Hobbs, *SPIE* **1435,** 298 (1991).
M. Haverlag, E. Stoffels, W. W. Stoffels, G. M. W. Kroesen, and F. J. de Hoog, *J. Vac. Sci. Technol. A* **12,** 3102 (1994).
G. A. Hebner, K. P. Killeen, and R. M. Biefeld, *J. Cryst. Growth* **98,** 293 (1989a).
G. A. Hebner, R. M. Biefeld, and K. P. Killeen, *Mater. Res. Soc. Symp. Proc.* **144,** 73 (1989b).
E. S. Hellman and J. S. Harris, Jr., *J. Cryst. Growth* **81,** 38 (1987).
D. Herriott, H. Kogelnick, and R. Kompfner, *Appl. Opt.* **3,** 523 (1964).

J. F. Hershberger, J. Z. Chou, G. W. Flynn, and R. E. Weston, Jr., *Chem. Phys. Lett.* **149,** 51 (1988).
G. Herzberg, *Atomic Spectra and Atomic Structure.* Dover, New York, 1937.
G. Herzberg, *Molecular Spectra and Molecular Structure. II. Infrared and Raman Spectra of Polyatomic Molecules.* Van Nostrand-Reinhold, New York, 1945.
G. Herzberg, *Molecular Spectra and Molecular Structure. I. Spectra of Diatomic Molecules.* Van Nostrand-Reinhold, New York, 1950.
G. Herzberg, *Molecular Spectra and Molecular Structure. III. Electronic Spectra and Electronic Structure of Polyatomic Molecules.* Van Nostrand-Reinhold, New York, 1966.
G. W. Hills, *Magn. Reson. Rev.* **9,** 15 (1984).
E. Hirota, *High Resolution Spectroscopy of Transient Molecules.* Springer-Verlag, Berlin, 1985.
E. Hirota and K. Kawaguchi, *Annu. Rev. Phys. Chem.* **36,** 53 (1985).
P. C. D. Hobbs, *SPIE* **1376,** 216 (1990).
P. C. D. Hobbs, *Opt. Photon. News,* April, p. 17 (1991).
G. R. Howes, C. J. Gamsky, and J. W. Taylor, *J. Vac. Sci. Technol. B* **12,** 3868 (1994).
K. P. Huber and G. Herzberg, *Molecular Spectra and Molecular Structure. IV. Constants of Diatomic Molecules.* Van Nostrand-Reinhold, New York, 1979.
D. E. Ibbotson, D. L. Flamm, and V. M. Donnelly, *J. Appl. Phys.* **54,** 5974 (1983).
N. Itabashi, K. Kato, N. Nishiwaki, T. Goto, C. Yamada, and E. Hirota, *Jpn. J. Appl. Phys.* **27,** L1565 (1988).
M. E. Jacox, *J. Phys. Chem. Ref. Data* **13,** 945 (1984).
M. E. Jacox, *J. Phys. Chem. Ref. Data* **17,** 269 (1988).
M. E. Jacox, *J. Phys. Chem. Ref. Data* **19,** 1387 (1990).
M. E. Jacox, *Vibrational and Electronic Energy Levels of Polyatomic Transient Molecules,* Monograph No. 3, *J. Phys. Chem. Ref. Data* (1994).
J. M. Jasinski, E. A. Whittaker, G. C. Bjorklund, R. W. Dreyfus, R. D. Estes, and R. E. Walkup, *Appl. Phys. Lett.* **44,** 1155 (1984).
G. M. Jellum and D. B. Graves, *J. Appl. Phys.* **67,** 6490 (1990).
S. R. Johnson, C. Lavoie, T. Tiedje, and J. A. Mackenzie, *J. Vac. Sci. Technol. B* **11,** 1007 (1993).
S. R. Johnson, C. Lavoie, E. Nodwell, M. K. Nissen, T. Tiedje, and J. A. Mackenzie, *J. Vac. Sci. Technol. B* **12,** 1225 (1994).
R. F. Karlicek, Jr. and A. Bloemeke, *J. Cryst. Growth* **73,** 364 (1985).
R. F. Karlicek, Jr., J. A. Long, and V. M. Donnelly, *J. Cryst. Growth* **68,** 123 (1984).
R. F. Karlicek, Jr., B. Hammarlund, and J. Ginocchio, *J. Appl. Phys.* **60,** 794 (1986).
D. S. Katzer and B. V. Shanabrook, *J. Vac. Sci. Technol. B* **11,** 1003 (1993).
T. Kawahara, A. Yuuki, and Y. Matsui, *Jpn. J. Appl. Phys.* **31,** 2925 (1992).
K. K. King, V. Tavitian, D. B. Geohegan, E. A. P. Cheng, S. A. Piette, F. J. Scheltens, and J. G. Eden, *Mater. Res. Soc. Symp. Proc.* **75,** 189 (1987).
D. Kirillov and R. A. Powell, U.S. Patent 5,118,200 (1992).
M. E. Klausmeier-Brown, J. N. Eckstein, I. Bozović, and G. F. Virshup, *Appl. Phys. Lett.* **60,** 657 (1992).
J. C. Knights, J. P. M. Schmitt, J. Perrin, and G. Guelachvili, *J. Chem. Phys.* **76,** 3414 (1982).
N. Kobayashi, H. Goto, and M. Suzuki, *J. Appl. Phys.* **69,** 1013 (1991).
N. Kobayashi, Y. Nakamura, H. Goto, and Y. Homma, *J. Appl. Phys.* **73,** 4637 (1993).

References

R. R. Krchnavek, H. H. Gilgen, J. C. Chen, P. S. Shaw, T. J. Licata, and R. M. Osgood, Jr., *J. Vac. Sci. Technol. B* **5**, 20 (1987).

T. D. Kunz, R. F. Menefee, B. D. Krenek, L. G. Fredin, and M. J. Berry, *High Temp. Sci.* **27**, 459 (1990).

Y. A. Kuritsyn, in *Laser Analytical Spectrochemistry* (V. S. Letokhov, ed.), Chapter 4, p. 152. Adam-Hilger, Bristol, 1986.

K. Kyuma, S. Tai, T. Sawada, and M. Nunoshita, *IEEE J. Quantum Electron.* **QE-18**, 676 (1982).

J. E. Lawler, E. A. Den Hartog, and W. N. G. Hitchon, *Phys. Rev. A* **43**, 4427 (1991).

W. S. Lee, G. W. Yoffe, D. G. Schlom, and J. S. Harris, Jr., *J. Cryst. Growth* **111**, 131 (1991).

G. W. Loge, N. Nereson, and H. Fry, *Appl. Opt.* **33**, 3161 (1994).

S. K. Loh and J. M. Jasinski, *J. Chem. Phys.* **95**, 4914 (1991).

C. Lu and Y. Guan, *J. Vac. Sci. Technol. A* **13**, 1797 (1995).

A. W. Mantz, *Microchem. J.* **50**, 351 (1994).

A. W. Mantz, *Spectrochim. Acta Part A* **51**, 2211 (1995).

D. Mazzarese, A. Tripathi, W. C. Conner, K. A. Jones, L. Calderon, and D. W. Eckart, *J. Electron. Mater.* **18**, 369 (1989).

A. R. McKellar, *Faraday Discuss. Chem. Soc.* **71**, 63 (1981).

D. C. Miller, J. J. O'Brien, and G. H. Atkinson, *J. Appl. Phys.* **65**, 2645 (1989).

A. C. G. Mitchell and M. W. Zemansky, *Resonance Radiation and Excited Atoms*. Cambridge, Univ. Press, Cambridge, UK, 1971.

R. Mitzner, A. Rosenfeld, and R. König, *Appl. Surf. Sci.* **69**, 180 (1993).

MKS Instruments, Inc., product information, 1995.

C. E. Moore, *Atomic Energy Levels,* NBS Circular 467, Vol. I (1949); Vol. II (1952); Vol. III (1958).

C. E. Moore and P. W. Merrill, *Partial Grotrian Diagrams of Astrophysical Interest,* National Standard Reference Data Series, NSRDS-NBS 23, 1968.

J. N. Moore, P. A. Hansen, and R. M. Hochstrasser, *Chem. Phys. Lett.* **138**, 110 (1987).

W. M. Moreau, *Semiconductor Lithography: Principles, Practices, and Materials.* Plenum, New York, 1988.

J. A. Mucha and J. Washington, *Mater. Res. Soc. Symp. Proc.* **334**, 31 (1994).

K. Nakamoto, *Infrared and Raman Spectra of Inorganic and Coordination Compounds,* 4th ed. Wiley, New York, 1986.

T. M. Niemczyk, B. Wangmaneerat, and D. M. Haaland, *J. Vac. Sci. Technol. A* **12**, 835 (1994).

J. Nishizawa, *J. Cryst. Growth* **56**, 273 (1982).

J. Nishizawa and N. Hayasaka, *Thin Solid Films* **92**, 189 (1982).

J. Nishizawa and H. Nihira, *J. Cryst. Growth* **45**, 82 (1978).

J. Nishizawa and T. Kurabayashi, *J. Electrochem. Soc.* **130**, 413 (1983).

J. Nishizawa and M. Saito, *J. Cryst. Growth* **52**, 213 (1981).

J. Nishizawa, H. Shimawaki, and Y. Sakuma, *J. Electrochem. Soc.* **133**, 2567 (1986).

J. Nishizawa, T. Kurabayashi, and J. Hoshina, *J. Electrochem. Soc.* **134**, 502 (1987a).

J. Nishizawa, H. Shimawaki, and Y. Sakuma, *J. Electrochem. Soc.* **134**, 3155 (1987b).

J. Nishizawa, H. Shimawaki, and Y. Sakuma, *J. Electrochem. Soc.* **135**, 1813 (1988).

J. Nishizawa, K. Aoki, S. Suzuki, and K. Kikuchi, *J. Electrochem. Soc.* **137**, 1898 (1990).

K. Nishikawa, K. Ono, M. Tuda, T. Oomori, and K. Namba, *Symp. Dry Process., 16th,* Tokyo, *1994,* III-4 (1994).

J. J. O'Brien and G. H. Atkinson, *Chem. Phys. Lett.* **130,** 321 (1986).

D. B. Oh, A. C. Stanton, H. M. Anderson, and M. P. Splichal, *J. Vac. Sci. Technol. B* **13,** 954 (1995).

H. Okabe, *Photochemistry of Small Molecules.* Wiley, New York, 1978.

J. A. O'Neill and J. Singh, *J. Appl. Phys.* **76,** 5967 (1994).

J. A. O'Neill and J. Singh, *J. Appl. Phys.* **77,** 497 (1995).

J. A. O'Neill, J. Singh, and G. S. Gifford, *J. Vac. Sci. Technol. A* **8,** 1716 (1990).

J. A. O'Neill, M. L. Passow, and T. J. Cotler, *J. Vac. Sci. Technol. A* **12,** 839 (1994).

J. F. Osmundsen, C. C. Abele, and J. G. Eden, *J. Appl. Phys.* **57,** 2921 (1985).

A. W. Ott, A. C. Dillon, H. K. Eaton, S. M. George, and J. D. Way, in *Microphysics of Surfaces: Nanoscale Processing,* OSA Tech. Dig. Ser., Vol. 5, p. 29. Optical Society of America, Washington, DC., 1995.

J. Park, Y. Lee and G. W. Flynn, *Chem. Phys. Lett.* **186,** 441 (1991).

R. W. B. Pearse and A. G. Gaydon, *The Identification of Molecular Spectra,* 4th ed. Chapman & Hall, London, 1976.

H. U. Poll, D. Hinze, and H. Schlemm, *Appl. Spectrosc.* **36,** 445 (1982).

W. J. Price, *Analytical Atomic Absorption Spectrophotometry.* Heyden, London, 1972.

A. D. Richards and H. H. Sawin, *J. Appl. Phys.* **62,** 799 (1987).

A. D. Richards, B. E. Thompson, K. D. Allen, and H. H. Sawin, *J. Appl. Phys.* **62,** 792 (1987).

J. A. Roth, T. J. de Lyon, and M. E. Adel, *Mater. Res. Soc. Symp. Proc.* **324,** 353 (1994).

M. Rothschild, in *Laser Microfabrication: Thin Film Processes and Lithography* (D. J. Ehrlich and J. Y. Tsao, eds.), Chapter 3, p. 163. Academic Press, Boston, 1989.

H. F. Sakeek, T. Morrow, W. G. Graham, and D. G. Walmsley, *Appl. Phys. Lett.* **59,** 3631 (1991).

S. Salim, K. F. Jensen, and R. D. Driver, *J. Cryst. Growth* **145,** 28 (1994).

S. Salim, C. K. Lim, and K. F. Jensen, *Chem. Mater.* **7,** 507 (1995).

S. Salim, C. A. Wang, R. D. Driver, and K. F. Jensen, *J. Cryst. Growth* (1996, in press).

D. W. Setser, ed., *Reactive Intermediates in the Gas Phase: Generation and Monitoring.* Academic Press, New York, 1979.

B. V. Shanabrook, J. R. Waterman, J. L. Davis, R. J. Wagner, and D. S. Katzer, *J. Vac. Sci. Technol. B* **11,** 994 (1993).

R. S. Sillmon, N. Bottka, J. E. Butler, and D. K. Gaskill, *J. Cryst. Growth* **77,** 73 (1986).

J. A. Silver, *Appl. Opt.* **31,** 707 (1992).

A. C. Stanton and C. E. Kolb, *J. Chem. Phys.* **72,** 6637 (1980).

A. J. Stirling and W. D. Westwood, *J. Appl. Phys.* **41,** 742 (1970).

J. C. Sturm and C. M. Reaves, *IEEE Trans. Electron Devices* **ED-39,** 81 (1992).

J. C. Sturm, P. V. Schwartz, and P. M. Garone, *Appl. Phys. Lett.* **56,** 961 (1990).

J. C. Sturm, P. M. Garone, and P. V. Schwartz, *J. Appl. Phys.* **69,** 542 (1991).

S. N. Suchard, ed., *Spectroscopic Data,* Vol. 1, Heteronuclear Diatomic Molecules, Parts A and B. IFI/Plenum, New York, 1975.

References

S. N. Suchard and J. E. Melzer, eds., *Spectroscopic Data,* Vol. 2. Homonuclear Diatomic Molecules. IFI/Plenum, New York, 1976.

H. C. Sun and E. A. Whittaker, *Appl. Opt.* **31,** 4998 (1992).

H. C. Sun and E. A. Whittaker, *Appl. Phys. Lett.* **63,** 1035 (1993).

H. C. Sun, E. A. Whittaker, Y. W. Bae, C. K. Ng, V. Patel, W. H. Tam, S. McGuire, B. Singh, and B. Gallois, *Appl. Opt.* **32,** 885 (1993).

H. C. Sun, V. Patel, B. Singh, C. K. Ng, and E. A. Whittaker, *Appl. Phys. Lett.* **64,** 2779 (1994).

K. Tachibana, H. Tadokoro, H. Harima, and Y. Urano, *J. Phys. D* **15,** 177 (1982).

L. L. Tedder, J. E. Crowell, and M. A. Logan, *J. Vac. Sci. Technol. A* **9,** 1002 (1991).

B. A. Thrush, *Acc. Chem. Res.* **14,** 116 (1981).

Y. P. Varshni, *Physica* **34,** 149 (1967).

P. L. G. Ventzek, R. M. Gilgenbach, C. H. Ching, and R. A. Lindley, *J. Appl. Phys.* **72,** 1696 (1992).

R. C. Wamsley, J. E. Lawler, J. H. Ingold, L. Bigio, and V. D. Roberts, *Appl. Phys. Lett.* **57,** 2416 (1990).

M. Watts, T. Perera, B. Ozarski, D. Meyers, and R. Tan, *Solid State Technol.* **31**(7), 59 (1988).

M. K. Weilmeier, K. N. Colbow, T. Tiedje, T. van Buuren, and L. Xu, *Can. J. Phys.* **69,** 422 (1991).

E. B. Wilson, Jr., J. C. Decius, and P. C. Cross, *Molecular Vibrations: The Theory of Infrared and Raman Vibrational Spectra.* Dover, New York, 1980.

R. C. Woods, R. L. McClain, L. J. Mahoney, E. A. Den Hartog, H. Persing, and J. S. Hamers, *SPIE* **1594,** 366 (1991).

J. Wormhoudt, *J. Vac. Sci. Technol. A* **8,** 1722 (1990).

J. Wormhoudt, A. C. Stanton, and J. Silver, *SPIE* **452,** 88 (1983).

J. Wormhoudt, A. C. Stanton, A. D. Richards, and H. H. Sawin, *J. Appl. Phys.* **61,** 142 (1987).

M. Xi, S. Salim, K. F. Jensen, and D. A. Bohling, *Mater. Res. Soc. Symp. Proc.* **334,** 169 (1994).

CHAPTER 9

Reflection

Reflection is a relatively simple, yet powerful, probe of films on a substrate and of the substrate itself. It can probe the interface between the substrate and the ambient medium above it, which is usually a gas, and in some configurations it is very sensitive to monolayers of adsorbates on a substrate. Also, reflection can be used to measure film thicknesses very accurately through optical interference in the film. Reflection probes can be easily installed on a wide range of thin film reactors. They are immune to the light collection issues in optical emission spectroscopy (OES), laser-induced fluorescence (LIF), and photoluminescence (PL) because the incident and reflected beams are well collimated. Still, some reflection-based probes, such as ellipsometry, are sensitive to other experimental factors, including wafer movement and the optical properties of reactor windows. Many reflection-based diagnostics can be employed in rotating-disk deposition systems.

The optical sources and detectors used in reflection are robust, straightforward to implement, relatively inexpensive, and involve established technologies. The optics of reflection have been well known for many decades. In simpler applications such as interferometry, data interpretation is trivial when the complex indices of refraction of all materials are known at the operating temperatures. In more demanding applications such as ellipsometry, where the goal may be to obtain thickness and composition information with monolayer-level sensitivity, data interpretation requires careful modeling of the material.

Within the context of this chapter, films and interfaces are assumed to be specular (smooth); i.e., the dimensions of any roughness are $\ll \lambda_{probe}$. Diffuse reflection due to surface roughness is described in Section 11.2.1 in the discussion of laser light scattering.

This chapter begins with a discussion of the optics of reflection. Then reflection diagnostics will be presented in three stages. First, methods that are mostly sensitive to the bulk properties of materials, thin films, and heterostructures will be discussed, such as reflection at an interface, interferometry, and photoreflectance. Then, reflection diagnostics that are specifically sensitive to surfaces are presented; these include surface infrared spectroscopies and surface differential spectroscopies, such as surface photoabsorption (SPA) and reflectance-difference (or anisotropy) spectroscopy (RDS/RAS). Finally, ellipsometry, a technique that senses bulk properties but has sufficient sensitivity to probe surfaces and interfaces, is discussed.

9.1 Optics of Reflection

The reflection of light waves at flat interfaces and the optics of thin films have been treated by Born and Wolf (1970), Klein and Furtak (1986), Hecht (1987), Knittl (1976), Macleod (1989), Potter (1985), Heavens (1964, 1965), Berning (1963), and others; Abelès (1971) has specifically addressed reflection from metallic films. Figure 9.1 illustrates the light beams involved in a reflection experiment at a single interface. The angle of incidence in medium 1 will be labeled θ_1 or just θ. This is the standard notation in much of the optics community, and is consistent with the notation in elastic light scattering (Chapter 11). In the ellipsometry community, this angle is often called ϕ (Azzam and Bashara, 1977). The incident (denoted by i) and reflected (r) beams make the same angle θ_1 with the normal to the interface. The transmitted (t) beam is refracted, and its propagation is described by the angle θ_2, given by Snell's law $\tilde{n}_1 \sin \theta_1 = \tilde{n}_2 \sin \theta_2$, as is illustrated in Figure 9.1. Here $\tilde{n}_1$ and $\tilde{n}_2$ are the respective complex indices of refraction of the materials; sometimes n_2/n_1 is denoted by n_{21}. For transparent materi-

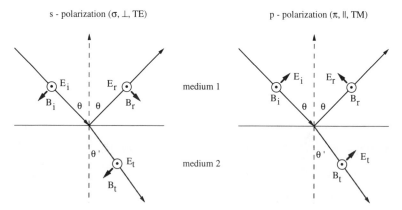

Figure 9.1 Incident (i), reflected (r), and transmitted (t) beams in reflection experiments at a single interface, with the electric (E) and magnetic (B) field vectors as indicated, in which the electric fields are either s- or p-polarized.

9.1 Optics of Reflection

als, the $\tilde{n}$ ($= n + ik$) are real ($k = 0$) and θ_2 is the angle of refraction. If medium 2 is absorbing ($k \neq 0$), the direction of the beam, i.e., the normal to the surface of constant real phase, is not exactly θ_2 (Klein and Furtak, 1986). The cited wavelength λ is that in vacuum unless otherwise stated. For most thin film diagnostics applications, it makes little difference whether the wavelength in the vacuum or the gas is used within the gas phase.

In most applications, it is usually a fine approximation to assume that each of the three beams is a plane wave. The wavevectors of the three beams are all in the same plane, called the place of incidence, that is defined by the normal to the interface and the wavevector of the incident beam. If $n_2 > n_1$, as in propagating into glass ($n_2 \approx 1.5$) from air ($n_1 \approx 1$), there is a transmitted beam for each θ. For $n_2 < n_1$, as from glass to air, there is a transmitted beam only for angles $\theta < \theta_c = \arcsin(n_2/n_1)$, where θ_c is known as the critical angle. When $\theta > \theta_c$, the beam is totally reflected, which is the basis for attenuated total internal reflection spectroscopy (Section 9.7.2).

For s (or σ, $\perp$, or transverse electric TE) linear polarization, the electric field vector is normal to the plane of incidence (points out of the paper in Figure 9.1); and for p (or π, $\|$, or transverse magnetic TM) polarization, it is in the plane of incidence (Figure 9.1).

The incident intensities with p and s polarizations are I_p^i and I_s^i; the electric field amplitudes of these beams are E_p^i and E_s^i, respectively. Similarly, the reflected beams are described by I_p^r, I_s^r, E_p^r, and E_s^r, and the transmitted beams by I_p^t, I_s^t, E_p^t, and E_s^t. There is no need to keep track of the transmitted beam when considering reflection at a single interface. For multiple interfaces, reflection from the structure involves the multiple transmissions and reflections that occur at each interface, which produce interference effects.

At each interface, the reflected fields can be related to the incident fields by $E_p^r = r_p E_p^i$ and $E_s^r = r_s E_s^i$. The coefficient for each polarization, r, will be called the *reflection coefficient*. Similarly, the transmitted fields are $E_p^t = t_p E_p^i$ and $E_s^t = t_s E_s^i$, where t is the *transmission coefficient*. For an arbitrary multilayer system, the reflection coefficients r_p and r_s can be calculated from known optical parameters and can also be measured. The coefficient r depends on the details of the structure of the material (including the temperature), the probe wavelength λ, angle of incidence θ and polarization.

Since $I \propto E^2$, the reflected intensities are related to the incident intensities by $I_p^r = R_p I_p^i$ and $I_s^r = R_s I_s^i$ where $R_p = |r_p|^2$ and $R_s = |r_s|^2$; $R_{p,s}$ is also the fraction of power that is reflected. $R_{p,s}$ will be called the *reflectance* for that polarization. The fraction of light that is reflected from a perfect interface between vacuum and a semiinfinite medium is the reflectivity for that medium. Similarly for the transmitted beams, $I_p^t = T_p I_p^i$ and $I_s^t = T_s I_s^i$, where $T_p = |t_p|^2$ and $T_s = |t_s|^2$. $T_{p,s}$ is the *transmittance* for each polarization. Clearly, $R_p + T_p = R_s + T_s = 1$ when there is no absorption. The appendix to this chapter describes the different names often used for r, t, R, and T, and the nuances in their usage.

The reflection from an optical element can be expressed by a 2×2 reflection matrix composed of reflection coefficients:

$$\begin{pmatrix} E_s \\ E_p \end{pmatrix}_{\text{reflected}} = \begin{pmatrix} r_{ss} & r_{sp} \\ r_{ps} & r_{pp} \end{pmatrix} \begin{pmatrix} E_s \\ E_p \end{pmatrix}_{\text{incident}} \quad (9.1)$$

For isotropic media $r_{sp} = r_{ps} = 0$ and r_{ss} and r_{pp} are the parameters r_s and r_p defined above.

There is no standard convention for the signs of the imaginary terms in $\tilde{\varepsilon}$ and $\tilde{n}$. The expressions presented here assume that the dielectric constant is written as $\tilde{\varepsilon} = \varepsilon' + i\varepsilon''$ and $\tilde{n} = n + ik$, for the reasons given in Chapters 2 and 4. (Very often, $\tilde{\varepsilon}$ is expressed as $\varepsilon_1 + i\varepsilon_2$, as is seen in the figures reproduced in this chapter from the literature. This notation is not used here in equations because it could lead to confusion with the subscript denoting the medium.) Some expressions in the literature differ from those presented here because they assume different sign conventions, which are sometimes unstated. Since data interpretation depends on this sign, consistency in notation is essential.

9.1.1 *Reflection for Simple Structures*

9.1.1.1 Two Semiinfinite Media

Consider the reflection of a beam at the interface between semiinfinite media labeled 1 and 2 (Figure 9.2). Both media are assumed to be isotropic, so the dielectric functions are scalars.

For *s*-polarization incident on this interface, the reflection and transmission coefficients are

$$r_s = \frac{\tilde{n}_1 \cos\theta_1 - \tilde{n}_2 \cos\theta_2}{\tilde{n}_1 \cos\theta_1 + \tilde{n}_2 \cos\theta_2} = -\frac{\sin(\theta_1 - \theta_2)}{\sin(\theta_1 + \theta_2)} \quad (9.2a)$$

$$t_s = \frac{2\tilde{n}_1 \cos\theta_1}{\tilde{n}_1 \cos\theta_1 + \tilde{n}_2 \cos\theta_2} = \frac{2\sin\theta_2 \cos\theta_1}{\sin(\theta_1 + \theta_2)} \quad (9.2b)$$

For *p*-polarization, they are

$$r_p = \frac{\tilde{n}_2 \cos\theta_1 - \tilde{n}_1 \cos\theta_2}{\tilde{n}_2 \cos\theta_1 + \tilde{n}_1 \cos\theta_2} = \frac{\tan(\theta_1 - \theta_2)}{\tan(\theta_1 + \theta_2)} \quad (9.3a)$$

$$t_p = \frac{2\tilde{n}_1 \cos\theta_1}{\tilde{n}_2 \cos\theta_1 + \tilde{n}_1 \cos\theta_2} = \frac{2\sin\theta_2 \cos\theta_1}{\sin(\theta_1 + \theta_2)\cos(\theta_1 - \theta_2)} \quad (9.3b)$$

The angle of incidence θ_1 in medium 1 and that angle θ_2 in medium 2 are related by Snell's law, and $\tilde{n}_1$ and $\tilde{n}_2$ can be complex. These equations are known as the Fresnel relations.

Later in this chapter, the polarization labels will be dropped when the type of polarization is understood. When applied to an arbitrary interface,

9.1 Optics of Reflection

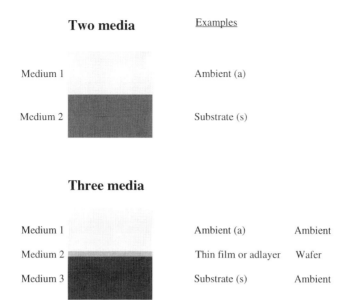

Figure 9.2 Illustration of two- and three-media material systems encountered in reflection experiments. [Sometimes these are denoted as two- and three-phase models (Section 9.11).]

say between media j and k, the reflection coefficient will be expressed as r_{jk} and the reflectance $R = |r|^2$ will be expressed as R_{jk}. Consequently, Equations 9.2a and 9.3a are r_{12} for the respective polarizations. In some instances, the two media will be called the ambient (subscript a) and substrate (subscript s), respectively, to conform with the notation used in some communities. In general, the angle of incidence in the ambient will be called θ instead of θ_1.

The reflection coefficients are sometimes expressed in a compact form that combines Equations 9.2 and 9.3 with Snell's law. For the ambient (a)/substrate (s) system,

$$r_s^0 = \frac{\tilde{n}_{a\perp} - \tilde{n}_{s\perp}}{\tilde{n}_{a\perp} + \tilde{n}_{s\perp}} \tag{9.4}$$

$$r_p^0 = \frac{\tilde{\varepsilon}_s \tilde{n}_{a\perp} - \tilde{\varepsilon}_a \tilde{n}_{s\perp}}{\tilde{\varepsilon}_s \tilde{n}_{a\perp} + \tilde{\varepsilon}_a \tilde{n}_{s\perp}} \tag{9.5}$$

where $\tilde{n}_{j\perp} = (\tilde{\varepsilon}_j - \tilde{\varepsilon}_a \sin^2\theta)^{1/2}$ and θ is the angle of incidence in the ambient. The superscript 0 on r indicates the absence of an overlayer.

Figure 9.3 plots the reflectance R_{12} from an air/GaAs interface where $\tilde{n}_{GaAs}$ is $\approx 4 + 2i$ at 400 nm. R increases monotonically with angle θ for s-polarization. For p-polarization, R decreases to a minimum at the pseudo-

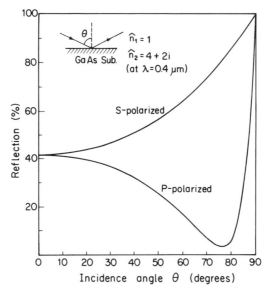

Figure 9.3 Calculated reflectance of GaAs versus angle of incidence at 400 nm (from Kobayashi and Horikoshi, 1990).

Brewster angle and then increases at larger angles. This is also seen for metals (Figure 9.4a). For lossless dielectrics ($k = 0$), $R = 0$ at Brewster's angle θ_B, where $\tan \theta_B = n_2/n_1$, as is seen in Figure 9.5a. The reflection coefficient r changes sign at θ_B. When the angle of incidence is at or near the pseudo-Brewster angle, the magnitude of the reflected beam is very sensitive to small changes at the interface, such as the appearance of very thin absorbing layers at a gas/solid interface. This is the basis of surface photoabsorption (SPA) and p-polarized reflectance spectroscopies described in Section 9.9.

At normal incidence, the reflectance between two media is

$$R_{12} = \frac{(n_2 - n_1)^2 + (k_2 - k_1)^2}{(n_2 + n_1)^2 + (k_2 + k_1)^2} \tag{9.6}$$

There is no distinction between s- and p-polarization. At the interface of vacuum and a material with $\tilde{n} = n + ik$, this becomes

$$R_{12} = \frac{(n - 1)^2 + k^2}{(n + 1)^2 + k^2} \tag{9.7}$$

which is also the reflectivity for that material. Note that the difference in the signs of the reflection coefficients in Equations 9.2a and 9.3a at $\theta = 0°$ is due to the different conventions for E_i and E_r in Figure 9.1.

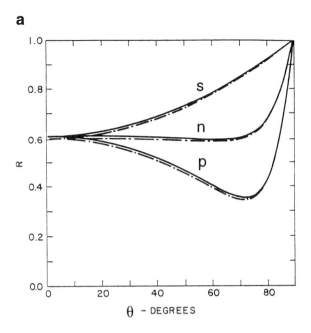

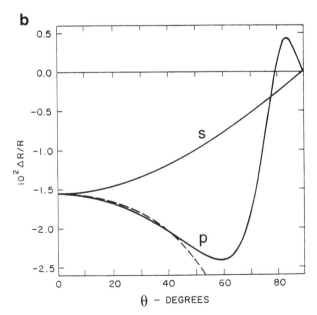

Figure 9.4 (a) The reflectance of a bare metal substrate ($n_3 = 2.0$, $k_3 = 4.0$; solid lines) in a liquid ambient ($n_1 = 1.333$, $k_1 = 0$) and this substrate covered by one monolayer (10^{-3} λ thick) of an absorbing film ($n_2 = 3.0$, $k_2 = 1.5$; dot-dashed line), for s and p polarizations. (b) The normalized change of the reflectance $\Delta R/R$ due to this film, with the dashed curve being $1/\cos\theta$ (from McIntyre and Aspnes, 1971). (For a gas ambient the reflectance would be the same for media 2 and 3 with optical constants divided by 1.333—i.e., $n_3 = 1.500$, $k_3 = 3.001$, $n_2 = 2.251$, and $k_2 = 1.125$—and with d/λ increased by 1.333 to 1.333×10^{-3}.)

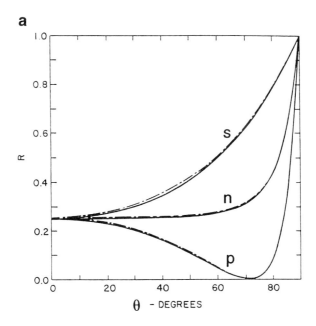

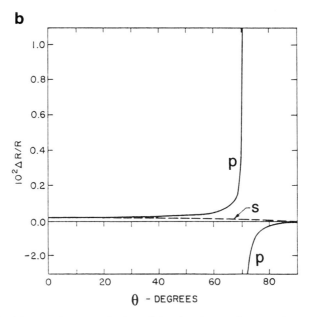

Figure 9.5 (a) The reflectance of a bare dielectric substrate ($n_3 = 4.0$, $k_3 = 0$; solid lines) in a liquid ambient ($n_1 = 1.333$, $k_1 = 0$) and this substrate covered by one monolayer ($10^{-3} \lambda$ thick) of an absorbing film ($n_2 = 3.0$, $k_2 = 1.5$; dot-dashed line) for s and p polarizations. (b) The normalized change of the reflectance $\Delta R/R$ due to this film (from McIntyre and Aspnes, 1971). (For a gas ambient, the reflectance would be the same for media 2 and 3 with optical constants divided by 1.333—i.e., $n_3 = 3.001$, $k_3 = 0$, $n_2 = 2.251$, and $k_2 = 1.125$—and with d/λ increased by 1.333 to 1.333×10^{-3}.)

9.1 Optics of Reflection

Azzam and Bashara (1977) give the reflection matrices (Equation 9.1) for an isotropic ambient (medium 1) and uniaxially or biaxially anisotropic substrates (medium 2).

9.1.1.2 Three Media: A Film on a Semiinfinite Substrate or Wafer in an Ambient

With the media labeled 1, 2, and 3 (Figure 9.2), the reflection coefficient is

$$r_{123} = \frac{r_{12} + r_{23}e^{i\beta_2}}{1 + r_{12}r_{23}e^{i\beta_2}} \quad (9.8)$$

where r_{12} and r_{23} come from Equation 9.2a (or 9.4) for s-polarization and from Equation 9.3a (or 9.5) for p-polarization. (r_{23} is the complex reflectance inside medium 2, just above the 2/3 interface.) For a film on a substrate, the ambient is medium 1, the film is medium 2, and the substrate is medium 3. No interference is assumed to occur in the substrate, which means that it is either semiinfinite or optically thick. The film thickness is d_2; when there is only one film in the problem the thickness will be called d. To conform with notation used by some, at times parameters in medium 1 will be labeled a (for ambient), medium 2 will have unlabeled parameters (n, d, etc.), and medium 3 will be labeled s (for substrate) [as in Equations 9.4 and 9.5]. When examining a wafer in an ambient, interference in the substrate (medium 2) is of interest. Since 1 and 3 both refer to the ambient in this case, the label 3 is changed to 1. Note that then $r_{23} = r_{21} = -r_{12}$.

The phase shift in the film (or substrate) for a round trip is

$$\beta_2 = \frac{2\pi}{\lambda}(2\tilde{n}_2 d_2 \cos\theta_2) = \frac{2\pi}{\lambda}(2d_2\sqrt{\tilde{n}_2^2 - \tilde{n}_1^2\sin^2\theta}) \quad (9.9)$$

where θ is the angle of incidence in the ambient medium. At normal incidence $\beta_2 = 4\pi\tilde{n}_2 d_2/\lambda$. (In some references, the phase shift β_2 is defined as half this quantity, so "2β" appears in Equations 9.8, 9.10, and 9.11 instead of β_2.)

The reflectance from this structure is

$$R_{123} = \frac{|r_{12}|^2 + |r_{23}|^2 e^{[i(\beta_2 - \beta_2^*)]} + r_{12}^* r_{23} e^{i\beta_2} + r_{12} r_{23}^* e^{-i\beta_2^*}}{1 + |r_{12}|^2|r_{23}|^2 e^{[i(\beta_2 - \beta_2^*)]} + r_{12} r_{23} e^{i\beta_2} + r_{12}^* r_{23}^* e^{-i\beta_2^*}} \quad (9.10a)$$

or

$$R_{123} = \frac{r_{12}^2 + r_{23}^2 + 2r_{12}r_{23}\cos\beta_2}{1 + r_{12}^2 r_{23}^2 + 2r_{12}r_{23}\cos\beta_2} \quad \text{for } k_i = 0 \quad (9.10b)$$

where for Equation 9.10b it has been assumed that all three media are transparent.

The $\cos\beta_2$ term is due to interference in the film (or wafer). When $\tilde{n}$, λ, d_2, or θ is changed monotonically, the reflectance oscillates. A new fringe occurs whenever β_2 changes by 2π, which means that $2\tilde{n}_2 d_2 \cos\theta_2/\lambda$ changes by 1. This gives the change in $\tilde{n}$, λ, d_2, or θ between adjacent maxima or adjacent minima, and twice the optical distance between adjacent maxima and minima. These interference patterns can be used to measure temperature and film thickness in reflectometry (Section 9.5) and ellipsometry (Section 9.11).

In terms of the reflectances at an interface between media 1 and 2 (R_{12}) and between 2 and 3 (R_{23}), and the corresponding phase upon reflection δ_{12} and δ_{23} respectively [$r_{jk} = |r_{jk}|\exp(i\delta_{jk})$], the reflectance for the three-phase system can also be expressed as (Born and Wolf, 1970; McIntyre and Aspnes, 1971)

$$R_{123} = \frac{R_{12} + R_{23}\, e^{-2\mathrm{Im}(\beta_2)} + 2\sqrt{R_{12} R_{23}}\, e^{-\mathrm{Im}(\beta_2)} \cos[\delta_{12} - \delta_{23} - \mathrm{Re}(\beta_2)]}{1 + R_{12} R_{23}\, e^{-2\mathrm{Im}(\beta_2)} + 2\sqrt{R_{12} R_{23}}\, e^{-\mathrm{Im}(\beta_2)} \cos[\delta_{12} + \delta_{23} + \mathrm{Re}(\beta_2)]} \quad (9.11)$$

For transparent materials ($k_i = 0$, $\delta_{ij} = 0$), the extremum values for R occur when $\cos\beta_2 = \pm 1$. For normal incidence these values are

$$R_+ = \left(\frac{n_1 - n_3}{n_1 + n_3}\right)^2 \quad (9.12)$$

and

$$R_- = \left(\frac{n_1 n_3 - n_2^2}{n_1 n_3 + n_2^2}\right)^2 \quad (9.13)$$

The maximum reflectance $R_{\max}$ occurs when $r_{12} r_{23} \cos\beta_2 = |r_{12} r_{23}|$, and the minimum value $R_{\min}$ occurs when $r_{12} r_{23} \cos\beta_2 = -|r_{12} r_{23}|$. When $r_{12} r_{23} > 0$, i.e., when n_2 is between n_1 and n_3, $R_{\max} = R_+$ and $R_{\min} = R_-$. When $r_{12} r_{23} < 0$, i.e., when n_2 is not between n_1 and n_3, $R_{\max} = R_-$ and $R_{\min} = R_+$.

The film index n_2 can be determined from $R_{\max}$ and $R_{\min}$ by

$$n_2 = \sqrt{n_1 n_3}\left(\frac{\dfrac{n_1 + n_3}{n_1 - n_3} \mp \sqrt{R_-/R_+}}{\dfrac{n_1 + n_3}{n_1 - n_3} \pm \sqrt{R_-/R_+}}\right)^{1/2} \quad (9.14)$$

$R_-/R_+ = R_{\min}/R_{\max}$ when $r_{12} r_{23} > 0$ and $R_{\max}/R_{\min}$ when $r_{12} r_{23} < 0$. The upper signs are used when $(n_1 n_3 - n_2^2)/(n_1 - n_3)$ is positive and the lower signs are used when it is negative. Again, this equation assumes that there is no absorption (real n).

This analysis has assumed that the film has a uniform refractive index and that the interfaces are sharp. Modifications are necessary when the film has a continuously varying refractive index (Jacobsson, 1966).

9.1.1.3 Three Media—A Very Thin Isotropic Film on a Semiinfinite Substrate

The limit of very thin overlayers, with thickness $d \ll \lambda$, is important in differential reflectometry (Section 9.8) and surface photoabsorption spec-

9.1 Optics of Reflection

troscopy (Section 9.9). McIntyre and Aspnes (1971) have shown how the above expressions for a film of arbitrary thickness simplify for thin isotropic layers ($\mu_i = 1$).

For s polarization (θ is the angle of incidence in medium 1),

$$r_s(d) \approx r_s(0)\left(1 + \frac{4\pi i d n_1 \cos\theta}{\lambda}\left(\frac{\tilde{\varepsilon}_2 - \tilde{\varepsilon}_3}{\varepsilon_1 - \tilde{\varepsilon}_3}\right)\right) \tag{9.15}$$

$$R_s(d) \approx R_s(0)\left(1 - \frac{8\pi d n_1 \cos\theta}{\lambda}\,\text{Im}\left(\frac{\tilde{\varepsilon}_2 - \tilde{\varepsilon}_3}{\varepsilon_1 - \tilde{\varepsilon}_3}\right)\right) \tag{9.16}$$

For p polarization,

$$r_p(d) \approx r_p(0)\left(1 + \frac{4\pi i d n_1 \cos\theta}{\lambda}\left(\frac{\tilde{\varepsilon}_2 - \tilde{\varepsilon}_3}{\varepsilon_1 - \tilde{\varepsilon}_3}\right)\left(\frac{1 - (\varepsilon_1/\tilde{\varepsilon}_2\tilde{\varepsilon}_3)(\tilde{\varepsilon}_2 + \tilde{\varepsilon}_3)\sin^2\theta}{1 - (1/\tilde{\varepsilon}_3)(\varepsilon_1 + \tilde{\varepsilon}_3)\sin^2\theta}\right)\right) \tag{9.17}$$

$$R_p(d) \approx R_p(0)\left(1 - \frac{8\pi d n_1 \cos\theta}{\lambda}\,\text{Im}\left(\left(\frac{\tilde{\varepsilon}_2 - \tilde{\varepsilon}_3}{\varepsilon_1 - \tilde{\varepsilon}_3}\right)\left(\frac{1 - (\varepsilon_1/\tilde{\varepsilon}_2\tilde{\varepsilon}_3)(\tilde{\varepsilon}_2 + \tilde{\varepsilon}_3)\sin^2\theta}{1 - (1/\tilde{\varepsilon}_3)(\varepsilon_1 + \tilde{\varepsilon}_3)\sin^2\theta}\right)\right)\right) \tag{9.18}$$

At normal incidence, Equations 9.16 and 9.18 reduce to

$$R_n(d) \approx R_n(0)\left(1 - \frac{8\pi d n_1}{\lambda}\,\text{Im}\left(\frac{\tilde{\varepsilon}_2 - \tilde{\varepsilon}_3}{\varepsilon_1 - \tilde{\varepsilon}_3}\right)\right) \tag{9.19}$$

These equations assume that ε_1 is real.

Equations 9.16, 9.18, and 9.19 are often presented in terms of $\Delta R/R$, where $\Delta R = R(d) - R(d = 0)$; this fraction then equals the second term in the brackets in the respective expressions. [In surface infrared reflectometry (Section 9.7), the convention is $\Delta R = R(0) - R(d)$.] As in Section 9.1.1.2, region 1 is often labeled the ambient (subscript a), region 2 the overlayer (o), and region 3 the substrate (s). Figures 9.4a and 9.5a show how a monolayer of an absorbing film changes the reflectance of a metal and a transparent dielectric, respectively.

For s-polarization, $|\Delta R/R|$ varies as $\cos\theta$, so it decreases monotonically with increasing angle, as shown in Figure 9.4b for a metal substrate. For p-polarization, $\Delta R/R$ is more complicated and depends on the dielectric functions of the three media (McIntyre and Aspnes, 1971). For a strongly absorbing film on a metal substrate, with $|\varepsilon_3'| \gg \varepsilon_1$ and $|\varepsilon_2'| \gg \varepsilon_1$, $|\Delta R|/R$ varies as $1/\cos\theta$ for small angles. As seen in Figure 9.4b, it increases with θ and then decreases to zero, and then $\Delta R/R$ changes sign and becomes zero at grazing incidence. Maximum sensitivity is generally attained for p-polarized light at an angle of incidence between 45° and the pseudo-Brewster angle. Figure 9.5b plots $|\Delta R/R|$ for a lossless dielectric.

For normal incidence, $\Delta R/R$ for a weakly absorbing film on a strongly absorbing (metallic) substrate is small and largely determined by the properties of the substrate. For an absorbing film on a transparent substrate, Equation 9.19 reduces to

$$\frac{\Delta R}{R} = -\frac{4n_1 n_2 \alpha_2 d}{n_1^2 - n_3^2} \qquad (9.20)$$

where $\alpha_2 = 4\pi k_2/\lambda$, so $\Delta R/R$ is proportional to the absorption coefficient of the film α_2, and is positive if $n_3 > n_1$ and negative if $n_3 < n_1$ (McIntyre and Aspnes, 1971).

9.1.1.4 Three Media—a Very Thin Anisotropic Film on a Semiinfinite Substrate

If there are surface anisotropies, this thin overlayer has biaxial, not cubic, symmetry. This feature is probed in reflection-difference spectroscopy (RDS/RDA), which is discussed in Section 9.10. The principal dielectric tensor components of this layer are ε_{xx}, ε_{yy}, and ε_{zz}, which are referenced to the crystal coordinate system and are in general complex and unequal. The laboratory coordinate normal to the surface is z and that along the intersection of the surface and the plane of incidence is x. In cases of interest, the z axes of the crystal and laboratory frame are generally the same. α'' is defined as the Eulerian angle between the crystal and laboratory x axes. To simplify notation, the overlayer (medium 2) parameters are not subscripted.

The reflection coefficients are (Hingerl et al., 1993)

$$r_{ss}(\theta,\alpha'') = r_s^0 \left(1 + \frac{4\pi i d n_{a\perp}}{\lambda} \frac{\tilde{\varepsilon}_s - \hat{\varepsilon} - \Delta\varepsilon \cos(2\alpha'')}{\tilde{\varepsilon}_s - \varepsilon_a}\right) \qquad (9.21)$$

$$r_{sp}(\theta,\alpha'') = -r_{ps}(\theta,\alpha'') = \frac{4\pi i d n_{a\perp}}{\lambda} \frac{n_a \tilde{n}_{s\perp} \Delta\varepsilon \sin(2\alpha'')}{(\tilde{\varepsilon}_s n_{a\perp} + \varepsilon_a \tilde{n}_{s\perp})(n_{a\perp} + \tilde{n}_{s\perp})} \qquad (9.22)$$

$$r_{pp}(\theta,\alpha'') = r_p^0 \left(1 + \frac{4\pi i d n_{a\perp}}{\lambda} \frac{\tilde{\varepsilon}_s - \hat{\varepsilon} - \left(\frac{\tilde{\varepsilon}_s}{\varepsilon_{zz}} - \frac{\hat{\varepsilon}}{\tilde{\varepsilon}_s}\right)\varepsilon_a \sin^2\theta + \frac{\Delta\varepsilon}{\tilde{\varepsilon}_s}\tilde{n}_{s\perp}^2 \cos(2\alpha'')}{(\tilde{\varepsilon}_s - \varepsilon_a)(\cos^2\theta - \frac{\varepsilon_a}{\tilde{\varepsilon}_s}\sin^2\theta)}\right) \qquad (9.23)$$

The first subscript refers to the linear polarization of the incident light and the second to that of the detected beam (p is parallel and s is perpendicular to the plane of incidence). $\hat{\varepsilon} = (\varepsilon_{xx} + \varepsilon_{yy})/2$ and $\Delta\varepsilon = (\varepsilon_{xx} - \varepsilon_{yy})/2$, which can be complex. It is assumed that ε_a are n_a are real. r_s^0 and r_p^0 are the reflection coefficients for the bare substrate, given by Equations 9.4 and 9.5, and $\tilde{n}_{j\perp}$ is defined after these equations. With no film present ($d = 0$), it is seen that $r_{ss} = r_s^0$, $r_{pp} = r_p^0$, and $r_{sp} = r_{ps} = 0$.

9.1.1.5 Multiple Films

The interference effects in reflection from a substrate overlaid by more than one film are easily handled by a recursive procedure. For example,

9.1 Optics of Reflection

consider a given multilayer system, called the "old" system, and a "new" system, consisting of an overlayer of thickness d and refractive index $\tilde{n}$ grown on top of the old system. A probing beam propagates at an angle θ' in this new layer, related to the angle of incidence θ by Snell's law. The reflection coefficient for a two-medium system consisting of semiinfinite ambient and overlayer material media is r_{ao} (Equations 9.2–9.5). If the reflection coefficient from the old system for a beam beginning in the overlayer (and not in the ambient) is called r_{old}, then the reflection coefficient from the new multilayer system with the overlayer is r_{new}:

$$r_{new} = \frac{r_{ao} + r_{old}Z}{1 + r_{ao}r_{old}Z} \qquad (9.24a)$$

where $Z = \exp(i\beta)$, with $\beta = 4\pi\tilde{n}d \cos\theta'/\lambda = 2ik_{oz}d$, as in Equation 9.9. This recursion expression is equivalent to Equation 9.8 with $r_{123} \to r_{new}$, $r_{12} \to r_{ao}$, and $r_{23} \to r_{old}$. (r_{old} is the "virtual" reflection coefficient defined inside the material, and not the r from the ambient into the "old" system.)

The recursion formula Equation 9.24a can be expressed in a slightly different way (W. G. Breiland, private communication, 1995). Consider a system with layer $m-1$ atop layer m, which is atop layer $m+1$, and so on. The (virtual) reflection coefficient $\tilde{r}_{m,m+1}$ is that for light impinging (from above) in medium m just above the interface between media m and $m+1$. With $r_{m-1,m}$ being the reflection coefficient between two semiinfinite media (as in Equations 9.2 to 9.5) and $Z_m = \exp(2ik_{mz}d_m)$

$$\tilde{r}_{m-1,m} = \frac{r_{m-1,m} + \tilde{r}_{m,m+1}Z_m}{1 + r_{m-1,m}\tilde{r}_{m,m+1}Z_m} \qquad (9.24b)$$

An alternative recursion relation (W. G. Breiland, private communication, 1995) tracks the effective reflection coefficient $\hat{r}_{m,m+1}$, which is that measured in medium $m+1$ just below the interface between media m and $m+1$:

$$\hat{r}_{m-1,m} = \frac{r_{m,m+1} + \hat{r}_{m,m+1}}{1 + r_{m,m+1}\hat{r}_{m,m+1}} Z_m \qquad (9.24c)$$

(Some researchers prefer to reverse the numbering of the layers, with the substrate being layer 1, the first film being layer 2, and so on.)

Using Equations 9.24a and 9.24b, along with Equations 9.8 and 9.9, for two films (media 2 and 3) on a semiinfinite substrate (medium 4), the reflection coefficient is

$$r_{1234} = \frac{r_{12} + r_{23}\,e^{i\beta_2} + r_{34}\,e^{i(\beta_2+\beta_3)} + r_{12}r_{23}r_{34}\,e^{i\beta_3}}{1 + r_{12}r_{23}\,e^{i\beta_2} + r_{12}r_{34}\,e^{i(\beta_2+\beta_3)} + r_{23}r_{34}\,e^{i\beta_3}} \qquad (9.25)$$

If it assumed that an overlayer of thickness d atop a substrate has a constant dielectric function that is nearly equal to that of the substrate, then $|r_{23}| \ll 1$ and Equations 9.8 and 9.24a can be approximated by the exponential spiral

$$r_{123}(d) \approx r_{12} + [r_{123}(0) - r_{12}]\exp(2ik_{oz}d) \tag{9.26a}$$

If the overlayer is very thin ($|2k_{oz}d| \ll 1$), then Equation 9.26a can be approximated by

$$r_{123}(d) \approx r_{123}(0) + 2ik_{oz}d[r_{123}(0) - r_{12}] \tag{9.26b}$$

Equations 9.24a and b have been used to interpret data in reflection and ellipsometric measurements made during deposition because they minimize errors during the rapid analysis needed for closed-loop feedback control. They are particularly useful in controlling small changes in the composition and refractive index. In these analysis algorithms, the overlayer is not necessarily a new layer, but can be merely the uppermost part of the growing film. Then d in Equations 9.26 is a parameter that is set to analyze the film periodically, and the details of the films and substrate below it need not be known. Many approaches have been developed that utilize various forms of the Fresnel equations for efficient analysis of multilayer systems in real time, such as the virtual interface model. These are detailed in Section 9.11.1. (Also see Figure 19.1.)

Repeated application of Equations 9.24a and b to analyze multilayered structures is computationally straightforward. Reflection from multiple films may arise because of the complex nature of the structure being grown (a semiconductor heterostructure, dielectric mirror, etc.) or because modeling requires dividing the film into several layers (e.g., rough surface/oxide layer/transition film/main film/substrate).

An alternative (and mathematically equivalent) approach to the use of recursion relations is the matrix approach, which is similar to the Jones vector method from Section 2.3 (Azzam and Bashara, 1977; Born and Wolf, 1970; Klein and Furtak, 1986; Jellison, 1993a). The elements of the electric field vector of the beams at one position z' in the optical system [$E^+(z')$ traveling in the $+z$ direction and $E^-(z')$ traveling in the $-z$ direction] are related to those at z'' [$E^+(z'')$ traveling in the $+z$ direction and $E^-(z'')$ traveling in the $-z$ direction]. This is done separately for each polarization. Then

$$\begin{pmatrix} E^+(z'') \\ E^-(z'') \end{pmatrix} = \begin{pmatrix} M_{++} & M_{+-} \\ M_{-+} & M_{--} \end{pmatrix} \begin{pmatrix} E^+(z') \\ E^-(z') \end{pmatrix} \tag{9.27}$$

or

$$\mathbf{E}(z'') = \mathbf{M}\mathbf{E}(z') \tag{9.28}$$

In the Abelès matrix method, each layer is represented by a 2×2 transfer matrix, while in the Hayfield and White method the multilayer system is divided into a series of interfaces and layers, each described by its own $\mathbf{M}$ (Jellison, 1993a). Following this latter approach, when z' and z'' are on either side of an interface between layers i and j, $\mathbf{M}$ is composed of reflection and transmission coefficients ($\mathbf{I}_{ij}$). When z' and z'' are near the two interfaces in one layer (i), $\mathbf{M}$ describes the phase due to propagation in that layer ($\mathbf{L}_i$).

The multilayer system is described by the product of such matrices, so the overall scattering matrix $\mathbf{S}$ for m layers (followed by a substrate),

$$\mathbf{S} = \mathbf{I}_{01}\mathbf{L}_1\mathbf{I}_{12}\mathbf{L}_2 \ldots \mathbf{L}_m\mathbf{I}_{m(m+1)} \tag{9.29}$$

relates the beams before and after the multilayer:

$$\mathbf{E}(\text{after}) = \mathbf{S}\mathbf{E}(\text{before}) \tag{9.30}$$

The reflection coefficient of the multilayer system is easily obtained from the 2×2 matrix $\mathbf{S}$. Details can be found in the above-cited references.

Least-squares fits are often used to determine layer thicknesses from these expressions. However, this procedure can be complicated and time-consuming for multilayer systems. Several algorithms have been devised that accelerate this analysis and improve the accuracy of the analysis. Swart and Lacquet (1990) and Swart (1994) have developed an algorithm based on fast Fourier analysis for spectroscopic reflectance of multilayered structures. It is designed for transparent and dispersionless films, and therefore works best in the infrared, away from absorption features.

9.2 Reflectometry, Ellipsometry, and Polarimetry

Reflection measurements can be made with a reflectometer, ellipsometer, or polarimeter (Hauge, 1980; Aspnes, 1985a). When performed as a function of wavelength, such reflection experiments are called spectroscopic reflectometry (SR), spectroscopic ellipsometry (SE), etc.

In reflectometry, only intensities are measured, and they are measured to determine the reflectances $R_p = |r_p|^2$ and $R_s = |r_s|^2$. Such measurements are made by a reflectometer, which is a type of photometer since it measures intensities. Reflectometry is performed at any angle of incidence, including normal incidence where the reflectance is $R_n = |r_n|^2$.

In ellipsometry, only the polarization state χ (see Equation 9.31' below in Section 9.11.1) is of interest, and absolute intensities are not measured. The ellipsometer determines the reflection coefficient ratio

$$\rho = r_p/r_s = \tan \Psi e^{i\Delta} \tag{9.31}$$

For isotropic media, ellipsometry is conducted at angles other than normal incidence. An ellipsometer keeps track of the change of phase of the electric field upon reflection.

Polarimetry is more general than reflectometry and ellipsometry, and includes features of both. It measures intensities and polarization states and is also concerned with depolarized or unpolarized light. A complete polarimeter is a polarization state detector that measures the four components of the Stokes vector. In incomplete or partial polarimeters, S_0, the intensity, may not be measured absolutely.

The merits of near-normal reflectometry and ellipsometry can be compared by examining how each can be used to obtain the complex dielectric constant $\tilde{\varepsilon}$ of an unknown two-medium system. Consider a light beam that is incident from a gas with index of refraction 1 onto a medium with complex index of refraction $\tilde{n} = n + ik$ and dielectric constant $\tilde{\varepsilon}$, with $\varepsilon' = n^2 - k^2$ and $\varepsilon'' = 2nk$. The reflection coefficient $r = |r|\exp(i\delta)$, where $|r| = \sqrt{R}$. For the near-normal incidence reflectometry experiment, Equations 9.2–9.5 can be used to show that

$$n = \frac{1 - R}{1 - 2\sqrt{R}\cos\delta + R} \tag{9.32}$$

$$k = \frac{2\sqrt{R}\sin\delta}{1 - 2\sqrt{R}\cos\delta + R} \tag{9.33}$$

The phase δ at this particular wavelength is not measured directly, so n and k are not known.

However, ε' and ε'' are the real and imaginary parts of an analytic function, so they are related by the Kramers–Kronig relations (such as Equations 4.12 and 4.13) to give

$$\delta(\omega) = -\frac{\omega}{\pi} \mathsf{P} \int_0^\infty \frac{\ln R(\omega')}{\omega'^2 - \omega^2} d\omega' \tag{9.34}$$

where P is the Cauchy principal value. Consequently, ε' and ε'' (and therefore n and k) can be obtained indirectly only if measurements are made over a sufficiently large frequency range (Smith, 1985). Two independent parameters, R_s and R_p, however, are available at nonnormal incidence.

In contrast, ellipsometry gives both n and k for this two-phase system at any wavelength. Because two parameters are determined for each measurement, Ψ and Δ, two optical parameters can be determined. With a measurement made at an angle θ to the normal for the two-medium problem,

$$\tilde{\varepsilon} = \sin^2\theta \left(1 + \tan^2\theta \left(\frac{1-\rho}{1+\rho}\right)^2\right) \tag{9.35}$$

There has been much activity in extracting surface information from kinetic reflectometric and ellipsometric probes for real-time control. Two extreme examples illustrate the range of use of these reflection probes. If the uppermost layers are at least partially transparent, both reflectometry and ellipsometry can provide real-time information about film thickness, optical constants, and temperature (as a function of λ and θ) through interference effects. If the uppermost film is optically thick, i.e., $\alpha(\lambda)d \gg 1$ where α is the absorption coefficient and d is its thickness, reflection probes only the region near the surface and is insensitive to any buried layers.

Ellipsometry has been used to determine the dielectric constant ε of many bulk materials by examining this region near the surface. Moreover, it is sensitive to surface overlayers, roughness, and porosity, which makes it a valuable probe of the surface during film processing. In reflectometry, intensity fluctuations and surface roughness are experimental problems, while in ellipsometry, the former is not an experimental concern (Equation 9.31) and the latter is amenable to analysis.

Ellipsometry is more complicated experimentally than reflectometry, but can yield more information. Both are heavily used in optical diagnostics. Polarimetry is even more complicated than ellipsometry and less widely used, although it can give more information. As will be seen in Section 9.11, in some experimental regimes ellipsometry can give ambiguous results, while polarimetry does not.

Humphreys-Owen (1961) and Hunter (1965) have compared various methods for extracting optical constants from reflectometry experiments performed at various combinations of incident angle and polarization.

9.3 Optical Dielectric Functions

Use of reflection methods to analyze and control thin film processes requires knowing the real and imaginary parts of either the dielectric function or the index of refraction. These parameters are sometimes measured within the processing apparatus with the aim of repeatable and precise diagnostic measurements. More often, published parameters are used. Such information is available for many of the materials common in thin film processing, often as a function of wavelength and sometimes as a function of temperature. The sources listed in Table 9.1 include several original papers and comprehensive compilations by Palik (1985, 1991) and Wolfe (1978). Much of these optical data on bulk materials were obtained by ellipsometry; the accuracy of these data depends on how well the surface region was modeled. Also, high doping levels can modify the optical constants. Figure 9.6 plots ε' and ε'' for several common semiconductors (Irene, 1993).

Optical data for alloys, such as ternary and quaternary alloys of III–V semiconductors, are limited (Table 9.1). Optical constants for alloys sometimes interpolate linearly between the corresponding limits, such as the limiting binaries for ternary alloys; but sometimes there are strong nonlinearities, such as bowing effects (Böer, 1990). Some of the dielectric functions reported in the book edited by Palik (1991) are presented for alloys at various compositions, such as $Al_xGa_{1-x}As$, Si_xGe_{1-x}, and $Hg_{1-x}Cd_xTe$.

Section 18.4.2 discusses the change of optical dielectric functions with temperature, which occurs in part due to the temperature variation of critical-point energies (also see Thomas, 1991). Table 9.1 lists several

Table 9.1 Sources for the Wavelength Dependence of $\tilde{\varepsilon}$ and $\tilde{n}$

Source	Reference
Selected primary sources	
Ambient temperature	
AlAs	Herzinger et al. (1995)
Si, Ge, GaP, GaAs, GaSb, InP, InAs, InSb	Aspnes and Studna (1983)
Si(100), (111), (110)	Jellison (1992a)
Various glasses	Jellison and Sales (1991)
GeSi	Jellison et al. (1993); Carline et al. (1994)
GaAs(100), (111), GaP(100), Ge(100)	Jellison (1992b)
GaInP	Schubert et al. (1995)
InAlAs, InAlGaAs	Dinges et al. (1993)
Variable temperature	
Si	Jellison and Modine (1983), Vuye et al. (1993), Sampson et al. (1994a)
GeSi	Humlíček and Garriga (1993); Pickering et al. (1995)
GaAs	Aspnes et al. (1990a,b), Yao et al. (1991), Kuo et al. (1994); Maracas et al. (1995a)
AlGaAs	Kuo et al. (1995)
Ge, Si, InAs, GaAs, InP, GaP, CdTe, ZnSe, ZnS (15–35°C)	Bertolotti et al. (1990)
Si, GaAs, InP (1.15, 1.31, 1.53, and 2.39 μm)	McCaulley et al. (1994)
Secondary sources	
Al, Ag, Au, Ba, Be, Bi, Ca, Cd, Ce, Co, Cr, Cs, Cu, Fe, Ga, Ge, Hg, In, Ir, K, La, Mg, Mn, Mo, Na, Nb, Nd, Ni, Pb, Pd, Pt, Pu, Rb, Re, Rh, Sb, Se, Si, Sn, Sr, Ta, Th, Ti, V, W, Zn, Zr	Hass and Hadley (1972)
Metals: Al, Ag, Au, Cu, Ir, Mo, Ni, Os, Pt, Rh, W	Palik (1985)
Semiconductors: CdTe, GaAs, GaP, Ge, InAs, InSb, InP, PbSe, PbS, PbTe, c-Si, a-Si, SiC, ZnS	
Insulators: As_2Se_3, As_2S_3, C(diamond), LiF, $LiNbO_3$, KCl, SiO_2 (type-α crystalline, glass), SiO, Si_3N_4, NaCl, TiO_2 (rutile)	
Metals: Be, Co, C(graphite), Cr, Fe, Hg, K, Li, Na, Nb, Pd, Ta, V	Palik (1991)
Semiconductors: AlAs, AlSb, $Al_xGa_{1-x}As$, CdSe, CdS, GaSb, Si_xGe_{1-x}, PbSnTe, $Hg_{1-x}Cd_xTe$, Se, β-SiC, Te, SnTe, ZnSe, ZnTe	
Insulators: Al_2O_3, AlON (spinel), $BaTiO_3$, BeO, a-diamondlike C films, CaF_2, $(C_2H_4)_n$ (polyethylene), CsI, Cu_2O/CuO, H_2O, KBr, KH_2PO_4 (KDP), $MgAl_2O_4$ (spinel), MgF_2, MgO, NaF, $SrTiO_3$, ThF_4, Y_2O_3	

(continues)

Table 9.1 (*Continued*)

Source	Reference
GaAs	Aspnes (1990)
AlGaAs	Adachi (1993)
InGaAs	Bhattacharya (1993)
Assorted materials	Wolfe (1978), Tompkins (1993)

sources that tabulate these temperature variations. This change with temperature is illustrated in Figure 9.6 (also see Figure 9.96).

9.4 Reflection at the Interface with a Semiinfinite Material

Reflection at a single interface is a good model of a structure when the top layer or the substrate itself is opaque at the probing wavelength, or when the substrate is "semiinfinite." Changes in the dielectric constant of this top layer are monitored by measuring R by using Equations 9.2–9.5. In many diagnostics applications, changes in the reflectance are monitored either versus time or relative to a standard because absolute measurements are more difficult and often are not necessary. Such experiments are therefore similar to differential reflectometry measurements (Sections 9.8 and 9.9). However, the diagnostics described in this section focus on the properties of either the bulk sample or the top many monolayers ($\geqslant 100$ Å) in the sample, while those described in Sections 9.8 and 9.9 focus more on the nature of the surface. While all kinetic measurements of thin film processes clearly vary with time, only those of relatively fast processes ($\ll 1$ sec) are usually called time-resolved reflectivity (TRR). (It is actually a measurement of the time-resolved reflectance.)

9.4.1 Thermometry

The temperature dependence of the reflectance arises from the temperature dependence of n and k (Table 9.1). Guidotti and Wilman (1992a,b) made use of this dependence in their optical beam thermometer (OBT) which monitors *in situ* the reflectance of a Si wafer in a processing reactor. They devised a nulling optical bridge to reduce noise to the shot noise limit of the detector, which is diagrammed in Figure 9.7. The optical beam was linearly polarized, chopped (for lock-in detection), and then split by a polarizing beamsplitter into two beams: one that reflects from the wafer before detection and one that is detected directly, perhaps after reflecting from a silicon reference mirror. The difference in detected signals drives

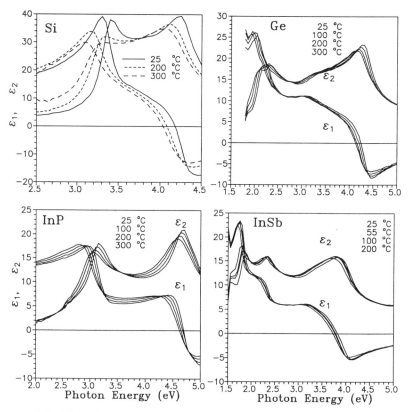

Figure 9.6 Dielectric functions of Si, Ge, InP, and InSb at several temperatures up to 300°C (from Irene, 1993). (The curves for the latter three can be identified by the decrease in peak energies with increasing temperature.)

the rotation angle of the beamsplitter to equalize the signals. A change in wafer temperature is indicated by a change in beamsplitter angle. A resolution of 0.2°C is possible under ideal conditions. This OBT is a precise *in situ* monitor if the surface does not change, i.e., as long as no films are deposited upon it (oxides, nitrides), it remains smooth, its composition (doping) does not change, etc. If the surface changes in a well characterized manner, a temperature calibration curve can be determined to correct for these changes (as in pyrometry in Chapter 13). This OBT is relatively inexpensive, robust, and compact.

The temperature dependence of the reflectance can also be used to monitor processes where the temperature changes very fast (TRR). One such application is thermometry during local heating of surfaces by focused beams that are scanned across the surface. For example, England *et al.* (1991) measured the temporal evolution of the surface temperature of a silicon wafer as an

9.4 Reflection at the Interface with a Semiinfinite Material

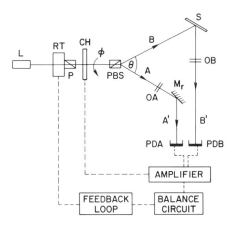

Figure 9.7 Schematic of the null-optical bridge for reflectance-based temperature determination, with a light source (L), rotator (RT) for a polarizer (P), a polarizing beamsplitter (PBS), and photodetectors (PDA, PDB) (from Guidotti and Wilman, 1992a).

electron beam was swept across it by monitoring the reflectance, as is shown in Figure 9.8. Examples of TRR that involve interference effects are described in Section 9.5. Kempkens *et al.* (1990) used TRR (632.8 nm) to monitor the electrode temperature during the starting of an arc discharge lamp, with resolution sufficient to resolve AC (60-Hz) electrode heating.

Another thermometric technique involving reflection senses thermal expansion of the wafer. This novel imaging method, called optical micrometry or wafer diameter extensometry, has been detailed by Peuse *et al.* (1992), Snow (1992), and Peuse and Rosekrans (1993, 1994), and has been described in the overview by Peters (1991). As diagrammed in Figure 9.9, reflection references the edge of the wafer to the focal plane of an optical system by using the Foucault knife edge test. With proper wafer mounting, the wafer edge moves as the temperature changes, because of thermal expansion, by ~0.8 μm/°C (for an 8-inch Si wafer, using the linear coefficient for thermal expansion of Si: 4×10^{-6}/°C). When temperature changes, the signal detected by the photocell decreases and feedback to the optomechanical servo system returns the signal to its former value. This technique is insensitive to thin overlayers and senses only the average wafer temperature. The temperature resolution is reported to be 0.3°C, based on the 0.25-μm resolution in measuring the expansion of an 8-inch wafer (from 100 to 1300°C). Temperature repeatability to within 1% (3-σ) was demonstrated on 6-inch wafers (Peuse and Rosekrans, 1993).

9.4.2 Monitoring Phase Changes and Annealing

Time-resolved reflectivity is also an effective real-time monitor of "surface" processing, though its sensitivity is not limited to the region near the surface.

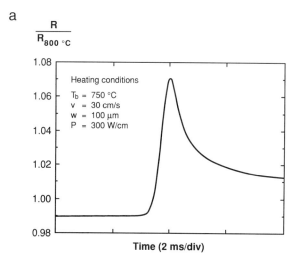

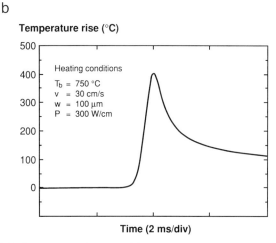

Figure 9.8 (a) Reflectance transient and (b) the deduced temperature profile during the sweep of an electron line beam over a focused laser-probe spot on a silicon wafer (from England et al., 1991).

In early work, Auston et al. (1978) used TRR at 632.8 nm (He–Ne laser) to probe melting near the surface during laser annealing of ion-implanted Si by a Q-switched Nd:glass laser. Lowndes et al. (1983) used TRR to determine the onset and duration of melting in silicon for a comparison with thermal melting models. While the reflectivity of silicon increases relatively slowly with increasing T from ambient temperature (~34.7% at 632.8 nm for normal incidence) to the melting point (~43.5%, 1410°C), it increases abruptly when it melts (~72%). This change in reflectivity is

9.4 Reflection at the Interface with a Semiinfinite Material

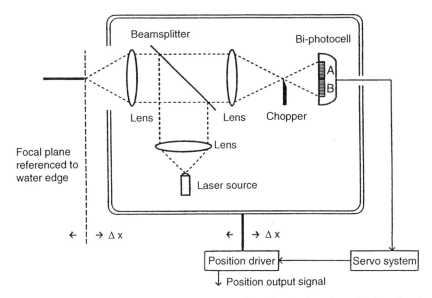

Figure 9.9 Measurement of wafer temperature by reflection-based monitoring of wafer thermal expansion. [From B. Peuse, A. Rosekrans, and K. Snow, *In-situ* temperature control for RTP via thermal expansion measurement, in *Rapid Thermal and Laser Processing, SPIE* **1804,** 45 (1993).]

dramatic, so monitoring the reflected laser clearly indicates changes from solid to liquid and vice versa.

Several more recent studies have used laser reflection measurements, sometimes in combination with such other methods as transient conductance, to determine the onset and duration of laser melting of amorphous silicon films (Lowndes *et al.*, 1984; Thompson *et al.*, 1984; Im *et al.*, 1993). While the only feature seen by monitoring the reflectance during laser melting of crystalline silicon is the increase in R during melting, followed by a decrease after crystallization, there are some reports of additional structure during laser melting of a-Si. The inset in Figure 9.10 shows the reflectance during excimer laser crystallization of amorphous silicon films (Im *et al.*, 1993), in which R oscillates before melting because of the explosive crystallization formation of fine-grained polycrystalline silicon. The width of the main hump denotes the duration of melting. The abrupt change in slope of the melt duration versus energy density in the main part of this figure signals the transition from partial to complete melting. Related examples of TRR that involve interferometry are described in Section 9.5.

Dilhac *et al.* (1989) monitored the reflectance at 632.8 nm during platinum silicide formation upon heating a Pt film on a Si wafer. At 580°C, the film reflectance dropped from that for Pt (~74%) when Pt_2Si was formed (to

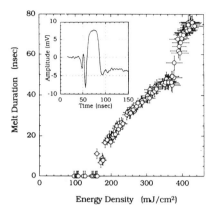

Figure 9.10 Melt duration as a function of laser energy density during crystallization of amorphous silicon films by an excimer laser, with an inset showing a representative reflectance trace (from Im *et al.*, 1993).

~56%) and then again when PtSi was formed (to ~38%); it remained constant thereafter. This TRR measurement served as an endpoint detector for the thin film process. In another study, Dilhac *et al.* (1990) used TRR to calibrate a thermocouple in a rapid thermal processor by using the increased reflectance when a germanium film melts.

In molecular beam epitaxy, wafer thermometry is sometimes referenced to a visually observed change in reflectance associated with the melting temperature of InSb (525°C) or the eutectic temperature of Si and Au (370°C) or Si and Al (577°C) (Chow, 1991; Panish and Temkin, 1993; Maracas *et al.*, 1995a). This calibration procedure is intrusive, and not recommended for manufacturing systems.

9.4.3 Stress Analysis

In almost every application of specular reflection in this section, the light beam is used to probe the dielectric function of the material. In a quite different application, specular reflection has been used as an *in situ* probe of film and wafer stress by monitoring the change in the angles of incidence and reflection, and hence the deflection angle, when the wafer warps due to differential strain in the film and wafer. The stress in a film on the wafer σ_f is related to the curvature in the wafer by Stoney's equation:

$$\sigma_f = \frac{E_s h^2}{6d(1-\nu_s)}\left(\frac{1}{\mathcal{R}} - \frac{1}{\mathcal{R}_0}\right) \quad (9.36)$$

where E_s is Young's modulus, ν_s is Poisson's ratio, and h is the thickness

9.4 Reflection at the Interface with a Semiinfinite Material

of the substrate, d is the film thickness, and $\mathcal{R}_0$ and $\mathcal{R}$ are the wafer curvatures before and after the process that creates the stressed film. The curvature is determined in a differential manner by comparing the angular deviation of two parallel beams that are incident on the wafer at symmetric positions about the center, and then reflected. This is depicted in Figure 9.11.

Kobeda and Irene (1986) developed this method for *in situ* analysis of the oxidation of silicon to separate the effects of intrinsic stress, due to the expanded molar volume of SiO_2 *vis-à-vis* Si, and extrinsic stress, due to thermal expansion. Sensitivity was increased by propagating the reflected beams a long distance before analysis. Teal and Muraka (1987) used this method to study the stress in $TaSi_x$ films on Si as a function of temperature. Leusink *et al.* (1992) improved the sensitivity of this method by propagating the reflected beams from a mirror before detection. This group then used this method to study *in situ* the kinetics of C49 $TiSi_2$ formation from sputter-deposited amorphous Ti–Si multilayers (Jongste *et al.*, 1993) and the stresses that develop during the chemical vapor deposition (CVD) of tungsten films on Si by the hydrogen reduction of WF_6 (Leusink *et al.*, 1993).

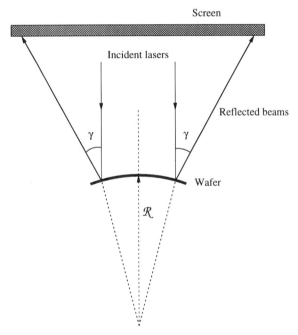

Figure 9.11 Schematic of the experimental apparatus used to monitor the wafer curvature $\mathcal{R}$ by reflectometry. The deflection angle γ is determined on the screen or by two location-sensitive detectors. To ensure ready separation of the incident and reflected beams, the incident beams usually do not hit the wafer at normal incidence. The wafer curvature and deflection angle have been greatly exaggerated in this figure. (Adapted from Leusink *et al.*, 1992).

9.5 Interferometry

All reflection spectroscopies are sensitive to optical interference effects arising from multiple reflections in a probed film, say of thickness d and refractive index n (Figure 9.12). As seen in Equations 9.10 and 9.11, this interference produces a term in the reflectance R (and transmittance) that is a sinusoidal function of $\beta = 4\pi nd \cos\theta_2/\lambda = 4\pi d(\tilde{n}^2 - \sin^2\theta)^{1/2}/\lambda$, where θ is the angle of incidence in the ambient medium, which has refractive index 1. Here β is the normalized optical path length of a beam during a roundtrip cycle in the film; R depends on the angle of incidence, θ, and on how n depends on λ and temperature T. When this optical path changes because d, T, λ, or θ is varied, the reflectance changes. R returns to its original value when β changes by 2π. The maximum and minimum values of R are given by Equations 9.12 and 9.13. For interference to occur, the wavelength must be transmitted in the layer, i.e., $\alpha d \lesssim 1$.

Fringes occur in the reflected intensity as β changes, and at normal incidence a new fringe occurs whenever $\beta = 4\pi nd/\lambda$ changes by 2π; this fringe pattern is called an interferogram. At constant temperature, a new fringe occurs during growth or etching when the thickness d changes by $\lambda/2n$. When the material in the film does not change, fringes can also appear when the thickness d and index n change because of variations in temperature. Consequently, film thickness and refractive indices can be measured by using interferometry, as has been reviewed by Pliskin and Zanin (1983) [also see Tarof *et al.* (1989)], and temperature can also be determined; each of these parameters can be monitored *in situ* and in real time during processing. This section focuses on measuring film thickness and substrate temperature by reflection interferometry with one polarization. Interferometry is usually performed at or near normal incidence, where polarization effects are unimportant, although this is not absolutely necessary. Interfer-

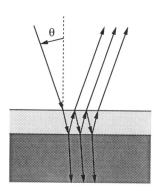

Figure 9.12 Schematic of reflection interferometry.

9.5 Interferometry

ence effects are also important in reflection ellipsometry, which is discussed in Section 9.11.

In some applications, reflection monitoring at a single wavelength is sufficient, while in others multiple-wavelength monitoring can improve precision and supply additional data (optical constants, etc.); sometimes, this is absolutely necessary. Multiple laser lines can be used or, more simply, dispersed white light can be employed. While data acquisition may be too slow for real-time multiwavelength monitoring when the wavelength is scanned with a monochromator, acquisition with multichannel detection is usually fast enough. When rapid turnaround is required, perhaps at many points on the wafer, data analysis may involve comparison with data in look-up tables (of previously analyzed cases) rather than real-time analysis using the approaches described in Section 9.1. Multiwavelength reflection interferometry has sometimes been called CARIS—constant-angle reflection interference spectroscopy. Similar information is obtained by probing at multiple angles of incidence, which is sometimes called VAMFO—variable-angle monochromatic fringe observation. This latter method is more suitable for *ex situ* characterization than for *in situ* observation in a thin film reactor.

Interference effects are monitored using a laser or a broadband optical source and can be observed in the reflected and transmitted beams. Observation in reflection is discussed here, although interference effects will also be seen in transmission as long as the wafer is partially transparent. Collection of transmitted light is difficult in some reactors. Analogous interference effects can be monitored by using light from the reactor itself, such as atomic and molecular optical emission (optical emission interferometry, Section 6.3.1.3) or thermal radiation (pyrometric interferometry, Section 13.4), or by other processes, such as spontaneous Raman scattering (Section 12.4.2).

One disadvantage of interferometry is that the reflected signal must be monitored throughout processing and needs to be referenced to a known starting condition. Ellipsometry often does not have this disadvantage; moreover, it provides much more information than reflection interferometry, including more detail about the surface and interfacial layers. It is, however, more complex and more expensive; it is also sensitive to the polarization-altering properties of reactor windows and requires careful calibration. Often, simple single- or multiple-wavelength reflection interferometry provides sufficient information for process development and real-time monitoring and control. Simpler, cheaper, and more robust than ellipsometry, interferometry is certainly a useful diagnostic for closed-loop control during manufacturing.

The next two sections focus on the use of reflection interferometry for metrology and thermometry. Reflection spectroscopy can also be employed to determine the composition of films. For example, Franke *et al.* (1993,

1994) used infrared reflection spectroscopy to determine the phosphorus concentration in phosphosilicate glass (PSG) films, and they assessed its potential application for at-line process monitoring and quality control during microelectronics manufacturing. Such methods can also involve interferometric effects.

9.5.1 Interferometric Metrology

Single- and multiple-wavelength interference has become a common tool for real-time thickness monitoring and endpoint detection in many thin film processes, including deposition, etching, annealing, and photoresist processing. Multiwavelength reflection spectroscopy using white-light sources is widely used for *ex situ* measurements of film thickness (Tencor Instruments, 1990; Keenan *et al.*, 1991; Prometrix Corporation, 1991; Kondo *et al.*, 1992). Using a wide range of wavelengths eliminates the ambiguity that can occur when just one wavelength is used and improves accuracy. Off-line commercial instruments can monitor and map film thicknesses from ~20 Å to 5 μm by using near-UV to near-IR light, sometimes with spot sizes down to 3 μm. The reflected light can be detected with a scanning monochromator or, for faster analysis, with a multichannel array detector. Fast software is used to compute the thickness for relatively thin films (0.1–8 μm), while fast Fourier analysis is used for thicker films (2–150 μm); alternatively, comparison can be made to previously calculated cases (lookup tables). In very early work on real-time monitoring, Konnerth and Dill (1975) used spectroscopic (interferometric) reflectance measurements to monitor etching and photoresist development *in situ*. Real-time monitoring of film growth and etching by interferometry is an example of time-resolved reflectivity, albeit one with modest temporal requirements.

Single-wavelength interferometry has been used by several groups to monitor deposition. For example, Jensen *et al.* (1988) used laser reflectometry with a He–Ne laser (632.8 nm) to monitor excimer laser-assisted, photolytic metalorganic (organometallic) vapor phase epitaxy (MOVPE/ OMVPE) of CdTe layers on GaAs substrates and HgTe epilayers on a CdTe buffer layer atop GaAs. Using the same wavelength, Sankur *et al.* (1991) monitored the reflected light for closed-loop control of thickness and composition during the OMVPE deposition of AlGaAs layers. Similarly, Farrell *et al.* (1991) and J. V. Armstrong *et al.* (1992) monitored the metalorganic molecular beam epitaxy (MOMBE) of AlGaAs on GaAs, dubbing the technique "dynamic optical reflectivity" (DOR). They showed that both the growth rate and refractive index could be determined from the reflectance trace. Bajaj *et al.* (1993) showed that interferometry senses the interdiffusion between alternating growing CdTe and HgTe layers during metalorganic chemical vapor deposition (MOCVD). Figure 9.13b shows the reflection interferogram during the growth of this multilayer pattern

9.5 Interferometry

depicted in Figure 9.13a, while Figure 9.13c shows the details during one period, where it is clear that interdiffusion must be included in the model. Irvine et al. (1994) integrated laser reflectometry and other probes to monitor and control the growth of II–VI layers, such as HgCdTe. Stoner et al. (1992) and C.-H. Wu et al. (1993) monitored the CVD of diamond films with real-time reflection interferometry. Severin and Severijns (1990) have developed an *in situ* sensor of film growth based on interferometry.

In situ optical monitoring and control is routinely used in the deposition of thin film optical coatings (Macleod, 1989), and reflection/transmission interferometry is one of the most important methods. In optoelectronics, optical coatings are often used to make high-reflectance (HR) mirrors or antireflective (AR) coatings. High-reflectance mirrors designed for a vacuum wavelength λ consist of alternating quarter-wavelength layers of materials 1 and 2 with respective thickness $\lambda/4n_1$ and $\lambda/4n_2$. The peak reflectances required for vertical cavity surface emitting lasers are >98%. Antireflective coatings can be made of a single layer of thickness $\lambda/(4\sqrt{n_1n_2})$, between materials with indices n_1 and n_2, or multiple layers (Macleod, 1989). Reflection coefficients must be $\lesssim 10^{-3}$ in optical amplifiers for telecommunications and as low as $\sim 10^{-5}$ for external cavity modelocking.

Babić et al. (1991, 1992) have integrated an *in situ* normal incidence reflectometer into a magnetron reactive sputtering system for the deposition of silicon/silicon oxynitride multilayer coatings for optoelectronics applications. A circular hole was placed in the anode to introduce the laser through it and onto the wafer at near-normal incidence. This modification of the anode actually improved wafer uniformity near the edge of the wafer. The deposition rate was determined by monitoring the interference fringes, while the index of refraction of the deposited material was monitored by comparing the maximum and minimum reflectance in the fringes by using Equations 9.12–9.14. Using this process monitoring method and feedback control, they made silicon/silicon nitride mirrors with peak reflectances of ~99%.

Most of these reflection studies used near-normal-incidence light to probe film growth, often because it is quite convenient experimentally. Only one reactor port is needed to conduct reflectometry at normal incidence. Also, normal-incidence experiments are insensitive to the polarization properties of reactor windows. Still, several interferometry investigations have been conducted at other angles. Takano et al. (1993) used *p*-polarized light reflectance at an incident angle of 70° (670 nm, laser diode) to monitor the thickness and optical constants during the plasma-assisted CVD of a-Si:H thin films. Dietz et al. (1995) employed *p*-polarized reflectance spectroscopy (PRS) at an incident angle near the pseudo-Brewster angle to monitor GaP epitaxy on Si(100). For pulsed chemical beam epitaxy (atomic layer epitaxy, ALE), they also observed that the reflected intensity contained a periodic oscillation superimposed on the usual interferometry

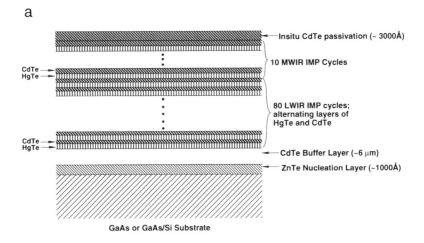

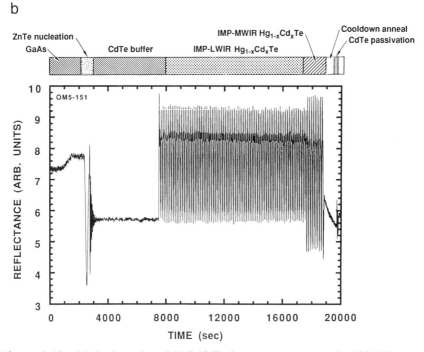

Figure 9.13 (a) Configuration of HgTe/CdTe heterostructure grown by MOCVD on a GaAs/Si substrate and (b) real-time reflectance monitoring during the growth of this heterostructure at 632.8 nm. (c) The reflectance during two cycles of the growth of the structure in (b), showing *in situ* monitored data (solid line), the theoretical profile assuming interdiffusion between the CdTe and HgTe layers (dashed line), and the theoretical profile assuming no interdiffusion (dotted line) (from Bajaj *et al.*, 1993). (© 1993 IEEE.)

9.5 Interferometry   357

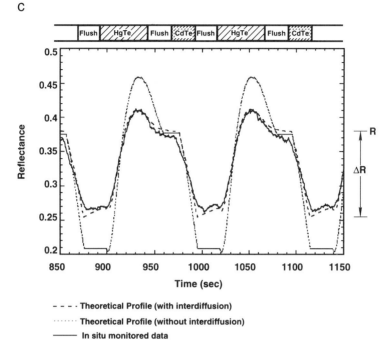

Figure 9.13 (*Continued*)

trace (see Figure 9.61a). This additional modulation is due to the surface modifications occurring with the alternating pulses of reactants to the surface, and is discussed further in Section 9.9.

Multiwavelength interferometry is a powerful real-time sensor of film deposition. Killeen and Breiland (1994) used near-normal, spectroscopic interferometry to monitor the growth of AlAs, GaAs, and AlGaAs films in a rotating-disk MOCVD reactor. Light from a tungsten–halogen lamp was delivered to the substrate with fiber optics, as is illustrated in Figure 9.14. The reflected light was delivered to a spectrometer by fiber optics, and a 1024-element Si photodiode array recorded the spectral region from 385 to 995 nm with a 1-sec integration time. Interference patterns are observable as long as the coherence length of the light is greater than the optical path length of the film. The coherence length is $\lambda^2/\Delta\lambda$, where $\Delta\lambda$ is the spectral resolution of the spectrometer/diode array detection system. For $\lambda = 385$ nm, the coherence length was >60 μm in vacuum or >12 μm in AlGaAs ($n < 5$), which is much greater than the film thicknesses. Examples of this type of monitoring are shown as the gray-scale images in Figure 9.15, which probes the growth of a thick AlAs layer on GaAs,

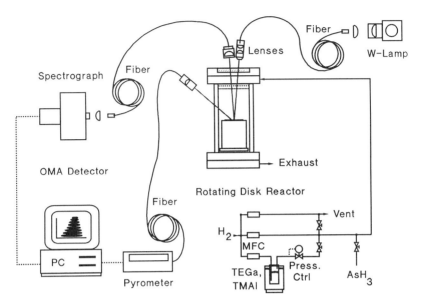

Figure 9.14 Experimental apparatus for *in situ* spectroscopic reflectance monitoring in a rotating-disk MOCVD reactor (from Killeen and Breiland, 1994). (© 1994 IEEE.)

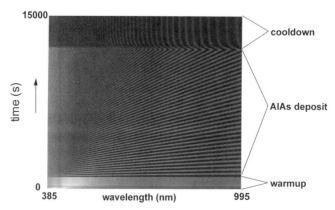

Figure 9.15 Gray-scale data visualization of spectroscopic reflectance during the growth of a thick layer of AlAs on a GaAs substrate by MOCVD at 650°C, followed by cooling after growth. The gray-scale intensities are proportional to the reflectance, with white being the maximum reflectance and black the minimum (from Killeen and Breiland, 1994). (© 1994 IEEE.)

9.5 Interferometry

and Figure 9.16, which probes the growth of a 15-period stack of AlAs/Al$_{0.5}$Ga$_{0.5}$As, which is a high-reflectance distributed Bragg reflector centered at 628 nm.

This multiwavelength method gathers much more information than single-wavelength interferometry, with little added complication and expense. Film thicknesses are often overdetermined. It is more versatile than single-wavelength methods because one or several optimal wavelengths can be chosen for a specific application. In particular, interference effects can be absent during the monitoring of a given layer with a single wavelength for one of two reasons: (1) strong absorption in that layer (450 nm, for the GaAs layer in Figure 9.17) or (2) interference effects in the underlying layer when that layer is an even number of quarter-wavelengths optically thick (AlAs for 673 nm in Figure 9.17, so the GaAs layer becomes an "absentee layer"). When the underlying layer is an odd number of quarter-wavelengths "thick," as is the AlAs layer in Figure 9.17 with 805-nm probing, the fringe visibility is maximized in the (GaAs) layer. Although spectroscopic ellipsometry (SE) gathers even more optical information, spectroscopic reflectometry collects sufficient information for rapid real-time monitoring and control of multi-monolayer films, and does not require the careful control of the polarization effects of optics that is necessary in SE.

Bacher *et al.* (1992) and Chalmers and Killeen (1992) both used spectroscopic reflectance measurements on partially completed vertical-cavity surface-emitting lasers (VCSELs) to modify the growth recipe for the remainder of the MBE growth process, thereby correcting for errors that occurred during the first phases of growth. The measurements were con-

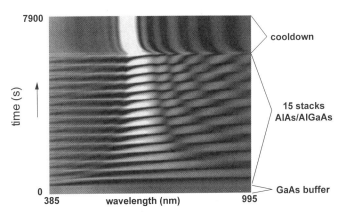

Figure 9.16 Gray-scale spectral reflectance during the growth of a 15-period stack of AlAs/Al$_{0.5}$Ga$_{0.5}$As by MOCVD to form a high-reflectance distributed Bragg reflector (DBR) centered at 628 nm. The gray scale is described in the caption to Figure 9.15 (from Killeen and Breiland, 1994). (© 1994 IEEE.)

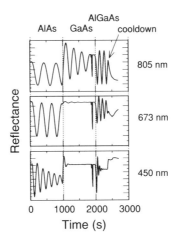

Figure 9.17 The reflectance monitored versus time at three wavelengths during the consecutive growth of AlAs, GaAs, and AlGaAs layers by MOCVD, followed by cooldown. The transient at 1950 sec was caused by an experimental artifact (from Killeen and Breiland, 1994). (© 1994 IEEE.)

ducted in line after the sample was repositioned within the MBE–reactor complex, without breaking vacuum, and allowed to cool to room temperature; such measurements can also be performed with the sample *in situ*.

Interferometry has been used to monitor etch rates during dry and wet etching of layered structures. For example, Busta *et al.* (1979) used interferometry (He–Ne laser, 632.8 nm) to monitor etch rates and to find the endpoint during plasma etching of Si_3N_4, SiO_2, and polysilicon on SiO_2. They also used the change in reflectance to determine the endpoint during etching of metals [tantalum on gold] (Equations 9.2–9.5). Vodjdani and Parrens (1987) measured etch depths *in situ* during reactive ion etching (RIE) of GaAs in $Cl_2/CH_4/H_2/Ar$ mixtures (632.8 nm). Hayes *et al.* (1990) monitored the CH_4/H_2 RIE of masked and unmasked InP/InGaAsP heterostructures by using He–Ne lasers operating at 1.15 μm and, for better depth resolution, 632.8 nm. They observed interference between reflections from the surface, buried heterointerfaces, and the polished back surface of the wafer, and used it to determine the etch rate, to identify heterolayers, and to determine the endpoint. This is illustrated by Figure 9.18, which shows the RIE etching of a laser base structure, consisting of (from top to bottom) p-$In_{0.72}Ga_{0.28}As_{0.6}P_{0.4}(Q_1)$/p-InP/$In_{0.72}Ga_{0.28}As_{0.6}P_{0.4}(Q_2)$/$In_{0.84}Ga_{0.16}As_{0.33}P_{0.67}(Q_3)$/grating in the InP substrate. Hicks *et al.* (1994) also used reflection interferometry to monitor the dry etching of InP/InGaAsP and GaAs/AlAs multilayer structures. Pope *et al.* (1994) have described the integration of laser interferometry into manufacturing tools for real-time control of silicon dioxide etching (by remote plasmas) of the slope portion of contacts and vias for integrated circuits.

9.5 Interferometry

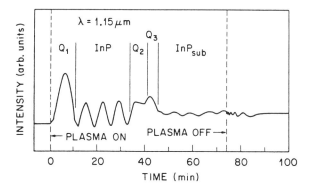

Figure 9.18 Reflectance interferogram taken at 1.15 μm during the reactive ion etching of a diode laser base structure. The solid lines denote heterointerfaces and the Q_i identify different InGaAsP quaternaries (from Hayes et al., 1990).

Collot et al. (1991) demonstrated how OES and laser interferometry can be used to monitor reactive ion etching of III–V heterostructures, including RIE etching of GaAs/AlGaAs heterostructures by CCl_2F_2/He and $SiCl_4$/He mixtures (top trace in Figure 6.17), and InGaAs/InP heterostructures in a CH_4/H_2 mixture. Derivative signals are seen to determine the interface more clearly than the signal itself. Vernon et al. (1992) studied thermal etching of InP by Cl_2 by infrared laser interferometry, with 1.52 μm from a He–Ne laser. The interval between fringes determined the etch rate, while the contrast and intensity of the interferogram gave qualitative information about the surface morphology during etching. Maynard and Hershkowitz (1993) proposed novel ways to present interferometric data during RIE. Wipiejewski and Ebeling (1993) used reflection interferometry (laser diode, 776 nm) to determine and control the etch depth and rate during the wet chemical etching of multilayer AlGaAs structures. Figure 9.19 shows how the etch rate increases when the liquid etchant is agitated. Böbel et al. (1994) have developed an instrument that simultaneously measures thickness, by reflection interferometry using LEDs at 630 and 950 nm, and temperature, by pyrometric interferometry (Section 13.4) at 950 nm; the pyrometry analysis uses data from the reflectance measurements. (Also, consult other references to this work that are cited in Section 13.4.)

Economou and co-workers have used multichannel interferometry to monitor plasma etching in real time at various radial positions, as is diagrammed in Figure 9.20 (Economou et al., 1991). This was combined with tomographic analysis of optical emission (Section 6.3.6) to assess the uniformity of etching processes. Aydil and Economou (1991) showed that in one state of the parallel-plate reactor the polysilicon etch rates (with Cl_2) are radially uniform, while in another state, where a plasmoid forms, the etching

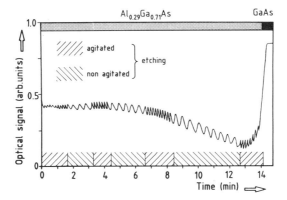

Figure 9.19 Reflectance monitoring at 776 nm (laser diode) during the wet etching (20 NH$_4$OH: 2 H$_2$O$_2$: 100 H$_2$O) of an Al$_{0.29}$Ga$_{0.71}$As layer either with or without agitation (from Wipiejewski and Ebeling, 1993).

rates are much greater in the center, as is seen in Figure 9.21. Aydil and Economou (1993) used spatially resolved interferometry to model plasma etching of polysilicon by chlorine.

Reflection interferometry can be used to follow thin film processes other than the deposition and etching of films. One important application is time-

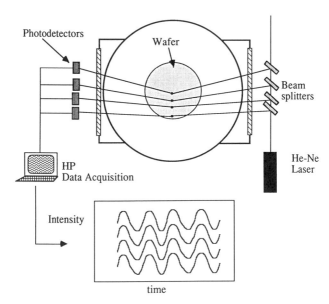

Figure 9.20 Schematic of spatially resolved laser interferometry apparatus (from Aydil and Economou, 1991).

9.5 Interferometry

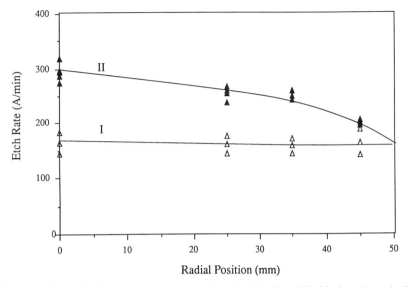

Figure 9.21 Polysilicon etch rates as a function of radius in an RF chlorine plasma in the absence (I) and presence (II) of a plasmoid, measured using the apparatus of Figure 9.20 (from Aydil and Economou, 1991).

resolved reflectometry to follow the solid-phase recrystallization of a-Si films in real time, as has been reviewed by Olson and Roth (1988). During the solid-phase epitaxy (SPE) of a-Si on a c-Si substrate, the a-Si film gets thinner and the reflectance oscillates because of interference, as is illustrated in Figure 9.22. The film thickness changes by $\lambda/4n_{a\text{-}Si}$ between adjacent interference maxima and minima, which is 32.6 nm for 632.8-nm light from a He–Ne laser. The a-Si film must be optically thin to perform such measurements. The time-dependent reflectance is compared to the known reflectance as a function of film thickness to obtain a plot of film depth versus time. This is converted into a SPE rate versus film depth. The SPE rate depends on temperature in an Arrhenius-like manner $\propto \exp(-\mathcal{E}_{act}/k_B T)$, with $\mathcal{E}_{act} \approx 2.7$ eV, and ranges from $\sim 10^{-10}$ to 1 cm/sec for $T = 500$–$1300°C$. Therefore, for $T > 1000°C$ the reflectance varies on time scales faster than 1 msec. Studies involving TRR include that by Dilhac *et al.* (1990), who used TRR to monitor the solid-phase epitaxial growth of As$^+$-implanted Si wafers in a rapid thermal processor, and that by England *et al.* (1993), who used electron beam heating for crystallization. Random nucleation and growth (RNG) can also be monitored with TRR when epitaxy does not occur (Olson and Roth, 1988). One example is the study by Im *et al.* (1993), who recorded reflectance oscillations due to explosive crystallization during laser melting (Figure 9.10).

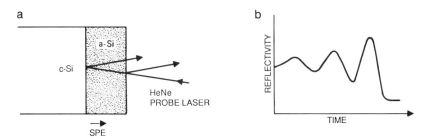

Figure 9.22 Measurement of the SPE (solid-phase epitaxy) of an amorphous silicon layer by time-resolved reflectivity (from Olson and Roth, 1988).

Interferometry has been used to probe the processing of polymer films, including those used as photoresist; this has been reviewed by Saenger and Tong (1990, 1991), Moreau (1988), and Watts et al. (1988). The film thickness and/or refractive index of a resist depend on the volume and density of its components (sensitizer, resin, and solvent) and how these components are bonded together, and can change due to solvent uptake (swelling) or dissolution, film shrinking during drying and curing, temperature changes, and dry etching. Usually interference in the polymer/substrate system is monitored, sometimes with an overlying solvent as the ambient. This gives the optical path length, i.e., nd. The film thickness d can be obtained directly by placing a plate over the polymer film and monitoring interference in the air gap where the film has been locally removed. In real-time monitoring, wavelengths must be chosen that do not develop the photoresist.

The Fresnel expressions (Equations 9.2–9.5) assume sharp interfaces between media with different refractive indices. In the transition layer between layers i and $i + 1$, the index may vary linearly from n_i to n_{i+1}. For example, a three-medium model with a transition layer would represent solvent/transition layer/dry polymer/substrate, while a four-medium model with a transition layer would represent solvent/saturated polymer/transition layer/dry polymer/substrate. Krasicky et al. (1988) corrected the three-layer model for a diffused layer of thickness d_t at the 1/2 interface by replacing r_{12} by fr_{12} and β_2 by $\beta_2 + \beta_t$ in Equations 9.8 and 9.10. For a linearly varying index,

$$f = \frac{|\sin \gamma/2|}{\gamma/2} \tag{9.37}$$

where

$$\beta_t = \frac{\gamma}{2} = \frac{2\pi}{\lambda} \langle n_t \rangle d_t \tag{9.38}$$

and $\langle n_t \rangle = (n_1 + n_2)/2$. Jacobsson (1966) has detailed the reflectance of films with continuously varying refractive index.

9.5 Interferometry

Saenger and Tong (1987) monitored thickness changes of freshly spun 18.3% solution of polyamic acid in NMP (*N*-methyl-2-pyrrolidone) by interferometry (632.8 nm) as it was dried and cured on a silicon wafer to form the polymer polyimide. Tong *et al.* (1989) used this method to study the transport of strong swelling agents in thin polyimide films. Krasicky *et al.* (1987a,b, 1988) used interferometry to monitor the dissolution of polymer films, which included the modeling of the transition layers. Konnerth and Dill (1975) employed multiwavelength interferometry to follow dissolution. The measurement of the rate of plasma etching of transparent polymers by interferometry is now routine. For example, Davies *et al.* (1990) used interferometry to monitor photoresist thickness and uniformity during plasma ashing (etching).

In situ monitoring by multiwavelength interferometry can be used to optimize the steps of the photoresist coating process to obtain the desired resist thickness (Metz *et al.*, 1991, 1992). Such control is needed because of tight process tolerances. A system similar to that in Figures 5.9d (bifurcated) or 9.14 can be used. A quartz halogen lamp, filtered to prevent exposure of the resist, is delivered to the wafer in a resist spin coater module through a bifurcated fiber optic cable. This same fiber optic collects the light, which is dispersed by a spectrometer and detected with a photodiode array (450–950 nm). The use of these many wavelengths increases the accuracy of the film thickness measurement. This interferometric method can be used for process development or real-time control, as is demonstrated in Figure 9.23. Figure 9.23a shows that the thickness changes mostly during the first 10 sec during spinning, while Figure 9.23b shows that desolvation during baking occurs when the wafer reaches the critical temperature, here ~5 sec into the bake cycle. The off-line scan of film thickness across the wafer after baking in Figure 9.23c shows a 144-Å thickness variation across the wafer, with the greatest thickness at the edges; this measurement can also be made in-line.

Interferometry can also be used for *in situ* endpoint control during photoresist development, as is schematically represented in Figure 9.24. Using this method, Thomson (1990) showed that the endpoint times can differ significantly even in a wafer fabrication facility that is in excellent statistical control, as seen in Figure 9.25. Early work by Lauchlan *et al.* (1985) examined the critical dimension (CD) (linewidth) of developed photoresist versus exposure dose, resist thickness, and softbake (prebake) temperature. Light from a broadband source was passed through a yellow filter to prevent exposure of the photoresist film. Reflected light was analyzed after passing through narrowband interference filters with central wavelengths optimized for the film under the resist. The total development time was determined relative to the control point, which is the last extremum in the detected reflected signal. Because the developed region in the photoresist is a small fraction (5–10%) of the probed area (diameter ~10 mm), the visibility of the fringes is small; i.e., the difference of the reflectances at the fringe

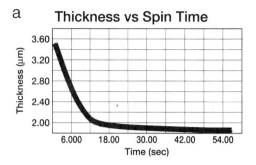

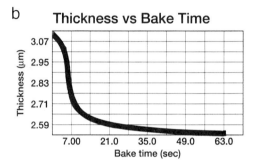

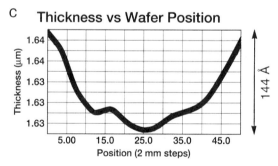

Figure 9.23 Reflectance monitoring of photoresist thickness versus (a) spin time, (b) bake time, and (c) wafer position, obtained with a spectrometer/photodiode array assembly (450–950 nm) and a fiber optic cable delivery system, as in Figure 5.9d (from Metz et al., 1992).

maxima and minima is a small fraction of the average reflectance. They developed smoothing techniques needed to improve the signal/noise ratio for resist exposed through a contact mask with 6% exposed area. Sautter and Batchelder (1987) extended this work by delivering light to many different parts of the wafer by using fiber optics. Reid and Sautter (1992) presented and analyzed the factors that affect the reliability of interferometry in endpoint detection of photoresist development. Novembre et al.

9.5 Interferometry

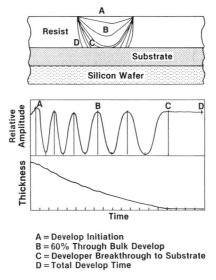

Figure 9.24 Schematic of the resist profile, reflectance, and local film thickness during the develop cycle for endpoint detection. [From A. G. Reid and K. M. Sautter, Develop endpoint detection—a manufacturer's perspective, in *Integrated Circuit Metrology, Inspection, and Process Control VI, SPIE* **1673,** 284 (1992).]

(1989) used interferometry to monitor the development rate of electron-beam-exposed resists on photomask substrates, and thereby to control the feature size.

Watts *et al.* (1988) have described how interferometry can be used to monitor the optical thickness of resist films during coating, softbaking, postexposure baking, and developing. They have also shown how resists atop wafers can be monitored by absorption and how the results of reflection interferometry and absorption measurements can be used together. For example, if after the softbake the thickness [which depends on the amounts of photoactive compound (PAC), resin, and solvent] is monitored by interferometry and if absorption in the film [which depends on the amount of PAC], is also measured, then variations in the PAC due to the coating step can be distinguished from the amount of solvent that is retained after the softbake. Absorption is usually measured by measuring the fraction of light reflected from the resist/wafer interface, which makes two passes in the film, rather than by transmission. This technique is described in Section 8.3.

Interferometry can also be used to follow surface modifications, such as oxidation. For example, Baufay *et al.* (1987) monitored the reflectance at 632.8 and 514.5 nm in real time during the oxidation of copper induced by continuous-wave (cw) argon-ion laser heating. The oscillations in the reflectance gave the thickness of the copper oxide layers as a function of time.

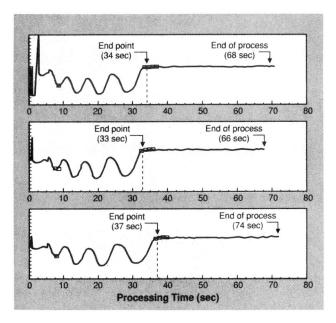

Figure 9.25 Real-time reflectance monitoring shows that the endpoint time for photoresist development varied for these three patterned wafers, even though the process was in excellent statistical control (from Thomson, 1990).

9.5.2 Interferometric Thermometry

Interferometric reflectometry is a simple, inexpensive, and noninvasive method of measuring temperature. As applied to measuring wafer temperature, it has been called wafer thermometry (Donnelly and McCaulley, 1990; Donnelly et al., 1992). In addition to applications in process development, interferometric thermometry has great promise for real-time monitoring during manufacturing.

Donnelly and McCaulley (1990) reported the use of interferometry to measure the temperature of Si, GaAs, and InP wafers using a 1.15-μm He–Ne laser. A full interference cycle occurred, corresponding to a change in optical path length (nd) of $\lambda/2$, for every ~3 K temperature change with these ~500-μm-thick wafers; consequently, temperature changes of ± 0.2 K are easily detected. Sankur and Gunning (1990) then demonstrated the use of 3.39-μm radiation from a He–Ne laser to monitor the temperature of a GaAs wafer. In monitoring semiconductor wafers, the photon energy must be below the fundamental band gap to ensure transmission and multiple internal reflections, but near enough the band gap so that the refractive index depends strongly on temperature.

9.5 Interferometry

Interferometry had been used in earlier work to monitor the temperature of glass plates. Hacman (1968) sensed the thermal expansion of glass plates interferometrically using 632.8 nm from a He–Ne laser. Bond *et al.* (1981) monitored the temperature change of a glass substrate in a plasma reactor in a similar way. Saenger (1988) used this method to monitor glass substrates that were transiently heated by an excimer laser.

One application of this technique is demonstrated in Figure 9.26 from Donnelly *et al.* (1992). Figure 9.26a shows the interferogram of a silicon wafer recorded by using a 1.15-μm He–Ne laser during heating by a N_2 plasma. The interferogram does not have the sinusoidal pattern that is obtained from single-wavelength interference, but has an additional interference beating pattern, because this laser emitted two closely spaced wavelengths, at 1.1526 μm (83%) and 1.1605 μm (17%). (This two-wavelength scheme has some advantages, but is seldom used.) Figure 9.26b follows the temperature changes in the plasma-heated wafer as a function of time.

For a substrate of thickness h (or film of thickness d), the interference phase β at temperature T is referenced relative to that at T_0, β_0. The number of fringes is $\Delta F = (\beta - \beta_0)/2\pi$. Also (Saenger and Gupta, 1991)

$$\frac{d\beta}{dT} = \frac{2\pi}{\lambda} 2nh(\tilde{\alpha} + \tilde{\beta}) \tag{9.39}$$

where $\tilde{\alpha} = (1/h)(dh/dT)$ is the linear coefficient of thermal expansion and $\tilde{\beta} = (1/n)(dn/dT)$ is the fractional change in refractive index with temperature of the wafer or film. (Both $\tilde{\alpha}$ and $\tilde{\beta}$ can depend on temperature.) When a new fringe appears, the temperature has changed by

$$\Delta T/\text{fringe} = \left(\frac{d\beta/dT}{2\pi}\right)^{-1} = \frac{\lambda}{2nh(\tilde{\alpha} + \tilde{\beta})} \tag{9.40}$$

Therefore, the probe becomes more sensitive to small changes in temperature when h is large and when $\tilde{\alpha} + \tilde{\beta}$ is large. So it is a more sensitive probe when measuring interference in a (thick) wafer than in a thin film on the wafer. The relative size of $\tilde{\alpha}$ and $\tilde{\beta}$ depends on the material and T. For semiconductors, typically $\tilde{\beta} \gg \tilde{\alpha}$. For silicon, $\tilde{\alpha} \approx 2.5 \times 10^{-6}/°C$ and $\tilde{\beta} \approx 50 \times 10^{-6}/°C$ at 1.5 μm. While this relation also holds for some insulators—for example, for fused silica $\tilde{\alpha} \approx 0.6 \times 10^{-6}/°C$ and $\tilde{\beta} \approx 6 \times 10^{-6}/°C$ at 633 nm—in others thermal expansion can be dominant, such as for soda lime glass where thermal expansion is very large, $\tilde{\alpha} = 9.5 \times 10^{-6}/°C$ (Bond *et al.*, 1981).

In general, more information is available about the temperature dependence of $\tilde{\alpha}$ (McCaulley *et al.*, 1994). $\tilde{\beta}$ depends on λ through the wavelength dependence of n. This can be very important for photon energies near the band gap. Choosing the probe energy near or above the band gap may seem preferable because $\tilde{\beta}$ would likely be very large; however, this can be done only if the concomitant absorption is weak ($\alpha d \ll 1$), to ensure

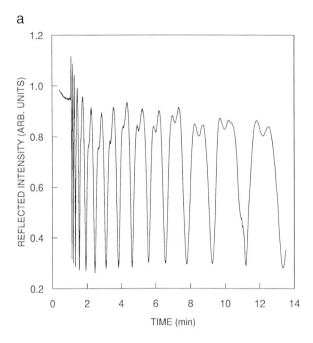
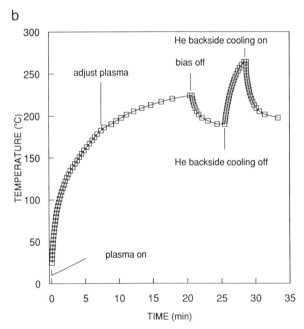

Figure 9.26 (a) Reflectance monitoring of a silicon wafer during N_2 plasma-induced heating. The interferogram has a double-peaked structure because two wavelengths near 1.15 μm were used: 1.1526 and 1.1605 μm. (b) Tracking the temperature of a silicon wafer during N_2 plasma heating by using interferometry, as in (a) (from Donnelly *et al.*, 1992).

9.5 Interferometry

multiple reflections in the film. In wafer thermometry, it is usually wise to choose the photon energy low enough below the room-temperature band gap so that the photon energy still is smaller than the band gap at the highest probed T, since the gap usually decreases with T (Section 18.4.2).

Experimentally, interferometric thermometry can be applied by using calibration data (a_i) that relate the number of fringes (ΔN) that are recorded as the temperature is changed to T, as referenced to a temperature T_0 (Donnelly and McCaulley, 1990). Using a polynomial fit,

$$T(K) = a_0 + a_1\left(\frac{\Delta N}{h}\right) + a_2\left(\frac{\Delta N}{h}\right)^2 + \cdots + a_5\left(\frac{\Delta N}{h}\right)^5 \quad (9.41)$$

where $a_0 = T_0$. Equation 9.41 can be used to determine any T from the measured number of fringes if the starting temperature is T_0. If a starting temperature $T_i \neq T_0$, an offset number of fringes $\Delta N'$ must be calculated from

$$\Delta N' = h(a_0' + a_1' T_i + a_2' T_i^2 + \cdots + a_6' T_i^6) \quad (9.42)$$

$\Delta N'$ is added to the number of fringes counted from T_i to T, and then this sum is the ΔN used in Equation 9.41 to obtain T.

All measurements must be made at the same wavelength and angle of incidence (usually 0°) as those used in obtaining the calibration constants. Donnelly and McCaulley (1990) give a_i and a_i' (which are determined from a_i) for Si, InP, and GaAs at 1.15 μm by using a calibration run referenced to a thermocouple. Saenger et al. (1992) have determined similar calibration coefficients for Si and GaAs at 1.523 μm, and have discussed how these coefficients scale for small changes in wavelength.

Direct calculation of the relation between ΔN and T is possible by using Equation 9.40 if $\tilde{\alpha} + \tilde{\beta}$ is accurately known. Furthermore, such calculations are needed when conducting measurements away from normal incidence because the propagation angle in the wafer depends on n (Equation 9.9).

Since experimental values for $\tilde{\alpha}$ are easier to find than those for $\tilde{\beta}$, McCaulley et al. (1994) used laser interferometry to measure $\tilde{\beta}(T)$ of Si, GaAs, and InP at 1.15, 1.31, 1.53, and 2.39 μm. Figure 9.27 plots $\tilde{\beta}$ in GaAs versus temperature for several infrared wavelengths. Polynomial expressions for experimentally determined $\tilde{\beta}(T)$ were given that have an estimated uncertainty of 5% from 77 to 900 K. They noted that although theory makes fairly good predictions of the absolute values and temperature dependences of $\tilde{\beta}$, the calculated values do not have the accuracy needed for interferometric thermometry; the experimental fits must be used.

Saenger and Gupta (1991) measured $\tilde{\alpha} + \tilde{\beta}$ for five materials for potential use as thin film interferometric temperature sensors (25–870°C, 632.8 nm): c-axis Al_2O_3 (corundum), MgO $\langle 100 \rangle$ (periclase), $MgAl_2O_4$ $\langle 111 \rangle$ (spinel), $Y_2O_3ZrO_2$ (yttria-stabilized zirconia) and SiO_2 (suprasil-fused silica). Of

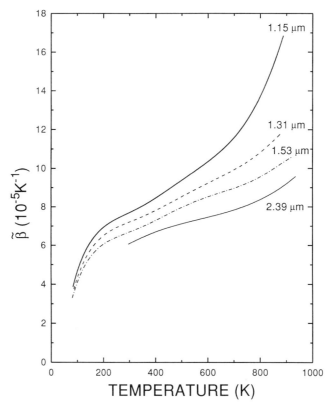

Figure 9.27 Fitted value of $\tilde{\beta}(T)$, the fractional change in the refractive index $(1/n)(dn/dT)$, for GaAs at four wavelengths (from McCaulley *et al.*, 1994).

these, MgO has the highest $\tilde{\alpha} + \tilde{\beta}$, 17.9 and 26.0 × 10^{-6}/°C at 25 and 800°C, respectively, which is smaller than that for Si.

Two potential problems can occur in the practical implementation of wafer interferometric thermometry, which relate to the required smoothness of the wafer and the direction of the temperature change. Early studies were conducted on wafers that were polished on both sides. Donnelly (1993a) has shown that interferometric thermometry can also be performed on wafers that are polished only on one side, which is standard for semiconductor wafers used in manufacturing. The interference pattern is the same as for wafers with both sides polished, although the contrast of the fringes is lower (Figure 9.28). Using 1.55-μm light from a distributed-feedback InGaAsP semiconductor laser, the contrast in the interferogram for a Si wafer polished on both sides was 0.97, where contrast is defined as $(I_{max} - I_{min})/I_{max}$ and I_{min} and I_{max} are the minimum and maximum fringe

9.5 Interferometry

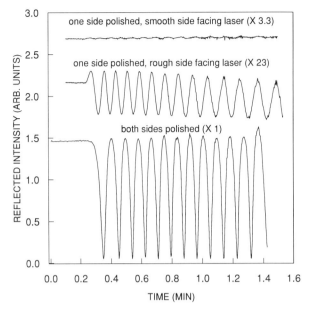

Figure 9.28 Interferograms at near-normal angle of incidence (1.55 μm, distributed-feedback InGaAsP semiconductor laser) of 0.6-mm-thick silicon wafers heated from room temperature to ~60°C. The wafer is either polished on only one side, with the laser incident from either side, or on both sides (from Donnelly, 1993a).

intensities. For the wafer polished on one side, the contrast was 0.22 with the rough side facing the laser beam and only ~0.006 with the polished side facing the beam; the contrast in the latter situation is too low to be useful. Using longer wavelengths decreases scattering at the rough surface and improves the fringe visibility.

The other disadvantage of the described implementation of interferometric thermometry is its inability to determine the sign of a temperature change. In monitoring temperature T versus time t, a change in sign of dT/dt gives a characteristic cusp in the R versus t trace when the reflectance is not at an extremum. However, the sign of dT/dt is not known at a fringe maximum or minimum. While it is not necessary to know this sign during heating and cooling cycles, it is essential during steady-state operation when temperature control is important—the sense of the change must be known to correct for a temperature drift. This problem can be solved by examining how the interference pattern changes with thickness or wavelength. Donnelly and McCaulley (1990) and Donnelly (1993b) have exploited the thickness nonuniformity that is common in wafers by comparing the interferograms at nearby points. These interferograms change with T differently because of their different thicknesses; consequently, the direction of temperature change can be inferred with a high degree of confidence. This principle is illustrated in Figure 9.29 for an interferometer with a wedge-shaped sensor, which approximates a wafer locally. Donnelly (1993b) has accomplished this by comparing light reflected into the four different sectors of a Ge quadrant photodiode detector (Figure 9.30). When the reflected beam is centered on the detector, each quadrant samples a contiguous region, presumably, with a slightly different thickness.

Saenger *et al.* (1992) have proposed to solve this problem by examining reflectance at one point by using a wavelength-modulated semiconductor laser. When the laser wavelength is modulated over a range corresponding to a fraction of a fringe $\Delta\lambda \ll \lambda^2/2nh$, the magnitude of the reflectance oscillations is proportional to $dR/d\lambda$. If the phase β increases with T, as in Si near 1.5 μm, then when R increases (decreases) in time, $dR/d\lambda$ is negative (positive) during heating and positive (negative) during cooling. They used an InGaAsP distributed-feedback laser $\lambda \sim 1.55$ μm that was wavelength-modulated by $\pm \sim 0.12$ Å at 1 kHz by modulating the bias current. This corresponds to a total change of 1/25 fringes for a ~ 500-μm-thick Si wafer. The reflected signal was measured with a Ge photodiode, which was then normalized by the output power of the diode laser and sent to a lock-in detector. The resulting signal gives the magnitude and sign of the reflectance modulation, which gives $dR/d\lambda$. Figure 9.31 compares the temperature obtained from interferometry with that from a noncontact fluoroptic probe (Section 14.3). dT/dt is uniquely determined, with the transition from heating to cooling near 280 sec noted, even though the reflectance is at a maximum.

A related, yet simpler, method by Kikuchi *et al.* (1994) and Kurosaki *et al.* (1995) determines dT/dt from the overshooting or undershooting of transmitted (or reflected) wavelength-chirped pulses from a pulsed infrared diode laser.

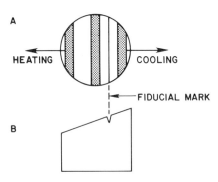

Figure 9.29 Schematic of wedge interferometer from the (A) top and (B) side. The dark bands, representing minima in reflectance, move in the direction indicated during heating and cooling (from Saenger *et al.*, 1992).

9.5 Interferometry

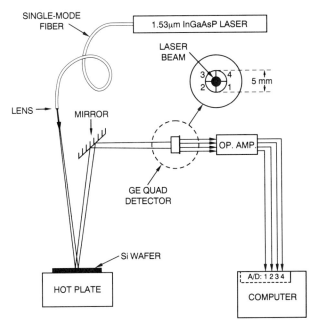

Figure 9.30 Schematic diagram of the interferometric thermometry setup for determining the direction of the wafer temperature change, using a Ge quadrant detector (from Donnelly, 1993b).

Saenger *et al.* (1992) have proposed alternative wavelength-modulation schemes using either full-fringe wavelength modulation or two laser wavelengths whose spacing corresponds to a quarter of a fringe. Donnelly (1993b) has suggested that wavelength modulation can be used to calibrate the thickness nonuniformity method, which would then be employed. Using two laser wavelengths (Figure 9.26a) often removes the ambiguity that can occur with a single wavelength.

Wafer interferometry has been used in several applications in thin film processing. Donnelly *et al.* (1992) employed it to monitor wafer temperature in a plasma reactor (Figure 9.26). McCrary *et al.* (1992) used a 1.53-μm distributed-feedback laser to monitor the temperature profile on InP substrates placed in different positions in a horizontal, low-pressure MOCVD reactor. With the 2-inch substrate placed in the central well susceptor, the temperature across the wafer was uniform within ± 2–$4°C$ for pressures from 20 to 1000 mbar, as is illustrated in Figure 9.32. When it was placed in the downstream position on the two-well susceptor, temperature varied by as much as $\pm 17°C$ across the wafer. This explains the better compositional uniformity in InGaAsP layers grown in the former position. Hayes *et al.* (1990) used interferome-

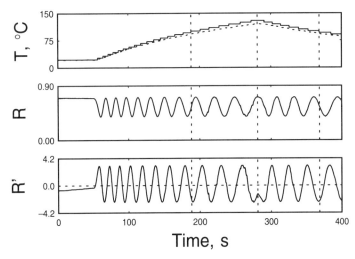

Figure 9.31 Experimental results showing temperature T (stepped solid line from interferometry and dotted line from a noncontact fluoroptic probe), reflectance R, and the wavelength derivative R' of a Si wafer during thermal cycling (from Saenger *et al.*, 1992).

try to monitor the changes in wafer thickness and temperature during reactive ion etching of InP wafers.

Interferometry can be used to monitor very rapid changes of temperature due to a pulsed or scanning heating source, which is another example of time-resolved reflectivity (TRR). Applications of this type of heating include recrystallization and annealing using lasers and beams, and rapid thermal processing. In one example of TRR, Murakami *et al.* (1984) monitored the reflectance of a He–Ne laser (632.8 nm) nearly normally incident on a Si(001) film on a sapphire substrate (SOS) during heating by a shuttered krypton-ion laser (647 nm, 676 nm; $\theta \approx 60°$). Interference in the silicon layer depends on the temperature dependence of n_{Si} and on thermal expansion. Since both lasers are focused, this method provides time- and space-resolved measurement of temperature. Timans *et al.* (1988) used TRR to probe the time-dependent temperature changes in silicon-on-insulator (SOI) structures during heating by a scanning electron beam. Although interference in the silicon layer in the studied structure—capping oxide film/silicon film/isolating oxide film/silicon substrate—is the most important effect, interference in the oxide layers must be included in the analysis.

9.6 Photoreflectance

In photoreflectance (PR), an amplitude-modulated optical beam is incident on a medium (Figure 9.33), which modulates the carrier concentra-

9.6 Photoreflectance

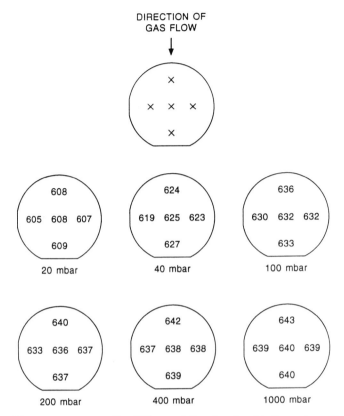

Figure 9.32 Temperature profiles (°C) on an InP wafer placed in the center well susceptor of a LP MOCVD reactor, at the indicated pressures, as measured by reflection interferometry at 1.53 μm (from McCrary et al., 1992).

tion near the surface and therefore also the built-in surface electric field. This causes a modulation in the reflectance ΔR, which is measured as a function of wavelength. Critical-point energies are then obtained by curve fitting. Photoreflectance is a contactless form of electroreflectance (ER), where an external electric field is applied directly; ER is less useful as a real-time probe.

Photoreflectance is a type of differential reflectometry. While the aim of the reflection-based spectroscopies discussed in Sections 9.8–9.10 is to probe the surface, PR uses modulation-induced changes near the surface or an interface to probe the bulk-like features of the wafer, thin film, or heterostructure. Reviews of photoreflectance and related modulation reflectance spectroscopies can be found in Aspnes (1980), Pollak (1981), Shen et al. (1987), Bottka et al. (1988), Pollak and Shen (1990), Glembocki (1990), and Glembocki and Shanabrook (1992). These modulation methods

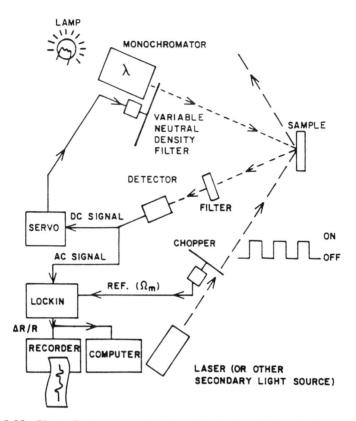

Figure 9.33 Photoreflectance apparatus (from Glembocki and Shanabrook, 1992).

are used extensively in *ex situ* materials characterization to measure alloy composition, subband energies in quantum well structures, and the presence and homogeneity of built-in electric fields in MODFET (modulation doped field-effect transistor) structures.

The reflectance changes can be related to the derivatives of the dielectric function (Equation 4.40). This can be generally expressed as

$$\frac{\Delta R}{R} = \alpha_s \Delta \varepsilon' + \beta_s \Delta \varepsilon'' \tag{9.43}$$

where $\Delta \tilde{\varepsilon} = \Delta \varepsilon' + i\Delta \varepsilon''$ is the differential change induced by the modulation and α_s and β_s are the Seraphin coefficients, which depend on the unperturbed ε' (ε_1) and ε'' (ε_2) (Seraphin and Bottka, 1965).

In the low-electric field regime, the PR lineshape for band-to-band transitions is related to the third derivative of ε, with $\Delta \varepsilon \approx (\hbar\Theta)^3 \partial^3(\varepsilon \mathcal{E}^2)/\partial \mathcal{E}^3$. $\mathcal{E}$ is the photon energy and $\hbar\Theta$ is the electro-optic energy parameter equal

9.6 Photoreflectance

to $(e^2\hbar^2E^2/2\mu_\parallel)^{2/3}$. E is the dominant electric field and $\mu_\parallel$ is the reduced interband effective mass in the direction of this field. E may represent built-in DC or externally applied AC or DC fields. The Aspnes low-field derivative form (Pollak, 1981) is often used for fitting lineshapes:

$$\frac{\Delta R}{R} = \text{Re}[Ae^{i\phi_{ex}}(\mathcal{E} - \mathcal{E}_i + i\Gamma_{\mathcal{E}})^{-n}] \qquad (9.44)$$

where $\mathcal{E}$ is the photon energy, $\mathcal{E}_i$ is the critical-point feature such as the fundamental band gap, $\Gamma_{\mathcal{E}}$ is the broadening factor, and ϕ_{ex} is a phase factor. The parameter n is related to the critical-point type (n' in Equation 4.40) and the particular derivative. For GaAs, low-field spectra are fit for a three-dimensional critical point, with $n = 5/2$. This PR lineshape feature is sharper than that obtained by photoluminescence and ellipsometry because PR measures a derivative of the dielectric function. Consequently, it is easier to obtain critical-point energies (Figure 9.34) and to resolve shifts and splittings due to anisotropic strain.

At high electric fields, the absorption edge shifts to lower energies because of the Franz–Keldysh (FK) effect. Franz–Keldysh oscillations are seen in $\Delta R/R$ for photon energies above the critical-point energy, with the extrema given by

$$m\pi = \phi + \tfrac{4}{3}[(\mathcal{E}_m - \mathcal{E}_{bg})/\hbar\Theta]^{3/2} \qquad (9.45)$$

where the photon energy of the mth extremum is $\mathcal{E}_m$, $\mathcal{E}_{bg}$ is the band gap, and ϕ is an arbitrary phase factor (points 1–5 in Figure 9.35).

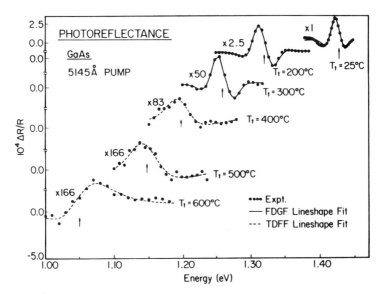

Figure 9.34 *In situ* photoreflectance spectra of GaAs at various OMVPE reactor temperatures (from Capuder *et al.*, 1990). (© 1990 IEEE.)

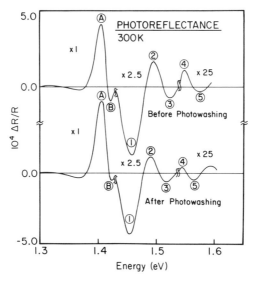

Figure 9.35 Photoreflectance spectrum of bare n-GaAs(100) at 300 K before and after photowashing, showing the Franz–Keldysh oscillations [features 1–5] (from Pollak and Shen, 1990). (© 1990 IEEE.)

One potential application of *in situ* PR is thermometry. The dependence of band gap energy on temperature $\mathcal{E}_{bg}(T)$ has been measured by using photoreflectance (Shen *et al.*, 1988; Shen, 1990), as well as by other techniques, and modeled by using the low-field PR lineshapes (Equation 9.44) and the Varshni equation or the phonon-dependence expression for $\mathcal{E}_{bg}(T)$ (Equations 18.7 and 18.8, respectively) (see Table 18.2). In turn, PR can be used to measure the temperature of semiconductors *in situ* by using this known dependence of the band gap energy on T. Capuder *et al.* (1990) measured the PR spectrum of GaAs and AlGaAs in an OMVPE reactor to 650°C to demonstrate the potential of this thermometric probe (Figure 9.34). With an error of ±5 meV for determining the fundamental gap energy from PR, the accuracy of the temperature measurement is no better than ±10°C. Given current uncertainties in the Varshni parameters, the uncertainty in temperature measurements by PR is actually much larger.

Pollak and Shen (1990) used PR to examine the decrease in the built-in potential in GaAs due to a surface treatment often called photowashing. This treatment unpins the Fermi level and reduces the surface electric field, as seen by the closer FK oscillations in Figure 9.35 after photowashing. This unpinning is thought to lower the surface state density, which can also be monitored with photoluminescence. (See Section 14.2, and Figures 14.3 and 14.4.) Although photoreflectance was not used as an *in situ* probe in this case, it demonstrates another potential real-time application of PR.

9.7 Surface Infrared Reflectometry

Nakanishi and Wada (1993) used PR to probe subsurface damage in GaAs *in situ* after Ar plasma etching, but not in real time because of the background plasma emission.

9.7 Surface Infrared Reflectometry

Surface infrared reflectometry (SIRR) is the most widely used surface infrared spectroscopy (SIRS). In SIRR, adsorbates on a surface are probed by the loss of reflected light due to the absorption by adsorbate vibrational modes. SIRR probes adsorbate identity, concentration, bonding configurations, and orientations, and is perhaps the best optical method for analyzing adsorbates on surfaces. Probing adsorbates on surfaces can reveal many things about a process: the fundamental chemical steps of etching and deposition, surface contamination, and the stability of a surface after cleaning. SIRR is widely used for fundamental studies and has been increasingly used in process development studies. However, it is not readily adaptable for real-time monitoring and control.

For strongly absorbing substrates, such as metals, only the external reflection mode shown in Figure 9.36a can be used, where the beam is incident

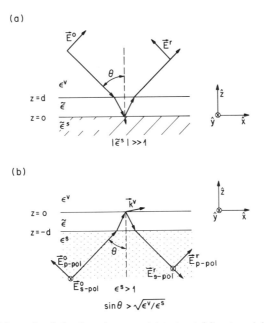

Figure 9.36 Schematic of the three-layer model to treat (a) external (IRRAS) and (b) internal (ATR) reflection (from Chabal, 1988).

from the gas ambient. This is called infrared reflection-absorption spectroscopy (IRRAS or IRAS) or reflection–absorption infrared spectroscopy (RAIRS). For transparent substrates, such as semiconductors within specific windows in the infrared, internal reflection can be used (Figure 9.36b), as well as external reflection. The internal reflection mode is called attenuated total internal reflection (ATR) spectroscopy. Chabal (1988, 1994) has reviewed all types of SIRS, including IRRAS and ATR spectroscopy, and Hoffman (1983) has reviewed IRRAS. Characteristic vibrational mode absorption frequencies are given by Nakamoto (1986) and in Table 8.1.

9.7.1 External Reflection Mode: Infrared Reflection–Absorption Spectroscopy

For detecting adlayers on metals, the external mode (IRRAS) is most sensitive at glancing angles, because adsorbate absorption is greatest when θ is near the pseudo-Brewster angle. This angle approaches 90° for conductors. This enhancement occurs for p-polarization, as is seen by using Equation 9.18 and expanding to the lowest-order terms of $\varepsilon_a/\tilde{\varepsilon}_m \ll 1$ [where a is the ambient (medium 1), m is the metal (medium 3)] (Chabal, 1988):

$$\left(\frac{\Delta R}{R}\right)_p \approx \frac{8\pi d n_a^3}{\lambda} \frac{\sin^2\theta}{\cos\theta} \mathrm{Im}\left(\frac{-1/\tilde{\varepsilon}}{1 - (\varepsilon_a/\tilde{\varepsilon}_m)\tan^2\theta}\right) \quad (9.46)$$

using the common IRRAS convention that $\Delta R = R(0) - R(d)$. The unsubscripted dielectric function is that of the adlayer, or more precisely that for the adsorbate response normal to the interface. $\Delta R/R$ peaks at an angle slightly less than the pseudo-Brewster angle, $\theta_B = \tan^{-1}\sqrt{|\tilde{\varepsilon}_m|/\varepsilon_a}$ (Greenler, 1975), which is ~87° for Ni at 6 μm. Since experimental limits for grazing incidence are usually ~80–85°, the denominator in the Equation 9.46 bracket can often be approximated by 1. Equation 9.16 shows that there is no analogous $1/\cos\theta$ enhancement for s-polarization.

For semiconductors, the external mode is less sensitive than the internal mode, as is illustrated in Figure 9.37; however, it is easier to implement. One way to increase the sensitivity of external mode reflection from semiconductors is to use the external reflection mode with enhancement by a buried metal layer (BML) (Bermudez, 1992). The semiconductor layer atop the metal or metal silicide is optically thin but chemically thick. Calculations show that the buried layer can increase the change in reflectance due to an adsorbate by an order of magnitude when p-polarized light is used (Figure 9.38). For example, a 3-Å-thick layer of H_2O on a Si wafer changes the reflectance at 6.1 μm (H_2O bend) by $\Delta R \sim +0.35 \times 10^{-4}$ near an 85° angle of incidence (not shown in the figure), while for this same layer on ~1000-Å-thick Si atop Ni, the change is $\sim -6 \times 10^{-4}$ near 83°.

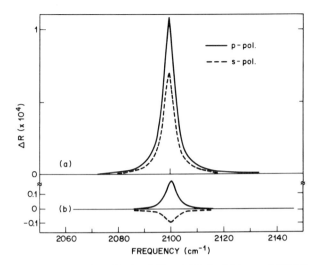

Figure 9.37 Model calculation of $\Delta R = R(0) - R(d)$ for (a) internal (ATR) and (b) external (IRRAS) reflection for a surface with a weakly absorbing, isotropic monolayer (from Chabal, 1988).

Koller et al. (1988) used the improved sensitivity of double-modulation Fourier transform methods for in situ external infrared reflection-absorption spectroscopy of plasma enhanced CVD (PECVD) SiO_2 films deposited on HgCdTe, Si, and Al substrates. Kawahara et al. (1991) used IRRAS to monitor the deposition of SiO_2 films on Si by SiH_4/O_2 low-pressure CVD. Kobayashi et al. (1993) monitored the surface reactions of the selective CVD of tungsten from $WF_6 + SiH_4$ by Fourier-transform infrared (FTIR) reflection spectroscopy of the surface. Absorption both by the adsorbates on the surface and gas-phase species can be seen. By redirecting the mirrors that steer the IR beam, absorption by gas-phase species alone can be monitored. While no adsorbed species were seen when WF_6 alone dosed W or SiO_2 surfaces, WF_6 was seen to react readily with Si-containing adsorbates produced by SiH_4 dosing of a W surface. Figure 9.39 shows the spectrum when the tungsten substrate at 150°C is exposed to silane. Features due to gas-phase silane are seen, along with a peak near 1160 cm^{-1} that represents the growth of a silicon-based adlayer. Nishikawa et al. (1994) used external FTIR reflection spectroscopy (80° angle of incidence) to probe the surface of Si during etching by an electron cyclotron resonance (ECR) plasma, and found evidence of $SiCl_x$, with $x = 1-4$.

Cumpston et al. (1995) used IRRAS to examine the curing of polyimide thin films spin coated onto metallized (Al, Cr, or Cu) silicon wafers (in situ, after cooling) and, to study bulk and interfacial reactions important

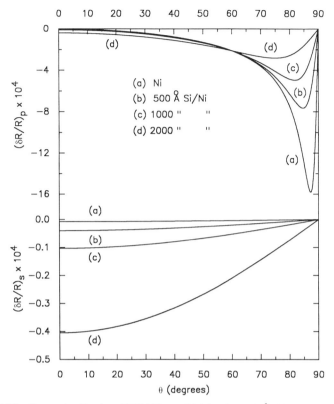

Figure 9.38 External-reflection (IRRAS) sensitivities for a 3 Å-thick dielectric (H$_2$O, $\tilde{n} = 1.311 + 0.132i$) on a dielectric layer atop a metal substrate (or thick buried layer) at 6.1 μm (from Bermudez, 1992).

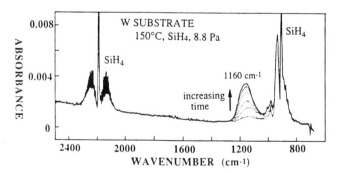

Figure 9.39 FTIR external reflectance absorption spectrum of a W substrate measured every 2 min after exposure to SiH$_4$ (from Kobayashi *et al.*, 1993).

9.7 Surface Infrared Reflectometry

to electroluminescent polymers, the photo- and electric field-induced degradation of polymers on Al.

9.7.2 Internal Reflection Mode: Attenuated Total Internal Reflection Spectroscopy

9.7.2.1 Theory

When a beam travels from a more optically dense medium (the substrate, with refractive index n_s) to one that is less dense (the ambient, $n_a < n_s$), the incident beam is totally reflected when $\theta > \theta_c = \arcsin(n_a/n_s)$, the critical angle. This is known as total internal reflection (TIR). Although 100% of the incident light intensity is reflected, there is still an evanescent, exponentially decaying electromagnetic wave that enters the less dense medium. The electric field and the intensity in the ambient decay as

$$E_i^a(z) = E_i^a(z=0)\exp(-\gamma_{TIR}z) \tag{9.47}$$

$$I^a(z) = I^a(z=0)\exp(-2\gamma_{TIR}z) \tag{9.48}$$

where

$$\gamma_{TIR} = \frac{2\pi(\sin^2\theta - n_{as}^2)^{1/2}}{\lambda_s} \tag{9.49}$$

z is the normal to the interface, $n_{as} = n_a/n_s$, and $\lambda_s = \lambda_{vacuum}/n_s$. E_i^a and I^a represent the field inside the medium, which are related to E_{inc} (for each polarization) and I_{inc}.

Using the conventions for TIR, with s and p polarizations as shown in Figure 9.36b, the field in the ambient (vacuum) is related to the incident field, with magnitude E_{inc}, as follows: For s polarization:

$$E_y^a(z=0) = \frac{2\cos\theta}{(1-\varepsilon_{as})^{1/2}} E_{inc} \tag{9.50}$$

and for p polarization:

$$E_x^a(z=0) = \frac{2(\sin^2\theta - \varepsilon_{as})^{1/2}}{(1-\varepsilon_{as})^{1/2}[(1+\varepsilon_{as})\sin^2\theta - \varepsilon_{as}]^{1/2}} \cos\theta\, E_{inc} \tag{9.51}$$

$$E_z^a(z=0) = \frac{2\cos\theta}{(1-\varepsilon_{as})^{1/2}[(1+\varepsilon_{as})\sin^2\theta - \varepsilon_{as}]^{1/2}} \sin\theta\, E_{inc} \tag{9.52}$$

where $\varepsilon_{as} = \varepsilon_a/\varepsilon_s = (n_a/n_s)^2$ is assumed to be real (transparent media) and <1.

Perfect TIR occurs only in the ideal two-medium limit. When there are absorbing adsorbates on the surface, the reflectance R is <100%, so this spectroscopy is instead called attenuated total internal reflection (ATR or ATIR) or internal reflection spectroscopy (IRS); IRS is the "formally" accepted nomenclature (Mirabella, 1993). Since FTIR spectroscopy (Sections 5.1.3.2 and 8.2.1.2; Back, 1991) is usually used to probe internal reflection, FTIR is sometimes appended to the technique acronym, as in ATR-FTIR.

For these very thin adsorbate layers, Equations 9.50–9.52 are still accurate for the field in the adlayer. With s and p polarizations, there are evanescent electric fields in the x, y, and z directions. Since dipoles interact with parallel electric fields ($-\mu \cdot \mathbf{E}$), each component of the surface dipoles can be probed. For $|\varepsilon_s| \gg \varepsilon_a$, each field component approaches $2 \cos \theta\, E_{inc}$, so each dipole component can be probed with equal sensitivity. ATR methods have been reviewed by Harrick (1967), Hoffmann (1983), Chabal (1988), and Mirabella (1993).

The changes in reflectance due to adsorbates are

$$\Delta R_s = \frac{2\pi}{\lambda n_s \cos \theta} |r_s^0|^2\, \mathcal{I}_y \varepsilon_y'' d_y \tag{9.53}$$

$$\Delta R_p = \frac{2\pi}{\lambda n_s \cos \theta} |r_p^0|^2 \left(\mathcal{I}_x \varepsilon_x'' d_x + \mathcal{I}_z \frac{\varepsilon_a^2}{\varepsilon_z'^2 + \varepsilon_z''^2} \varepsilon_z'' d_z \right) \tag{9.54}$$

where

$$\mathcal{I}_y = \frac{4\varepsilon_s \cos^2\theta}{\varepsilon_s - \varepsilon_a} \tag{9.55}$$

$$\mathcal{I}_z = \frac{4\varepsilon_s \cos^2\theta}{\varepsilon_s - \varepsilon_a} \frac{(\varepsilon_s/\varepsilon_a)\sin^2\theta}{[(\varepsilon_s + \varepsilon_a)/\varepsilon_a]\sin^2\theta - 1} \tag{9.56}$$

$$\mathcal{I}_x = \frac{4\varepsilon_s \cos^2\theta}{\varepsilon_s - \varepsilon_a} \frac{(\varepsilon_s/\varepsilon_a)\sin^2\theta - 1}{[(\varepsilon_s + \varepsilon_a)/\varepsilon_a]\sin^2\theta - 1} \tag{9.57}$$

and ΔR is defined as $R(0) - R(d)$ and r^0 is r with no adlayer. Here ε_i'' and d_i are the imaginary part of the dielectric function and the thickness of the adlayer associated with coordinate i (Chabal, 1988).

9.7.2.2 Instrumentation

Chabal (1988) has demonstrated that internal-reflection (ATR) signals for semiconductor surfaces (per reflection) are much larger than those for external reflection, as is seen in Figure 9.37. Because ΔR per internal reflection is very small for submonolayer and monolayer coverages, such coverages are difficult to monitor except when hydrogen is the adsorbate. Multimonolayer thin films are relatively easy to detect.

Multiple reflections are necessary to probe weak absorptions. Many internal reflections (~10–40) are possible in transparent media using the

9.7 Surface Infrared Reflectometry

configurations shown in Figures 9.40 and 9.41. In this case, ATR spectroscopy is also called multiple internal reflection (MIR), MIR infrared spectroscopy (MIRIRS), or ATR–MIR. Several examples of MIR element configurations are given in Figure 9.40. Often in processing, interesting events occur only on the "top" surface, and only the internal reflections from the top surface contribute to the signal (Figure 9.40a). In some cases, absorptions at the top and bottom of the element are probed (Figure 9.40b), as during the study of gas–surface interactions at varying temperatures or wet processing.

The MIR element must be highly transparent, which depends on the element and the temperature range to be studied. For semiconductor MIR elements, infrared absorption from free carriers (intrinsic and from dopants) and from vibrational transitions from impurities can limit the range of studies for which ATR is possible. For example, semi-insulating GaAs is a good MIR element in the near-infrared below 400°C, but at higher temperatures free-carrier absorption becomes strong and the element is opaque above 450°C (Annapragada and Jensen, 1991). Such high-temperature absorption can limit the use of ATR for real-time investigation of high-temperature processes, such as CVD. Furthermore, at higher temperatures blackbody radiation from the hot element is another problem; it must be shielded from the path of the IR beam. These elements can be resistively heated by passing current through them, as J. R. Creighton (private communication, 1994) has demonstrated with doped-GaAs ele-

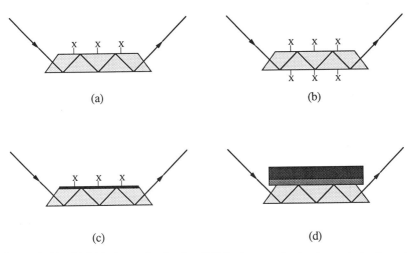

Figure 9.40 Multiple internal reflection (MIR) elements for *in situ* analysis of species adsorbed (a) on one surface of the element, (b) on both surfaces, and (c) on (an absorbing) film predeposited on the element; (d) *ex situ* analysis of a wafer placed in contact with the element.

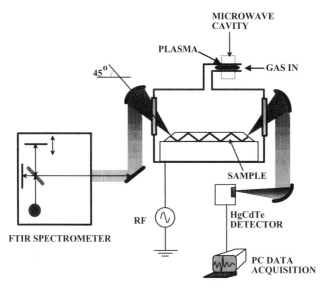

Figure 9.41 Schematic of apparatus used for hybrid microwave-RF plasma cleaning of Si with real-time monitoring by ATR (from Zhou *et al.*, 1993).

ments. However, this technique has a tradeoff: at low doping densities it is difficult to heat the sample, while at higher dopant densities free-carrier absorption becomes significant.

Usually the element is chosen of a given material and cut in a given manner to probe processes occurring on a specific surface of that material. Absorption can preclude this even at ambient temperature, as is the case in Si where multiphonon absorption makes it opaque below 1500 cm^{-1} (Higashi and Chabal, 1993). This precludes the study of lower-energy vibrational modes, such as those due to oxides and halides on the surface. To overcome this limitation, ATR studies on silicon have at times been conducted using a Ge element overlaid with a Si film (Kawamura *et al.*, 1991), as is illustrated in Figure 9.40c; Ge is transparent above 650 cm^{-1}.

Figure 9.40d shows yet another mode for ATR studies for opaque substrates, albeit one that is more suitable for *ex situ* analysis than for real-time *in situ* studies (Harrick, 1967; Olsen and Shimura, 1989; Sawara *et al.*, 1992). The top of the substrate to be examined is placed in mechanical contact with a transparent MIR element. Olsen and Shimura (1989), Ling *et al.* (1993), and Bjorkman (1994, 1995) studied the oxide formed on a silicon wafer and the results of other surface passivation and cleaning procedures on Si by using a Ge MIR element. The signal was enhanced by 1500× relative to external reflection, due to both the multiple reflections and the electric field enhancement in the oxide that is sandwiched between the two high-refractive-index materials (Si and Ge). This technique is not complicated by the interference effects that can affect external reflection probes. Yet another way to overcome semiconductor substrate absorption is to use the external reflection mode that is enhanced by a buried metal layer, as described above (Bermudez, 1992). Also, Richardson *et al.* (1989) examined IR absorption at the surface by transmission at Brewster's angle; this has higher sensitivity than external reflection, but is less sensitive than ATR-MIR.

ATR studies are usually conducted with a FTIR spectrometer (Johnston, 1993; Section 5.1.3.2) equipped with mirrors for external analysis (ATR-FTIR). When emission from the MIR element is a concern, the element should be placed between the interferometer stage and the detector, as illustrated in Figure 9.41.

9.7.2.3 Applications

ATR has been used to examine a variety of processes on surfaces, including passivation (such as oxidation), cleaning (including the removal of oxides), etching, and deposition. Probing hydrogen on surfaces is important for understanding cleaning procedures and deposition processes. ATR is well suited to the study of hydrogen on surfaces, because the modes are strong and the mode frequencies are transmitted through most MIR elements of interest.

Chabal, Higashi, and co-workers have studied the appearance of monohydrides and dihydrides on Si(100) and (111) for a variety of passivation and cleaning procedures (Chabal, 1988, 1993, 1994; Chabal *et al.*, 1989; Higashi *et al.*, 1990; Jakob and Chabal, 1991; Jakob *et al.*, 1991). Dihydrides dominate the Si(100) surface after dilute aqueous HF etching (Chabal *et al.*, 1989). Ideal hydrogen termination of the Si(111) surface is observed after buffered aqueous HF etching in basic solutions (Chabal *et al.*, 1989). This treatment removes the surface oxide and terminates (passivates) the surface with hydrogen, as seen in the FTIR spectrum in Figure 9.42 (top). The sharp, highly polarized (p) peak at 2083.7 cm^{-1} indicates ideal termination with silicon monohydride ($\equiv$Si$-$H) oriented normal to the surface (Higashi *et al.*, 1990). With different surface treatments, more complex spectra are obtained; these can be resolved by using different incident polarizations, protonated/deuterated solutions, and mode calculations. Figure 9.42 (bottom) shows the result of treatments that leave a trihydride (T) and uncoupled monohydride (M') normal to the surface, and a dihydride (D') and coupled monohydride (M) tilted relative to the surface normal. Similar work has been performed by Bjorkman *et al.* (1995). Higashi and Chabal (1993) have reviewed the use of ATR and other surface-sensitive methods in analyzing the chemical composition and morphology of silicon surfaces after cleaning and passivation.

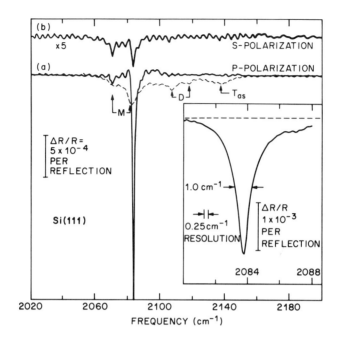

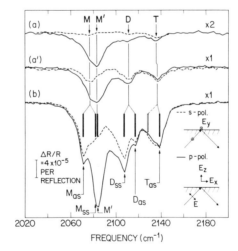

Figure 9.42 (Top) Internal reflection spectra of HF-treated Si(111) surfaces [solid curves: pH-modified, buffered HF (pH = 9–10); dashed curves: dilute HF] (from Higashi *et al.,* 1990). (Bottom) Polarized IR internal reflection spectra of silicon–hydrogen vibrations on Si(111) for (a) 10% H:90% D, (a′), 25% H:75% D, and (b) 100% H hydrogen fluoride treatment, with the monohydride (M, ~2077 cm^{-1}), dihydride peaks (D, ~2111 cm^{-1}), and trihydride (T, ~2137 cm^{-1}) -related peaks (from Chabal *et al.,* 1989).

9.7 Surface Infrared Reflectometry

Gottscho, Chabal, and co-workers probed the plasma passivation of GaAs and Si surfaces in real time using *in situ* ATR. These studies were conducted at room temperature in an environment suitable for integration into cluster tools. In one study, the surface of a GaAs wafer was monitored by ATR and photoluminescence simultaneously during passivation by a remote NH_3 (or H_2) discharge (Aydil *et al.*, 1993; 1995). The ATR spectrum was taken with a FTIR spectrometer purged with N_2 (Figure 9.41). The infrared beam from the FTIR spectrometer internally reflected from the exposed surface 38 times, and was then detected with a liquid-nitrogen-cooled HgCdTe detector. Figure 9.43 displays the ratio of the ATR after passivation to that before. The transmission at a frequency increases when the surface concentration of a resonant adsorbate decreases. This is seen for the As—O stretch (~850 cm^{-1}) and the CH_x (x = 2,3) stretches (2850, 2912 cm^{-1}), so the concentrations of As—O and CH_x both decrease when GaAs is exposed to this hydrogen-containing plasma effluent. Conversely, transmission decreases when the density of a surface species increases. In Figure 9.43, this is observed for the H_2O scissor (~1600 cm^{-1}), the As—H stretch (~2100 cm^{-1}), and the H—OH stretch (~3200 cm^{-1}). Therefore, the concentrations of both AsH and H_2O increase during the passivation. The H_2O-related peaks are also seen with a H_2 discharge and shift to lower frequencies, as expected, when ND_3 or D_2 is used.

These absorptions were also monitored in real time by Aydil *et al.* and compared to the efficiency of photoluminescence (PL). Figure 14.7 shows that As—O is removed from the surface, as monitored by the decrease in absorption near 850 cm^{-1}, well before PL increases and that As—H

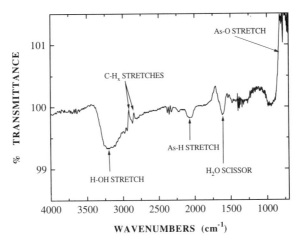

Figure 9.43 Ratio of ATR transmission after GaAs plasma passivation to that before passivation (from Aydil *et al.*, 1993).

is formed on the surface, as seen by the increase in absorption near 2100 cm^{-1}, simultaneously with the increase in PL. Arsenic oxide is removed as fast as H atoms from the discharge can be transported to the surface, while the As–H passivation layer is formed later, perhaps after diffusion of H through the native oxide layer.

This same group studied silicon surface cleaning by remote H_2 and NH_3 plasmas using real-time ATR spectroscopy (Zhou et al., 1993). As above, the FTIR spectra were referenced to those before processing. Figure 9.44 shows that the H_2 plasma removes the oxide from Si(100) (increase of the SiO–H band) and produces Si–H (decrease of the Si–H band), as does the comparison spectrum taken after treating the wafer with aqueous HF. However, the plasma treatment is more effective in reducing hydrocarbon contamination, as seen by the increase in the C–H bands near 2900 cm^{-1}.

ATR has been used to examine several other types of surface processing. For example, Niwano et al. (1994a) studied the initial stages of ultraviolet/ozone oxidation of Si(100) and (111) surfaces using ATR. Figure 9.45 shows the decrease of SiH and SiH$_2$ features during such treatment of HF-passivated Si(100) surfaces, with the subsequent growth and decrease of SiH$_x$O$_y$ absorption peaks. In related work, Niwano et al. (1994b) investigated the initial stages of the oxidation of hydrogen-terminated Si(100) and (111) stored in air.

Qi et al. (1994) and Gee and Hicks (1992) studied the orientation of hydrogen adsorbed on (2×4)–$c(2 \times 8)$ and (2×6) surface reconstructions on GaAs(100) by using polarized ATR. Absorption by s- and p-polarized light

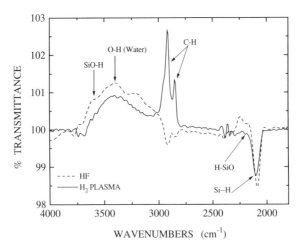

Figure 9.44 ATR spectra of a Si surface treated either by a H_2 plasma or a HF dip, each ratioed to the spectrum before treatment (from Zhou et al., 1993).

9.7 Surface Infrared Reflectometry

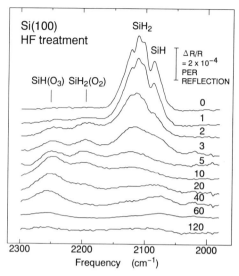

Figure 9.45 Si—H stretching vibration spectra of HF-treated Si(100) surfaces for different exposure times of UV ozone (listed in seconds) obtained by ATR (from Niwano *et al.*, 1994a).

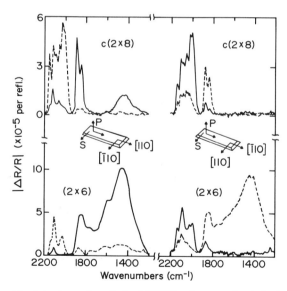

Figure 9.46 Absolute change in polarized (ATR) reflectance due to hydrogen adsorption on GaAs(100) with the noted surface reconstructions and polarizations (*s:* dashed lines; *p:* solid lines) (from Qi *et al.*, 1994).

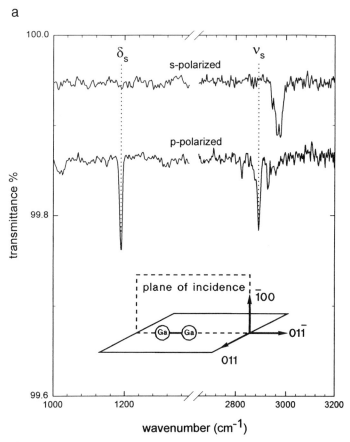

Figure 9.47 Polarized ATR spectra of Ga-rich GaAs(100)–(1 × 2)–CH$_3$ with the plane of incidence either (a) parallel or (b) perpendicular to the Ga—Ga dimers. The features near 1191 and 2889 cm^{-1} correspond to the symmetric CH$_3$ bending (δ_s) and stretching (ν_s) modes, respectively (from Creighton, 1994).

was examined for two GaAs(110) MIR elements, one with the long (propagating) axis along [110] and the other with the long axis along [1̄10]; the hydrogen dosed only one of the reflecting surfaces in both cases. As seen in Figure 9.46 for both reconstructions, the As—H vibrations from 1950 to 2190 cm^{-1} are primarily s-polarized for the crystal oriented along [110] and p-polarized for the element aligned along [1̄10] (Qi et al., 1994). These two measurements are consistent with each other, and with the interpretation that the As—H bonds orient along [1̄10], parallel to the As-dimer bonds. Similarly, the Ga—H vibrations from 1200 to 1900 cm^{-1}, are mostly p-polarized for the [110]-aligned crystal and s-polarized for the [1̄10] crystal. The Ga—H bonds orient along [110], which is perpendicular to the As dimers (and parallel to

9.7 Surface Infrared Reflectometry

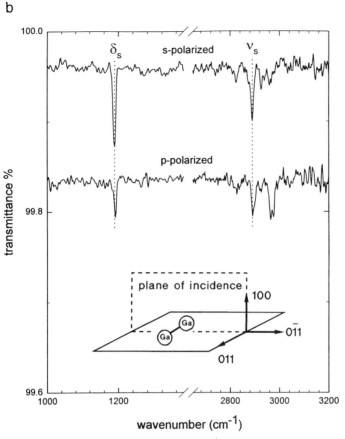

Figure 9.47 (*Continued*)

the Ga dimers). These As—H and Ga—H stretching vibrations would not be polarized along the crystal axes, as observed here, if there were substantial intermixing of the As and Ga atoms at the surface.

Similar information about Ga-dimer orientation was obtained by Creighton (1994), who studied the orientation of Ga—CH$_3$ bonds on Ga-rich GaAs(100)–(1 × 2)-CH$_3$ surfaces. The GaAs MIR element was dosed by trimethylgallium on both reflecting surfaces. Figure 9.47 shows the different orientations of the Ga dimers on the top and bottom surfaces of the element, as determined by alternately using *s*- and *p*-polarized light to probe the absorption by CH$_3$.

Crowell *et al.* (1991) studied the dissociative adsorption of disilane and digermane on Ge(111) by ATR using a Ge MIR element. Figure 9.48 shows the Si—H (~2100 cm^{-1}) and Ge—H stretches (~1960 cm^{-1}) of adsorbed

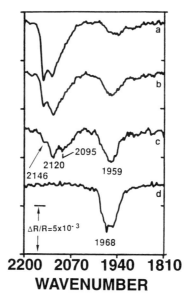

Figure 9.48 Multiple internal reflection infrared spectroscopy (ATR) spectrum in the Si−H and Ge−H stretching region obtained at 150 K after Ge(111) was exposed to Si_2H_6 at 150 K and heated to (a) 200 K, (b) 300 K, and (c) 400 K; or (d) to H atoms at 400 K (from Crowell et al., 1991).

species after disilane dosing at 150 K, heating to (a) 200 K, (b) 300 K, and (c) 400 K and cooling to 150 K. As temperature is increased, SiH_3 is seen to successively convert into SiH_2, which converts into SiH, and GeH is formed at the same time. Spectrum (d) is a comparison with H-atom-dosed Ge. In another deposition study, Gladfelter et al. (1994) studied the deposition of aluminum from dimethylethylamine alane on a native-oxide-covered Si MIR element.

Real-time ATR studies of MOCVD are hindered by substrate absorption at the high temperatures needed for deposition. When necessary, these studies are performed after processing, but still in situ. Annapragada and Jensen (1991) studied the fundamental steps in GaAs MOCVD using a MIR GaAs element that allowed 12 internal reflections. Because of strong element absorption at the temperatures needed for reaction (∼500–600°C), the samples were dosed at higher temperatures and analyzed <400°C. They examined Ga-rich, H- and As-treated GaAs surfaces at 380°C, after dosing with tert-butylarsine at 600°C. Only the dosed Ga-rich surface showed strong alkyl stretching bands, near 2950 cm^{-1}. Tripathi et al. (1989) used in situ ATR to probe the thermal reactions of trimethylgallium and ammonia on a GaAs element, which is a reaction that is important in the MOCVD of GaN.

9.7 Surface Infrared Reflectometry

Xi *et al.* (1996) used ATR to study the chemisorption and decomposition of the MOCVD and MOMBE precursor *tris*(dimethylamino)phosphine (DMAP) on GaAs (100)-c(8 × 2). Figure 9.49 shows the IR spectrum with DMAP continuously impinging onto the surface at 375 K, which looks like the spectrum observed when several monolayers of DMAP are first deposited at 140 K and are then annealed (and decompose) at higher temperatures. This figure also shows the spectrum of dimethylamine (DMA) after deposition at 140 K and subsequent heating to 310 K to remove the weakly bound physisorbed layers. The lack of similarity between these two spectra suggests that DMA is not formed during the decomposition of DMAP on GaAs surfaces.

Miyazaki (1995) have used ATR to probe surface reactions during the plasma-enhanced CVD of silicon carbide (CH_4 + SiH_4 plasma) and silicon (a-Si or polysilane, SiH_4 plasma). For the early stages of silicon growth, they found that SiH_2 is the major species on the surface when the substrate is at 200°C, but SiH_3 and $(SiH_2)_n$ chains are dominant when the substrate is at room temperature.

ATR of metal alkyls can be useful in understanding photon-assisted growth of metals, as well as in semiconductor CVD. Sanchez *et al.* (1988) and O'Neill *et al.* (1989) studied the surface photochemistry of dimethylcadmium (DMCd) on silicon, using a Si MIR element. Figure 9.50 shows the

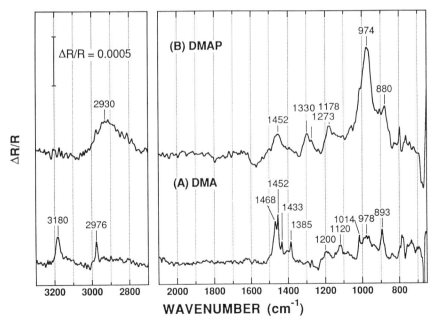

Figure 9.49 Comparison of the ATR spectra of (A) dimethylamine (DMA) adsorbed on GaAs(100) at 140 K and annealled to 310 K and (B) *tris*(dimethylamino)phosphine (DMAP) impinging on GaAs(100) at 375 K (from Xi *et al.*, 1996).

differences in the profiles of the FTIR spectra of gas-phase DMCd and physisorbed and chemisorbed DMCd on passivated Si in the region of CH stretches. Adsorption on oxide-passivated Si surfaces (as shown in this figure) was seen to be stronger than that on hydrogen-passivated surfaces. Excimer-laser decomposition of chemisorbed DMCd was observed by monitoring these bands when using 193 nm exposure, but not with 248 nm. Of course, absorption by the deposited metal limits this application of ATR to the study of monolayers, as here, or other very thin films.

ATR spectroscopy can be used to monitor liquid–solid processes, as well as the dry processes at gas–solid interfaces detailed above. For example, Yota and Burrows (1991) used this method in real-time to study the passivation of silicon and germanium surfaces by hydrofluoric acid, which was buffered with ammonium fluoride. This treatment leaves a residue that slowly corrodes the surface unless the surface is rinsed with water after treatment. Niwano *et al.* (1994c) have also examined the surface of Si(100)

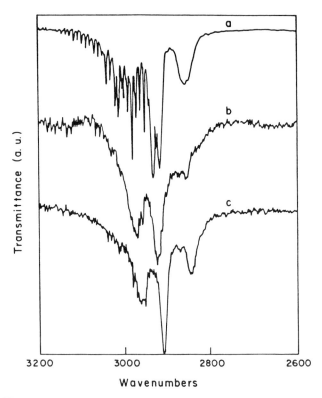

Figure 9.50 ATR absorption spectra of dimethylcadmium (a) in the gas phase, (b) physisorbed on silicon, and (c) chemisorbed on silicon, with arbitrary transmittance scales for each (from Sanchez *et al.*, 1988).

during immersion in HF solutions. Rappich *et al.* (1993) used ATR spectroscopy to monitor wet etching of Si(111). They showed that the Si surface remains covered with Si—H bonds during etching by NaOH solutions even though the etch rate is very fast. Yota and Burrows (1993) examined chemical and electrochemical treatments of GaAs(100) by Na_2S and $(NH_4)_2S$ solutions with ATR.

In a quite different application of ATR, Schnitzer *et al.* (1990) discussed monitoring the processing of photoresist thin films by the absorption of tunable diode lasers transmitted through photoresist-coated silver halide optical fibers. The fiber itself serves as the MIR element and absorption occurs by ATR at the photoresist infrared absorption resonances. (See Section 8.3 for related diagnostics.)

Section 11.2.1.4 describes another application of the TIR optical configuration, the use of total internal reflection microscopy for the *in situ* determination of defects during deposition.

9.8 Differential Reflectometry

Differential reflectometry (DR) compares the surfaces, interfaces, or thicknesses of two samples by measuring the difference in their reflectances. In a broader definition, DR also includes measuring the small changes in the reflectance of a single surface during processing as a function of time. Since DR is particularly sensitive to changes on the surface, it has also been called differential surface reflectance (DSR) or surface differential reflectivity (SDR); DSR has been reviewed by Selci *et al.* (1987). The electronic structure of the surface can be studied, such as that of the adsorption layer during thin film growth, when the probe wavelength is scanned (200–800 nm), which is spectral differential reflectometry (also called SDR). The term DR will be used to refer to differential reflectometry, which will be assumed to monitor surface changes, and SDR will be used to denote spectroscopic DR. DR is a form of reflectance modulation spectroscopy.

The change in reflectance due to an isotropic surface layer ($d \ll \lambda$) can be obtained from Equations 9.16 and 9.18 for *s*- and *p*-polarized light, respectively (McIntyre and Aspnes, 1971). For normal incidence, they reduce to Equation 9.19 to give the fractional change in reflectance:

$$\frac{\Delta R}{R(0)} = -\frac{8\pi d n_a}{\lambda} \operatorname{Im} \frac{\tilde{\varepsilon}_s - \tilde{\varepsilon}}{\tilde{\varepsilon}_s - \varepsilon_a} \qquad (9.19')$$

where ε_a, $\tilde{\varepsilon}$, and $\tilde{\varepsilon}_s$ are the respective complex dielectric constants of the ambient, surface layer, and substrate; n_a is the ambient index of refraction; and λ is the probe wavelength in vacuum. $\Delta R = R(d) - R(0)$, where $R(0)$ is the reflectance of the bare substrate.

For visible light, $\Delta R/R(0) \approx -1\%$ $\text{Im}[(\tilde{\varepsilon}_s - \tilde{\varepsilon})/(\tilde{\varepsilon}_s - \varepsilon_a)]$ per monolayer. If ε_a and $\tilde{\varepsilon}$ are real and $\text{Re}(\varepsilon_s) \gg \varepsilon_a$, this imaginary term is $\sim \text{Im}(\varepsilon_s)[(\varepsilon - \varepsilon_a)/|\varepsilon_s|^2]$. Therefore, this change in reflectance can be large when the substrate is strongly absorbing, and so $\text{Im}(\varepsilon_s)$ is large, as it is for metals and semiconductors. Kleint and Merkel (1989) have discussed changes in reflectance spectra due to surface adsorbates in terms of surface band structure features.

Two other types of differential reflection diagnostics are presented in the next two sections: surface photoabsorption, which follows the reflectance of p-polarized light near Brewster's angle, and reflectance difference (anisotropy) spectroscopy, which compares the reflection coefficient at normal incidence for two different polarizations on one surface.

Differential reflectance can be used in several modes. (1) Spectra taken during processing of the surface can be compared with reference spectra taken before processing. (2) The processed sample can be compared with a controlled *ex situ* reference. (3) DR can be performed on the sample and an adjacent *in situ* reference. (4) Process conditions can be modulated periodically, and the synchronous changes in reflectance are then detected.

The experimental arrangement for SDR in this third mode is shown in Figure 9.51 (Pahk *et al.,* 1991; also see Holbrook and Hummel, 1973; Hummel, 1983). Radiation from a lamp, such as a 600-W Xe high-pressure arc lamp, is dispersed by a monochromator and focused onto the sample. The incident and reflected beams are nearly normal to the sample surface. The incident beam is directed to the surface by an oscillating mirror which rapidly rasters the probing beam back and forth between the two adjacent samples at a frequency ω. The reflected light is collected and detected by a photomultiplier (or other light detector), which is preceded by a diffuser to average out spatial inhomogeneities in the detector response. This signal is sent to a lock-in amplifier synchronized at ω to give the difference in reflected beam intensities $S_{\text{lock-in}}$, which is proportional to the difference of the reflectances, $\Delta R = R_1 - R_2$, and to a low-pass filter, which gives a signal $S_{\text{low pass}}$ that is proportional to the average reflectance $\langle R \rangle = (R_1 + R_2)/2$. The differential reflectance is $\Delta R/\langle R \rangle = S_{\text{lock-in}}/S_{\text{low pass}}$. This normalization is needed because $S_{\text{lock-in}}$ and $S_{\text{low pass}}$ both depend linearly on the spectral intensity of the source, which can vary greatly with λ.

This mode of spectral differential reflectance must be performed on surfaces that do not change appreciably during the measurement. Time-resolved studies are typically performed with a fixed-wavelength laser (as in Sections 9.9 and 9.10), chosen to maximize the reflectance changes for the electronic structure of the given surface. Although spectral, time-resolved DR can be performed with parallel detection of the dispersed white light, little work has been done in this area.

Even though DR does not provide as much information as ellipsometry (Section 9.11), it is still very sensitive to near-surface optical properties and

9.8 Differential Reflectometry

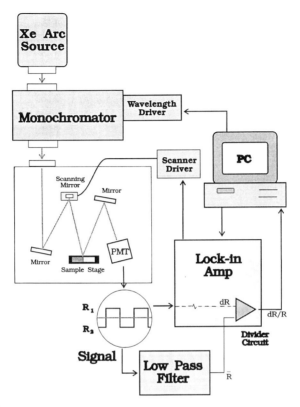

Figure 9.51 Experimental apparatus for differential reflectance spectroscopy (from Burns et al., 1993).

has some practical advantages *vis-à-vis* ellipsometry for *in situ* studies. The DR optical system is simpler, and optical alignment is less critical. While ellipsometry provides absolute measurements, DR is sensitive to the small differential changes between the sample and reference that may occur, for example, due to the changing composition of an alloy in CVD, surface modifications from ion bombardment, or the etching of thin oxide layers on a film. Therefore, a controllable reference sample is required in DR. The differential nature of DR can be used to cancel out uncontrolled changes in the experimental environment. As such, it is well suited for *in situ* process studies, as well as *ex situ* characterization. Under certain conditions, it can be useful for real-time control.

Pahk *et al.* (1991) probed the etching of SiO_2 films on Si substrates by liquid HF/H_2O both *ex situ* and *in situ* using the apparatus in Figure 9.51. In the *ex situ* experiment, two samples with the same initial oxide thickness were compared, one in which the oxide had been etched by the solution and the other being the unetched reference sample. In the *in situ* experiment, DR was measured as two samples, with different initial oxide thicknesses, were etched by the solution. The measured $\Delta R/\langle R \rangle$ was compared with that calculated using Equation 9.10 for a given oxide thickness on the substrate.

The high sensitivity of DR to small changes on the surface is demonstrated by the use of spectral differential reflectometry by Chongsawangvirod and Irene (1991) to measure the very small differences (<1%) in $\Delta R/R$ for (100), (110), (111), (311), and (511) Si surfaces, as seen in Figure 9.52—which is corrected for differences in the oxide thickness. The bulk contributions are the same in each case. Burns *et al.* (1993) used DR *ex situ* to examine the level of surface damage after ion implantation in Si wafers, and after rapid thermal annealing. Hummel *et al.* (1988) conducted similar studies of damage due to ion implantation using DR.

Brewer *et al.* (1990) used reflectance to monitor changes in the surface composition of CdTe(100) due to KrF laser irradiation. The reflectance was monitored at 632.8 nm continuously as a train of pulses impinged on the surface. When a train of pulses with fluence $\gg 40$ mJ/cm^2 irradiated the surface, the reflectance slowly increased, eventually by ~1%. This indicated an increasingly Te-rich surface layer. If a series of pulses with fluence $\ll 40$ mJ/cm^2 then hit the surface, the reflectance slowly decreased to its initial value, which indicated that the surface slowly returned to its initial stoichiometric composition.

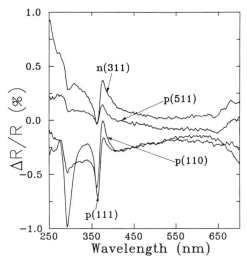

Figure 9.52 Differential reflectance spectra of several Si surface orientations (and doping types), compared to the (100) orientation. These spectra have been corrected for differences in the thickness of the surface oxides (from Chongsawangvirod and Irene, 1991).

9.9 Surface Photoabsorption and p-Polarized Reflectance Spectroscopies

Reflection experiments are particularly sensitive near the pseudo-Brewster angle for p-polarized light because this configuration most effectively probes the surface and the bulk. Time-resolved reflectometry (or reflectivity; TRR) experiments using p-polarized light to probe the surface has been called by different names. Surface photoabsorption (SPA) or photoabsorption (PA) is surface differential reflectometry performed with p-polarized light, well above the absorption edge and incident near the pseudo-Brewster angle, to probe strongly absorbing surface layers. When light that is below the absorption edge is used, this method has been called p-polarized reflectance spectroscopy (PRS) or Brewster angle reflectance spectroscopy (BARS). This section describes the use of all such reflectometries to probe absorbing or transparent surface layers as a function of time by using p-polarized light (for $\theta \gg 0°$). Such methods are very useful diagnostics of the surface for *in situ* fundamental studies and process development. The simplicity and low cost of these diagnostics make them appropriate for use as a real-time diagnostic during manufacturing.

This sensitivity of SPA can be understood from Equations 9.17 and 9.18. The change in reflectance due to the adsorbate can be obtained by replacing d by $d_0\Theta$, where d_0 is the thickness of a monolayer of the adsorbate and Θ is the fractional surface coverage of the adsorbate. This assumes that the nonlinear coverage dependence of the optical properties of the adsorption layer and the substrate can be neglected. As illustrated in Figure 9.4b and seen from Equation 9.58 (below), $|\Delta R|/R$ becomes particularly large for p-polarization and reaches a maximum between 45° and the pseudo-Brewster angle (McIntyre and Aspnes, 1971). SPA is particularly sensitive to small changes in the imaginary part of the index of refraction of the surface k, which accounts for its name since k is responsible for absorption. This SPA signal can be quite large, as much as several percent during the growth of a monolayer, and can be about an order of magnitude larger than reflectance-difference spectroscopy signals (Section 9.10). Alternatively, sometimes it may be better to monitor ΔR, whose magnitude generally peaks near grazing incidence.

Minimizing noise is very important in these TRR experiments since $\Delta R/R(0)$ is at most ~1% in such experiments. Therefore, it is necessary to choose light sources with low amplitude noise and to signal average. Light-source noise is of less concern in DR where the light source alternates between two samples (Figure 9.51). Figure 9.53 shows the experimental implementation of SPA.

From Equation 9.18, the fractional change in reflectance of p-polarized light for an isotropic surface layer is

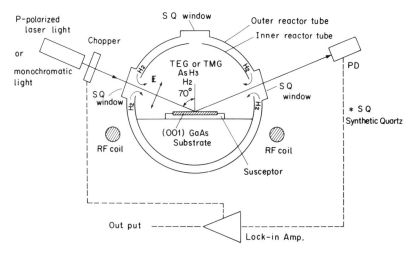

Figure 9.53 Horizontal MOVPE reactor with real-time SPA monitoring (from Kobayashi et al., 1991a).

$$\left.\frac{\Delta R_p}{R_p}\right|_{\text{SPA}} = -\frac{8\pi d n_{a\perp}}{\lambda} \text{Im} \left(\frac{\tilde{\varepsilon}_s - \tilde{\varepsilon} - (\tilde{\varepsilon}_s^2 - \tilde{\varepsilon}^2) \frac{\varepsilon_a}{\tilde{\varepsilon}\tilde{\varepsilon}_s} \sin^2\theta}{(\tilde{\varepsilon}_s - \varepsilon_a)(\cos^2\theta - \frac{\varepsilon_a}{\tilde{\varepsilon}_s} \sin^2\theta)} \right) \quad (9.18')$$

When probing the growth of semiconductors $|\tilde{\varepsilon}_s|, |\tilde{\varepsilon}| \gg \varepsilon_a$, and this equation can usually be approximated by

$$\left.\frac{\Delta R_p}{R_p}\right|_{\text{SPA}} = -\frac{8\pi d n_{a\perp}}{\lambda} \text{Im} \left(\frac{\tilde{\varepsilon}_s - \tilde{\varepsilon}}{(\tilde{\varepsilon}_s - \varepsilon_a)(\cos^2\theta - \frac{\varepsilon_a}{\tilde{\varepsilon}_s} \sin^2\theta)} \right) \quad (9.58)$$

Since $\tan^2\theta_B = \tilde{\varepsilon}_s/\varepsilon_a$, the denominator on the right-hand side of this approximate expression is small and the fractional change is large. So p-polarized light couples well to the bulk and is a good probe of the bulk and the surface.

The dielectric function components important in SPA can be seen by examining the fractional change in the reflection coefficient $[r_{pp}(\theta,\alpha'',d) - r_p^0(d=0)]/r_p^0(d=0)$ from Equation 9.23, which is the general form for anisotropic layers. When probing the growth of semiconductors $|\varepsilon_s| \sim |\varepsilon_{ii}| \gg \varepsilon_a$, so the terms involving ε_{zz} are very small. Consequently, the main contribution of SPA comes from the component of the dielectric tensor that lies in the surface and the plane of incidence (Hingerl et al., 1993).

Hingerl et al. (1993) have shown that surface photoabsorption and reflectance difference spectroscopies (RDS, Section 9.10) give similar spectral information, as is formally discussed in Section 9.10.2. They are both differ-

9.9 Surface Photoabsorption and p-Polarized Reflectance Spectroscopies

ential spectroscopies that can give real-time spectroscopic information. The SPA signals are typically larger, and SPA experiments are less sensitive to polarization perturbations caused by the optics. RDS is more versatile when experimental conditions are changed because it compares two optical signals on a sample at the same time, while SPA compares signals before and after a process-induced change. For example, consider the analysis of a surface at different substrate temperatures. The dielectric function of the bulk substrate varies with temperature, so reflection contributions due to the bulk change with T. Variations in temperature always cancel out in RDS, but can possibly change the baseline reflectance in SPA.

Kobayashi and Horikoshi (1989), Makimoto et al. (1990) and Kobayashi et al. (1991a) first developed SPA for the *in situ* study of GaAs growth with alternating deposition of Ga and As on GaAs. In this early work the 325-nm line from a He–Cd laser was used, along with several argon-ion and He–Ne laser lines, and the angle of incidence was 20°; in later work by this group an angle of incidence of ~70° was used, near the pseudo-Brewster angle. Figure 9.54a shows that the *p*-polarized signal strongly tracks the sequence of AsH_3 and triethylgallium (TEGa). This is seen in more detail in Figure 9.54b. Kobayashi and Horikoshi (1989) showed that the SPA signal tracks with RHEED (reflection high-energy electron diffraction) measurements. Kobayashi and Horikoshi (1990) examined the spectral dependence (2.0–3.7 eV, dispersed white light) of the SPA signal ($\Delta R/R$) for Ga- and As-terminated surfaces. They examined this for *p*- and *s*-polarized light and incidence azimuths along [100], [110], and [$\bar{1}$10]. As seen in Figure 9.55, the signal was largest for *p*-polarized light near 2.6 eV. For light along [$\bar{1}$10], this peak value was ~2%, while for [100] and [110] it was ~1%.

Kobayashi et al. (1991b) used SPA to probe the steps in atomic layer epitaxy (ALE). Figure 9.56 shows the $\Delta R/R_{As}$ of an As-stabilized GaAs(001) surface (470°C) monitored with 470-nm light during a 1-sec pulsed supply of Ga organometallics and afterwards during a H_2 purge ($\theta = 70°$, [$\bar{1}$10] azimuth). R_{As} is the reflectance of the arsenic-stabilized surface. $\Delta R/R_{As}$ increased immediately by 3×10^{-2} when the triethylgallium (TEGa) was incident (left side of Figure 9.56), and did not change during the H_2 purge. This indicates that a Ga layer was deposited immediately. In contrast, it increased by 7×10^{-3} when a pulse of trimethylgallium (TMGa) was incident and then slowly increased to 3×10^{-2} during the H_2 purge (right side). This demonstrates that the surface layer formed by the TMGa exposure is modified during the H_2 purge to form the same Ga layer seen with the TEGa pulse. Figure 9.57 plots the initial reflectance change during the TMGa pulse and shows that SPA is a good sensor of growth that terminates after one monolayer. Stable growth of a single monolayer occurs for a wide range of TMGa doses up to 485°C, but atomic-layer limited growth does not occur at 520°C.

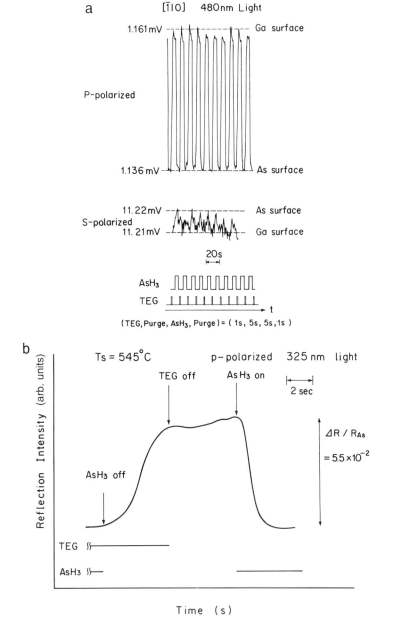

Figure 9.54 (a) Typical surface photoabsorption (SPA) signal traces during atomic layer epitaxy of GaAs at 560°C (from Kobayashi and Horikoshi, 1990). (b) Real-time SPA monitoring during a deposition sequence (from Kobayashi et al., 1991a).

9.9 Surface Photoabsorption and p-Polarized Reflectance Spectroscopies

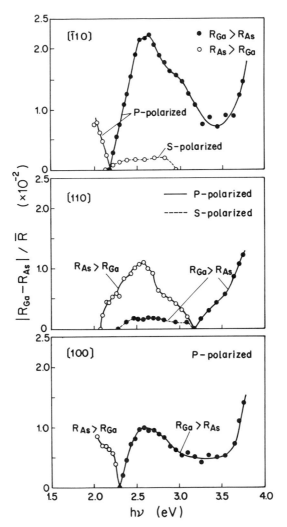

Figure 9.55 Spectral dependence of SPA for different incidence azimuths during GaAs atomic layer epitaxy at 560°C (from Kobayashi and Horikoshi, 1990).

The alternating deposition of monolayers of group III and V elements in the growth of other III–V semiconductors has also produced large oscillations in the SPA reflectance (Kobayashi and Kobayashi, 1992). Figure 9.58 shows the ALE growth of five monolayers of InAs on an InP barrier by using trimethylindium (TMIn) and AsH_3. Then another InP barrier was grown by flowrate-modulation epitaxy (FME) by alternating triethylindium (TEIn), which decomposes to form an In surface, and PH_3.

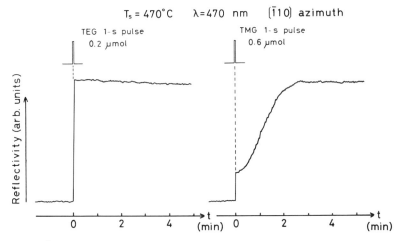

Figure 9.56 SPA reflectance changes during H_2 purge after TEGa and TMGa pulses (from Kobayashi et al., 1991b).

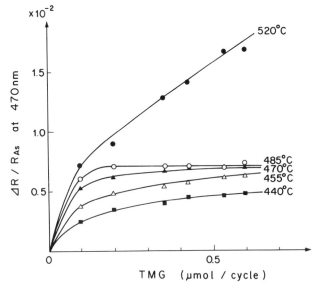

Figure 9.57 Initial SPA reflectance change as a function of the amount of TMGa in the pulse, at several temperatures (from Kobayashi et al., 1991b).

9.9 Surface Photoabsorption and *p*-Polarized Reflectance Spectroscopies

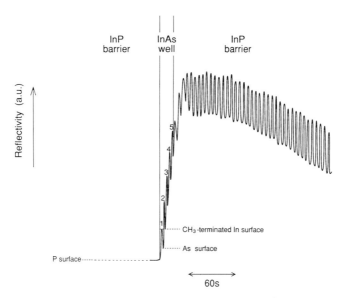

Figure 9.58 SPA reflectance change for atomic layer epitaxial growth of five InAs monolayers followed by the growth of an InP barrier (from Kobayashi and Kobayashi, 1992).

Nishi *et al.* (1992, 1993) monitored the surface by SPA during GaAs ALE, using alternating GaCl and AsH_3 supplies, and InP ALE, with alternating InCl and *tert*-butyl phosphine (TBP). (The GaCl comes from thermally decomposed diethylgallium chloride and the InCl from the reaction between trimethylindium and HCl.) They demonstrated that for GaAs ALE, the difference spectrum during the GaCl and AsH_3 injection cycles depends on the plane of the incident *p*-polarized light source.

Ohshita and Hosoi (1993) monitored the Si(100) surface by SPA during Si vapor-phase epitaxy (VPE) from SiH_2Cl_2/H_2, by using a 488-nm argon-ion laser incident at 80°. When the reactant delivery to the substrate was turned on, the SPA signal increased due to the $SiCl_2$ adsorbate on the Si(100) surface; and when it was turned off, it decreased due to $SiCl_2$ desorption and reaction. The rates of these processes were monitored by SPA.

J. Eng, H. Fang, S. Vemuri, I. P. Herman, and B. E. Bent (unpublished, 1995) used SPA to monitor the composition of the surface during the etching of GaAs(100) by pulses of HCl gas. Figure 9.59 shows that at 840 K the reflectance at 488 nm decreases when the HCl flow is turned on and it increases when the HCl is turned off. With HCl incident, GaCl and As_2 both desorb from the surface during steady-state etching. When the flow is off, As_2 continues to desorb at this temperature, making the surface more Ga-rich and increasing the reflectance (as in Figure 9.54).

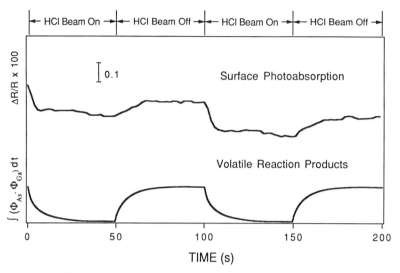

Figure 9.59 SPA traces using 488 nm during the etching of GaAs(100) [840 K] by pulses of HCl, compared to the flux of desorbing GaCl and As_2 (from J. Eng, H. Fang, S. Vemuri, I. P. Herman, and B. E. Bent, unpublished, 1995).

Weegels et al. (1994) monitored reflectance changes during electron cyclotron resonance (ECR) plasma processing of GaAs. Surface damage during plasma etching with Ar and plasma cleaning with H_2 was spectroscopically monitored using p-polarized light incident at 61°.

Eres and Sharp (1992, 1993) and Sharp and Eres (1992) employed a slightly different differential reflectometry scheme to investigate the kinetics of surface-limited CVD. Specifically, they used a pulsed molecular beam to modulate digermane delivery to a heated substrate and studied the deposition of germanium on Si and Ge substrates. This probed the surface during and after impingement with a p-polarized 632.8-nm He–Ne laser with 0.02% root-mean-square (RMS) noise directed at a 37.5° angle of incidence. The reflected light was detected with a silicon photodiode, amplified, and then digitized by a fast analog-to-digital converter for signal averaging. This TRR experiment is very similar to SPA, but the angle of incidence differs from the near pseudo-Brewster angle often used in SPA.

Figure 9.60 displays the differential reflectance signal during the digermane pulse, which shows the chemisorption of the digermane layer (~1 msec, see inset), and the slower (~100 msec) desorption of the reaction products from the chemisorption event (with the pulse off). This could be the desorption of molecular hydrogen, which would leave behind a new Ge layer, or the desorption of hydridic species, which would indicate that only a fraction of the chemisorbed digermane is converted into a Ge film. Single-

9.9 Surface Photoabsorption and *p*-Polarized Reflectance Spectroscopies

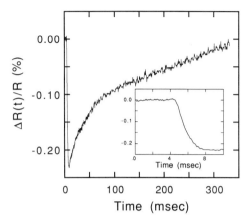

Figure 9.60 Transient differential reflectance during Ge epitaxy using a pulsed molecular jet of digermane, showing the fast change during the pulse (also see inset) and the much slower recovery process (from Sharp and Eres, 1992).

wavelength TRR cannot distinguish between these two routes, but the kinetic model suggests the former. This is an example in which spectroscopic TRR, perhaps with two or three wavelengths, could help identify the adsorbate through its characteristic optical properties. Interferometric effects were observed in the reflectance during the growth of thicker films, and periodic oscillations in the reflectance were seen in some regimes during pulsed impingement due to oscillations in the hydrogen coverage on the surface.

In monitoring GaP ALE growth on Si(001) by *p*-polarized reflectance spectroscopy (PRS) [632.8 nm, θ ~ 75°], Dietz *et al.* (1995) noticed a modulation in the reflectance associated with the periodic sequence of reactant pulses that was superimposed upon the usual reflection interferometry trace seen during heteroepitaxy (Section 9.5.1), as seen in Figure 9.61a. This differs from SPA in several ways. The energy of the probing photons is below the band gap, so film growth can be probed, and yet surface-related structure can still be seen. This fine structure in the reflectance trace that tracks the pulse sequence is also amplitude modulated. In GaP homoepitaxy, this amplitude-modulated fine structure is still seen (Figure 9.61b), albeit without the underlying structure due to interference in the growing film. This fine structure can be used to calibrate layer thickness during homoepitaxy. Laser light scattering (LLS, Section 11.2.1) was also monitored during hetero- and homoepitaxy; for homoepitaxy it was seen to be cross-correlated with the PRS signal, as seen in Figure 9.61b. Since the amplitude modulation in the fine structure was observed by both PRS and LLS, it was thought to be due to cyclic changes in surface structure. Related work is described by Bachmann *et al.* (1995).

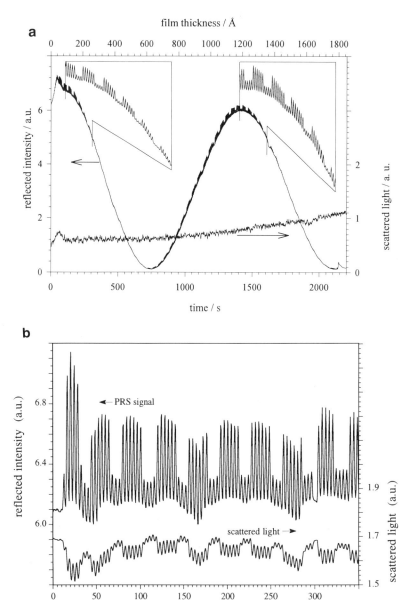

Figure 9.61 Real-time p-polarized reflectance and laser light scattering monitoring of (a) the heteroepitaxial ALE of GaP on Si(001) and (b) the homoepitaxial ALE of GaP on GaP(001) (from Dietz et al., 1995).

Dietz and Lewerenz (1993) and Dietz *et al.* (1994) used this reflectometry method, which they called Brewster-angle reflectance spectroscopy (BARS), for *in situ* characterization of film growth or etching of transparent films on Si. They found that reflectance measurements are sensitive to the growth of ~10–20 Å of a dielectric overlayer at the Si pseudo-Brewster angle $\theta_B = 75.637°$ at 632.8 nm ($\varepsilon' = 15.25$, $\varepsilon'' = 0.17$). They also monitored the growth of alternating SiO_2/Si_3N_4 dielectric multilayers on Si, using remote plasma CVD, by taking the time derivative of R_p; this results in a discontinuity at each new interface. They termed this method Brewster angle reflectance differential spectroscopy (BARDS).

The ultimate sensitivity of BARS as an *in situ* probe during the growth of very thin layers is limited by the depolarization caused by the chamber windows, which increases I_s/I_p after the Glan–Thompson polarizing prism from ~10^{-6} to ~10^{-5}, and by the angular divergence of the laser beam. For growth on Si, the effects of depolarization and beam divergence lead to large errors in determining reflectance for overlayers <10 Å thick when $\theta = \theta_B$, but this error is relatively small (5–10%) for these thin overlayers if θ is smaller than θ_B by 0.3°. Offsetting the angle of incidence slightly from Brewster's angle also helps determine unambiguously whether the index of refraction of the growing film is larger or smaller than that of the substrate.

Lewerenz and Dietz (1993) have used Brewster angle spectroscopy to identify defects in semiconductors.

9.10 Reflectance-Difference (Anisotropy) Spectroscopy

The interaction of light with uniaxial and biaxial materials depends on the relative orientation of the light polarization and the crystal axes because of anisotropy. The reflection coefficients from semiinfinite media composed of such anisotropic materials generally depend on polarization for all angles of incidence, including normal incidence. In isotropic materials, such as cubic semiconductors, contributions due to local anisotropies cancel in the bulk and, based on this bulk contribution, the reflection coefficient r is independent of the angle between the electric field vector and any crystal axis. However, the anisotropy of the surface cannot be neglected because it leads to small, yet measurable, differences in the reflection coefficient for different polarizations, even at normal incidence.

The reflection probe that senses this surface anisotropy at normal incidence is called reflectance-difference spectroscopy (RDS) or reflectance-anisotropy spectroscopy (RAS); it is a normal-incidence version of ellipsometry. As such, RDS is also a type of surface differential reflectometry. It can be used for more fundamental studies of surface structure, chemistry, and dynamics, and for real-time process monitoring. It is capable of monitoring the surface during most stages of MOCVD and MBE growth, including cleaning (deoxidation), pregrowth conditioning, and homoepitaxial or heteroepitaxial growth. RDS has been instrumental in process development studies and is a prime candidate for real-time monitoring and control during manufacturing. The use of RDS as a surface monitor has been reviewed by Aspnes (1993c, 1994, 1995a), Ploska *et al.* (1994), Drévillon and Yakovlev (1994), and Pickering (1994).

The surface anisotropy can be due to the different geometric ordering of atoms in two orthogonal directions parallel to the surface, which can cause anisotropies in surface absorption and refractive index. The (001) growth surfaces of III–V and II–VI semiconductors are terminated by cation and anion dimer bonds oriented along the [110] and [$\bar{1}$10] directions. Figure 9.62 illustrates this for As-terminated GaAs.

The RDS response appears to arise from both a *structural* component due to the geometric periodicity of the surface unit cell and a resonant *chemical* component involving the surface metal dimer states. In some growth systems, there also appears to be a *morphological* component due to anisotropic 3D growth. RDS can be understood theoretically by a continuum model, where the optical properties of the surface layer are described by a general complex dielectric tensor, or by the discrete dipole model, where the atoms near the surface are described as dipoles with specific polarizability (Wentink *et al.*, 1992). RDS is not sensitive to long-range order, but to the averaged anisotropies within the probed area.

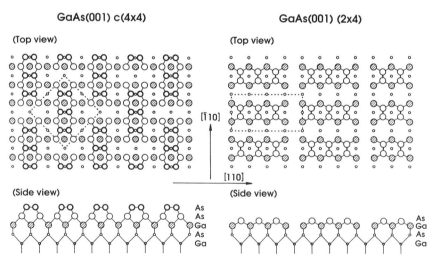

Figure 9.62 Surface structures for two different surface reconstructions of GaAs(100) (from Figure 4 in Richter, 1993 and from Biegelsen *et al.*, 1990).

9.10 Reflectance-Difference (Anisotropy) Spectroscopy

In the dielectric model, the RDS signal is related to the difference in the dielectric response in perpendicular directions on the surface. This is equivalent to the difference in normal-incidence reflectances for $\alpha = 0°$ and $90°$ in Equations 9.21–9.23 (Hingerl et al., 1993)

$$\left.\frac{\delta r}{r}\right|_{\text{RDS}} = \left.\frac{\delta r}{r}\right|_{\text{magnitude}} + i\delta\tilde{\theta} = \frac{r_x - r_y}{\langle r \rangle} \qquad (9.59a)$$

$$= \frac{4\pi i d n_a}{\lambda} \frac{\varepsilon_{xx} - \varepsilon_{yy}}{\varepsilon_s - \varepsilon_a} \qquad (9.59b)$$

where r_x and r_y are the reflection coefficients for x- and y-linear polarization and $\langle r \rangle$ is their average. [The difference in optical response due to different incident polarizations, as in RDS, is denoted by δ, while that due to differences in surface properties, as in DR and SPA, is denoted by Δ. This differs from the notation used in Hingerl et al. (1993).]

9.10.1 Instrumentation

Aspnes et al. (1988a,b) and Acher and Drévillon (1992) have detailed the construction and operation of instruments capable of performing RDS. Figure 9.63 shows the experimental apparatus for RDS using a photoelastic modulator (PM), while Figure 9.64 details the alignment of the optical elements used for RDS. For spectroscopic measurements, a white-light source (Xe lamp) is collimated and transmitted through a polarizer that linearly polarizes the beam at $0°$ to bisect [110] and [$\bar{1}$10]. Making these axes the x and y directions in Equation 9.59 maximizes the signal. The reflected beam propagates through a photoelastic modulator (at frequency $\sim$50 kHz) and an analyzer at $45°$ before dispersion by a spectrometer and photomultiplier detection. Thus this instrument is similar to the polarization-modulated ellipsometer (PME) described in Section 9.11.2. This operates as a near-null optical system when the reflection coefficients for linear polarization along [110] (r_{110}) and [$\bar{1}$10] ($r_{\bar{1}10}$) are equal.

The modulation in the detected signal S is (Aspnes et al., 1988b)

$$\frac{\delta S(t)}{S} = 2(-\text{Im}(\delta r/r) + \delta_1 \cos 2\bar{\theta}_1 + \delta_2 \cos 2\bar{\theta}_2 - 2a_P) J_1(\delta_C) \sin \omega t \qquad (9.60)$$
$$+ 2(\text{Re}(\delta r/r) + 2 \Delta P + 2 \Delta C) J_2(\delta_C) \cos 2\omega t$$

The first terms in the two sets of brackets contain the important spectral information, while the other terms are optical corrections. $\delta r = r_{110} - r_{\bar{1}10}$ and $r = (r_{110} + r_{\bar{1}10})/2$, so the term varying at ω gives the imaginary part of the fractional anisotropy of the reflection coefficient $\delta r/r$, while that varying at 2ω gives the real part. This latter term is related to the anisotropy in reflectance $\delta R/R = (R_{110} - R_{\bar{1}10})/R = 2 \text{Re}(\delta r/r)$. The reference azimuths for the polarizer, modulator (compensator), and analyzer are $-45° + \Delta P$,

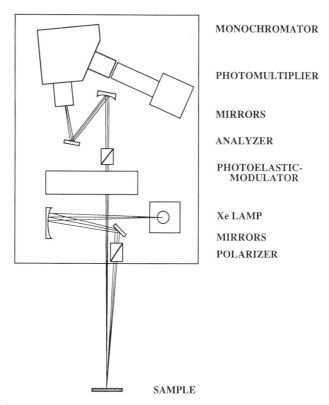

Figure 9.63 Schematic of a reflectance-difference spectrometer (from Kamiya et al., 1992c).

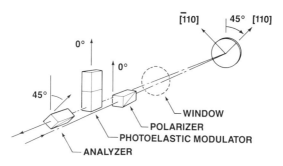

Figure 9.64 The polarization-sensitive optical elements in a photoelastic-modulator reflection-difference spectrometer (from Aspnes et al., 1988b).

9.10 Reflectance-Difference (Anisotropy) Spectroscopy

$+45°+\Delta C$, and ΔA, relative to 45° in Figure 9.64, and those denoting the incident- and reflected-beam window strain are $\bar{\theta}_1$ and $\bar{\theta}_2$. Retardation due to window strain affecting the incident beam, window strain affecting the reflected beam, and the modulator are δ_1, δ_2, and δ_C, respectively. J_1 and J_2 are Bessel functions.

This PM version is standard, and has a sensitivity for $\delta R/R = 2\,\mathrm{Re}(\delta r/r)$ of $\sim 5 \times 10^{-5}$. Nonuniform window strain can make measurement of $\mathrm{Im}(\delta r/r)$ less sensitive (Aspnes *et al.*, 1988a,b). Kamiya *et al.* (1992c) have demonstrated real-time RDS with ~ 80 spectral points between 1.5 and 5.5 eV in 25 sec. Other types of RD instruments have also been used for *in situ* monitoring. In earlier RDS instruments, the crystal x and y axes were exchanged either by rotating the sample (Aspnes, 1985b) or by rotating a polarizer (Aspnes *et al.*, 1987). Later, Aspnes *et al.* (1988b) described a rotating analyzer system. These instruments are slower, determine only $\mathrm{Re}(\delta r/r)$, and have relatively poor sensitivity *vis-à-vis* PM systems. Aspnes *et al.* (1990c) have demonstrated a double-modulation version with PM modulation of the incident polarization (50 kHz) and a slow rotation of the sample (0.1 Hz) to eliminate systematic errors. Other variations and developments in RDS instrumentation may be found in Acher and Drévillon (1992) and Pickering (1994).

9.10.2 Comparing RDS with SPA and Ellipsometry

Although the signals in RDS are smaller than those in SPA and polarization-related calibration is much more involved in RDS, RDS has the advantage that it measures differences on the surface at one time; SPA tracks the surface as a function of time and is therefore subject to drifts. Furthermore, the spectroscopic capability of RDS is a demonstrated strength of the method; the spectroscopic capability of SPA has been investigated less comprehensively. When RDS is used to monitor a fast process, only a single wavelength has been examined—one that is most sensitive to surface changes, as in SPA.

Hingerl *et al.* (1993) have demonstrated that surface photoabsorption and RDS can give similar optical information. Using Equation 9.23, the normalized difference in reflection coefficients of two films a and b, both with thickness d, is

$$\left.\frac{\Delta r_{pp}}{r_{pp}}\right|_{\mathrm{SPA}} = \frac{4\pi i d n_{a\perp}}{\lambda} \frac{(\varepsilon_{xx}^b - \varepsilon_{xx}^a)\left(1 - \frac{\varepsilon_a}{\tilde{\varepsilon}_s}\sin^2\theta\right) + \left(\frac{1}{\varepsilon_{zz}^b} - \frac{1}{\varepsilon_{zz}^a}\right)\varepsilon_a\tilde{\varepsilon}_s\sin^2\theta}{(\tilde{\varepsilon}_s - \varepsilon_a)\left(\cos^2\theta - \frac{\varepsilon_a}{\tilde{\varepsilon}_s}\sin^2\theta\right)} \quad (9.61)$$

where ε_{ii} for the two films can be complex. If $|\tilde{\varepsilon}_s|, |\varepsilon| \gg \varepsilon_a$, after neglecting higher-order terms,

$$\left.\frac{\Delta R_p}{R_p}\right|_{SPA} = \frac{8\pi d n_{a\perp}}{\lambda} \text{Im}\left(\frac{\varepsilon_{xx}^a - \varepsilon_{xx}^b}{\tilde{\varepsilon}_s \cos^2\theta - \varepsilon_a}\right) \qquad (9.62)$$

If film b is the same anisotropic film as film a, but rotated by 90°, then Equations 9.59b and 9.62 are seen to be very similar. Hingerl et al. (1993) were able to simulate SPA data by using the data base they obtained from RDS experiments.

Ellipsometry at a fixed angle is not suited for detecting anisotropic effects in the surface plane because r_p and r_s are different for $\theta \neq 0°$, even for isotropic samples. Wentink et al. (1992) pointed out the close connection between RDS and variable-angle ellipsometry. They made RDS measurements at normal incidence (n) on Ge(001) 2 × 1 by using a modified rotating-analyzer ellipsometer. The reflection coefficient ratio ρ, which compares p- and s-polarized light in ellipsometry (Equation 9.31), then compares the reflection coefficients for light polarized along the ($\bar{1}10$) and (110) axes and becomes the complex anisotropic reflectance ratio

$$\rho_n = \frac{r_{\bar{1}10}}{r_{110}} = \tan \Psi \, e^{i\Delta} \qquad (9.63)$$

9.10.3 Applications

Early work on RDS was performed by Aspnes and Studna (1985), Aspnes (1985b), and Aspnes et al. (1987). Richter (1993), Aspnes (1993c, 1994, 1995a), and Pickering (1994) have reviewed more recent developments.

Much work has centered on the growth of GaAs. The four primary surface reconstructions of GaAs are (4 × 2) {Ga-rich, terminated by Ga dimers along [110]}, (2 × 4) {As-rich, As dimers along [$\bar{1}$10]}, c(4 × 4) {highly As-rich, double layer of As atoms with the outer layer as As dimers along [110]}, and d(4 × 4) {partially disordered c(4 × 4) phase due to an excess of adsorbed As} (Kamiya et al., 1992a,b). (See Figure 9.62. Also, note that every fourth dimer is missing because of local charge neutrality.) The RD spectra for these reconstructions are given in Figure 9.65. Kamiya et al. (1992c) used RDS to explore in detail GaAs(100) surfaces under UHV for a variety of processing conditions.

Spectroscopic RDS clearly demonstrates the differences between surface features (Figure 9.65). Features at 1.9 eV are due to transitions on Ga dimers (4 × 2), while those at 2.6 and 4.2 eV are due to transitions on As dimers (2 × 4) (Kamiya et al., 1992b). This information provides fundamental understanding of surface processes and specifies the wavelength where anisotropies are largest. This latter point is important because real-time control is best performed with a single wavelength, and this wavelength should be chosen to maximize the anisotropy. For example, the maximum surface anisotropy between (2 × 4) As-terminated and (4 × 2) Ga (Al) -terminated surfaces of GaAs (AlAs) occurs at 2.0–2.5 eV (3.5 eV),

9.10 Reflectance-Difference (Anisotropy) Spectroscopy

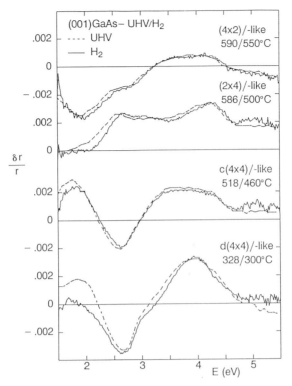

Figure 9.65 RD spectra of the primary reconstructions of GaAs(100) in UHV and atmospheric pressure H_2 at selected sample temperatures (from Kamiya et al., 1992a).

as is seen in Figure 9.65 for GaAs. This figure shows the GaAs(100)–(2 × 4) RDS spectrum has a camelback profile, while the GaAs(100)–c(4 × 4) has a deep minimum at 2.6 eV, which is attributed to electronic transitions on the As dimers. The change in sign of $\delta r/r$ is due to the change in orientation of the dimers, from along [$\bar{1}$10] for (2 × 4) to [110] for c(4 × 4). This is seen in Figure 9.62.

These GaAs reconstructions have been observed with RDS in UHV MBE systems (Aspnes et al., 1990c; Resch et al., 1993) and also at higher pressures, as in a MOVPE reactor with a 1 atm hydrogen ambient (Kamiya et al., 1992a,b,c), as seen in Figure 9.65. For both MOVPE and UHV conditions, RD shows a progression from d(4 × 4) to c(4 × 4) to (2 × 4)-like reconstruction as temperature is increased, and under analogous conditions, the surface reconstructions and RDS spectra are similar for both types of epitaxial growth (also see Figure 9.66). The (2 × 4) reconstruction is a less arsenic-rich surface and is favored at higher temperatures because

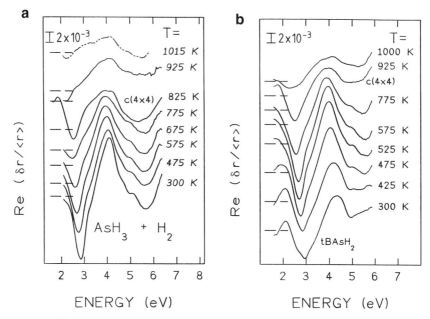

Figure 9.66 RD spectra of GaAs(100) in a MOVPE reactor at different temperatures with H_2 and (a) AsH_3 or (b) $tBAsH_2$ (from Figures 7 and 8 in Richter, 1993).

of the loss of As from the surface. RDS measurements have shown that conversion can be prevented up to successively higher temperatures by adding AsH_3 to the H_2 in the reactor (Reinhardt et al., 1993), and to even higher temperatures by replacing the arsine with the more-arsenic-rich *tert*-butylarsine (Richter, 1993), as is seen in Figure 9.66 (compare to Figure 9.65). Addition of trimethylgallium (TMGa) in an MOVPE reactor tends to lessen the As-dimer features at 2.6 and 4.0 eV.

Reflection high-energy electron diffraction (RHEED) can monitor monolayer-by-monolayer epitaxial growth, seen as oscillations in the signal (see Joyce et al., 1986; also see Figure 9.68 below). These oscillations are observed only when the growth is layer-by-layer (2D growth); they are washed out for continuous growth due to surface roughness and at higher temperatures where growth occurs primarily at surface steps, instead of on the terraces, due to enhanced surface diffusion. However, RHEED can be used only during ultrahigh vacuum (UHV) processes, such as MBE, because it probes with electrons. RDS is a promising optical monitor of 2D growth for all pressures, and as such is a real-time probe of the growth rate (in terms of monolayers) during MOVPE. While RHEED is sensitive to long-range order in surface structure, with the oscillations due to the periodic cycle between smooth and rough (2D island growth) surfaces, RD usually

9.10 Reflectance-Difference (Anisotropy) Spectroscopy

senses the relative surface concentrations of anion and cation dimer bonds and is therefore sensitive to the local electronic structure.

Evidence for RDS growth oscillations has been seen by several groups during GaAs epitaxy. They were first observed during (UHV) MBE growth (Aspnes et al., 1987; Scholz et al., 1992), then during MOVPE at lower pressures during growth (Jönsson et al., 1990 [10^{-3} Torr]; Samuelson et al., 1991; Jönsson et al., 1992 [10^{-3} to 0.3 mbar, = $\sim 10^{-3}$ to 0.2 Torr]), and most recently in the standard high-pressure (10^4 Pa, = 0.1 atm) MOVPE environment (Reinhardt et al., 1993; Richter, 1993). Since RHEED can be, and is routinely, used during MBE, these earliest observations are of moderate interest; however, since RHEED cannot be used at higher pressures, the observations of RD oscillations during higher-pressure (MOVPE/MOCVD) growth are very significant. These RD oscillations are clear in Figure 9.67a (at 2.6 eV) during MOVPE of GaAs with 10^4 Pa H_2, 70 Pa AsH_3, and various partial pressures of TMGa. These traces show an increase in the growth rate with increasing TMGa partial pressure. Figure 9.67b shows the growth rate evaluated using these oscillations. Variations in the growth rate with changing carrier-gas pressure were also observed. Such time-resolved studies are easiest to perform at the wavelength most sensitive to the As_2 dimer features.

Harbison *et al.* (1988) compared RHEED, RDS, and reflectance signals during the growth of AlAs on As-stabilized (2 × 4) AlAs using 3.5 eV, and found that RHEED and RDS have the same oscillation period, as is seen

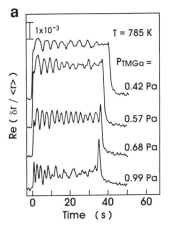
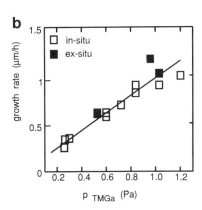

Figure 9.67 (a) RD signal at 2.6 eV during homoepitaxy (MOVPE) on GaAs(100) for fixed H_2, fixed AsH_3, and different TMGa partial pressures, showing growth oscillations (from Figure 10 in Richter, 1993). (b) Dependence of the growth rate of GaAs by MOVPE on the partial pressure of TMGa evaluated by the time period of RD oscillations (open squares) and from postgrowth thickness analysis (closed squares) (from Reinhardt et al., 1993).

in Figure 9.68 (also see Colas *et al.,* 1991). S. R. Armstrong *et al.* (1992) also found the same oscillation period for RHEED and RDS during the MBE growth of AlAs on GaAs(001), with the RD oscillations continuing after the RHEED oscillations were damped. These studies suggest that one RD oscillation period corresponds to the growth of a single atomic bilayer.

Several other investigations have used RDS to follow and control surface reactions relevant to GaAs growth. Maa and Dapkus (1991) studied the surface reactions of trimethylgallium and *tert*-butylarsine GaAs(100) surfaces by using real-time RDS. Aspnes *et al.* (1992b) and Kamiya *et al.* (1992b) have used RDS to simulate atomic layer epitaxy of GaAs(100) by cyclic and noncyclic exposures of reactant gases, as is seen in Figure 9.69. Related studies on pseudo–atomic-layer epitaxial growth of GaAs have been made by Armstrong *et al.* (1993). Paulsson *et al.* (1991, 1993) studied the kinetics of GaAs MOVPE by using various Ga- and As-bearing precursors. Morris *et al.* (1995) used RDS to examine the surfaces of $Al_xGa_{1-x}As$ layers (x = 0.0, 0.25, 0.50, 0.75, 1.0) grown by MBE on GaAs(001). Berkovits and Paget (1993) used RDS to study Na_2S-passivated GaAs during annealing. Jönsson *et al.* (1993) followed surface anisotropy during the reaction of H_2S on GaAs with kinetic RDS. Armstrong *et al.* (1994a) used RDS to

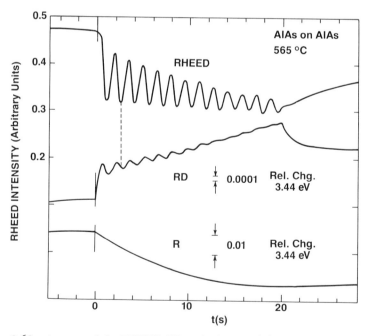

Figure 9.68 Averages of nine RHEED, RD, and reflectance (R) traces during AlAs growth on an As-stabilized (2 × 4) AlAs surface (from Harbison *et al.,* 1988).

9.10 Reflectance-Difference (Anisotropy) Spectroscopy

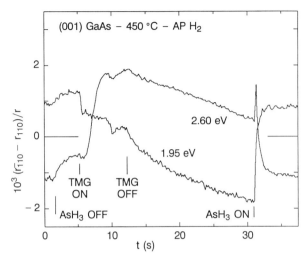

Figure 9.69 Transient RD traces during atomic layer (homo)epitaxy cycles on GaAs(001) monitored at two different photon energies (from Aspnes *et al.*, 1992b).

study n-GaAs(001) electrodes during anodic oxidation in aqueous KCl solutions. Rumberg *et al.* (1995) used RDS to monitor AlAs growth by MOMBE.

The epitaxial growth of other materials has been studied as well. Koch *et al.* (1991) followed the RDS transient during the growth of several heterostructures, including InP/GaInAs superlattices. Similar studies of dimer formation on the surface of InP and the growth of InP have been conducted by Ploska *et al.* (1994). Figure 9.70 shows that the different surfaces in various InGaAsP materials have distinctive RDS spectra, so RDS can be used for submonolayer monitoring and control of surface concentrations for group III and group V elements (Ploska *et al.*, 1994; Jönsson *et al.*, 1994). In addition to the homoepitaxial growth [GaP on GaP(100)], this figure shows examples of heteroepitaxial growth [InGaAs grown lattice matched to InP(100)] and surface modifications ("GaAsP" obtained by exposing GaAs to phosphine and "InAsP" obtained by exposing InP to arsine). González *et al.* (1995) probed the atomic layer MBE growth of $GaAs_{1-x}P_x$ layers with RDS to investigate the incorporation of P_2. Armstrong *et al.* (1994b) examined the homoepitaxial growth of InP on InP(001). Scholz *et al.* (1992) employed RDS to monitor the MBE of InAs on GaAs(100). Esser *et al.* (1994) studied ordered Sb layers on GaAs(110) and InP(110) with RDS. Surface anisotropies of group IV surfaces have also been investigated by using RDS, including Ge(001) 2×1, by Wentink *et al.* (1992), and Si(100), by Müller *et al.* (1993).

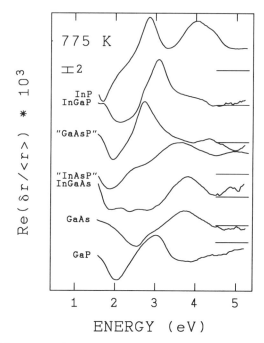

Figure 9.70 RD spectra under hydrogen flow for several III–V (100) epilayers (InP, InGaP, InGaAs, GaAs, and GaP) and surfaces ("GaAsP"—GaAs surfaces exposed to phosphine; "InAsP"—InP surfaces exposed to arsine) (from Ploska et al., 1994).

Typical fractional anisotropies in these cited growth studies are ~0.1–1% (for $|\delta r/r|$). Acher et al. (1990) saw much larger reflectance anisotropies during the low-pressure MOCVD, with $|\delta r/r|$ exceeding 5% for InAs on InP (Figure 9.71) and 100% for InP on GaAs/Si. This was attributed to anisotropic roughness during the 3D growth of these lattice-mismatched depositions. Scholz et al. (1992) made similar observations for InAs coverages larger than four monolayers on GaAs(100).

RDS has also been used to follow surface preparation before growth. Substrate deoxidation has been followed for GaAs wafers under AsH_3/H_2 flow [as temperature was increased] (Reinhardt et al., 1993) and InP wafers under PH_3/H_2 flow (Ploska et al., 1994). In both cases, RDS features due to the oxide are seen to disappear above a certain temperature and are replaced by a characteristic RDS spectrum of the surface of the substrate material. The RDS signal of InP has a peak near 2.6 eV, while the oxide has a minimum near 4.7 eV and a maximum near 5.5–6.1 eV. Ploska et al. (1994) followed the removal of the oxide layer from InP by kinetic RDS at 2.6 eV, which is indicated by the signal increase at 2.6 eV at 755 K in Figure 9.72.

9.11 Ellipsometry

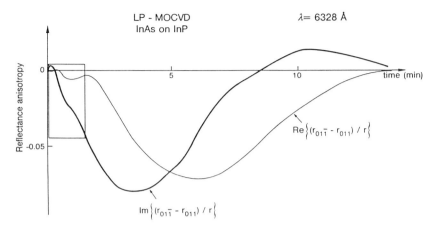

Figure 9.71 RD record of the MOCVD growth of an InAs layer on an InP substrate at 632.8 nm, showing very large anisotropies related to the anisotropic three-dimensional growth of this layer (from Acher *et al.*, 1990).

9.11 Ellipsometry

Ellipsometry is the measurement of the state of polarization of the light beam after it has interacted with a medium to obtain information about

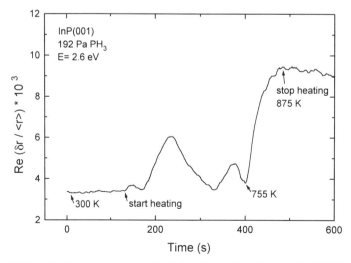

Figure 9.72 RD kinetic trace at 2.6 eV during the heating of an oxidized InP(100) wafer in hydrogen/phosphine mixtures showing the removal of the oxide layer (from Ploska *et al.*, 1994).

that medium. The polarization can be modified by reflection, transmission (and refraction), and scattering. Reflection ellipsometry, which compares the Fresnel coefficients for p- and s-polarized light, is commonly used to study thin films, surfaces, and interfaces. Transmission ellipsometry (sometimes called polarimetry, which is not to be confused with an alternative meaning given in Section 9.2) is commonly used to measure natural and induced optical anisotropies in bulk samples, which may be refractive or absorptive; it is sometimes used to examine films by analysis of the measured Fresnel coefficients. Scattering ellipsometry is used to measure particle sizes and their distribution (Azzam and Bashara, 1977). This section discusses the widespread use of reflection ellipsometry, which will simply be called ellipsometry, as an *in situ* and real-time diagnostic.

As such, ellipsometry examines the change in polarization of light upon reflection from a specular surface. In spectroscopic ellipsometry (or, alternatively, spectral ellipsometry or spectroellipsometry, SE) these changes are measured as a function of wavelength λ, typically from 200 to 800 nm. In this wavelength range, SE depends on electronic transitions from the valence to conduction bands through the real and imaginary parts of the dielectric function $\varepsilon'(\lambda)$ and $\varepsilon''(\lambda)$ (or ε_1 and ε_2), and it provides a macroscopically averaged probe of electronic structure, crystallinity, alloy concentrations, and molecular binding on the surface. Single-wavelength ellipsometry (SWE) is commonly used for *ex situ* film analysis when only two parameters are unknown, such as the thickness d and index of refraction n of one given film. SWE is much simpler than SE, yet far less powerful. For example, only SE can be used to analyze a multilayer system to obtain the composition and thicknesses of buried layers. This is demonstrated in the *ex situ* SE analysis in Figure 9.73 (Vedam *et al.*, 1985).

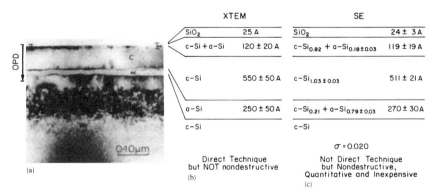

Figure 9.73 Analysis of a multilayer by cross-sectional transmission electron microscopy [XTEM, (a,b)] and spectroscopic ellipsometry [SE, (c)] (from Vedam *et al.*, 1985).

Although ellipsometry probes essentially bulk properties, with proper modeling and careful application of experimental ellipsometric corrections, the high precision of SE can provide surface and interfacial characterization and analysis with submonolayer resolution (~0.01 nm). Consequently, SE can be a very good surface probe during processing. The use of SE to determine the optical properties of the material near the surface, as is needed during epitaxial growth, can be termed near-surface dielectric function spectroscopy (NSDFS). Moreover, microscopic roughness on the surface, surface contamination, and crystalline damage can be differentiated by their different ellipsometric spectra. As with other reflection-based probes, ellipsometry can monitor the interference of waves from reflections within multilayer structures to determine the optical path length of films, nd, from which the film thickness d can be determined. Use of SE as an *in situ* probe depends on the reliability of the model of the material and the speed of data acquisition and analysis. There is a sizable data base of optical functions for use in SE (Table 9.1).

Azzam and Bashara (1977) is an authoritative overview of the basic principles of ellipsometry. Recent reviews on real-time spectroscopic ellipsometry and instrumentation for thin film processing include those by Aspnes and Chang (1989), Collins (1990), Irene (1993), Jellison (1993a), Drévillon (1993), Collins *et al.* (1994), Nguyen *et al.* (1994), and Pickering (1994, 1995). Seminal papers on ellipsometry have been collected by Azzam (1991). Tompkins (1993) details practical ellipsometry techniques and reviews several *in situ* and *ex situ* applications of ellipsometry. Applications in electrochemistry, which are sometimes important in thin film processing, are surveyed by Greef (1993); they will not be discussed here. The proceedings volume on spectroscopic ellipsometry edited by Boccara *et al.* (1993) is another excellent reference. Other discussions of the optics and instrumentation of ellipsometry include Muller (1969), Hauge (1980), Hauge *et al.* (1980), Aspnes (1985a), and Riedling (1987). These sources can also be consulted for details concerning the standard (Nebraska convention) notation used in ellipsometry. With minor exceptions, this notation will be followed here. One such exception is made to conform with other conventions followed in this volume; specifically, θ is used to denote the angle of incidence instead of the usual ellipsometric designation ϕ. The complex index of refraction is $\tilde{n} = n + ik$ and the dielectric function is $\tilde{\varepsilon} = \varepsilon' + i\varepsilon''$. To avoid confusion with medium indices, $\tilde{\varepsilon}$ is not expressed as $\varepsilon_1 + i\varepsilon_2$, a form which is common in the literature (and is used in several of the presented figures).

9.11.1 *Theory and Modeling*

Figure 9.1 depicts the optical beams in ellipsometry. With the p- and s-polarized components of the electric field labeled as E_p and E_s, the incident field is characterized by E_p^i and E_s^i and the reflected field by E_p^r and E_s^r. The polarization state incident on the surface is defined as $\chi_i = E_s^i/E_p^i$ and that reflected from the surface is $\chi_r = E_s^r/E_p^r$. For isotropic media,

$$E_p^r = r_p E_p^i \quad \text{and} \quad E_s^r = r_s E_s^i \tag{9.64}$$

where r_p and r_s are the (complex amplitude) reflection coefficients and $r_{p,s} = |r_{p,s}|e^{i\delta_{p,s}}$. Cartesian axes are sometimes chosen so that the beam propagates along the z axis, with p polarization along the x axis and s polarization along the y axis, and as such the electric fields have components $E_x(z,t) = \text{Re}\{E_{0x} e^{-i(\omega t - kz + \delta_x)}\}$ and $E_y(z,t) = \text{Re}\{E_{0y} e^{-i(\omega t - kz + \delta_y)}\}$ (Equations 2.2 and 2.3). When this is done, the p and s labels can be replaced by x and y, respectively.

A reflection coefficient ratio ρ is often defined and then related to the ellipsometric angles Ψ and Δ:

$$\rho = \chi_i/\chi_r = r_p/r_s = \tan \Psi e^{i\Delta} \tag{9.31'}$$

Since only ratios of measurable quantities are important in ellipsometry, accurate measurements of intensities are not needed; this is one reason why ellipsometry can be more accurate than reflectometry. Each of the parameters in Equation 9.31' is a function of wavelength λ, angle of incidence θ, and the thickness and dielectric function of each probed layer.

The interpretation of the ellipsometric angles becomes clearer by examining the case where $E_p^i = E_s^i$. Then $\chi_i = 1$ and

$$\tan \Psi = E_{0x}/E_{0y} \tag{9.65}$$

$$\Delta = \delta_x - \delta_y \tag{9.66}$$

where the electric field amplitudes and phases are those of the reflected beam. In terms of the Stokes parameters of the reflected beam,

$$\cos 2\Psi = -S_1/S_0 \tag{9.67}$$

$$\tan \Delta = -S_3/S_2 \tag{9.68}$$

Sometimes it is preferable to express the ellipsometric angles in terms of the associated ellipsometry parameters:

$$N_e = \cos 2\Psi \tag{9.69}$$

$$S_e = \sin 2\Psi \sin \Delta \tag{9.70}$$

$$C_e = \sin 2\Psi \cos \Delta \tag{9.71}$$

where $N_e^2 + S_e^2 + C_e^2 = 1$ for polarized light and <1 for light partially depolarized by reflection.

Several different ellipsometric configurations can be used to measure these parameters. They differ in their ability to determine the three ellipsometric parameters. For example, in some configurations the ellipsometer

9.11 Ellipsometry

may be insensitive to small changes in an ellipsometric angle or to its sign. Once the ellipsometric measurement has determined the angles Ψ and Δ for each incident λ, this information can be used along with a model of the material structure to determine the thicknesses and dielectric functions of the probed layers. Using the characteristic features in the wavelength dependence of $\tilde{\varepsilon}$ for the materials in the different layers, SE can determine many more than the two parameters (among the thickness d, ε', and ε'' for each layer) obtainable by SWE.

Perhaps the most critical stage of SE is developing a realistic model of ε for the different layers and, particularly, for the interfaces and the overlayers at the surface (Aspnes and Chang, 1989; Jellison, 1993a,b). Media can be approximated by two-phase models (ambient/substrate), three-phase models (ambient/film/substrate), and higher-order multiphase models (ambient/multiple films/substrate), as illustrated in Figures 9.2 and 9.74a.

In the two-phase model, the interface between material 1 with complex dielectric function $\tilde{\varepsilon}_1$ and material 2 with complex dielectric function $\tilde{\varepsilon}_2$ is sharp and both media are homogeneous. By inserting the Fresnel reflection coefficients, Equations 9.2–9.5, into Equation 9.31', the ratio of the dielectric functions at a particular wavelength can be expressed in terms of the quantity ρ, which is itself determined from the experimentally determined values for Ψ and Δ:

$$\frac{\tilde{\varepsilon}_2}{\tilde{\varepsilon}_1} = \sin^2\theta \left(1 + \tan^2\theta \left(\frac{1-\rho}{1+\rho}\right)^2\right) \qquad (9.35')$$

If the ambient medium is vacuum or a gas, then $\tilde{\varepsilon}_1 = 1$ and the right-hand side is the dielectric function of the medium $\tilde{\varepsilon}_2$. $\tilde{\varepsilon}_2$ can also be expressed in terms of the associated ellipsometric parameters

$$\varepsilon_2' = \sin^2\theta \left(1 + \tan^2\theta \left(\frac{N_e^2 - S_e^2}{(1+C_e)^2}\right)\right) \qquad (9.72)$$

$$\varepsilon_2'' = -2\sin^2\theta \tan^2\theta \frac{N_e S_e}{(1+C_e)^2} \qquad (9.73)$$

Multiphase models reduce to this limit when the uppermost layer is optically thick. One common way of presenting ellipsometric data for arbitrary multiphase media is to use Equation 9.35' for analysis, as if it were a two-phase system consisting of a uniform substrate with a clean and smooth interface. The resulting dielectric function for the "substrate" is called the pseudodielectric function $\langle\tilde{\varepsilon}\rangle = \langle\varepsilon'\rangle + i\langle\varepsilon''\rangle$. $\langle\tilde{\varepsilon}\rangle$ can then be used to evaluate the parameters $\tilde{\varepsilon}_i$ and d_i of the individual layers in the multiphase model.

The expressions for the three-phase model, with an overlayer (o) on the substrate (s) in an ambient (a), are most easily evaluated numerically (Aspnes and Chang, 1989). However, they are relatively simple when the film thickness $d \ll \lambda$. Inserting Equations 9.15 and 9.17 into Equation 9.31' gives

$$\rho(d) = \rho(0)\left(1 + \frac{4\pi i d n_a \cos\theta}{\lambda} \frac{\tilde{\varepsilon}_s(\tilde{\varepsilon}_s - \tilde{\varepsilon}_o)(\tilde{\varepsilon}_o - \varepsilon_a)}{\tilde{\varepsilon}_o(\tilde{\varepsilon}_s - \varepsilon_a)(\tilde{\varepsilon}_s \cot^2\theta - \varepsilon_a)}\right) \quad (9.74)$$

where $\rho(0)$ is the reflection coefficient ratio for the two-phase (ambient/substrate) case and $\varepsilon_a = n_a^2$ is assumed to be real. Using this in Equation 9.35', the pseudodielectric function is

$$\langle\tilde{\varepsilon}\rangle \sim \tilde{\varepsilon}_s + \frac{4\pi i d n_a}{\lambda} \frac{\tilde{\varepsilon}_s(\tilde{\varepsilon}_s - \tilde{\varepsilon}_o)(\tilde{\varepsilon}_o - \varepsilon_a)}{\tilde{\varepsilon}_o(\tilde{\varepsilon}_s - \varepsilon_a)} (\tilde{\varepsilon}_s/\varepsilon_a - \sin^2\theta)^{1/2} \quad (9.75)$$

This is useful in studying thin overlayers and surface roughness.

When $|\tilde{\varepsilon}_s| \gg |\tilde{\varepsilon}_o| \gg |\varepsilon_a|$,

$$\langle\tilde{\varepsilon}\rangle \sim \tilde{\varepsilon}_s + \frac{4\pi i d n_a}{\lambda} \tilde{\varepsilon}_s^{3/2} \quad (9.76)$$

In this limit measurements depend on the thickness d of the overlayer film, but not on its optical properties.

Expressions for r and ρ for multilayer systems are obtained by repeated application of Equation 9.24 or by using the recursive and matrix methods described earlier in this chapter. Within each layer the material can be isotropic, uniaxial, or biaxial; have uniform or nonuniform/graded thickness or refractive index; or be homogeneous or inhomogeneous (Jellison, 1993a,b; Chindaudom and Vedam, 1994).

If a phase is homogeneous, its dielectric function can be obtained from the literature (Table 9.1). Dielectrics are often simulated by using a Lorentz oscillator model, such as Equation 4.18 (Jellison, 1993a,b). Microscopically inhomogeneous phases are often modeled using the effective medium approximation (EMA) (Aspnes, 1982; Aspnes and Chang, 1989; Jellison, 1993a,b). In the most widely used EMA, the effective dielectric constant $\tilde{\varepsilon}$ is given by the Bruggerman expression

$$0 = f_a \frac{\tilde{\varepsilon}_a - \tilde{\varepsilon}}{\tilde{\varepsilon}_a + K\tilde{\varepsilon}} + f_b \frac{\tilde{\varepsilon}_b - \tilde{\varepsilon}}{\tilde{\varepsilon}_b + K\tilde{\varepsilon}} \quad (9.77)$$

This assumes two randomly mixed phases with dielectric constants $\tilde{\varepsilon}_a$ and $\tilde{\varepsilon}_b$ with fractions f_a and f_b, respectively ($f_a + f_b = 1$), and is straightforwardly extended for more than two phases. $K = 2$ assumes spherical microstructure. The Maxwell–Garnet EMA is an alternative approach that assumes layered phases.

Figure 9.74b shows that $\tilde{\varepsilon}$ is different for crystalline and amorphous Si. ε_{c-Si} peaks near 3.4 and 4.2 eV due to the E_1 and E_2 peaks, respectively, while ε_{a-Si} has a single peak near 3.6 eV. In a partially amorphous silicon film parts of ε would be attributed to c-Si and a-Si, as long as the dimensions of the local structure were $\ll \lambda$, and possibly also to a third phase, i.e., voids

9.11 Ellipsometry

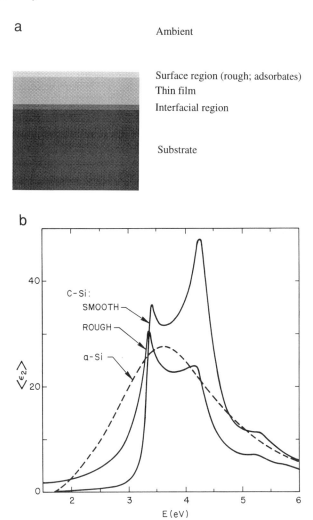

Figure 9.74 (a) Multiphase model for ellipsometric analysis of a film atop a substrate, with a surface overlayer region (which could represent roughness and/or adsorbates), the film, an interfacial region between the film and substrate (which could be a graded region), and the substrate. The model parameters are the dielectric function for each region and the thickness of the top three regions. (b) Measured pseudodielectric function $\langle \varepsilon'' \rangle$ for a smooth c-Si wafer, a microscopically rough c-Si wafer, and a smooth a-Si film (from Aspnes, 1982).

(Aspnes, 1982). Similarly, polycrystalline silicon can be modeled as partly due to c-Si and partly to voids. When a medium X has voids that occupy a fraction $f_{voids} \ll 1$, then Equation 9.77 gives $\tilde{\varepsilon} \approx \tilde{\varepsilon}_X(1 - 3f_{voids}/2)$ if $|\tilde{\varepsilon}_X| \gg 1$. Microscopically rough surfaces, which are commonly formed during plasma etching, can be modeled as an overlayer of the substrate with $f_{voids} = 0.4$–0.6.

Ellipsometry monitors specularly reflected beams and is useful for determining roughness when the typical dimensions are $\ll \lambda$. When the dimensions are $> \lambda$, the collected light signal is partially lost to diffuse scattering for both polarizations, and so will not affect the ellipsometric analysis. This more macroscopic roughness can be analyzed by using laser light scattering (LLS, Section 11.2.1).

The evolution of a thin film process monitored by ellipsometry is often represented by the trajectory of the ellipsometric angles as an implicit function of time. For example, as a uniform absorbing film is deposited on a substrate, the Δ–Ψ trajectory starts at (Δ,Ψ) for the substrate and then spirals to (Δ,Ψ) for the bulk material of the film (see Figure 9.79, p. 445). Alternatively, the locus of the imaginary and real parts of the pseudodielectric function $\langle\varepsilon''\rangle$ vs $\langle\varepsilon'\rangle$ can be plotted. Then, during this deposition this locus evolves from the $\langle\varepsilon''\rangle,\langle\varepsilon'\rangle$ coordinate of the initial substrate to that of the optically thick film by an exponential spiral (Figure 9.75).

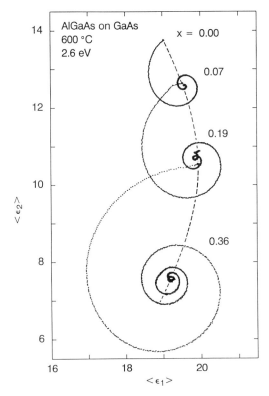

Figure 9.75 Locus of the pseudodielectric function at 2.6 eV for the sequential MOMBE depositions of $Al_xGa_{1-x}As$ films for the indicated Al fractions (from Aspnes et al., 1990a).

9.11 Ellipsometry

During deposition these kinetic ellipsometric data can be analyzed to determine the dielectric function of the most recently deposited material by using an approach based on direct use of the Fresnel relations, differentiation of ρ or $\langle \tilde{\varepsilon} \rangle$ with respect to thickness, or the virtual interface (VI) technique. These NSDFS methods have been reviewed by Aspnes (1995b, also private communication, 1995). (See Section 9.1.1.5.)

The classical Fresnel approach utilizing the three-phase model gives only the average refractive index of the overlayer film (o); it has been applied assuming either that the film is transparent ($k_o = 0$) or that k_o is not necessarily zero (Urban et al., 1992; Urban and Comfort, 1994). More complex n-phase models (with k_{phase} not necessarily zero) can be constructed by sequentially building on previously deposited layers and curve-fitting the Fresnel relations to provide $\tilde{\varepsilon}$ for the outermost layer (Theeten, 1980; Hottier and Laurence, 1981; I.-F. Wu et al., 1993). To avoid the instabilities inherent in the classical Fresnel approach, the outer layer usually should not be thinner than ~ 50 Å.

In the derivative approach, the near-surface properties are obtained either from the ρ or $\langle \tilde{\varepsilon} \rangle$ trajectory with increasing thickness during growth or from their derivatives with respect to thickness. In the exponential spiral approximation (ESA) (Theeten et al., 1979; Hottier et al., 1980; Hartley, 1988; Aspnes et al., 1990b; Hartley et al., 1992a,b), when an overlayer with thickness d and dielectric constant $\tilde{\varepsilon}_o$ is deposited, to first order the pseudodielectric constant $\langle \tilde{\varepsilon} \rangle$ locus evolves as

$$\langle \tilde{\varepsilon}(d) \rangle \approx \tilde{\varepsilon}_o + [\langle \tilde{\varepsilon}(0) \rangle - \tilde{\varepsilon}_o]\exp(2ik_{oz}d) \qquad (9.78)$$

where $ck_{oz}/\omega = \tilde{n}_{oz} = (\tilde{\varepsilon}_o - \tilde{\varepsilon}_a \sin^2\theta)^{1/2}$. (This relation assumes that the "back-reflected" wave is weak, and is similar to Equation 9.26a.) This is an exponential spiral that converges on $\tilde{\varepsilon}_o$ (the focus) when the layer becomes optically thick. Two properties of this spiral were used in early attempts to control growth, which do not require knowledge of the substructure (as described in Aspnes, 1993a,b): (1) Any line drawn between any point on this spiral and the focus makes a constant angle with the tangent at that point. (2) If the growth occurs at a fixed rate, the length of this line is proportional to the spacing between data points. $\tilde{\varepsilon}_o$ of a film can then be determined during growth, even for optically thin layers, $|2k_{oz}d| \ll 1$, by using this exponential spiral approach or, equivalently, by solving for $\tilde{\varepsilon}_o$ in Equation 9.78:

$$\tilde{\varepsilon}_o \approx \langle \tilde{\varepsilon}(0) \rangle + [\langle \tilde{\varepsilon}(0) \rangle - \langle \tilde{\varepsilon}(d) \rangle]/(2ik_{oz}d) \qquad (9.79)$$

This relation gives a running-value measurement of the overlayer dielectric function (Aspnes et al., 1990b), assuming a small optical path, and is similar to Equation 9.26b.

Although the ESA is simple, it is not a very good approximation and is best utilized when the optical properties of the overlayer and substrate are similar. A superior derivative-like method is based on the linearized virtual-interface (VI) equations.

The VI model determines the optical properties of the uppermost part of the film structure, such as the most recently deposited part—with arbitrary composition and thickness d—by using the reflectance from the structure before the growth of this last part of the film (Aspnes, 1993a,b, 1994, 1995b, also private communication, 1995). d is a parameter that is set to analyze the film periodically. The optical contribution of the entire underlying region of the sample, the virtual substrate, is represented by a single virtual reflectance at a virtual interface, which is illustrated in Figure 19.1. Other details about this underlying region are not needed. This virtual interface remains at the same distance below the surface as processing, such as deposition, continues. VI approaches now dominate the real-time analysis and control of film processes using ellipsometry (Aspnes, 1995b, also private communication, 1995).

Linearized virtual interface theory uses the value and derivative of the ellipsometric data, as does the ESA; however, it is more exact because it uses the linearized form of the exact Fresnel relations. Because the virtual reflectances are generally different for s and p polarizations, an assumption is often needed to enable analysis of the ellipsometric data. As described by Aspnes (1993b) and used by Aspnes *et al.* (1992a), in the virtual-substrate approximation (VSA) it is assumed that the s and p reflections come from the same virtual substrate. [This assumption was initially called the common-pseudosubstrate approximation (CPA) (Aspnes, 1993b)]. With linearized VI theory and the VSA assumption, results are often accurate to 1 part in 10^3 (which is much better than that using the ESA), the speed of analysis is fast enough for closed-loop feedback control, and the analysis is mathematically stable (in contrast to the conventional Fresnel approach). One weakness of this linearized approach is that the deposition rate must be known in order to determine the near-surface dielectric function. If knowledge of the dielectric function is assumed, the VSA can be used to determine the growth rate.

Urban and Tabet (1993, 1994) have curve-fit the exact Fresnel equations to obtain an exact virtual interface theory that provides the two virtual reflectances, the thickness and dielectric function. *A priori* knowledge of the deposition rate is not needed with this approach. Because analysis is slower with this model than with linearized theory, Urban and Tabet used neural network techniques (Section 19.2.2) to accelerate the data analysis, to enable real-time analysis in some cases.

Figure 9.75 shows the three spirals formed in the successive growth of $Al_xGa_{1-x}As$ layers with $x = 0.07, 0.19$, and 0.36 on a GaAs substrate. Time is an implicit variable along this locus. Plotted in this manner, the reflection coefficient evolves in a similar way. The reflectance oscillations seen in

9.11 Ellipsometry

interferometry (Section 9.5) are one-dimensional projections of these spirals plotted versus time (Aspnes, 1994).

Equation 9.31' assumes isotropic media and it must be modified for anisotropic media. For example, for the rotating-analyzer system with polarizer angle P (to be discussed in the next section) the apparent ρ is (Hingerl et al., 1993)

$$\langle \rho \rangle = \frac{r_{pp} + r_{sp} \tan P}{r_{ss} + r_{ps} \cot P} \tag{9.80}$$

The reflection coefficients from a very thin anisotropic layer on an isotropic substrate are given by Equations 9.21–9.23. Parikh and Allara (1994) have discussed the effects of optical anisotropy on SE of thin films and surfaces.

9.11.2 *Instrumentation*

Figures 9.76–9.78 illustrate the basic elements of the most common types of ellipsometers. The incident light source passes through a polarizer (P, such as a Rochon polarizer) and sometimes another optical element (X), such as a compensator (C) that converts this linearly polarized light to elliptically polarized light, before it hits the sample (S). The reflected beam passes through an analyzer (A, such as a Glan–Taylor polarizer) to determine the state of polarization before measurement by the detector. The analyzer is sometimes preceded (Y) or followed (Z) by another optical element. The ellipsometer arrangement is denoted by these elements. For example, when a polarizer and compensator are before the sample and an analyzer is after it, the arrangement is labeled PCSA. If an optical element automatically rotates, its designation is subscripted, as in P_{rot}. The azimuthal angles for the polarizer, analyzer, and compensator (often designated $\theta_{p,a,c}$) are measured in a positive sense counterclockwise from the plane of incidence when looking into the beam.

In *null ellipsometry,* the azimuthal settings of the optical elements (P, C, A), the phase retardation of the compensator (δ_c), and the angle of incidence are set to reduce the detected signal to zero. In *photometric ellipsometry,* the light flux through the system is monitored as these optical parameters are varied. These readings are performed for several fixed settings of these optical parameters in a *static photometric ellipsometer,* while in a *dynamic photometric ellipsometer* one or more of the parameters P, C, A, and δ_c is periodically varied and the signal is analyzed by Fourier analysis. Dynamic photometric ellipsometers are typically used for real-time diagnostics.

In single-wavelength ellipsometry, a laser, such as a He–Ne laser at 632.8 nm, or filtered white light is used as the light source. In spectroscopic ellipsometry, a Xe arc lamp that is dispersed by a spectrometer either immediately after the lamp or just before the detector is used as the light

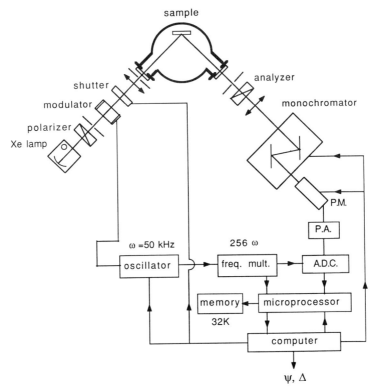

Figure 9.76 Schematic of a polarization-modulated spectroscopic ellipsometer (from Drévillon *et al.*, 1982, as modified by Collins, 1990).

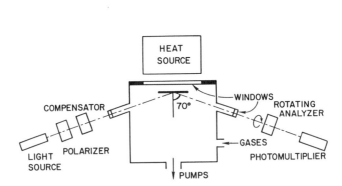

Figure 9.77 Schematic of an automatic rotating-analyzer ellipsometer mounted on a rapid-thermal processing chamber for *in situ* analysis (from Sampson *et al.*, 1994b).

9.11 Ellipsometry

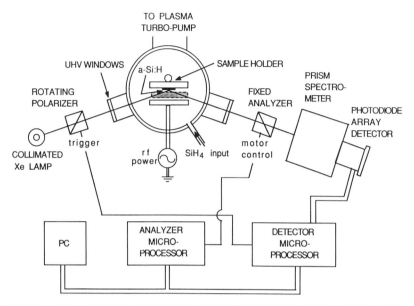

Figure 9.78 Schematic of a rotating-polarizer ellipsometer, with multichannel detection for spectroscopic analysis, installed on a reactor used for plasma-enhanced CVD of amorphous silicon and used for *in situ* analysis (from An *et al.*, 1992).

source. In either case, the light is collimated before passing through the optical elements. Grating and prism spectrometers can be used. The latter has poorer spectral resolution but no interference from multiple orders; the overlapping of orders from a diffraction grating can be a serious problem for wide wavelength scans. In early instruments, a rotating interference filter was used instead of a spectrometer, which gave relatively poor spectral resolution. Photomultipliers are typically used as the detector (Figures 9.76 and 9.77). Analysis of SE data can be accelerated by using multichannel detectors, such as photodiode arrays (PDAs) and charge-coupled device (CCD) arrays, to detect light dispersed by monochromators at the exit of the ellipsometer (Figure 9.78).

In the classical null ellipsometer (PCSA), the compensator azimuthal angle is set at $\pm 45°$ and the polarizer is manually rotated so that the elliptically polarized incident beam is linearly polarized upon reflection. This is ascertained by nulling the transmission through the analyzer. The azimuths of the incident and reflected beams determine ρ. While it is very accurate, this slow manual method has been supplanted by three types of automatic (dynamic) ellipsometers: self-compensating (or nulling), polarization-modulated (or phase-modulated), and rotating-element ellipsometers. Although each of these "automated" ellipsometers is capable of the same precision, each has distinct advantages for use in SWE and SE.

The automatic self-compensating ellipsometer has a PXSYA arrangement where the polarizer and analyzer are fixed and X and Y are Faraday or Pockels cells that null the signal. SWE measurements can be made in ~1 msec and SE scans in ~3 sec. Nulling ellipsometers are the most accurate of the ellipsometer configurations. Still, the automatic self-compensating ellipsometer is rarely used for real-time analysis.

The optical arrangement in a polarization-modulated ellipsometer (PME) is PXSA, where X is a birefringent phase modulator, such as a Pockels cell or piezobirefringent plate, that dynamically elliptically polarizes the light incident on the sample (Figure 9.76) (Drévillon, 1993). Since the retardance of this element can be varied at 100 kHz, the measurement time can be as fast as 10 μsec when the detected light is analyzed by phase-sensitive detection or Fourier analysis. Fourier analysis is performed after fast analog-to-digital conversion (Drévillon *et al.*, 1982). Higher precision is obtained with longer measurement times (~5 msec). The configuration of Drévillon *et al.* (1982) permits evaluation of two associated ellipsometric parameters in a single scan. That of Jellison and Modine (1990) uses a Wollaston prism as the analyzer to split the reflected beam into two orthogonally polarized beams and thereby determine all three parameters, N_e, S_e, and C_e, in a single scan. Therefore, Ψ and Δ can be obtained with good sensitivity and no ambiguity in the sign. PME is suited for high-speed, real-time spectroscopic ellipsometry because it is the fastest automatic ellipsometric technique; however, it is not fully achromatic and requires extra calibration. This method is also called phase-modulated ellipsometry.

In the rotating-element ellipsometer at least one of the optical elements rotates, typically at 50–100 Hz (Collins, 1990). This is the most common automatic ellipsometer due to its relative simplicity. The most common rotating model is the rotating analyzer ellipsometer (RAE, PSA_{rot}) introduced by Aspnes and Studna (1975) (Figure 9.77). Other typical optical arrangements for real-time spectroscopic use include the common $P_{rot}SA$ (rotating-polarizer ellipsometer, RPE; Figure 9.78), and also $PSA_{rot}A$ and $PP_{rot}SA$; at least two of the three normalized Stokes parameters can be obtained with these instruments. The RAE system has been adapted by Bu-Abbud *et al.* (1986) to permit variable-angle spectroscopic ellipsometry (VASE). With the relatively complex $PSC_{rot}A_{rot}$ arrangement, all three normalized Stokes parameters can be determined. The modulated signal is processed by Fourier analysis, and the minimum time resolution is ~5 msec. Since these instruments are relatively slow, real-time SE measurements have been limited to slow processes, such as depositions with < 10 Å/sec growth rates.

RAE (PSA_{rot}) and RPE ($P_{rot}SA$) systems have advantages in the spectroscopic mode because they are achromatic when they are used without a compensator; however, they have disadvantages. For example, without a compensator, RAE determines only N_e and C_e, as can be seen below from

Equations 9.81–9.87. Since $C_e \propto \cos \Delta$, an RAE cannot determine the sign of Δ and is insensitive to Δ when Δ is near 0 or 180°. This is not an issue for continuous real-time analysis of etching or deposition from a known initial condition, but is a concern for analysis of a film when no additional information is available (particularly when the extinction coefficient k of the material is small). Moreover, RAE is most precise when circularly polarized light is incident on the rotating analyzer (Aspnes, 1974), which means that for best results the reflection-induced phase shift needs to be near 90° if no compensator is used. (This makes it relatively insensitive for small k, such as near fundamental band gaps.) These weaknesses disappear if a compensator is used, but problems with achromaticity then become significant. In *ex situ* RAE measurements, Ψ is typically measured without a compensator and Δ is measured both with and without a compensator.

Rotating-compensator systems, such as $PSC_{rot}A$, are the best rotating-element ellipsometers for analysis at a single wavelength because they can determine Δ better than rotating polarizer and analyzer systems, especially when $|\cos \Delta| \approx 1$, and because the handedness of the reflected light, given by the sign of Δ, is certain. However, extensive wavelength corrections are needed when used in the SE mode.

Instrumental polarization effects decide the relative merits of the rotating analyzer and polarizer instruments. In SE the monochromator is generally placed after the ellipsometer optical elements (and before the detector) to eliminate background light from the plasma, etc. The monochromator must be placed before the system for RAE, because the monochromator throughput depends on polarization, but may be placed afterwards for RPE; therefore rotating-polarizer systems are preferred when any residual light in the reactor must be filtered out. Rotating-analyzer systems are preferred when reactor light is not important. These systems are immune to background light because the DC term is not needed for data analysis, as is seen below in Equation 9.81.

While visible/near ultraviolet SE (~ 0.2–1 μm) is sensitive to the different electronic structure in materials, infrared spectroscopic ellipsometry (~ 2–16 μm) is sensitive to differences in vibrational structure. This is important in identifying hydrocarbonated molecules in lithography resists and silicon oxides, nitrides, and oxynitrides. A glowbar is used as the light source, and germanium plates at Brewster's angle or etched metallic grids on BaF_2 are used as polarizers and analyzers. These instruments use a grating spectrometer or, more recently, an FTIR spectrometer for spectral analysis (Ferrieu, 1989; Zalczer *et al.*, 1993; Röseler, 1993; Drévillon, 1993).

Data analysis in an automatic SE is illustrated by considering the rotating-analyzer configuration $PCSA_{rot}$. With the analyzer rotating at a rate ω, the detected signal is

$$S(t) = \eta_d I_0[1 + \alpha_e \cos(2\omega t) + \beta_e \sin(2\omega t)] \tag{9.81}$$

where I_0 is the average intensity, α_e and β_e are the normalized Fourier coefficients, η_d is the detector efficiency, and $t = 0$ has been specified relative to a reference phase determined in calibration.

This can be related to the polarization states by

$$\chi_r = [\beta_e \pm i(1 - \alpha_e^2 - \beta_e^2)^{1/2}]/(1 + \alpha_e) \tag{9.82}$$

$$\chi_i = [\tan C' + \rho_c \tan(P' - C')]/[1 - \rho_c \tan C' \tan(P' - C')] \tag{9.83}$$

where $\rho_c = T_c \exp(-i\delta_c)$ is the relative amplitude transmittance of the compensator, and P' and C' are reference azimuths for the polarizer and compensator.

With an ideal compensator $T_c = 1$ and $C' = 0$, the ellipsometry angles are given by

$$\tan \Psi = [(1 + \alpha_e)/(1 - \alpha_e)]^{1/2} |\tan P'| \tag{9.84}$$

$$\cos(\Delta + \delta_c) = \beta_e/(1 - \alpha_e^2)^{1/2} \tag{9.85}$$

where $0° \leq \Psi \leq 90°$ and $-180° < \Delta \leq 180°$. With $P' = 90°$ and the retardation of the compensator $\delta_c = 0°$,

$$\Psi = \cos^{-1}\left(\sqrt{(1 - \alpha_e)/2}\right) = \tfrac{1}{2}\cos^{-1}(-\alpha_e) \tag{9.86}$$

$$\Delta = \cos^{-1}\left(\frac{\beta_e}{\sqrt{1 - \alpha_e^2}}\right) \tag{9.87}$$

Details about data analysis can be found in the references cited in this chapter, including Aspnes and Chang (1989), Collins (1990), and Jellison (1993a). Fourier analysis is used to analyze data collected in single-wavelength ellipsometry. When a multichannel detector is used, the Fourier components are determined by integrating the signal four times a cycle, then processing the data from the four quadrants with the Hadamard summation (Kim *et al.*, 1990; An and Collins, 1991). Details on the rapid data acquisition analysis for real-time monitoring may be found in Kim *et al.* (1990), An and Collins (1991), An *et al.* (1992), Johs *et al.* (1993), and Piel *et al.* (1993). In simpler problems, such as analysis of two homogeneous phases, the model equations (Equations 9.31', 9.72, and 9.73) are solved directly to obtain the dielectric functions. For more complicated problems, numerical analysis is required, and a least-squares regression analysis (Aspnes, 1981) is used to determine the model parameters that best fit the data. The above-mentioned virtual interface algorithm can also be used. In general, numerical analysis is needed to solve multiphase models.

The fit of the model (ρ_{calc}) with the spectroscopic ellipsometry data (ρ_{exp}) is judged by using a figure-of-merit estimator of the form

9.11 Ellipsometry

$$\chi_{fit}^2 = \frac{1}{n-m-1} \sum_{i=1,n} \frac{|\rho_{exp}(\lambda_i) - \rho_{calc}(\lambda_i, \mathbf{z})|^2}{X} \quad (9.88)$$

where measurements are made at n wavelengths and the vector $\mathbf{z}$ has m components representing model parameters, such as film thicknesses, constituent fractions, etc. An unbiased estimator is generally used, with $X = 1$. Jellison (1991) has shown that using a biased estimator, with $X = |\delta\rho(\lambda_i)|^2$ where the error in each data point is $\delta\rho(\lambda_i)$, weights the fit more accurately. Very closely related to the unbiased estimator is regression analysis performed to minimize the mean square (MSE) function, defined as

$$\text{MSE} = \frac{1}{N} \sum_{i,j} \{[\Psi_{exp}(\lambda_i, \theta_j) - \Psi_{calc}(\lambda_i, \theta_j)]^2 + [\Delta_{exp}(\lambda_i, \theta_j) - \Delta_{calc}(\lambda_i, \theta_j)]^2\} \quad (9.89)$$

where there are N data points at different λ and incident angles θ.

Careful calibration and correction for errors, along with physically reasonable modeling, are essential to obtain meaningful results from ellipsometry (Azzam and Bashara, 1977; Collins, 1990). The Aspnes residual calibration procedure is commonly used for rotating-element ellipsometers (Aspnes and Studna, 1975; Johs, 1993). Correction for polarization changes due to windows is essential in real-time applications in vacuum chambers. Errors in Ψ and Δ of less than 0.01° are possible with proper calibration (Jellison, 1993a), but 0.01–0.02° is more typical for a well aligned ellipsometer and quite optimistic for a real-time monitoring ellipsometer (see below). The potential for submonolayer accuracy is illustrated by comparing this measurement accuracy with the ellipsometric sensitivity to an overlayer with $n = 1.5$ on Si, such as SiO_2 (632.8 nm, $\theta = 70°$). Calculations show that Δ decreases by ~0.3° per 0.1 nm increase in overlayer thickness for ultrathin films (Irene, 1993).

The typical lateral spatial resolution in ellipsometry is ~0.1–1 mm. Recent studies have tried to improve this resolution either by focusing the beam to ~10 μm (Erman and Theeten, 1986; Barsukov et al., 1988) or by imaging the surface (~2 μm resolution) (Stiblert and Sandström, 1983; Beaglehole, 1988; Cohn et al., 1988).

Several types of automatic spectroscopic ellipsometers are available commercially, some of which are easily adaptable for use in real time. Spectrally they typically range from ~0.25 to 1.0 μm, but may extend to ~1.7 μm. Infrared versions (2–16 μm) are also available. SEs designed for on-line use from ~0.3 to 1.0 μm employ silicon photodiode arrays to obtain a 128-point spectrum in ~0.02–0.1 sec. Software packages for data analysis are readily available from vendors, although they must be used with careful examination of the problem at hand. Analysis can be improved by optimizing the angle of incidence and by using several angles of incidence; however, this latter method is less practical for real-time analysis. Typical angles of incidence are ~60–75°, usually near Brewster's angle. In process development studies, samples should be mounted on *xyz* translation stages in the chamber.

The implementation of ellipsometry in a film chamber for accurate *in situ* monitoring requires care. Maracas *et al.* (1992, 1995b) have described how to install SE onto an MBE chamber. Strain-free windows must be used, as they must for all applications of ellipsometry *in situ* in a vacuum chamber. The actual angle of incidence must be known to at least 0.1°, or else there can be large errors in the determined thickness, especially for buried layers; also, the plane of incidence must be normal to the substrate. Both of these conditions must also hold when the wafer is being rotated, since the goal of 1% thickness uniformity across the wafer can be attained only with wafer rotation. The wafer must be mounted rigidly enough for optical analysis, but not so rigidly that slip planes are formed that destroy the layer. The manipulator, upon which the sample is mounted, must be very stable, especially to temperature changes.

The typical time scales for data acquisition and analysis for real-time SE control of most thin film processes is on the order of a second. However, time-resolved ellipsometry, usually using a single wavelength and with less accuracy, can be conducted on times scales of nanoseconds and shorter to examine fast processes, such as those utilizing pulsed excimer lasers (Jellison and Lowndes, 1985a,b; Jellison, 1993a).

9.11.3 *Comparison with Other Reflection Methods*

Spectroscopic ellipsometry is the most general of reflection spectroscopies used for diagnostics. [Spectroscopic polarimetry is more general, but seldom used (Section 9.2).] It can be used to determine any information obtainable from any other reflection method. For example, interferometric effects are easily obtained from SE. The main contribution to ellipsometry and surface photoabsorption comes from the component of the dielectric tensor that lies in the surface and the plane of incidence. SE data obtained at different angles of incidence also contain information obtainable from reflection-difference spectroscopy (Hingerl *et al.,* 1993; Wentink *et al.,* 1992). Although a variable-angle spectroscopic ellipsometer can be adapted to conduct any reflection experiment (RDS, SPA), the instruments developed for these other spectroscopies are simpler and more readily applied than are ellipsometers.

The spectroscopic reflection interferometry methods described in Section 9.5 provide less information than SE, and particularly less information about interfacial grading. While SE is often the preferred method, such reflectance probes sometimes supply sufficient information about layer compositions and thicknesses in multilayer systems, without the experimental complexities and added cost of SE. In general, the accuracy of optical reflectometry experiments is usually >1%, while the accuracy of ellipsometric measurements can be <1%.

9.11.4 Applications in Real-Time Analysis

Automated ellipsometry is being used increasingly as a real-time probe during a wide range of thin film processes. Overviews have been written by Aspnes and Chang (1989) for plasma-assisted etching, Herman and Sitter (1989) and Maracas *et al.* (1992) for molecular beam epitaxy, Pickering (1994, 1995) for epitaxy by CVD and MBE, and Irene (1993) for oxidation and related processes. Historically, single-wavelength ellipsometry has been used to monitor fast process (<1 sec), while spectroscopic ellipsometry has been used for *in situ* post-processing analysis and to monitor slow processes. Submonolayer real-time characterization of the surface region is desired on a time scale that is short compared to the time needed to grow a monolayer (~1 sec) (Aspnes, 1993c, 1994, Pickering, 1994, 1995).

With developments in rapid acquisition and analysis (An *et al.*, 1992) for SE analysis of film thicknesses and, for deposition, composition, SE can be used for real-time (kinetic) monitoring of even fast (<1 sec) processes. Improvements in speed and cost have made ellipsometry a strong candidate for real-time monitoring and control during manufacturing, as well as a valuable tool in process development (Aspnes, 1993c, 1994; Pickering, 1994). Such real-time monitoring is also called kinetic ellipsometry.

Spectroscopic ellipsometry increases the power of such measurements, particularly in determining the composition of alloys, multilayer thicknesses (including buried layers; Figure 9.73), crystallinity, porosity, and medium inhomogeneity. However, it also increases the analysis time, so there is a tradeoff between the number of wavelengths examined (and the spectral resolution) and analysis time. Sometimes, monitoring is needed at only a few wavelengths, so data interpretation for real-time control can be faster than if a full spectral analysis were necessary. SE capability permits an optimal choice of these wavelengths on the basis of layer thicknesses and the dielectric function critical points of the overlayer, films and substrate, and temperature changes. Wavelengths (or wavelength for SWE) can be chosen to enable a measurement particularly sensitive to some of these parameters and insensitive to others. For example, the dielectric functions plotted in Figure 9.6 illustrate that for some photon energies, ε_1 and ε_2 (ε' and ε'') are very sensitive to temperature, making them good choices for monitoring temperature changes; but for others ε_1 and ε_2 change little with temperature, making them useful for measurements that are sensitive to properties other than temperature (Irene, 1993). The choice of wavelength determines the sampling depth, which is due to absorption. The compilation of dielectric functions for materials of interest as a function of wavelength and temperature has been crucial to the processing applications of SE (Table 9.1).

Ellipsometry has been used extensively to monitor the growth of thin films of the older, more established deposition methods of sputtering and evaporation. In one very early use of real-time ellipsometry, Yamaguchi et al. (1969) monitored optical constants and thickness during the deposition of Ag films by evaporation (546.1 nm, Hg spectral line). More recently, in situ ellipsometry has been extensively used to monitor epitaxial growth by MOCVD and MBE. In one such early application, Theeten et al. (1979) followed the VPE growth of $Ga_{1-x}Al_xAs$ layers and the GaAlAs/GaAs interfacial regions in real time by using single-wavelength ellipsometry (632.8 nm). While single-wavelength probing is still being used in some applications, spectroscopic ellipsometry is fast becoming a standard technique for real-time probing, especially during the epitaxial growth of compound semiconductors. Maracas et al. (1992, 1995b) have discussed design factors in adapting a spectroscopic ellipsometer to an MBE chamber, as well as the application of SE to thermometry and in determining the growth rate, film thickness, and alloy concentration in thick films and quantum-well structures.

The growth of silicon films has been monitored by ellipsometry, often with the aim of investigating the nucleation step of growth. In one early study, Hottier and Theeten (1980) and Hottier and Cadoret (1982) monitored low-pressure CVD of Si on Si_3N_4 by using real-time ellipsometry and observed that the Δ–Ψ trajectory calculated assuming three-dimensional growth differs from that observed (Figure 9.79), particularly during the early stages of deposition, because of nucleation and coalescence. The trajectory follows the calculation when the nucleation step is included; this is also seen in the below-cited studies by Drévillon and co-workers and Collins and co-workers.

Drévillon et al. (1982) monitored the deposition of hydrogenated amorphous silicon (a-Si:H) in a low-pressure silane glow discharge by high-speed polarization-modulated ellipsometry. A single wavelength was used, 350 nm, corresponding to the absorption peak of the amorphous silicon film (from Figure 9.74b). The growth was on a transparent fused-silica substrate ($\Delta = 360°$), with a ~12 Å/sec growth rate. Data were taken every 5 msec during this run of ~4 sec.

Collins and Cavese (1987) compared the nucleation and growth of a-Si:H films grown by a glow discharge (SiH_4 or SiH_4/H_2) and by sputtering in Ar/H_2 mixtures. In the first 50 Å of the glow-discharge growth, the amorphous silicon had bulk density which led to subsequent layer-by-layer growth. In contrast, the first 200–400 Å of the sputtered film was found to have a density much below that of the bulk. An et al. (1990, 1991) used real-time ellipsometry to determine the dielectric functions of magnetron-sputtered a-Si, remote plasma-enhanced CVD of hydrogenated a-Si, and a-Si chemically modified by atomic H exposure.

9.11 Ellipsometry

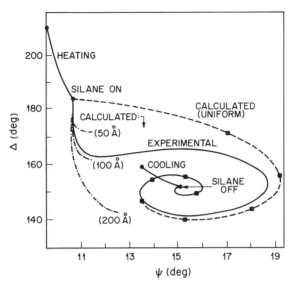

Figure 9.79 *In situ* ellipsometric monitoring of the chemical vapor deposition of amorphous Si on Si$_3$N$_4$, compared to predictions of deposition models assuming nucleation or two-dimensional (uniform) growth (from Hottier and Theeten, 1980, as presented by Donnelly, 1989).

The time evolution of the bonded-hydrogen content was estimated, along with that of the bandgap, in this last process, which has been studied further by An *et al.* (1993). As seen in Figure 9.80, Collins (1990) was able to monitor the growth of amorphous Si on Mo by hemispherical nucleation with nuclei separated by 40 Å. This group has also reviewed real-time spectroscopic ellipsometry during the growth of tetrahedrally bonded materials (Collins *et al.*, 1994) and aluminum (Nguyen *et al.*, 1994), including their own research.

Collins *et al.* (1993) have reviewed the use of a multichannel rotating-polarizer spectroscopic ellipsometer for real-time studies of the nucleation and growth of Al, a-Si:H, and a-SiC:H (Figure 9.78). This spectrometer can acquire ~50 point spectra from 1.3 to 4.3 eV in 16 msec, and has a precision of 0.04° for Ψ and 0.12° for Δ for a Si surface at 2.5 eV. Li *et al.* (1992) have detailed this application to the thin film coalescence during a-Si:H growth. Other applications of this spectrometer have been reviewed by An *et al.* (1992). Figure 9.81 shows the multichannel acquisition of the pseudodielectric function during the growth of a-Si:H by PECVD.

Drévillon and co-workers (Kumar *et al.*, 1986; Drévillon *et al.*, 1987; Layadi *et al.*, 1993) have monitored the nucleation and growth of microcrystalline and amorphous Si by using (visible/ultraviolet) SE, while Blayo and

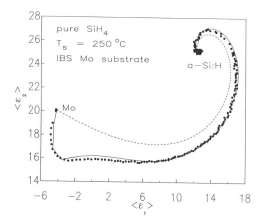

Figure 9.80 Real-time ellipsometric monitoring of the growth of hydrogenated amorphous silicon on Mo by plasma-enhanced CVD at 3.4 eV, expressed in terms of the pseudodielectric function (circles). This is compared to simulations assuming layer-by-layer growth (dashed curve) and hemispherical nucleation with nuclei initially on a 40-Å square grid (continuous curve) (from Collins, 1990).

Drévillon (1990) used infrared (phase-modulated) ellipsometry (IRPME) to monitor the interfacial region formed during the growth of a-Si:H and a-SiC:H on glass. IRPME allows direct identification of chemical species incorporated into the film. The interaction of the film with the substrate is indicated by the presence of the BO vibration group near 1480 cm^{-1}. Ossikovski et al. (1993) extended this work to the *in situ* study of a-Si:H/a-SiN$_x$ and p-doped a-Si:H/intrinsic a-Si:H interfaces.

Collins (1989) monitored the growth, surface modification, and etching of amorphous carbon films by ellipsometry. In later work by the same group, Hayashi et al. (1992) monitored the growth of diamond films in a microwave plasma-enhanced CVD reactor (5–10% CO/H$_2$) with near-infrared ellipsometry (1.55 μm). This growth occurs by the formation and subsequent coalescence of nuclei. Larger grains and nuclei can be studied with this longer wavelength than with the usual 200–800-nm range in visible/near-ultraviolet ellipsometry. Figure 9.82 illustrates the model of film growth, while Figure 9.83a plots the expected Δ–Ψ trajectory for diamond growth on Si for different nucleation densities. Figure 9.83b shows the experimental trajectory for wet-cleaned substrates. The nucleation densities were determined to be 2×10^9/cm^2 and 3×10^{10}/cm^2 for wet- and dry-cleaned substrates, respectively.

With a fixed flow of reactants, the composition of a deposited alloy film can change during deposition because the nature of the surface changes, and so real-time monitoring is required to control alloy composition. In a

9.11 Ellipsometry

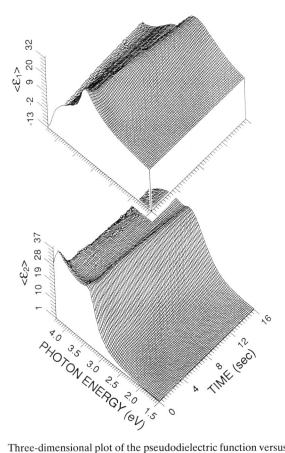

Figure 9.81 Three-dimensional plot of the pseudodielectric function versus photon energy and time acquired during the growth of a-Si : H by plasma enhanced CVD (from An *et al.*, 1992).

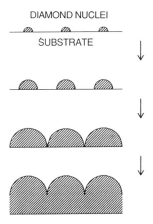

Figure 9.82 Thin film nucleation model for the CVD of diamond (from Hayashi *et al.*, 1992).

a

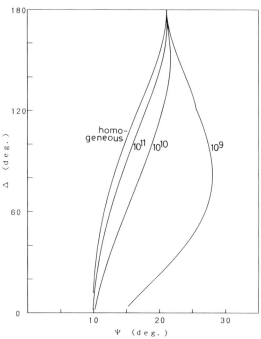

b

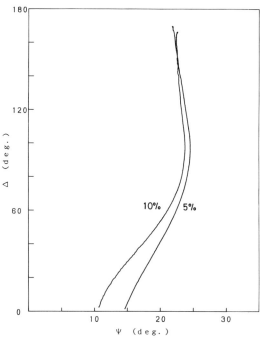

9.11 Ellipsometry

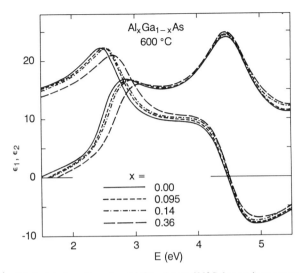

Figure 9.84 Dielectric functions for $Al_xGa_{1-x}As$ at 600°C for various compositions (from Aspnes et al., 1990b).

series of investigations, Aspnes and co-workers have devised and demonstrated SE methods to control the epitaxial growth of AlGaAs alloys by metalorganic molecular beam epitaxy (MOMBE). Aspnes *et al.* (1990a,b) showed how single-wavelength ellipsometry with a rotating polarizer can be used to determine the composition near the surface of $Al_xGa_{1-x}As$ films on GaAs in real time during MOMBE and to control film composition to a specific value of x. Measurements were made with photon energies near 2.6 eV (~4750 Å) because ε'' is very sensitive to the alloy fraction x in $Al_xGa_{1-x}As$ at the 600°C growth temperature, as is seen in Figure 9.84. The flow of the triethylaluminum [$Al(C_2H_5)_3$] was regulated to maintain the target value of x, while the flows of the triethylgallium [$Ga(C_2H_5)_3$] and cracked arsine (AsH_3) sources were kept constant. Figure 9.85 plots the different trajectories of $\langle \varepsilon'' \rangle$ versus $\langle \varepsilon' \rangle$ for regulated and unregulated flow, with time (or film thickness) as the running variable. x was maintained to a precision of ±0.001 when the flow was regulated. Figure 9.86 shows

---

Figure 9.83 (a) Simulated trajectory of the ellipsometry parameters during the nucleation and growth of diamond on silicon (at 1.55 μm) assuming homogeneous growth or nucleation with the indicated nuclei density (in cm^{-2}). (b) Experimental trajectory obtained in real time during diamond film growth by microwave plasma-enhanced CVD, with the indicated CO/H_2 ratios in the gas mixture (from Hayashi et al., 1992).

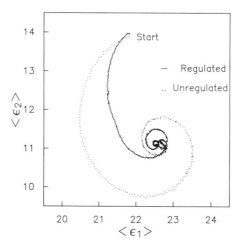

Figure 9.85 Pseudodielectric function during regulated (closed-loop control) and unregulated growth of AlGaAs by MOMBE (from Aspnes et al., 1990b).

the trajectory during continued regulated and unregulated growth. (Note the finer scales in this figure.) Aspnes et al. (1992a) deposited $Al_xGa_{1-x}As$ parabolic wells by using closed-loop ellipsometric control. The target values for x were followed within 0.02, as is seen in Figure 9.87. These control studies used linearized virtual interface (VI) theory with the virtual substrate approximation (VSA), as is described in Section 9.11.1 (Aspnes, 1993a,b,c, 1994, 1995a,b, also private communication, 1995), because it minimizes analysis errors and enables analysis that is rapid enough for feedback control. Aspnes et al. (1994a,b) also used VI/VSA theory to investigate the ellipsometry-monitored modification of InP(001) surfaces during the capture of As.

I.-F. Wu et al. (1993) developed a model based on classical Fresnel analysis (which is also described in Section 9.11.1), which they have dubbed the "adaptive multilayer model." They used it to control the growth of SiO_x antireflection coatings on traveling-wave semiconductor laser amplifiers (AlGaAs, InGaAsP) by monitoring the surface in real time with a rotating-analyzer ellipsometer. The refractive index of SiO_x varies from 1.9 to 1.45 as x is adjusted between 1 and 2; x was so tuned by varying the O_2 pressure in the evaporation chamber. The averaged refractive index of the film was controlled within ± 0.01 and amplifier facet reflectances $\sim 10^{-4}$ were obtained reproducibly.

Maracas et al. (1992) have used SE to monitor layer thickness, alloy composition, and wafer temperature during the MBE of AlGaAs and GaAs epitaxial layers (250–1000 nm). In related work, Droopad et al. (1994)

9.11 Ellipsometry

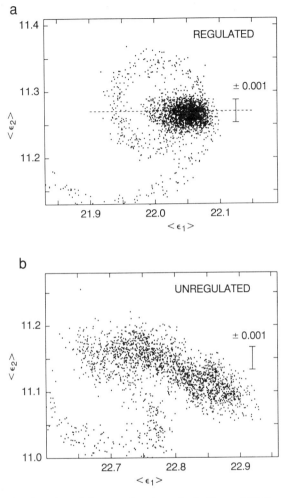

Figure 9.86 Continuation of the growth trajectories in Figure 9.85 (from Aspnes *et al.*, 1990b).

tracked SE data at three wavelengths (380, 465, and 550 nm) to determine growth rates and alloy composition during the MBE of an AlGaAs/GaAs/AlAs structure. Unlike RHEED, this method can be used to monitor the entire growth process, even with wafer rotation, even though wobble increases the noise in the data. Eyink *et al.* (1993a,b) used *in situ* ellipsometry to grow reproducibly low-temperature GaAs (LT-GaAs) layers by MBE by using an ellipsometric signature based on arsenic capping of the surface; thus they were also able to elucidate the process regions of LT-GaAs growth. LT-GaAs is a high-resistivity buffer used for subsequent growth of epitaxial device layers. Celii *et al.* (1995) monitored the growth of pseudo-

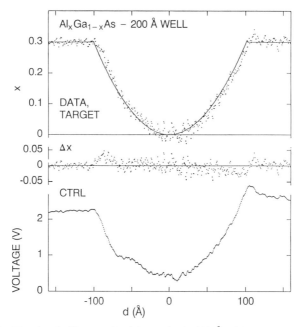

Figure 9.87 The chemical beam epitaxial growth of a 200-Å Al$_x$Ga$_{1-x}$As parabolic quantum well with real-time feedback control of composition. This shows (top) the value of x for the outer running 3.1 Å determined from $\langle \varepsilon'' \rangle$, (center) the difference between the measured and target values of x, and (bottom) the control voltage (from Aspnes et al., 1992a).

morphic AlAs/In$_{0.53}$Ga$_{0.47}$As/InAs resonant tunneling diodes (RTDs) by using *in situ* spectroscopic ellipsometry and compared the structural information obtained by SE with the device characteristics of the RTDs.

Composition control is a serious concern during the growth of mercury cadmium telluride heterostructures. Demay and co-workers used real-time SE to adjust the composition during the MBE growth of HgCdTe layers (Demay et al., 1987a) and examine interdiffusion during the annealing of thin CdTe layers on HgTe (Demay et al., 1987b). Hartley et al. (1992a,b) used real-time SE to control composition, growth rates, interdiffusion, and the formation of surface-related growth defects during the MBE growth of Cd$_{0.2}$Hg$_{0.8}$Te layers and CdTe/HgTe superlattices deposited on CdTe and Cd$_{0.96}$Zn$_{0.04}$Te(100). Figure 9.88 shows increased Hg incorporation with decreasing temperature during the MBE of HgCdTe. Johs et al. (1993) monitored the rate of MOVPE growth of CdTe on GaAs by using SE, by accumulating accurate ellipsometric data at 12 wavelengths in <1 sec. The growth rate and film thickness determined in real time (with feedback control) are shown in Figure 9.89. In this figure, the growth rate was decreased just before 5 minutes.

9.11 Ellipsometry

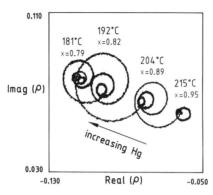

Figure 9.88 Real-time ellipsometric monitoring and control of the MBE growth of $Hg_{1-x}Cd_xTe$ epilayers on CdTe(100) with successively larger Hg fractions (from Hartley et al., 1992a,b).

Early work on monitoring Si epitaxy was performed by Hottier and Cadoret (1983) who monitored Si homoepitaxy at 546.1 nm. Pickering et al. (1987) and Pickering (1991) used dual-wavelength ellipsometry, with 364 and 488 nm from an argon-ion laser, along with laser light scattering (Section 11.2.1), to monitor in situ UHV VPE of Si/GeSi multilayers and related processing steps, such as cleaning. For wafers cleaned with the RCA procedure, the ellipsometric angle Δ was observed to increase after the oxide layer atop silicon was removed by heating because of the removal of this lower index layer. Δ decreased during the prebake and early stages of homoepitaxial growth of Si on Si because of surface roughening, and later returned to the value expected for microscopically smooth epitaxial silicon. No such change was seen with UV ozone cleaning. Figure 11.18 shows a minimal change of Δ and Ψ during optimized oxide removal and homoepitaxial silicon growth. While this dual-wavelength method has more versatility than SWE, it is still limited. In more recent work by this group (Pickering et al., 1993, 1995), Si/GeSi growth has been monitored by SE. Figure 11.24 shows the ellipsometric angles at 364 and 488 nm, along with the intensity of scattered laser light, during the CVD of GeSi and Si epilayers. Oscillations are seen only with 488 nm because of the limited penetration depth at 364 nm. Period-to-period reproducibility in multiple quantum well structures was ±1 monolayer (also see Pickering, 1994).

In addition to these studies of semiconductor deposition, ellipsometry has been used to monitor the deposition of other materials, such as oxide insulators and other insulators. Chang et al. (1983) used in situ ellipsometry to monitor the deposition of films by plasma-enhanced beam deposition. For example, SiO_2 and Al_2O_3 were deposited by reacting beams of Si and

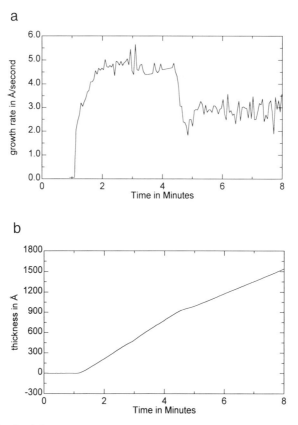

Figure 9.89 Real-time, multiwavelength ellipsometric determination of the (a) growth rate and (b) film thickness of a CdTe layer grown on GaAs by MOVPE, with feedback control based on these measurements. The growth rate was changed at ~4.5 minutes (from Johs et al., 1993).

Al atoms, respectively, with atomic oxygen formed in the plasma discharge. An and Sugimoto (1991, 1992) used single-wavelength ellipsometry (546.1 nm) to follow the MOCVD growth of Ta_2O_5 films from different reactants and to follow the dissolution of these as-deposited and annealed films in HCl and HF solutions. The goal of these studies was to optimize the corrosion characteristics of the Ta_2O_5 films. Similar ellipsometric MOCVD growth and corrosion-resistance studies have been performed by the same group to follow the MOCVD of TiO_2 (Toyoda and Sugimoto, 1990) and Nb_2O_5 (Hara et al., 1994) films. Rivory et al. (1993) used ellipsometry to monitor the evaporative deposition of CaF_2 on Si and on SiO_2 films atop Si. The void fraction was found to decrease with increased thickness, and for growth on Si a CaF_2/Si mixture formed at the interface. Logothetidis et al.

9.11 Ellipsometry

(1995) used SE (1.5 to 5.5 eV) to monitor the growth of TiN$_x$ films by DC reactive magnetron sputtering of Ti in N$_2$. Trolier–McKinstry *et al.* (1994) evaluated work on characterizing ferroelectric films by SE, including *in situ* annealing studies.

Control of run-to-run reproducibility and homogeneity are the two main problems during the growth of tin-doped indium oxide In$_2$O$_3$:Sn (ITO) by reactive DC magnetron sputtering in Ar/O$_2$ atmospheres. Fukarek and Kersten (1994) have used single-wavelength ellipsometry (632.8 nm) to monitor this deposition. Alternative ways to monitor the growth of this transparent conducting oxide optically are by *in situ* monitoring by optical emission (Section 6.3.3) and optical transmittance in the UV (Section 8.3).

Houdy (1988) used single-wavelength ellipsometry (632.8 nm) to control the RF diode sputtering of soft x-ray optics, including multilayer stacks of C/W (200 layers, 40–80 Å periods) and Si/W (100 layers, 20–30 Å periods). Typically in such multilayer, soft x-ray optics, thicknesses have to be controlled better than 0.1 Å and interface roughness has to be <3 Å. Yamamoto and Arai (1993) extended this technique to ultrathin Au and Mo layers and to Au/C multilayers.

Ellipsometry is also proving to be an important monitor of etching. Haverlag *et al.* (1989) used single-wavelength ellipsometry (632.8 nm) to follow RF plasma etching of SiO$_2$ films on Si. (See Figure 9.93 below.) In addition to measuring the etch rate, they established that a ~200 Å layer was present on top of the oxide during etching, probably due to roughening of the SiO$_2$ film, and that at higher pressures a polymer layer forms on the Si after the oxide layer has been completely removed. Kroesen *et al.* (1993) and Zhang *et al.* (1993) monitored the reactive ion etching (RIE) of GeSi alloys in real time using an automatic rotating-compensator ellipsometer operating at a single wavelength (632.8 nm). Using the measured complex index of refraction, the Ge concentration can be determined in real time with this method. Plasma emission did not affect the observations since Ψ and Δ were determined using the second and fourth harmonics of the compensator rotation frequency. Oehrlein (1993) used ellipsometry to follow the SiF$_x$ layer that is formed on the surface during RIE and plasma etching of Si by CF$_4$ and SF$_6$. The thickness of this surface layer varied between ~0.5 and 1.7 nm depending on sheath voltage, and the layer was stable after the discharge was turned off, thus validating postplasma XPS measurements.

Blayo *et al.* (1994) monitored the surface during silicon etching in a helicon high-density plasma by using real-time ellipsometry at 2.9 eV. In both HBr and Cl$_2$ plasmas, Ψ and Δ each rapidly changed to a new steady-state value that depended on RF bias, and then rapidly returned to a value near the pre-etch value when the plasma was turned off. This suggested that there is an overlayer during plasma exposure that is stable while the plasma is on, but desorbs after plasma extinction. N. Layadi, J. T. C. Lee, and V. M. Donnelly (private communication, 1996) have instead attributed this effect to the delay time between extinguishing the source and bias power. In similar kinetic ellipsometry monitoring, Oehrlein (1993) did not see a volatile overlayer during plasma etching with fluorinated etchants, which justified his use of in-line XPS for post-processing analysis of adsorbed surface layers. [Volatile layers were not seen during silicon etching by chlorinated and brominated species in a helical resonator plasma by using laser-induced thermal desorption (see Section 15.1).]

Ellipsometry is also sensitive to the damage to the surface that can occur during processing with charged particles. Aspnes and Chang (1989) have surveyed the use of ellipsometry to monitor damage due to plasmas, and Irene (1993) has reviewed its use to monitor ion beam damage and etching. As an example, Haverlag et al. (1992) monitored the damage to the silicon surface in real time during etching in CF_4 and SF_6 ECR plasmas. Figure 9.90 shows that the damaged, fluorinated layer near the surface is expected to increase Ψ, while the CF_x overlayer is expected to decrease Δ and increase Ψ. Figure 9.91 follows these parameters during etching.

Haverlag and Oehrlein (1992) found that ellipsometry can be used for endpoint detection during etching of patterned surfaces, but only under

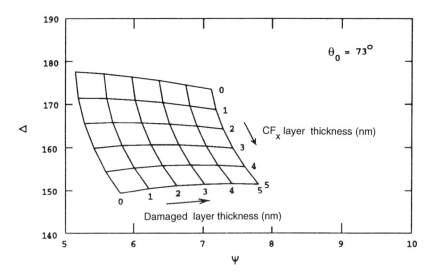

Figure 9.90 Simulated changes in the ellipsometric parameters Δ and Ψ as a function of the damaged silicon thickness and fluorocarbon layer thickness during ECR plasma etching of Si by CF_4 (from Haverlag et al., 1992).

9.11 Ellipsometry

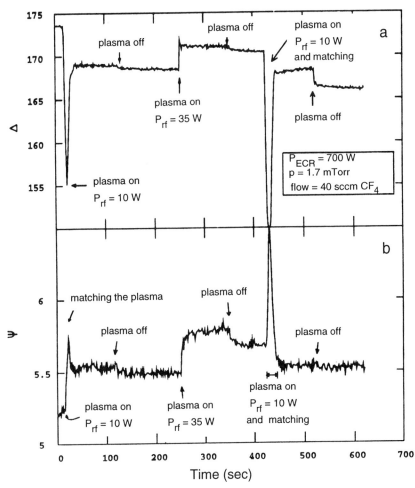

Figure 9.91 Changes in Δ and Ψ during the etching of a Si wafer in an RF biased ECR plasma of CF$_4$ (from Haverlag et al., 1992).

limited conditions. The endpoint of oxide etching was recognized as a sudden change in the direction of the Δ–Ψ trajectory only when the laser was parallel to the mask pattern lines. When it was normal to them, etching of the mask, and not the trench, was probed by the ellipsometer laser. In contrast, the time derivative of the reflected light intensity, in normal incidence reflectometry, was seen to be a good indicator of endpoint for all process conditions. Blayo et al. (1995) also examined the use of ellipsometry to monitor etching of patterned surfaces, and observed significant changes in Δ and Ψ even when the patterned lines were perpendicular to the plane of incidence. These changes, which may be useful for endpoint detection, cannot be explained by geometric models, but seem to be due to the coupling between the beam and the submicron surface features, and are semiquantitatively explained by a polarization scattering loss model. In this investigation, they also noted how underlying topography can affect ellipsometric measurements.

Henck (1992) monitored RIE of a polysilicon film by CF_4, HCl, and HCl/HBr etchants in real time by using a rotating-compensator single-wavelength ellipsometer. The polysilicon film was grown on a silicon wafer covered with silicon dioxide, and the *in situ* measurement of the thickness of this film was reproducible within 2–4 Å. The software and hardware were optimized for fast determination of the film thickness. The birefringent plate compensator rotated once every 0.008 sec, Ψ and Δ were measured every 0.08 sec, and the film thickness was obtained every 0.15 sec. In related work, Henck *et al.* (1992) used rapid SE for real-time thickness measurement and control during plasma etching. The thickness of the layers in a three-layer-stack of silicon nitride/poly-Si/SiO_2 on a Si wafer could be determined rapidly enough for control, in 3 sec for 40 wavelengths. Duncan *et al.* (1994) further detailed rapid real-time SE, using a photodiode array and rapid regression analysis. Stefani and Butler (1994) used real-time SWE ellipsometric measurements conducted at a single point on the wafer during plasma etching of silicon in combination with response surface methodology (Section 19.2.1) to infer plasma etch uniformity across the entire wafer. This group also integrated SE on a reactor for real-time tool control (Maung *et al.,* 1994) and used *in situ* ellipsometry for robust endpointing, fault detection and Run-to-Run control (Butler and Stefani, 1994; Butler *et al.,* 1994) (also see Section 19.3).

Real-time monitoring of the etching of more complex structures, such as multilayer structures with ternary semiconductors and materials with different types of crystallinity, may require spectroscopic ellipsometry. Heyd *et al.* (1991, 1992) monitored the argon ion etching of a ~300 Å Si-doped GaAs/~500 Å $Al_{0.23}Ga_{0.77}As$/0.5 μm undoped GaAs/doped GaAs (substrate) heterostructure with a rotating polarizer configuration. In addition to measuring film thickness and the AlGaAs composition, this analysis monitored the native GaAs oxide, the damaged (amorphized) GaAs or AlGaAs overlayer, and the temperature near the surface. A Si photodiode array provided 80 points from 1.5 to 4.0 eV and data were averaged over 40 polarizer rotations. Within this 3.2-sec acquisition time, ~3.8 Å of GaAs is etched. With ~13 sec for calibration and data analysis, the cycle is repeated every ~16 sec. Nafis *et al.* (1993) used *in situ* (and *ex situ*) SE to monitor ECR plasma etching of bulk Si, GaAs, and InP; SiO_2 layers on Si; and GaAs/AlGaAs and InP/InGaAs heterostructures. Etching-induced damage near the surface was modeled as a crystalline/amorphous semiconductor mixture.

9.11 Ellipsometry

Ellipsometry has also been used to monitor wet etching. In their study of ellipsometric monitoring of aqueous HF etching of Si, Gould and Irene (1988) found evidence for the formation of a fluorocarbon film on the Si following etching. Yao et al. (1993) used in situ SE to investigate the effect of HF cleaning of Si(100) surfaces. Flack et al. (1984) and Manjkow et al. (1987) used ellipsometry to follow the dissolution of resists.

The formation of insulators on semiconductors and conductors, by deposition or surface reactions, is readily monitored by ellipsometry. Several examples of deposition have already been described above, including that by I.-F. Wu et al. (1993), where they used ellipsometry to control the growth of SiO_x antireflection coatings, and the deposition of films by plasma-enhanced beam deposition by Chang et al. (1983). Irene (1993) has reviewed the well-studied use of ellipsometry to monitor the thermal oxidation of Si to form SiO_2 overlayers. More recent work in this area includes the study of ultrathin silicon oxide formation in an RF discharge by Kuroki et al. (1994) and that on rapid thermal oxidation, which is discussed below. Rauh et al. (1993) has monitored the oxidation of thin copper films by using automated spectroscopic ellipsometry.

Theeten et al. (1980) examined the plasma oxidation of GaAs in real time using an instrument capable of spectroscopic measurements and a variable angle of incidence. Kinetic measurements were conducted at a fixed wavelength and a fixed angle of incidence, and more complete ellipsometric analyses of the surface oxide were made with the oxidation processes temporarily interrupted. The formation of a uniform and transparent oxide at a constant rate would be indicated by perfectly periodic behavior. Such behavior was almost observed (at 300 nm), as is seen in Figure 9.92. The peak of tan Ψ decreased in each cycle, which indicated that the film was weakly absorbing; this was attributed to the presence of 2% a-As in the oxide. The slight decrease in cos Δ in later cycles indicated nonuniformity of the thickness of the oxide layer within the probed region. Hu et al. (1991) and Joseph et al. (1992) have used real-time ellipsometry to follow ECR plasma oxidation of silicon, and found that the process could be modeled with a three-phase model with a pure silicon dioxide film/interface layer/silicon wafer.

Such processes are also related to the surface and interface sensitivity of SE. Yakovlev et al. (1992) studied the evolution of the SiO_2/Si interface during high-temperature annealing (750–1100°C) by variable-angle SE with the sample immersed in a transparent liquid ambient having optical properties very similar to the SiO_2 overlayer. This spectroscopic immersion ellipsometry approach increased the sensitivity to the interface properties, and enabled observation of a rapid reduction in interfacial microroughness, which was followed by a slow decay of the interfacial suboxide during annealing.

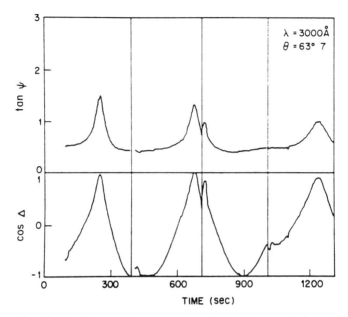

Figure 9.92 Kinetic ellipsometric measurements during plasma oxidation of GaAs(100) surfaces, interrupted at the vertical lines (from Theeten *et al.*, 1980).

Ellipsometric measurements at elevated temperatures are important because they provide optical data needed for thin film processing (Table 9.1), such as deposition, and because surfaces undergo significant changes at high temperatures, even in the absence of reactants. For example, Yao and Snyder (1991) used SE to probe GaAs(100) in real time and observed native oxide desorption at ~577°C and associated surface degradation; desorption of As from an arsenic-capped GaAs at ~350°C produced a clean and smooth surface. [Native oxide desorption on GaAs has also been investigated by Yao *et al.* (1991).] One early suggestion of using ellipsometry to probe wafer temperature was made by Tomita *et al.* (1986).

Kroesen *et al.* (1991) used *in situ* ellipsometry (632.8 nm, $\theta = 74.14°$) to follow Si wafer temperature, with and without an oxide, after an RF plasma had been turned off. Cooling from ~100°C to ambient was compared with measurements by a fluoroptic probe (Section 14.3) connected to the wafer. Oxide thicknesses were chosen so that $\Delta = 0°$ or $360°$ and $\Psi \sim 70°$. In this regime, Δ is very sensitive to oxide thickness, while Ψ is very sensitive to the refractive index of bulk oxide but insensitive to the thickness. Figure 9.93 shows the real-time evolution of the ellipsometric parameters during etching.

Sampson and Massoud (1993) used *in situ* ellipsometry (RAE) (see Figure 9.77) to determine the temperature of a silicon wafer in a rapid

9.11 Ellipsometry

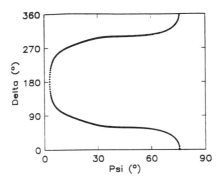

Figure 9.93 Ψ–Δ plot measured during the plasma etching of a SiO$_2$ film on Si with a rotating-analyzer ellipsometer at 632.8 nm. When the film thickness is such that Δ ≈ 0° or 360° (Ψ ~ 70°), Ψ is insensitive to thickness changes, whereas Δ is very sensitive (~1°/0.2 nm); however, Ψ is very sensitive to refractive index changes. With the plasma turned off, either the Δ (thickness) or Ψ (refractive index) change can be used to determine temperature (for Δ ~ 0°) (from Kroesen *et al.*, 1991).

thermal processor by using data from Jellison and Modine (1983). The variation of the ellipsometric angles for a bare Si wafer with temperature is shown in Figure 9.94. Δ and Ψ were determined within 0.01° in this high-

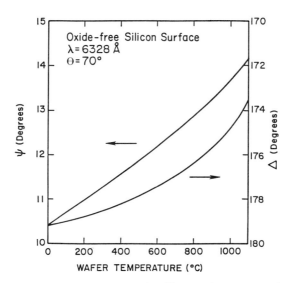

Figure 9.94 Temperature dependence of the ellipsometric parameters for a bare silicon wafer (from Sampson and Massoud, 1993).

resolution ellipsometer (632.8 nm). Figure 9.95 shows the rapid ellipsometric measurement of temperature during rapid thermal heating. Using 413.3 nm (xenon lamp filtered by a monochromator), the maximum error is estimated to be ±1.4°C from 0 to 700°C. At 632.8 nm, the worst-case error is larger, ±10°C. This improvement with shorter wavelength occurs because of the increased dependence of ε'' on temperature as the photon energy increases from 2.0 to 3.0 eV (Figure 9.96); ε' shows less dependence on temperature. Since the index of refraction n monotonically increases with T to 900°C for photon energies up to ~2.8 eV, but not for higher energies, the choice of 442.8 nm (2.8 eV) seems wisest [Sampson *et al.* (1994a)].

In related work, Sampson *et al.* (1993) monitored and controlled rapid thermal oxidation of silicon; both the thickness of the SiO_2 layer and the wafer temperature were monitored. Conrad *et al.* (1993) demonstrated closed-loop control of both silicon wafer temperature and silicon oxide thickness during rapid thermal oxidation (RTO) by using single-wavelength ellipsometry (632.8 nm). Doping of the silicon substrate (type and density) was found to have very little effect on the dielectric constant for photon energies <3.4 eV, and not to affect temperature measurements made by single-wavelength ellipsometry at 632.8 nm (Sampson *et al.*, 1994b; Viña and Cardona, 1984). Surface roughness levels affected the measured ellipsometric parameters at 632.8 nm only for >50 nm peak-to-valley roughness (Sampson *et al.*, 1994b). Since roughness in semiconductor device processing

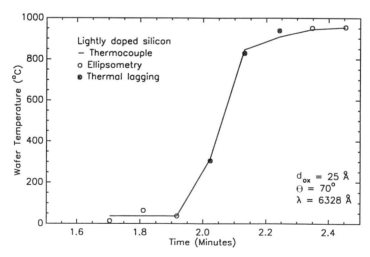

Figure 9.95 Comparison of the temperature of a silicon wafer measured in real time using ellipsometry to that using a thermocouple for a single-step increase in lamp power in a rapid thermal processor (from Sampson et al., 1993).

9.11 Ellipsometry

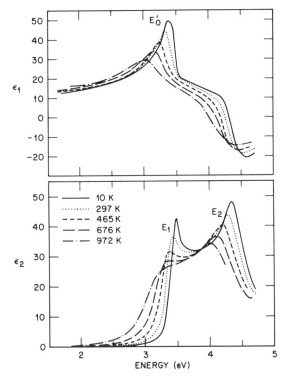

Figure 9.96 Dielectric function of silicon at several temperatures, showing peaks associated with several critical points (from Jellison and Modine, 1983).

is much less than 50 nm, it is not a practical concern in ellipsometric thermometry of silicon.

Maracas *et al.* (1992, 1995b) have addressed the use of SE to determine wafer temperature during the MBE growth of GaAs/AlGaAs heterostructures. In related work, Kuo *et al.* (1994) and Maracas *et al.* (1995a) modeled high-temperature SE data with the Lorentz oscillator model to obtain ε' and ε'' for GaAs. Thermometry in which the emissivity is determined by ellipsometry (RAE) and then used in the pyrometric determination of temperature is described in Section 13.4, where it is termed ellipsometric pyrometry (Hansen *et al.*, 1988, 1989). (Work by Cardona and co-workers on ellipsometric determination of critical-point parameters is cited in Section 18.4.2.)

Because ellipsometry is sensitive to interfacial regions, it can be used to monitor surface preparation procedures, such as cleaning, and to monitor monolayer films on surfaces. Aspnes (1994) has characterized the application of ellipsometry to the probing of adsorbates as the ellipsometric equivalent of surface differential reflectance. For example, Figure 9.97 shows the evolution of the pseudodielectric constant during cleaning of an air-grown natural overlayer on a c-Si wafer, as probed at 4.25 eV (Aspnes and Studna, 1981; Aspnes, 1983; Pickering, 1995). Gautard *et al.* (1985) studied the preparation of InP(100) surfaces by chemical cleaning treatments by using ellipsometry. Other interfacial phenomena can be monitored by ellipsometry. Andrieu and d'Avitaya (1989) analyzed the adsorption of antimony on Si(111) using ellipsometry (0.05 monolayer detection limit) to elucidate processes that are important for depositing *n*-type doped Si by MBE; later, this group made similar studies of Ga adsorption on Si(111) (Andrieu

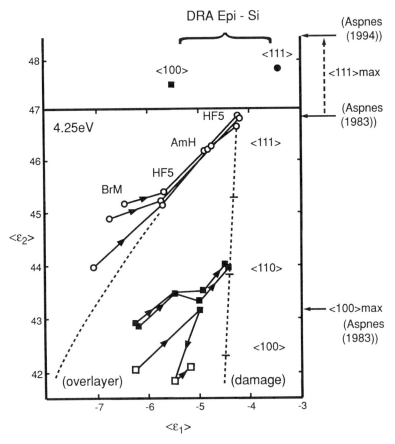

Figure 9.97 Variation of $\langle \varepsilon \rangle$ at 4.25 eV of air-exposed Si after different chemical treatments of the surface, as obtained by ellipsometry. This figure demonstrates how optical data are sensitive to the condition of the surface and how these data "vary" with time as sample preparation procedures improve. [From Pickering, 1995, as adapted from D. E. Aspnes, 1994, 1995 (private communication), which is an updated version from Aspnes, 1983.]

and d'Avitaya, 1991). Habraken *et al.* (1980) compared the ellipsometric response of metal surfaces that are clean with those having submonolayer and monolayer films. Gendry *et al.* (1990) used high-spatial-resolution SE (10-μm spot size) to probe the very thin sulfide layers grown on InP by thermal or plasma treatment of the surface. Yakovlev *et al.* (1993) monitored the interaction of silane with a Pd film at 250°C using ellipsometry. The surface processes depend on the silane flux, and can produce Pd_2Si, palladium hydrides, and porosity. "Fast" (~msec) monitoring was performed by using SWE, while more detailed analysis of slower processes was performed by using SE.

Tompkins (1993) has reviewed other *in situ* applications of single-wavelength ellipsometry, including the dissolution and swelling of thin polymer films, dry oxidation of metals, electrochemical oxidation of metals, and the formation of fluoropolymers on silicon during RIE, as well as the *ex situ* characterization of many materials.

APPENDIX
Terminology

There is no consistent nomenclature for r and R, even among standard optics texts, as is seen in the Table 9A.1. The exact nomenclature sometimes depends on the nuance in usage.

Table 9A.1 Nomenclature in Reflection Experiments[a]

Source	r	R
Azzam and Bashara (1977)	Complex-amplitude reflection coefficient	Reflectance, intensity reflectance
Born and Wolf (1970)	Reflection coefficient	Reflectivity
Guenther (1990)	Amplitude reflection coefficient	Reflectivity, reflectance[b]
Hecht (1987)	Amplitude reflection coefficient	Reflectance
Holm (1991)	Fresnel amplitude reflection coefficient	Reflectance, reflectivity[b]
Klein (1970)	Reflection coefficient	Reflectivity
Klein and Furtak (1986)	Reflection coefficient	Reflectance
Saleh and Teich (1991)	Complex-amplitude reflectance	Power reflectance
This work	Reflection coefficient	Reflectance

[a] Terms for transmission are similar.
[b] See text.

The names given to r, which is the ratio of the reflected electric field amplitude to that incident for a given polarization, include the "reflection coefficient," the "amplitude reflection coefficient," and the "complex amplitude reflectance," among others. Analogously, t, which is the ratio of the transmitted electric field amplitude to that incident, has been called the "transmission coefficient," the "amplitude transmission coefficient," and the "complex amplitude transmittance." The nomenclature adopted here is the *reflection coefficient* for r and the *transmission coefficient* for t.

The differences in nomenclature for the fraction of power or intensity that is reflected $R = |r|^2$ or transmitted $T = |t|^2$ are more subtle and diverse. R has been called the "reflectivity," "reflectance," and "power (or intensity) reflectance," among other terms, for a specific polarization. Guenther (1990) and Holm (1991) use the term "reflectance" to mean the ratio of the measured reflected flux to incident flux, specifically in radiometric units (according to Guenther); it is therefore defined to be an experimental parameter. Holm defines "reflectivity" as the fundamental or characteristic property of a material, i.e., a theoretical reflectance, while Guenther defines it as the ratio of reflected and incident Poynting vector magnitudes for a wave of known frequency and phase.

In this volume, the term *reflectivity* will be used in the limited sense defined by Holm, so that the reflectivity of a given type of material is the fraction of light (power or intensity) that is reflected from an ideal interface between the vacuum and a semiinfinite medium of that material. This is given by Equation 9.7. Reflectivity is therefore a fundamental property of a material. The *reflectance* is defined as the fraction of light (power or intensity) that is reflected from any structure, as determined by measurement. It is also the fraction of light reflected from a structure more complex than a semiinfinite medium, even as determined by calculation or theory. Similarly the *transmittance* is the fraction of light that is transmitted through a structure.

These considerations are also important in other chapters. In Chapter 8, the absorbed fraction should be called the *absorbance*. In the discussion of emitted thermal radiation in Chapter 13, the ratio of the thermal radiation emitted from a real object to that from a blackbody is usually called the "emissivity," although this term more precisely corresponds to this ratio for an ideal homogeneous material. To be consistent with the nomenclature adopted here, this measured ratio should really be called the *emittance;* however, the term *emissivity* is adopted in this book because it is in such widespread use.

References

F. Abelès, *Phys. Thin Films* **6,** 151 (1971).
O. Acher and B. Drévillon, *Rev. Sci. Instrum.* **63,** 5332 (1992).
O. Acher, S. M. Koch, F. Omnes, M. Defour, M. Razeghi, and B. Drévillon, *J. Appl. Phys.* **68,** 3564 (1990).
S. Adachi, ed., *Properties of Aluminum Gallium Arsenide,* INSPEC EMIS Data Rev. Ser. No. 7. INSPEC, London, 1993.
C. H. An and K. Sugimoto, *J. Jpn. Inst. Met., Sendai* **55,** 58 (1991).
C. H. An and K. Sugimoto, *J. Electrochem. Soc.* **139,** 1956 (1992).

References

I. An and R. W. Collins, *Rev. Sci. Instrum.* **62,** 1904 (1991).
I. An, H. V. Nguyen, N. V. Nguyen, and R. W. Collins, *Phys. Rev. Lett.* **65,** 2274 (1990).
I. An, Y. M. Li, C. R. Wronski, H. V. Nguyen, and R. W. Collins, *Appl. Phys. Lett.* **59,** 2543 (1991).
I. An, Y. M. Li, H. V. Nguyen, and R. W. Collins, *Rev. Sci. Instrum.* **63,** 3842 (1992).
I. An, R. W. Collins, H. V. Nguyen, K. Vedam, H. S. Witham, and R. Messier, *Thin Solid Films* **233,** 276 (1993).
S. Andrieu and F. A. d'Avitaya, *Surf. Sci.* **219,** 277 (1989).
S. Andrieu and F. A. d'Avitaya, *J. Cryst. Growth* **112,** 146 (1991).
A. V. Annapragada and K. F. Jensen, *Mater. Res. Soc. Symp. Proc.* **204,** 53 (1991).
J. V. Armstrong, T. Farrell, T. B. Joyce, P. Kightley, T. J. Bullough, and P. J. Goodhew, *J. Cryst. Growth* **120,** 84 (1992).
S. R. Armstrong, R. D. Hoare, M. E. Pemble, I. M. Povey, A. Stafford, A. G. Taylor, P. Fawcette, B. A. Joyce, D. R. Klug, J. Neave, and J. Zhang, *J. Cryst. Growth* **124,** 37 (1992).
S. R. Armstrong, R. D. Hoare, I. M. Povey, M. E. Pemble, A. Stafford, A. G. Taylor, and D. R. Klug, *Appl. Surf. Sci.* **69,** 46 (1993).
S. R. Armstrong, M. E. Pemble, and A. R. Turner, *Surf. Sci.* **307–309,** 1028 (1994a).
S. R. Armstrong, G. H. Fan, M. E. Pemble, H. H. Abdul Ridha, and A. R. Turner, *Surf. Sci.* **307–309,** 1051 (1994b).
D. E. Aspnes, *J. Opt. Soc. Am.* **64,** 639 (1974).
D. E. Aspnes, in *Handbook of Semiconductors* (M. Balkanski, ed.), Vol. 2, Chapter 4A, p. 109. North-Holland Publ., Amsterdam, 1980.
D. E. Aspnes, *SPIE* **276,** 188 (1981).
D. E. Aspnes, *Thin Solid Films* **89,** 249 (1982).
D. E. Aspnes, *J. Phys. Colloq. C10,* **44,** 3 (1983).
D. E. Aspnes, in *Handbook of Optical Constants of Solids* (E. D. Palik, ed.), Chapter 5, p. 89. Academic Press, Boston, 1985a.
D. E. Aspnes, *J. Vac. Sci. Technol. B* **3,** 1498 (1985b).
D. E. Aspnes, ed. *Properties of Gallium Arsenide,* 2nd ed., INSPEC EMIS Data Rev. Ser. No. 2. INSPEC, London, 1990.
D. E. Aspnes, *Appl. Phys. Lett.* **62,** 343 (1993a).
D. E. Aspnes, *J. Opt. Soc. Am. A* **10,** 974 (1993b).
D. E. Aspnes, *Thin Solid Films* **233,** 1 (1993c).
D. E. Aspnes, *Surf. Sci.* **307–309,** 1017 (1994).
D. E. Aspnes, *Mater. Sci. Eng. B* **30,** 109 (1995a).
D. E. Aspnes, *IEEE J. Sel. Top. Quantum Electron.* **1,** 1054 (1995b).
D. E. Aspnes and R. P. H. Chang, in *Plasma Diagnostics* (O. Auciello and D. L. Flamm, eds.), Vol. 2, Chapter 3, p. 67. Academic Press, Boston, 1989.
D. E. Aspnes and A. A. Studna, *Appl. Opt.* **14,** 220 (1975).
D. E. Aspnes and A. A. Studna, *SPIE* **276,** 227 (1981).
D. E. Aspnes and A. A. Studna, *Phys. Rev. B* **27,** 7466 (1983).
D. E. Aspnes and A. A. Studna, *Phys. Rev. Lett.* **54,** 1956 (1985).
D. E. Aspnes, J. P. Harbison, A. A. Studna, and L. T. Florez, *Phys. Rev. Lett.* **59,** 1687 (1987).
D. E. Aspnes, J. P. Harbison, A. A. Studna, and L. T. Florez, *Appl. Phys. Lett.* **52,** 957 (1988a).

D. E. Aspnes, J. P. Harbison, A. A. Studna, and L. T. Florez, *J. Vac. Sci. Technol. A* **6,** 1327 (1988b).
D. E. Aspnes, W. E. Quinn, and S. Gregory, *Appl. Phys. Lett.* **56,** 2569 (1990a).
D. E. Aspnes, W. E. Quinn, and S. Gregory, *Appl. Phys. Lett.* **57,** 2707 (1990b).
D. E. Aspnes, Y. C. Chang, A. A. Studna, L. T. Florez, H. H. Farrell, and J. P. Harbison, *Phys. Rev. Lett.* **64,** 192 (1990c).
D. E. Aspnes, W. E. Quinn, M. C. Tamargo, M. A. A. Pudensi, S. A. Schwarz, M. J. S. P. Brasil, R. E. Nahory, and S. Gregory, *Appl. Phys. Lett.* **60,** 1244 (1992a).
D. E. Aspnes, I. Kamiya, H. Tanaka, and R. Bhat, *J. Vac. Sci. Technol. B* **10,** 1725 (1992b).
D. E. Aspnes, M. C. Tamargo, M. J. S. P. Brasil, R. E. Nahory, and S. A. Schwarz, *Appl. Phys. Lett.* **64,** 3279 (1994a).
D. E. Aspnes, M. C. Tamargo, M. J. S. P. Brasil, R. E. Nahory, and S. A. Schwarz, *J. Vac. Sci. Technol. A* **12,** 1180 (1994b).
D. H. Auston, C. M. Surko, T. N. C. Venkatesan, R. E. Slusher, and J. A. Golovchenko, *Appl. Phys. Lett.* **33,** 437 (1978).
E. S. Aydil and D. J. Economou, *J. Appl. Phys.* **69,** 109 (1991).
E. S. Aydil and D. J. Economou, *J. Electrochem. Soc.* **140,** 1471 (1993).
E. S. Aydil, Z. Zhou, K. P. Giapis, Y. Chabal, J. A. Gregus, and R. Gottscho, *Appl. Phys. Lett.* **62,** 3156 (1993).
E. S. Aydil, Z. H. Zhou, R. A. Gottscho, and Y. J. Chabal, *J. Vac. Sci. Technol. B* **13,** 258 (1995).
R. M. A. Azzam, ed., *Selected Papers on Ellipsometry.* SPIE, Bellingham, WA, 1991.
R. M. A. Azzam and N. M. Bashara, *Ellipsometry and Polarized Light.* North-Holland Publ., Amsterdam, 1977.
D. I. Babić, J. J. Dudley, M. Shirazi, E. L. Hu, and J. E. Bowers, *J. Vac. Sci. Technol. A* **9,** 1113 (1991).
D. I. Babić, T. E. Reynolds, E. L. Hu, and J. E. Bowers, *J. Vac. Sci. Technol. A* **10,** 939 (1992).
K. Bacher, B. Pezeshki, S. M. Lord, and J. S. Harris, Jr., *Appl. Phys. Lett.* **61,** 1387 (1992).
K. J. Bachmann, N. Dietz, A. E. Miller, D. Venables, and J. T. Kelliher, *J. Vac. Sci. Technol. A* **13,** 696 (1995).
D. M. Back, *Phys. Thin Films* **15,** 265 (1991).
J. Bajaj, S. J. C. Irvine, H. O. Sankur, and S. A. Svoronos, *J. Electron. Mater.* **22,** 899 (1993).
D. O. Barsukov, G. M. Gusakov, and A. A. Komarnitskii, *Opt. Spectrosc. (Engl. Transl.)* **64,** 782 (1988).
L. Baufay, F. A. Houle, and R. J. Wilson, *J. Appl. Phys.* **61,** 4640 (1987).
D. Beaglehole, *Rev. Sci. Instrum.* **59,** 2557 (1988).
V. L. Berkovits and D. Paget, *Thin Solid Films* **233,** 9 (1993).
V. M. Bermudez, *J. Vac. Sci. Technol. A* **10,** 152 (1992).
P. H. Berning, *Phys. Thin Films* **1,** 69 (1963).
M. Bertolotti, V. Bogdanov, A. Ferrari, A. Jascow, N. Nazorova, A. Pikhtin, and L. Schirone, *J. Opt. Soc. Am. B* **7,** 918 (1990).
P. Bhattacharya, ed., *Properties of Lattice-Matched and Strained Indium Gallium Arsenide,* INSPEC EMIS Data Rev. Ser. No. 8. INSPEC, London, 1993.

D. K. Biegelsen, R. D. Bringans, J. E. Northrup, and A. Swartz, *Phys. Rev. B* **41**, 5701 (1990).

C. H. Bjorkman, M. Fukuda, T. Yamazaki, S. Miyazaki, and M. Hirose, *Jpn. J. Appl. Phys. (Part 1)* **34**, 722 (1995).

C. H. Bjorkman, T. Yamazaki, S. Miyazaki, and M. Hirose, *J. Appl. Phys.* **77**, 313 (1995).

N. Blayo and B. Drévillon, *Appl. Phys. Lett.* **57**, 786 (1990).

N. Blayo, I. Tepermeister, J. L. Benton, G. S. Higashi, T. Boone, A. Onuoha, F. P. Klemens, D. E. Ibbotson, J. T. C. Lee, and H. H. Sawin, *J. Vac. Sci. Technol. B* **12**, 1340 (1994).

N. Blayo, R. A. Cirelli, F. P. Klemens, and J. T. C. Lee, *J. Opt. Soc. Am. A* **12**, 591 (1995).

F. G. Böbel, H. Möller, B. Hertel, G. Ritter, and P. Chow, *Solid State Technol.* **37**(8), 55 (1994).

A. C. Boccara, C. Pickering, and J. Rivory, eds., *Thin Solids Films* **233**, 1 (1993); **234**, 307 (1993).

K. W. Böer, *Survey of Semiconductor Physics*. Van Nostrand-Reinhold, New York, 1990.

R. A. Bond, S. Dzioba, and H. M. Naguib, *J. Vac. Sci. Technol.* **18**, 335 (1981).

M. Born and E. Wolf, *Principles of Optics,* 4th ed. Pergamon, Oxford, 1970.

N. Bottka, D. K. Gaskill, R. S. Sillmon, R. Henry, and R. Glosser, *J. Electron. Mater.* **17**, 161 (1988).

P. D. Brewer, J. J. Zinck, and G. L. Olson, *Appl. Phys. Lett.* **57**, 2526 (1990).

G. H. Bu-Abbud, N. M. Bashara, and J. A. Woollam, *Thin Solid Films* **137**, 27 (1986).

T. M. Burns, E. A. Irene, S. Chevacharoenkul, and G. E. McGurie, *J. Vac. Sci. Technol. B* **11**, 78 (1993).

H. H. Busta, R. E. Lajos, and D. A. Kiewit, *Solid State Technol.* **22**(2), 61 (1979).

S. W. Butler and J. A. Stefani, *IEEE Trans. Semicond. Manuf.* **SM-7**, 193 (1994).

S. W. Butler, J. Stefani, M. Sullivan, S. Maung, G. Barna, and S. Henck, *J. Vac. Sci. Technol. A* **12**, 1984 (1994).

K. Capuder, P. E. Norris, H. Shen, Z. Hang, and F. H. Pollak, *J. Electron. Mater.* **19**, 295 (1990).

R. T. Carline, C. Pickering, D. J. Robbins, W. Y. Leong, A. D. Pitt, and A. G. Cullis, *Appl. Phys. Lett.* **64**, 1114 (1994).

F. G. Celii, Y.-C. Kao, A. J. Katz, and T. Moise, *J. Vac. Sci. Technol. A* **13**, 733 (1995).

Y. J. Chabal, *Surf. Sci. Rep.* **8**, 211 (1988).

Y. J. Chabal, in *Internal Reflection Spectroscopy: Theory and Applications* (F. M. Mirabella, Jr., eds.), Chapter 8, p. 191. Dekker, New York, 1993.

Y. J. Chabal, in *Handbook on Semiconductors,* Vol. 2 (M. Balkanski, ed.). Elsevier, New York, 1994.

Y. J. Chabal, G. S. Higashi, K. Raghavachari, and V. A. Burrows, *J. Vac. Sci. Technol. A* **7**, 2104 (1989).

S. A. Chalmers and K. P. Killeen, *Appl. Phys. Lett.* **62**, 1182 (1993).

R. P. H. Chang, S. Darack, E. Lane, C. C. Chang, D. Allara, and E. Ong, *J. Vac. Sci. Technol. B* **1**, 935 (1983).

P. Chindaudom and K. Vedam, *Phys. Thin Films* **19**, 191 (1994).

S. Chongsawangvirod and E. A. Irene, *J. Electrochem. Soc.* **138**, 1748 (1991).

P. P. Chow, in *Thin Film Processes II* (J. L. Vossen and W. Kern, eds.), Chap. II-3, p. 133. Academic Press, Boston, 1991.

R. F. Cohn, J. W. Wagner, and J. Kruger, *Appl. Opt.* **27**, 4664 (1988).

E. Colas, D. E. Aspnes, R. Bhat, A. A. Studna, J. P. Harbison, L. T. Florez, M. A. Koza, and V. G. Keramidas, *J. Cryst. Growth* **107**, 47 (1991).

R. W. Collins, *J. Vac. Sci. Technol. A* **7**, 1378 (1989).

R. W. Collins, *Rev. Sci. Instrum.* **61**, 2029 (1990).

R. W. Collins and J. M. Cavese, *J. Appl. Phys.* **62**, 4146 (1987).

R. W. Collins, I. An, H. V. Nguyen, and Y. Lu, *Thin Solid Films* **233**, 244 (1993).

R. W. Collins, I. An, H. V. Nguyen, Y. Li, and Y. Lu, *Phys. Thin Films* **19**, 49 (1994).

P. Collot, T. Diallo, and J. Canteloup, *J. Vac. Sci. Technol. B* **9**, 2497 (1991).

K. A. Conrad, R. K. Sampson, H. Z. Massoud, and E. A. Irene, *J. Vac. Sci. Technol. B* **11**, 2096 (1993).

J. R. Creighton, *Appl. Surf. Sci.* **82/83**, 171 (1994).

J. E. Crowell, G. Lu, and B. M. H. Ning, *Mater. Res. Soc. Symp. Proc.* **204**, 253 (1991).

B. H. Cumpston, J. P. Lu, B. G. Willis, and K. F. Jensen, *Mater. Res. Soc. Symp. Proc.* **385**, 103 (1995).

J. T. Davies, T. Metz, R. N. Savage, and H. Simmons, *SPIE* **1392**, 551 (1990).

Y. Demay, J. P. Gailliard, and P. Medina, *J. Cryst. Growth* **81**, 97 (1987a).

Y. Demay, D. Arnoult, J. P. Gailliard, and P. Medina, *J. Vac. Sci. Technol. A* **5**, 3139 (1987b).

N. Dietz and H. J. Lewerenz, *Appl. Surf. Sci.* **69**, 350 (1993).

N. Dietz, D. J. Stephens, G. Lucovsky, and K. J. Bachmann, *Mater. Res. Soc. Symp. Proc.* **324**, 27 (1994).

N. Dietz, A. Miller, and K. J. Bachmann, *J. Vac. Sci. Technol. A* **13**, 153 (1995).

J.-M. Dilhac, C. Ganibal, and T. Castan, *Appl. Phys. Lett.* **55**, 2225 (1989).

J.-M. Dilhac, C. Ganibal, N. Nolhier, and L. Amat, *SPIE* **1393**, 349 (1990).

H. W. Dinges, H. Burkhard, R. Lösch, H. Nickel, and W. Schlapp, *Thin Solid Films* **233**, 145 (1993).

V. M. Donnelly, in *Plasma Diagnostics* (O. Auciello and D. L. Flamm, eds.), Chapter 1, p. 1. Academic Press, Boston, 1989.

V. M. Donnelly, *Appl. Phys. Lett.* **63**, 1396 (1993a).

V. M. Donnelly, *J. Vac. Sci. Technol. A* **11**, 2393 (1993b).

V. M. Donnelly and J. A. McCaulley, *J. Vac. Sci. Technol. A* **8**, 84 (1990).

V. M. Donnelly, D. E. Ibbotson, and C.-P. Chang, *J. Vac. Sci. Technol. A* **10**, 1060 (1992).

B. Drévillon, *Prog. Cryst. Growth Charact.* **27**, 1 (1993).

B. Drévillon and V. Yakovlev, *Phys. Thin Films* **19**, 1 (1994).

B. Drévillon, J. Perrin, R. Marbot, A. Violet, and J. L. Dalby, *Rev. Sci. Instrum.* **53**, 969 (1982).

B. Drévillon, C. Godet, and S. Kumar, *Appl. Phys. Lett.* **50**, 1651 (1987).

R. Droopad, C. H. Kuo, S. Anand, K. Y. Choi, and G. N. Maracas, *J. Vac. Sci. Technol. B* **12**, 1211 (1994).

W. M. Duncan, S. A. Henck, J. W. Kuehne, L. M. Loewenstein, and S. Maung, *J. Vac. Sci. Technol. B* **12**, 2779 (1994).

D. Economou, E. S. Aydil, and G. Barna, *Solid State Technol.* **34**(4), 107 (1991).

J. M. C. England, N. Zissis, P. J. Timans, and H. Ahmed, *J. Appl. Phys.* **70**, 389 (1991).

J. M. C. England, P. J. Timans, C. Hill, P. D. Augustus, and H. Ahmed, *J. Appl. Phys.* **73**, 4332 (1993).
G. Eres and J. W. Sharp, *Appl. Phys. Lett.* **60**, 2764 (1992).
G. Eres and J. W. Sharp, *J. Vac. Sci. Technol. A* **11**, 2463 (1993).
M. Erman and J. B. Theeten, *J. Appl. Phys.* **60**, 859 (1986).
N. Esser, R. Hunger, J. Rumberg, W. Richter, R. Del Sole, and A. I. Shkrebtii, *Surf. Sci.* **307–309**, 1045 (1994).
K. G. Eyink, Y. S. Cong, M. A. Capano, T. W. Haas, R. A. Gilbert, and B. G. Streetman, *J. Electron. Mater.* **22**, 1387 (1993a).
K. G. Eyink, Y. S. Cong, R. Gilbert, M. A. Capano, T. W. Haas, and B. G. Streetman, *J. Vac. Sci. Technol. B* **11**, 1423 (1993b).
T. Farrell, J. V. Armstrong, and P. Kightley, *Appl. Phys. Lett.* **59**, 1203 (1991).
F. Ferrieu, *Rev. Sci. Instrum.* **60**, 3212 (1989).
W. W. Flack, J. S. Papanu, D. W. Hess, D. S. Soong, and A. T. Bell, *J. Electrochem. Soc.* **131**, 2200 (1984).
J. E. Franke, L. Zhang, T. M. Niemczyk, D. M. Haaland, and J. H. Linn, *J. Electrochem. Soc.* **140**, 1425 (1993).
J. E. Franke, T. M. Niemczyk, and D. M. Haaland, *Spectrochim. Acta, Part A* **50A**, 1687 (1994).
W. Fukarek and H. Kersten, *J. Vac. Sci. Technol. A* **12**, 523 (1994).
D. Gautard, J. L. LaPorte, M. Cadoret, and C. Pariset, *J. Cryst. Growth* **71**, 125 (1985).
P. E. Gee and R. F. Hicks, *J. Vac. Sci. Technol. A* **10**, 892 (1992).
M. Gendry, J. Durand, M. Erman, J. B. Theeten, L. Nevot, and B. Pardo, *Appl. Surf. Sci.* **44**, 309 (1990).
W. L. Gladfelter, M. G. Simmonds, L. A. Zazzera, and J. F. Evans, *Mater. Res. Soc. Symp. Proc.* **334**, 273 (1994).
O. J. Glembocki, *SPIE* **1286**, 2 (1990).
O. J. Glembocki and B. V. Shanabrook, *Semicond. Semimetals* **36**, 221 (1992).
Y. González, L. González, and F. Briones, *J. Vac. Sci. Technol. A* **13**, 73 (1995).
G. Gould and E. A. Irene, *J. Electrochem. Soc.* **135**, 1535 (1988).
R. Greef, *Thin Solid Films* **233**, 32 (1993).
R. G. Greenler, *J. Vac. Sci. Technol.* **12**, 1410 (1975).
R. D. Guenther, *Modern Optics.* Wiley, New York, 1990.
D. Guidotti and J. G. Wilman, *Appl. Phys. Lett.* **60**, 524 (1992a).
D. Guidotti and J. G. Wilman, *J. Vac. Sci. Technol. A* **10**, 3184 (1992b).
F. H. P. M. Habraken, O. L. J. Gijzeman, and G. A. Bootsma, *Surf. Sci.* **96**, 482 (1980).
D. Hacman, *Optik* **28**, 115 (1968).
G. P. Hansen, S. Krishnan, R. H. Hauge, and J. L. Margrave, *Metallu. Trans. A* **19A**, 1889 (1988).
G. P. Hansen, S. Krishnan, R. H. Hauge, and J. L. Margrave, *Appl. Opt.* **28**, 1885 (1989).
N. Hara, E. Takahashi, J. H. Yoon, and K, Sugimoto, *J. Electrochem. Soc.* **141**, 1669 (1994).
J. P. Harbison, D. E. Aspnes, A. A. Studna, L. T. Florez, and M. K. Kelly, *Appl. Phys. Lett.* **52**, 2046 (1988).
N. J. Harrick, *Internal Reflection Spectroscopy.* Wiley (Interscience), New York, 1967; 2nd printing by Harrick Scientific Corporation, Ossining, NY, 1979.

R. H. Hartley, U.S. Patent 4,770,895 (1988).
R. H. Hartley, M. A. Folkard, D. Carr, P. J. Orders, D. Rees, I. K. Varga, V. Kumar, G. Shen, T. A. Steele, H. Buskes, and J. B. Lee, *J. Vac. Sci. Technol.* B **10,** 1410 (1992a).
R. H. Hartley, M. A. Folkard, D. Carr, P. J. Orders, D. Rees, I. K. Varga, V. Kumar, G. Shen, T. A. Steele, H. Buskes, and J. B. Lee, *J. Cryst. Growth* **117,** 166 (1992b).
G. Hass and L. Hadley, in *American Institute of Physics Handbook* (D. E. Gray, ed.), 3rd ed., Chapter 6g, pp. 6–118. McGraw Hill, New York, 1972.
P. S. Hauge, *Surf. Sci.* **96,** 108 (1980).
P. S. Hauge, R. H. Muller, and C. G. Smith, *Surf. Sci.* **96,** 81 (1980).
M. Haverlag and G. S. Oehrlein, *J. Vac. Sci. Technol.* B **10,** 2412 (1992).
M. Haverlag, G. M. W. Kroesen, C. J. H. de Zeeuw, Y. Creyghton, T. H. J. Bisschops, and F. J. de Hoog, *J. Vac. Sci. Technol.* B **7,** 529 (1989).
M. Haverlag, D. Vender, and G. S. Oehrlein, *Appl. Phys. Lett.* **61,** 2875 (1992).
Y. Hayashi, W. Drawl, R. W. Collins, and R. Messier, *Appl. Phys. Lett.* **60,** 2868 (1992).
T. R. Hayes, P. A. Heimann, V. M. Donnelly, and K. E. Strege, *Appl. Phys. Lett.* **57,** 2817 (1990).
O. S. Heavens, *Phys. Thin Films* **2,** 193 (1964).
O. S. Heavens, *Optical Properties of Thin Solid Films.* Dover, New York, 1965.
E. Hecht, *Optics,* 2nd ed. Addison-Wesley, Reading, MA, 1987.
S. A. Henck, *J. Vac. Sci. Technol.* A **10,** 934 (1992).
S. A. Henck, W. M. Duncan, L. M. Lowenstein, and J. Kuehne, *SPIE* **1803,** 299 (1992).
M. A. Herman and H. Sitter, *Molecular Beam Epitaxy.* Springer-Verlag, Berlin, 1989.
C. M. Herzinger, H. Yao, P. G. Snyder, F. G. Celii, Y.-C. Kao, B. Johs, and J. A. Woollam, *J. Appl. Phys.* **77,** 4677 (1995).
A. R. Heyd, I. An, R. W. Collins, Y. Cong, K. Vedam, S. S. Bose, and D. L. Miller, *J. Vac. Sci. Technol.* A **9,** 810 (1991).
A. R. Heyd, R. W. Collins, K. Vedam, S. S. Bose, and D. L. Miller, *Appl. Phys. Lett.* **60,** 2776 (1992).
S. E. Hicks, W. Parkes, J. A. H. Wilkinson, and C. D. W. Wilkinson, *J. Vac. Sci. Technol.* B **12,** 3306 (1994).
G. S. Higashi and Y. J. Chabal, in *Handbook of Semiconductor Wafer Cleaning Technology* (W. Kern, ed.), Chapter 10, p. 433. Noyes, Park Ridge, NJ, 1993.
G. S. Higashi, Y. J. Chabal, G. W. Trucks, and K. Raghavachari, *Appl. Phys. Lett.* **56,** 656 (1990).
K. Hingerl, D. E. Aspnes, I. Kamiya, and L. T. Florez, *Appl. Phys. Lett.* **63,** 885 (1993).
F. M. Hoffmann, *Surf. Sci. Rep.* **3,** 107 (1983).
J. A. Holbrook and R. E. Hummel, *Rev. Sci. Instrum.* **44,** 463 (1973).
R. T. Holm, in *Handbook of Optical Constants of Solids II* (E. D. Palik, ed.), Chapter 2, p. 21. Academic Press, Boston, 1991.
F. Hottier and R. Cadoret, *J. Cryst. Growth* **56,** 304 (1982).
F. Hottier and R. Cadoret, *J. Cryst. Growth* **61,** 245 (1983).
F. Hottier and G. Laurence, *Appl. Phys. Lett.* **38,** 863 (1981).

F. Hottier and J. B. Theeten, *J. Cryst. Growth* **48,** 644 (1980).
F. Hottier, J. Hallais, and F. Simondet, *J. Appl. Phys.* **51,** 1599 (1980).
P. Houdy, *Rev. Phys. Appl.* **23,** 1653 (1988).
Y. Z. Hu, J. Joseph, and E. A. Irene, *Appl. Phys. Lett.* **59,** 1353 (1991).
J. Humlíček and M. Garriga, *Appl. Phys. A* **56,** 259 (1993).
R. E. Hummel, *Phys. Status Solidi A* **76,** 11 (1983).
R. E. Hummel, W. Xi, P. H. Holloway, and K. A. Jones, *J. Appl. Phys.* **63,** 2591 (1988).
S. P. F. Humphreys-Owen, *Proc. Phys. Soc., London* **77,** 949 (1961).
W. R. Hunter, *J. Opt. Soc. Am.* **55,** 1197 (1965).
J. S. Im, H. J. Kim, and M. O. Thompson, *Appl. Phys. Lett.* **63,** 1969 (1993).
E. A. Irene, *Thin Solid Films* **233,** 96 (1993).
S. J. C. Irvine, J. Bajaj, and R. V. Gil, *J. Electron. Mater.* **23,** 167 (1994).
R. Jacobsson, *Prog. Opt.* **5,** 247 (1966).
P. Jakob and Y. J. Chabal, *J. Chem. Phys.* **95,** 2897 (1991).
P. Jakob, P. Dumas, and Y. J. Chabal, *Appl. Phys. Lett.* **59,** 2968 (1991).
G. E. Jellison, Jr., *Appl. Opt.* **30,** 3354 (1991).
G. E. Jellison, Jr., *Opt. Mater.* **1,** 41 (1992a).
G. E. Jellison, Jr., *Opt. Mater.* **1,** 151 (1992b).
G. E. Jellison, Jr., in *Characterization of Optical Materials* (G. J. Exarhos, ed.), Chapter 2, p. 27. Butterworth-Heinemann, Boston, 1993a.
G. E. Jellison, Jr., *Thin Solid Films* **234,** 416 (1993b).
G. E. Jellison, Jr. and D. H. Lowndes, *Appl. Opt.* **24,** 2948 (1985a).
G. E. Jellison, Jr. and D. H. Lowndes, *Appl. Phys. Lett.* **47,** 718 (1985b).
G. E. Jellison, Jr. and F. A. Modine, *Phys. Rev. B* **27,** 7466 (1983).
G. E. Jellison, Jr. and F. A. Modine, *Appl. Opt.* **29,** 959 (1990).
G. E. Jellison, Jr. and B. C. Sales, *Appl. Opt.* **30,** 4310 (1991).
G. E. Jellison, Jr., T. E. Haynes, and H. H. Burke, *Opt. Mater.* **2,** 105 (1993).
J. E. Jensen, P. D. Brewer, G. L. Olson, L. W. Tutt, and J. J. Zinck, *J. Vac. Sci. Technol. A* **6,** 2808 (1988).
S. Johnston, *Fourier Transform Infrared: A Constantly Evolving Technology.* Ellis Horwood, New York, 1991.
B. Johs, *Thin Solid Films* **234,** 395 (1993).
B. Johs, D. Doerr, S. Pittal, I. B. Bhat, and S. Dakshinamurthy, *Thin Solid Films* **233,** 293 (1993).
J. F. Jongste, P. F. A. Alkemade, G. C. A. M. Janssen, and S. Radelaar, *J. Appl. Phys.* **74,** 3869 (1993).
J. Jönsson, K. Deppert, S. Jeppesen, G. Paulsson, L. Samuelson, and P. Schmidt, *J. Appl. Phys.* **56,** 2414 (1990).
J. Jönsson, K. Deppert, and L. Samuelson, *J. Cryst. Growth* **124,** 30 (1992).
J. Jönsson, K. Deppert, and L. Samuelson, *J. Appl. Phys.* **74,** 6146 (1993).
J. Jönsson, F. Reinhardt, M. Zorn, K. Ploska, W. Richter, and J. Rumberg, *Appl. Phys. Lett.* **64,** 1998 (1994).
J. Joseph, Y. Z. Hu, and E. A. Irene, *J. Vac. Sci. Technol. B* **10,** 611 (1992).
B. A. Joyce, P. J. Dobson, J. H. Neave, K. Woodbridge, J. Zhang, P. K. Larsen, and B. Boelger, *Surf. Sci.* **186,** 423 (1986).
I. Kamiya, D. E. Aspnes, H. Tanaka, L. T. Florez, J. P. Harbison, and R. Bhat, *Phys. Rev. Lett.* **68,** 627 (1992a).

I. Kamiya, D. E. Aspnes, H. Tanaka, L. T. Florez, J. P. Harbison, and R. Bhat, *J. Vac. Sci. Technol. B* **10,** 1716 (1992b).
I. Kamiya, D. E. Aspnes, L. T. Florez, and J. P. Harbison, *Phys. Rev. B* **46,** 15894 (1992c).
T. Kawahara, A. Yuuki, and Y. Matsui, *Jpn. J. Appl. Phys.* **30,** 431 (1991).
K. Kawamura, S. Ishizuka, H. Sakaue, and Y. Horiike, *Jpn. J. Appl. Phys.* **30,** 3215 (1991).
W. A. Keenan, M. E. Keefer, P. F. Gise, and L. A. Thornquist, *Microelectron. Manuf. Technol.,* February, p. 19 (1991).
H. Kempkens, W. W. Byszewski, P. D. Gregor, and W. P. Lapatovich, *J. Appl. Phys.* **67,** 3618 (1990).
J. Kikuchi, R. Kurosaki, S. Fujimura, and H. Yano, *SPIE* **2336,** 111 (1994).
K. P. Killeen and W. G. Breiland, *J. Electron. Mater.* **23,** 179 (1994).
Y.-T. Kim, R. W. Collins, and K. Vedam, *Surf. Sci.* **233,** 341 (1990).
M. V. Klein, *Optics.* Wiley, New York, 1970.
M. V. Klein and T. E. Furtak, *Optics,* 2nd ed., Wiley, New York, 1986.
Ch. Kleint and M. Merkel, *Surf. Sci.* **213,** 657 (1989).
Z. Knittl, *Optics of Thin Films.* Wiley, London, 1976.
N. Kobayashi and Y. Horikoshi, *Jpn. J. Appl. Phys.* **28,** L1880 (1989).
N. Kobayashi and Y. Horikoshi, *Jpn. J. Appl. Phys.* **29,** L702 (1990).
N. Kobayashi and Y. Kobayashi, *J. Cryst. Growth* **124,** 525 (1992).
N. Kobayashi, T. Makimoto, Y. Yamauchi, and Y. Horikoshi, *J. Cryst. Growth* **107,** 62 (1991a).
N. Kobayashi, Y. Yamauchi, and Y. Horikoshi, *J. Cryst. Growth* **115,** 353 (1991b).
N. Kobayashi, Y. Nakamura, H. Goto, and Y. Homma, *J. Appl. Phys.* **73,** 4637 (1993).
E. Kobeda and E. A. Irene, *J. Vac. Sci. Technol. B* **4,** 720 (1986).
S. M. Koch, O. Acher, F. Omnes, M. Defour, B. Drévillon, and M. Razeghi, *J. Appl. Phys.* **69,** 1389 (1991).
K. B. Koller, W. A. Schmidt, and J. E. Butler, *J. Appl. Phys.* **64,** 4704 (1988).
N. Kondo, N. Fujiwara, and A. Abematsu, *SPIE* **1673,** 392 (1992).
K. L. Konnerth and F. H. Dill, *IEEE Trans. Electron Devices* **ED-22,** 452 (1975).
P. D. Krasicky, R. J. Greole, and F. Rodriguez, *Chem. Eng. Commun.* **54,** 279 (1987a).
P. D. Krasicky, R. J. Greole, J. A. Jubinsky, F. Rodriguez, Y. M. N. Namaste, and S. K. Obendorf, *Polym. Eng. Sci.* **27,** 282 (1987b).
P. D. Krasicky, R. J. Greole, and F. Rodriguez, *J. Appl. Polym. Sci.* **35,** 641 (1988).
G. M. W. Kroesen, G. S. Oehrlein, and T. D. Bestwick, *J. Appl. Phys.* **69,** 3390 (1991).
G. M. W. Kroesen, G. S. Oehrlein, E. de Frésart, and M. Haverlag, *J. Appl. Phys.* **73,** 8017 (1993).
S. Kumar, B. Drévillon, and C. Godet, *J. Appl. Phys.* **60,** 1542 (1986).
C. H. Kuo, S. Anand, R. Droopad, K. Y. Choi, and G. N. Maracas, *J. Vac. Sci. Technol. B* **12,** 1214 (1994).
C. Kuo, S. Anand, H. Fathollahnejad, R. Ramamurti, R. Droopad, and G. Maracas, *J. Vac. Sci. Technol. B* **13,** 681 (1995).
H. Kuroki, K. G. Nakamura, I. Kamioka, T. Kawabe, and M. Kitajima, *J. Vac. Sci. Technol. A* **12,** 1431 (1994).
R. Kurosaki, J. Kikuchi, Y. Kobayashi, Y. Chinzei, S. Fujimura, and Y. Horiike, *SPIE* **2635,** 224 (1995).

L. Lauchlan, K. Sautter, T. Batchelder, and J. Irwin, *SPIE* **539,** 227 (1985).
N. Layadi, P. Roca i Cabarrocas, V. Yakovlev, and B. Drévillon, *Thin Solid Films* **233,** 281 (1993).
G. J. Leusink, T. G. M. Oosterlaken, G. C. A. M. Janssen, and S. Radelaar, *Rev. Sci. Instrum.* **63,** 3143 (1992).
G. J. Leusink, T. G. M. Oosterlaken, G. C. A. M. Janssen, and S. Radelaar, *J. Appl. Phys.* **74,** 3899 (1993).
H. J. Lewerenz and N. Dietz, *J. Appl. Phys.* **73,** 4975 (1993).
Y. M. Li, I. An, H. V. Nguyen, C. R. Wronski, and R. W. Collins, *Phys. Rev. Lett.* **68,** 2814 (1992).
L. Ling, S. Kuwabara, T. Abe, and F. Shimura, *J. Appl. Phys.* **73,** 3018 (1993).
S. Logothetidis, I. Alexandrou, and A. Papadopoulos, *J. Appl. Phys.* **77,** 1043 (1995).
D. H. Lowndes, R. F. Wood, and R. D. Westbrook, *Appl. Phys. Lett.* **43,** 258 (1983).
D. H. Lowndes, R. F. Wood, and J. Narayan, *Phys. Rev. Lett.* **52,** 561 (1984).
B. Y. Maa and P. D. Dapkus, *Appl. Phys. Lett.* **58,** 2261 (1991).
H. A. Macleod, *Thin-Film Optical Filters,* 2nd ed. McGraw-Hill, New York, 1989.
T. Makimoto, Y. Yamauchi, N. Kobayashi, and Y. Horikoshi, *Jpn. J. Appl. Phys.* **29,** L207 (1990).
J. Manjkow, J. S. Papanu, D. W. Hess, D. S. Soane (Soong), and A. T. Bell, *J. Electrochem. Soc.* **134,** 2003 (1987).
G. N. Maracas, J. L. Edwards, K. Shiralagi, K. Y. Choi, R. Droopad, B. Johs, and J. A. Woolam, *J. Vac. Sci. Technol. A* **10,** 1832 (1992).
G. N. Maracas, C. H. Kuo, S. Anand, and R. Droopad, *J. Appl. Phys.* **77,** 1701 (1995a).
G. N. Maracas, C. H. Kuo, S. Anand, R. Droopad, G. R. L. Sohie, and T. Levola, *J. Vac. Sci. Technol. A* **13,** 727 (1995b).
S. Maung, S. Banerjee, D. Draheim, S. Henck, and S. W. Butler, *IEEE Trans. Semicond. Manuf.* **SM-7,** 184 (1994).
H. L. Maynard and N. Hershkowitz, *IEEE Trans. Semicond. Manuf.* **SM-6,** 373 (1993).
J. A. McCaulley, V. M. Donnelly, M. Vernon, and I. Taha, *Phys. Rev. B* **49,** 7408 (1994).
V. R. McCrary, V. M. Donnelly, S. G. Napholtz, T. R. Hayes, P. S. Davisson, and D. C. Bruno, *J. Cryst. Growth* **125,** 320 (1992).
J. D. E. McIntyre and D. E. Aspnes, *Surf. Sci.* **24,** 417 (1971).
T. E. Metz, R. N. Savage, and H. O. Simmons, *SPIE* **1594,** 146 (1991).
T. E. Metz, R. N. Savage, and H. O. Simmons, *Semicond. Int.,* February, p. 68 (1992).
F. M. Mirabella, Jr., ed., *Internal Reflection Spectroscopy: Theory and Applications.* Dekker, New York, 1993.
S. Miyazaki, H. Shin, Y. Miyoshi, and M. Hirose, *Jpn. J. Appl. Phys. (Part 1)* **34,** 787 (1995).
W. M. Moreau, *Semiconductor Lithography: Principles, Practices, and Materials.* Plenum, New York, 1988.
S. J. Morris, J.-Th. Zettler, K. C. Rose, D. I. Westwood, D. A. Woolf, R. H. Williams, and W. Richter, *J. Appl. Phys.* **77,** 3115 (1995).
A. B. Müller, F. Reinhardt, U. Resch, W. Richter, K. C. Rose, and U. Rossow, *Thin Solid Films* **233,** 19 (1993).
R. H. Muller, *Surf. Sci.* **16,** 14 (1969).

K. Murakami, Y. Tohmiya, K. Takita, and K. Masuda, *Appl. Phys. Lett.* **45,** 659 (1984).
S. Nafis, N. J. Ianno, P. G. Snyder, W. A. McGahan, B. Johs, and J. A. Woollam, *Thin Solid Films* **233,** 253 (1993).
K. Nakamoto, *Infrared and Raman Spectra of Inorganic and Coordination Compounds,* 4th ed. Wiley, New York, 1986.
H. Nakanishi and K. Wada, *Jpn. J. Appl. Phys.* **32,** 6206 (1993).
H. V. Nguyen, I. An, and R. W. Collins, *Phys. Thin Films* **19,** 127 (1994).
K. Nishi, A. Usui, and H. Sakaki, *Appl. Phys. Lett.* **61,** 31 (1992).
K. Nishi, A. Usui, and H. Sakaki, *Thin Solid Films* **225,** 47 (1993).
K. Nishikawa, K. Ono, M. Tuda, T. Oomori, and K. Namba, *Symp. Dry Process., 16th,* Tokyo, *1994,* III-4 (1994).
M. Niwano, J.-I. Kageyama, K. Kinashi, N. Miyamoto, and K. Honma, *J. Vac. Sci. Technol. A* **12,** 465 (1994a).
M. Niwano, J.-I. Kageyama, K. Kurita, K. Kinashi, I. Takahashi, and N. Miyamoto, *J. Appl. Phys.* **76,** 2157 (1994b).
M. Niwano, Y. Kimura, and N. Miyamoto, *Appl. Phys. Lett.* **65,** 1692 (1994c).
A. E. Novembre, W. T. Tang, and P. Hsieh, *SPIE* **1087,** 460 (1989).
G. S. Oehrlein, *J. Vac. Sci. Technol. A* **11,** 34 (1993).
J. E. Olsen and F. Shimura, *J. Appl. Phys.* **66,** 1353 (1989).
G. L. Olson and J. A. Roth, *Mater. Sci. Rep.* **3,** 1 (1988).
J. A. O'Neill, E. Sanchez, and R. M. Osgood, Jr., *J. Vac. Sci. Technol. A* **7,** 2110 (1989).
Y. Ohshita and N. Hosoi, *J. Cryst. Growth* **131,** 495 (1993).
R. Ossikovski, H. Shirai, and B. Drévillon, *Thin Solid Films* **234,** 363 (1993).
U.-S. Pahk, S. Chongsawangvirod, and E. A. Irene, *J. Electrochem. Soc.* **138,** 308 (1991).
E. D. Palik, *Handbook of Optical Constants of Solids.* Academic Press, Boston, 1985.
E. D. Palik, *Handbook of Optical Constants of Solids II.* Academic Press, Boston, 1991.
M. B. Panish and H. Temkin, *Gas Source Molecular Beam Epitaxy.* Springer-Verlag, Berlin, 1993.
A. N. Parikh and D. L. Allara, *Phys. Thin Films* **19,** 279 (1994).
G. Paulsson, K. Deppert, S. Jeppesen, J. Jönsson, L. Samuelson, and P. Schmidt, *J. Cryst. Growth* **111,** 115 (1991).
G. Paulsson, B. Junno, and L. Samuelson, *J. Cryst. Growth* **127,** 1014 (1993).
L. Peters, *Semicond. Int.* **14**(9), 56 (1991).
B. Peuse and A. Rosekrans, *Mater. Res. Soc. Symp. Proc.* **303,** 125 (1993).
B. W. Peuse, A. Rosekrans, and K. A. Snow, *SPIE* **1804,** 45 (1993).
B. Peuse and A. Rosekrans, *SPIE* **2091,** 301 (1994).
C. Pickering, *Thin Solid Films* **206,** 275 (1991).
C. Pickering, in *The Handbook of Crystal Growth* (D. T. J. Hurle, ed.), Vol. 3, p. 817. North-Holland Publ., Amsterdam, 1994.
C. Pickering, in *Photonic Probes of Surfaces; Electromagnetic Waves: Recent Developments in Research* (P. Halevi, ed.), Vol. 2, Chapter 1. Elsevier, Amsterdam, 1995.
C. Pickering, D. J. Robbins, I. M. Young, J. L. Glasper, M. Johnson, and R. Jones, *Mater. Res. Soc. Symp. Proc.* **94,** 173 (1987).

C. Pickering, R. T. Carline, D. J. Robbins, W. Y. Leong, D. E. Gray, and R. Greef, *Thin Solid Films* **233,** 126 (1993).

C. Pickering, D. A. O. Hope, R. T. Carline, and D. J. Robbins, *J. Vac. Sci. Technol. A* **13,** 740 (1995).

J.-P. Piel, J.-L. Stehle, and O. Thomas, *Thin Solid Films* **233,** 301 (1993).

W. A. Pliskin and S. J. Zanin, in *Handbook of Thin Film Technology* (L. I. Maissel and R. Glang, eds.), Chapter 11. McGraw-Hill, New York, 1983.

K. Ploska, W. Richter, F. Reinhardt, J. Jönsson, J. Rumberg, and M. Zorn, *Mater. Res. Soc. Symp. Proc.* **334,** 155 (1994).

F. H. Pollak, *SPIE* **276,** 142 (1981).

F. H. Pollak and H. Shen, *J. Electron. Mater.* **19,** 399 (1990).

J. Pope, Jr., R. Woodburn, J. Watkins, R. Lachenbruch, and G. C. Viloria, *SPIE* **2091,** 185 (1994).

R. F. Potter, in *Handbook of Optical Constants of Solids* (E. D. Palik, ed.), Chapter 2, p. 11. Academic Press, Boston, 1985.

Prometrix Corporation, *Solid State Technol.* **34**(9), 57 (1991).

H. Qi, P. E. Gee, and R. F. Hicks, *Phys. Rev. Lett.* **72,** 250 (1994).

J. Rappich, H. J. Lewerenz, and H. Gerischer, *J. Electrochem. Soc.* **140,** L187 (1993).

M. Rauh, P. Wissmann, and M. Wölfel, *Thin Solid Films* **233,** 289 (1993).

A. G. Reid and K. M. Sautter, *SPIE* **1673,** 284 (1992).

F. Reinhardt, W. Richter, A. B. Müller, D. Gutsche, P. Kurpas, K. Ploska, K. C. Rose, and H. Zorn, *J. Vac. Sci. Technol. B* **11,** 1427 (1993).

U. Resch, S. M. Scholz, U. Rossow, A. B. Müller, and W. Richter, *Appl. Surf. Sci.* **63,** 106 (1993).

H. H. Richardson, H.-C. Chang, C. Noda, and G. E. Ewing, *Surf. Sci.* **216,** 93 (1989).

W. Richter, *Philos. Trans. R. Soc. London, Ser. A* **344,** 453 (1993).

K. Riedling, *Ellipsometry for Industrial Applications.* Springer, New York, 1987.

J. Rivory, S. Fisson, V. Van Nguyen, G. Vuye, Y. Wang, F. Abelès, and K. Yu-Zhang, *Thin Solid Films* **233,** 260 (1993).

A. Röseler, *Thin Solid Films* **234,** 307 (1993).

J. Rumberg, F. Reinhardt, W. Richter, T. Farrell, and J. Armstrong, *J. Vac. Sci. Technol. B* **13,** 88 (1995).

K. L. Saenger, *J. Appl. Phys.* **63,** 2522 (1988).

K. L. Saenger and J. Gupta, *Appl. Opt.* **30,** 1221 (1991).

K. L. Saenger and H. M. Tong, *J. Appl. Polym. Sci.* **33,** 1777 (1987).

K. L. Saenger and H. M. Tong, in *New Characterization Techniques for Polymer Thin Films* (H. M. Tong and L. T. Nguyen, eds.), Chapter 4, p. 95. Wiley, New York, 1990.

K. L. Saenger and H. M. Tong, *Polym. Eng. Sci.* **31,** 432 (1991).

K. L. Saenger, F. Tong, J. S. Logan, and W. M. Holber, *Rev. Sci. Instrum.* **63,** 3862 (1992).

B. E. A. Saleh and M. C. Teich, *Fundamentals of Photonics.* Wiley, New York, 1991.

R. K. Sampson and H. Z. Massoud, *J. Electrochem. Soc.* **140,** 2673 (1993).

R. K. Sampson, K. A. Conrad, E. A. Irene, and H. Z. Massoud, *J. Electrochem. Soc.* **140,** 1734 (1993).

R. K. Sampson, K. A. Conrad, H. Z. Massoud, and E. A. Irene, *J. Electrochem. Soc.* **141,** 539 (1994a).

R. K. Sampson, K. A. Conrad, H. Z. Massoud, and E. A. Irene, *J. Electrochem. Soc.* **141,** 737 (1994b).
L. Samuelson, K. Deppert, S. Jeppesen, J. Jönsson, G. Paulsson, and P. Schmidt, *J. Cryst. Growth* **107,** 68 (1991).
E. Sanchez, P. Shaw, J. A. O'Neill, and R. M. Osgood, Jr., *J. Vac. Sci. Technol. A* **6,** 765 (1988).
H. Sankur and W. Gunning, *Appl. Phys. Lett.* **56,** 2651 (1990).
H. Sankur, W. Southwell, and R. Hall, *J. Electron. Mater.* **20,** 1099 (1991).
K. M. Sautter and T. Batchelder, *Microelectron. Manuf. Testing,* September, p. 36 (1987).
K. Sawara, T. Yaskaka, S. Miyazaki, and M. Hirose, *Proc. Int. Workshop Sci. Technol. Surf. React. Processes,* Tokyo, *1992,* p. 93 (1992).
I. Schnitzer, A. Katzir, U. Schiessl, H. Bottner, and M. Tacke, *SPIE* **1228,** 246 (1990).
S. M. Scholz, A. B. Müller, W. Richter, D. R. T. Zahn, D. I. Westwood, D. A. Woolf, and R. H. Williams, *J. Vac. Sci. Technol. B* **10,** 1710 (1992).
M. Schubert, V. Gottschalch, C. M. Herzinger, H. Yao, P. G. Snyder, and J. A. Woollam, *J. Appl. Phys.* **77,** 3416 (1995).
S. Selci, F. Ciccacci, G. Chiarotti, P. Chiaradia, and A. Cricenti, *J. Vac. Sci. Technol. A* **5,** 327 (1987).
B. O. Seraphin and N. Bottka, *Phys. Rev.* **139,** A560 (1965).
P. J. Severin and A. P. Severijns, *J. Electrochem. Soc.* **137,** 1306 (1990).
J. W. Sharp and D. Eres, *J. Cryst. Growth* **125,** 553 (1992).
H. Shen, *SPIE* **1286,** 125 (1990).
H. Shen, P. Parayanthal, Y. F. Liu, and F. H. Pollak, *Rev. Sci. Instrum.* **58,** 1429 (1987).
H. Shen, S. H. Pan, Z. Hang, J. Leng, F. H. Pollak, J. N. Woodall, and R. N. Sacks, *Appl. Phys. Lett.* **53,** 1080 (1988).
D. Y. Smith, in *Handbook of Optical Constants of Solids* (E. D. Palik, ed.), Chapter 3, p. 35. Academic Press, Boston, 1985.
K. A. Snow, WO Patent 92/21011 (1992).
J. Stefani and S. W. Butler, *J. Electrochem. Soc.* **141,** 1387 (1994).
L. Stiblert and T. Sandström, *J. Phys. (Paris)* **44,** C10–79 (1983).
B. R. Stoner, B. E. Williams, S. D. Wolter, K. Nishimura, and J. T. Glass, *J. Mater. Res.* **7,** 257 (1992).
P. L. Swart, *Mater. Res. Soc. Symp. Proc.* **324,** 75 (1994).
P. L. Swart and B. M. Lacquet, *J. Electron. Mater.* **19,** 1383 (1990).
A. Takano, M. Kawasaki, and H. Koinuma, *J. Appl. Phys.* **73,** 7987 (1993).
L. E. Tarof, C. J. Miner, and A. J. Springthorpe, *J. Electron. Mater.* **18,** 361 (1989).
V. L. Teal and S. P. Muraka, *J. Appl. Phys.* **61,** 5038 (1987).
Tencor Instruments, *Solid State Technol.* **33**(7), 50 (1990).
J. B. Theeten, *Surf. Sci.* **96,** 275 (1980).
J. B. Theeten, F. Hottier, and J. Hallais, *J. Cryst. Growth* **46,** 245 (1979).
J. B. Theeten, R. P. H. Chang, D. E. Aspnes, and T. E. Adams, *J. Electrochem. Soc.* **127,** 378 (1980).
M. E. Thomas, in *Handbook of Optical Constants of Solids II* (E. D. Palik, ed.), Chapter 8, p. 177. Academic Press, Boston, 1991.
M. O. Thompson, G. J. Galvin, J. W. Mayer, P. S. Peercy, J. M. Poate, D. C. Jacobson, A. G. Cullis, and N. G. Chew, *Phys. Rev. Lett.* **52,** 2360 (1984).

References

M. Thomson, *Solid State Technol.* **33**(5), 171 (1990).
P. J. Timans, R. A. McMahon, and H. Ahmed, *Appl. Phys. Lett.* **53**, 1844 (1988).
T. Tomita, T. Kinosada, T. Yamashita, M. Shiota, and T. Sakurai, *Jpn. J. Appl. Phys.* **25**, L925 (1986).
H. G. Tompkins, *A User's Guide to Ellipsometry.* Academic Press, Boston, 1993.
H. M. Tong, K. L. Saenger, and C. J. Durning, *J. Polym. Sci., Part B: Polym. Phys.* **27**, 689 (1989).
M. Toyoda and K. Sugimoto, *J. Jpn. Inst. Met., Sendai* **54**, 925 (1990).
A. Tripathi, D. Mazzarese, W. C. Conner, and K. A. Jones, *J. Electron. Mater.* **18**, 45 (1989).
S. Trolier-McKinstry, P. Chindaudom, K. Vedam, and R. E. Newnham, *Phys. Thin Films* **19**, 249 (1994).
F. K. Urban, III and J. C. Comfort, *Thin Solid Films* **253**, 262 (1994).
F. K. Urban, III and M. F. Tabet, *J. Vac. Sci. Technol. A* **11**, 976 (1993).
F. K. Urban, III and M. F. Tabet, *J. Vac. Sci. Technol. A* **12**, 1952 (1994).
F. K. Urban, III, D. C. Park, and M. F. Tabet, *Thin Solid Films* **220**, 247 (1992).
G. A. Vawter, J. F. Klem, and R. E. Leibenguth, *J. Vac. Sci. Technol. A* **12**, 1973 (1994).
K. Vedam, P. J. McMarr, and J. Narayan, *Appl. Phys. Lett.* **47**, 339 (1985).
M. Vernon, T. R. Hayes, and V. M. Donnelly, *J. Vac. Sci. Technol. A* **10**, 3499 (1992).
L. Viña and M. Cardona, *Physica B (Amsterdam)* **117**, 356 (1984).
N. Vodjdani and P. Parrens, *J. Vac. Sci. Technol. B* **5**, 1591 (1987).
G. Vuye, S. Fisson, V. Nguyen Van, Y. Wang, J. Rivory, and F. Abelès, *Thin Solid Films* **233**, 166 (1993).
M. Watts, T. Perera, B. Ozarski, D. Meyers, and R. Tan, *Solid State Technol.* **31**(7), 59 (1988).
L. M. Weegels, T. Saitoh, H. Oohashi, and H. Kanbe, *Appl. Phys. Lett.* **64**, 2661 (1994).
D. J. Wentink, H. Wormeester, P. De Boeij, C. Wijers, and A. van Silfhout, *Surf. Sci.* **274**, 270 (1992).
T. Wipiejewski and K. J. Ebeling, *J. Electrochem. Soc.* **140**, 2028 (1993).
W. L. Wolfe, in *Handbook of Optics* (W. G. Driscoll, ed.), Chapter 7, p. 7-1. McGraw Hill, New York, 1978.
C.-H. Wu, W. H. Weber, T. J. Potter, and M. A. Tamor, *J. Appl. Phys.* **73**, 2977 (1993).
I.-F. Wu, J. B. Dottellis, and M. Dagenais, *J. Vac. Sci. Technol. A* **11**, 2398 (1993).
M. Xi, S. Salim, and K. F. Jensen, *J. Vac. Sci. Technol. A* (1996, submitted).
V. A. Yakovlev, Q. Liu, and E. A. Irene, *J. Vac. Sci. Technol. A* **10**, 427 (1992).
V. A. Yakovlev, B. Drévillon, N. Layadi, P. R. i Cabarrocas, *J. Appl. Phys.* **74**, 2535 (1993).
T. Yamaguchi, S. Yoshida, and A. Kinbara, *Jpn. J. Appl. Phys.* **8**, 559 (1969).
M. Yamamoto and A. Arai, *Thin Solid Films* **233**, 268 (1993).
H. Yao and P. G. Snyder, *Thin Solid Films* **206**, 283 (1991).
H. Yao, P. G. Snyder, and J. A. Woollam, *J. Appl. Phys.* **70**, 3261 (1991).
H. Yao, J. A. Woollam, and S. A. Alterovitz, *Appl. Phys. Lett.* **62**, 3324 (1993).
J. Yota and V. A. Burrows, *Mater. Res. Soc. Symp. Proc.* **204**, 345 (1991).
J. Yota and V. A. Burrows, *J. Vac. Sci. Technol. A* **11**, 1083 (1993).
G. Zalczer, O. Thomas, J.-P. Piel, and J.-L. Stehlé, *Thin Solid Films* **234**, 356 (1993).
Y. Zhang, G. S. Oehrlein, E. de Frésart, and J. W. Corbett, *J. Vac. Sci. Technol. A* **11**, 2492 (1993).
Z.-H. Zhou, E. S. Aydil, R. A. Gottscho, Y. J. Chabal, and R. Reif, *J. Electrochem. Soc.* **140**, 3316 (1993).

CHAPTER **10**

Interferometry and Photography

This chapter is an overview of the use of interferometry and photographic imaging in optical diagnostics. Since optical interferometry and imaging play important roles in many of the optical diagnostics presented in other chapters—and have been addressed in those chapters—this chapter describes in detail only those applications that do not overlap these other probes.

These techniques are especially useful for visualizing flow (Lauterborn and Vogel, 1984; Merzkirch, 1987; Hesselink, 1988; Yang, 1989; Adrian, 1991; Freymuth, 1993), as is particularly important in chemical vapor deposition (CVD). Interferometry (Section 10.1) and shadowgraphy and schlieren photography (Section 10.2) are based on index-of-refraction inhomogeneities, and so they provide information about local densities and temperature. [Elastic scattering (Section 11.1.2.2) is also used for flow visualization.] Interferometry is also an important monitor of the wafer surface. For example, speckle photography and interferometry, which are useful for metrology and thermometry of the surface, are covered in Section 11.3.

10.1 Interferometry

Interferometry can be important in diagnostics in three ways. (1) Stand-alone interferometers can serve as spectroscopic instruments that analyze light collected from a reactor. (2) Interference effects can occur within a physical structure, such as a thin film or multiple films, and this interference can serve as the basis of a diagnostic. (3) The gas-phase medium in a reactor can be placed within an arm of an interferometer, as is done in CVD fluid dynamics measurements of density and temperature. In each of these three cases, interferometry compares the optical path length integrated along a ray in the reactor or film, $\int n\, dx$ (where n is the refractive index), either to the optical length along a reference path or to an absolute value of phase (such as π or $\pi/2$).

Figure 10.1 illustrates four common types of interferometers. Details about the optics of these interferometers can be found in Born and Wolf (1970), Klein and Furtak (1986), and Hecht (1987). Philbert *et al.* (1989b) have described the use of several of these instruments for flow visualization. Breiland and Ho (1993) have reviewed the use of interferometry for *in situ* density and temperature measurements in chemical vapor deposition reactors. Stamper *et al.* (1986) have discussed the use of interferometry for probing laser-produced plasmas.

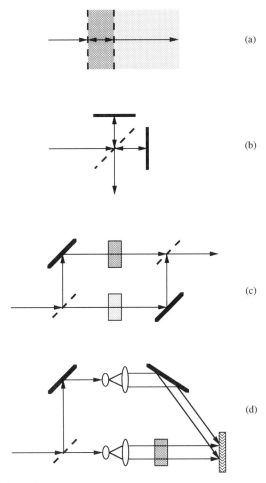

Figure 10.1 Schematic depiction of (a) Fabry–Perot, (b) Michelson, (c) Mach–Zehnder, and (d) holographic interferometers.

10.1 Interferometry

Fabry–Perot interferometry is schematically depicted in Figure 10.1a. As a stand-alone interferometer, it consists of two parallel mirrors separated by a distance L, and resolves frequency components within the free spectral tbrange $c/2L$ when used with plane-parallel mirrors. Fabry–Perot interferometers are used for high-resolution spectroscopy, and have been used occasionally in optical emission spectroscopy (OES, Chapter 6) and laser-induced fluorescence (LIF, Chapter 7) diagnostics. A solid optical element with two parallel surfaces, an etalon, serves the same function, and is used intracavity to a laser to narrow its frequency output. The interference effect in thin films, illustrated in Figures 9.12 and 10.1a, is actually Fabry–Perot interferometry. Many optical spectroscopies described in this volume utilize this interference to monitor film thickness, dielectric properties, and temperature, including OES (Section 6.3.1.3), reflectometry (Section 9.5), ellipsometry (Section 9.11), Raman scattering (Section 12.4.2), and pyrometry (Section 13.4).

The Michelson interferometer is illustrated in Figure 10.1b. As a stand-alone unit, it can be used to perform Fourier transform spectroscopy (FTS). The principles of this method are described in Section 5.1.3.2 (Figure 5.8) and applications of FTS are presented in several sections, including Sections 8.2.1, 9.7, and 13.4. The Michelson interferometer can also be used to measure the optical path length of a medium, such as the gaseous region of a thin film reactor, by placing it in one of the two arms.

Walkup *et al.* (1986) measured the plasma electron density in the plume during pulsed laser ablation by placing the plume in one arm of a Michelson interferometer. The index of refraction n of free electrons, as in a plasma, is <1 (as obtained by using Equation 4.19 with $\nu_0 = 0$), while that due to a neutral gas is >1 (for probe frequencies far below resonance, as in the middle visible or infrared); consequently, the sign of $n - 1$ provides a measure of the relative density of electrons and neutrals. This method is much more sensitive to free electrons than to neutral atoms at the 632.8-nm and 1.15-μm wavelengths available from a He–Ne laser. Walkup *et al.* saw the interferometer signal decrease during the ablation of Al_2O_3, showing that $n - 1 < 0$ and that many electrons were formed, as is seen in Figure 10.2; the interferometer signal also tracked Al* emission. In contrast, this signal increased ($n - 1 > 0$) during the ablation of organic polymers, showing that very few electrons were formed; also, the interferometry time profile was very different from that of C_2* optical emission—it began later and decayed much slower.

Bollmann and Haberger (1992) have developed an interferometric method, which is similar to Michelson interferometry, that determines the average temperature of a wafer by thermal expansion. Two small grating-like structures are fabricated on a Si(100) wafer by using a KOH etch, so the grooved surfaces are in the (111) plane. These two structures have parallel grooves, and are placed on opposite sides of the wafer to maximize

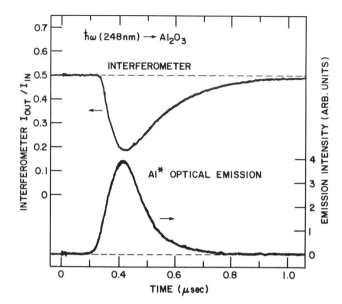

Figure 10.2 Michelson interferometer signal (upper curve) and 396.2-nm Al* optical emission (lower curve) after 248-nm laser ablation of Al_2O_3 (from Walkup et al., 1986).

sensitivity. As shown in Figure 10.3, a He–Ne laser is split into two beams, each of which is directed normal to these (111) facets. The light retroreflected from these two structures is recombined to form a series of interference stripes on the charge-coupled device (CCD) detector, which migrate on the detector as changes in temperature alter the distance between the two structures. One disadvantage of this method for real-time monitoring of temperature during manufacturing is the need to fabricate the grating

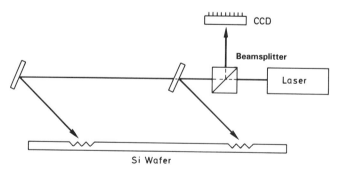

Figure 10.3 Measurement of average wafer temperature by interferometric sensing of thermal expansion (from Bollmann and Haberger, 1992).

10.1 Interferometry

patterns. Sections 11.2.2.3 and 11.3 describe other applications of interferometry to wafer thermometry.

The Mach–Zehnder interferometer depicted in Figure 10.1c compares the optical paths in the two arms, one containing the medium of interest and the other usually containing optics that compensate for path length differences with the first arm. One application of this interferometer is in hook spectroscopy, which makes use of the (anomalous) dispersion near the electronic transitions of atoms (Equation 4.19). In this method, the (fringe) output of a Mach–Zehnder interferometer is focused into a stigmatic grating spectrograph. The wavelength-dispersed fringe pattern at the exit focal plane has hook-like features because of dispersion. The wavelength spread between the hooks gives the density if the oscillator strength is known, and vice versa. Sappey *et al.* (1993) used hook spectroscopy to determine the absolute copper-atom density in laser-ablated copper plasmas, as is depicted in Figure 10.4.

In conventional inteferometry, such as Michelson and Mach–Zehnder interferometry, the two beams are combined at one time and place to

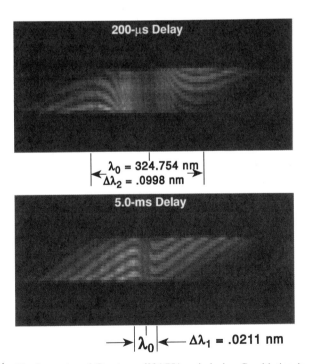

Figure 10.4 Hook spectra of Cu atoms (324.754 nm) during Cu ablation in 25 Torr He taken at delay times of 200 μsec (top) and 5.0 msec (bottom). The hook separation is much smaller in the later spectrum because the Cu density has decreased by a factor of ~35 (from Sappey *et al.*, 1993).

provide optical path information. In interference holography, or holographic interferometry, (Vest, 1979), depicted in Figure 10.1d, comparisons are made of conditions that occur at different times. After splitting a laser into two paths and expanding both, the beam passing through the medium (sample beam) interferes with the reference beam on the holographic plate. The phase and amplitude information stored in the plate is used to compare the medium under different experimental conditions for which the path lengths are slightly different.

In the "double-exposure" method, holograms are made on the same plate under two different conditions, such as with the reactor gas at ambient and normal operating temperatures. The plate is then developed, repositioned, and illuminated with the same reference beam. The reconstructed beam is that of the cell, with an interference pattern representing the difference in optical path lengths between the two conditions. In the closely related "real-time" method, a hologram is made at one condition, say at ambient temperature. It is then developed and put back in place. Illumination with the same reference beam gives the reconstructed original sample beam, while illumination with this reference beam and the new sample beam, with new experimental conditions, gives a real-time interference pattern relative to the first condition. While this provides the same information as with the double exposure method, changes due to changing conditions can be monitored in real time. However, very careful repositioning of the developed holographic plate is necessary in the real-time method to achieve good accuracy. In fact, the film is often developed in place to avoid repositioning problems.

Another development in real-time holography is the use of a photoconducting plastic plate, which can create a phase grating, instead of photographic film. An electric charge is placed on the plastic plate before exposure. When the reference hologram is taken, charge migrates between the light and dark regions in the photoconducting plastic. The plate is then "developed" in place with a large electric field that creates strains that alter the refractive index. This produces phase shifts that are proportional to the hologram light intensity. The pattern can be erased after use by heating the plastic to smooth out the strains.

Giling (1982a,b) used both holographic methods to examine flow in several types of horizontal CVD cells. In an air-cooled cylindrical cell, Giling observed a stable pattern of parallel fringes above the susceptor for low flow rates of H_2 and He, which indicated stable laminar flow. For faster flow rates, concentric circles developed (H_2, Figure 10.5a), indicating that there was a cold axial region above the susceptor; the flow pattern was still stable. For N_2 and Ar, the pattern became very complicated when the flow rate was fast in this cell (N_2, Figure 10.5b) and the real-time pattern fluctuated; the spiral patterns in Figure 10.5b demonstrate the combined effect of forward and convective motions. In a rectangular CVD cell, this flow condition led to

10.1 Interferometry

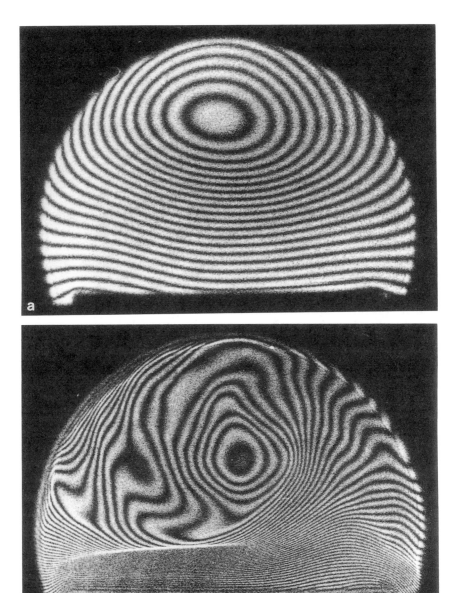

Figure 10.5 Interference pattern in an air-cooled cylindrical reactor ($T_{susceptor}$ = 1350 K) obtained by interference holography with (a) H_2 flow at high flow velocity (70 cm/sec) and (b) N_2 flow at high flow velocity (18 cm/sec) (from Giling, 1982a).

turbulent flow. Williams and Peterson (1986) conducted a similar study in a vertical metalorganic (MO) CVD reactor. Benet *et al.* (1981) used interference holography to measure temperature profiles in a stagnation-flow CVD reactor. They examined paths across the radially symmetric reactor and used Abel inversion methods to invert the data (Section 5.2.1.3), as is shown in Figure 10.6. More recently, Chehouani *et al.* (1993) modified the apparatus used in the Benet *et al.* study to achieve faster and more accurate temperature measurements in a large cold wall CVD reactor.

Breiland and Ho (1993) have concluded that interference holography is better than the other interference and photography methods discussed in this chapter for investigating flows in CVD reactors because it is the least sensitive to aberrations in optics, such as those due to the reactor windows. While the tracer method gives the flow streamlines directly, interference methods, such as interference holography, have the advantage that they do not require such intrusive additives. However, the three-dimensional temperature profiles and flow patterns must be deduced from the measured optical path length differences with these methods, which is difficult to do quantitatively.

This technique has also been used to probe the plume formed during laser ablation. For example, Ehrlich and Wagner (1991) employed double-

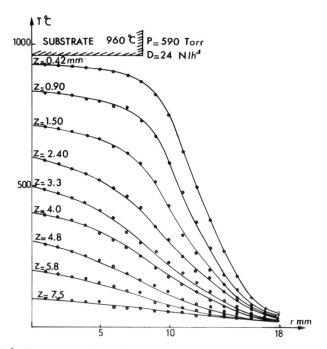

Figure 10.6 Temperature in a vertical reactor at several heights (z) as a function of radius (r), determined by interference holography (from Benet *et al.*, 1981).

exposure pulsed holography to probe the plume during pulsed laser deposition. Lindley *et al.* (1993) used resonant holographic interferometry to measure absolute-line density profiles of neutral aluminum atoms in the plume formed during KrF laser ablation from an aluminum target. This measurement was made specific and very sensitive to this species alone by tuning the dye laser used in the holography near the 394.401-nm Al transition from the ground state.

Mostovych *et al.* (1991) used polarization (Nomarski) interferometry to probe the plasma formed after laser ablation of aluminum. This type of interferometry has been reviewed by Stamper *et al.* (1986).

10.2 Photography, Imaging, and Microscopy

Photography and other imaging techniques can be used in conjunction with many of the optical spectroscopies described in the other chapters. Light that is emitted by a medium (optical emission, Chapter 6; laser-induced fluorescence, Chapter 7; photoluminescence, Chapter 14), transmitted through a medium (Chapter 8), specularly reflected from a surface (Chapter 9), or scattered from a medium (Chapters 11 and 12) can be imaged to obtain a 2D or 3D picture. Section 5.2.1 describes several ways to image light from a medium, including direct two-dimensional imaging.

This section describes photography diagnostics that do not cleanly fit into the topics of any of these cited chapters and imaging and microscopy methods that are not presented in other chapters. One imaging method described elsewhere in this volume is the use of the light-sheet technique (Section 5.2.1.2) for 2D imaging of LIF, which is described in Chapter 7. It can also be used for flow illumination by using nonresonant processes, such as elastic scattering (Philbert *et al.*, 1989c). One topic in this section is high speed photography, which has been reviewed by Courtney-Pratt (1986), Pfeiffer (1986), and Hugenschmidt (1986). Several of the interferometric techniques described in Section 10.1 overlap with the high-speed, as well as other, photographic methods described in this section.

Shadowgraphy and schlieren photography are often used for flow visualization (Philbert *et al.*, 1989a); they have not been widely used for thin film processing. Both methods are based on refractive-index inhomogeneities in transparent media that cause beam deflections. A shadowgraph records the transmission through the uniformly illuminated medium, as illustrated in Figure 10.7a. In schlieren (Figure 10.7b), the medium is illuminated by light passing through the schlieren diaphragms (D1—primary or entrance, D2—secondary or exit) that are placed in the focal planes of the field lenses L1 and L2. The diaphragms, say a square aperture (D1) and a knife-edge (D2), are aligned in a homogeneous medium so that half the rays from the imaged region (A) are transmitted past D2 and are detected at

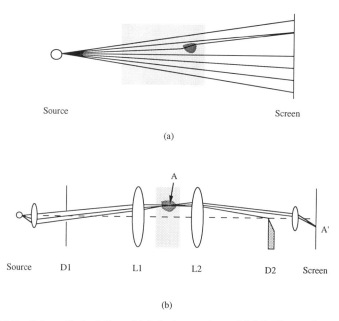

Figure 10.7 Schematic depiction of (a) shadowgraphy and (b) Schlieren photography.

A'. When, due to the process, there are inhomogeneities in the refractive index in that imaged region A, the fraction of light that is transmitted past the knife-edge and is imaged at A' changes.

These techniques have been combined with fast imaging methods to follow the plume formed during pulsed laser deposition. Gupta *et al.* (1991) used ultrafast photography (shadowgraphy) with a synchronized dye laser to probe the evolution of fragments expelled during excimer laser ablation of $YBa_2Cu_3O_{7-x}$, as used for pulsed laser deposition. Srinivasan *et al.* (1989, 1990), Braren *et al.* (1991), and Kelly *et al.* (1992) used this method to study the ablation of polymers, including polymethylmethacrylate (PMMA) and polyethyleneterephthalate (PET). Figure 10.8 shows time-resolved images of PMMA ablation at 248 nm, which depict absorption and elastic scattering of the probe laser by the fragments and the effects of the resulting shock wave. Ventzek *et al.* (1990, 1992) reported the use of schlieren photography and shadowgraphy to examine hydrodynamic phenomena in the plumes of ablated polymers and aluminum. Dyer and Sidhu (1988) and Dyer *et al.* (1990) used streak photography to show prompt emission from atomic and ionic species during ablation from polymer and $YBa_2Cu_3O_{7-x}$ targets. Kim *et al.* (1988) and Zyung *et al.* (1989) developed an ultrafast imaging method, with imaging that provides very good spatial resolution as well as temporal resolution, to study regions damaged or ablated by a laser.

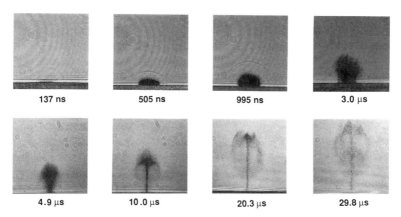

Figure 10.8 Ablation of PMMA by a single 248-nm (XeCl laser) pulse that is normal to the surface, with the dye laser probe (596 nm) delayed and parallel to the surface and illuminating the plume from behind. The widths of the photos correspond to 1.8 mm in the upper row and 2.9 mm in the lower row (from B. Braren, K. G. Casey, and R. Srinivasan, unpublished; also see Srinivasan *et al.*, 1989).

Use of Michelson and holographic interferometry to probe laser ablation is described earlier in this chapter (Section 10.1). Plume dynamics can also be followed by OES (Section 6.3.4), LIF (Section 7.2.2), absorption (Section 8.2.2.1.2), and pyrometry (Section 13.5). Among the techniques described is fast imaging of plume emission with an intensified charge-coupled device (ICCD) array camera (Figures 6.29 and 6.30).

Microscopy is widely used for wafer inspection. Imaging the processed wafer onto a vidicon with a microscope is often used for real-time observation during processing with localized laser beams (direct laser writing) (Herman, 1989). Figure 5.11 depicts an apparatus that can be used for laser writing, sample observation by a microscope, and for Raman and LIF diagnostics in backscattering configuration. One of the thermometric methods described in Section 9.4.1 (Figure 9.9), which is based on thermal expansion of the wafer, utilizes optical microscopy.

References

R. J. Adrian, *Annu. Rev. Fluid Mech.* **23**, 261 (1991).

S. Benet, R. Bergé, S. Brunet, S. Charar, B. Armas, and C. Combescure, in *Proceedings of the Eighth International Conference on Chemical Vapor Deposition* (J. M. Blocher, Jr., G. E. Vuillard, and G. Wahl, eds.), p. 188. Electrochemical Society, Pennington, NJ, 1981.

D. Bollmann and K. Haberger, *Microelectron. Eng.* **19**, 383 (1992).

M. Born and E. Wolf, *Principles of Optics,* 4th ed. Pergamon, Oxford, 1970.

B. Braren, K. G. Casey, and R. Kelly, *Nucl. Instrum. Methods B* **58,** 463 (1991).
W. G. Breiland and P. Ho, in *Chemical Vapor Deposition* (M. L. Hitchman and K. F. Jensen, eds.), Chapter 3, p. 91. Academic Press, Boston, 1993.
H. Chehouani, S. Bénet, S. Brunet, and B. Armas, *J. Phys. IV, Colloque C3, Supplément* II **3,** 83 (1993).
J. S. Courtney-Pratt, in *Fast Electrical and Optical Measurements,* (J. E. Thompson and L. H. Luessen, eds.), Vol. II, p. 595. Martinus Nijhoff, Dordrecht, 1986.
P. E. Dyer and J. Sidhu, *J. Appl. Phys.* **64,** 4657 (1988).
P. E. Dyer, A. Issa, and P. H. Key, *Appl. Phys. Lett.* **57,** 186 (1990).
M. J. Ehrlich and J. W. Wagner, *Appl. Phys. Lett.* **58,** 2883 (1991).
P. Freymuth, *Rev. Sci. Instrum.* **64,** 1 (1993).
L. J. Giling, *J. Electrochem. Soc.* **129,** 634 (1982a).
L. J. Giling, *J. Phys.* (*Paris*) **43,** C5-235 (1982b).
A. Gupta, B. Braren, K. G. Casey, B. W. Hussey, and R. Kelly, *Appl. Phys. Lett.* **59,** 1302 (1991).
E. Hecht, *Optics,* 2nd ed. Addison–Wesley, Reading, MA, 1987.
I. P. Herman, *Chem. Rev.* **89,** 1323 (1989).
L. Hesselink, *Annu. Rev. Fluid Mech.* **20,** 421 (1988).
M. Hugenschmidt, in *Fast Electrical and Optical Measurements* (J. E. Thompson and L. H. Luessen, eds.), Vol. II, p. 643. Martinus Nijhoff, Dordrecht, 1986.
R. Kelly, A. Miotello, B. Braren, A. Gupta, and K. Casey, *Nucl. Instrum. Methods B* **65,** 187 (1992).
H, Kim, J. C. Postlewaite, T. Zyung, and D. D. Dlott, *J. Appl. Phys.* **64,** 2955 (1988).
M. V. Klein and T. E. Furtak, *Optics,* 2nd ed. Wiley, New York, 1986.
W. Lauterborn and A. Vogel, *Annu. Rev. Fluid Mech.* **16,** 223 (1984).
R. A. Lindley, R. M. Gilgenbach, and C. H. Ching, *Appl. Phys. Lett.* **63,** 888 (1993).
W. Merzkirch, *Flow Visualization,* 2nd ed. Academic Press, Orlando, FL, 1987.
A. N. Mostovych, K. J. Kearney, J. A. Stamper, and A. J. Schmitt, *Phys. Rev. Lett.* **66,** 612 (1991).
W. Pfeiffer, in *Fast Electrical and Optical Measurements* (J. E. Thompson and L. H. Luessen, eds.), Vol. II, p. 609. Martinus Nijhoff, Dordrecht, 1986.
M. Philbert, J. Surget, and C. Véret, in *Handbook of Flow Visualization* (W.-J. Yang, ed.), Chapter 12, p. 189. Hemisphere, New York, 1989a.
M. Philbert, J. Surget, and C. Véret, in *Handbook of Flow Visualization* (W.-J. Yang, ed.), Chapter 13, p. 203. Hemisphere, New York, 1989b.
M. Philbert, J. Surget, and C. Véret, in *Handbook of Flow Visualization* (W.-J. Yang, ed.), Chapter 14, p. 211. Hemisphere, New York, 1989c.
A. D. Sappey, T. K. Gamble, and D. K. Zerkle, *Appl. Phys. Lett.* **62,** 564 (1993).
R. Srinivasan, B. Braren, K. G. Casey, and M. Yeh, *Appl. Phys. Lett.* **55,** 2790 (1989).
R. Srinivasan, K. G. Casey, B. Braren, and M. Yeh, *J. Appl. Phys.* **67,** 1604 (1990).
J. A. Stamper, E. A. McLean, S. P. Obenschain, and B. H. Ripin, in *Fast Electrical and Optical Measurements* (J. E. Thompson and L. H. Luessen, eds.), Vol. II, p. 691. Martinus Nijhoff, Dordrecht, 1986.
P. L. G. Ventzek, R. M. Gilgenbach, J. A. Sell, and D. M. Heffelfinger, *J. Appl. Phys.* **68,** 965 (1990).
P. L. G. Ventzek, R. M. Gilgenbach, C. H. Ching, and R. A. Lindley, *J. Appl. Phys.* **72,** 1696 (1992).

References

C. M. Vest, *Holographic Interferometry.* Wiley, New York, 1979.
R. E. Walkup, J. M. Jasinski, and R. W. Dreyfus, *Appl. Phys. Lett.* **48,** 1690 (1986).
J. E. Williams and R. W. Peterson, *J. Cryst. Growth* **77,** 128 (1986).
W.-J. Yang, ed., *Handbook of Flow Visualization.* Hemisphere, New York, 1989.
T. Zyung, H. Kim, J. C. Postlewaite, and D. D. Dlott, *J. Appl. Phys.* **65,** 4548 (1989).

CHAPTER 11

Elastic Scattering and Diffraction from Particles and Nonplanar Surfaces (Scatterometry)

This chapter describes elastic scattering of light from particles and from surfaces that have random or periodic deviations from a perfectly smooth plane. Since the frequencies of the incident and scattered light are the same in elastic scattering, optical analysis is not concerned with frequency shifts. Instead, the angular pattern of the scattering profile, the absolute intensity of the scattering, and the dependence on the incident and scattered electric field polarizations are analyzed by electromagnetic diffraction theory. Sometimes, losses due to absorption also affect the scattering profile. Elastic scattering is applicable to process studies and to real-time monitoring and control. Elastic light scattering is also widely called laser light scattering (LLS) or scatterometry. Speckle photography and interferometry are other types of diagnostic measurements based on elastic scattering.

Elastic scattering from particles in a gas or liquid or on a surface is important in thin film processing, as is that from surfaces that are either rough or have periodically etched or deposited features. While both can be explained by diffraction theory, they involve different approximations and are treated separately in this chapter. Elastic scattering is useful for real-time control because instrumentation costs are low. Scattering profiles, which are often called scattergrams, are compared to theoretical predictions or to scattergrams from reference samples. It is straightforward to collect light that is scattered into the complete range of angles needed for detailed analysis in *ex situ* studies, but it is more difficult to do so during *in situ* investigations. In particular, the angular range of collected light can be severely limited by constraints in production reactors. Still, even a limited angle range can be useful for many applications.

Elastic or Rayleigh scattering is distinguished from the inelastic scattering processes described in Chapter 12—Raman and Brillouin scattering—in which the scattered photon energy is different from that incident. Technically, Rayleigh scattering is quasi-elastic scattering by nonpropagating density changes in gases, liquids, and solids caused by entropy fluctuations. Elastic scattering by particles that are much smaller than the wavelength of light is sometimes called Rayleigh scattering. In quasi-elastic scattering, the scattered light has the same frequency as the incident light but can have a broader spectral width. Such scattering was used by Hori *et al.* (1996) to measure electron density and temperature (Thomson scattering, $\omega_0 = 0$ in Equations 4.18 and 4.23) and neutral atom density (Rayleigh scattering, $\omega \ll \omega_0$) in processing plasmas.

11.1 Detection of Particles

The appearance of particulates in thin film processing usually indicates trouble. Whether they originate from solvents, from interactions with the walls of the processing chamber and connecting tubes, or from the gas-phase chemistry occurring in the processing chamber, particulates on the surface of a wafer can lead to defects. Particles as small as 0.08 μm (and smaller) and particle densities $\gtrsim 0.1/cm^2$ constitute a major problem, as is described in the volumes edited by Tolliver (1988) and Donovan (1990) on particle identification and control in semiconductor manufacturing, and the roadmap by the Semiconductor Industry Association (1994), as well as in overviews by Tullis (1988), Peters (1992), Selwyn (1993), and Menon and Donovan (1993). Use of IR and Raman spectroscopies for *ex situ* particle analysis has been described by Fisher and Davidson (1990).

The major source of particles in microelectronics fabrication is formation during processing (O'Hanlon and Parks, 1992). Particulates can produce defects because their physical presence affects a thin film process locally. Alternatively, particles can decompose during later stages of processing, such as by reacting with water vapor, to produce surface contamination. Inputs to the reactor and process, including reaction gases, solvents, etc., were once the dominant sources of particulates. Particle levels from these sources and from people and the environment have been reduced, so they now contribute only a small faction of the total particle count.

Elastic light scattering is useful in detecting particles in the gas phase, on surfaces, and in liquids, and has been used increasingly to monitor particles *in situ* during several phases of thin film processing. In fact, because this technique is both straightforward and sensitive, in some investigations particles are purposely introduced into a flowing gas, in chemical-vapor deposition (CVD) systems and more general studies of flow dynamics, to examine the streamlines.

11.1 Detection of Particles

Use of elastic scattering to monitor particles *in situ* in process chambers and exhaust lines of plasma etchers, ion implanters, CVD tools, and resist strippers has dramatically reduced the cost of tool ownership and maintenance (Peters, 1992). Processes and cleaning procedures have been optimized, tool cleanliness has improved, and maintenance needs have decreased, resulting in increased tool availability. Use of particle monitors lessens the need for costly "testing" monitor wafers; moreover, particle sensors often yield more accurate information than that obtained by using these monitor wafers. Several vendors offer sensors that detect particles ≥ 0.2 μm in diameter.

The mechanisms for particle formation and transport have been much scrutinized, as have optimal particle measurement methodologies, and have at times, been the subject of some controversy (Selwyn, 1993). In particular, the formation of particles can be very important in plasma processing where particles can form heterogeneously and homogeneously, the latter resulting in a more highly monodispersed particle size distribution. Understanding particle transport in plasmas is essential; it is influenced by viscous forces, radiometric forces, gravity, turbulence, etc. *In situ* measurement of particles in plasmas is now preferred to post-processing analysis on monitor wafers or downstream measurements, although particle sensors are still often installed on the roughing lines of vacuum systems.

In situ particle monitors are also used to optimize and monitor processes other than plasma etching (Peters, 1992). In CVD, they are used to adjust gas flow rates to eliminate gas-phase nucleation and to signal the need for electrode replacement. Particle counting helps detect incomplete stripping of resists in photoresist ashing and to determine photoresist problems during ion implantation. During sputtering, flaking of films from clips or targets can be observed and cathode arcing can be detected with particle monitors. With *in situ* monitoring, tools can be cleaned only when needed, e.g., when particle counts exceed allowable levels, thereby optimizing tool availability.

Ideally, it would be nice if light scattering could identify the composition of the particle (or the different compositions if several types of particles are present), determine their shape, and measure the distribution of the density of particles as a function of size. It would also be nice if these measurements could be made with sufficient spatial resolution and speed so that these parameters could be mapped in three dimensions. Although the examination of the scattered light at different scattering angles, for different polarization configurations, and with several laser wavelengths can provide much information, it is still not sufficient to describe the system fully. Usually, several simplifying assumptions about the size distribution and the particle shape must be made. Although the direct problem of determining the scattering profile of a known particle distribution is relatively straightforward, the inverse problem of determining the distribution from the scattering profile has neither a simple nor unique solution.

11.1.1 *Theory*

A detailed theoretical treatment of the scattering of light by particles has been given by van de Hulst (1957), Kerker (1969), Bohren and Huffman (1983), and Bohren (1995). Leith (1990) has written a brief overview of the subject. Conditions in film reactors usually permit two simplifying assumptions: (1) The particle density is low enough so that each particle scatters light independently; (2) each incident photon is scattered at most once (single scattering event). It will be further assumed that there is one type of particle, which is homogeneous and identifiable by its composition and shape.

The general theory of light scattering treats the incident field E_i as a plane wave and the scattered beam E_s as a spherical wave. The notation of Bohren and Huffman (1983) is adopted here; they used the same phase factor for the plane wave as in Equations 2.2 and 2.3. The scattered wave is observed a distance r from the particle, at a scattering angle θ from the direction of the incident laser and at an azimuthal angle ϕ (Figure 11.1). The polarizations of these fields are defined as being either parallel ($\parallel$) or normal ($\perp$) to the scattering plane, which is defined by the wavevectors of the incident and scattered beams.

The components of the scattered electric field E_s are related to the incident field E_i by the amplitude scattering matrix S_i as

$$\begin{pmatrix} E_{\parallel s} \\ E_{\perp s} \end{pmatrix} = \frac{e^{ikr-ikz}}{-ikr} \begin{pmatrix} S_2 & S_3 \\ S_4 & S_1 \end{pmatrix} \begin{pmatrix} E_{\parallel i} \\ E_{\perp i} \end{pmatrix} \qquad (11.1)$$

In general S_i depends on the scattering angle θ and the azimuthal angle ϕ.

Alternatively, the scattering problem can be posed in terms of the Stokes parameters. In scattering theory, the Stokes parameters are labeled as I, Q, U, and V or as s_1, s_2, s_3 and s_4, instead of the ellipsometric notation (S_i) presented in Section 2.3, to avoid confusion with the amplitude scattering matrix. The Stokes parameters of the scattered and incident beams are related by

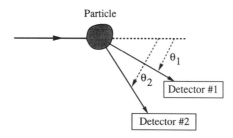

Figure 11.1 Schematic of elastic scattering from particles.

11.1 Detection of Particles

$$\begin{pmatrix} I_s \\ Q_s \\ U_s \\ V_s \end{pmatrix} = \frac{1}{k^2 r^2} \begin{pmatrix} S_{11} & S_{12} & S_{13} & S_{14} \\ S_{21} & S_{22} & S_{23} & S_{24} \\ S_{31} & S_{32} & S_{33} & S_{34} \\ S_{41} & S_{42} & S_{43} & S_{44} \end{pmatrix} \begin{pmatrix} I_i \\ Q_i \\ U_i \\ V_i \end{pmatrix} \quad (11.2)$$

where the scattering matrix $S_{ij}(\theta,\phi)$ can be expressed in terms of the amplitude scattering matrix $S_i(\theta,\phi)$ (Bohren and Huffman, 1983).

The total power removed from the laser beam by the particle, integrated over θ and ϕ, is $\mathcal{P}_s = C_{ext} I_i$, where C_{ext} is the cross-section (which is an area) and I_i is the incident intensity. This total removal of photons, termed extinction, is due to scattering and absorption by the particles, with $C_{ext} = C_{sca} + C_{abs}$. (While σ is commonly used to denote a cross-section, and is used elsewhere in this volume, C_i is common in particle scattering communities and will be used here.)

Maxwell's equations have been solved for scattering by transparent and absorbing particles that are spheres, cylinders, etc. Three parameters are used in assessing the approximations that are useful in analyzing scattering. (1) $x = 2\pi a/\lambda$ is the normalized size of the particle, where a is the characteristic dimension, which is the radius for a sphere, and λ is the wavelength of light in the ambient (λ_{vac}/n_2; λ_{vac} is the wavelength in vacuum). (2) m is the relative index of refraction of the two media with $m = (n_1 + ik_1)/(n_2 + ik_2)$, where the particle is medium 1 and the ambient is medium 2. For scattering silica particles in a gas in the visible region, m is real, while for scattering aluminum particulates m is complex. (3) $2x(n_1 - n_2)$ is the phase shift incurred by light passing through the particle.

When $x \ll 1$ ($a \ll \lambda$), scattering is called Rayleigh scattering and is independent of the shape of the particle. Mie (or Lorenz–Mie) scattering is the term often used to describe all types of scattering from homogeneous spheres. When $n_1 \sim n_2$, or more precisely when $n_1 - n_2$ and $x(n_1 - n_2)$ are small, scattering from spheres is termed Rayleigh–Gans scattering. When $x \gg 1$ ($a \gg \lambda$), geometric optics can be used to describe scattering.

Particulates formed during thin film processing may be irregularly shaped. Still, in most studies they have been approximated as spheres to simplify analysis; this is assumed here. Scattering is independent of the azimuthal angle ϕ for spherical particles. Also then $S_3 = S_4 = 0$, so that if the incident light is linearly polarized either normal ($I_{\perp i}$) or parallel ($I_{\parallel i}$) to the scattering plane, the scattered light is similarly polarized [$I_{\perp s}(\theta)$ and $I_{\parallel s}(\theta)$, respectively]. The total scattered intensity is $I(\theta) = I_{\perp s}(\theta) + I_{\parallel s}(\theta)$.

$I_{\perp s}$ and $I_{\parallel s}$ can be expressed as

$$I_{\perp s}(\theta) = \frac{i_1(\theta)}{k^2 r^2} I_{\perp i} \quad (11.3)$$

$$I_{\parallel s}(\theta) = \frac{i_2(\theta)}{k^2 r^2} I_{\parallel i} \quad (11.4)$$

where $i_1 = |S_1(\theta)|^2$ and $i_2 = |S_2(\theta)|^2$. The cross-section for scattering into a solid angle $d\Omega = \sin\theta d\theta d\phi$ is then $(\lambda^2/4\pi)(|S(\theta)|^2)d\Omega$, where $|S(\theta)|^2 = |S_1(\theta)|^2$ for $\perp$-polarized light, $|S_2(\theta)|^2$ for $\|$-polarized light, and $(|S_1(\theta)|^2 + |S_2(\theta)|^2)/2$ for unpolarized incident light. For spheres, $S_1(0°) = S_2(0°) = S(0°)$.

By solving Maxwell's equations, the amplitude scattering functions for a sphere can be shown to be

$$S_1(\theta) = \sum_{n=1}^{\infty} \frac{2n+1}{n(n+1)} [a_n \pi_n(\cos\theta) + b_n \tau_n(\cos\theta)] \quad (11.5)$$

$$S_2(\theta) = \sum_{n=1}^{\infty} \frac{2n+1}{n(n+1)} [a_n \tau_n(\cos\theta) + b_n \pi_n(\cos\theta)] \quad (11.6)$$

where π_n and τ_n are related to the associated Legendre polynomials $P_n^{(1)}(\cos\theta)$ by

$$\pi_n(\cos\theta) = \frac{1}{\sin\theta} P_n^{(1)}(\cos\theta) \quad (11.7)$$

$$\tau_n(\cos\theta) = \frac{d}{d\theta} P_n^{(1)}(\cos\theta) \quad (11.8)$$

and the scattering coefficients are

$$a_n = \frac{m\psi_n(mx)\psi_n'(x) - \psi_n(x)\psi_n'(mx)}{m\psi_n(mx)\xi_n'(x) - \xi_n(x)\psi_n'(mx)} \quad (11.9)$$

$$b_n = \frac{\psi_n(mx)\psi_n'(x) - m\psi_n(x)\psi_n'(mx)}{\psi_n(mx)\xi_n'(x) - m\xi_n(x)\psi_n'(mx)} \quad (11.10)$$

where $\psi_n(x)$ and $\xi_n(x)$ are Ricatti–Bessel functions. $\psi_n(x) = xj_n(x)$, where j_n is the spherical Bessel function, and $\xi_n(x) = xh_n^{(1)}(x)$, where $h_n^{(1)}$ is the spherical Hankel function. The scattering coefficients can be determined from Equations 11.9 and 11.10, and can also be found in tables provided by Gumprecht and Sliepcevich (1951) and Denman *et al.* (1966).

The extinction cross-section can be expressed as

$$C_{ext} = \frac{\lambda^2}{2\pi} \sum_{n=1}^{\infty} (2n+1)\text{Re}(a_n + b_n) \quad (11.11)$$

or, equivalently, as

$$C_{ext} = \frac{4\pi}{k^2} \text{Re}(S(0°)) \quad (11.12)$$

The scattering cross-section is

11.1 Detection of Particles

$$C_{sca} = \frac{\lambda^2}{2\pi} \sum_{n=1}^{\infty} (2n+1) (|a_n|^2 + |b_n|^2) \tag{11.13}$$

The scattering efficiency can be defined for each cross-section $Q = C/\pi a^2$, where a is the particle radius.

Several of these results simplify for small particles. The first terms for the first three scattering coefficients are (Bohren and Huffman, 1983)

$$a_1 = -\frac{2}{3} i \frac{m^2-1}{m^2+2} x^3 - \frac{2}{5} i \frac{(m^2-2)(m^2-1)}{(m^2+2)^2} x^5 + \frac{4}{9} \left(\frac{m^2-1}{m^2+2}\right)^2 x^6 + O(x^7) \tag{11.14a}$$

$$a_2 = -\frac{1}{15} i \frac{m^2-1}{2m^2+3} x^5 + O(x^7) \tag{11.14b}$$

$$b_1 = -\frac{1}{45} i (m^2-1) x^5 + O(x^7) \tag{11.14c}$$

$$b_2 = O(x^7) \tag{11.14d}$$

The first few terms of the expansions for the different cross-sections (and the associated efficiencies) can be obtained by using Equations 11.14. In particular,

$$Q_{ext} = -4i \frac{m^2-1}{m^2+2} x - \frac{4}{15} i \frac{(m^2-1)^2(m^4+27m^2+38)}{(m^2+2)^2(2m^2+3)} x^3 + \frac{8(m^2-1)^2}{3(m^2+2)^2} x^4 + \ldots \tag{11.15}$$

Further, $S(0°)$ can be obtained by using Equation 11.12. For very small x,

$$Q_{abs} = 4x \, \text{Im}\left(\frac{m^2-1}{m^2+2}\right) \tag{11.16}$$

which is $\propto 1/\lambda$.

For very small particles ($x \ll 1$, $a \ll \lambda$), only the a_1 term is important in Equations 11.5 and 11.6. With $\pi_1(\cos\theta) = 1$ and $\tau_1(\cos\theta) = \cos\theta$, Equations 11.3 and 11.4 then give the Rayleigh formula. The scattered intensities for a single particle using polarized light are

$$I_{\perp s}(\theta) = \frac{8 I_{\perp i} \pi^4 a^6}{r^2 \lambda^4} \left|\frac{m^2-1}{m^2+2}\right|^2 \tag{11.17a}$$

$$I_{\parallel s}(\theta) = \frac{8 I_{\parallel i} \pi^4 a^6}{r^2 \lambda^4} \left|\frac{m^2-1}{m^2+2}\right|^2 \cos^2\theta \tag{11.17b}$$

m can be complex, but x, $x\,\text{Re}(m)$, and $x\,\text{Im}(m)$ are each $\ll 1$. The scattering efficiency is

$$Q_{sca} = \frac{8}{3} x^4 \left|\frac{m^2-1}{m^2+2}\right|^2 \tag{11.17c}$$

Each of these factors is $\propto 1/\lambda^4$.

These expressions for Rayleigh scattering are valid for all particle shapes. $I_{\|s}$ differs from $I_{\perp s}$ only by the $\cos^2\theta$ factor. The scattered intensity shows the $1/\lambda^4$ factor ($\propto \omega^4$) characteristic of many scattering phenomena. Since the scattering probability per photon $\propto 1/\lambda^3$, shorter wavelengths scatter with greater probability than longer wavelengths. The scattering rate increases very rapidly with particle radius $\propto a^6$.

In Rayleigh scattering ($a \ll \lambda$), $I_{\perp s}$ is independent of angle, while $I_{\|s} \propto \cos^2\theta$ and is smaller than $I_{\perp s}$, as is seen in Figure 11.2. Figures 11.3 and 11.4a illustrate how the magnitude and angular dependence of the scattering cross-section change as the radius a increases. In particular, $I_{\|s}$ and $I_{\perp s}$ begin to peak in the forward direction ($\theta = 0°$) as the particle size increases, and both eventually develop a second, but smaller, peak for backscattering ($\theta = 180°$). Figure 11.3 shows that $I_{\|s}$ and $I_{\perp s}$ vary rapidly with θ for larger particles, though this variation is somewhat weaker for $I_{\|s}$, and that, in contrast to Rayleigh scattering, $I_{\|s}$ is usually larger than $I_{\perp s}$. Figure 11.4b depicts the rapid increase in scattering cross-section with particle radius and the oscillatory dependence of scattering intensity on size when the particle has a complex index of refraction.

These scattering formulas apply to a single particle. The total intensity of scattered light is

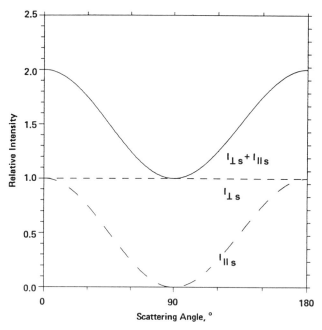

Figure 11.2 Relative intensity of polarized elastic scattering versus scattering angle using 632.8 nm for a sphere smaller than 0.05 μm, which is in the Rayleigh region. Reprinted from Leith (1990), p. 51, by courtesy of Marcel Dekker, Inc.

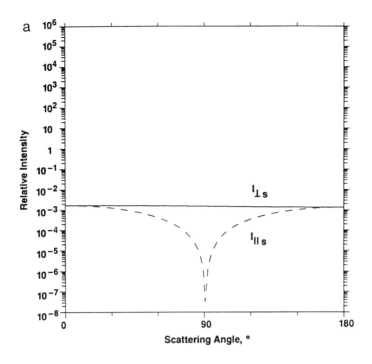

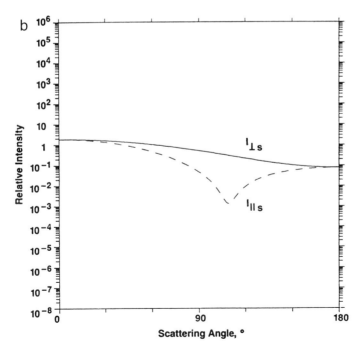

Figure 11.3 Relative intensity of polarized elastic scattering versus scattering angle using 632.8 nm for a sphere of diameter (a) 0.10 μm, (b) 0.30 μm, (c) 1.0 μm, and (d) 7.0 μm. Reprinted from Leith (1990), pp. 52–54, by courtesy of Marcel Dekker, Inc. (*Continues*)

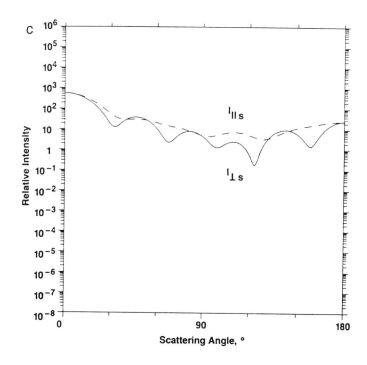

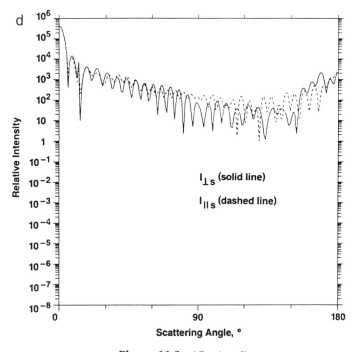

Figure 11.3 (*Continued*)

11.1 Detection of Particles

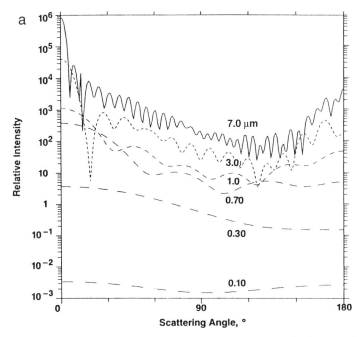

Figure 11.4 (a) Relative intensity of $I_{\|s} + I_{\perp s}$ versus scattering angle for scattering from spheres with different diameters, at 632.8 nm; (b) relative intensity for $I_{\|s} + I_{\perp s}$ versus particle size for spheres made of glass ($m = 1.54$) and carbon ($m = 1.59 + 0.66i$) in air, at a 0° scattering angle (632.8 nm). Reprinted from Leith (1990), p. 56, by courtesy of Marcel Dekker, Inc. (*Continues*)

$$I_{\text{tot}}(\theta) = \int \mathcal{N}(a) I_s(\theta, a) da \qquad (11.18)$$

where $\mathcal{N}(a)$ is the number of scatterers with radius a in the scattering volume and I_s is the scattered intensity for a given polarization from one particle with radius a. If there are several types of particles with different compositions or shapes, the contribution for each one must be summed.

Figures 11.1 and 11.5 show typical experimental setups for elastic scattering of particles.

11.1.2 Particles in Gases

11.1.2.1 Plasma Reactors

Particulate formation in plasma reactors is a potentially serious problem that has been seen in deposition, sputtering, and etching plasmas. Often the particles are localized in electrostatic traps. Process development studies have been geared to minimizing both the rate of particle formation and the impact of particles on the processing on the substrate.

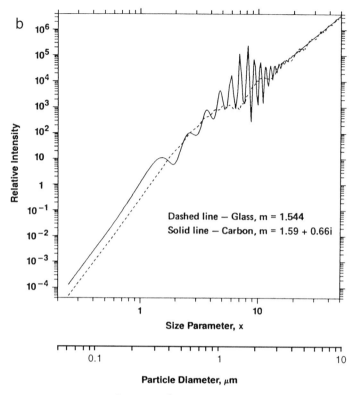

Figure 11.4 (*Continued*)

The simplest way to detect particles is by using a continuous-wave (cw) laser, such as a He–Ne laser (632.8 nm), and detectors placed at one or, preferably two scattering angles (Figure 11.1). One of the scattering angles is usually near 0°, because most of the scattering is at small angles when the particles are larger than λ. The laser can be scanned slowly, say between the electrodes, to monitor density variations. A charge-coupled device (CCD) camera can be placed near θ = 0°, with the laser following a fixed trajectory in the reactor (Selwyn *et al.*, 1990b) or one that is rastered across the chamber by reflecting it from a rotating or vibrating mirror before it enters the chamber (Selwyn *et al.*, 1991) (Figure 11.5). Figure 11.6 is a photograph of particles detected in an argon plasma with this system. Other laser wavelengths can be used as well (488 nm from an argon-ion laser; 647 nm from a krypton-ion laser), and more than one wavelength can be used at once. In more qualitative studies the volume in the reactor can be illuminated by a white-light source, with detection by a CCD camera (Dorier *et al.*, 1992).

In early work, Spears *et al.* (1986, 1988) and Spears and Robinson (1988) examined particle formation and spatial distribution in silane RF discharges

11.1 Detection of Particles

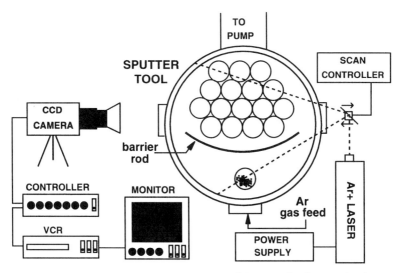

Figure 11.5 Schematic diagram of rastered laser light scattering in a production quartz sputtering deposition tool (from Selwyn, 1991).

by using 532 nm from a frequency-doubled pulsed Nd:YAG laser; they also used laser-induced fluorescence (LIF) to examine gas-phase species in the discharge. Another early study on the use of laser scattering in processing reactors is noted here, which was not performed in a discharge. Olson and Kibbler (1986) found that a light-scattering aerosol is sometimes formed in the thermal boundary layer above the substrate during metalorganic chemical-vapor deposition (MOCVD), such as during the growth of GaAs with trimethylgallium (TMGa) and AsH_3. Particle formation can be serious in thermal CVD, as well as in plasma-enhanced CVD (PECVD) and plasma etching.

Jellum and Graves (1990a,b) used angular dissymmetry, i.e., measurements at different scattering angles, to estimate particle sizes in DC and RF sputtering Ar discharges with Al electrodes. This technique is useful when the particle circumference ($2\pi a$) is of the same order as the scattering wavelength; this condition was met for ~100–200-nm particles by using 514.5 nm from an excimer laser pumped dye laser. The particles were assumed to be pure aluminum with $n = 0.82 + 6.25i$ and two scattering angles were used, 12° and 90°. Figure 11.7 shows the relative scattered intensity at these two angles, along with the inferred particle radius and density, as a function of distance from the upper (powered) electrode. In addition to this continuous elastic scattering, occasionally "speckled" scattering was seen, corresponding to macroscopic, ~1-mm-long rods. Figure 11.8 shows that the intensity of optical emission at the 810.5 nm line of Ar, monitored at the 90° port, depends on the distance from the electrodes and decreases when particles are detected (He–Ne laser, 632.8 nm). Related

Figure 11.6 Photograph of particle traps over a square ceramic substrate in an Ar plasma with a graphite electrode obtained by using the laser light scattering system depicted in Figure 11.5 (from Selwyn, 1991).

studies by this group, detailed in Section 7.2.1.1, demonstrate how LIF measurements of atoms in plasmas can be affected by the presence of particulates.

Jellum *et al.* (1991) used elastic light scattering (632.8 nm, $\theta = 170°$) to show that the carbon particles formed in a methane discharge tend to congregate near the cooled electrode and away from the heated electrode, as is seen in Figure 11.9. Several other investigations cited in this section have similarly demonstrated that particles tend to segregate in different regions because of electrostatic traps.

Selwyn *et al.* (1989) studied plasma etching of Si wafers by chlorine using laser light scattering to detect particles and two-photon LIF (Section 7.2.1.2) to detect chlorine atoms. Pulsed 233.3-nm radiation was used for both diagnostics. Chlorine-containing negative ions were also detected because this high-fluence laser detached electrons and then excited the Cl atoms. Spatial measurements of light scattering showed that particles are trapped at the sheath boundaries, while comparison of the elastic scattering and two-photon LIF results indicated that particles form and grow in the layer of negative ions.

11.1 Detection of Particles

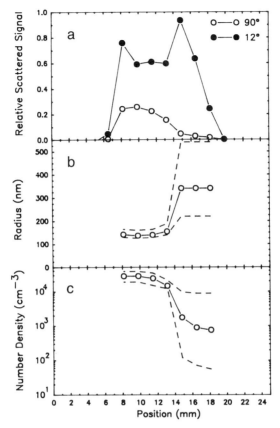

Figure 11.7 (a) Relative scattering intensity at 12° and 90° as a function of position from the lower grounded electrode (at 25 mm) and upper powered electrode (at 0 mm) for an Ar discharge. (b) Estimate of particle radius from scattering theory, with error bounds (dashed lines). (c) Corresponding estimate of number density, with error bounds [upper (lower) dashed lines in (b) correspond to the lower (upper) dashed lines here] (from Jellum and Graves, 1990b).

Selwyn *et al.* (1990a, 1991) used rastered laser light to study light scattering, as shown in Figure 11.5. This study suggested that particles are negatively charged and electrostatically trapped at the plasma/sheath boundary. Selwyn (1991) and Selwyn and Patterson (1992) used rastered laser light scattering to evaluate and modify reactive ion etching (RIE) and sputter deposition tools to minimize the effect of particle formation. The existence of particle traps is clearly seen in Figure 11.6. In this study, the topographies of electrodes were designed to trap particles intentionally in specified regions of the plasma, away from the wafer. Geha *et al.* (1992) combined these methods with Langmuir probing to examine electrostatic traps in plasmas containing silicon, graphite, aluminum, and stainless steel wafers and underlying disks.

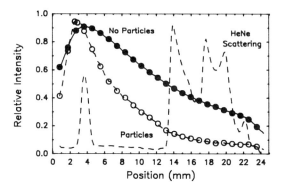

Figure 11.8 Comparison of plasma-induced optical emission profiles in an argon plasma with aluminum electrodes when particles are present (open circles) and not present (closed circles). The dashed line is scattering at 170° from a He–Ne laser (632.8 nm), which indicates particle density as a function of distance from the powered electrode (from Jellum and Graves, 1990b).

Using laser light scattering, O'Neill et al. (1990) showed that the appearance time of particles during plasma etching of Si by CF_nCl_{4-n} monotonically increases with the fluorination of the halocarbon. Because the delay in particle appearance did not correlate inversely with the etch rate, they deduced that particle formation depended on the chemical nature of the reactant and not merely the amount of available (etched) silicon.

In other studies, Bouchoule et al. (1991) investigated particle formation in silane/argon discharges with continuous and pulsed RF excitation. Dorier et al. (1992) examined the threshold for powder formation in RF silane plasmas as a function of input power, RF frequency, and electrode conditions (Figure 11.10). Anderson et al. (1990) studied particle generation in silane/ammonia RF discharges. Selwyn et al. (1990b) monitored particles during SiO_2 sputtering. Yoo and Steinbrüchel (1992) investigated particle formation in an Ar sputtering plasma with a Si substrate and during RIE of Si in CCl_2F_2/Ar plasmas.

In most of these studies, scattered light was collected from many particles and background scattering from the reactor walls was minimized. Hobbs (1995) has described a tracking system based on light scattering that can be used to detect and map single submicron particles in plasma chambers, flow pipes, or other processing environments with minor sensitivity to background scattering. This *in situ* scanning coherent lidar (laser radar) [ISICL] system is depicted in Figure 11.11. It uses a modified Michelson interferometer and homodyne detection, runs at the quantum limit of detection because of noise cancellation, and is insensitive to misalignment. ISICL is immune to stray

11.1 Detection of Particles

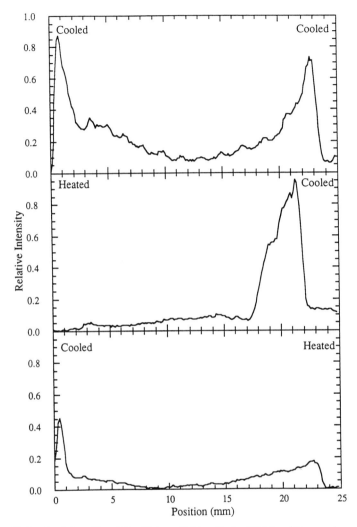

Figure 11.9 Spatial profiles of He–Ne laser light scattering (632.8 nm) from carbon particles in an argon glow discharge with parallel-plate electrodes that are either heated or cooled. (The particles were initially formed in a methane discharge and then were studied with only argon flowing.) (From Jellum et al., 1991.)

light, such as (Lambertian) scattered light from the chamber walls that is $10^6\times$ stronger than the signal from the scattering particle, because of spatial filtering, polarization analysis, and electronic analysis of the signal. This system can track the trajectory of the targeted particle and can also be used to detect debris peeling off from the walls of the processing chamber.

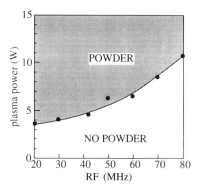

Figure 11.10 RF frequency dependence of plasma power threshold for the onset of powder formation in a silane plasma with parallel-plate electrodes (from Dorier *et al.*, 1992).

11.1.2.2 Flow Visualization and Velocimetry

Although particle formation is usually monitored because it is deleterious, it can be beneficial in some cases. Sometimes, particles are purposely introduced into a reactor because their locations can be readily monitored by elastic scattering. This is a method of flow visualization that is used in CVD reactors; another method of flow visualization relies on interferometry (optical path-length methods) and is described in Chapter 10. Optical detec-

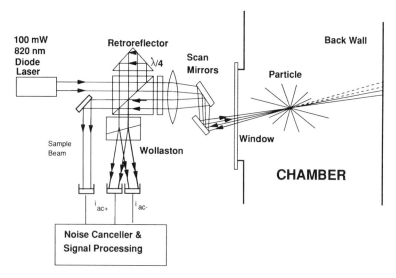

Figure 11.11 Apparatus for a scanning coherent lidar (laser radar) system for detecting and mapping isolated submicron particles in a chamber, which needs only one access window (from Hobbs, 1995).

11.1 Detection of Particles

tion of tracers is also useful in measuring the flow velocities in CVD reactors, i.e., velocimetry. Lauterborn and Vogel (1984), Merzkirch (1987), Hesselink (1988), Yang (1989), Adrian (1991), and Freymuth (1993) have reviewed methods of flow visualization for the study of three-dimensional streamlines. Breiland and Ho (1993) have detailed the use of such optical methods for flow visualization and velocimetry in CVD.

In flow visualization, the tracer is usually smoke consisting of micrometer-sized particles. The smoke can be made by a chemical reaction (e.g., $TiCl_4$ reacting with water vapor to form TiO_2 or kerosene dripped on a hot wire in air to form white smoke), a solid powder source, or an atomizer. The injected tracer follows the streamlines if the particles are sufficiently small, and the streamlines can then be seen by elastic light scattering. These flow patterns can be measured by photography or by electronic image recording with CCD arrays. Figure 11.12 illustrates this flow visualization in a rotating-disk reactor, where Breiland and Evans (1991) used smoke generated from kerosene dropped on a hot wire in air to see the streamlines with 514.5 nm from an argon-ion laser. The streamlines remain straight until they reach the boundary layer and then begin to spiral for the faster inlet volume flow rate in (a). Looping, recirculating paths are seen for the slower flow rate in (b).

A dark layer is often seen above the susceptor in the flow visualization of CVD reactors. This layer was initially interpreted as evidence for the existence of a stagnant layer above the surface; however, it is now understood to be caused by thermophoretic forces that are exerted on the tracer particles in a thermal gradient. Therefore, the particles do not follow the streamlines when the thermal gradient is large, and tracer methods give only qualitative information about flow in CVD reactors. This flow of particles from hot to cold regions has the benefit of forcing particles, which may form due to homogeneous chemistry, away from the growing surface

 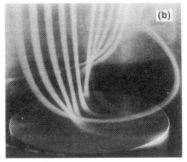

Figure 11.12 Flow visualization of streamlines above a room-temperature disk spinning at 100 rpm by light scattering from injected smoke (620 Torr He, 514.5 nm), for inlet volume flow rates (a) 11 slm (standard liters/minute) and (b) 3 slm (from Breiland and Evans, 1991).

in a horizontal CVD reactor. Fotiadis and Jensen (1990) have reviewed and detailed the effect of thermophoretic transport on flow visualization in CVD reactors. This problem does not occur when phosphorescing molecules are used as the tracer, as is described in Section 7.2.1.1.

Laser velocimetry, or laser Doppler velocimetry (LDV) or anemometry, measures gas flow velocities (Drain, 1980; Adrian, 1991) and has been used to make such measurements in CVD reactors (Breiland and Ho, 1984; Ho et al., 1984; Koppitz et al., 1984). A laser is split and then recombined at an angle (2ϕ) to set up an interference pattern, as is seen in Figure 11.13. The particle elastically scatters light when it passes through interference maxima, leading to a "Doppler burst." The velocity component normal to the interference fringes, v_j, is determined by

$$v_j = \frac{\lambda \nu}{2 \sin \phi} \qquad (11.19)$$

where ν is the measured frequency of the periodic pattern of the burst. By rotating the two beams into a different plane, by say 90°, the velocity component v_k normal to v_j (and to the laser symmetry axis) can be obtained. Koppitz et al. (1984) determined the velocity profiles for H_2 and N_2 in a metalorganic CVD (MOCVD) reactor.

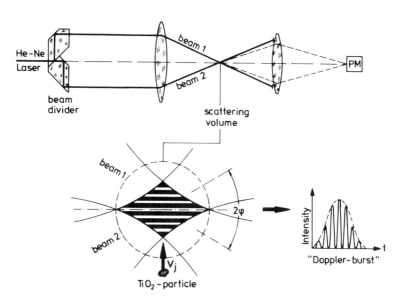

Figure 11.13 Schematic diagram of the dual-beam laser velocity technique. [From M. Koppitz, W. Richter, R. Bahnen, and M. Heyen, in *Laser Processing and Diagnostics* (D. Bäuerle, ed.), Light scattering diagnostics of gas-phase epitaxial growth (MOCVD-GaAs), page 532, Figure 3, 1984. Copyright © Springer-Verlag.]

11.1.3 *Particles in Liquids*

Knollenberg (1988) has reviewed the monitoring systems that are commonly used to detect particulates in the fluids in semiconductor processing; light scattering is the dominant method employed. Particulates in liquids can be measured *in-line*, where the monitoring sensor is mounted on a line through which all the liquid is monitored; *on-line*, where a small portion of the main flow is diverted through the line with the sensor; or in an *extractive-batch* mode, where a sample is extracted off-line for batch analysis. In *"in situ"* in-line instruments the sensor measures particles in a small (<1%), though representative, sample of the flow, while in *volumetric* instruments the sensor analyzes all the liquid in the flow. Solid particles in liquids scatter less light than those in gases because of the lower index ratio m in the liquid ($n_{air} = 1$, $n_{water} = 1.33$). Moreover it is more strongly scattered in the forward direction, which can make detection easier. Polystyrene latex spheres (PLS) are usually used for calibration in optical particle counters.

In volumetric sampling, a sheet of light formed by a cylindrical lens cuts a slice across the flow line (632.8 nm). Commercial sensors can measure the transmitted light, which measures beam extinction, or scattered light (Knollenberg, 1988). Particle sizes are determined from pulse height analysis. As shown in Figure 11.4b, scattering signals vary as $\propto a^2$ for larger particles, while they vary as $\propto a^6$ for submicron particles.

In commercial *"in situ"* instruments, light passing through the line is collected with a lens, which has a dump spot in the center to remove the unscattered laser beam. In this dark-field arrangement, only light scattered at small angles is collected. The region that is sampled is defined by an aperturing scheme that determines particle trajectories or by comparing scattered light by p- and s-polarized beams that are collinearly focused, but to different foci (Knollenberg, 1988).

Most particle analysis in gases and liquids is performed in dark field with detectors placed at scattering angles $\gg 0°$, often at 90° for convenience. The goal is usually counting particles, and not determining the refractive index of the scatterer, even though the index can provide clues about the source of the particles. However, particle counting can be inaccurate if the refractive index is not measured because the assumed index can be wrong; in particular, bubbles in liquids can be incorrectly counted as particles. Taubenblatt and Batchelder (1991, 1992a,b) and Batchelder and Taubenblatt (1992) have used a Nomarski interferometer to monitor single particles in transmission in fluid flow by examining the amplitude and phase of the light scattered in the forward direction by one particle. This arrangement gives a unique value of the particle size and its refractive index, which identifies the composition of that specific particle because the real and imaginary parts of $S(0°)$ in Equations 11.12 and 11.15 have unique signatures that depend on a, n_1, and k_1. Particles with different a and n_1 can have the same phase shift $[4\pi a(n_1 - n_2)/\lambda]$, but will have different scattering amplitudes.

Figure 11.14 is a schematic of this transmission Nomarski interferometer. Circularly polarized light from a laser, with a gaussian transverse profile, is separated into s- and p-polarized beams. When focused in the flow cell, the foci for the two polarizations are laterally separated, forming the two arms of the interferometer; at a given time a particle may be in one focus but not in the other. The transmitted light is then polarization-analyzed to detect the relative phase and extinction of these two paths by separate differential measurements; the difference of $s + p$ and $s - p$ resolved light gives I_{phase}, while the difference of s and p resolved light gives $I_{extinction}$. For small phase shifts, the relative phase shift of the two paths $\Delta\phi \propto iS(0°)$. Since $I_{phase} \propto \text{Re}(\Delta\phi)$ and $I_{extinction} \propto \text{Im}(\Delta\phi)$, particle parameters can be determined by using Equation 11.15. This differential arrangement ($I_{extinction}$) eliminates common mode noise.

Figure 11.15 shows the calculated x versus y trajectories of five particles with different n and k (and also bubbles), with particle size as the implicit variable in each curve. The measured data are seen to fall on the trajectory calculated for the specific particle type, which shows that this method uniquely identifies the particle type as well as its size (Taubenblatt and Batchelder, 1991, 1992b).

Another advantage of this bright-field Nomarski transmission method for detecting particles in liquids relates to the magnitude of the refractive-index mismatch. Particle detectors placed at 90° detect scattered light when the particle and ambient have very different refractive indices (wide angle scattering). When the indices of the particle and fluid nearly match, scattering is peaked in the forward direction and a 90° detector will not pick it up, but the transmission Nomarski detector will. This technique is also reviewed by Blackford and Grant (1993), Grant (1993), and Batchelder and Taubenblatt (1993).

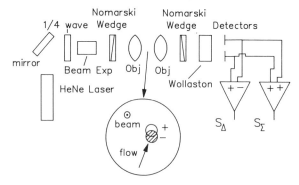

Figure 11.14 Experimental apparatus based on Nomarski optics for the measurement of both the phase shift and extinction in forward scattering by a single particle (from Taubenblatt and Batchelder, 1991).

11.1 Detection of Particles

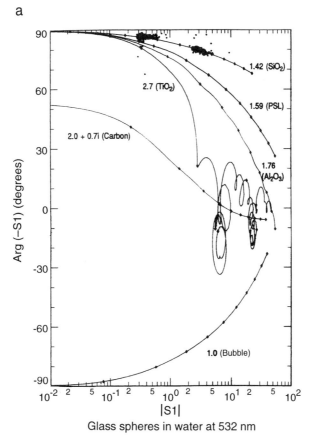

Glass spheres in water at 532 nm

Figure 11.15 Calculated trajectory of arg(-$S1$) versus |$S1$| for spherical particulates of refractive index $n + ik$ in water with particulate diameter an implicit variable along the trajectory, using 532 nm [$S1 = S(0°)$]. Experimental points obtained by using the apparatus in Figure 11.14 are shown for (a) glass spheres, (b) PSL (polystyrene latex) spheres, (c) india ink, and (d) white latex paint in water, demonstrating the ability to identify the type of particle and determine its diameter (from Taubenblatt and Batchelder, 1992b). (*Continues*)

11.1.4 Particles on Surfaces

Gise (1988) has reviewed the technology of detecting particles on surfaces and other surface defects. Surface defects can be classified into three categories, each of which can be detected by light scattering: (1) Scratches, cracks, and other specular defects, whose characteristic dimensions (width, depth) are large compared to the wavelength λ of the probing light. These scatter preferentially into specific directions. (2) Particles on the surface. (3) Surface irregularities and fine particles ≪λ over large fractions of the surface.

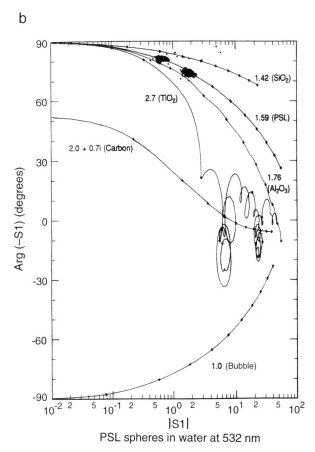

Figure 11.15 (*Continued*)

This results in a haze on the surface that scatters light over a wide angle. This surface roughness is described in detail in Section 11.2.1.

Wafers are often inspected for particles after they have been cleaned. Sometimes, spikes and pits appear on the surface as a result of the cleaning procedure, and scattering from these features often cannot be distinguished from that due to particles. Spikes can form when the cleaning steps involve the removal of a material from the surface, e.g., by sequential oxidation/etching steps, and when the local presence of oxide or bubbles, which may arise because of metal contaminants, prevents the cleaning of the underlying wafer. Pits, often called by the misnomer "crystal originated particles" or COPs, are formed by the etching of crystal defects. Morita *et al.* (1994) decreased the scattering contribution from COPs, *vis-à-vis* real particles, by using a laser that illuminated the surface at grazing incidence rather than the more common perpendicular illumination.

11.1 Detection of Particles

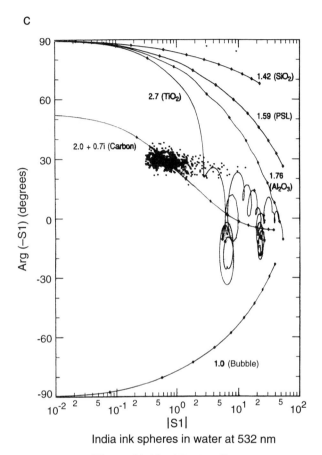

Figure 11.15 (*Continued*)

To first order, the scattering from a particle on a surface can be understood by considering first the scattering from that particle in air and then the effect of the surface. With light incident normal to the surface, backscattered light ($90° < \theta < 180°$) is unaffected by the surface, while forward-scattered light ($0° < \theta < 90°$) is reflected by the surface. For particles much larger than λ, forward scattering near $0°$ dominates, which is seen as backscatter, attenuated by the reflectance $R(\theta = 0°)$.

In typical inspection systems a focused laser (632.8 nm) is scanned along a line on the surface by reflection from a rotating polygonal (multifaceted) mirror or an oscillating mirror. The wafer is translated in the direction normal to this line between scans, for two-dimensional coverage. Scattered light is collected by a linear-array fiber optic bundle. Particles are sized from the height and width of the detected pulses. In an alternative method, the laser is focused on the wafer while the substrate is scanned. One way

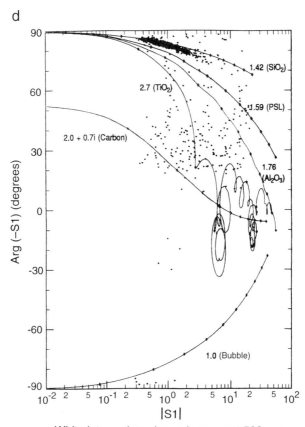

White latex paint spheres in water at 532 nm

Figure 11.15 (*Continued*)

is to rotate the wafer at a constant frequency (~1500 rpm) and translate it at a constant speed (~0.75 mm/sec) to trace out a spiral. These systems are made for *ex situ* inspection stations.

While such methods work well with unpatterned wafers, particle detection is also needed on patterned substrates. Because it is difficult to distinguish scattering from the pattern and that from particulates and other defects, the methods used for unpatterned wafers must be modified. Particle scattering from a repetitively patterned surface, such as a DRAM (dynamic random access memory) repeat, is typically performed by examining the diffraction pattern and blocking out the main diffraction peaks from the repetitive pattern with a spatial filter. The remaining features are then attributed to defects and particulates on the surface. Miyazaki *et al.* (1992) have shown how this method can be used to inspect 16M DRAMs with 0.3-μm resolution. This

11.1 Detection of Particles

type of detection is usually done in bright field. Taubenblatt and Batchelder (1992c) have examined the diffraction pattern in dark field to lessen the background, i.e., by blocking out the main diffraction bright spots that are now dimmer (Figure 11.16). In converting to the dark-field configuration, chang-

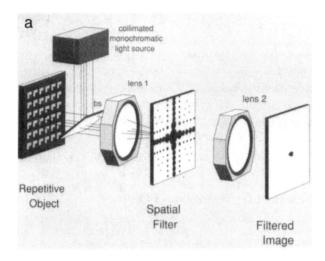

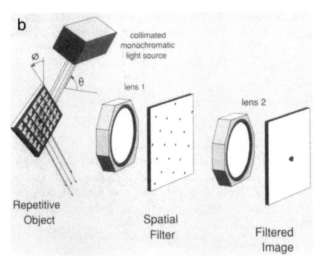

Figure 11.16 Integrated inspection of repetitively patterned semiconductor wafers for defects and particulates by using a spatial filtering system consisting of an illumination source, beam splitter (bs), a Fourier-transform lens, a spatial filter, and an inverse Fourier-transform lens. Part (a) shows a conventional spatial filter system with bright-field illumination, while (b) shows a spatial filter system with dark-field illumination, with oblique illumination and 45° azimuthal orientation of the wafer, which improves performance (from Taubenblatt and Batchelder, 1992c).

ing θ in this figure helps, but changing φ helps even more because the observation is then off the main grid lines. Nakata and Akiyama (1989) have described a related method for detecting particles on surfaces by using diagonal illumination and spatial frequency filtering.

Another technique for particle detection makes use of the depolarizing nature of light scattering from particles (Koizumi and Akiyama, 1985). s-Polarized light impinges the wafer at a grazing angle ($\theta \sim 89°$). Light scattered to the surface normal (0°) is collected, then analyzed to let only p-polarized light through—which is due mostly to particles—and then detected.

Hattori and Koyata (1991) have described an automated particle detection and identification system for very large-scale integration (VLSI) processing that uses light scattering to detect particles, and optical (fluorescence) and nonoptical techniques for identification.

Analogous methods are used to inspect reticles for particles and defects (Gise, 1988).

11.2 Diffraction from Surface Features

Elastic light scattering from patterns on surfaces can be understood by vector diffraction theory. Use of these techniques for *in situ* and *ex situ* analysis of surface topography is commonly called laser light scattering (LLS) or scatterometry. While it is convenient to subdivide such applications by the magnitude of the vertical variations and the lateral periodicity of the topography, all of these probes are essentially diffraction experiments. Two categories are considered here. One is scattering from relatively smooth surfaces where the vertical deviations are typically ~ 1 Å ($\ll \lambda$) and there is no periodicity (surface roughness). The other is scattering from periodic features where the vertical deviations can be ~ 1 μm ($\sim \lambda$), such as those formed during patterning. Sometimes the former is called laser light scattering and the latter is called scatterometry, although both terms are really applicable to either category. Scattering from periodic surfaces can also be called diffractometry. These scattering methods are noncontact and nondestructive, and often can be used for fast, *in situ* monitoring. Overviews of this type of elastic scattering have been written by McNeil *et al.* (1992a,b, 1993a,b). Whitehouse (1994) details optical and nonoptical techniques for the metrology of surface topography.

11.2.1 *Laser Light Scattering (Scatterometry) from Random Surface Features*

Laser light scattering (LLS) can be used to analyze surface roughness and, in effect, to probe the atomic-level deviations from planarity of the surface. The basics of scattering from microscopically rough surfaces, where the

11.2 Diffraction from Surface Features

vertical deviations are $\ll \lambda$, have been provided in a series of papers by Church and Zavada (1975) and Church et al. (1977, 1979). Although the goal of their investigations was to assess the roughness of diamond-turned optics, many of their results can be used for real-time laser light scattering from surfaces during thin film processing. Their work was based on earlier studies by Barrick (1970) on the scattering of radio waves from "rough" surfaces, which is important for radar. Bennett and Bennett (1967) have discussed the use of elastic light scattering to measure surface roughness of thin film optics. Beckmann (1967) presented another early review of scattering from rough surfaces. More recently, Jacobson et al. (1992) have reviewed angle-resolved scattering measurements in the context of polished optics. Stover (1975, 1990) provides a comprehensive treatment of visible light scattering from rough surfaces in particular, and of scatterometry in general. Other valuable reviews and surveys of surface scattering and scatterometry have been presented by Eastman (1978), Bennett and Mattsson (1989), Pickering (1994, 1995), Church and Takacs (1995a), and Stover (1995).

The degree of surface roughness of films can indicate the nature of film growth (3D versus 2D) [Tong and Williams, 1994] and the introduction of defects at the surface. Furthermore, surface roughness can affect (and be affected by) wafer cleaning and adhesion, degrade the breakdown voltage of thin gate oxides in microelectronics, and decrease the reflectance of metal layers in optoelectronics. Surface roughness can be evaluated by using laser light scattering, atomic-force microscopy (AFM), Nomarski microscopy, and interferometric profilometry, as has been discussed by Malik et al. (1993), and by x-ray scattering, as has been discussed by Als-Nielsen (1986), Stearns (1992), Chason et al. (1994), and Church and Takacs (1995b). While each of these methods is well suited for in-line or *ex situ* analysis, LLS is the technique that is best suited for real-time surface monitoring. It should be emphasized that LLS experiments produce only a statistical understanding of the surface topography. Although this is usually satisfactory for real-time probing, it is a much more modest understanding of the surface than that obtainable by using other techniques, such as AFM. While both LLS and AFM can measure sub-Å level variations in vertical topography, the lateral resolution in LLS is $\sim\lambda$, while that in AFM is much better, $\sim$Å.

The goal of LLS is to determine the surface topography, which can be described by the surface profile $Z(\mathbf{r},t)$. Consider a laser with intensity I_i, wavelength λ, and wavevector $\mathbf{k}_i$ that is incident on the surface at an angle θ_i to the surface normal. On a smooth surface there is a specular reflection, which is also in the plane of incidence but is symmetrically at this same angle on the other side of the surface normal. For rough surfaces, there is diffuse reflection in addition to specular reflection. This diffusely scattered light has intensity I_s and can be detected in a solid angle $d\Omega$ in the direction θ_s, ϕ_s, where the scattered wavevector is $\mathbf{k}_s$. ϕ_s is the azimuthal angle of the detected scattered beam relative to the plane of incidence (Figure 11.17). Because this is elastic

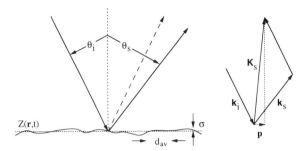

Figure 11.17 Illustration of the beams involved in laser light scattering from a surface with vertical topography described by $Z(\mathbf{r},t)$, with characteristic vertical displacement σ and average transverse length parameter d_{av}. The incident angle is θ_i, the incident wavevector is $\mathbf{k}_i$, the scattered angle is θ_s, and the scattered wavevector is $\mathbf{k}_s$. The component of the change in wavevector $\mathbf{K}_s$ parallel to the surface and resolved in the plane of incidence is $\mathbf{p}$. As depicted in this figure, the azimuthal angle ϕ_s describing $\mathbf{k}_i$ is zero, and the component $\mathbf{K}_s$ parallel to the surface and resolved normal to the plane of incidence, $\mathbf{q}$, is also zero.

scattering, $k = k_i = k_s = 2\pi/\lambda$. The change in wavevector, the scattering wavevector $\mathbf{K}_s = \mathbf{k}_s - \mathbf{k}_i$, has components parallel to the surface; these can be resolved into components that are parallel, $p = (2\pi/\lambda)(\sin\theta_s \cos\phi_s - \sin\theta_i)$, and normal, $q = (2\pi/\lambda)\sin\theta_s \sin\phi_s$, to the plane of incidence.

Although both LLS and reflection ellipsometry (Section 9.11) are sensitive to surface roughness, they are complementary methods. LLS experiments probe only roughness with lateral features $> \lambda/2$, although the vertical features are usually $\ll \lambda$. Ellipsometry senses only roughness $\ll \lambda$ because it determines the ratio of the complex reflectances of specularly reflected light with different polarizations.

11.2.1.1 General Concepts

The main features of LLS can be understood by considering a sinusoidal diffraction grating, with period d and peak-to-valley amplitude $2a$. The surface profile for the grating is

$$Z(x) = a \sin(2\pi x/d) \qquad (11.20)$$

The scattering angles θ_s are the diffraction angles θ_m given by the grating equation

$$\sin\theta_m = \sin\theta_i + m\frac{\lambda}{d} \qquad (11.21)$$

where m is the grating order $= 0, \pm 1, \pm 2, \pm 3, \ldots$. The vector theory of diffraction shows that the scattering intensity varies as

$$I_s \propto (ka)^{2|m|} \qquad (11.22)$$

11.2 Diffraction from Surface Features

For monolayer-level roughness, $a \ll \lambda$ ($ka \ll 1$), this equation suggests that, in addition to specular reflection ($m = 0$), only the ± 1 orders are important. Consequently, a one-to-one correspondence between diffraction pattern and surface topography is possible. Moreover, the scattering intensity increases as the square of the standard deviation of the roughness ($\propto a^2$) and becomes larger when short wavelengths are used ($\propto 1/\lambda^2$).

For small angle scattering ($m = \pm 1$) at near-normal incidence, Equation 11.21 becomes

$$|\theta_s - \theta_i| \sim \lambda/d \tag{11.23}$$

This shows that the angular spread of the scattered beam is determined by the reciprocal of the characteristic lateral measure of roughness (normalized by λ), while Equation 11.22 demonstrates that the scattered intensity varies as the square of the characteristic vertical amplitude of the roughness (normalized by λ).

The choice of laser wavelength determines the range of the lateral correlation distances probed by LLS. For $d \gg \lambda$, Equation 11.23 shows that the angular spread is very small and therefore it is difficult to obtain useful information. The minimum feature size that can be probed by LLS is $d = \lambda/2$, which is seen from Equation 11.21 in first order with $\theta_i = -\theta_m = 90°$. (Diffraction from these features into $m = \pm 1$ is not possible for a smaller d.) For a given θ_i, θ_m, and λ, scattering is most sensitive to lateral features with d given by this equation.

11.2.1.2 Theory

In measuring the angle-resolved scatter (ARS), the scattered intensity I_s is

$$\frac{dI_s(t)}{d\Omega} = 4\ k^4 \cos\theta_i \cos^2\theta_s\ QW(p,q,t)\ I_i \tag{11.24}$$

The "surface factor" $W(p,q,t)$ is the surface power spectral density (PSD), which can also be expressed as $W(\mathbf{K}_s,t)$. It contains all the information on how elastic scattering depends on surface topography, and is the parameter determined from the measurement. This equation is valid in the smooth surface approximation for $(k\sigma)^2 \ll 1$, where σ is the root-mean-square (RMS) vertical surface roughness. For scattering with visible light, this means that $\sigma \ll 100$ Å, and so this expression is certainly valid for surface disruptions up to several monolayers. The differential scattering cross-section is $d\sigma/d\Omega = (dI_s/d\Omega)/I_i$, which is also called the scatter function or the cosine-corrected BRDF (bidirectional reflection distribution function). Sometimes angle-resolved scattering is expressed in terms of the BRDF, which is $(dI_s/d\Omega)/(I_i \cos\theta_s)$.

The "optical factor" Q is related to the surface reflectance. Expressions for Q are given by Church et al. (1977, 1979). If the scattered radiation is measured with a polarization insensitive detector, then

$$Q = \sum_{\beta} Q_{\alpha\beta} \qquad (11.25a)$$

for incident light with α polarization and

$$Q = \frac{1}{2} \sum_{\alpha\beta} Q_{\alpha\beta} \qquad (11.25b)$$

for unpolarized incident light. α and β are indices that represent the incident and scattered polarization, and can be either s- or p-polarized. $Q_{\alpha\beta}$ are presented in Table 11.1. In the limit of small scattering angles, where $\theta_s \to \theta_i$ and $\phi_s \to 0$, $Q_{ss} \to R_s(\theta_i)$, Q_{sp}, $Q_{ps} \to 0$, and $Q_{pp} \to R_p(\theta_i)$, where R are the Fresnel reflectances given in Section 9.1.1.1.

W is the Fourier transform of the surface profile:

$$W(p,q,t) = \frac{1}{A}\left|\frac{1}{2\pi}\int \exp(i\mathbf{K}_s\cdot\mathbf{r})Z(\mathbf{r},t)d\mathbf{r}\right|^2 \qquad (11.26a)$$

$$= \frac{1}{A}\left|\frac{1}{2\pi}\int dx \int dy \, \exp(ipx + iqy)\, Z(x,y,t)\right|^2 \qquad (11.26b)$$

where r is the coordinate in the plane and A is the area of a square with length $2L$.

These expressions for the scattering intensity and W simplify in special limits. For small scattering angles $QI_i = R(\theta_i)I_i$, which is the reflected intensity I_r, and $\theta_s \approx \theta_i$ so

Table 11.1 The $Q_{\alpha\beta}$ Factors*

$$Q_{ss} = |\varepsilon - 1|^2 \left|\frac{\cos\phi_s}{(\cos\theta_i + \sqrt{\varepsilon - \sin^2\theta_i})(\cos\theta_s + \sqrt{\varepsilon - \sin^2\theta_s})}\right|^2$$

$$Q_{sp} = |\varepsilon - 1|^2 \left|\frac{\sqrt{\varepsilon - \sin^2\theta_s}\sin\phi_s}{(\cos\theta_i + \sqrt{\varepsilon - \sin^2\theta_i})(\varepsilon\cos\theta_s + \sqrt{\varepsilon - \sin^2\theta_s})}\right|^2$$

$$Q_{ps} = |\varepsilon - 1|^2 \left|\frac{\sqrt{\varepsilon - \sin^2\theta_i}\sin\phi_s}{(\varepsilon\cos\theta_i + \sqrt{\varepsilon - \sin^2\theta_i})(\cos\theta_s + \sqrt{\varepsilon - \sin^2\theta_s})}\right|^2$$

$$Q_{pp} = |\varepsilon - 1|^2 \left|\frac{\sqrt{\varepsilon - \sin^2\theta_i}\sqrt{\varepsilon - \sin^2\theta_s}\cos\phi_s - \varepsilon\sin\theta_i\sin\theta_s}{(\varepsilon\cos\theta_i + \sqrt{\varepsilon - \sin^2\theta_i})(\varepsilon\cos\theta_s + \sqrt{\varepsilon - \sin^2\theta_s})}\right|^2$$

where $\varepsilon = \tilde{\varepsilon}_2/\tilde{\varepsilon}_1$ for light that travels in medium 1 and scatters off medium 2.

* From Church et al. (1977, 1979).

11.2 Diffraction from Surface Features

$$\frac{dI_s(t)}{d\Omega} = 4k^4\cos^3\theta_i W(p,q,t)I_r \quad (11.27)$$

For the limiting case of a quasi–one-dimensional, grating-like surface topography, these expressions can be integrated over y to give

$$\frac{dI_s(t)}{d\theta_s} = 4k^3\cos\theta_i \cos^2\theta_s Q W_1(p,t)I_i \quad (11.28)$$

where

$$W_1(p,t) = \frac{1}{2L}\frac{1}{2\pi}\left|\int_{-L}^{+L} \exp(ipx)\, Z(x,t)dx\right|^2 \quad (11.29)$$

with p and Q evaluated at $\phi_s = 0$.

For an isotropically rough surface, Equation 11.26 becomes

$$W_2(u,t) = \frac{1}{\pi R^2}\left|\int_0^R J_0(ux)Z(x,t)x\, dx\right|^2 \quad (11.30)$$

where $u^2 = p^2 + q^2$, J_0 is the Bessel function of the first kind of order 0, and x is really a radial coordinate.

W_1 and W_2 are interrelated, and both are related to the autocovariance function of the profile $Z(x,t)$, which is given by (Church et al., 1979):

$$C(v,t) = \frac{1}{2L}\int_{-L}^{+L} Z(x,t)Z(x+v,t)dx \quad (11.31)$$

Slightly different terminology is used in some more recent sources (Church and Takacs, 1995a), with the spatial frequency expressed in (ordinary) frequency f_x and f_y (Hz), instead of the radial frequency units used here, p and q, and the one- and two-dimensional power spectral density functions called $S_1(f_x)$ and $S_2(f_x,f_y)$, instead of W_1 and W (and W_2).

Many quantities of interest can be expressed in terms of the moments of the PSD W. These moments can be expressed as

$$m_n = \int_{-\infty}^{+\infty} p^n W_1(p,t)dp \quad (11.32)$$

(n even), or alternatively in terms of W_2 or C. The RMS profile height is $\sigma = m_0^{1/2}$, the RMS slope is $m_{sl} = m_2^{1/2}$, and the RMS curvature is $c = m_4^{1/2}$. The average transverse length parameter is $d_{av} = 2\pi(m_0/m_2)^{1/2} = 2\pi\sigma/m_{sl}$. [The dimensionless shape parameter $\bar{\alpha} = m_0 m_4/m_2^2 = (\sigma c/m_{sl}^2)^2$ can also be defined.] The total integrated scatter (TIS) is $4k^2 m_0 = (2k\sigma)^2$. The angular spread of the scattering spectrum (angular radius of gyration) about the specular direction is $\Delta\theta = (m_2/m_0)^{1/2}/k = \lambda/d_{av}$. These expressions of TIS and $\Delta\theta$ are precise mathematical formulations of Equations 11.22 and 11.23, respectively.

This expression for total integrated scatter is the low scatter limit of a more exact expression derived from that for specular reflectance at normal incidence by Bennett and Porteus (1961) (see also Pickering, 1995):

$$\text{TIS} = 1 - e^{-(4\pi\sigma/\lambda)^2} \sim (4\pi\sigma/\lambda)^2 = 4k^2\sigma^2 \tag{11.33}$$

Equation 11.32 can be generalized to include instrumental effects. For arbitrary lateral structure, the RMS roughness is given by

$$\sigma^2 = \int_{p_{min}}^{p_{max}} \int_{q_{min}}^{q_{max}} M(p,q)W(p,q)dq\,dp \tag{11.34}$$

where $M(p,q)$ is the modulation transfer function (MTF) and the instrumental bandwidth is given by the integration limits (McNeil et al., 1992a, 1993a).

Different models of the surface topography can be considered. A normal Markov process of random defect scatterers would suggest an exponential autocovariance function:

$$C(v) = \sigma^2 e^{-|v|/l} \tag{11.35}$$

where σ is the vertical RMS roughness and l is the transverse autocorrelation length of the roughness (Church et al., 1977, 1979). For a one-dimensional shot model, this leads to

$$W_1(p) = \frac{1}{\pi} \frac{\sigma^2 l}{1 + (lp)^2} \tag{11.36}$$

In the small scattering angle approximation (in the plane of incidence), this becomes

$$W_1(p) \sim \frac{1}{\pi} \frac{\sigma^2}{k^2 l} \frac{1}{(\theta_s - \theta_i)^2} \tag{11.37}$$

For a two-dimensional shot model, this leads to

$$W_2(u) = \frac{1}{2\pi} \frac{\sigma^2 l^2}{[1 + (lu)^2]^{3/2}} \tag{11.38}$$

In the small scattering angle approximation, this becomes

$$W_2(u) \sim \frac{1}{2\pi} \frac{\sigma^2}{k^3 l} \frac{1}{|\theta_s - \theta_i|^3} \tag{11.39}$$

This means that these 1D and 2D models predict scattering that falls off as the second and third power, respectively, of the scattering angle from the specular direction. There has been little quantitative analysis of small angle scattering for wafers roughened during processing. However, "$1/\theta^2$" scattering is very common from highly polished mirrors and lenses.

Church (1988) has pointed out that the finish of highly polished optical surfaces is frequently fractal-like with $W_1(p) = K_n/p^n$ for the 1D or corrugated surface and $W_2(u) = \{\Gamma[(n+1)/2]/(2\Gamma(1/2)\Gamma(n/2))\} K_n/u^{n+1}$ for 2D isotropically rough surfaces, with $1 < n < 3$. (Γ is the gamma function and K_n is a constant.) With $n = 2$, which is called the Brownian fractal, these expressions correspond to Equations 11.36 and 11.38, respectively, in the small angle approximation. The intrinsic surface parameters for such surfaces are K_n and n and not σ and l.

11.2.1.3 Instrumentation

Laser light scattering of rough surfaces is usually measured with a single fixed-frequency laser, such as a He–Ne laser (632.8 nm), argon-ion laser (488 nm), or He–Cd laser (325 nm), incident at a fixed angle θ_i and light collected at an angle θ_s. A filter that transmits only the laser line can be placed before the detector to limit the collection of stray light. Such a setup is most sensitive to lateral structure given by d in the grating equation (Equation 11.21) in first order. Sometimes multiple scattering angles are used. Anisotropies in roughness can be observed by rotating the plane of the incident and scattered wavevectors, which can be accomplished by rotating the sample or the optics. LLS can also be used with rotating samples; any surface anisotropy is indicated by an oscillation in the scattered intensity (see Figure 11.21 below). LLS is very sensitive to polarization.

Surface morphological features less than about 1 nm high are routinely detected. Although submonolayer sensitivity, which requires careful exclusion of other sources of light, has not been common, it is certainly possible. To date, most LLS data obtained during thin film processing have been analyzed only in a semiquantitative manner. In process development studies, it has been common for researchers to analyze the surface after the process with more sensitive *ex situ* spectroscopies, such as atomic-force microscopy (AFM), to assist the interpretation of LLS data acquired *in situ*.

11.2.1.4 Applications in Real-Time Probing

The simplicity of LLS makes it a very attractive tool for real-time monitoring; it is often implemented in conjunction with other optical probes. LLS has been used as a real-time monitor during substrate cleaning and during both homoepitaxial and heteroepitaxial growth. In one early study, Olson and Kibbler (1986) used LLS to monitor the onset of lattice-mismatched growth during the MOCVD heteroepitaxy of GaInAs on GaAs, which was noted by the large increase in laser scatter. They also monitored LLS during the growth of GaInAs/GaAs superlattices on GaAs. During the MOCVD of GaP on Si, they saw oscillations in the LLS signal due to interference in the film.

In another early investigation, Robbins *et al.* (1987a,b) and Pickering *et al.* (1987) monitored the surface topography during oxide removal from Si substrates and subsequent growth of Si by molecular-beam epitaxy (MBE); this was followed by later work by this group (Pidduck et al., 1989a,b; Pickering, 1991; Pickering *et al.*, 1994). In their earlier work there was surface roughening during oxide removal, especially when there was carbon contamination, which was indicated by an increase in LLS. The LLS signal decreased during Si homoepitaxy because the surface became smoother. Figure 11.18 shows the results in later studies with a cleaner system, where there is no perturbation during the growth of the silicon epilayer. *In situ* ellipsometry traces are also shown in this figure (Section 9.11).

Smith *et al.* (1993) used real-time LLS to monitor the growth of GaAs films on GaAs by MBE. LLS showed that the thicker the oxide, the rougher the surface after the *in situ* oxide desorption step that preceded growth. During the homoepitaxial growth, the laser used for LLS was incident both along the

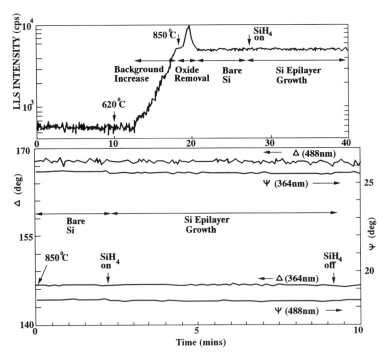

Figure 11.18 (a) Laser light scattering trace (488 nm) monitoring oxide removal from a Si surface at 850°C and subsequent epitaxial growth (UHV-VPE) of Si; (b) ellipsometry traces during the Si epitaxial growth phase (from Pickering, 1991).

[110] and [1̄10] crystal azimuths. The LLS intensity initially decreased for both directions, indicating smoothing. Then it remained essentially constant for the [110] direction, but increased monotonically for the [1̄10] direction. *Ex situ* AFM showed an anisotropic ridge topology, which was sensed by LLS. The LLS intensity decreased during post-deposition annealing, which indicated recovery to the equilibrium condition.

In similar work, Lavoie *et al.* (1992) also noticed roughening (i.e., increased LLS) during thermal desorption of the oxide and smoothing (decreased LLS) during MBE growth of a homoepitaxial GaAs buffer layer (001). In later work, they conducted measurements at three different scattering angles, with either *s*- or *p*-polarized light incident along either the [110] or [11̄0] directions along the surface (457 nm, argon-ion laser), and found that high-spatial frequency roughness (small d) decreases during growth (where the initial surface roughness was caused by the oxide desorption), while low-spatial frequency roughness (large d) increases (Lavoie *et al.*, 1994). In another study involving *in situ* treatment of the GaAs surface, Rouleau and Park (1993) found that GaAs surfaces cleaned by H atoms at elevated temperature are smoother than those prepared by thermal desorption.

As noted by Olson and Kibbler (1986), an increase in LLS can be used to monitor the onset of dislocations during lattice-mismatched growth. In films thinner than the critical thickness d_c, the in-plane lattice constant of the film adjusts to that of the bulk substrate, and the commensurate layer is elastically strained and dislocation-free. When the film thickness exceeds d_c, misfit dislocations form in the film and it relaxes to become incommensurate and strain-free. These dislocations cause a surface roughness of ~0.8–1.2 nm that increases the scattered signal. Celii *et al.* (1993a,b,c) have used LLS as an *in situ* probe of surface roughness during the MBE growth of strained InGaAs layers on GaAs and of strained layer InGaAs/GaAs superlattices on GaAs, with the experimental setup depicted in Figure 11.19. They saw a dramatic increase in laser light scattering when the deposit attained the critical thickness (d_c) in both cases, as is seen in Figure 11.20. The LLS signal was not isotropic and diffuse, which would indicate random surface roughness. Instead, it peaked in specific directions, which indicates a preferred direction of the rough features. Figure 11.21 shows the LLS signal as a function of azimuthal angle. The strong peaks at 90° and 270° are due to scattering from α dislocations, which run along [011̄], while the weaker peaks at 0° and 180° are due to scattering from the rarer β dislocations, which run along [011]. Figure 11.22 demonstrates that the LLS intensity tracks the dislocation density. By monitoring small scattering angles, Yang *et al.* (1993) found that during MBE growth of InAs on GaAs, the surface begins to roughen after two monolayers have been deposited. They also monitored the slip-line generation during the growth of InGaAs/GaAs superlattices on vicinal GaAs substrates.

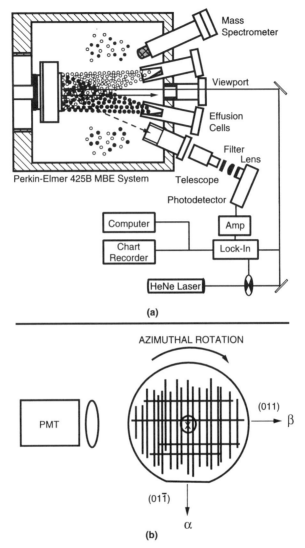

Figure 11.19 (a) Schematic of an MBE system with capability of laser light scattering and *in situ* mass spectrometry. (b) Azimuthal rotation of the wafer changes the detected LLS signal due to preferential scattering of misfit dislocations (from Celii *et al.*, 1993c).

Bertness *et al.* (1994) used LLS to monitor the growth of InP layers by organometallic vapor-phase epitaxy (OMVPE) on InP substrates cut with misorientation from the (001) plane. Such miscuts can enhance step flow and lead to smoother morphologies. Figure 11.23 compares the LLS signal for four such substrate cuts: no misorientation (0°), miscut 2° toward [0$\bar{1}$1] (2°), and miscut 4° toward [$\bar{1}\bar{1}$1] (4°A) or toward [1$\bar{1}$1] (4°B).

11.2 Diffraction from Surface Features

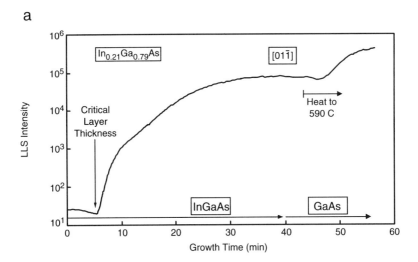

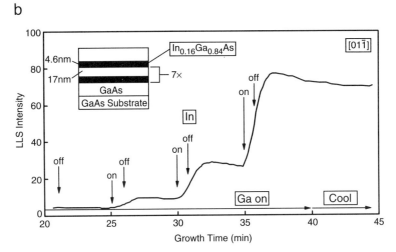

Figure 11.20 Laser light scattering signal (543 nm, He–Ne laser, sensitivity to α dislocations) during the MBE growth and relaxation of (a) an $In_{0.21}Ga_{0.79}As$ layer on GaAs and (b) an $In_{0.16}Ga_{0.84}As$/GaAs strained layer superlattice on GaAs, with α dislocations first observed during the deposition of the sixth InGaAs layer (from Celii et al., 1993a).

Only the 2° sample showed an increase in surface roughness during the outgas phase (PH_3 flow), which was due either to the desorption of surface oxides or to the formation of indium clusters. The films on the singular (001) (0°) substrate showed the least increase in diffuse scattering; consequently, they were the smoothest films. The roughness of the films grown on the misoriented (vicinal) substrates depended on deposition temperature.

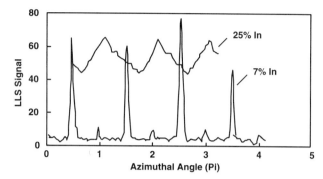

Figure 11.21 Laser light scattering azimuthal scans obtained during the MBE growth of $In_{0.07}Ga_{0.93}As$ and $In_{0.25}Ga_{0.75}As$, showing highly anisotropic and isotropic scattering, respectively (from Celii et al., 1993b).

Dietz et al. (1995) used LLS together with p-polarized reflectance spectroscopy (PRS) to probe GaP atomic layer epitaxy on Si(001) and GaP(001) substrates. This is shown in Figure 9.59 and discussed further in Section 9.9.

The heteroepitaxy of II–VI and group IV semiconductors has also been monitored by using LLS. In monitoring the MBE growth of ZnSe on GaAs (0.25% mismatch), Rouleau and Park (1993, 1994) found a rapid increase in diffusely scattered light, presumably when the film attained the critical thickness. With continued growth, the diffuse scatter periodically varied because of optical interference within the ZnSe film. They also saw similar behavior during the growth of ZnTe on GaAs (7.9% mismatch). Pickering (1991) monitored the CVD of $Si_{0.77}Ge_{0.23}$ epilayers on Si by using LLS and

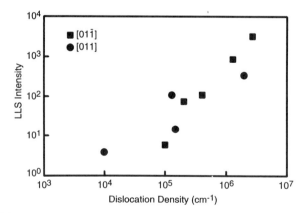

Figure 11.22 Correlation between the LLS signal (543 nm, He–Ne laser) from α or β dislocations in relaxed InGaAs layers grown by MBE with dislocation densities determined ex situ by TEM (from Celii et al., 1993c).

11.2 Diffraction from Surface Features

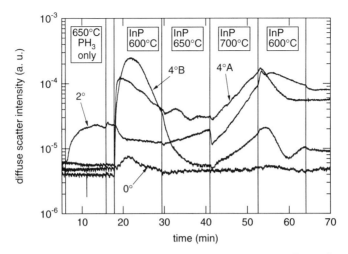

Figure 11.23 Laser light scattering (632.8 nm) during the InP OMVPE growth at different temperatures on singular and vicinal InP(001) substrates: singular (001) (labeled 0°), miscut 2° toward [0$\bar{1}$1] (2°), miscut 4° toward [$\bar{1}\bar{1}$1] (4°A), or miscut 4° toward [1$\bar{1}$1] (4°B) (from Bertness et al., 1994). (© 1994 IEEE.)

spectroscopic ellipsometry (Section 9.11). As seen in Figure 11.24, LLS increases during the growth of the alloy—presumably due to the roughening of the surface—and decreases when silicon is regrown on top—due to smoothing.

Elastic scattering from randomly "rough" surfaces has also been used in other systems and for other processes. For example, Al-Jumaily et al. (1987) used LLS to examine the roughness of sputter-deposited Cu films. Although this study was conducted *ex situ*, such analysis could also be implemented *in situ*.

In a quite different application, Johnson et al. (1993) used LLS to help monitor the wafer temperature during MBE growth. The GaAs wafers were polished on one side (for subsequent thin film processing) and textured on the other side (to promote diffuse scattering). The transmission of near-band-gap light from a white-light source was monitored through a roundtrip in the wafer by measuring the diffuse reflectance from the textured back surface. This is actually an absorption probe, very similar to the transmission technique that is described in Section 8.4. It has two advantages *vis-à-vis* monitoring by transmission: (1) the measured absorption is due to a double pass through the wafer rather than a single pass, and (2) access to only one side of the wafer is required rather than to both sides. In this technique, diffuse reflection of the white-light source from the front surface is a potential background signal. Johnson et al. (1993) observed that this background level increased after oxide desorption, which roughened the front surface. This thermometric technique is detailed further in Section 8.4.

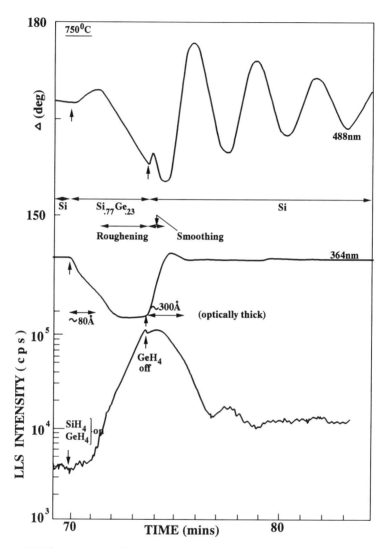

Figure 11.24 Real-time ellipsometry (364 nm, 488 nm) and LLS (488 nm) measured simultaneously during $Si_{0.77}Ge_{0.23}$/Si multilayer growth (UHV-VPE) (from Pickering, 1991).

Total internal reflection microscopy (TIRM) is a closely related diagnostic, in which light is coupled through the leg of a prism and totally reflected from the hypotenuse; TIR is discussed in Section 9.7.2. The elastic scattering of light from a film atop the hypotenuse face can indicate the presence of defects in the film. Using an argon-ion laser and a long working distance microscope, Williams *et al.* (1992) were able to examine *in situ* ZrO_2 films

11.2 Diffraction from Surface Features

that were deposited on a prism by electron beam evaporation. They observed point defects and film crystallites in these films.

Another scattering probe of surface roughness, based on a different physical mechanism, is very sensitive to surface composition as well as to roughness. Eagen and Weber (1979) detected monolayer levels of adsorbed oxygen on metals via surface-plasmon resonances that were measured by elastic scattering induced by the surface roughness; the spectrum of the inelastically scattered light can also be examined. While this method can be used *in situ,* it is invasive since it uses a prism to couple light to the surface and is limited to high-reflectivity metals.

11.2.2 *Scatterometry from Periodic Surface Features*

The periodic surface structure shown in Figure 11.25a is fabricated by using a patterning process; it could represent a distributed-feedback reflector in a semiconductor laser, a small test-pattern region on a large wafer that probes the etching process (Figure 11.25a1), or the latent image after patterned photoresist exposure (Figure 11.25a2). Diffraction from this structure produces a characteristic multiple-slit interference/diffraction pattern, as seen in Figures 11.26 and 11.27. This type of scatterometry has been used both as an *in situ* monitoring probe as well as an *ex situ* characterization tool. Naqvi *et al.* (1994) have reviewed several such applications of scatterometry.

As diagrammed, the pattern in Figure 11.25a1 (or 2) repeats every d, which gives the pitch. The etched pattern is a wide and c deep, and the unetched pattern is b wide; width $d = a + b$. The diagram shows a sidewall angle ξ, but the walls of the etched feature can be more complex. A parallel beam at wavelength λ incident on this structure produces a multiple-slit interference pattern which can determine the etch width, depth, and sidewall angle. Determination of the repeat distance d from the rapid oscillations in Figures 11.26 and 11.27 is trivial and usually not interesting. It is the envelope that provides information about the profile.

The measured scattergram can be related to the profile by using either diffraction calculations or a data set of scattergrams from previously measured profiles. Petit (1980, 1990) has reviewed the integral and differential electromagnetic theory of diffraction from gratings. Moharam and Gaylord (1982, 1986) have developed a rigorous coupled-wave theory that has been used to determine scattergrams from such periodic structures in several studies.

Figure 11.25b shows the arrangement of a typical scatterometer used for *ex situ* characterization (McNeil *et al.,* 1992a); it can be used *in situ* only when the angle tuning fits the reactor requirements. Use of s-polarized light avoids surface plasma wave coupling in conducting samples, although both s- and p-polarized light are used in some applications. Two-

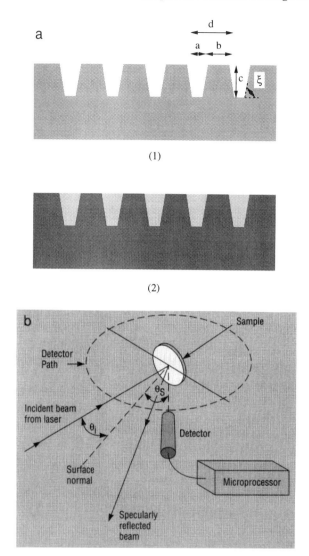

Figure 11.25 (a) Periodic structure in (1) topography after etching and (2) refractive index after photoresist exposure (latent image). Experimental arrangement of (b) a "standard" scatterometer, (c) a scatterometer for two-dimensional scatter patterns, and (d) a 2Θ scatterometer arrangement used for short-pitch, small-CD structures (from McNeil et al., 1992a).

dimensional scatter patterns can be viewed by placing a hemispherical frosted dome about the sample (Figure 11.25c); the dome has a hole in it to allow the laser to irradiate the sample. A camera views the scattering pattern, which is then analyzed. In the "2Θ Scatterometer" the angle of incidence is tuned by rotating the sample, and the detector is continuously

11.2 Diffraction from Surface Features

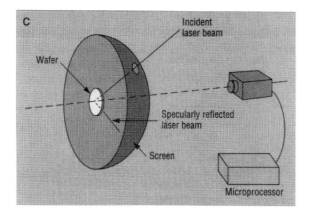

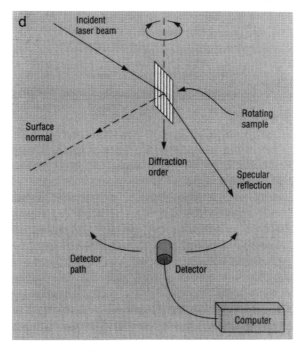

Figure 11.25 (*Continued*)

moved to collect light from one order (Figure 11.25d). This method provides a large data set for analysis, and is often used for *ex situ* analysis of short-pitch [small-CD (critical dimension)] structures.

11.2.2.1 Etch Profiles

One goal of these studies is to use a diffraction pattern that is fabricated by lithography somewhere on the film to monitor the etch rate of the film

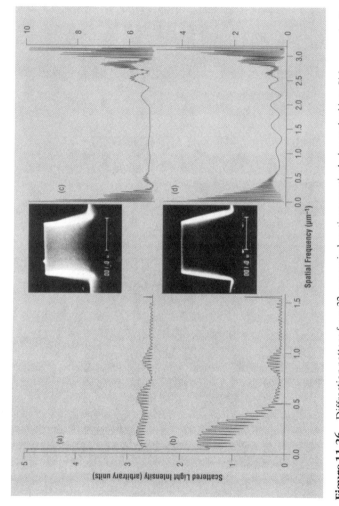

Figure 11.26 Diffraction pattern from 32-μm period gratings reactively-ion etched into Si for a nominally vertical profile [top SEM, (a) and (c)—baselines shifted up] and an overcut profile [bottom SEM, (b) and (d)], both etched to the same depth (λ = 632.8 nm). The patterns were obtained with either normal [(a) and (b)] or grazing [(c) and (d)] incidence (from Giapis et al., 1991).

11.2 Diffraction from Surface Features

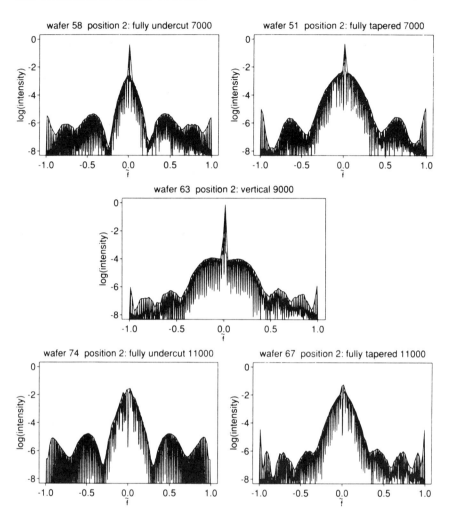

Figure 11.27 Logarithm of scattered light intensity from etched profiles as a function of dimensionless spatial frequency for different types of profiles (as denoted). The raw data and corresponding envelope functions are shown for each (from Krukar *et al.*, 1993).

in real time. In early work, Kleinknecht and Meier (1978) monitored both wet and plasma etching of SiO_2 and Si_3N_4 gratings on Si by measuring the intensity of a diffraction order (Equation 11.21). The goal was to determine the etch depth of the film and to describe lateral underetching by the etching process, which was detected by a decrease in the first-order diffraction signal. In related work, Sternheim *et al.* (1983) monitored the oscillations in the zero-order mode during the etching of isolation areas in silicon for oxide-isolated bipolar devices. In a series of studies, Braga *et al.* (1983) and Mendes *et al.* (1984, 1985) determined the endpoint in etching of a film by monitoring both the zero- and first-order peaks of a grating pattern test area. Damar *et al.* (1984) explored procedural and instrumentation aspects of such monitoring.

Later work has concentrated on using light scattering to detect submicron differences in etch profiles for critical dimension and linewidth control. Grimard *et al.* (1989) evaluated the use of scattering for real-time monitoring during plasma etching by comparing experiment and theory for simple etch geometries. The theory involved simple scalar diffraction analysis, which is valid for the $d/\lambda \gg 1$ (they assumed $d/\lambda \geq 10$). Similar work was done by Bishop *et al.* (1991) using a coupled-wave approach, and by Chapados and Paranjpe (1992).

In a comprehensive study, Giapis *et al.* (1991) and Krukar *et al.* (1993) measured the scattergrams over 180° from Si wafers with photolithographically defined lines and spaces produced by RIE (632.8 nm). A wide range of etch depths and sidewall profiles was generated by varying etching conditions. Figure 11.26 shows the diffraction patterns for vertical and overcut patterns, measured alternately at normal and grazing incidence. Figure 11.27 plots the scattered intensity, along with the envelope functions, for features that were etched to the same depth, but which had fully undercut, fully tapered, or vertical sidewalls. The envelope functions are clearly distinctive in each case. Giapis *et al.* (1991) used principal-components analysis (Section 19.1) to classify the scattergrams according to etch profile and depth. Krukar *et al.* (1993) used alternately linear discriminant analysis and neural network training to catalog scattergrams, and found that both were fast enough for in-line inspection. They noted that discriminant analysis is capable of classifying scattergrams with more subtle differences than the principal-component analysis used in the earlier study. In this *ex situ* study, etch features were classified with >95% accuracy with discriminant analysis, which bodes well for the use of scatterometry as a diagnostic for real-time monitoring and control during etching. Although careful *in situ* control during etching is possible, it is constrained by optical access in the reactor, which can limit the collection of all diffracted orders. One disadvantage of this technique for real-time control is the need for a grating-like test pattern. This consideration affects microelectronics applications, but not optical device applications where gratings are normally present. More recent work by this group has involved the use of ultraviolet lasers for analysis of submicron features. In related work, Chapados (1994) has described an *in situ* sensor that uses scatterometry with a He-Cd laser at 441.5 nm to determine the critical dimensions of gratings in dielectric and metal films on semiconductor wafers. Scatterometry can also be used to monitor the planarization of thin films atop grating structures.

11.2.2.2 Latent Images in Photoresist

Control of the critical dimension (CD) of circuit features requires control of the CD of the resist profile. Scatterometry can be used to control this

11.2 Diffraction from Surface Features

by probing the latent image formed in photoresist when it is exposed, as well as before baking and development. Since the photoactive compound (PAC) concentration in the photoresist changes on local exposure, a grating on the photomask generates a periodic PAC profile with a periodic refractive index that diffracts light. This process can be modeled by using photolithography simulation tools (Crisalle *et al.*, 1989), such as PROLITH (Mack, 1985) and SAMPLE (Oldham *et al.*, 1979), which predict PAC concentrations as a function of exposure, resist, and substrate conditions. The index of refraction of the photoresist depends linearly on the PAC concentration. Rigorous coupled-wave theory (Moharam and Gaylord, 1982) can be used to predict the diffraction pattern from this latent image phase grating.

Use of scatterometry to monitor the intensity of light diffraction from the exposed, undeveloped resist into first (or second) order provides quantitative feedback for real-time closed-loop control. For tight control of CDs, this is superior to the use of test wafers to optimize conditions and open-loop control. Such control is needed because the required exposure dose depends on the optical properties of the underlying film. Yoon *et al.* (1992) have described other methods of characterizing latent images that are based on microscopy.

Scatterometry has been used to optimize exposure (Hickman *et al.*, 1992) and focus (Adams, 1991; Milner *et al.*, 1993). Figure 11.28a shows the how the first-order diffraction efficiency increases with exposure time (dose), while Figure 11.28b shows that there is a unique diffraction efficiency (before baking and development, as in Figure 11.28a) for each CD of the resist grating, as measured after development (Hickman et al., 1992). The resist must not be sensitive to the chosen laser wavelength, so 632.8 nm (He–Ne laser) was chosen to probe the Hunt 204 photoresist, which is sensitive to the 436-nm Hg g line. CD variations were reduced by almost a factor of 4 by monitoring diffraction from the latent image. The intensity of the first-order diffraction peak always appears to be a maximum with optimal focusing conditions; for the second-order peak, optimization depends on the layer beneath the resist (Milner *et al.*, 1993). The focus position can be controlled with a precision of 0.1 μm, and this method offers the possibility of even finer control.

Scatterometry is also very important for monitoring post-exposure bake (PEB) for chemically amplified resists. Other applications for scatterometry are for accurate linewidth metrology for photomask gratings, the measurement of random and periodic structure on line edges in masks, and the accurate measurement of phase shifts in phase shift masks (McNeil *et al.*, 1992a).

11.2.2.3 Optical Diffraction Thermometry

Diffraction can be used to measure temperature by monitoring changes in length due to thermal expansion, even though this fractional change is

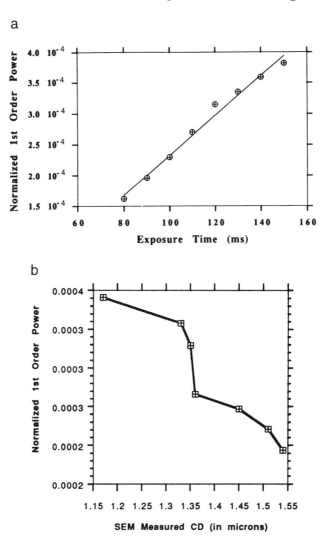

Figure 11.28 Normalized first-order diffracted power (632.8 nm) from a latent image grating in photoresist on a silicon wafer versus (a) exposure time and (b) the critical dimension (CD) measured in these gratings after development (from Hickman et al., 1992).

very small. For Si, $\tilde{\alpha} \sim 3 \times 10^{-6}/°C$, so a 3°C temperature rise increases the length by 1 part in 10^5. One way to conduct such a measurement is projection moiré interferometry, in which the change of the period of a grating δ due to thermal expansion is made by comparison to a standard that does not vary with temperature; this change is too small to measure it absolutely. Other optical diagnostics that utilize interferometry to sense the change in

11.2 Diffraction from Surface Features

distance between two widely separated points on a wafer caused by thermal expansion are described in Sections 10.1 and 11.3.

11.2.2.3.1 *Projection Moiré Interferometry*

By using projection moiré interferometry, Zaidi *et al.* (1992), Brueck *et al.* (1993), and Lang *et al.* (1994) have compared the grating period $\delta(T)$ with the spacing of an interferometric pattern made by two lasers, $\delta_{virtual}$, and have made noncontact temperature measurements on Si with a resolution of ±0.05°C. Projection moiré interferometry (Post, 1982) is a variation of conventional moiré interferometry (Guild, 1956; Theocaris, 1969; Durelli and Parks, 1970; Kafri and Glatt, 1990), in which patterns are produced by the superposition of two gratings that have slightly different pitch (period) or angle (orientation).

The experimental arrangement for the temperature measurement is shown in Figure 11.29 (Zaidi *et al.*, 1992). Two laser beams at the same wavelength λ (632.8 nm), which can be split from the same laser, are symmetrically incident on the substrate grating, each at an angle of incidence θ_i, to form an interferometric pattern with period $\delta_{virtual} = \lambda/(2 \sin \theta_i)$. θ_i is chosen so that diffraction into the $m = -1$ order from each laser is nearly normal to the surface, so $\theta_{-1} \sim 0°$. Using the grating equation (Equation 11.21),

$$\sin \theta_{-1} = \sin \theta_i - \lambda/\delta \tag{11.40}$$

The two diffracted beams interfere to form a moiré interference pattern with period

$$\delta_{moiré} = \lambda/(2 \sin \theta_{-1}) \tag{11.41}$$

Using Equation 11.40, this can also be expressed as

$$1/\delta_{moiré} = 1/\delta_{virtual} - 2/\delta \tag{11.42}$$

Since this is most sensitive when $\theta_{-1} \sim 0°$ and $\delta_{moiré}$ is very large, $\delta_{virtual}$ is chosen to be $\sim \delta_0/2$, where δ_0 is the period at ambient temperature.

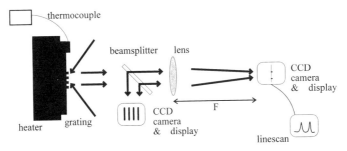

Figure 11.29 Schematic of apparatus used for temperature measurements by projection moiré interferometry (from Zaidi *et al.*, 1992).

Using these expressions and the linear coefficient of thermal expansion $\tilde{\alpha}\,[=(1/\delta)(d\delta/dT)]$,

$$d\delta_{\text{moiré}}/dT = 2\tilde{\alpha}(\delta_{\text{moiré}})^2/\delta \qquad (11.43)$$

Since $[d\delta_{\text{moiré}}/dT]/\delta_{\text{moiré}} = [2\delta_{\text{moiré}}/\delta]\tilde{\alpha}$, this intereference method can amplify the effect of thermal expansion by $2\delta_{\text{moiré}}/\delta \sim 10^3{-}10^4$.

This moiré pattern can be viewed by a CCD camera, as is shown in Figure 11.30, and processed digitally by a fast Fourier transform to obtain T. Alternatively, the pattern can be projected through a lens of focal length f and imaged onto a CCD camera in the focal plane, which is a distance f away (Figure 11.29). This gives two spots, which is the two-dimensional Fourier transform of the moiré pattern, as is seen in Figure 11.31. The spacing of the two spots is $L = \lambda f/\delta_{\text{moiré}}$, which changes with temperature as $dL/dT = 2\lambda f\tilde{\alpha}/\delta_0$. This separation is more simply understood by individual consideration of the two diffracted beams, which are paraxial rays ($\theta_{-1} \sim 0°$) in the geometric optics limit. A parallel beam focused by a thin lens with focal length f is imaged a distance $f\phi$ below the optic axis, where ϕ is the angle the beam makes with the optic axis of the lens. Therefore the separa-

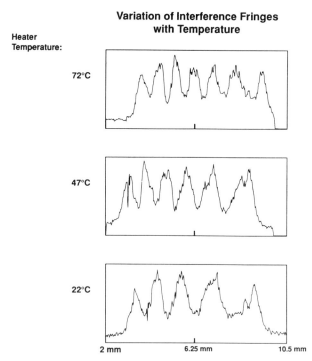

Figure 11.30 Observed moiré fringes from a Si grating illuminated with a laser at 632.8 nm at different temperatures (from Zaidi et al., 1992).

11.2 Diffraction from Surface Features

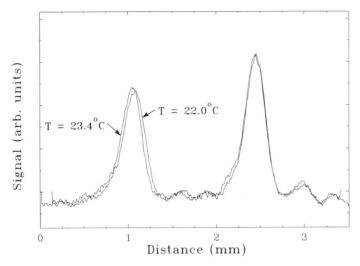

Figure 11.31 Linescans across the Fourier transform image of the moiré pattern, which show that the peak separations change with temperature (from Zaidi et al., 1992).

tion of the two spots is $L = 2f\theta_{-1}$, which, using Equation 11.41, agrees with the previous expression.

Figure 11.31 shows that changes in the peak separation can be seen with ~1°C temperature variations. The minimum temperature difference that can be measured with this method depends inversely on the number of grooves that are covered by the lasers. Lang et al. (1994) have described an automated data acquisition system for continuous monitoring of two diffraction order angles. Using image processing techniques, relative temperature changes can be calculated from the order separation.

Projection moiré interferometry can be used for *in situ*, real-time measurement of *T*, relative to a standard temperature, with high precision ~ 1°C. It is inexpensive, noninvasive, and allows rapid determination of *T*. This method is independent of tilt to first order, and is relatively insensitive to warping of the wafer as the substrate temperature changes and to thin films that can appear on the grating during some processing steps. Effects of all wafer rigid body motions can be eliminated to all orders by tracking the spot centroids and the angle of the line between the spot (Lang et al., 1994). However, this technique requires that the substrate have a grating patterned on it, which could be part of a test area; therefore, it cannot be used to measure temperature at arbitrary points. Furthermore, this grating test pattern would use valuable "real estate" on the wafer during manufacturing.

Pichot and co-workers (Pichot and Guillaume, 1982; Durandet et al., 1990) have developed another contactless temperature probe based on the thermal expansion of a grating, but one in which this expansion is compared to a reference grating at a fixed temperature (Figure 11.32), rather than to the grating formed by the interference of two lasers. Durandet *et al.* (1990) used this method to monitor the temperature of the surface of a silicon wafer during electron cyclotron resonance (ECR) plasma etching. Figure 11.33 shows that the temperature increases for the first 10 min in an argon plasma and then remains constant. Since the same temperature rise is seen in a nitrogen plasma when an identical amount of power is deposited, this heating was attributed to ion bombardment. When a small amount of SF_6 is added at 10 min (Figure 11.33), the temperature increases again, which they attributed to the exothermic chemical reaction of silicon etching by fluorine.

11.3 Speckle Photography and Interferometry

Speckle metrology, including speckle photography and interferometry, involves scattering of light from a rough surface to measure in-plane displace-

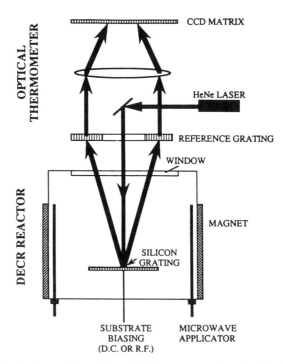

Figure 11.32 Implementation of the optical thermometer based on the thermal expansion of a diffraction grating referenced to an external grating in a distributed ECR (DECR) reactor (from Durandet *et al.,* 1990).

11.3 Speckle Photography and Interferometry

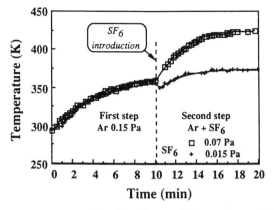

Figure 11.33 The temperature rise of a silicon wafer in a plasma, which is first heated by an Ar plasma and then heated and etched by an SF_6/Ar mixture, monitored by the system described in Figure 11.32 (from Durandet et al., 1990).

ment or deformation (Françon, 1979; Dainty, 1984). It has also been used to probe inhomogeneous fluid media, such as fluids seeded with scattering particles (speckle velocimetry) and fluids with refractive index gradients; as such, it has been used extensively in fluid dynamics (Vest, 1979; Lauterborn and Vogel, 1984; Merzkirch, 1989; Shakher and Nirala, 1994).

In speckle photography, light scattered from the medium is imaged to produce a speckle pattern with speckles of size $d_s = 1.2\lambda_s(M + 1)f_\#$, where λ_s is the wavelength of the scattered light and M is the magnification of the imaging system consisting of a lens with $f_\# = f$ (its focal length)/D (its diameter). Successive images superimposed on the same photographic plate (specklegram) display movement of the speckle pattern, which can indicate particle flow (and velocity) or in-plane translation and deformation of the surface of a solid. For example, if the particle moves a distance d_p between speckle images, then the speckle pattern on the specklegram has patterns that are separated by $d_s = Md_p$. One way to extract this information optically is by illuminating the specklegram with coherent light (wavelength λ_f) and imaging the interference pattern with a lens of focal length f_f (Lauterborn and Vogel, 1984; Merzkirch, 1989). The light scattered by the two patterns interferes to produce Young's interference fringes separated by $d_f = \lambda_f f_f/d_s = \lambda_f f_f/Md_p$, from which the displacement d_p can be determined.

Light scattered from the roughened backside of a wafer exhibits a speckle pattern that can change with temperature because the distance between the topographical features responsible for scattering vary due to thermal expansion. Such speckle interferometry could be used anywhere, and perhaps at several places on the backside of the wafer, where there is big, relatively unused "real estate." Probing the backside of the wafer makes this approach for real-time monitoring in production tools more attractive than other methods that require prepatterning on the frontside of the wafer.

Burckel *et al.* (1994, 1995) have described different approaches to speckle interferometry. One technique is electronic speckle-pattern interferometry (ESPI) in which two lasers (He–Ne or diode) impinge on the surface at the same place. The scattered light from both beams is imaged onto a lens and the speckle field is monitored. The cross-term in the interference signal can be processed to sense differential translational motion as the temperature changes, i.e., thermal expansion. Analysis is improved by controlling the phase shift of one of the beams (phase shift interferometry, PSI).

One weakness of full-field methods such as ESPI and projection moiré interferometry is the processing requirement for analyzing the image, which may involve ~1/3 Mpixels. This makes it difficult to achieve accuracies of ~1°C in the ~10 msec that is sometimes needed in rapid thermal processing (RTP). The speckle technique introduced by Burckel *et al.* (1995), subfeature speckle interferometry (SUFSI), requires only a few detector elements and is therefore more compatible with the bandwidth requirements needed for real-time control. It relies on amplitude interference, rather than the intensity cross-correlations between speckle intensity patterns in ESPI. Figure 11.34 shows two ways to implement SUFSI, both of which utilize a

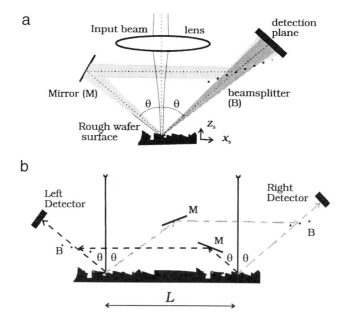

Figure 11.34 Experimental arrangement for subfeature speckle interferometry (SUFSI) to measure (a) in-plane translation and (b) differential translation (strain) and tilt (from Burckel *et al.*, 1995).

11.3 Speckle Photography and Interferometry

single laser focused to a spot size w_f. The speckle pattern has a size (or correlation length) $d_{\text{speckle},i} \sim \lambda R_i/w_f$, where R_i is the path length from the sample to the particular detector. Interference from the spherical waves emanating from each scattering region leads to a fringe pattern because the wavefronts from each have different radii of curvature. If these radii are R_r and R_l, then the fringes have characteristic separation $d_{\text{fringe}} \sim \sqrt{2\lambda R_r R_l /|R_r - R_l|}$. In many ESPI measurements $R_r \sim R_l$ so $d_{\text{fringe}} \to \infty$; with this geometry, d_{fringe} can be set $\leq d_{\text{speckle}}$.

In the single-point geometry in Figure 11.34a, speckle fields are collected from symmetrically placed solid angles. The two collected speckle fields have opposite phase shifts for horizontal translations (x_s), leading to a net phase shift of $2k\Delta x_s \sin\theta$, and so this geometry is sensitive to this in-plane translation (x_s) [$k = 2\pi/\lambda$]. The phases are independent of the other horizontal translation y_s and have a common phase shift for vertical translation (z_s), which cancel.

The two-point geometry in Figure 11.34b is sensitive to differential translation (strain), such as that caused by thermal expansion, and to tilt. The phase shift at each detector ($i = l, r$) is $\mp k(\Delta x_s^{(1)} - \Delta x_s^{(2)})\sin\theta$. The effect of thermal expansion can be separated from that of tilt by analyzing both the sum and difference of the phase shifts from the two detectors. Figure 11.35 shows that the deviation between temperature measurements of a Si wafer by SUFSI and a thermocouple is very small (RMS deviation of 0.92°C). In this case the tilt correction was small and could be neglected, so the phase difference between the two detectors (left − right) varied with temperature as $-2kL \sin\theta[\tilde{\alpha} \Delta T]$. $\tilde{\alpha}$ is the linear coefficient of thermal expansion and L is the distance between the two scattering points.

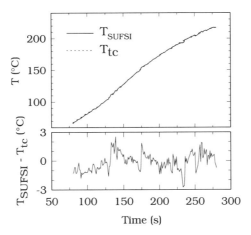

Figure 11.35 Temperature obtained by SUFSI, T_{SUFSI}, and a thermocouple, T_{tc} (from Burckel et al., 1995).

Voorhes and Hall (1991) have likened the two-beam interference speckle interferometry technique, in which a laser is split into two beams that are focused at the same angle of incidence to the same point, to laser Doppler velocimetry (LDV) [Figure 11.13]. In this version of laser extensometry, changes in interference fringe patterns from diffusely reflecting surfaces are detected. They have also described a simpler procedure to measure strain due to thermal expansion, in which light scattered from a single beam is collected by a lens, transmitted through two laterally separated apertures, and detected by a single detector. They cite a 0.1°C resolution and rapid response.

These examples of speckle thermometry are similar to the technique described by Bollmann and Haberger (1992), presented in Section 10.1, which uses interferometry from features across the wafer to measure thermal expansion and therefore sense temperature. However, that study utilized light diffracted from features patterned on different places on the wafer, while this work used that from the unpatterned roughened wafer backside.

References

T. E. Adams, *SPIE* **1464,** 294 (1991).
R. J. Adrian, *Annu. Rev. Fluid Mech.* **23,** 261 (1991).
G. A. Al-Jumaily, S. R. Wilson, L. L. Dehainaut, J. J. McNally, and J. R. McNeil, *J. Vac. Sci. Technol. A* **5,** 1909 (1987).
J. Als-Nielsen, in *Structure and Dynamics of Surfaces* (W. Schommers and P. von Blanckenhagen, eds.), Chap. 5. Springer-Verlag, Berlin, 1986.
H. M. Anderson, R. Jairath, and J. L. Mock, *J. Appl. Phys.* **67,** 3999 (1990).
D. E. Barrick, in *Radar Cross Section Handbook,* Chapter 9. Plenum, New York, 1970.
J. S. Batchelder and M. A. Taubenblatt, *Solid State Technol.* **35**(10), S1 (October, 1992).
J. S. Batchelder and M. A. Taubenblatt, *Microcontamination,* April, p. 35 (1993).
P. Beckmann, *Prog. Opt.* **6,** 53 (1967).
H. E. Bennett and J. M. Bennett, in *Phys. Thin Films* **4,** 1 (1967).
H. E. Bennett and J. O. Porteus, *J. Opt. Soc. Am.* **51,** 123 (1961).
J. M. Bennett and L. Mattsson, *Introduction to Surface Roughness and Scattering.* Optical Society of America, Washington, DC, 1989.
K. A. Bertness, C. Kramer, J. M. Olson, and J. Moreland, *J. Electron. Mater.* **23,** 195 (1994).
K. P. Bishop, S. M. Gaspar, L. M. Milner, S. S. H. Naqvi, and J. R. McNeil, *SPIE* **1545,** 64 (1991).
D. B. Blackford and D. C. Grant, *Microcontamination,* February, p. 27 (1993).
C. F. Bohren, in *Handbook of Optics* (M. Bass, ed.), 2nd ed., Vol. 1, Chapter 6, p. 6.1. McGraw-Hill, New York, 1995.

References

C. F. Bohren and D. R. Huffman, *Absorption and Scattering of Light by Small Particles.* Wiley, New York, 1983.

D. Bollmann and K. Haberger, *Microelectron. Eng.* **19,** 383 (1992).

A. Bouchoule, A. Plain, L. Boufendi, J. P. Blondeau, and C. Laure, *J. Appl. Phys.* **70,** 1991 (1991).

E. S. Braga, G. F. Mendes, J. Frejlich, and A. P. Mammana, *Thin Solid Films* **109,** 363 (1983).

W. G. Breiland and G. H. Evans, *J. Electrochem. Soc.* **138,** 1806 (1991).

W. G. Breiland and P. Ho, in *Chemical Vapor Deposition* (M. L. Hitchman and K. F. Jensen, eds.), Chapter 3, p. 91. Academic Press, Boston, 1993.

W. G. Breiland and P. Ho, in *Proceedings of the Ninth International Conference on Chemical Vapor Deposition* (McD. Robinson, C. H. J. van den Brekel, G. W. Cullen, J. M. Blocher, and P. Rai-Choudhury, eds.), p. 44. Electrochemical Society, Pennington, NJ, 1984.

S. R. J. Brueck, S. H. Zaidi, and M. K. Lang, *Mater. Res. Soc. Symp. Proc.* **303,** 117 (1993).

D. Burckel, S. H. Zaidi, M. K. Lang, A. Frauenglass, and S. R. J. Brueck, *Mater. Res. Soc. Symp. Proc.* **342,** 17 (1994).

D. Burckel, S. H. Zaidi, A. Frauenglass, M. Lang, and S. R. J. Brueck, *Opt. Lett.* **20,** 315 (1995).

F. G. Celii, E. A. Beam, III, L. A. Files-Sesler, H.-Y. Liu, and Y. C. Kao, *Appl. Phys. Lett.* **62,** 2705 (1993a).

F. G. Celii, L. A. Files-Sesler, E. A. Beam, III, and H.-Y. Liu, *J. Vac. Sci. Technol. A* **11,** 1796 (1993b).

F. G. Celii, Y. C. Kao, H.-Y. Liu, L. A. Files-Sesler, and E. A. Beam, III, *J. Vac. Sci. Technol. B* **11,** 1014 (1993c).

P. Chapados, Jr. *SPIE* **2091,** 323 (1994).

P. Chapados, Jr. and A. Paranjpe, *SPIE* **1803,** 283 (1992).

E. Chason, T. M. Mayer, B. K. Kellerman, D. T. McIlroy, and A. J. Howard, *Phys. Rev. Lett.* **72,** 3040 (1994).

E. L. Church, *Appl. Opt.* **27,** 1518 (1988).

E. L. Church and P. Z. Takacs, in *Handbook of Optics* (M. Bass, ed.), 2nd ed., Vol. 1, Chapter 7, p. 7.1. McGraw-Hill, New York, 1995a.

E. L. Church and P. Z. Takacs, *Opt. Eng.* **34**(2), 353 (1995b).

E. L. Church and J. M. Zavada, *Appl. Opt.* **14,** 1788 (1975).

E. L. Church, H. A. Jenkinson, and J. M. Zavada, *Opt. Eng.* **16,** 360 (1977).

E. L. Church, H. A. Jenkinson, and J. M. Zavada, *Opt. Eng.* **18,** 125 (1979).

O. D. Crisalle, S. R. Keifling, D. E. Seborg, and D. A. Mellinchamp, *SPIE* **1185,** 171 (1989).

J. C. Dainty, ed., *Laser Speckle and Related Phenomena.* Springer-Verlag, Berlin, 1984.

H. S. Damar, F. P. Chan, T. T. Wu, and A. R. Neureuther, *SPIE* **470,** 157 (1984).

H. H. Denman, W. Heller, and W. J. Pangonis, *Angular Scattering Functions for Spheres.* Wayne State Univ. Press, Detroit, MI, 1966.

N. Dietz, A. Miller, and K. J. Bachmann, *J. Vac. Sci. Technol. A* **13,** 153 (1995).

R. P. Donovan, ed., *Particle Control for Semiconductor Manufacturing.* Dekker, New York, 1990.

J.-L. Dorier, C. Hollenstein, A. A. Howling, and U. Kroll, *J. Vac. Sci. Technol. A* **10**, 1048 (1992).

L. E. Drain, *The Laser Doppler Technique*. Wiley, New York, 1980.

A. Durandet, O. Joubert, J. Pelletier, and M. Pichot, *J. Appl. Phys.* **67**, 3862 (1990).

A. J. Durelli and V. J. Parks, *Moiré Analysis of Strain*. Prentice–Hall, Englewood Cliffs, NJ, 1970.

C. F. Eagen and W. H. Weber, *Phys. Rev. B* **19**, 5068 (1979).

J. M. Eastman, *Phys. Thin Films* **10**, 167 (1978).

W. G. Fisher and J. M. Davidson, in *Particle Control for Semiconductor Manufacturing* (R. P. Donovan, ed.), Chapter 20, p. 341. Dekker, New York, 1990.

D. I. Fotiadis and K. F. Jensen, *J. Cryst. Growth* **102**, 743 (1990).

N. Françon, *Laser Speckle and Applications in Optics* (translated by H. H. Arsenault). Academic Press, New York, 1979.

P. Freymuth, *Rev. Sci. Instrum.* **64**, 1 (1993).

S. G. Geha, R. N. Carlile, J. F. O'Hanlon, and G. S. Selwyn, *J. Appl. Phys.* **72**, 374 (1992).

K. P. Giapis, R. A. Gottscho, L. A. Clark, J. B. Kruskal, D. Lambert, A. Kornblit, and D. Sinatore, *J. Vac. Sci. Technol. A* **9**, 664 (1991).

P. Gise, in *Handbook of Contamination Control in Microelectronics* (D. L. Tolliver, ed.), Chapter 12, p. 383. Noyes, Park Ridge, NJ, 1988.

D. C. Grant, *Microcontamination*, March, p. 37 (1993).

D. S. Grimard, F. L. Terry, Jr., and M. E. Elta, *SPIE* **1185**, 234 (1989).

J. Guild, *The Interference System of Crossed Diffraction Gratings*. Oxford Univ. Press (Clarendon), Oxford, 1956.

R. O. Gumprecht and C. M. Sliepcevich, *Tables of Light-Scattering Functions for Spherical Particles*. Edwards, Ann Arbor MI, 1951.

T. Hattori and S. Koyata, *Solid State Technol.* **34**(9), S1 (September 1991).

L. Hesselink, *Annu. Rev. Fluid Mech.* **20**, 421 (1988).

K. C. Hickman, S. M. Gaspar, K. P. Bishop, S. S. H. Naqvi, J. R. McNeil, G. D. Tipton, B. R. Stallard, and B. L. Draper, *J. Vac. Sci. Technol. B* **10**, 2259 (1992).

P. Ho, M. E. Coltrin, and W. G. Breiland, in *Laser Processing and Diagnostics* (D. Bäuerle, ed.), p. 515. Springer-Verlag, Berlin, 1984.

P. C. D. Hobbs, *Appl. Opt.* **34**, 1579 (1995).

T. Hori, M. D. Bowden, K. Uchino, K. Muraoka, and M. Maeda, *J. Vac. Sci. Technol. A* **14**, 144 (1996).

R. D. Jacobson, S. R. Wilson, G. A. Al-Jumaily, J. R. McNeil, J. M. Bennett, and L. Mattsson, *Appl. Opt.* **31**, 1426 (1992).

G. M. Jellum and D. B. Graves, *J. Appl. Phys.* **67**, 6490 (1990a).

G. M. Jellum and D. B. Graves, *Appl. Phys. Lett.* **57**, 2077 (1990b).

G. M. Jellum, J. E. Daugherty, and D. B. Graves, *J. Appl. Phys.* **69**, 6923 (1991).

S. R. Johnson, C. Lavoie, T. Tiedje, and J. A. Mackenzie, *J. Vac. Sci. Technol. B* **11**, 1007 (1993).

O. Kafri and I. Glatt, *The Physics of Moiré Metrology*. Wiley, New York, 1990.

M. Kerker, *The Scattering of Light and Other Electromagnetic Radiation*. Academic Press, New York, 1969.

H. P. Kleinknecht and H. Meier, *J. Electrochem. Soc.* **125**, 798 (1978).

R. G. Knollenberg, in *Handbook of Contamination Control in Microelectronics* (D. L. Tolliver, ed.), Chapter 8, p. 257. Noyes, Park Ridge, NJ, 1988.

M. Koizumi and N. Akiyama, *SPIE* **538,** 239 (1985).
M. Koppitz, W. Richter, R. Bahnen, and M. Heyen, in *Laser Processing and Diagnostics* (D. Bäuerle, ed.), p. 530. Springer-Verlag, Berlin, 1984.
R. Krukar, A. Kornblit, L. A. Clark, J. Kruskal, D. Lambert, E. A. Reitman, and R. Gottscho, *J. Appl. Phys.* **74,** 3698 (1993).
M. K. Lang, G. W. Donohoe, S. H. Zaidi, and S. R. J. Brueck, *Opt. Eng.* **33,** 3465 (1994).
W. Lauterborn and A. Vogel, *Annu. Rev. Fluid Mech.* **16,** 223 (1984).
C. Lavoie, S. R. Johnson, J. A. Mackenzie, T. Tiedje, and T. van Buuren, *J. Vac. Sci. Technol. A* **10,** 930 (1992).
C. Lavoie, M. K. Nissen, S. Eisebitt, S. R. Johnson, J. A. Mackenzie, and T. Tiedje, *Mater. Res. Soc. Symp. Proc.* **324,** 119 (1994).
D. Leith, in *Particle Control for Semiconductor Manufacturing* (R. P. Donovan, ed.), Chapter 4, p. 47. Marcel Dekker, New York, 1990.
C. Mack, *SPIE* **538,** 207 (1985).
I. J. Malik, S. Pirooz, L. W. Shive, A. J. Davenport, and C. M. Vitus, *J. Electrochem. Soc.* **140,** L75 (1993).
J. R. McNeil, S. S. H. Naqvi, S. M. Gaspar, K. C. Hickman, K. P. Bishop, L. M. Milner, R. H. Krukar, and G. A. Petersen, *Microlithogr. World,* November/December, p. 16 (1992a).
J. R. McNeil, S. S. H. Naqvi, S. M. Gaspar, K. C. Hickman, and S. R. Wilson, in *Encyclopedia of Materials Characterization* (C. R. Brundle, C. A. Evans, and S. Wilson, eds.). Butterworth–Heinemann, Boston, 1992b.
J. R. McNeil, S. S. H. Naqvi, S. M. Gaspar, K. C. Hickman, K. P. Bishop, L. M. Milner, R. H. Krukar, and G. A. Petersen, *Solid State Technol.* **36**(3), 29 (March 1993a).
J. R. McNeil, S. S. H. Naqvi, S. M. Gaspar, K. C. Hickman, K. P. Bishop, L. M. Milner, R. H. Krukar, and G. A. Petersen, *Solid State Technol.* **36**(4), 53 (April 1993b).
G. F. Mendes, L. Cescato, J. Frejlich, E. S. Braga, and A. P. Mammana, *Thin Solid Films* **117,** 107 (1984).
G. F. Mendes, L. Cescato, J. Frejlich, E. S. Braga, and A. P. Mammana, *J. Electrochem. Soc.* **132,** 190 (1985).
V. B. Menon and R. P. Donovan, in *Handbook of Semiconductor Wafer Cleaning Technology* (W. Kern, ed.), Chapter 9, p. 379. Noyes, Park Ridge, NJ, 1993.
W. Merzkirch, *Flow Visualization,* 2nd ed. Academic Press, Orlando, FL, 1987.
W. Merzkirch, in *Handbook of Flow Visualization* (W.-J. Yang, ed.), Chapter 11, p. 181. Hemisphere, New York, 1989.
L. M. Milner, K. P. Bishop, S. S. H. Naqvi, and J. R. McNeil, *J. Vac. Sci. Technol. B* **11,** 1258 (1993).
Y. Miyazaki, H. Tanaka, N. Kosada, and T. Tomoda, *SPIE* **1673,** 515 (1992).
M. G. Moharam and T. K. Gaylord, *J. Opt. Soc. Am.* **72,** 1385 (1982).
M. G. Moharam and T. K. Gaylord, *J. Opt. Soc. Am. A* **3,** 1780 (1986).
E. Morita, H. Okuda, and F. Inoue, *Semicond. Int.* **17**(8), 156 (1994).
T. Nakata and N. Akiyama, *Jpn. J. Appl. Phys.* **28,** 2396 (1989).
S. S. H. Naqvi, S. H. Zaidi, S. R. J. Brueck, and J. R. McNeil, *J. Vac. Sci. Technol. B* **12,** 3600 (1994).
J. F. O'Hanlon and H. G. Parks, *J. Vac. Sci. Technol. A* **10,** 1863 (1992).

W. G. Oldham, S. N. Nandgaonker, A. R. Neureuther, and M. O'Toole, *IEEE Trans. Electron Devices* **ED-26,** 717 (1979).

J. M. Olson and A. Kibbler, *J. Cryst. Growth* **77,** 182 (1986).

J. A. O'Neill, J. Singh, and G. S. Gifford, *J. Vac. Sci. Technol. A* **8,** 1716 (1990).

L. Peters, *Semicond. Int.* **15**(12), 52 (November, 1992).

R. Petit, *Electromagnetic Theory of Gratings,* Top. Curr. Phys. Springer-Verlag, Berlin, 1980.

R. Petit, *SPIE* **1266,** 2 (1990).

M. Pichot and M. Guillaume, U. S. Patent 4,525,066 (1982).

C. Pickering, *Thin Solid Films* **206,** 275 (1991).

C. Pickering, in the *Handbook of Crystal Growth,* (D. T. J. Hurle, ed.), Vol. 3, p. 817. North-Holland, Amsterdam, 1994.

C. Pickering, in *Photonic Probes of Surfaces; Electromagnetic Waves: Recent Developments in Research, Vol. 2* (P. Halevi, ed.). Elsevier, Amsterdam, 1995.

C. Pickering, D. J. Robbins, I. M. Young, J. L. Glasper, M. Johnson, and R. Jones, *Mater. Res. Soc. Symp. Proc.* **94,** 173 (1987).

C. Pickering, D. A. O. Hope, W. Y. Leong, D. J. Robbins, and R. Greef, *Mater. Res. Soc. Symp. Proc.* **324,** 53 (1994).

A. J. Pidduck, D. J. Robbins, A. G. Cullis, D. B. Gasson, and J. L. Glasper, *J. Electrochem. Soc.* **136,** 3083 (1989a).

A. J. Pidduck, D. J. Robbins, D. B. Gasson, C. Pickering, and J. L. Glasper, *J. Electrochem. Soc.* **136,** 3088 (1989b).

D. Post, *Opt. Eng.* **21,** 458 (1982).

D. J. Robbins, I. M. Young, A. J. Pidduck, C. Pickering, J. L. Glasper, and D. B. Gasson, *Mater. Res. Soc. Symp. Proc.* **94,** 167 (1987a).

D. J. Robbins, A. J. Pidduck, A. G. Cullis, N. G. Chew, R. W. Hardeman, D. B. Gasson, C. Pickering, A. C. Daw, M. Johnson, and R. Jones, *J. Cryst. Growth* **81,** 421 (1987b).

C. M. Rouleau and R. M. Park, *J. Vac. Sci. Technol. A* **11,** 1792 (1993).

C. M. Rouleau and R. M. Park, *Mater. Res. Soc. Symp. Proc.* **324,** 125 (1994).

G. S. Selwyn, *J. Vac. Sci. Technol. B* **9,** 3487 (1991).

G. S. Selwyn, *Semicond. Int.* **16**(3), 72 (March, 1993).

G. S. Selwyn and E. F. Patterson, *J. Vac. Sci. Technol. A* **10,** 1053 (1992).

G. S. Selwyn, J. Singh, and R. S. Bennett, *J. Vac. Sci. Technol. A* **7,** 2758 (1989).

G. S. Selwyn, J. E. Heidenreich, and K. L. Haller, *Appl. Phys. Lett.* **57,** 1876 (1990a).

G. S. Selwyn, J. S. McKillop, K. L. Haller, and J. J. Wu, *J. Vac. Sci. Technol. A* **8,** 1726 (1990b).

G. S. Selwyn, J. E. Heidenreich, and K. L. Haller, *J. Vac. Sci. Technol. A* **9,** 2817 (1991).

Semiconductor Industry Association, *The National Technology Roadmap for Semiconductors,* Semiconductor Industry Association, San Jose, CA, 1994.

C. Shakher and A. K. Nirala, *Appl. Opt.* **33,** 2125 (1994).

G. W. Smith, A. J. Pidduck, C. R. Whitehouse, J. L. Glasper, and J. Spowart, *J. Cryst. Growth* **127,** 966 (1993).

K. G. Spears, T. J. Robinson, and R. M. Roth, *IEEE Trans. Plasma Sci.* **PS-14,** 179 (1986).

K. G. Spears and T. J. Robinson, *J. Phys. Chem.* **92,** 5302 (1988).

K. G. Spears, R. P. Kampf, and T. J. Robinson, *J. Phys. Chem.* **92,** 5297 (1988).

References

D. G. Stearns, *J. Appl. Phys.* **71,** 4286 (1992).

M. Sternheim, W. van Gelder, and A. W. Hartman, *J. Electrochem. Soc.* **130,** 655 (1983).

J. C. Stover, *Appl. Opt.* **14,** 1796 (1975).

J. C. Stover, *Optical Scattering: Measurement and Analysis.* McGraw Hill, New York, 1990.

J. C. Stover, in *Handbook of Optics* (M. Bass, ed.), 2nd ed., Vol. 2, Chapter 26, p. 26.1. McGraw-Hill, New York, 1995.

M. A. Taubenblatt and J. S. Batchelder, *Appl. Opt.* **30,** 4972 (1991).

M. A. Taubenblatt and J. S. Batchelder, *SPIE* **1821,** 152 (1992a).

M. A. Taubenblatt and J. S. Batchelder, presented at 1992 Optical Society Meeting (1992b).

M. A. Taubenblatt and J. S. Batchelder, *Appl. Opt.* **31,** 3354 (1992c).

P. S. Theocaris, *Moiré Fringes in Strain Analysis.* Pergamon, London, 1969.

D. L. Tolliver, ed., *Handbook of Contamination Control in Microelectronics.* Noyes, Park Ridge, NJ, 1988.

W. M. Tong and R. S. Williams, *Annu. Rev. Phys. Chem.* **45,** 401 (1994).

B. Tullis, in *Handbook of Contamination Control in Microelectronics* (D. L. Tolliver, ed.), Chapter 13, p. 410. Noyes, Park Ridge, NJ, 1988.

H. C. van de Hulst, *Light Scattering by Small Particles.* Wiley, New York, 1957; reprinted by Dover, New York, 1981.

C. M. Vest, *Holographic Interferometry.* Wiley, New York, 1979.

D. W. Voorhes and D. M. Hall, *SPIE* **1595,** 61 (1991).

D. J. Whitehouse, *Handbook of Surface Metrology.* Institute of Physics Publishing, Bristol, 1994.

F. L. Williams, G. A. Peterson, Jr., C. K. Carmiglia, and B. J. Pond, *J. Vac. Sci. Technol. A* **10,** 1472 (1992).

K. Yang, E. Mirabelli, Z.-C. Wu, and L. J. Schowalter, *J. Vac. Sci. Technol. B* **11,** 1011 (1993).

W.-J. Yang, ed., *Handbook of Flow Visualization.* Hemisphere, New York, 1989.

W. J. Yoo and C. Steinbrüchel, *J. Vac. Sci. Technol. A* **10,** 1041 (1992).

E. Yoon, R. W. Allison, Jr., R. P. Kovacs, and C. Dai, *SPIE* **1673,** 580 (1992).

S. H. Zaidi, S. R. J. Brueck, and J. R. McNeil, *J. Vac. Sci. Technol. B* **10,** 166 (1992).

CHAPTER **12**

Raman Scattering

Raman scattering refers to most types of inelastic scattering of electromagnetic radiation, which, in quantized form, consists of photons. Photons entering a medium can lose energy to the sample, which is referred to as Stokes scattering, or gain energy, which is anti-Stokes scattering (Figure 12.1). Raman scattering in solids is often described in terms of the excitation or de-excitation of quantized elementary excitations, such as optical phonons (vibrational Raman scattering) or magnons (magnetic Raman scattering), and the interaction with single electrons or collective excitations of electrons, plasmons (electronic Raman scattering). In molecules, Raman scattering leads to changes in the population of vibrational and rotational states.

Raman scattering has been widely used in the fundamental spectroscopic study of excitations in solids, liquids, and gases and has also been extensively used in materials characterization. Vibrational and rotational Raman scattering have been used for *in situ* diagnostics during thin film processing. Although magnetic and electronic Raman scattering are very important in fundamental studies of elementary excitations in solids and materials characterization (Herman, 1996), they are rarely used as real-time diagnostics and will not be covered here.

Overviews by Hayes and Loudon (1978), Long (1977), Cardona (1982), Mitra and Massa (1982), Ferraro and Nakamoto (1994), and Herman (1996) detail the theory and spectroscopic applications of Raman scattering. The Springer series on light scattering in solids (Cardona, 1975; Cardona and Güntherodt, 1982–1991) and gases and liquids (Weber, 1979) provide critical and detailed reviews of many types of inelastic light scattering. Eckbreth (1996) and Drake (1988) have reviewed the theory of Raman scattering and its use as a diagnostic in combustion and related gas-phase processes. The quantum mechanics and spectroscopy of Raman scattering have been reviewed in Sections 4.2.2 and 4.3.2.

All types of inelastic scattering can be classified as either Raman or Brillouin scattering. Figure 12.1 shows a simulated spectrum with elastic

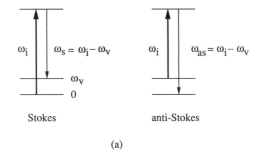

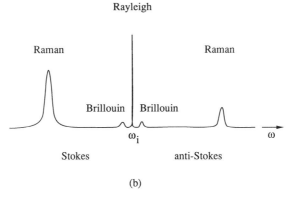

Figure 12.1 (a) Energy level diagrams for Stokes and anti-Stokes inelastic scattering, and (b) the spectrum of scattered light, showing the (elastic) Rayleigh peak and the Stokes and anti-Stokes peaks for both Brillouin and Raman scattering.

(Rayleigh) scattering (Chapter 11) and Stokes and anti-Stokes forms of Brillouin and Raman scattering. Scattering from optical phonons or any other excitation with energy >5 cm^{-1} is called Raman scattering, while that from density changes due to pressure fluctuations, which propagate as sound waves (or acoustic phonons), or any low-frequency excitation <5 cm^{-1} is called Brillouin scattering. Consequently, Raman and Brillouin scattering are also distinguished by the instruments needed for spectral analysis: grating monochromators, and sometimes Fourier-transform infrared (FTIR) spectrometers, are used for Raman scattering, while Fabry–Perot interferometers are needed for Brillouin scattering. Brillouin scattering is rarely used for real-time probing of film processing and will not be considered any further.

Raman scattering is a coherent, two-photon process in which the photon scatters only while the incident photon is present. It is formally different from the sequential absorption of a photon followed by fluorescence, although both types of processes can be important and hard to distinguish near resonance. Spontaneous Raman scattering, which is the version most commonly used in the study of materials and molecules, is very weak; only ~10^{-12} of the incident photons are scattered and detected. This is the main disadvantage of using Raman spectroscopy as an *in situ* probe. Several stimulated versions of Raman processes have been studied (Section 4.4). Of these, only coherent anti-Stokes Raman scattering (CARS) has proved to be particularly important in probing gases during thin film processing; CARS is described in Section 16.1. To distinguish spontaneous Raman scattering from these other more complex processes, it is sometimes called (in jest) "COORS"—common old ordinary Raman scattering.

Raman scattering is a versatile spectroscopy that is species-selective. One advantage of spontaneous Raman scattering as a diagnostic is that a single, fixed wavelength can be used to probe each molecule in the gas phase, even simultaneously, and to probe the substrate. Another advantage *vis-à-vis* transmission, is that Raman scattering can be used with reactors that have one port, while single-pass IR absorption requires input and output ports. The major disadvantage of Raman scattering is that signals are very weak.

Spontaneous vibrational Raman scattering can be used as a real-time diagnostic of the interfacial region near the surface during thin film processing, with monolayer sensitivity, such as during molecular beam epitaxy (MBE) (Nowak *et al.*, 1992). It is also a good probe of the composition and temperature of films, such as during chemical vapor deposition (CVD) (Herman, 1991). The gas above the substrate can be probed by vibrational and rotational Raman scattering; pure rotational Raman scattering can measure gas temperature by determining the relative population of rotational levels. Spontaneous Raman scattering has been used extensively in several other applications, including the remote sensing of the atmosphere, which is called Raman lidar (light detection and ranging) (She, 1990), and in combustion diagnostics (Lapp and Hartley, 1976; Eckbreth, 1996).

12.1 Kinematics and Dynamics of Spontaneous Raman Scattering

The frequency of the scattered light is determined by energy conservation: $\omega_s = \omega_i \pm \omega_v$ for Stokes ($-$) and anti-Stokes ($+$) scattering, where ω_i and ω_v are the frequencies of the incident light and the quantized excitation (Figure 12.1). With the conservation of momentum **q**, the scattered wavevector is $q_v = 2q_i \sin(\theta/2)$ when $\omega_i \gg \omega_v$, where θ is the scattering angle. The kinematics of vibrational and rotational Raman scattering in gases does not depend on the scattering angle. For optical phonons, only excitations near zonecenter (Γ) **q**$_v \sim$ **0** contribute to first-order scattering in crystalline and polycrystalline solids, where the incident photon interacts with only one excitation. This gives a sharp Raman peak for each allowed optic mode. In amorphous and microcrystalline materials, phonons far from zonecenter can also scatter, which can lead to a Raman peak that is asymmetric toward lower frequencies and one that is downshifted in frequency and broader vis-à-vis the corresponding peak in a crystal.

The power of the Raman scattered beam per unit solid angle in a medium is (Equation 5.13)

$$\tilde{\mathcal{P}}_{\text{Raman}} = N \left(\frac{d\sigma}{d\Omega}\right) I_{\text{inc}} v(\omega_s/\omega_i) \tag{12.1a}$$

$$= N \left(\frac{d\sigma}{d\Omega}\right) \mathcal{P}_{\text{inc}} l(\omega_s/\omega_i) \tag{12.1b}$$

where N is the number density of the scattering species, $d\sigma/d\Omega$ is the differential scattering cross-section (for photons), $\mathcal{P}_{\text{inc}}$ is the incident power, I_{inc} is the incident intensity, v is the scattering volume, and l is the length of the laser beam over which the scattered light is collected. The power observed by the detector is

$$\mathcal{P}_{\text{Raman}} = N \left(\frac{d\sigma}{d\Omega}\right) I_{\text{inc}} V_{\text{im}} (\omega_s/\omega_i) \Delta\Omega_{\text{eff}} \eta_o \tag{12.2}$$

where $\Delta\Omega_{\text{eff}}$ is the effective solid angle that is collected, V_{im} is the imaged volume ($<v$), and η_o is the optical collection efficiency (Section 5.2.1.1). The detected Raman signal $S_{\text{Raman}} = \eta_d \mathcal{P}_{\text{Raman}}$, where η_d is the detector efficiency. This means that $S_{\text{Raman}}/\hbar\omega_s$ counts per second are counted, which corresponds to the detected rate of scattering photons from ω_i to ω_s.

For scattering in gases $N = N_{vJ}$ (Equation 3.12), the density of scattering species in the initial vibrational–rotational energy level. Equations 12.1 and 12.2 give the scattered signal integrated over the spectral lineshape. The corresponding spectral distributions of the scattered light, such as $\mathcal{P}_{\text{Raman}}(\omega_s)$, are given by these expressions with the differential cross-section replaced by the spectral differential cross-section $d^2\sigma/d\Omega\, d\omega_s$. Sections 4.2.2 and 4.3.2 describe the physics of spontaneous Raman scattering and the Raman spectroscopy of molecules and solids.

Raman spectra are very sensitive to the polarization of the incident and detected scattered beams. The scattering arrangement is represented by **a(b,c)d**, where **a** and **d** are the directions of the incident and scattered beams, respectively, and **b** and **c** are their respective electric field polarizations. In backscattering geometry $\mathbf{d} = \bar{\mathbf{a}}$. In a "polarized" spectrum $\mathbf{c} \parallel \mathbf{b}$, while in a "depolarized" spectrum $\mathbf{c} \perp \mathbf{b}$. For example, the light scattered from a laser traveling in the z direction that is polarized in the x direction, with electric field amplitude E_x^{inc}, can be observed in the y direction, with either x or z polarization. The former configuration is $z(x,x)y$, which is polarized scattering and has scattered intensity $I_{xx} \propto \mu_x^2$ where $\mu_x = \hat{\alpha}_{xx} E_x^{\text{inc}}$; the latter configuration is $z(x,z)y$, which is depolarized scattering and has scattered intensity $I_{zx} \propto \mu_z^2$ where $\mu_z = \hat{\alpha}_{zx} E_x^{\text{inc}}$. $\hat{\alpha}$ is the polarizability (Section 4.1).

The depolarization ratio D is defined as the ratio of depolarized to polarized scattering intensities. It is a function of scattering angle θ and can be analyzed for incident polarization either parallel or normal to the scattering plane, which is defined by the wavevectors of the incident and scattered light. After rotational averaging, $D(\theta = 90°) = 3\gamma^2/(45a^2 + 4\gamma^2) \leq 3/4$ for incident light that is linearly polarized normal to the scattering plane, and it is 1 for light polarized in the scattering plane. a and γ are the isotropic and anisotropic parts of the polarizability, defined by Equations 4.46–4.48. If the incident light is nonpolarized, $D(\theta = 90°) = 6\gamma^2/(45a^2 + 7\gamma^2) \leq 6/7$. Depolarization ratios from randomly oriented species for different incident and scattered polarizations have been tabulated by Hirschfeld (1973) and Long (1977).

Vibrational Raman scattering, i.e., scattering from optical phonons, is the most common type of inelastic scattering in probing solids and is detailed in Sections 4.2.2.2 and 4.3.2.2. Both vibrational–rotational and pure rotational Raman scattering are possible in gases and are detailed in Sections 4.2.2.1 and 4.3.2.1. Pure rotational scattering is generally less widely used than vibrational–rotational scattering because the Raman shifts are relatively small, which makes it necessary to minimize light scattered at the laser frequency. (H_2 is an exception because rotational shifts in this molecule are quite large.) Identification is difficult in multicomponent mixtures, so rotational Raman scattering is best used when only one species is dominant. One advantage of pure rotational scattering is that the scattering cross-section for a single rotational transition is typically an order of magnitude larger than that of an entire vibrational branch in vibrational–rotational scattering (Eckbreth, 1996).

12.2 Instrumentation

Raman spectroscopy requires a light source, optics to deliver the light to the sample, optics to collect the scattered light, an instrument to spectrally disperse the scattered light and to remove stray light, and a photon detector. Chase (1994) has reviewed the state-of-the-art Raman instrumentation. Since the spontaneous scattering cross-section is small, high-intensity light sources are often needed in Raman scattering. Either continuous-wave (cw) lasers (ion lasers, such as argon-ion and krypton-ion lasers, and dye lasers) or pulsed lasers can be used because only the number of incident photons is important. However, pulsed lasers are preferred if gating is needed to remove background light, such as during plasma processing. The Raman scattering intensity, line shift, and/or linewidth can be measured. When the Raman lineshape is probed, the linewidth of the laser must be narrower than the linewidth of the elementary excitation, which for phonons is typically ~3 cm^{-1} at ambient temperature.

In Raman analysis of solids, the laser is often focused to a circular spot on the sample, and in absorbing samples, significant laser heating must be avoided. With cw laser irradiation the temperature rise at the surface of a strongly absorbing sample is given by Equation 4.71; this consideration puts a practical upper limit on P_{inc}. Because the scattered light is imaged onto the slit of a monochromator, the power of the incident laser can be increased (with no increase in heating) by focusing the laser to a line with a cylindrical lens, rather than to a circular spot.

The scattering efficiency of a material can be defined as $\mathcal{S} = N(d\sigma/d\Omega) \sum_{j=1,\ldots,l} |\mathbf{e}_s \cdot \overleftrightarrow{\mathbf{R}}_j \cdot \mathbf{e}_i|^2$, where $d\sigma/d\Omega$ is the scattering cross-section per molecule or unit cell, N is the density of molecules or units cells, $\mathbf{e}_i$ and $\mathbf{e}_s$ are the polarization unit vectors of the incident and scattered light, and $\overleftrightarrow{\mathbf{R}}_j$ is the Raman tensor for the vibrational mode of degeneracy l (Tsang, 1989). The scattering efficiencies for diamond ($\omega_p = 1332$ cm^{-1}), silicon (521 cm^{-1}), and germanium (301 cm^{-1}) at 514.5 nm are ~5.5, ~600, and 8670×10^{-7} cm^{-1}sr^{-1}, respectively (Tsang, 1989). Equations 12.1 and 12.2 can be used with $N(d\sigma/d\Omega)$ replaced by $\mathcal{S}$.

Raman scattering of optically thick samples or thin films is usually performed in backscattering configuration, with a scattering angle $\theta \approx 180°$ (Figure 12.2a). The scattered power inside the material is given by $\mathcal{S}^*P_{inc}$, where $\mathcal{S}^*$ is the effective internal scattering efficiency:

$$\mathcal{S}^* = \mathcal{S}\frac{1 - \exp[-(\mathcal{S} + \alpha_i + \alpha_s)d]}{\mathcal{S} + \alpha_i + \alpha_s} \quad (12.3)$$

where α_j is the absorption coefficient at ω_j ($j = $ i, s) and d is the sample thickness. When $\mathcal{S} \ll \alpha_i, \alpha_s$, as is common, the effective depth of the Raman probe is $d_{eff} \sim 1/(\alpha_i + \alpha_s)$ for optically thick samples, and the scattering volume v in Equation 4.25 is $\sim w^2/(\alpha_i + \alpha_s)$. This effective depth depends on the material, laser wavelength, and temperature. The lateral spatial resolution is given by the spot size at the focus w, Equation 2.16, which can approach the far-field diffraction limit for visible light, ~0.5 μm. Raman spectroscopy with such tightly focused lasers is called Raman microprobe scattering, and such analysis can be performed in backscattering configuration with the apparatus shown in Figure 5.11.

Using the above values for $\mathcal{S}$ and Equations 12.1 and 12.3, the Raman signal (inside the material) p_{Raman} when a monolayer ($d = 0.3$ nm) of diamond, Si or Ge is probed by 100 mW at 514.5 nm is 3.9×10^3, 4×10^5 or 6.8×10^6 counts per second (cps)/sr, respectively (Equation 12.1b). The detected count rate is much smaller because of several factors. A fraction of the incident and scattered light is each reflected at the interface, which decreases the signal by $(1 - R_i)(1 - R_s)$, where R is the reflectance; this factor is sometimes included in the definition of $\mathcal{S}^*$ in Equation 12.3. If

12.2 Instrumentation

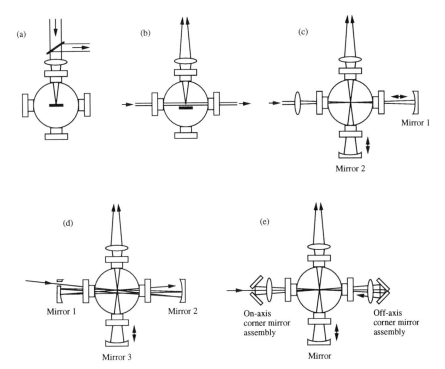

Figure 12.2 Experimental arrangements for Raman scattering external to the laser cavity, with (a) backscattering, as in Figure 5.11, and (b)–(e) 90° collection. An alternative backscattering scheme uses a small mirror to direct the laser to the sample, as in Figure 5.12. In (c)–(e), a curved mirror placed at −90° essentially doubles light collection. The signal can also be increased by retroreflecting the laser for either (c) a double pass or a multipass arrangement, the latter by using (d) an ellipsoidal light trap (Hill and Hartley, 1974) or (e) a roof top mirror design (Hill et al., 1977). Note that the lenses that focus the laser in (c) and (e) can be planoconvex lenses with the flat side to the reactor. [Adapted, in part, from Eckbreth (1996).]

the optical system collects a solid angle $\Delta\Omega$, only light scattered within an angle $\Delta\Omega_{\text{eff}} = \Delta\Omega/n^2$ in the semiconductor (of refractive index n) is collected because of refraction from the semiconductor to the ambient medium. Only a fraction of the collected photons $\eta_t = \eta_o\eta_d$ are detected because of optical transmission losses and the detector efficiency, as is discussed in Section 5.2.1.

Therefore, the signal S_{Raman} (cps) $= S^*(1 - R_i)(1 - R_s) (\Delta\Omega/n^2) \eta_t(\mathcal{P}_{\text{inc}}/\hbar\omega_{\text{inc}})d$. Assuming typical experimental conditions and no losses due to absorption by overlayers, Tsang (1989) has estimated that the detected Raman signal is ~0.36, 36, and 630 cps for a monolayer of diamond, Si, and Ge, respectively (100 mW, 514.5 nm). When bulk Si and Ge are analyzed with this laser [$d_{\text{eff}} \sim 1/(\alpha_i + \alpha_s)$], the signal is approximately 43,000 and 17,500 cps, respectively. Since each of these count rates is integrated over the Raman lineshape, the number of counts for a measurement at a given wavelength is much smaller.

With transparent media, alternative geometries include side scattering ($\theta \sim 90°$) (Figures 12.2b–12.2e) and in-line scattering ($\theta \sim 0°$), and the laser may not be focused at all (Figure 12.2b). In any case, light is collected by small $f_\#$ optics and directed, sometimes by optical fibers, to a monochromator that must disperse the scattered light and reject any collected light from the incident laser. The ultimate spatial resolution is limited by practical matters, such as the distance between the chamber window and the sample.

It is common to collect scattered light at $\theta \sim 90°$ when studying gases and liquids. In Raman scattering of gases, in cells that are external to the laser cavity, the exciting laser beam is often focused and a collecting mirror is put on the side opposite the collecting lens so light scattered to $\theta \sim -90°$ can also be collected (Figures 12.2c–12.2e). To increase the incident laser intensity, either a double pass (Figure 12.2c) or multipass (Figures 12.2d and 12.2e) arrangement is used, or the scattering cell is placed inside the laser cavity (intracavity) (Weber, 1979; Eckbreth, 1996).

Triple-stage spectrometers, which have a subtractive double stage for rejecting the laser line followed by a single-stage for dispersion, and double monochromators, which have two additive stages, are commonly used for dispersion (Section 5.1.3.1). An alternative approach uses a single-stage spectrometer preceded by a narrow line filter that rejects the laser line (Chase, 1994). This is very attractive as it involves less expensive equipment and can lead to greater throughput because it involves fewer diffraction gratings and mirrors. Improvements in the laser line filters, notably holographic notch filters, have diminished one previous disadvantage of this approach, i.e., the high transmission loss for light scattered within ~ 200 cm^{-1} of the laser line; however, the lack of the tunability of these filters can sometimes be a drawback.

The classic mode of photon detection in spontaneous Raman scattering is to scan the monochromator and to count photons with a photomultiplier (PMT). Parallel detection of the dispersed radiation by multichannel detectors, such as intensity-enhanced photodiode arrays, mepsicrons, and most notably, charge-coupled-device (CCD) detectors, has greatly improved the signal/noise ratio in Raman spectra (Section 5.1.2; and Chang and Long, 1982; Acker *et al.*, 1988; Tsang, 1989; Chase, 1994).

Spontaneous Raman spectra can be improved by increasing the flux of incident photons, improving photon collection, and by using better detectors. Since the Raman cross-section $\propto \omega^4$, the use of shorter wavelengths improves signal strength. Other ways to improve signal strength include tuning ω_i to a resonance in the cross-section (Cardona, 1982), the use of interference effects on specially designed substrates (Connell *et al.*, 1980), and the use of surface-enhanced Raman scattering (SERS) on roughened metal surfaces (Otto, 1984; Campion, 1985a,b; Jha, 1985). SERS, which can increase Raman signals by $\sim 10^4$–10^6, is caused by electromagnetic resonances and by other effects related to chemisorption. Because Raman processes usually are not surface-selective, the measurement of monolayers (Campion, 1985a,b) and very thin films (Tsang, 1989) on surfaces requires these signal-enhancing methods, as well as other techniques, such as the careful subtraction of the Raman signal from the substrate.

Since Raman scattering signals are very small, interference from photoluminescence in solids (or fluorescence in gases) must be kept to a minimum. Photoluminescence is not a concern when the photon energy is less than the fundamental gap $\mathcal{E}_0$ ($= \mathcal{E}_{bg}$, band gap) or in materials that do not fluoresce at all, such as indirect semiconductors. Although scattering near the band gap $\mathcal{E}_0$ is inadvisable in direct semiconductors, Raman scattering can be resonantly enhanced at the associated spin–orbit split gap $\mathcal{E}_0 + \Delta$ or at larger gaps, such as $\mathcal{E}_1$, without creating a strong photoluminescence background near the Raman peak. In some materials, particularly organic compounds, the fluorescence intensity tends to increase with ω_i, so the ω^4 improvement in the Raman signal is of little value. Near-infrared light, such as that from Nd/YAG lasers (1.06 μm), is sometimes used to avoid fluorescence. Because optical detectors in the near-infrared perform poorly, especially when employed with single-channel detection, Fourier transform (FT) Raman spectroscopy (Hendra *et al.*, 1991) is often used instead of dispersive spectroscopy. In FT Raman spectroscopy, spectral analysis and parallel detection are accomplished by using a scanning Michelson interferometer (Section 5.1.3.2).

12.3 Thermometry and Density Measurements in Gases

Early work on the use of Raman scattering to measure gas temperatures was performed by Sedgwick *et al.* (1975) and Sedgwick and Smith (1976). They profiled the temperature in a Si CVD reactor by using the ratio of the intensities of the Stokes and anti-Stokes lines in vibrational scattering from H_2 (Section 4.3.2.1.2). [They, along with Smith and Sedgwick (1977), also performed early work on identifying gas-phase species during CVD by using Raman scattering.] More recent studies have instead measured temperature by determining the dependence of Raman line strength on the rotational quantum number J, using Equation 4.39. Eckbreth (1996) can be consulted for more details on the use of Raman scattering for gas-phase thermometry.

Leyendecker *et al.* (1983) measured the temperature profile in a filament-heated CVD reactor, operating with a CH_4/H_2 mixture for carbon deposition. They used a cw argon-ion laser (488 nm) as the light source and photon counting for detection. Temperature was determined by comparing the pure rotational Raman transitions $J = 1 \rightarrow 3$ (Stokes shift of 587 cm^{-1}) and $J = 3 \rightarrow 5$

(1034 cm^{-1}) in H$_2$. For H$_2$, corrections due to deviations from the rigid rotor model had to be included (Cheung *et al.*, 1981). Because pure rotational Raman transitions are inactive in the spherical-top molecule CH$_4$, they also determined the temperature from the ratio of the vibrational Raman intensities of the CH$_4$ hot-band peak $v_1 + v_4 - v_4$ (near 2913 cm^{-1} in Figure 12.3) to the fundamental peak (at 2917 cm^{-1}). This ratio is $A \exp(-hv_4/k_BT)$, where A was obtained by using the temperature from H$_2$ in calibration runs. This measurement was necessary when no H$_2$ was present. Figure 12.4 shows the measured temperature profile for different CH$_4$/H$_2$ mixtures.

Doppelbauer *et al.* (1984) used similar methods to measure temperature and species density during the catalytic hydrogenation of acetylene (C$_2$H$_2$) to ethylene (C$_2$H$_4$) in C$_2$H$_2$/H$_2$ mixtures. They determined temperatures by using rotational Raman scattering in H$_2$, and the concentrations of the hydrocarbons by the intensity of the Q-branch peak in vibrational–rotational scattering, v_2 in C$_2$H$_2$ and v_1 in C$_2$H$_4$, relative to that for $J = 1 \rightarrow 3$ in H$_2$. They concluded that it was too difficult to determine concentration on the basis of the assignment of the pure-rotational Raman peaks in C$_2$H$_2$ and C$_2$H$_4$ (which do not overlap those of H$_2$). These last two studies were further detailed by Leyendecker *et al.* (1984).

Breiland *et al.* (1986) used Raman scattering and laser-induced fluorescence (LIF) diagnostics to compare *in situ* observations during Si CVD from SiH$_4$/H$_2$/He mixtures with model predictions. They used pulsed-laser Raman spectroscopy to profile the silane density and the temperature. The silane density was determined by integrating the v_1 vibrational Q-branch

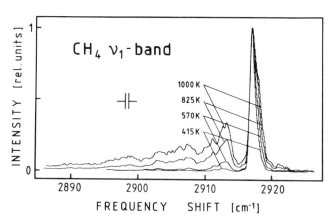

Figure 12.3 Raman spectrum of the v_1 band in CH$_4$ at different temperatures, with the fundamental peak seen at 2917 cm^{-1}. The ratio of the intensity of the hot band peak $v_1 + v_4 - v_4$ near 2913 cm^{-1} to that of the fundamental can be used to determine temperature. [From G. Leyendecker, J. Doppelbauer, and D. Bäuerle, *Appl. Phys. A* **30**, Raman diagnostics of CVD systems: Determination of local temperatures, page 240, Figure 3, 1983. Copyright © Springer-Verlag.]

12.3 Thermometry and Density Measurements in Gases

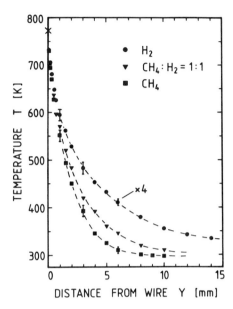

Figure 12.4 Temperature profiles in a model CVD reactor, with a central wire heated to a temperature of 773 K (to induce CVD) and a reactor pressure of 1000 mbar for different H_2/CH_4 mixtures, determined by Raman scattering in H_2 and/or CH_4. [From G. Leyendecker, J. Doppelbauer, and D. Bäuerle, *Appl. Phys. A* **30**, Raman diagnostics of CVD systems: Determination of local temperature, page 240, Figure 4, 1983. Copyright © Springer-Verlag.]

signal at 2187 cm^{-1}. These profiles are compared with model predictions in Figure 12.5. This was based on earlier work by Breiland and Kushner (1983), who used a frequency-tripled Nd:YAG laser (355 nm) to probe silane and nitrogen in a CVD reactor.

Breiland *et al.* (1986) and Breiland and Evans (1991) also determined the local gas temperature from the integrated intensity of 4–6 peaks in the rotational Raman spectrum of H_2, as is illustrated in Figure 12.6a for two different heights in a rotating-disk CVD reactor. The local temperature T was determined from the dependence of line strength on the rotational quantum number J by using Equation 4.39. This was corrected for alternating line strengths (g_I is 1 for even J and 3 for odd J) and for nonrigidity, which is important in H_2 (Cheung *et al.*, 1981). The temperature accuracy in T was estimated to be 10–25°C. Figure 12.6b compares the temperature profiles in a rotating-disk reactor obtained experimentally, by using Raman scattering or a thermocouple, and by calculation. The Raman measurements are clearly superior. In related work depicted in Figure 12.7, Breiland and Ho (1984) showed how rotational Raman scattering in nitrogen can also be used to obtain T_{rot}.

Koppitz *et al.* (1984a,b) used rotational Raman scattering to obtain the temperature profile of H_2 and N_2 in metalorganic CVD (MOCVD) reactors

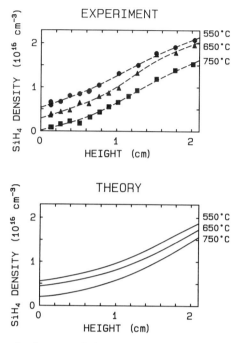

Figure 12.5 Silane density versus height above the susceptor in a CVD reactor during silicon deposition from silane in helium, with the experimental profiles (a) obtained by vibrational Raman scattering and the theoretical profiles (b) from a CVD model that includes gas-phase fluid mechanics and chemical kinetics considerations (from Breiland et al., 1986).

designed for GaAs deposition, as is illustrated in Figure 12.8. Koppitz et al. (1984b) also monitored the vibrational Raman scattering of Ga(CH$_3$)$_3$ (TMGa) (~520 cm^{-1}) and AsH$_3$ (two modes near 2120 cm^{-1}) as a function of position from the substrate—and hence as a function of temperature. Figure 12.9 shows the decrease of the TMGa signal with AsH$_3$ present. Monteil et al. (1986) used rotational Raman scattering (argon-ion laser, 488 nm) to profile the temperature distribution of H$_2$ and N$_2$ during metalorganic vapor-phase epitaxy (MOVPE) of GaAs. They also followed the Raman spectra of the gas-phase TMGa/AsH$_3$ mixture, and noted the production of gas-phase CH$_4$. In later work, Monteil et al. (1988) monitored the thermal decomposition of AsH$_3$ (2114 cm^{-1}) in H$_2$ in an atmospheric pressure MOVPE system. Broadening of this As—H stretching-vibration band of AsH$_3$ suggested the formation of AsH$_2$, or more probably AsH. Fotiadis et al. (1990) also spatially profiled temperature in CVD reactors by using these Raman techniques and compared their experimental results with predictions from finite-element models of flow and heat transfer.

Oosterlaken et al. (1994a,b) used rotational Raman scattering from H$_2$ (excimer laser pumped dye laser) to map the temperature profile in a cold-

12.3 Thermometry and Density Measurements in Gases

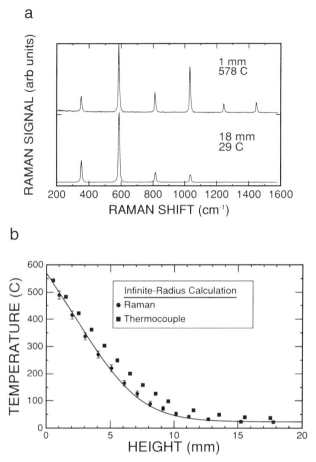

Figure 12.6 (a) Rotational Raman spectrum of hydrogen at two different heights (and therefore temperatures) in a rotating-disk CVD reactor, with the disk maintained at 635°C, as obtained with a laser at 354.7 nm; the spectra have been scaled so the 587-cm^{-1} peak heights are equal. (b) Comparison of the temperature profiles in a rotating-disk reactor (625 Torr H$_2$, 570°C disk, 1000 rpm disk) obtained experimentally (thermocouple—squares; Raman scattering from H$_2$—circles) and by calculation (solid lines) (from Breiland and Evans, 1991).

wall LPCVD reactor with flows of H$_2$/N$_2$/CF$_4$ and H$_2$/WF$_6$ mixtures. The partial pressure of H$_2$ was also determined this way, while the partial pressures of other species were determined by using vibrational Raman scattering. With the H$_2$/N$_2$/CF$_4$ mixtures, the local partial pressure was determined with a standard deviation of 4% at 100 Pa partial pressure (vibrational shift of 2331 cm^{-1} for N$_2$, 908 cm^{-1} for CF$_4$); the detection limits were 6 Pa for H$_2$ and 5 Pa for N$_2$ at room temperature. With H$_2$/WF$_6$ mixtures, the detection limit for WF$_6$ (771 cm^{-1}) was 0.3 Pa and for

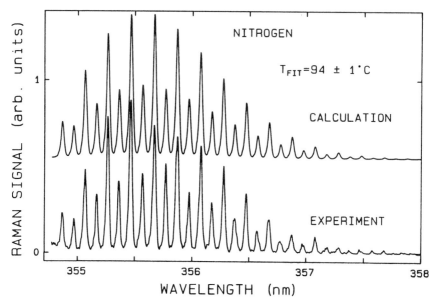

Figure 12.7 Rotational Raman spectrum of N_2 obtained with a laser at 354.7 nm, comparing the calculated and experimental spectra at 94°C (from Breiland and Ho, 1984).

HF (3962 cm^{-1}), a potential product, it was 20 Pa; no HF was detected during tungsten CVD.

12.4 Real-Time Raman Probing of Solids and Surfaces

12.4.1 *Surfaces and Interfaces*

While Raman scattering is not an inherently surface-sensitive spectroscopy, it is specifically sensitive to surfaces and interfaces in some instances. (1) The growth of a very thin layer, approximately a monolayer, can produce a phonon peak due to the new material that is shifted from the Raman feature of the substrate; this peak can sometimes be resonantly enhanced by suitable choice of laser wavelength. (2) The incident and detected polarizations can be configured to make scattering from the substrate forbidden. Raman scattering from the very thin surface layer can be allowed, perhaps because it is not bulk-like (while scattering from a thick film of this same material would be forbidden), or perhaps because of a resonance. (3) Modes at the interface of two films can have frequencies very different from those of either film, e.g., GaAs modes at a GaSb/InAs interface connected by As bonds (Shanabrook and Bennett, 1994). Tsang (1989) has reviewed early work on the Raman analysis of surface layers

12.4 Real-Time Raman Probing of Solids and Surfaces

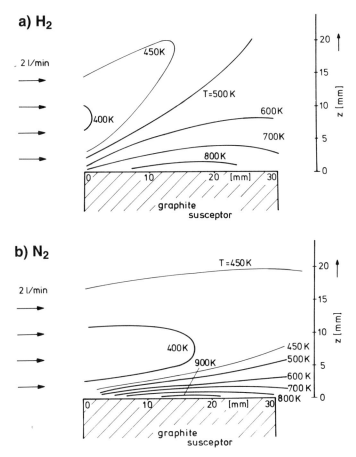

Figure 12.8 Temperature profiles (isotherms) for H_2 and N_2 in a metalorganic chemical-vapor deposition (MOCVD) reactor obtained by rotational Raman scattering (from Koppitz et al., 1984a).

and interfaces. Geurts and Richter (1987), McGilp (1990), and Richter (1993) have surveyed more recent work on monolayer-sensitive Raman probing in real time during deposition. Geurts (1993) has reviewed the use of Raman spectroscopy to analyze electronic band bending at III–V semiconductor interfaces.

Efficient collection and detection of scattered light is essential in such studies. Often CCD array detectors are employed for efficient parallel detection. Ways to enhance the scattering probability must also be considered. For example, choosing the laser wavelength near an electronic resonance can increase the Raman signal by more than an order of magnitude (Equation 4.28).

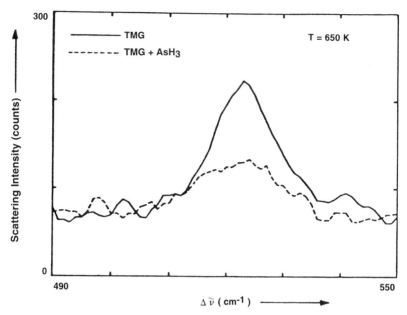

Figure 12.9 Vibrational Raman spectrum of trimethylgallium (TMGa), when it is either alone in an MOCVD reactor or in the presence of AsH$_3$. [From M. Koppitz, W. Richter, R. Bahnen, and M. Heyen, in *Laser Processing and Diagnostics* (D. Bäuerle, ed.), Light scattering diagnostics of gas-phase epitaxial growth (MOCVD-GaAs), page 534, Figure 9, 1984. Copyright © Springer-Verlag.]

Surface specificity can be achieved because of a relaxation of polarization selection rules for scattering. For example, first-order Raman scattering of longitudinal-optical (LO) phonons from (110)-oriented surfaces is symmetry-forbidden for zincblende semiconductors, while it is allowed for transverse optical (TO) phonons (Table 4.2). These selection rules are relaxed in electric fields (E), at impurities, for phonons with finite q vectors (very thin films), and near resonances (due to strong electron/phonon Fröhlich interactions—another finite q effect) (Table 4.2) (Pollak, 1991). Smit *et al.* (1989) used electric-field induced Raman scattering to probe *in situ* the electric field that forms at the surface of InAs(110) when chlorine and oxygen are adsorbed on it. They saw that the intensity of the LO phonon peak is $\propto E^2$, while that of TO phonons is independent of E.

In situ Raman scattering has been used to probe the early stages of growth during MBE. Zahn *et al.* (1991) examined the growth of CdS on InP(110) with Raman scattering performed *in situ* after each deposition step, as seen in Figure 12.10. The Raman spectra were taken with [001] incident and scattered polarizations and with the 457.9-nm argon-ion laser line, which is nearly resonant with an electronic transition in CdS. The features near 305 and 610 cm^{-1} are due to LO and 2LO scattering in the CdS layer. The absence of the other peaks that would be expected for

12.4 Real-Time Raman Probing of Solids and Surfaces

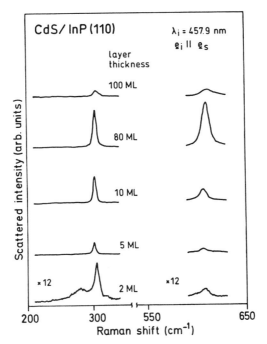

Figure 12.10 Resonant Raman spectra obtained *in situ* after the MBE growth of a series of CdS layers on InP(110) using a parallel polarization configuration and 457.9 nm (from Zahn et al., 1991).

wurtzite CdS, which is the preferred structure for bulk CdS, suggests that the layers are zincblende. Both TO (246 cm^{-1}) and LO scattering are forbidden based on crystal symmetry for zincblende crystals (Table 4.2); however, LO scattering becomes allowed and the overtone 2LO becomes strong near resonance because of the strong Fröhlich electron–phonon interaction. Raman scattering from the InP substrate is still forbidden because there is no resonant enhancement.

The CdS layer is still observable in the two-monolayer (ML) film in Figure 12.10. In this spectrum there is a weak and broad feature near 280 cm^{-1} that may be due to scattering from an In$_2$S$_3$-rich layer at the interface during the early stages of growth. The sudden decrease in the LO and 2LO intensities and the increase in their linewidths for the 100-monolayer film suggest that the crystal quality has worsened. This may be due to the formation of defects after the layer exceeded the critical thickness for lattice-matched growth.

Interfacial reactions are stronger in other systems. For example, Richter (1993) has observed strong interfacial reactions during the MBE growth of CdTe on InSb(110). The sequence of on-line Raman spectra in Figure 12.11 shows evidence for the formation of In$_2$Te$_3$, due to the strong reaction

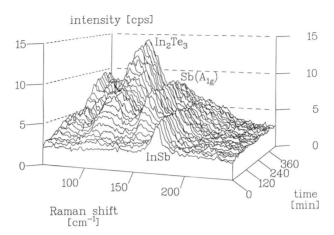

Figure 12.11 Raman spectra of CdTe grown by MBE on InSb(110). These spectra show evidence of In_2Te_3 formation and Sb liberation from the surface, but no indication of CdTe (from Figure 15 in Richter, 1993).

of Te with In, and the concomitant liberation of Sb as CdTe is deposited on the surface.

Nowak et al. (1992) studied the MBE growth of ZnSe on GaAs(110), again with the 457.9-nm argon-ion laser line, which is close to the direct band gap of ZnSe and leads to a Raman resonance. Figure 12.12a shows the Fröhlich-induced LO (252.3 cm^{-1}) and 2LO (504.6 cm^{-1}) scattering in a 400-Å-thick ZnSe layer for polarized scattering, with [001] incident and detected polarizations; they saw such signals for ZnSe thickness as thin as ~20 Å. These same two peaks, which are also Fröhlich-induced, are seen for [1$\bar{1}$0] polarized scattering, as shown in Figure 12.12b. In addition, they saw the symmetry-allowed TO phonon scattering (Table 4.2) from the GaAs substrate (268.2 cm^{-1}) and from the ZnSe film (207.5 cm^{-1}) with this latter configuration. The last peak is very weak, which demonstrates that the symmetry-forbidden Fröhlich-induced LO peak is much stronger than the symmetry-allowed deformation-potential-induced TO peak with this resonance. The very weak peak near 160 cm^{-1} may be due to Ga_2Se_3 at the ZnSe/GaAs interface. For substrate temperatures >670 K they saw no Raman features attributable to ZnSe; instead Raman features were seen, presumably due to Ga_2Se_3.

Richter and co-workers (Hünermann et al., 1991; Richter et al., 1992; Esser et al., 1993; Hunger et al., 1994; Resch-Esser et al., 1994) used *in situ* resonance Raman scattering to probe the electronic structure of Sb monolayers on InP(110), GaAs(100) and GaAs(110), and Si(111). They achieved submonolayer sensitivity, and observed mode frequencies and resonance behavior that were distinct from those of bulk Sb and the substrate.

12.4 Real-Time Raman Probing of Solids and Surfaces

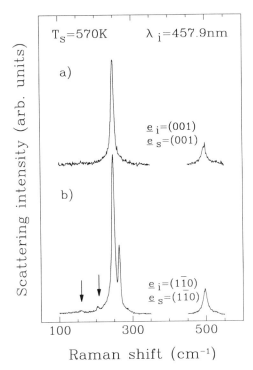

Figure 12.12 *In situ* resonant Raman spectra of 40 nm of ZnSe deposited by MBE on GaAs(110) taken with 457.9 nm and the indicated polarizations for the incident and scattered photons. Features in (a) are due to LO (252.3 cm^{-1}) and 2LO (504.6 cm^{-1}) phonon scattering in ZnSe. Also seen in (b) are TO phonon scattering in the GaAs substrate (268.2 cm^{-1}) and the ZnSe film (207.5 cm^{-1}, right arrow), and a weak peak near 160 cm^{-1} (left arrow) that may be due to Ga$_2$Se$_3$ at the ZnSe/GaAs interface (from Nowak *et al.*, 1992).

In early work, Nemanich *et al.* (1983) used interference-enhanced Raman scattering on specially prepared substrates to monitor the formation of a platinum silicide interfacial region at Pt/a-Si interfaces. Interfacial regions ~ 20 Å wide were detected. In similar *in situ* studies, Nemanich and Doland (1985) noticed the formation of a crystalline silicide with as little as 10 Å of Pd deposited on Si(111).

Hamilton and Anderson (1985) used *in situ* Raman scattering to monitor the oxidation of Fe–18Cr–3Mo. The observed signal-to-noise ratio indicated that Raman scattering should be able to characterize a monolayer oxide on the surface.

Raman scattering can also be used to probe surface adsorbates during surface passivation. Hines *et al.* (1993a,b) used polarized and angle-resolved Raman scattering to probe the orientation of hydrogen on hydrogen-terminated Si(111). This provides information that can complement infrared absorption studies by attenuated total internal reflection (ATR) spectroscopy, which is described in Section 9.7.2.

12.4.2 Thin Films and Substrates

Vibrational and electronic Raman scattering are widely used in *ex situ* materials characterization, including the analysis of semiconductors (Pollak, 1991), polymers (Bulkin, 1991), and catalysts (Stencel, 1990; Mehičić and Grasselli, 1991).

Many properties of semiconductors can be probed, including damage resulting from processing such as ion-implantation and reactive-ion etching, disorder, strain, interface quality, grain size and orientation, and doping levels. Many such studies have been performed *ex situ*, as has been reviewed by Pollak (1991) and Perkowitz (1993). For example, Tsang *et al.* (1985) used *ex situ* Raman scattering to characterize surface modifications after reactive ion etching of Si(100). In vibrational Raman scattering, substitutional dopant atoms can be detected by their local vibrational modes (when they are lighter than the host) and by their effects on the host modes due to disorder. Disorder caused by doping or other types of "damage" can lead to mode softening and broadening, and the relaxation of symmetry and polarization selection rules. For example, in backscattering from a (001) surface, TO modes may be seen in addition to the normally allowed LO modes, and disorder-activated modes, such as the DATA, disorder-activated transverse acoustic (TA) modes, may appear. Free carriers due to dopants can scatter by single-particle and collective effects, and sometimes couple to phonons (Pollak, 1991; Herman, 1996). To see these bulk $L_\pm$ modes [LO phonons coupled to (collective) plasmons] in *n*-type zincblende semiconductors, the Raman probe depth must exceed the width of the surface depletion layer, in which only LO phonons are seen. Electronic Raman scattering of inter-subband transitions in semiconductor quantum wells can be used to determine band offsets (Menendez and Pinczuk, 1988). Strain in a medium can also be determined by using Equations 4.62a and b.

The study by Bauer *et al.* (1993) illustrates several diagnostic capabilities of real-time monitoring of thin films by Raman scattering. They studied the growth of ZnSe on GaAs(001) by atomic layer epitaxy *in situ* with Raman scattering performed during growth interruptions. The polarization used allowed the scattering of LO phonons. Figure 12.13 shows the relatively large ZnSe LO phonon signal with 457.9 nm (argon-ion laser), which is resonant with the E_0 gap in ZnSe, and the smaller signal with 488 nm, which is below the gap. The ZnSe Raman shift increased from 251.7 to 253 cm^{-1} as the film thickness increased to 50 nm and beyond (to the critical thickness). This was attributed to an increase in compressive strain (see Equation 4.62a) in the epilayer when growth changed from the 3D island to the 2D layer-by-layer mode. When probing with 488 nm, which is not

12.4 Real-Time Raman Probing of Solids and Surfaces

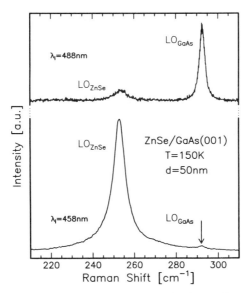

Figure 12.13 Raman spectra of 1LO phonon scattering from a 50-nm-thick ZnSe epilayer grown by atomic-layer epitaxy (ALE) on GaAs(100) and measured at 150 K. The LO features in GaAs and ZnSe can be selectively increased by using the 488.0- and 457.9-nm argon-ion laser lines, respectively (from Bauer *et al.*, 1993).

absorbed by the ZnSe, the Raman intensity is expected to increase linearly with thickness. Superimposed on top of this general trend, they saw an oscillation due to interference in the film, as seen in Figure 12.14. The

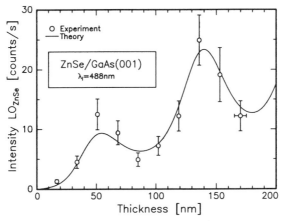

Figure 12.14 The intensity of Raman scattering from ZnSe LO phonons as a function of ZnSe epilayer thickness, showing oscillations due to interference effects in the layer superimposed upon the monotonic increase with increasing layer thickness (from Bauer *et al.*, 1993).

intensity of the LO phonon peak due to GaAs at 292 cm^{-1} also oscillated with ZnSe thickness, as expected, with no increase or decrease when averaged over a cycle. This type of interference-enhanced Raman scattering (IERS) has been described by Ramsteiner et al. (1989) and is similar to the interference effects described in Sections 6.3.1.3, 9.5, and 13.4.

Sugiura et al. (1995) used in situ Raman scattering to follow the strain in GaP layers grown on Si(111) by vapor phase epitaxy at 700°C, as they were cooled to room temperature. Using Equation 4.62b, they concluded that the tensile strain present at room temperature is due to the difference in thermal expansion coefficients of the film and substrate. In such studies it is important to differentiate between phonon shifts due to strain and those due to temperature changes (Section 12.4.2.1).

In a different application of this diagnostic, Nakamura and Kitajima (1992) performed in situ real-time Raman measurements of Si(111) during the irradiation by argon ions with either 3 or 5 keV energy. They found that the TO line at 521 cm^{-1} decreases with time during ion bombardment, indicating an increase in disorder, and that it decreases faster at 5 keV than at 3 keV.

Rosman et al. (1995) probed the growth of a diamond film in real time in a DC plasma jet CVD reactor by using Raman scattering. They used a pulsed laser (532 nm from a frequency-doubled Nd:YAG laser) and gated detection to overcome the large OE background. The effect of stress and temperature changes on the Raman shift were used to analyze the film and determine film temperature (Section 12.4.2.1).

Raman microprobe scattering has been used in real time to probe thin film processes that are stimulated by a tightly focused laser ($\sim$1 μm) to form laterally localized surface modifications (direct laser writing). The processing laser can also be used as the source of photons for Raman scattering, with the same $\sim$1-μm spot size. (Such fine resolution in Raman analysis is obtainable in specially designed chambers; $\sim$ <10 μm is achievable in conventional thin film reactors.) If the same laser is used for both purposes and is scanned across the surface, perhaps to deposit, etch, or dope a line feature, then Raman scattering probes the steady-state reaction front in real time. In the first such study, Magnotta and Herman (1986) probed during local laser CVD of Si from silane. They observed a Raman peak with a shift near 495 cm^{-1}, as is seen in Figure 12.15, in addition to the $\sim$300-cm^{-1} shift of the Ge substrate. This indicated the formation of solid silicon, which was at a temperature near the melting point. In a somewhat different application, van Duyne et al. (1986) used surface-enhanced Raman scattering (SERS) as an in situ probe of localized deposition of Ag films from solid and liquid precursors. Raman microprobe scattering can also be used during conventional pattering processes (Figure 7.17).

During etching, products are formed on the surface and then they desorb. In some cases the desorption step is slow and products remain on the

12.4 Real-Time Raman Probing of Solids and Surfaces

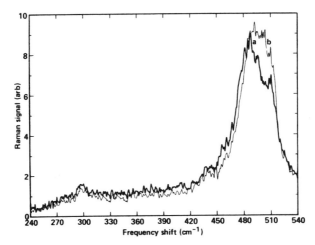

Figure 12.15 Raman spectra obtained with 514.5 nm (a) in real time during direct laser writing (local laser CVD) of silicon from silane on a germanium substrate and (b) *in situ* after deposition. Under these conditions the probed silicon was near its melting temperature (from Magnotta and Herman, 1986).

surface, which can then be monitored with Raman scattering. Tang and Herman (1990) monitored the formation of the nondesorbing products cupric chloride ($CuCl_2$) and cuprous chloride (CuCl) during the "etching" of Cu films by local laser heating in Cl_2 (Figure 12.16). To prevent the thermal reaction of Cu with Cl_2 at room temperature, they oxidized the Cu film to form a thin overlayer of Cu_2O. This layer was observed by

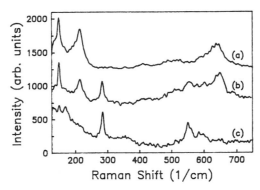

Figure 12.16 The Raman spectrum (a) before, (b) in real-time during, and (c) *in situ* after local laser etching by Cl_2 of a copper film on glass that was overlaid by a thin Cu_2O film. The features in (a), due to Cu_2O, are replaced in (b) and (c) by Raman peaks due to nondesorbing copper chlorides $CuCl_2$ (166, 283, and 563 cm^{-1}) and CuCl (broad peak below 200 cm^{-1}) (from Herman *et al.*, 1991, as adapted from Tang and Herman, 1990).

resonant Raman scattering before reaction (Figure 12.16a). Cu_2O has a strong allowed Raman peak at 218 cm^{-1} due to the two-phonon mode $2\Gamma_{12^-}$ and several other weaker peaks (154, 515, 635, and 665 cm^{-1}), some of which are forbidden. In crystalline Cu_2O the "forbidden" lines are observable due to resonance with the 1s exciton (such as with 488 nm); these lines are no longer formally forbidden in the films grown here because of the high degree of disorder in the film. The Raman spectrum taken in real time during "etching" (Figure 12.16b) shows an increase in $CuCl_2$ peaks near 166, 283, and 563 cm^{-1} and a decrease in the Cu_2O peaks. The spectrum obtained *in situ* after "etching" (Figure 12.16c) shows the $CuCl_2$ peaks, no evidence of copper oxide, and a broad feature below 200 cm^{-1} due to CuCl. In another example of enhanced Raman resonances in oxide films, Otto *et al.* (1992) used Raman scattering to follow the reduction of palladium oxide.

Burke *et al.* (1989) used *ex situ* Raman scattering to determine the composition of GeSi films deposited by thermal and excimer-laser CVD. GeSi has three Raman peaks, due to Ge−Ge, Ge−Si and Si−Si modes, near 300, 400, and 500 cm^{-1}, respectively, Stoichiometries determined by using the Si−Si mode near 500 cm^{-1} were found to agree very well with those determined by x-ray photoelectron spectroscopy (XPS) and Auger electron spectroscopy (AES). However, unlike these ultrahigh vacuum (UHV) methods, this Raman method is easy to implement for real-time probing during CVD. In some experimental regimes, the Raman peaks were observed to be very broad, indicating that an amorphous film was deposited. Laser annealing was seen to narrow these peaks, due to the transformation of the film from amorphous to polycrystalline GeSi, as was monitored by *in situ* Raman scattering.

Real-time changes in polarization of the scattered light can indicate a phase change during processing. Tang and Herman (1992) saw such a change during laser-assisted melting of Si(001) and Ge(001) and the laser-etching of these materials by Cl_2. (Laser-assisted etching of Si becomes fast only when the laser melts the surface.) In the crystal, $z(x,y)\bar{z}$ is allowed and $z(x,x)\bar{z}$ is forbidden [$x \parallel (100)$, $y \parallel (010)$], as is seen by the Si peak near 510 cm^{-1} in Figure 12.17b where hot crystalline Si was being probed. In partially molten material, this selection rule is still valid within each of the crystallites. However, since these crystallites rotate, both $z(x,y)\bar{z}$ and $z(x,x)\bar{z}$ appear to be allowed in the laboratory frame, as is seen by the peak near 482 cm^{-1} in Figure 12.17a,b. (There is no Raman scattering from molten Si.)

Raman scattering has also been used to determine doping concentration after ion implantation (Pollak, 1991; Perkowitz, 1993). The presence of the dopants induces disorder, which "relaxes" wavevector conservation. Although this method is not commonly used as an *in situ* monitor of dopant implantation (after annealing), it has found some use in processing. For example, Ashby and Myers (1991) used this as a pre-etch diagnostic for selective photochemical etching of GaAs. The etch rate in this process

12.4 Real-Time Raman Probing of Solids and Surfaces

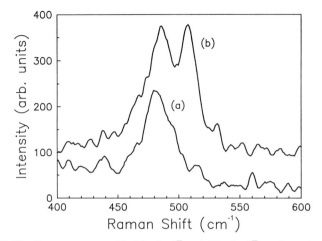

Figure 12.17 Raman spectra with (a) $z(x,x)\bar{z}$ and (b) $z(x,y)\bar{z}$ configurations [$x \parallel (100)$, $y \parallel (010)$] (514.5 nm, laser heating) of partially molten region in silicon, seen near 480 cm^{-1} in (a) and (b), surrounded by crystalline silicon [Si(001)], seen near 510 cm^{-1} only in (b) (from Tang and Herman, 1992).

depends on the concentration of photogenerated free carriers, which in turn depends on the recombination rate of these carriers. This recombination rate depends on the density of implantation-induced defects, which is determined from the LO-phonon Raman lineshape.

12.4.2.1 Thermometry

There are two ways to determine the temperature of a solid *in situ* during processing by using spontaneous vibrational Raman scattering: (1) from the Raman profile of either the Stokes or anti-Stokes peak and (2) from the relative amplitude of the Stokes and anti-Stokes peaks.

Vibrational Raman scattering studies have shown that zonecenter ($q = 0$) optical phonon frequencies ω_p generally decrease with temperature T, while the linewidths Γ_p increase with T. This is due to anharmonicity, as can be shown by perturbation theory. The coupling of the scattered optical phonon to two and three phonons of lower energy leads to changes in ω_p and Γ_p that vary as $\sim T$ and $\sim T^2$, respectively, above the Debye temperature (Balkanski *et al.*, 1983). Thermal volume expansion also contributes to the decrease of ω_p with temperature, as $\sim T$.

The temperature dependence of the Raman shift can be modeled as

$$\omega_p(T) = \omega_0 + \Delta^{(1)}(T) + \Delta^{(2)}(T) \qquad (12.4)$$

where ω_0 is the harmonic frequency of the optical mode, $\Delta^{(1)}(T)$ is due to thermal expansion, and $\Delta^{(2)}(T)$ is due to phonon–phonon coupling.

The thermal expansion contribution is

$$\Delta^{(1)}(T) = \omega_0 \left[\exp\left\{ -3\gamma_G \int_0^T \tilde{\alpha}(T')dT' \right\} - 1 \right] \quad (12.5)$$

where $\tilde{\alpha}(T)$ is the coefficient of linear thermal expansion and γ_G is the mode Grüneisen parameter, which describes how ω_p changes with volume V: $\gamma_G = -d \ln \omega_p / d \ln V$.

The anharmonic coupling term $\Delta^{(2)}(T)$ arises from cubic, quartic, and higher-order anharmonic deviations from harmonic behavior. It can be approximated as

$$\Delta^{(2)}(T) = A_1 \left\{ 1 + \sum_{j=1}^{2} \frac{1}{(e^{x_j} - 1)} \right\} + A_2 \left\{ 1 + \sum_{k=1}^{3} \left(\frac{1}{(e^{y_k} - 1)} + \frac{1}{(e^{y_k} - 1)^2} \right) \right\} \quad (12.6)$$

where $x_1 + x_2 = y_1 + y_2 + y_3 = \hbar\omega_0/k_B T$. The first term couples the optical phonon to two lower-energy phonons (three-phonon coupling), while the second term represents higher-order processes that couple it to three phonons (four-phonon coupling). $x_1 = x_2 = \hbar\omega_0/2k_B T$ and $y_1 = y_2 = y_3 = \hbar\omega_0/3k_B T$ has been assumed sometimes (Balkanski et al., 1983). However, $x_1 = 0.35 \, \hbar\omega_0/k_B T$, $x_2 = 0.65 \, \hbar\omega_0/k_B T$ works better for Si and Ge (Menendez and Cardona, 1984).

In the high-temperature limit ($k_B T \gg \hbar\omega_0$), $\Delta^{(1)}(T)$ and the first term in $\Delta^{(2)}(T)$ vary linearly with T, while the second term in Equation 12.6 varies as T^2. In this regime Equation 12.6 can be approximated by

$$\omega_p(T) = \omega_0 + aT + bT^2 \quad (12.7)$$

Similarly, the temperature dependence of the Raman linewidth can be modeled as

$$\Gamma_p(T) = B_1 \left\{ 1 + \sum_{j=1}^{2} \frac{1}{(e^{x_j} - 1)} \right\} + B_2 \left\{ 1 + \sum_{k=1}^{3} \left(\frac{1}{(e^{y_k} - 1)} + \frac{1}{(e^{y_k} - 1)^2} \right) \right\} \quad (12.8)$$

The first term corresponds to three-phonon processes (decay into two phonons) and the second term to four-phonon processes (decay into three phonons). As with the Raman shift, at high temperatures the width can also be modeled as

$$\Gamma_p(T) = \Gamma_0 + cT + dT^2 \quad (12.9)$$

The temperature dependence of the phonon shift and linewidth has been studied in detail for Si (Balkanski et al., 1983; Menendez and Cardona, 1984; Tang and Herman, 1991), Ge (Menendez and Cardona, 1984; Tang and Herman, 1991), GeSi alloys (Burke and Herman, 1993), and diamond (Anastassakis et al., 1971; Herchen and Cappelli, 1991; Dai et al., 1992; also

12.4 Real-Time Raman Probing of Solids and Surfaces

see the discussion of Rosman *et al.* in Section 12.4.2). Figure 12.18 shows the decrease of the peak frequencies in Si, Ge, and GeSi alloys up to 900 K. Similar studies have been conducted for several III–V and II–VI semiconductors, including GaAs, AlGaAs, and GaInP (Shealy and Wicks, 1987); AlSb (Raptis and Anastassakis, 1990); and ZnS, ZnO, and α-SiC (wurtzite-ZnO structure) (Mead and Wilkinson, 1977).

Since $d\omega_p/dT \sim -0.03$ cm^{-1}/K in c-Si (Figure 12.18) and ω_p can be measured well to 0.1 cm^{-1}, Raman scattering can determine temperature within 3 K. This is easily achieved if the temperature is relatively constant

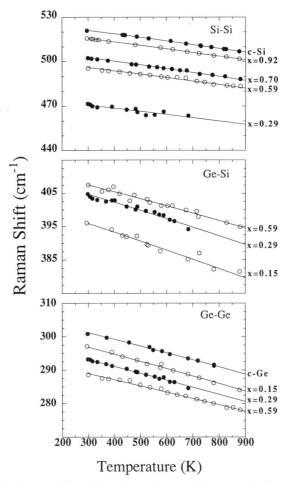

Figure 12.18 Raman shifts of the three polycrystalline Ge$_{1-x}$Si$_x$ alloy modes (Si—Si, Ge—Si, Ge—Ge) as a function of temperature, with the Si—Si and Ge—Ge modes also compared with the shifts in c-Si and c-Ge, respectively (from Burke and Herman, 1993).

over ~1 mm. Moreover, Raman scattering is capable of very fine lateral resolution. Pazionis *et al.* (1989) used this diagnostic to determine the temperature profile, with submicron resolution, of patterned Si films on silicon dioxide and sapphire heated by tightly focused lasers (~1 μm). An accuracy of <100 K near melting (1690 K) was reported for this application. Yamada *et al.* (1986) performed similar measurements for unpatterned Si films on sapphire, by using focused elliptical beams (~5 μm).

Temperature can also be determined from the ratio of the Stokes (S) and anti-Stokes (AS) scattering intensities (Compaan *et al.*, 1985). Using Equation 4.25 and the discussion in Section 4.2.2.2, this ratio is

$$\frac{I_S}{I_{AS}} = \left(\frac{\omega_s^3 n_s \chi'^2(\omega_i,\omega_s)}{\omega_a^3 n_a \chi'^2(\omega_i,\omega_a)}\right) \frac{n(\omega_p) + 1}{n(\omega_p)} = \left(\frac{\omega_s^3 n_s \chi'^2(\omega_i,\omega_s)}{\omega_a^3 n_a \chi'^2(\omega_i,\omega_a)}\right) \exp(\hbar\omega_p/k_B T) \quad (12.10)$$

where $\omega_s = \omega_i - \omega_p$ and $\omega_a = \omega_i + \omega_p$. Although the bracketed term is ~1, it cannot be set equal to 1. To determine T without significant error, this term must be determined exactly, especially near resonances, as in silicon (Jellison *et al.*, 1983; Kip and Meier, 1990). This correction is usually not important in diamond because Raman measurements are usually made far below the band gap (Anastassakis *et al.*, 1971; Herchen and Cappelli, 1991; Dai *et al.*, 1992). Because $(n(\omega_p) + 1)/n(\omega_p) \rightarrow 1$ at high T, this method is useful only when $k_B T \lesssim \hbar\omega_p$. Even higher temperatures can be measured by analyzing two-phonon scattering instead of this one-phonon process (Hayes and Loudon, 1978); however, this is not practical in film diagnostics because the signals are even weaker than in one-phonon scattering.

This Raman temperature diagnostic is applicable to all common semiconductors. Unlike several transmission probes (Section 8.4), this sensor requires access only on one side of a wafer. Unlike, interferometric probes (Sections 9.5.2, 10.1, 11.2.2.3, and 11.3), Raman scattering gives an absolute, not relative, measure of temperature and does not need any special surface preparation. It is relatively insensitive to (transparent) overlayers and can be calibrated for the effect of stress. Raman thermometry is also capable of submicron lateral spatial resolution.

One disadvantage of this sensor for real-time monitoring is its relatively high cost if typical research laboratory equipment is used. The cost of the lasers (water-cooled argon-ion lasers), spectrometers (double), and detectors (CCD array) makes this method too expensive. However, with better engineering, this versatile method could become practical. For example, a competitively priced integrated system could be designed to control temperature to a specific set point by using fiberoptic laser delivery and collection, excitation by an air-cooled argon-ion, He–Ne, or solid-state laser, spectral analysis with holographic notch filters and other filters, and detection by a photomultiplier or a modest array detector.

References

W. P. Acker, B. Yip, D. H. Leach, and R. K. Chang, *J. Appl. Phys.* **64,** 2263 (1988).
E. Anastassakis, H. C. Hwang, and C. H. Perry, *Phys. Rev. B* **4,** 2493 (1971).
C. I. H. Ashby and D. R. Myers, *J. Electron. Mater.* **20,** 695 (1991).
M. Balkanski, R. F. Wallis, and E. Haro, *Phys. Rev. B* **28,** 1928 (1983).
S. Bauer, H. Berger, P. Link, and W. Gebhardt, *J. Appl. Phys.* **74,** 3916 (1993).
W. G. Breiland and G. H. Evans, *J. Electrochem. Soc.* **138,** 1806 (1991).
W. G. Breiland and P. Ho, in *Proceedings of the Ninth International Conference on Chemical Vapor Deposition* (McD. Robinson, C. H. J. van den Brekel, G. W. Cullen, J. M. Blocher, and P. Rai-Choudhury, eds.), p. 44. The Electrochemical Society, Pennington, NJ, 1984.
W. G. Breiland and M. J. Kushner, *Appl. Phys. Lett.* **42,** 395 (1983).
W. G. Breiland, M. E. Coltrin, and P. Ho, *J. Appl. Phys.* **59,** 3267 (1986).
B. J. Bulkin, in *Analytical Raman Spectroscopy* (J. G. Grasselli and B. J. Bulkin, eds.), Chapter 7. Wiley, New York, 1991.
H. H. Burke and I. P. Herman, *Phys. Rev. B* **48,** 15016 (1993).
H. H. Burke, I. P. Herman, V. Tavitian, and J. G. Eden, *Appl. Phys. Lett.* **55,** 253 (1989).
A. Campion, *J. Vac. Sci. Technol. B* **3,** 1404 (1985a).
A. Campion, *Annu. Rev. Phys. Chem.* **36,** 549 (1985b).
M. Cardona, ed., *Light Scattering in Solids.* Springer-Verlag, Berlin, 1975.
M. Cardona, in *Light Scattering in Solids II* (M. Cardona and G. Güntherodt, eds.), Chapter 2. Springer-Verlag, Berlin, 1982.
M. Cardona and G. Güntherodt, eds., *Light Scattering in Solids II–VI.* Springer-Verlag, Berlin, 1982–1991.
R. K. Chang and M. B. Long, in *Light Scattering in Solids II* (M. Cardona and G. Güntherodt, eds.), Chapter 3. Springer-Verlag, Berlin, 1982.
B. Chase, *Appl. Spectrosc.* **48**(7), 14A (1994).
L. M. Cheung, D. M. Bishop, D. L. Drapcho, and G. M. Rosenblatt, *Chem. Phys. Lett.* **80,** 445 (1981).
A. Compaan, M. C. Lee, and G. J. Trott, *Phys. Rev. B* **32,** 6731 (1985).
G. A. N. Connell, R. J. Nemanich, and C. C. Tsai, *Appl. Phys. Lett.* **36,** 31 (1980).
S. Dai, J. P. Young, G. M. Begun, and G. Mamantov, *Appl. Spectrosc.* **46,** 375 (1992).
J. Doppelbauer, G. Leyendecker, and D. Bäuerle, *Appl. Phys. B* **33,** 141 (1984).
M. C. Drake, *Mater. Res. Soc. Symp. Proc.* **117,** 203 (1988).
A. C. Eckbreth, *Laser Diagnostics for Combustion Temperature and Species,* 2nd ed. Gordon & Breach, Luxembourg, 1996.
N. Esser, M. Köpp, P. Haier, A. Kelnberger, and W. Richter, *J. Vac. Sci. Technol. B* **11,** 1481 (1993).
J. R. Ferraro and K. Nakamoto, *Introductory Raman Spectroscopy.* Academic Press, Boston, 1994.
D. I. Fotiadis, M. Boekholt, K. F. Jensen, and W. Richter, *J. Cryst. Growth* **100,** 577 (1990).
J. Geurts, *Surf. Sci. Reports* **18,** 1 (1993).
J. Geurts and W. Richter, *Springer Proc. Phys.* **22,** 328 (1987).

J. C. Hamilton and R. J. Anderson, *High Temp. Sci.* **19,** 307 (1985).
W. Hayes and R. Loudon, *Scattering of Light by Crystals.* Wiley, New York, 1978.
P. Hendra, C. Jones, and G. Warnes, *Fourier Transform Raman Spectroscopy: Instrumentation and Chemical Applications.* Ellis Horwood, New York, 1991.
H. Herchen and M. A. Cappelli, *Phys. Rev. B* **43,** 11740 (1991).
I. P. Herman, *SPIE* **1594,** 298 (1991).
I. P. Herman, "Raman Scattering," in *The Encyclopedia of Applied Physics* (G. L. Trigg, ed.), Vol. 15, p. 587. VCH, New York, 1996.
I. P. Herman, H. Tang, and P. P. Leong, *Mater. Res. Soc. Symp. Proc.* **201,** 563 (1991).
R. A. Hill and D. L. Hartley, *Appl. Opt.* **13,** 186 (1974).
R. A. Hill, A. J. Mulac, and C. E. Hackett, *Appl. Opt.* **16,** 2004 (1977).
M. A. Hines, Y. J. Chabal, T. D. Harris, and A. L. Harris, *Phys. Rev. Lett.* **71,** 2280 (1993a).
M. A. Hines, T. D. Harris, A. L. Harris, and Y. J. Chabal, *J. Electron Spectrosc. Relat. Phenom.* **64/65,** 183 (1993b).
T. Hirschfeld, *Appl. Spectrosc.* **27,** 389 (1973).
M. Hünermann, J. Geurts, and W. Richter, *Phys. Rev. Lett.* **66,** 640 (1991).
R. Hunger, N. Blick, N. Esser, M. Arens, W. Richter, V. Wagner, and J. Geurts, *Surf. Sci.* **307–309,** 1061 (1994).
G. E. Jellison, Jr., D. H. Lowndes, and R. F. Wood, *Phys. Rev. B* **28,** 3272 (1983).
S. S. Jha, *Surf. Sci.* **158,** 190 (1985).
B. J. Kip and R. J. Meier, *Appl. Spectrosc.* **44,** 707 (1990).
M. Koppitz, O. Vestavik, W. Pletschen, A. Mircea, M. Heyen, and W. Richter, *J. Cryst. Growth* **68,** 136 (1984a).
M. Koppitz, W. Richter, R. Bahnen, and M. Heyen, in *Laser Processing and Diagnostics* (D. Bäuerle, ed.), p. 530. Springer-Verlag, Berlin, 1984b.
M. Lapp and D. L. Hartley, *Combust. Sci. Technol.* **13,** 199 (1976).
G. Leyendecker, J. Doppelbauer, D. Bärerle, P. Geittner, and H. Lydtin, *Appl. Phys. A* **30,** 237 (1983).
G. Leyendecker, J. Doppelbauer, and D. Bäuerle, in *Laser Processing and Diagnostics* (D. Bäuerle, ed.), p. 504. Springer-Verlag, Berlin, 1984.
D. A. Long, *Raman Spectroscopy.* McGraw Hill, New York, 1977.
F. Magnotta and I. P. Herman, *Appl. Phys. Lett.* **48,** 195 (1986).
J. F. McGilp, *J. Phys.: Condens. Matter* **2,** 7985 (1990).
D. G. Mead and G. R. Wilkinson, *J. Raman Spectrosc.* **6,** 123 (1977).
M. Mehičić and J. G. Grasselli, in *Analytical Raman Spectroscopy* (J. G. Grasselli and B. J. Bulkin, eds.), Chapter 10. Wiley, New York, 1991.
J. Menendez and M. Cardona, *Phys. Rev. B* **29,** 2051 (1984).
J. Menendez and A. Pinczuk, *IEEE J. Quantum Electron.* **QE-24,** 1698 (1988).
S. S. Mitra and N. E. Massa, in *Handbook of Semiconductors* (W. Paul, ed.), Chapter 3, p. 81. North-Holland Publ., Amsterdam, 1982.
Y. Monteil, M. P. Berthet, R. Favre, A. Hariss, J. Bouix, M. Vaille, and P. Gibart, *J. Cryst. Growth* **77,** 172 (1986).
Y. Monteil, R. Favre, P. Raffin, J. Bouix, M. Vaille, and P. Gibart, *J. Cryst. Growth* **93,** 159 (1988).
K. G. Nakamura and M. Kitajima, *J. Appl. Phys.* **71,** 3645 (1992).
R. J. Nemanich and C. M. Doland, *J. Vac. Sci. Technol. B* **3,** 1142 (1985).

R. J. Nemanich, M. J. Thompson, W. B. Jackson, C. C. Tsai, and B. L. Stafford, *J. Vac. Sci. Technol. B* **1,** 519 (1983).
C. Nowak, D. R. T. Zahn, U. Rossow, and W. Richter, *J. Vac. Sci. Technol. B* **10,** 2066 (1992).
T. G. M. Oosterlaken, G. J. Leusink, G. C. A. M. Janssen, S. Radelaar, K. J. Kuijlaars, C. R. Kleijn, and H. E. A. van den Akker, *J. Appl. Phys.* **76,** 3130 (1994a).
T. G. M. Oosterlaken, G. J. Leusink, G. C. A. M. Janssen, S. Radelaar, C. R. Kleijn, and K. J. Kuijlaars, *Proc. Conf. Adv. Metall. ULSI* Austin, TX, *1994* (1994b).
A. Otto, in *Light Scattering in Solids IV* (M. Cardona and G. Güntherodt, eds.), Chapter 6. Springer-Verlag, Berlin, 1984.
K. Otto, C. P. Hubbard, W. H. Weber, and G. W. Graham, *Appl. Catalysis B* **1,** 317 (1992).
G. D. Pazionis, H. Tang, and I. P. Herman, *IEEE J. Quantum Electron.* **QE-25,** 976 (1989).
S. Perkowitz, *Optical Characterization of Semiconductors: Infrared, Raman and Photoluminescence Spectroscopy.* Academic Press, San Diego, CA, 1993.
F. H. Pollak, in *Analytical Raman Spectroscopy* (J. G. Grasselli and B. J. Bulkin, eds.), Chapter 6. Wiley, New York, 1991.
M. Ramsteiner, C. Wild, and J. Wagner, *Appl. Opt.* **28,** 4017 (1989).
Y. S. Raptis and E. Anastassakis, *Solid State Commun.* **76,** 335 (1990).
U. Resch-Esser, U. Frotscher, N. Essser, U. Rossow, and W. Richter, *Surf. Sci.* **307–309,** 597 (1994).
W. Richter, *Philos. Trans. R. Soc. London, Ser. A* **344,** 453 (1993).
W. Richter, N. Esser, A. Kelnberger, and M. Köpp, *Solid State Commun.* **84,** 165 (1992).
N. Rosman, L. Abello, J. P. Chabert, G. Verven, and G. Lucazeau, *J. Appl. Phys.* **78,** 519 (1995).
T. O. Sedgwick and J. E. Smith, Jr. *J. Electrochem. Soc.* **123,** 254 (1976).
T. O. Sedgwick, J. E. Smith, Jr., R. Ghez, and M. E. Cowher, *J. Cryst. Growth* **31,** 264 (1975).
B. V. Shanabrook and B. R. Bennett, *Phys. Rev. B* **50,** 1695 (1994).
C. Y. She, *Contemp. Phys.* **31,** 247 (1990).
J. R. Shealy and G. W. Wicks, *Appl. Phys. Lett.* **50,** 1173 (1987).
K. Smit, L. Koenders, and W. Mönch, *J. Vac. Sci. Technol. B* **7,** 888 (1989).
J. E. Smith, Jr. and T. O. Sedgwick, *Thin Solid Films* **40,** 1 (1977).
J. N. Stencel, *Raman Spectroscopy for Catalysis.* Van Nostrand, New York, 1990.
M. Sugiura, M. Kishi, and T. Katoda, *J. Appl. Phys.* **77,** 4009 (1995).
H. Tang and I. P. Herman, *J. Vac. Sci. Technol. A* **8,** 1608 (1990).
H. Tang and I. P. Herman, *Phys. Rev. B* **43,** 2299 (1991).
H. Tang and I. P. Herman, *J. Appl. Phys.* **71,** 3492 (1992).
J. C. Tsang, in *Light Scattering in Solids V* (M. Cardona and G. Güntherodt, eds.), Chapter 6, p. 233. Springer-Verlag, Berlin, 1989.
J. C. Tsang, G. S. Oehrlein, I. Haller, and J. S. Custer, *Appl. Phys. Lett.* **46,** 589 (1985).
R. P. van Duyne, R. I. Altkorn, and K. L. Haller, *IEEE Circuits Devices* **2**(1) 61 (1986).

A. Weber, ed., *Raman Spectroscopy of Gases and Liquids.* Springer-Verlag, Berlin, 1979.

M. Yamada, K. Nambu, Y. Itoh, and K. Yamamoto, *J. Appl. Phys.* **59,** 1350 (1986).

D. R. T. Zahn, Ch. Maierhofer, A. Winter, M. Reckzügel, R. Srama, A. Thomas, K. Horn, and W. Richter, *J. Vac. Sci. Technol. B* **9,** 2206 (1991).

CHAPTER 13

Pyrometry

All materials in thermal equilibrium emit thermal radiation characterized by a spectrum that depends on the temperature T. Pyrometry or radiation thermometry is the determination of temperature by analyzing this emission spectrum. Depending on the wavelength range studied, this procedure is called either optical or infrared pyrometry and is performed with an instrument called a pyrometer.

The wafer temperature in thin film processing is often monitored by measuring the intensity of the emitted radiation in a narrow spectral band about one or several wavelengths by using a (photoelectric) automatic pyrometer; such an instrument is also called a narrow-band, "brightness," or "monochromatic" pyrometer. This method is often used in rapid thermal processing because of its fast response. In a wide-band or "total" radiation pyrometer, essentially all thermal radiation is collected on a radiation sensor; it is rarely used in thin film processing. Also less commonly used in these applications is the "disappearing-filament" optical pyrometer, which is a version of narrow-band interferometry that is useful at relatively high temperatures. In this instrument, based on visual comparison, the source of radiation is imaged through a tungsten ribbon filament of a "standard lamp." The current through the filament is changed until the brightness of the filament matches that of the measured object, whereupon it disappears. This procedure can be performed visually by the operator or automatically. Pyrometry is sometimes performed in comparison with a standard reference temperature, such as the normal melting point of gold, 1064.43°C, in the International Practical Temperature Scale.

The physics and technology of pyrometry have been reviewed in the book edited by DeWitt and Nutter (1988), and is more briefly covered in chapters in volumes on thermometry and related sensors by Norton (1989), Schooley (1986), Benedict (1984), and McGee (1988). Anderson (1990) and Peyton et al. (1990) have reviewed some of the applications of pyrometry in thin film processing; Peyton has also surveyed the costs of the various types of pyrometers. Roozeboom and Parekh (1990) and Roozeboom (1991, 1993a,b) have detailed the use of pyrometry in commercially available rapid thermal processing (RTP) reactors and have reviewed advances in pyrometry that are useful for RTP thermometry. Öztürk *et al.* (1991) have described how errors in temperatures determined by pyrometry affect manufacturing by rapid thermal chemical vapor deposition (RTCVD). Another important application of thermal emission is in measuring and controlling the thickness of thin film layers by a method called pyrometric interferometry.

13.1 Theoretical and Experimental Considerations

A general discussion of thermal radiation is provided by Franson *et al.* (1962), Eisberg and Resnick (1985), Nutter (1985), DeWitt and Incropera (1988), and McGee (1988). The Planck blackbody spectrum of the intensity emitted from a surface in thermal equilibrium at temperature T between frequencies v and $v + dv$ is

$$I(v,T)dv = \frac{2hv^3 n^2 \hat{\varepsilon}(v,T)}{c^2} \frac{1}{\exp(hv/k_B T) - 1} dv \tag{13.1}$$

where h is Planck's constant, k_B is Boltzmann's constant, c is the speed of light, and n is the refractive index (=1 for vacuum, ≈1.00028 for air).

The emissivity $\hat{\varepsilon} = 1$ for an ideal blackbody. (The emissivity is represented here by $\hat{\varepsilon}$ rather than the commonly used ε to avoid confusion with the dielectric function; ε is still used in the figures.) Technically, the term *emissivity* refers to this parameter for an ideal, homogeneous, semiinfinite material (see Chapter 9 appendix), while *emittance* refers to the measured value of this parameter and to the theoretically determined value of this parameter for composite structures. Even though emittance is usually the more technically correct nomenclature, the term *emissivity* is used here because it is so common in the literature.

In terms of wavelength, Planck's law is often expressed as

$$I(\lambda,T)d\lambda = \frac{c_1 \hat{\varepsilon}(\lambda,T)}{\lambda^5} \frac{1}{\exp(c_2/\lambda T) - 1} d\lambda \tag{13.2}$$

The first radiation constant $c_1 = 2hc^2/n^2 = 1.19089 \times 10^{-16}$ W m^2 (for $n = 1$), and the second radiation constant $c_2 = hc/nk_B = 0.0143883$ m K (for $n = 1$).

Wien's displacement law gives the wavelength at which $I(\lambda,T)$ peaks:

$$\lambda_{\max} T = 2.89787 \times 10^{-3} \text{ m K} \tag{13.3}$$

where the constant on the right-hand side is sometimes called the third radiation constant c_3.

13.1 Theoretical and Experimental Considerations

Using radiometric nomenclature (Section 2.1), the intensity given in Equations 13.1 and 13.2 represents the spectral radiant energy emitted in a solid angle $d\Omega$ normal to the surface from an area dA in a time dt, and is not strictly the radiant intensity. If viewed at an angle θ to the normal, the intensity is reduced according to the Lambert cosine law to $I_\theta(\lambda,T) = I(\lambda,T)\cos\theta$. When integrated over the solid angle of a hemisphere, the hemispherical spectral radiant intensity $W(\lambda,T) = \pi I(\lambda,T)$ is obtained, which is also known as the spectral radiant emittance or exitance $M(\lambda,T)$. Sometimes, $W(\lambda,T)$ is expressed in the form of Equation 13.2, with c_1 replaced by $c_1' = \pi c_1 = 3.7414 \times 10^{-16}$ W m². Integration of $W(\lambda,T)$ over all wavelengths gives the Stefan–Boltzmann law

$$W(T) = \sigma T^4 \tag{13.4}$$

for blackbodies ($\hat{\varepsilon} = 1$), where $\sigma = (\pi^4/15)(c_1/c_2^4) = 2\pi^5 k_B^4/15c^2h^3 = 5.6687 \times 10^{-8}$ W/K⁴ m² ($n = 1$) is the Stefan–Boltzmann constant.

The spectral radiance $L(\lambda,T)$ is the spectral radiant flux per unit solid angle $d\Omega$ per unit projected area, where the spectral radiant flux $\Phi(\lambda,T)$ is that radiant energy per unit time and wavelength that passes through the projected area element dA in the direction θ (which is $dA\cos\theta$). Since this is equivalent to the radiant emission $I_\theta\, dA$ per unit of projected area dA in the direction θ ($dA\cos\theta$), $L(\lambda,T) = I(\lambda,T)$.

Real materials are seldom perfect blackbodies, i.e., bodies that absorb all incident radiation. Thermal emission is corrected by the emissivity factor $\hat{\varepsilon}(\lambda,T)$ in Equations 13.1 and 13.2. From Kirchhoff's law of radiation, the emissivity is

$$\hat{\varepsilon}(\lambda,T) = 1 - R(\lambda,T) - T_{\text{trans}}(\lambda,T) \tag{13.5}$$

where R is the reflectance and T_{trans} is the transmittance of the emitter, so $\hat{\varepsilon}$ is the emitter absorbance. $R(\lambda,T)$ is really the total hemispherical reflectance, not just the specular reflectance. For blackbodies $\hat{\varepsilon}(\lambda,T) = 1$ for all λ and T. Graybodies are objects whose emissivity is equal at all wavelengths, $\hat{\varepsilon}(T)$. Figure 13.1 shows the spectral radiance of blackbody emission for several wavelengths. Errors of hundreds of degrees are possible in pyrometry if the emissivity is not accurately known, as is seen in Figure 13.2. Figure 13.2a shows that the flux near 3 μm is the same for a graybody at 1000°C with $\hat{\varepsilon} = 0.5$ and a blackbody at 800°C, while Figure 13.2b illustrates that the fluxes near 0.9–1.0 μm are equal for this same graybody and a blackbody at 930°C.

The reflectance and transmittance can include the effects of interference due to films on the surface and surface roughness. For a semiinfinite material, $\hat{\varepsilon}(\lambda,T) = 1 - R(\lambda,T)$. In general, $\hat{\varepsilon}$ can depend strongly on λ and film thickness when films are being processed, so the graybody assumption may be poor during film processing even if it is a good assumption for the bare wafer. Uncertainties in $\hat{\varepsilon}$ are responsible for much of the error in pyrometric

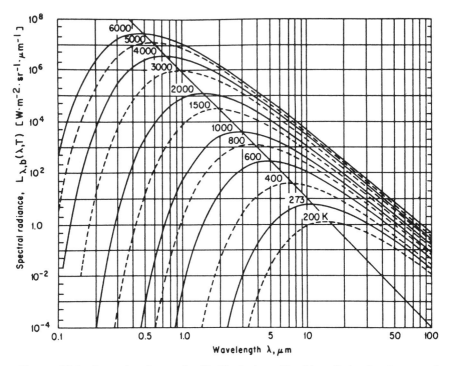

Figure 13.1 Spectral radiance of a blackbody from Planck's radiation law for several temperatures (from Nutter, 1985). (Copyright ASTM. Reprinted with permission.)

temperature measurements. Roozeboom (1993b) has classified three types of emissivities that are important in pyrometry during thin film processing. The *intrinsic emissivity* is that of the starting wafer, which, for optically-thick wafers, is the emissivity of the wafer material. The *extrinsic emissivity* is that including layers on top of the wafer or buried in it, which is technically the emittance. The *effective emissivity* also includes the effects of the optical properties in a reflective chamber due to its components.

When pyrometric measurements are made at photon energies higher than the peak in $I(\nu,T)$ ($\lambda < \lambda_{\max}$), as is common, the high-energy approximation of Planck's radiation law can be used. With $h\nu \gg k_B T$, the 1 in the denominator of Equations 13.1 and 13.2 can be ignored, giving Wien's approximation. Using this approximation, the true temperature T of the object with emissivity $\hat{\varepsilon}$ can be related to the spectral temperature T_λ, which is the temperature of a blackbody that emits the same spectral flux as this object at a given λ, by

$$\frac{1}{T} = \frac{1}{T_\lambda} + \frac{\lambda}{c_2} \ln \hat{\varepsilon}(\lambda, T) \tag{13.6a}$$

This gives a measure of how much T_λ underestimates T. More generally,

13.1 Theoretical and Experimental Considerations

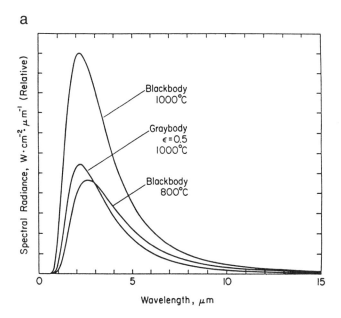

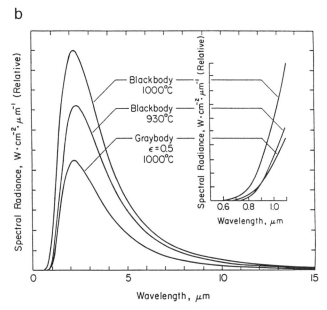

Figure 13.2 Spectral radiances for a blackbody and a graybody with emissivity of 0.5 at several temperatures, illustrating why accurate knowledge of the emissivity is essential. [From Tenney (1988) in *Theory and Practice of Radiation Thermometry* (D. P. DeWitt and G. D. Nutter, eds.). Copyright © 1988 John Wiley & Sons, Inc. Reprinted by permission of John Wiley & Sons, Inc.]

if a structure is assumed to have an emissivity $\hat{\varepsilon}_a$, so pyrometry determines a temperature T_a, but it really has an emissivity $\hat{\varepsilon}$, then the true temperature is T:

$$\frac{1}{T} = \frac{1}{T_a} + \frac{\lambda}{c_2} \ln\left(\frac{\hat{\varepsilon}(\lambda,T)}{\hat{\varepsilon}_a(\lambda,T)}\right) \qquad (13.6b)$$

The thermal radiation power $\mathcal{P}$ sensed by a pyrometer is

$$\mathcal{P} = Af(h)L(\lambda,T)\Delta\lambda \qquad (13.7)$$

where A is the area on the wafer imaged by the pyrometer, h is the distance between the wafer and the pyrometer, $f(h)$ is a factor describing the collection efficiency of the pyrometer, and $\Delta\lambda$ is the wavelength band measured by the pyrometer. In production reactors pyrometers usually sense thermal radiation from the backside of the wafer.

Measurements are usually made with quantum detectors, such as compound semiconductor diodes (Figure 13.3c), for semiconductor processing. Thermal detectors are not useful for many applications, such as rapid thermal processing (RTP), because of their long response times. Typical pyrometers have response times from 300 msec to 10 sec, but they can be customized to 1–10 msec, which is adequate for most RTP applications (Peyton *et al.*, 1990). Pyrometric measurements are usually made at a single wavelength (Section 13.2), although measurements at multiple wavelengths are sometimes made to correct for errors due to uncertain emissivities (Section 13.3). All optics, including reactor windows and lenses, must transmit well at the infrared wavelengths of interest. Sometimes, optical fibers and light pipes are used to assist light collection.

13.2 Single-Wavelength Pyrometry

Single-wavelength (one-color) pyrometry is by far the most common in thin film processing, where it is widely used in RTP and MBE (molecular-beam epitaxy), as well as in other cold-wall reactors. A wide range of customized optical pyrometers is available (Peyton *et al.*, 1990). Figure 13.4 shows schematics of the pyrometric measurement of a wafer in cold-, warm-, and hot-wall RTP chambers. The largest sources of experimental error in pyrometry are uncertainties in the emissivity and the collection of stray light, such as radiation from heating lamps reflecting from the wafer.

The wavelength of the observed light is chosen to (1) maximize thermal radiation (which favors operating near λ_{max} in Wien's law, Equation 13.3), (2) minimize light from the lamps that heat the wafer (which favors longer λ), and (3) maximize transmission of thermal radiation through reactor windows (which favors shorter λ). Note that at 400, 800, and 1200°C, λ_{max} is 4.3, 2.7, and 2.0 μm, respectively. The typical lamp emission and window

13.2 Single-Wavelength Pyrometry

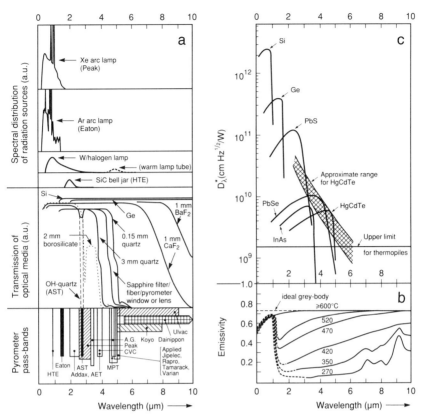

Figure 13.3 Spectral features of rapid thermal processing (RTP) systems: (a) Spectral properties of heat sources, optical media, and pyrometers; (b) the emissivity of Si (after Sato, 1967); (c) the detectivity for temperature sensors at room temperature [after Dimmock, 1972 (© 1972 IEEE)]; (from Roozeboom, 1991).

transmission spectra shown in Figure 13.3a illustrate this tradeoff in choosing the measured wavelength. Bandpass filters are often used to select the desired pyrometer range (Figure 13.3a). (Also see Adel *et al.*, 1994.)

Long-arc xenon lamps emit little radiation at $\lambda > 1.4$ μm—especially with a water-cooled wall. When these lamps are used for RTP heating, pyrometers typically are operated in the 1.4–1.5 μm or 2.6–3.5 μm ranges. Tungsten filament lamps deliver little radiation at wavelengths >4.5 μm when the light is transmitted through quartz filters or chamber walls. When these tungsten lamps are used, long-wavelength bands are chosen, such as ~4.5–5.0 μm or 5–9 μm. Operation at these longer wavelengths requires either the use of special materials for reactor windows or very thin quartz windows (Roozeboom and Parekh, 1990; Roozeboom, 1991).

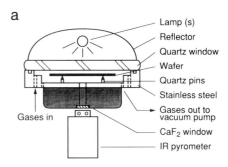
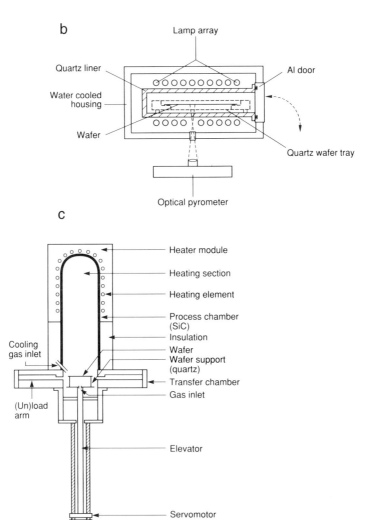

Figure 13.4 Schematic configuration of a typical (a) cold-wall, (b) warm-wall, and (c) hot-wall RTP chamber (from Roozeboom and Parekh, 1990).

13.2 Single-Wavelength Pyrometry

Another way to avoid sensing reflected lamp radiation is by transmitting the lamp radiation through a mirror that transmits most of the light, but rejects a very narrow bandwidth centered at, say, λ_c. Radiation from the wafer is collected and sent through a filter with a narrow-band transmission window at λ_c, so the pyrometer senses only thermal radiation (Accufiber citation in Roozeboom and Parekh, 1990). λ_c has been chosen to be 950 nm for silicon-wafer thermometry because the photon energy is above the band gap (1.1 μm) and the emissivity is not a function of temperature (Figure 13.3b).

Even when the spectral overlap of thermal radiation and other flux cannot be avoided, their individual contributions can sometimes be separated. One such approach, which is described in Section 13.2.1, uses internal light pipes to sense a modulation in the reflected flux. Loarer *et al.* (1990) describe a different approach, based on the photothermal effect (Section 15.2), in which a modulated laser (chopped cw [continuous-wave] or pulsed) modulates the temperature of the substrate locally. This modulates the thermal emission, but not the reflected flux (if the change in temperature affects blackbody emission much more than it does the optical properties of the substrate). Depending on the effect of this modulation on the thin film process, which is a function of the laser duty cycle and the actual temperature rise, this method can be somewhat intrusive.

The measurement of temperature hinges on accurate knowledge of the emissivity of the wafer for all process conditions. From Equation 13.5, the emissivity is seen to be sensitive to the properties of the wafer, the surface morphology, and films on the wafer. Uncertainties in $\hat{\varepsilon}$ can lead to large errors in pyrometric thermometry.

Sato (1967) made an extensive study of the emissivity of bulk Si as a function of temperature and wavelength (Figure 13.3b). More recently, Timans has measured the emissivity of GaAs (1992) and Si (1993). The emissivity of Si at wavelengths longer than the band edge ($\sim$1.1 μm, 300 K) is due to free-carrier absorption, and so silicon wafers are nearly transparent below the band gap at low T. Consequently, the emissivity of a Si wafer is a strong function of temperature, especially below 600°C. Above about 500°C the intrinsic carrier concentrations are high enough that Si wafers are typically opaque at usual pyrometric wavelengths, and $\hat{\varepsilon} \sim 0.7$ for $\lambda >$ 5 μm. Above 600°C, a bare silicon wafer behaves like a graybody with emissivity 0.7. Standard optical pyrometers cannot be used for silicon wafers below 400°C unless emissivity corrections are carefully analyzed.

Thin films of silicon dioxide, silicon nitride, and polysilicon can change the emissivity of a silicon wafer. Pettibone *et al.* (1986) have shown that oxide layers on silicon can cause temperature errors of 10–50°C. Öztürk *et al.* (1991) have demonstrated that there can be $\pm$50°C thermometry errors during polysilicon growth on a 1000-Å silicon dioxide layer atop a

Si wafer. Figure 13.5 shows how the calculated emissivity varies with silicon dioxide layer thickness for polysilicon/SiO$_2$/Si wafer structures. Wood *et al.* (1990) corrected temperature measurements by calculating the emissivity using known physical parameters.

Nulman *et al.* (1990) have examined how the emissivity of a Si wafer depends on temperature, surface roughness, and the nature of the film overlayer. Figure 13.6 plots the observed emissivity of a silicon wafer at 4.5 μm (below the band gap) for measurements made through a 800-nm-thick SiO$_2$ overlayer under three different conditions: from both the smooth device (D2) and roughened back (B2) sides of a wafer oxidized on both sides, and from the back side of a wafer oxidized only on the back side (B9). Below 600°C, the three emissivities are very different, but each increases rapidly with wafer temperature. For the 10-Ω cm resistivity of this wafer, free-carrier absorption is weak and transmission through the wafer at this below-band-gap photon energy is significant, so $\hat{\varepsilon} < 1 - R$. Although both wafers are oxidized on the back side, the emissivities measured from the back side differ at these low temperatures because of the different device sides—one is oxidized and the other is not.

As T increases, the free-carrier concentration increases and free-carrier absorption increases. Above 600°C, both wafers are opaque at 4.5 μm and both back-side measurements are identical. However, emissivities from the front and back oxidized surfaces of the first wafer still differ because the device surface is smooth while the back surface is rough. In this regime, each emissivity decreases with a slope of 8.89×10^{-5}/°C, mainly because

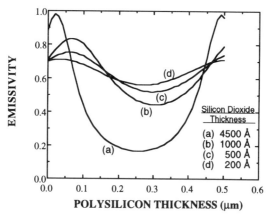

Figure 13.5 The calculated emissivity at 3.8 μm of a polysilicon/SiO$_2$/Si three-layer structure as a function of polysilicon thickness for different silicon dioxide thicknesses (from Öztürk *et al.*, 1991). This demonstrates that films on a surface can greatly change the emissivity of the material structure and therefore affect the accuracy of the pyrometric determination of temperature. (© 1991 IEEE.)

13.2 Single-Wavelength Pyrometry

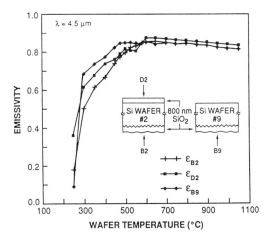

Figure 13.6 Measured emissivity at 4.5 μm as a function of temperature for Si wafers with 800 nm SiO$_2$ on both sides (#2) [observed from the smooth top (D2) and from the rough back side (B2)] and on the back side only (#9) [observed from the rough back side (B9)] (from Nulman et al., 1990).

of the temperature dependence of the index of refraction, and thus the reflectance changes. Vandenabeele and Maex (1992) have arrived at similar conclusions about the effect of surface roughness on wafer emissivity.

Moslehi et al. (1992) measured temperature in an RTP tool by using narrow-band infrared pyrometry centered at 5.35 μm. The back-side emissivity was determined using Equation 13.5, with the reflectance and transmittance measured with a pulsed CO laser lasing near 5.35 μm and chalcogenide optical fibers, which improved the accuracy of single-wavelength pyrometry. At 600°C, emissivities were within ±15% of model predictions. Dilhac et al. (1991) also used back-side reflectance measurements during rapid thermal processing to correct for emissivity changes. A more general, and more complicated, approach is ellipsometric pyrometry, in which the emissivity is obtained in real time from Equation 13.5 by using a reflection ellipsometer to determine the optical properties (dielectric functions, film thicknesses, etc.) of the substrate. Hansen et al. (1988, 1989) have used a rotating-analyzer ellipsometer to determine the optical constants of hot transition metals, and have used this method *in situ*, combined with pyrometry, to measure temperatures. Peyton et al. (1990) have described commercialized versions of this instrument. Ellipsometry can also be used to determine the temperature directly (Section 9.11). Murray (1967) has shown that measuring the polarization of thermally emitted radiation and reflected radiation can improve pyrometric measurements.

Sometimes the silicon wafer is coated with thick layers of silicon dioxide, as in some applications of RTP and in radiantly heated selective chemical vapor deposition (CVD). Delfino and Hodul (1992) have shown that pyrometric measurements can be made more accurately and at lower temperatures for such wafers than for uncoated Si wafers by using narrow-band infrared pyrometry centered near the Si—O absorption band, ~9.2 µm. Measurements were made with a narrow-band filter centered near this band (9.4 ± 0.3 µm) and a broad-band filter (11 ± 3 µm), which is more typical in infrared pyrometry. When there are thick films (1.3 µm) on both sides of the wafer, the emissivity with the 9.4-µm filter reaches its maximum of 0.41 at 270°C and remains at this value for higher T, while with the 11-µm filter the maximum emissivity of 0.66 occurs only at higher T, ≥ 430°C. Wafers with thinner oxide films also exhibit more uniform emissivity at lower T, particularly with the 9.4-µm based pyrometer.

Careful temperature analysis in RTP reactors is crucial because rapid thermal processing relies on high-temperature operation (for short times). Virtually runaway conditions can occur during growth with only small errors in temperature, because of the rapid increase in deposition rate with increasing temperature. The deposition rate in the kinetically controlled regime is given by the Arrhenius relation $A \exp(-\mathcal{E}_{act}/k_B T)$, where $\mathcal{E}_{act}$ is the activation energy, which is 1.6 eV for polysilicon CVD from silane. Using this relation and Equation 13.6b, Öztürk et al. (1991) show (Figure 13.7) that the deviation in the thickness of RTP-deposited polysilicon films from the targeted value of 4000 Å can be very large because of emissivity and temperature errors. This is also seen in Figure 18.1 where a fractional error in the temperature measurement translates into a much larger fractional error in film thickness. Consequently, temperature calibration of RTP pyrometers is crucial (Roozeboom and Parekh, 1990; Roozeboom, 1993a,b; Öztürk *et al.*, 1991). These pyrometers can be absolutely calibrated

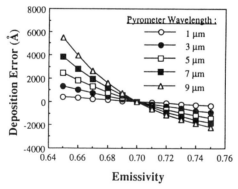

Figure 13.7 Polysilicon deposition error, relative to the desired film thickness of 4000 Å, that occurs in Si CVD in a rapid thermal processor, assuming an activation energy of 1.6 eV, when the surface emissivity deviates from the assumed value of 0.7, for various pyrometer wavelengths (from Öztürk *et al.*, 1991). (© 1991 IEEE.)

13.2 Single-Wavelength Pyrometry

by comparison with thermocouples and by measuring the extent of well-studied reactions, such as the rate of oxide growth and activation of dopant implants. Note that Figure 13.7 suggests that errors are smaller at shorter wavelengths.

In molecular-beam epitaxy (MBE), the pyrometer must be shielded from stray light, such as that from the MBE effusion cells, the substrate heater, and even light from control lights on electronics displays. During MBE film growth, coatings on the optical viewports can lead to time-varying errors in the measured thermal radiation. This can also be a problem in rapid thermal processing. Growth of a film with a band gap smaller than that of the wafer can affect lamp-heated MBE by increasing the energy absorbed by the substrate, thereby heating it to an even higher temperature, and by changing the emissivity, thereby affecting pyrometric measurements. Katzer and Shanabrook (1993), Shanabrook et al. (1993) and Nouaoura et al. (1995) have examined the use of pyrometry to monitor wafer temperature during MBE.

Pyrometry has also been used to measure temperature during other thin film processes. Patel et al. (1991) have used a charge-coupled device (CCD)-based camera with a PtSi Schottky barrier detector array to monitor thermal emission (3–5 μm) during plasma etching of Si layers atop SiO_2(film)/Si(wafer). This thermal imaging system, which is capable of detecting a noise-equivalent differential temperature of 0.1°C, was used to monitor temperature and emissivity changes and to serve as an etching endpoint detector. Rio and Moore (1995) used an IR camera (8–12 μm) to monitor the temperature of a silicon wafer during ion implantation. Vandenabeele et al. (1992) used the emissivity changes in pyrometry to detect the endpoint of the transformation of CoSi to $CoSi_2$.

Thermal imaging is also used to control temperature during sputter deposition of Co-alloy/Cr thin films (in magnetic disks for high-density recordings). Temperature control is important in this in-line sputtering system, in which the substrate moves through the chamber, because the deposition temperature affects the coercive force of these films. Fujita et al. (1992) measured temperature uniformity *in situ* in such a sputtering reactor by measuring thermal emission, with an InSb sensor (3–5.4 μm), through a duct designed to minimize stray IR light.

Thermal imaging has also been used in the "later" stages of electronics production, i.e., beyond the thin film processing stage. Linnander (1993) has reviewed the use of infrared thermography (usually in the 3–5 μm range) in the design, development, and production of printed circuit boards, hybrid circuits, and other electronic components, where temperature differences as small as 0.1°C can be detected in times as short as 1/30 sec.

Since the emissivity depends on the absorption properties of the surface (Equation 13.5), thermal radiation will peak at characteristic absorption peaks of adsorbates and films on a substrate. Under OMVPE [organometallic vapor-phase epitaxy] growth conditions (400–700°C), Mazzarese et al. (1992) noted enhanced thermal emission at the vibrational frequencies of adsorbates associated with the TMGa [trimethylgallium] and NH_3 reactants, by using FTIR [Fourier-transform infrared] spectroscopy. Wangmaneerat et al. (1992a,b) and Niemczyk et al. (1993, 1994) showed how to determine the temperature and composition of phosphosilicate glass (PSG) and borophosphosilicate glass (BPSG) thin films *in situ* by using multivariate statistics (chemometrics) to evaluate the infrared emission spectrum from these films atop silicon wafers. The spectral emissivity is sensitive to the characteristic absorption bands in the film, as is understood from Equation 13.5. Figure 13.8a shows how the emissivity changes with temperature, Figure 13.8b shows the partial least-squares (PLS) weight-loading vectors (which peak where the emissivity peaks) obtained in the multivariate analysis, and Figure 13.8c plots the temperatures determined from such spectra. These measurements were made using an off-axis ellipsoidal mirror and analyzed from ~550 to 1600 cm^{-1} with a commercial FTIR spectrometer (McGuire et al., 1992). Niemczyk et al. (1994) showed that the standard errors in analyzing BPSG films by using the monitor wafer data set was ~0.09 wt% for boron and phosphorus content, 36 Å in the film thickness, and 1.9°C in temperature. The errors were slightly larger in monitoring BPSG atop product wafers (0.13 wt%, 120 Å, and 5.9°C, respectively), in part because of the smaller set sampled. Franke et al. (1994) have reviewed the use of infrared spectroscopies, including emission, reflection, and absorption, to characterize dielectric thin films quantitatively.

In most applications, pyrometry is used to measure relatively uniform temperature profiles on a wafer. Thermal radiation from a locally heated spot can be imaged into a detector for micropyrometry analysis. Sedgwick (1981) adapted another method to measure the local temperature rise on a Si wafer during localized laser heating (by an argon-ion laser, focused to a 50–100 μm diameter) by modifying a standard "disappearing hot wire or filament" pyrometer with a long working distance magnifier (15×).

13.2.1 *Optical Fiber Thermometry*

Optical fibers can be used for *in situ* single-wavelength pyrometry measurements, although they are limited by the ~500–700°C softening temperature of most glasses (Vanzetti and Traub, 1988). Light pipes have been developed that can withstand even higher temperatures for use in pyrometry (Dils et al., 1986; Kreider, 1985). The light pipes collect and trap a wide angle of incident light, ~50°, by total internal reflection, which is later delivered to fiber optics (Dils et al., 1986). Sapphire light pipes have a flat response from the ultraviolet to the near-infrared. The light pipes are enclosed in a sheath for protection from reactor processes.

13.2 Single-Wavelength Pyrometry

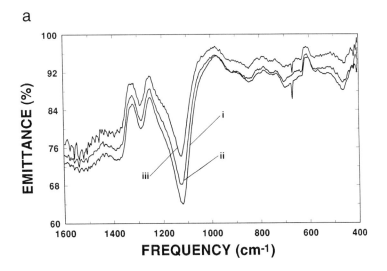

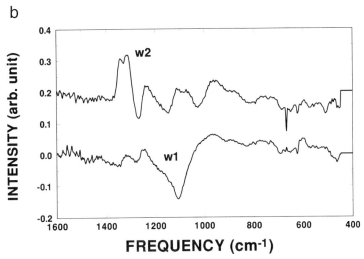

Figure 13.8 (a) Emittance (emissivity) spectra of phosphosilicate (PSG) glass at (i) 225°C, (ii) 166°C, and (iii) 119°C. (b) The first (W1) and second (W2) PLS (partial least-squares) weight-loading vectors for temperature determination, with W2 offset by +0.2. (c) Temperature determined from the emittance from several PSG samples (from Wangmaneerat *et al.*, 1992b).

(Continues)

In pyrometry, the detector must usually be shielded from light from the heating lamps, either physically or by a wise choice of the monitored wavelength. Such shielding is not necessary in the ripple-amplitude method of optical fiber thermometry described by Schietinger *et al.* (1991),

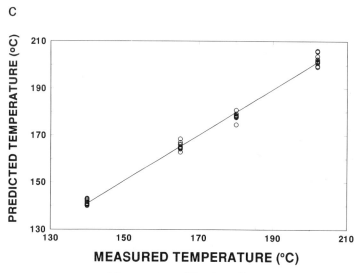

Figure 13.8 (*Continued*)

Schietinger and Adams (1992a,b) and Fiory *et al.* (1993), in which the effect of the detected lamp radiation is removed and the emissivity is determined under some conditions. In the ripple-amplitude method, light is collected by two light pipes placed in the reactor, as shown in Figure 13.9—one directed to the lamps and one to the wafer. The light collected from the lamps gives a signal $S_l(t) = S_{l,0} + \Delta S_l(t)$, where the second contribution is the 120-Hz modulated ripple from the 60-Hz power supply. The signal from the pipe collecting light from the wafer $S_w(t)$ has components from the

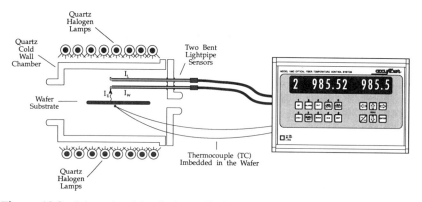

Figure 13.9 Schematic of the ripple-amplitude method for obtaining the emissivity for pyrometry measurements in an RTP reactor (adapted from Schietinger *et al.*, 1991).

13.2 Single-Wavelength Pyrometry

wafer thermal radiation E_w and reflected flux from the lamps $F_w(t)$ [with ripple component $\Delta F_w(t)$]. Since F_w has the same ripple fraction as does S_l, the wafer reflectance $R = \Delta F_w(t)/\Delta S_l(t)$ can be determined. From this $F_w(t) = RS_l(t)$, and so

$$E_w = S_w(t) - RI_l(t) = S_w(t) - \left(\frac{\Delta F_w(t)}{\Delta S_l(t)}\right) S_l(t) \qquad (13.8)$$

T can be determined from E_w and the expressions in Section 13.1. Moreover, if the wafer has zero transmittance, as at 1 μm where silicon photodetectors can be used, the emissivity is also known; $\hat{\varepsilon} = 1 - R$.

This technique is insensitive to surface roughness because of the large collection angle of the light pipes. In the commercial device (Accufiber, 1995; Peyton *et al.*, 1990) the temperature range is 500–1900°C, with repeatability of 0.05% and 1–50 readings per second. Fiory *et al.* (1993) report accuracies of at least ±5°C from 650 to 1050°C for pyrometry at 1 μm in RTP reactors. Detection at longer wavelengths may lower this lower temperature limit. This method involves *in situ* probes, which can be a disadvantage.

A related modulation method by Vandenabeele and Maex (1992a,b) (also see Roozeboom, 1993a,b, for more details) uses two pyrometers to monitor radiation at 2.95 μm. As seen in Figure 13.10, the wafer is heated from the top by two rows of lamps; 2.95 μm is filtered from the top row by a thick pyrex plate and the lower row is modulated at 4 Hz. One pyrometer detects only 2.95 μm from the lower bank of lamps. The other pyrometer senses 2.95-μm thermal radiation emitted by the wafer and 2.95 μm reflected off the wafer, originating from the modulated lower

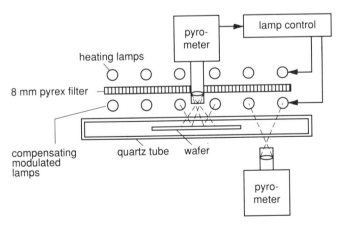

Figure 13.10 A modulation method for emissivity determination for RTP, similar to that in Figure 13.9 but without optical fibers or light pipes in the chamber (from Roozeboom, 1993a, based on work by Vandenabeele and Maex, 1992a,b).

bank of lamps. This method permits measurements down to 250°C without the use of fibers.

13.3 Dual-Wavelength Pyrometry

In dual-wavelength or two-color pyrometry, also called radiation ratio thermometry, measurements are performed at two different wavelengths λ_1 and λ_2 (Tenney, 1988). The ratio of the thermal radiation collected at these two wavelengths is given, in the Wien approximation of the Planck radiation law (Equation 13.2), by

$$\mathcal{R} = \frac{I(\lambda_1,T)}{I(\lambda_2,T)} = \frac{\hat{\varepsilon}(\lambda_1,T)}{\hat{\varepsilon}(\lambda_2,T)} \left(\frac{\lambda_2}{\lambda_1}\right)^5 \exp\left(-\frac{c_2}{T}\left(\frac{1}{\lambda_1} - \frac{1}{\lambda_2}\right)\right) \quad (13.9a)$$

The same ratio is measured for a blackbody at temperature T_r:

$$\mathcal{R} = \frac{I(\lambda_1,T_r)}{I(\lambda_2,T_r)} = \left(\frac{\lambda_2}{\lambda_1}\right)^5 \exp\left(-\frac{c_2}{T_r}\left(\frac{1}{\lambda_1} - \frac{1}{\lambda_2}\right)\right) \quad (13.9b)$$

The ratio temperature T_r is an analog of the spectral temperature T_λ given in Equation 13.6a. Equating Equations 13.9a and 13.9b gives

$$\frac{1}{T} = \frac{1}{T_r} + \frac{1}{c_2(\lambda_1^{-1} - \lambda_2^{-1})} \ln\left(\frac{\hat{\varepsilon}(\lambda_1,T)}{\hat{\varepsilon}(\lambda_2,T)}\right) \quad (13.10)$$

This means that the ratio temperature is the true temperature if the emissivity is the same at both wavelengths. This is automatically true for graybodies at all T. [Note that Si is approximately a graybody only for $T > 600°C$ (Figure 13.3b).] A weaker condition, but one that has a greater chance to approximate realistic conditions, is that $\hat{\varepsilon}(\lambda_1)/\hat{\varepsilon}(\lambda_2)$ is independent of temperature. Then by using the temperature-independent correction of this equation, T is always obtainable from T_r. For dual-wavelength pyrometry to be useful, $\hat{\varepsilon}(\lambda_1)/\hat{\varepsilon}(\lambda_2)$ should not change when the emissivity changes because of deposition, oxidation, etc. Because both absorption in the wafer and interference effects in films strongly depend on λ and T, this ratio may change during processing.

Nordine (1986) has compared the accuracy of one-, two-, and three-color pyrometry, and found that emissivity corrections for multiple-wavelength pyrometry exhibit a larger variation with temperature than those for single-wavelength pyrometry. Mordo *et al.* (1991a,b) and Gat and French (1992) have demonstrated improved temperature control and repeatability during RTP by using a modified version of two-wavelength pyrometry with pyrometry from the back side of the wafer. They established an empirical correlation curve between the emissivity (at one wavelength) with the apparent difference in the temperatures determined at the two wavelengths.

For calibration, the emissivity was measured at the two wavelengths [3.4 ± 0.1 μm and 4.77 ± 0.1 μm (4.3–4.7 μm in earlier work)] as a function of the oxide thickness on the wafer back side; an oscillatory dependence was seen at both wavelengths.

Dual-wavelength pyrometers require less calibration than single-wavelength versions. They are commercially available, but exhibit less flexibility than single-wavelength versions (Peyton et al., 1990).

13.4 Pyrometric Interferometry

Although the changes in emissivity due to interference and absorption in films overlying the substrate complicate temperature measurements, they can be used to measure the thickness and the absorption properties of these films. Thermal radiation is used to probe these films optically, much as a laser (or other optical source) is used in reflection interferometry (Section 9.5) and optical emission (OE) from the plasma is used for OE interferometry (Section 6.3.1.3).

The emissivity of a system consisting of a film (medium 2) of thickness d growing on an opaque substrate (3) in an ambient (1) is (Grothe and Boebel, 1993)

$$\hat{\varepsilon} = \frac{(1 - R_{12})(1 - R_{23}e^{-2\alpha d}) - 4\sqrt{R_{12}R_{23}}\,e^{-\alpha d}\sin(4\pi n_2 d/\lambda + \delta_{23})\sin\delta_{12}}{1 + R_{12}R_{23}e^{-2\alpha d} + 2\sqrt{R_{12}R_{23}}\,e^{-\alpha d}\cos(4\pi n_2 d/\lambda + (\delta_{23} + \delta_{12}))} \quad (13.11)$$

where

$$r_{ij} = \sqrt{R_{ij}}\exp(i\delta_{ij}) \quad (13.12)$$

Here R_{ij} is the reflectance and r_{ij} is the reflection coefficient at the ij interface, and $\alpha = 4\pi k_2/\lambda$ is the (intensity) absorption coefficient of the film. (See Section 9.1.) The emissivity oscillates with optical path length much like the reflectance, which is expected because they are related by Kirchhoff's law (Equation 13.5). For the assumed opaque substrate, $\hat{\varepsilon} = 1 - R_{123}$, where the reflectance R_{123} is given by Equation 9.11.

Very early work in using the interference effects of thermal radiation to determine process parameters was done by Dumin (1967), who monitored the growth of polycrystalline Si on sapphire. Using the same method, Kamins and Dell'Oca (1972) monitored the thickness of polycrystalline Si on oxidized silicon. More recently, SpringThorpe and Majeed (1990) reemphasized the potential application of these methods, by monitoring the oscillations in thermal radiation at 0.94 μm during the MBE growth of GaAs, $Ga_{1-x}Al_xAs$, and AlAs layers. Growth rates and alloy concentrations were determined from the oscillation periods. They also used this method in conjunction with reflection interferometry using a He–Ne laser (632.8 nm). Snail and Marks (1992) used pyrometric interferometry

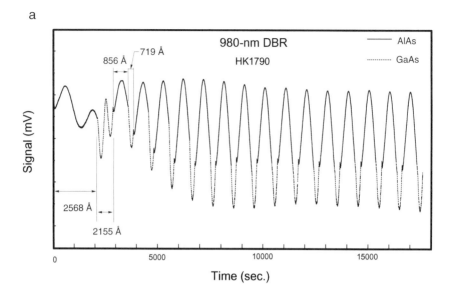

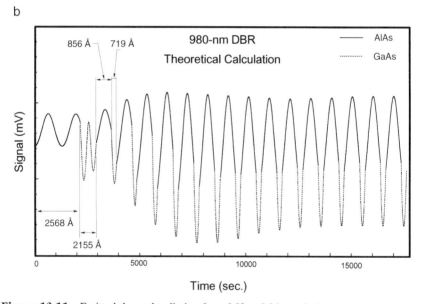

Figure 13.11 Emitted thermal radiation from 0.92 to 0.96 μm during the growth of a λ/4 AlAs/GaAs 980-nm distributed Bragg reflector (DBR) mirror stack by MBE: (a) Experimental data; (b) theoretical calculation (from Houng et al., 1994).

13.4 Pyrometric Interferometry

to measure the growth rate and temperature of diamond films as they were deposited in an oxygen–acetylene flame.

Boebel and Möller (1993) showed that film thickness and temperature can be monitored simultaneously by measuring interference in thermally emitted radiation at multiple wavelengths. High-resolution measurements are possible, to 0.1 nm for thickness and 0.025 K for temperature, with real-time evaluation. They used this method to monitor the thermal oxidation of silicon at 751.2 and 952.4 nm. Grothe and Boebel (1993) used pyrometric interferometry (920 nm) to control the optical thickness of GaAs and AlGaAs films during deposition by MBE. In particular, they monitored the growth of alternating $\lambda/4$ layers of $Al_{0.2}Ga_{0.8}As$ and GaAs, which is useful for growing Bragg reflectors for vertical-cavity surface emitting lasers. A temperature stability of 1 K is required to control the thickness to 2 nm. Phase-shift effects due to absorption are sometimes important. The Al content in the AlGaAs can also be determined from the interference pattern and known (or experimentally measured) refractive indices. The same group also found that single-wavelength monitoring at 950 nm gave good

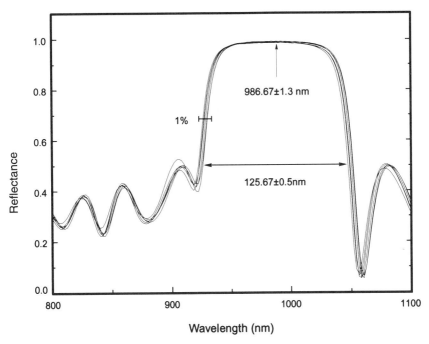

Figure 13.12 Reflectance spectra of six 980-nm DBR wafers grown by MBE with *in situ* feedback control by using the thermal radiation monitored in Figure 13.11. The center wavelength and FWHM of the DBRs are reproducible to ±0.2% with this feedback control (from Houng *et al.*, 1994).

control of thickness and temperature, with <100 msec evaluation time (Böbel *et al.*, 1994a). In related work, Houng *et al.* (1994) monitored the emitted thermal radiation from 0.92 to 0.96 μm for *in situ* real-time control of the MBE growth of AlAs/GaAs (980 nm) and AlAs/AlGaAs (780 nm) distributed Bragg reflectors. Figure 13.11 compares the experimental and calculated (oscillatory) thermal radiation emitted in this band during the growth of a 980-nm distributed Bragg reflector. Figure 13.12 demonstrates that the center wavelength and the FWHM (full width at half maximum) of the AlAs/GaAs stack are very reproducible (to ±0.2%) when the thermal radiation is used to provide process feedback.

Böbel *et al.* (1994b,c) have developed an instrument that simultaneously measures temperature and thickness by pyrometric interferometry at 950 nm and reflection interferometry using (chopped) LEDs (light emitting diodes) at 630 and 950 nm (Figure 13.13). In this reflection-supported pyrometric interferometry, which they called PYRITE, the reflectance determined by using the LED at 950 nm is utilized in the pyrometry measurements at that same wavelength (Equation 13.5, with $T_{\text{trans}} = 0$). Reflection measurements at 630 nm are needed only for thickness determination.

Infrared pyrometric interferometry can sometimes be used to monitor homoepitaxial growth. Visible-based optical methods are often insensitive

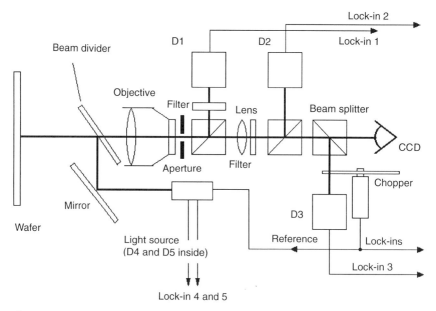

Figure 13.13 Experimental apparatus for reflection-supported pyrometric interferometry that simultaneously measures temperature and thickness by pyrometric interferometry at 950 nm and reflection interferometry using (chopped) LEDs at 630 and 950 nm (from Böbel *et al.*, 1994b).

13.4 Pyrometric Interferometry

to homoepitaxial growth because the dielectric functions of the film and substrate are the same in the visible. However, the refractive indices of the epilayer and substrate are very different in the infrared if they are doped differently owing to free-carrier effects. Infrared light will then interfere in the film, which can be used to monitor thickness. Use of infrared interference to measure the thickness of homoepilayers *ex situ*, using an external IR source, is a standard technique (Cox and Stalder, 1973; Cox, 1974; Pearce, 1988). Yu *et al.* (1992) and Zhou *et al.* (1994a,b) demonstrated how to measure the thickness of Si epitaxial films on Si wafers by monitoring thermal radiation in an ultrahigh-vacuum (UHV) single-wafer CVD reactor or a furnace. Interferograms were made with a scanning Michelson interferometer, so this is a type of Fourier-transform infrared spectroscopy (FTIRS,

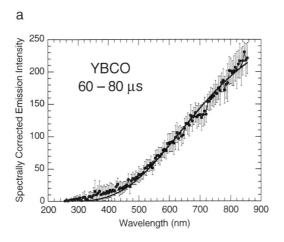

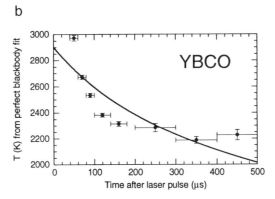

Figure 13.14 (a) Emission spectrum from particles 60–80 μsec after KrF-laser ablation of $YBa_2Cu_3O_{7-x}$ in vacuum, which is attributed to thermal radiation. This is compared to a blackbody spectrum, to give a temperature to 2670 K. (b) Particle temperatures as a function of time after the laser pulse, using the blackbody radiation analysis of (a) (from Geohegan, 1993).

Section 5.1.3.2). The accuracy of *in situ* measurement was 2%, compared to 0.3% for *ex situ* measurements. For lightly doped epilayers (0.5 Ω cm) the low-temperature limit for analysis was determined to be 150°C, while the upper limit was 900°C for more heavily doped substrates (n-type, 0.001–0.002 Ω cm) and 650°C for lightly doped substrates (n-type, 0.01–0.02 Ω cm). Zhou *et al.* (1994b) used emission FTIR to monitor growth rates and incubation times in real time during Si CVD.

13.5 Thermal Radiation during Pulsed-Laser Deposition

Long-lived, broad-band radiation has been observed in some studies of the plume in pulsed laser deposition (PLD), which has been attributed to thermal radiation from particulates in the plume. Rohlfing (1988) observed a long-lived continuum in PLD of graphite in helium when low laser fluences were used and also at higher fluences after atomic and molecular emission had ceased. This was attributed to thermal emission from carbon particles because the emission spectra fit the Planck blackbody distribution function, with $T = 2500–4000$ K, depending on conditions and measurement time. Estimates of particle size were also obtained by using the dependence of emissivity on particle diameter. Geohegan (1993) used ICCD [intensified charge-coupled device] camera, 2D-gated imaging to follow the emission from the plume of laser-ablated BN and $YBa_2Cu_3O_{7-x}$. All emission from 10 μsec to 1.5 msec after the laser pulse matched the blackbody radiation profile (Figure 13.14a). Particle temperatures were observed to decrease from ~3000 K at the first measurement to ~2200 K about 1500 μsec (450 μsec) after the pulse for BN ($YBa_2Cu_3O_{7-x}$) (Figure 13.14b).

References

Accufiber product information. Luxtron Corp., 1995.
M. E. Adel, S. Mangan, and Y. Ish-Shalom, *SPIE* **2091,** 311 (1994).
R. L. Anderson, *SPIE* **1392,** 437 (1990).
R. P. Benedict, *Fundamentals of Temperature, Pressure, and Flow Measurements.* Wiley, New York, 1984.
F. G. Boebel and H. Möller, *IEEE Trans. Semicond. Manuf.* **SM-6,** 112 (1993).
F. G. Böbel, H. Möller, A. Wowchak, B. Hertl, J. Van Hove, L. A. Chow, and P. P. Chow, *J. Vac. Sci. Technol. B* **12,** 1207 (1994a).
F. G. Böbel, H. Möller, B. Hertel, G. Ritter, and P. Chow, *Solid State Technol.* **37**(8), 55 (1994b).
F. G. Böbel, A. Wowchak, P. P. Chow, J. Van Hove, and L. A. Chow, *Mater. Res. Soc. Symp. Proc.* **324,** 105 (1994c).
P. F. Cox, *J. Electrochem. Soc.* **121,** 1233 (1974).
P. F. Cox and A. F. Stalder, *J. Electrochem. Soc.* **120,** 287 (1973).

References

M. Delfino and D. T. Hodul, *IEEE Trans. Electron Devices* **ED-39,** 89 (1992).
D. P. DeWitt and F. P. Incropera, in *Theory and Practice of Radiation Thermometry* (D. P. DeWitt and G. D. Nutter, eds.), Chapter 1, p. 21. Wiley, New York, 1988.
D. P. DeWitt and G. D. Nutter, eds., *Theory and Practice of Radiation Thermometry.* Wiley, New York, 1988.
J.-M. Dilhac, C. Ganibal, and N. Nolhier, *Mater. Res. Soc. Symp. Proc.* **224,** 3 (1991).
R. R. Dils, J. Geist, and M. L. Reilly, *J. Appl. Phys.* **59,** 1005 (1986).
J. O. Dimmock, *J. Electron. Mater.* **1,** 255 (1972).
D. J. Dumin, *Rev. Sci. Instrum.* **38,** 1107 (1967).
R. Eisberg and R. Resnick, *Quantum Physics of Atoms, Molecules, Solids, Nuclei, and Particles,* 2nd ed., p. 2. Wiley, New York, 1985.
A. T. Fiory, C. Schietinger, B. Adams, and F. G. Tinsley, *Mater. Res. Soc. Symp. Proc.* **303,** 139 (1993).
J. E. Franke, T. M. Niemczyk, and D. M. Haaland, *Spectrochim. Acta Part A* **50A,** 1687 (1994).
K. Franson, S. Katz, and E. Raisen, in *Temperature: Its Measurement and Control in Science and Industry* (F. G. Brickwedde, ed.), Vol. 3, Part 1, p. 529. Reinhold, New York, 1962.
E. Fujita, K. Furusawa, H. Kataoka, N. Tsumita, T. Yonekawa, and N. Shige, *J. Vac. Sci. Technol. A* **10,** 1657 (1992).
A. Gat and M. French, U.S. Patent 5,165,796 (1992).
D. B. Geohegan, *Appl. Phys. Lett.* **62,** 1463 (1993).
H. Grothe and F. G. Boebel, *J. Cryst. Growth* **127,** 1010 (1993).
G. P. Hansen, S. Krishnan, R. H. Hauge, and J. L. Margrave, *Metall. Trans. A* **19A,** 1889 (1988).
G. P. Hansen, S. Krishnan, R. H. Hauge, and J. L. Margrave, *Appl. Opt.* **28,** 1885 (1989).
Y. M. Houng, M. R. T. Tan, B. W. Liang, S. Y. Wang, and D. E. Mars, *J. Vac. Sci. Technol. B* **12,** 1221 (1994).
T. I. Kamins and C. J. Dell'Oca, *J. Electrochem. Soc.* **119,** 112 (1972).
D. S. Katzer and B. V. Shanabrook, *J. Vac. Sci. Technol. B* **11,** 1003 (1993).
K. G. Kreider, in *Applications of Radiation Thermometry* (J. C. Richmond and D. P. DeWitt, eds.), p. 151. ASTM, Philadelphia, 1985.
B. Linnander, *IEEE Circuits Devices* **9**(4), 35 (1993).
T. Loarer, J.-J. Greffet, and M. Huetz-Aubert, *Appl. Opt.* **29,** 979 (1990).
D. Mazzarese, K. A. Jones, and W. C. Conner, *J. Electron. Mater.* **21,** 329 (1992).
T. D. McGee, *Principles and Methods of Temperature Measurement.* Wiley, New York, 1988.
J. A. McGuire, B. Wangmaneerat, T. M. Niemczyk, and D. M. Haaland, *Appl. Spectrosc.* **46,** 178 (1992).
D. Mordo, Y. Wasserman, and A. Gat, *SPIE* **1595,** 52 (1991a).
D. Mordo, Y. Wasserman, and A. Gat, *Semicond. Int.,* **14**(11), 86 (October 1991b).
M. M. Moslehi, R. A. Chapman, M. Wong, A. Paranjpe, H. N. Najm, J. Kuehne, R. L. Yeakley, and C. J. Davis, *IEEE Trans. Electron Devices* **ED-39,** 4 (1992).
T. P. Murray, *Rev. Sci. Instrum.* **38,** 791 (1967).
T. M. Niemczyk, J. E. Franke, B. Wangmaneerat, C. S., Chen, and D. M. Haaland, in *Computer-Enhanced Analytical Spectroscopy* (C. L. Wilkins, ed.), Vol. 4, p. 165. Plenum, New York, 1993.

T. M. Niemczyk, J. E. Franke, S. Zhang, and D. M. Haaland, *Mater. Res. Soc. Symp. Proc.* **341,** 119 (1994).
P. C. Nordine, *High Temp. Sci.* **21,** 97 (1986).
H. N. Norton, *Handbook of Transducers.* Prentice-Hall, Englewood Cliffs, NJ, 1989.
M. Nouaoura, L. Lassabatere, N. Bertru, J. Bonnet, and A. Ismail, *J. Vac. Sci. Technol. B* **13,** 83 (1995).
J. Nulman, S. Antonio, and W. Blonigan, *Appl. Phys. Lett.* **56,** 2513 (1990).
G. D. Nutter, in *Applications of Radiation Thermometry* (J. C. Richmond and D. P. DeWitt, eds.) *ASTM Spec. Tech. Publ.* **STP 895,** p. 3. ASTM, Philadelphia, 1985.
M. C. Öztürk, F. Y. Sorrell, J. J. Wortman, F. S. Johnson, and D. T. Grider, *IEEE Trans. Semicond. Manuf.* **SM-4,** 155 (1991).
V. Patel, M. Patel, S. Ayyagari, W. F. Kosonocky, D. Misra, and B. Singh, *Appl. Phys. Lett.* **59,** 1299 (1991).
C. W. Pearce, in *VLSI Technology* (M. Sze, ed.), Chapter 2, p. 51. McGraw-Hill, New York, 1988.
D. W. Pettibone, J. R. Suarez, and A. Gat, *Mater. Res. Soc. Symp. Proc.* **52,** 209 (1986).
D. Peyton, H. Kinoshita, G. Q. Lo, and D. L. Kwong, *SPIE* **1393,** 295 (1990).
V. Rio and G. Moore, *Semicond. Int.* **18**(6), 131 (1995).
E. A. Rohlfing, *J. Chem. Phys.* **89,** 6103 (1988).
F. Roozeboom, *Mater. Res. Soc. Symp. Proc.* **224,** 9 (1991).
F. Roozeboom, *Mater. Res. Soc. Symp. Proc.* **303,** 149 (1993a).
F. Roozeboom, in *Rapid Thermal Processing Science and Technology* (R. B. Fair, ed.), Chapter 9, p. 349. Academic Press, Boston, 1993b.
F. Roozeboom and N. Parekh, *J. Vac. Sci. Technol. B* **8,** 1249 (1990).
T. Sato, *Jpn. J. Appl. Phys.* **6,** 339 (1967).
C. W. Schietinger and B. E. Adams, U.S. Patent 5,154,512 (1992a).
C. W. Schietinger and B. E. Adams, U.S. Patent 5,166,080 (1992b).
C. Schietinger, B. Adams, and C. Yarling, *Mater. Res. Soc. Symp. Proc.* **224,** 23 (1991).
J. F. Schooley, *Thermometry.* CRC, Boca Raton, FL, 1986.
T. O. Sedgwick, *Appl. Phys. Lett.* **39,** 254 (1981).
B. V. Shanabrook, J. R. Waterman, J. L. Davis, R. J. Wagner, and D. S. Katzer, *J. Vac. Sci. Technol. B* **11,** 994 (1993).
K. A. Snail and C. M. Marks, *Appl. Phys. Lett.* **60,** 3135 (1992).
A. J. SpringThorpe and A. Majeed, *J. Vac. Sci. Technol. B* **8,** 266 (1990).
A. S. Tenney, in *Theory and Practice of Radiation Thermometry* (D. P. DeWitt and G. D. Nutter, eds.), Chapter 6, p. 459. Wiley, New York, 1988.
P. J. Timans, *J. Appl. Phys.* **72,** 660 (1992).
P. J. Timans, *J. Appl. Phys.* **74,** 6353 (1993).
P. Vandenabeele and K. Maex, *J. Appl. Phys.* **72,** 5867 (1992a).
P. M. N. Vandenabeele and K. I. J. Maex, European Patent application, filed June, 1992b.
P. Vandenabeele, R. J. Schreutelkamp, K. Maex, C. Vermeiren, and W. Coppeye, *Mater. Res. Soc. Symp. Proc.* **260,** 653 (1992).
R. Vanzetti and A. C. Traub, in *Theory and Practice of Radiation Thermometry* (D. P. DeWitt and G. D. Nutter, eds.), Chapter 11, p. 625. Wiley, New York, 1988.

References

B. Wangmaneerat, J. A. McGuire, T. M. Niemczyk, D. M. Haaland, and J. H. Linn, *Appl. Spectrosc.* **46,** 340 (1992a).

B. Wangmaneerat, T. M. Niemczyk, and D. M. Haaland, *Appl. Spectrosc.* **46,** 1447 (1992b).

S. Wood, P. Apte, T.-J. King, M. Moslehi, and K. Saraswat, *SPIE* **1393,** 337 (1990).

F. Yu, Z.-H. Zhou, P. Stout, and R. Reif, *IEEE Trans. Semicond. Manuf.* **SM-5,** 34 (1992).

Z.-H. Zhou, S. Compton, I. Yang, and R. Reif, *IEEE Trans. Semicond. Manuf.* **SM-7,** 87 (1994a).

Z. H. Zhou, I. Yang, H. Kim, F. Yu, and R. Reif, *J. Vac. Sci. Technol. A* **12,** 1938 (1994b).

CHAPTER **14**

Photoluminescence

Photoluminescence (PL) is the solid-state analog of laser-induced fluorescence (LIF), which is discussed in Chapter 7. After the absorption of a photon induces a transition to an excited electronic state, the material relaxes radiatively (spontaneous emission), nonradiatively (thermal relaxation), or through both routes (Figure 14.1). This radiative decay is called photoluminescence. In some solid-state, and also gas-phase, systems, optical emission (OE) is further characterized as either fluorescence or phosphorescence. Fluorescence is the rapid emission that occurs when the transition between the higher and lower states is allowed. Phosphorescence is the relatively slow emission that occurs when the transition between the two states is "forbidden," as is that between spin triplet and singlet states.

Figure 14.1 shows that photoluminescence in solids can be the result of transitions between electronic states that are localized to dopants, impurities, etc., or transitions between (delocalized) band states. In phosphors (or fluorophors), which are of the former type of material, atomic ions, such as rare-earth ions, dope insulating hosts (Figure 14.1a). The electronic transitions due to these dopants are localized at the ion, but are still influenced by the host crystal via crystal field splittings, strain, etc. Also, the luminescence decay rate and efficiency of the material depend in part on the rate of energy transfer from the ion to the host crystal (γ_{nonrad}), which is a function of the material and of temperature. The emitted light can have a sharp or broad spectrum, depending on these dopant–host interactions.

In semiconductors (Figure 14.1b), absorption induces an electronic transition between the valence band and the conduction band, leaving a free electron in the conduction band and a hole (electron vacancy) in the valence band. The electron and hole rapidly thermally relax, the electron to the bottom of the conduction band and the hole to the top of the valence band. In direct semiconductors, such as GaAs and InP, the bottom of the conduction band and the top of the valence band correspond to electron states with the same wavevector (**k**). The electron and hole can recombine

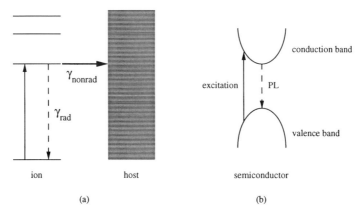

Figure 14.1 Schematic of the energy levels involved in the photoluminescence in solids due to transitions between (a) relatively sharp electronic states, perhaps due to dopants or defects, and (b) electronic bands.

radiatively—i.e., the electron relaxes from the conduction band to the valence band—and the material luminesces. This is known as band-to-band emission. In indirect semiconductors, such as Si and AlAs, the bottom of the conduction band and the top of the valence band correspond to carriers with different wavevectors. Electron/hole recombination must be accompanied by the absorption or emission of a phonon, and photoluminescence is usually very weak, even at cryogenic temperatures (Pankove, 1971). Consequently, PL can be a useful diagnostic of film processing only for direct gap semiconductors.

At low laser intensities, band-to-band (or band-edge) photoluminescence in direct semiconductors occurs very near the band gap energy $\mathcal{E}_{bg}$. In the parabolic-band approximation the spectral profile of luminescence is given by (Bebb and Williams, 1972)

$$\mathcal{P}(\hbar\omega) = N_n N_p C(\hbar\omega, T) \frac{2\pi}{(\pi k_B T)^{3/2}} \times \left\{ \frac{f_h^{3/2} + f_l^{3/2}}{m_h^{3/2} + m_l^{3/2}} (\hbar\omega - \mathcal{E}_{bg})^{1/2} \right.$$
$$\times \exp[-(\hbar\omega - \mathcal{E}_{bg})/k_B T] + \exp(-\Delta/k_B T) \frac{f_s^{3/2}}{m_h^{3/2} + m_l^{3/2}} \quad (14.1)$$
$$\left. \times (\hbar\omega - \mathcal{E}_{bg} - \Delta)^{1/2} \exp[-(\hbar\omega - \mathcal{E}_{bg} - \Delta)/k_B T] \right\}$$

where

$$C(\hbar\omega, T) = \frac{2e^2}{\hbar^2 m^2 c^3} \left(\frac{2\pi\hbar^2}{k_B T}\right)^{3/2} |p_{cv}|^2 n(\hbar\omega)\hbar\omega \quad (14.2)$$

$f_i = m_i/(m_c + m_i)$, N_n is concentration of electrons, N_p is the concentration of holes (light plus heavy), Δ is the spin–orbit splitting energy, p_{cv} are the transition matrix elements, $n(\hbar\omega)$ is the refractive index, and m_c, m_h, m_l, and m_s are the respective effective masses of electrons, heavy holes, light holes, and holes in the splitoff band. For large spin–orbit splitting, the second term in the bracket in Equation 14.1 can be neglected.

The band gap energy varies with material composition, as in ternary (and quaternary) semiconductors such as AlGaAs, so PL can be used to characterize composition. $\mathcal{E}_{bg}$ also varies with temperature, which can also be determined by PL (Section 14.3). At very high laser intensities, the PL spectrum can change, as is described below.

Electrons and holes in semiconductors can also bind to form hydrogen-atom-like states called excitons, which can be free and roam through the crystal or be bound to an impurity. Excitonic effects are important only at temperatures where the exciton binding energies are $>k_B T$, which means cryogenic temperatures in bulk crystals and sometimes up to room temperature in semiconductor heterostructures. Photoluminescence due to the recombination of free excitons occurs at an energy that is slightly lower than band-to-band PL, by virtue of the free exciton binding energy. Similarly, PL from bound excitons occurs at even lower energy because of the additional binding to the site, which is characteristic of the binding site. Spectral analysis of excitonic PL is useful in identifying the impurities that are present and in determining their concentrations. Excitonic PL is also sensitive to strain and confinement in films and heterostructures, and sometimes to the sharpness of interfaces. At low temperatures ($< \sim 50$ K), PL from excitons dominates over band-to-band emission in semiconductors and thus is a common and powerful tool for *ex situ* material characterization. It has been reviewed by Bebb and Williams (1972), Williams and Bebb (1972), Dean (1982), Schroder (1990), and Perkowitz (1993).

In situ optical probes of film processing are usually conducted at and above room temperature, where band-to-band PL usually dominates; consequently, much of the structural information obtainable from excitonic PL is unavailable. The presence of some types of defects near the surface can still affect the band-to-band PL intensity and decay rate (Section 14.2), although they have little effect on the PL spectrum. Such measurements are particularly significant in light-emitting materials, engineered for excitation by injection currents, because such defects also affect electroluminescence. The spatial profile of these impurities can be determined by varying the excitation depth of PL $\sim 1/\alpha(\lambda)$, which can be changed by tuning the laser wavelength. Band-to-band PL also has applications in thermometry (Section 14.3).

In principle, PL could also be used to monitor (undoped) insulators. However, laser wavelengths shorter than commonly available are needed to excite electrons in these very-large-band-gap materials. (The fluorophors and phosphors described in Section 14.3 are doped insulators.)

14.1 Experimental Considerations

Three spectroscopic features of PL can be used in diagnostics:

(1) Monitoring the wavelength of the PL features is sometimes useful. For band-to-band PL in semiconductors, the band gap energy, and therefore temperature, can be determined *in situ* during thin film processing by monitoring the peak of the PL spectrum (Section 14.3).

(2) The PL efficiency η_{PL} varies as

$$\eta_{PL} = \frac{\gamma_{rad}}{\gamma_{rad} + \gamma_{nonrad}} \qquad (14.3)$$

where γ_{rad} is the rate of radiative decay (electron/hole recombination in a semiconductor) and γ_{nonrad} is the nonradiative relaxation rate. Changes in γ_{nonrad} monitored by the PL efficiency can indicate changes in defect density or changes in temperature. Equation 14.3 applies both to PL during steady-state excitation and to time-integrated PL for pulsed excitation.

(3) The rate of decay of PL is γ_{PL}:

$$\gamma_{PL} = \gamma_{rad} + \gamma_{nonrad} \qquad (14.4)$$

Again, this rate is sensitive to defects and temperature through γ_{nonrad}. This rate can be monitored by using pulsed excitation and measuring the decay rate directly, or by using chopped (modulated) continuous-wave (cw) excitation and measuring the phase shift of the PL signal relative to that of the excitation light source.

Photoluminescence requires optical excitation within an absorption band. For semiconductor excitation, any fixed-line laser with photon energy above the band gap is sufficient. In selected applications, use of photon energies much above band gap may lead to undesired hot carrier effects or excessive thermal excitation (even with efficient PL). Often, it does not matter whether cw or pulsed optical sources are used, because photoluminescence is a linear spectroscopy; thus cw argon-ion lasers are commonly used. However, pulsed lasers or amplitude-modulated cw sources are needed in the presence of background light, such as the background glow during plasma processing, so the background can be removed by gated detection. While strong, sharp spectral features can often be resolved from plasma-induced emission (PIE) by dispersing the collected light in a spectrometer, the band-to-band PL profile is too broad to do this at processing temperatures; so cw excitation is not feasible. Pulsed and modulated sources are also needed to measure PL lifetimes (Section 14.3).

Photoluminescence is collected with a lens and is usually spectrally resolved by a spectrometer or transmitted by a band-pass filter before detection by a photomultiplier, semiconductor photovoltaic or photoconductive

14.2 Probing Defects and Damage

detector, or a multichannel detector. Figure 14.2 shows a typical apparatus for *in situ* PL. Photoluminescence probes $\sim 1/\alpha$ into the sample, where the absorption coefficient α depends on the material type, the probing wavelength, and T. Sometimes reabsorption of the emitted light can be significant. Excitation PL, the analog of excitation LIF where the excitation wavelength is tuned and nondispersed emission is detected, is sometimes used for spectroscopic analysis, particularly at cryogenic temperatures, but has rarely been used in diagnostics.

14.2 Probing Defects and Damage

The PL efficiency of a semiconductor is given by Equation 14.3, where γ_{rad} is the rate of radiative decay (electron/hole recombination) and γ_{nonrad} is the nonradiative relaxation rate. Nonradiative decay can be accelerated by defects in the bulk and on the surface. When absorption occurs very near the surface, as for photon energies that are much above the band gap in direct gap semiconductors, electrons and holes can rapidly diffuse to the surface and recombine nonradiatively. Photoluminescence has been very useful in assessing these defects in GaAs and other III–V semiconductors. Based on the body of earlier work, Yoon *et al.* (1992a,b) concluded that higher PL yield in GaAs can be correlated with increased Schottky barrier heights, reduced interface-state density, and reduced surface-recombination velocity. Elemental As, produced by the reaction of GaAs with the As_2O_3 native oxide, produces a mid-gap state that can pin the Fermi level and enhance γ_{nonrad}. Also, reactive ion etching (RIE) of semiconductors pro-

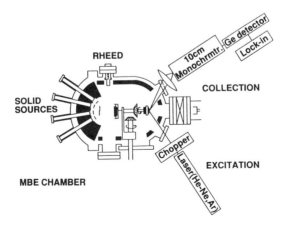

Figure 14.2 Apparatus for *in situ* PL measurements in an MBE chamber (from Sandroff *et al.*, 1991b). In this arrangement measurements were made with the wafer in the sample transfer position; however, with appropriately placed viewports, PL can be performed with the wafer in the normal growth position.

duces damage near the surface that can degrade device performance. Room-temperature PL has been employed *ex situ* in numerous studies to evaluate the RIE-induced damage in III–V semiconductors and to assess the reduction of this damage by surface passivation.

One prominent use of PL has been the *in situ* assessment of "photowashing" of GaAs by Offsey *et al.* (1986). Cleaned GaAs wafers were illuminated with visible light during washing with water and then dried. Photoluminescence was performed before and after each sequence, as illustrated in Figure 14.3. The increase of the PL intensity after photowashing has been attributed to the reduction of the surface-state density; this state reduction also unpinned the Fermi level. Figure 14.4 shows how band-edge PL varies after the unpinning treatment. Similar PL observations have been made and similar mechanistic conclusions have been drawn when the GaAs surface is coated with $Na_2S \cdot 9H_2O$ (Wilmsen *et al.*, 1988). Other studies utilizing real-time monitoring of GaAs by PL include those by Booyens *et al.* (1983), Haegel and Winnacker (1987), Guidotti *et al.* (1987), and Raja *et al.* (1988).

Yoon *et al.* (1992a,b) evaluated the reduction of such damage in GaAs caused by RIE in BCl_3/Cl_2 mixtures by monitoring PL *in situ* during attempted passivation in a H_2 plasma. Etching and passivation were performed in the same plasma reactor. This followed earlier work by this group on real-time monitoring during plasma processing by Mitchell *et al.* (1990) and Gottscho *et al.* (1990). In this study, a pulsed dye laser (500 nm) was used to excite PL, and gated detection was used to acquire the PL signal, thereby avoiding the effects of the background plasma glow. Moreover, this gated PL technique enabled Yoon *et al.* to use fluences

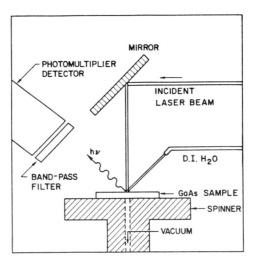

Figure 14.3 Experimental apparatus for unpinning GaAs(100) surfaces, with *in situ* PL analysis (from Offsey *et al.*, 1986).

14.2 Probing Defects and Damage

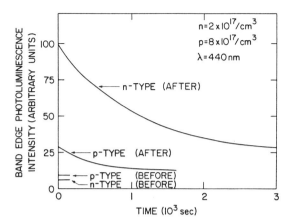

Figure 14.4 Band-edge PL from GaAs versus time either before or after the unpinning treatment, with 440-nm excitation (from Offsey et al., 1986).

low enough to avoid photodegradation of the GaAs surface. Figure 14.5 monitors GaAs PL during the RIE of an AlGaAs epilayer on GaAs. As the layer is etched away, more light is absorbed by the GaAs and the PL intensity increases. When the AlGaAs layer has been removed and etching of GaAs commences at about 4.2 min, PL decreases rapidly owing to RIE-induced damage to the GaAs. The results of successive 2-sec-long hydrogen

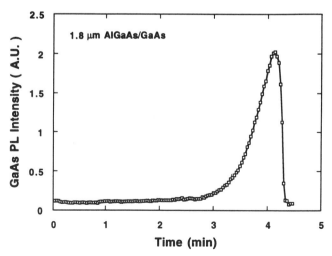

Figure 14.5 Variation in PL intensity from GaAs during RIE (BCl_3/Cl_2) of an $Al_{0.3}Ga_{0.7}As$/GaAs heterostructure, using a pulsed dye laser (500 nm) for PL excitation and synchronized gated integration for PL detection (from Yoon et al., 1992b).

plasma treatments to this damaged surface are shown in Figure 14.6. This treatment yields large PL signals, and presumably less damage; however, the PL signal repeatedly decays in ~20 min, possibly because of contamination in the chamber. Although the passivation was not entirely successful in this particular study, the *in situ* monitoring method was found to be sound.

In subsequent work by this group, Aydil *et al.* (1993a,b; 1995) used both PL and Fourier-transform IR methods to monitor the GaAs surface in real time during passivation with a remote NH_3 plasma. Figure 14.7 shows that PL increases simultaneously with the formation of As—H on the surface, as seen by the increase in absorption near 2100 cm^{-1}, and after As—O has been removed from the surface, as seen by the decrease in absorption near 850 cm^{-1}. Arsenic oxide is removed as fast as H atoms from the discharge can be transported to the surface, while the As—H layer is formed later, perhaps after diffusion of H through the native oxide layer. Passivation, as monitored by PL, apparently requires the formation of the As—H layer. Aydil *et al.* (1995) also examined GaAs passivation by a remote H_2 plasma.

Sandroff *et al.* (1991a) used *in situ* PL to compare recombination at (100) GaAs surfaces and interfaces immediately after the MBE growth of GaAs/AlGaAs single heterostructures and AlGaAs/GaAs/AlGaAs double heterostructures. They also used *in situ* PL to compare recombination at GaAs surfaces for different reconstructions and for the 2 × 4 surface modified by the deposition of thin layers of Se, Si, Zn and Te.

Room-temperature PL is an important tool for quality assessment over entire wafers of direct gap semiconductors, which are sometimes overlaid by thin films (Hovel and Guidotti, 1985). Such wafer-scale PL probing has

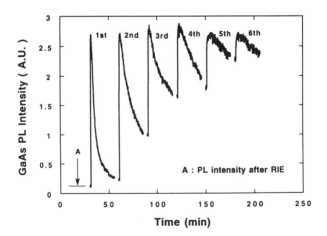

Figure 14.6 GaAs intensity change with repeated 2-sec hydrogen-plasma passivation treatments, after the RIE removal of an $Al_{0.3}Ga_{0.7}As$ overlayer (as in Figure 14.5) and subsequent purging of the reactor with hydrogen for 30 min (from Yoon *et al.*, 1992b).

14.2 Probing Defects and Damage

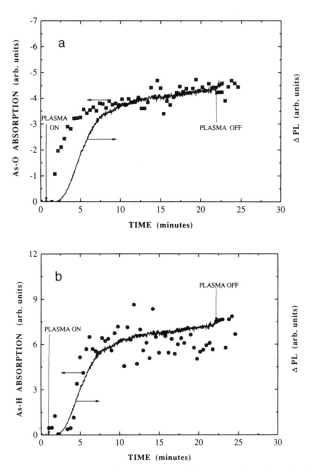

Figure 14.7 Real-time changes in (a) As—O and (b) As—H absorption obtained by ATR spectroscopy (Section 9.7.2) during the passivation of GaAs by a remote NH$_3$ microwave discharge, compared to changes in the PL intensity (from Aydil *et al.*, 1993b).

become a widespread *ex situ* characterization method; it can also be used for real-time, *in situ* applications. Several features of interest can be mapped quantitatively, even at ambient temperature: (1) defect concentrations (from the PL intensity), (2) composition of ternary compounds (from the PL wavelength), and (3) quantum-well thickness and interface sharpness in heterostructures.

Most PL mapping systems use a laser focused to 1 μm–2 mm and scan the wafer or the laser. These mappings are relatively slow, requiring several minutes to hours, and are not very suitable for *in situ* analysis. Livescu *et al.* (1990) have devised a real-time photoluminescence imaging system that is promising for *in situ* characterization and diagnostics. An argon-ion laser is expanded onto the substrate with a biconcave lens. Photoluminescence is collected with a lens, sent through a bandpass filter, and imaged onto a camera. A vidicon is used for visible light ($\lambda < 0.9$ μm), while a lead sulfide camera is used for infrared light (<2 μm). The camera signal is then processed to map the wafer features in two dimensions. This includes corrections for laser variations across the wafer due to the gaussian laser profile. The PL spectrum is mapped by using bandpass filters centered at different wavelengths. When only a few filters are needed (for defect analysis only a single filter is needed) this method is still very fast. An alternative approach for spectral mapping is to image the laser to a line and then image the PL onto a charge-coupled device (CCD) array. This gives the dispersed spectrum along the line, which is then scanned. While this would still not be a real-time probe, it does not require changing the filter. Carver *et al.* (1995) and Imler (1995) have presented other techniques for rapid PL scanning of semiconductors.

14.3 Thermometry

Changes in temperature affect band gap energies and relaxation rates (γ_{nonrad}). Applications of PL in thermometry involve monitoring both effects.

In thermal equilibrium, the luminescence spectrum of a direct band gap semiconductor is given by Equation 14.1. The maximum of band-to-band PL in a semiconductor occurs at an energy $\mathcal{E}_{bg}(T) + k_B T/2$, which exceeds the band gap because of the thermal distribution of free carriers. The change in the band gap with temperature is characterized by Equations 18.7 and 18.8. Sandroff *et al.* (1991b) used this band gap variation to determine the temperature of GaAs wafers *in situ* in an MBE [molecular beam epitaxy] chamber (Figure 14.2). The PL spectra from 25 to 450°C presented in Figure 14.8 show that both the peak PL energy and PL intensity decrease with increasing temperature. Figure 14.9 suggests that a 10°C change can be measured; an accuracy of ±8°C is reported up to 300°C. Use of this diagnostic during thermal processing is limited because the precision is low owing to the large linewidth of PL ($\sim k_B T$), especially at elevated temperatures.

Kirillov and Merz (1983) studied PL from cw-laser-heated GaAs and InP to determine the temperature from the peak of the PL emission and the known change in band gap $\mathcal{E}_{bg}(T)$. Figure 14.10 shows examples of these luminescence spectra and how the temperature within the laser-heated spot, determined by PL, changes with laser power. This diagnostic was also used in laser-heating studies of AlGaAs/GaAs structures by Salathé *et al.* (1981) and during the investigation of laser-heating-induced etching of GaAs in the presence of $SiCl_4$ by Takai *et al.* (1985). As and

14.3 Thermometry

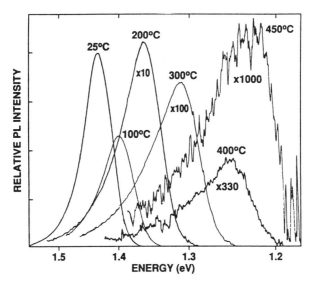

Figure 14.8 Photoluminescence spectra from GaAs in an AlGaAs/GaAs/AlGaAs heterostructure from 25 to 450°C, obtained in the MBE chamber of Figure 14.2 (from Sandroff *et al.*, 1991b).

Palmetshofer (1987) used PL to determine the temperature of CdTe during cw laser heating.

There are limitations to PL, other than large linewidth, when the sample is heated by a laser or is heavily doped. The Burstein–Moss shift (Pankove, 1971) is the increase in the band edge due to band filling, which occurs in heavily doped semiconductors; this can affect the temperature dependence of the PL energy. In particular, at high laser intensities, such as during cw laser heating, the electron/hole distributions can deviate from a thermal profile because there are more free carriers than expected at the "lattice" temperature. The PL then peaks at an energy above $\mathcal{E}_{bg} + k_B T/2$ because of this band filling. Band shrinkage due to the screening of the crystal field by laser-produced carriers is also possible (Casey and Panish, 1978). Kirillov and Merz (1983) concluded that these two electronic effects did not affect their temperature measurements.

The rate of nonradiative decay $\gamma_{nonrad}(T)$ in a phosphor increases with T, while the radiative rate γ_{rad} is insensitive to T. Therefore, the integrated or steady-state PL intensity of a phosphor decreases with increasing temperature, as seen from Equation 14.3. Also, the rate of luminescence decay in a phosphor typically increases with increasing T, as seen in Equation 14.4. Both of these effects have been used as temperature sensors, as has been detailed by Grattan and Zhang (1995). In some systems, the shift of spectral emission of a phosphor can change with temperature and can be used as a sensor.

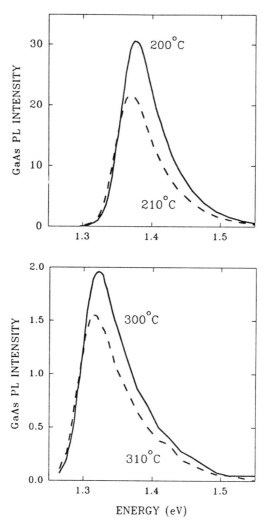

Figure 14.9 *In situ* PL spectra of GaAs (see Figures 14.2 and 14.8) at closely spaced temperatures (from Sandroff *et al.*, 1991b).

In the fluoroptic probe (Luxtron Corp., 1995; Egerton *et al.*, 1982), a small amount of phosphor is bonded to the tip of an optical fiber. Ultraviolet or blue/violet light travels down the fiber to excite the phosphor, and the visible luminescence is transmitted back along the same fiber and is detected. These probes determine temperature by measuring either the intensity or lifetime of the phosphor emission. Figure 14.11 shows the excitation and luminescence spectra of a europium-activated, rare-earth oxysulfide phosphor $(Gd_{0.99}Eu_{0.01})_2O_2S$ at the tip of a fiber, as measured by Egerton *et*

14.3 Thermometry

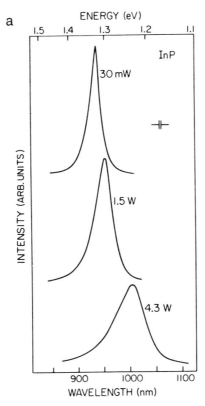

Figure 14.10 (a) Band-to-band luminescence spectra of InP using different values of laser power for simultaneous heating and PL excitation (beam diameter ~ 100 μm); (b) the temperature inside the laser-heated spot, as determined from these PL measurements (from Kirillov and Merz, 1983).

(Continues)

al. (1982), and illustrates the former method. The PL intensity of the observed line, denoted in the figure as c, generally decreases with increasing T, but the ratio of the intensities of lines c and a is a better (monotonic) monitor. Intensity-based systems are generally inferior to lifetime-based systems in terms of performance and cost.

The rate of PL decay can be monitored by using either a pulsed or an amplitude-modulated source. In the latter method, a modulated light source is used to excite the medium, e.g., a mechanically chopped ($f \sim 100$–1000 Hz) light source, such as the filtered output of a Hg arc lamp, or, increasingly, a modulated light-emitting diode (LED) or laser diode, and a lock-in amplifier is used to analyze the detected luminescence signal. The phosphor lifetime $\tau(T) = 1/(\gamma_{nonrad}(T) + \gamma_{rad})$ is determined by the phase lag in emission $\tan^{-1}(2\pi f \tau)$, measured by the lock-in amplifier.

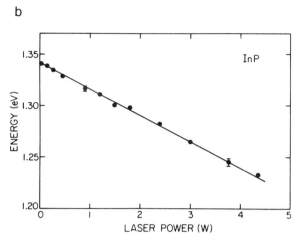

Figure 14.10 (*Continued*)

In the closely related phase-locked detection method, the signal is compared to the light source modulation driver, after it has been phase shifted. These methods filter out the background plasma light and are insensitive to changes in lamp intensity. The phase lag technique was used by McCormack (1981) to probe the temperature of a sensor—barium chlorofluoride activated by divalent samarium ($BaClF:Sm^{2+}$)—remotely with fibers. The 380–500-nm lamp radiation excited this material and 687.7-nm emission of the $^5D_0$–$^7F_0$ transition was detected. τ decreases from 3 msec at $\sim 20°C$ to 30 μsec at 350°C ($f = 150$ Hz).

A commercial fluoroptic probe has employed this phase-shift method to monitor UV-light-induced optical emission from phosphorus-doped $LaO_2S:Eu$ at the fiber tip. This instrument has a suitably fast response time (0.31 sec) and can be used from ~ 100 to 300°C. Hussla *et al.* (1987) have reported *in situ* temperature measurements of silicon wafers during plasma etching using this fluoroptic thermometry technology.

A more current fluoroptic technology (Wickersheim and Sun, 1987; Luxtron Corp., 1995) uses blue/violet light pulses to excite magnesium fluorogermanate activated with tetravalent manganese. The red phosphor emission decays exponentially, and the decay time is determined by multi-point digital integration of the emission decay curve. (Alternative analysis methods involve digital curve fitting or time constant analysis with data at two times.) The uncalibrated accuracy of this probe is $\pm 2°C$, the calibrated accuracy can approach $\pm 0.1°C$ very near the calibration temperature, and the precision is $\pm 0.1°C$. This technology can be used up to 450°C.

The fluoroptic probe can be placed on the wafer and attached with a thermally conducting aluminum-containing epoxy, or it can be put very near the wafer. Although contact with the wafer is needed for accurate

14.3 Thermometry

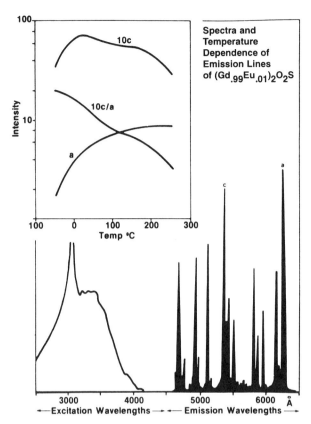

Figure 14.11 The absorption and emission characteristics of the Luxtron $(Gd_{0.99}Eu_{0.01})_2O_2S$ phosphor, along with the relative intensities of the noted a and c peaks (from Egerton *et al.*, 1982).

thermometry, it can lead to contamination during manufacturing. In the noncontact version, errors in temperature measurement can be large due to poor thermal contact through the gas. Still, this version has been used often in process studies of plasma etching. The high temperature limit of this sensor depends on the stability of the phosphor and the fibers.

The book by Grattan and Zhang (1995) details these and other sensors in fiber optic fluorescence thermometry. Other sensors they discussed that are potentially interesting for thin film diagnostics include ruby $(Al_2O_3:Cr^{3+})$, alexandrite $(BeAl_2O_4:Cr^{3+})$, $YVO_4:Eu$, $Y_2O_3:Eu$, and YAG:Tb. Figure 14.12 (from this volume) shows how the host affects the temperature dependence of the lifetimes in several Cr^{3+}-doped crystals. The regions where the lifetimes vary rapidly with temperature are useful for thermometry.

Bugos *et al.* (1989) have examined the potential thermometric application of temperature-dependent shifts in absorption, measured by excitation

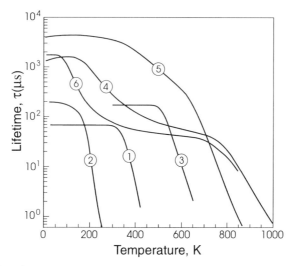

Figure 14.12 The temperature dependence of the fluorescence lifetimes for Cr^{3+} in various hosts: (1) LiSAF (LiSrAlF$_6$), (2) LiBAF (LiBaAlF$_6$), (3) LiCAF (LiCaAlF$_6$), (4) alexandrite, (5) ruby, and (6) emerald (from Figure 2.10 on page 53 in Grattan and Zhang, 1995).

emission, and emission profiles in europium-activated thermographic phosphors. The charge-transfer (CT) absorption peaks in both lanthanum phosphate (LaPO$_4$:Eu^{3+}) and lutetium phosphate (LuPO$_4$:Eu^{3+}) shift to the blue with increasing temperature (~15 nm for ~300°C rise), but the atomic lines seen in emission after CT band excitation did not shift.

In a very different application of such fluorescing and phosphorescing materials, Kolodner and Tyson (1982, 1983) spin-coated a sample with a thin film polymer heavily doped with a fluorophor whose PL intensity drops rapidly with temperature. In this rare-earth chelate europium thenoyl-trifluoroacetonate (EuTTA), the TTA ligand strongly absorbs UV light near 345 nm. This excitation is transferred to the $^5D_0$ level of the Eu^{3+} ion which fluoresces at 612 nm. However, there is a thermally activated transfer from the Eu^{3+} level to a triplet level in TTA that relaxes nonradiatively. As temperature increases, this transfer proceeds faster and the PL intensity at 612 nm decreases. Digital processing of the CCD array image showed that temperature could be measured with a resolution better than 0.01°C with 15-μm spatial resolution (Kolodner and Tyson, 1982). Further studies showed than 0.7-μm spatial resolution was possible with 0.08°C temperature resolution (Kolodner and Tyson, 1983).

In another study by the same group, Kolodner et al. (1983) followed the temperature of a glass resin doped with a luminescing material as it was etched by RIE. The PL efficiency of this phosphor, perdeutero-(*tris*-6,6,7,7,8,8,8-heptafluoro-2,2-dimethyl-3,5-octanedionato)europium (dEu-

FOD) changes more slowly with T than does EuTTA, but it is more robust in the RIE environment. The phase-shift method was used with a mechanical chopper (660 Hz) modulating the 300–350-nm filtered light output of a 100-W Hg arc lamp. Such methods in which the luminescing material must be on or part of the wafer are not generally applicable for real-time thermometry of thin film processing.

References

D. J. As and L. Palmetshofer, *J. Appl. Phys.* **62**, 369 (1987).
E. S. Aydil, K. P. Giapis, R. Gottscho, V. M. Donnelly, and E. Yoon, *J. Vac. Sci. Technol. B* **11**, 195 (1993a).
E. S. Aydil, Z. Zhou, K. P. Giapis, Y. Chabal, J. A. Gregus, and R. Gottscho, *Appl. Phys. Lett.* **62**, 3156 (1993b).
E. S. Aydil, Z. H. Zhou, R. A. Gottscho, and Y. J. Chabal, *J. Vac. Sci. Technol. B* **13**, 258 (1995).
H. B. Bebb and E. W. Williams, *Semicond. Semimetals* **8**, 181 (1972).
H. Booyens, J. H. Basson, A. W. R. Leitch, M. E. Lee, and C. M. Stander, *Surf. Sci.* **130**, 259 (1983).
A. R. Bugos, S. W. Allison, and M. R. Cates, *IEEE Proc. Southeastcon, 1989,* p. 361 (1989).
G. E. Carver, R. W. Heebner, and G. Astfalk, *IEEE J. Sel. Top. Quantum Electron.* **1**, 980 (1995).
H. C. Casey, Jr. and M. B. Panish, *Heterostructure Lasers,* Part A, Chapter 3. Academic, New York, 1978.
P. J. Dean, *Prog. Cryst. Growth Charact.* **5**, 89 (1982).
E. J. Egerton, A. Nef, W. Millikin, W. Cook, and D. Baril, *Solid State Technol.* **25**(8), 84 (1982).
R. A. Gottscho, B. L. Preppernau, S. J. Pearton, A. B. Emerson, and K. P. Giapis, *J. Appl. Phys.* **68**, 440 (1990).
K. T. V. Grattan and Z. Y. Zhang, *Fiber Optic Fluorescence Thermometry.* Chapman & Hall, London, 1995.
D. Guidotti, E. Hasan, H. J. Hovel, and M. Albert, *Appl. Phys. Lett.* **50**, 912 (1987).
N. M. Haegel and A. Winnacker, *Appl. Phys. A* **42**, 233 (1987).
H. J. Hovel and D. Guidotti, *IEEE Trans. Electron Devices* **ED-32**, 2331 (1985).
I. Hussla, K. Enke, H. Grünwald, G. Lorenz, and H. Stoll, *J. Phys. D* **20**, 889 (1987).
W. R. Imler, *IEEE J. Sel. Top. Quantum Electron.* **1**, 987 (1995).
D. Kirillov and J. L. Merz, *J. Appl. Phys.* **54**, 4104 (1983).
P. Kolodner and J. A. Tyson, *Appl. Phys. Lett.* **40**, 782 (1982).
P. Kolodner and J. A. Tyson, *Appl. Phys. Lett.* **42**, 117 (1983).
P. Kolodner, A. Katzir, and N. Hartsough, *Appl. Phys. Lett.* **42**, 749 (1983).
G. Livescu, M. Angell, J. Filipe, and W. H. Knox, *J. Electron. Mater.* **19**, 937 (1990).
Luxtron Corp., Product Information (1995).
J. S. McCormack, *Electron. Lett.* **17**, 630 (1981).
A. Mitchell, R. A. Gottscho, S. J. Pearton, and G. R. Scheller, *Appl. Phys. Lett.* **56**, 821 (1990).

S. D. Offsey, J. M. Woodall, A. C. Warren, P. D. Kirchner, T. I. Chappell, and G. D. Pettit, *Appl. Phys. Lett.* **48,** 475 (1986).

J. I. Pankove, *Optical Processes in Semiconductors.* Dover, New York, 1971.

S. Perkowitz, *Optical Characterization of Semiconductors: Infrared, Raman and Photoluminescence Spectroscopy.* Academic Press, San Diego, CA, 1993.

M. Y. A. Raja, S. R. J. Brueck, M. Osinski, and J. McInerney, *Appl. Phys. Lett.* **52,** 625 (1988).

R. P. Salathé, H. H. Gilgen, and Y. Rytz-Froidevaux, *IEEE J. Quantum Electron.* **QE-17,** 1989 (1981).

C. J. Sandroff, F. S. Turco-Sandroff, L. T. Florez, and J. P. Harbison, *J. Appl. Phys.* **70,** 3632 (1991a).

C. J. Sandroff, F. S. Turco-Sandroff, L. T. Florez, and J. P. Harbison, *Appl. Phys. Lett.* **59,** 1215 (1991b).

D. K. Schroder, *Semiconductor Material and Device Characterization,* Chapter 9. Wiley, New York, 1990.

M. Takai, H. Nakai, J. Tsuchimoto, K. Gamo, and S. Namba, *Jpn. J. Appl. Phys.* **24,** L705 (1985).

K. A. Wickersheim and M. H. Sun, *J. Microwave Power* **22**(2), 85 (1987).

E. W. Williams and H. B. Bebb, *Semicond. Semimetals* **8,** 321 (1972).

C. W. Wilmsen, P. D. Kirchner, and J. M. Woodall, *J. Appl. Phys.* **64,** 3287 (1988).

E. Yoon, R. A. Gottscho, V. M. Donnelly, and H. S. Luftman, *Appl. Phys. Lett.* **60,** 2681 (1992a).

E. Yoon, R. A. Gottscho, V. M. Donnelly, and W. S. Hobson, *J. Vac. Sci. Technol. B* **10,** 2197 (1992b).

CHAPTER **15**

Spectroscopies Employing Laser Heating

In most diagnostics, heating of the medium by the probing laser is avoided because it perturbs the medium (laser heating is discussed in Section 4.5). Such perturbations are particularly troublesome in thermometry because the goal is the measurement of temperature. Laser heating must be avoided even when the goal is measuring the rate of a process, because the rate often changes rapidly with an increase in temperature. In contrast, such incidental heating is of little concern in this chapter because laser heating is central to the optical diagnostics detection scheme discussed.

The use of lasers to heat gases and materials is widespread in materials processing, both to induce materials modifications and for spectroscopic analysis. Lasers evaporate material for physical vapor deposition, melt semiconductors for doping, and melt metals for welding. When laser heating is used for spectroscopic analysis, the heating induces a change that is monitored by a technique that may be optically or nonoptically based. In laser-induced thermal desorption, a pulsed laser desorbs adsorbates from a surface and these species are detected in the gas phase. In photothermal optical spectroscopy, a modulated laser heats a region near the surface or in the gas. This induces a periodic or pulsed change in temperature, small enough not to perturb the process, which is detected directly or indirectly through the change of a material property or the creation of another form of energy.

15.1 Laser-Induced Thermal Desorption

Thermal desorption of adsorbates by resistive heating of the substrate is commonly used in surface science studies of these surface adsorbates. In this method, called thermal desorption spectroscopy (TDS) or temperature-programmed desorption (TPD), different adsorbates desorb at distinct times, corresponding to different temperatures. The binding energy of each adsorbate can be determined from its desorption temperature. When pulsed lasers are used to heat a surface transiently and rapidly, this technique is called laser-induced thermal desorption (LITD) or laser desorption spectroscopy (LDS). Laser-induced thermal desorption has been used to study the fundamental chemistry at the interface in a thin film process and in process development, and has been surveyed by Stair and Weitz (1987) and George (1993).

Desorption temperatures are much higher in LITD than in resistively heated thermal desorption spectroscopy because the heating rates are much faster, $\sim 10^8$–10^{11} K/sec for LITD vis-à-vis ~ 1–100 K/sec in TDS. This rapid heating and the small duty cycle of pulsed-laser LITD are advantages in studying steady-state film processes. De Jong and Niemantsverdriet (1990) have shown how to obtain the correct activation energy, preexponential factor, and order of desorption as a function of adsorbate coverage from thermal desorption spectra (for TDS and LITD).

Laser-induced thermal desorption is similar to laser microanalysis, which is an *ex situ* method for chemical analysis of solids (Moenke-Blankenburg, 1989). The material to be characterized is positioned on a microscope stage to target the focused laser at the region of interest. The laser, usually a pulsed solid-state laser with pulse energy up to 1 J, is focused to 0.5–300 μm and evaporates 10^{-12}–10^{-6} g of material to a depth of 1–100 μm. This forms a microplasma whose atomic constituents are routinely identified by optical emission spectroscopy (OES, with micro- to nanogram sensitivity) or, for more sensitivity, by mass spectrometry (MS, with picogram sensitivity). Detection by atomic absorption and fluorescence is also common. Laser microvaporization has also been combined with excitation by an inductively coupled plasma to induce optical emission. Trace elements can be detected by this technique. Each of these materials-characterization methods has an analog with the more "gentle" LITD probing, which probes only about a monolayer at the surface.

Laser-induced thermal desorption leads to a transient increase of the desorbed species in the gas, which can be detected in several ways. Quadrupole mass spectrometry, sometimes in conjunction with time-of-flight methods, is the most common technique for analyzing the desorbed species in low-pressure environments, where the pressure is so low that collisions are infrequent. It is sensitive to all species and is well adapted to high-vacuum studies and processing. The related technique, Fourier-transform mass spectrometry, is also used. At higher pressures, monitoring of the desorbed species by optical methods, e.g. single and multiphoton ionization (including resonant-enhanced MPI) and laser-induced fluorescence (LIF), is often preferable, even though it may be difficult to detect all species with these techniques; George (1993) has reviewed these methods. Optical excitation

15.1 Laser-Induced Thermal Desorption

of these same desorbed species on the surface produces no fluorescence because of rapid relaxation by the surface. Laser-induced thermal desorption of surface adsorbates during plasma processing can also be monitored by a transient increase in optical emission (laser-desorption–plasma-induced emission, LD-PIE), as well as by LIF (LD-LIF) (Herman et al., 1994).

The temperature profile at the surface resulting from laser heating is obtained by solving the heat flow equation

$$\nabla \cdot [K \nabla T(\mathbf{r},t)] - \rho_d C \frac{\partial T(\mathbf{r},t)}{\partial t} = -f(\mathbf{r},t) \qquad (15.1)$$

where K is the thermal conductivity, ρ_d is the density, C is the specific heat, and f is the heat source per unit volume and time, which is due to laser heating here ($= \alpha(\mathbf{r},t) I(\mathbf{r},t)$, where α is the absorption coefficient and I is the laser intensity). In the most general case, numerical solution by finite-difference (or element) analysis is needed because of the temperature dependence of the optical and thermophysical properties, and the specific spatial and temporal profiles of the heating laser (Wood and Giles, 1981; Baeri and Campisano, 1982). Analytic solutions are possible when simplifying assumptions are made (Ready, 1971; Lin and George, 1983). The temperature profile and desorption kinetics during LITD have been investigated by several groups, including Brand and George (1986), Burgess et al. (1986), Philippoz et al. (1989), and George (1993).

In one approach (Brand and George, 1986; George, 1993), the temporal profile of the laser pulse intensity is approximated by a gaussian with FWHM pulse width t_p and the transverse profile is assumed to be a gaussian with (intensity) FWHM width r_p. The laser intensity profile is

$$I(r,t) = I_0 \exp[-4 \ln 2 \, (r/r_p)^2] \exp[-4 \ln 2 \, (t/t_p)^2] \qquad (15.2)$$

and so the pulse energy is

$$\mathcal{E} = \frac{\pi^{3/2}}{8(\ln 2)^{3/2}} I_0 r_p^2 t_p \qquad (15.3)$$

If the absorption depth is very small compared to the characteristic thermal diffusion distance during the pulse (Section 4.5), which in turn is much smaller than r_p, then the temperature at the surface at $r = 0$ is

$$T(t) = T_0 + \frac{(1-R)}{\sqrt{\pi K \rho_d C}} \int_0^t \frac{I(t-t')}{\sqrt{t'}} dt' \qquad (15.4)$$

where R is the reflectance and R, K, C, and ρ_d are assumed to be independent of T. (For many systems, such as semiconductors, these and earlier assumptions may not be valid.)

This equation provides two empirical results. The peak temperature occurs ~35% of the pulse length after the peak laser intensity ($t = 0$):

$$t(T_{max}) \sim 0.35 t_p \qquad (15.5)$$

and, with all parameters in the same units, this peak temperature is

$$T_{max} = T_0 + \frac{1.565(1-R)}{\sqrt{\pi K \rho_d C}} \sqrt{t_p}\, I_0 \qquad (15.6)$$

The desorption kinetics are described by

$$-\frac{d\Theta}{dt} = \nu_n \Theta^n \exp(-\mathcal{E}_d/k_B T) \qquad (15.7)$$

where Θ is the surface coverage, n is the order of the desorption, ν_n is the preexponential factor, and $\mathcal{E}_d$ is the desorption activation energy. Because of the fast heating rates in LITD, Equation 15.7 is dominated by desorption at T_{max}.

Laser-induced thermal desorption has been developed for use as a real-time, *in situ* probe of the surface layer during thin film processing in a series of studies by Herman *et al.* (1994), Cheng *et al.* (1994), Donnelly *et al.* (1994), and Cheng *et al.* (1995). Figure 15.1 shows the apparatus used to probe the adlayer during the etching of Si by Cl_2 and by Cl_2/O_2 and Cl_2/HBr mixtures in a helical resonator plasma. A XeCl laser (308 nm, 20-nsec pulse width, ~5 Hz) heated the substrate transiently and desorbed the steady-state adlayer. During etching by Cl_2, this resulted in transient increases in plasma-induced emission (LD-PIE) from Si and SiCl (Section 6.3.1, Figure 6.10) and LIF from desorbed SiCl (LD-LIF), which is reso-

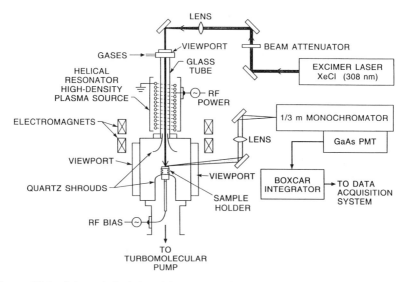

Figure 15.1 Schematic depiction of the apparatus for LITD optical detection during plasma etching (from Cheng *et al.*, 1994).

15.1 Laser-Induced Thermal Desorption

nantly excited during the same laser pulse (Section 7.2.2, Figures 6.10 and 7.25); with HBr in the gas mixture, LIF from desorbed SiBr was also observed (Figure 7.25).

This real-time monitor of the surface is important because etching involves two steps: (1) the formation of a "weakly" bound adlayer and (2) the removal of that adlayer. It is difficult to use other optical or nonoptical probes to monitor this layer accurately in the very reactive plasma environment. When combined with other probes—a Langmuir probe for plasma diagnostics and reflection interferometry (Section 9.5.1) for determining etching rates—LITD was shown to provide a fairly complete picture of the etching process; among other things, this can be used to determine which etching step is rate-limiting. For example, the thickness of the adsorbed layer during plasma etching by Cl_2 was found to be nearly independent of pressure, input RF power, and substrate bias voltage near normal etching conditions, and etching in this operating regime was seen to be limited by the flux of ions to the surface (Cheng *et al.*, 1994).

Figure 15.2 shows real-time monitoring of SiCl LD-LIF during Cl_2 etching of Si in the helical resonator, with the observation wavelength fixed and the laser fluence high enough to saturate the signal. When the discharge was turned on, the steady-state LD-LIF signal doubled, indicating that the presence of the plasma doubles the chlorine content of the surface layer, and it does so within one laser pulse (<200 msec). The laser repetition rate was chosen so that several monolayers were etched between laser

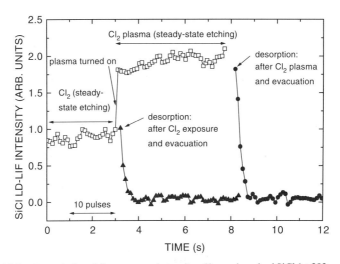

Figure 15.2 Laser-induced fluorescence intensity of laser-desorbed SiCl (at 293 nm) plotted for each laser pulse. The open symbols represent data taken during steady-state (plasma) etching of Si by Cl_2, while the closed symbols represent data taken after the process was terminated and after a time delay (not shown) for pumpdown (from Herman *et al.*, 1994).

pulses, and thus every laser pulse probed a steady-state surface layer. The thickness of this layer was found to be self-limited to about 1–2 monolayers. The stability of the surface layer was tested by blocking the laser, turning the plasma off (in the plasma-on case), and pumping away the reactive gas for several minutes (for both the plasma-on and plasma-off cases). When the laser pulse train was turned on again, a decaying LD-LIF was seen, and the signal from the first laser pulse was within ~10% of the steady-state value (for both cases). This demonstrated that for this particular system, in-line diagnostics, such as x-ray photoelectron spectroscopy (XPS), where the wafer is transferred in vacuum from the reactor to the diagnostics chamber, probe the same adlayer that is present during processing. In these studies, in-line XPS measurements were used to corroborate the LITD measurements.

Laser-induced thermal desorption is an accurate real-time monitor of surfaces during high-pressure and highly reactive processing, and has sub-monolayer sensitivity. When resonance and laser fluence requirements can be met, the use of the same laser for surface desorption and LIF excitation is preferable to using different lasers for the two processes. During plasma processing, the photon energy should be lower than the work function of the wafer to avoid the photoemission of electrons that could affect the plasma. Laser-induced thermal desorption is not a nonintrusive diagnostic because it entails the removal of the adlayer. However, it is a valuable probe for process development because it investigates the surface under steady-state conditions and may have selected applications in real-time monitoring.

The thermal interaction of chlorine with Si surfaces has also been studied with LITD by several groups. For example, Gupta *et al.* (1991) studied laser desorption following Cl_2 adsorption on Si(111) 7 × 7 by using mass spectrometry (ruby laser, 694.3 nm). De Jonge *et al.* (1986) used LIF to determine the vibrational and rotational distributions of SiCl ejected when a chlorine-exposed Si(111) surface was irradiated by a KrF laser (248 nm). Aliouchouche *et al.* (1993) employed LITD to desorb silicon chloride formed thermally on Si and LIF to probe products. Such laser desorption of a thermally produced adlayer is also called laser-induced etching. Strupp *et al.* (1992) desorbed In and Ga from Si(100) with the frequency-doubled 532-nm output from a Nd:YAG laser and used LIF to detect the desorbed In and Ga atoms. Figure 15.3 shows the depletion of In atoms from the surface and demonstrates that LITD obeys first-order desorption kinetics (A) and not half-order kinetics (B).

15.2 Thermal Wave Optical Spectroscopies

Light absorbed by a gas or a wafer relaxes in part by radiative decay, which produces photons that can be detected (PL, LIF), and in part by

15.2 Thermal Wave Optical Spectroscopies

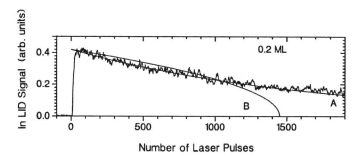

Figure 15.3 Laser-induced desorption of In with detection by LIF, compared with models of (A) first-order and (B) half-order desorption kinetics (from Strupp et al., 1992).

nonradiative recombination, which leads to a local temperature rise. While the optical diagnostics described in other chapters detect these emitted photons, thermal-wave optical (or photothermal) spectroscopies probe the local temperature rise caused by nonradiative recombination. This is accomplished either by sensing this rise directly or by detecting it indirectly by an effect induced by the temperature rise, such as the creation of an elastic wave or surface deformation or by a change in the index of refraction. As such, photothermal techniques are sensitive to the distinctive features and spatial variations in the optical, thermal, and acoustic properties of a structure. These methods are also nondestructive and noninvasive (as long as the temperature rise is too small to affect the process or the material).

Thermal-wave spectroscopies are often used in materials characterization to measure the absorption coefficient of films and for microscopic examination of features and defects, sometimes for integrated circuit inspection. Some photothermal methods require direct contact of a transducer to the wafer, while others do not. Much of the technological development in this field has been to develop such noncontact methods. Although these techniques have not been used frequently to probe thin film processes *in situ*, some believe that they will eventually be utilized more extensively.

Overviews of thermal wave spectroscopies have been given by Rosencwaig (1980), Tam (1986, 1987), Mandelis (1987), Sell (1989), Bein and Pelzl (1989), Bicanić (1991), and Christofides (1993). Collectively, these techniques are called thermal wave analysis (TWA), photothermal spectroscopy, or photoacoustic spectroscopies. Sometimes these terms are also used with more limited meanings. One potential source of confusion in assessing these techniques is the frequent use of different names to describe the same methods because there is no standard terminology in the literature. The terminology of Bein and Pelzl (1989) is adopted in most of this section.

Thermal waves are temporally and spatially oscillating temperature distributions, which are detected either directly or by other physical changes they cause. With a periodic heating source, $f(\mathbf{r},t) = f(\mathbf{r})\exp(-i\omega t)$, the heat flow equation (Equation 15.1) becomes

$$\nabla \cdot [K\nabla T(\mathbf{r})] + Kq^2 T(\mathbf{r}) = -f(\mathbf{r}) \tag{15.8}$$

where $q = (1 + i)(\omega \rho_d C/2K)^{1/2}$ is a characteristic thermal wave number (Favro et al., 1987). Because q is complex, the solution to this equation is a heavily damped wave, called a thermal wave, that damps in the characteristic thermal diffusion distance $\mu_d = (2K/\omega \rho_d C)^{1/2}$. Elastic waves are generated by the periodic thermal expansion of the thermal waves.

Thermal wave methods can probe wafers or gases (or liquids). Spectroscopic analysis is not needed when the main interest is mapping features such as defects in wafers; a laser at a fixed frequency is used to induce heating. In spectroscopic photothermal analysis, either spectrally dispersed white light from a Xe lamp or a tunable laser is used. In either case, the intensity is modulated with a mechanical chopper (to several kilohertz), an acousto-optic modulator (to several megahertz, for improved spatial resolution), or, more rarely, by electro-optic modulation. The detected signal is analyzed with a lock-in amplifier. In probing a wafer, tuning the modulation frequency ω permits sampling down to different depths, because the penetration depth of thermal waves decreases as ω increases. In some cases, the sample is excited by a pulsed laser.

Many detection methods can be used to sense absorption. In *photoacoustic spectroscopy (PAS)*, a thermal wave produced in a solid is transferred to the gas. This causes a pressure fluctuation in the gas that is detected by a microphone. A thermal wave generated by modulated absorption in the gas can also be detected this way. When heat diffusion plays a minor role, the term *optoacoustic* is sometimes used instead of photoacoustic.

In *thermoelastic response sensing,* the strains and stresses in the solid caused by the thermal wave are sensed. Low modulation frequencies fall within the *thermoelastic deformation regime,* while higher frequencies fall within the *ultrasonic regime.* In the ultrasonic regime, the thermal wave leads to the propagation of thermoelastically induced displacement at sound velocities, which can be detected by a piezoelectric transducer that is attached to the wafer. Elastic waves with frequencies up to several megahertz can be detected. This excellent time response leads to excellent spatial resolution. When the heating laser is strongly absorbed, detection of the ultrasonic signal provides a high-resolution evaluation of subsurface properties. All of these sensing methods are sometimes also called photoacoustic spectroscopy (PAS) or microscopy (PAM). These probes involve contact with a sensor, and so they have limited applications for real-time analysis during thin film processing.

In a related technique, *photothermal displacement spectroscopy,* slow surface displacements and deformations produced by the heating laser are sensed by reflecting a probe laser at the surface or by interferometry.

15.2 Thermal Wave Optical Spectroscopies

[Sometimes this is also called photothermal beam deflection (PBD), as is the next technique.]

In *photothermal beam deflection (PBD)*, a probe beam propagating in the gas above a surface senses the spatial and temporal changes in the index of refraction caused by thermal waves propagating from the heated surface. This results in a deflection of the probe beam, which is also known as the *mirage effect*. Figure 15.4 shows that the deflected beam is detected by a position-sensitive detector, such as a quadrant photodetector or, more primitively, by a photodetector that is preceded by an slit mounted slightly off from the peak of the unperturbed beam (also see Figure 15.8a).

Beam deflection is described by the paraxial equation

$$\frac{d}{ds}\left(n_0 \frac{dr_0}{ds}\right) = \nabla n(r,t) \tag{15.9}$$

where s is the light path, r_0 is the perpendicular displacement from the path for the uniform index (n_0) case, and the right-hand side is the gradient of the index of refraction normal to the path. If x is the direction of beam travel and y is the direction normal to the surface, the deflection angle ϕ is approximately

$$\phi(y) = \frac{1}{n_0}\int \frac{dn}{dy}\,dx \tag{15.10}$$

Photothermal beam deflection is also called photothermal deflection spectroscopy (PDS) or optical beam deflection (OBD) spectroscopy. As described below, it has been used as a real-time probe of laser ablation.

In *photothermal radiometry*, changes in thermal radiation (Chapter 13) due to the modulation in the heating laser are sensed by an external infrared detector, such as photoconducting HgCdTe and pyroelectric detectors; use of pyroelectric detectors is known as photopyroelectric (PPE) spectroscopy. In *microcalorimetry*, temperature changes are sensed by a small calorimeter

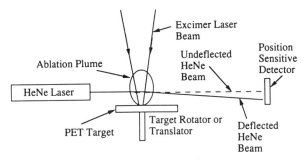

Figure 15.4 Experimental apparatus for photothermal beam analysis (from Figure 1, P. L. G. Ventzek, R. M. Gilgenbach, D. M. Heffelfinger, and J. A. Sell, *J. Appl. Phys.* **70**, 587 (1991)).

(thermocouple, a contact method). In thermal wave *photopyroelectric* characterization, a bonded pyroelectric detector measures the temperature rise. These techniques that directly measure temperature are sometimes collectively called *photothermal spectroscopy* (*PTS*). Temperature profiles can also be sensed by the *internal mirage effect,* which senses thermal-wave-induced gradients in the refractive index in the solid by laser beam deflection, and *thermal lensing* probing, which probes the thermally induced curvature in the index of refraction by a propagating laser.

In *modulated reflectance spectroscopy,* the reflection of a probe beam senses thermal waves and, in semiconductors, plasma waves produced by photocarriers. Rosencwaig (1987) has developed a "thermal wave" mapping method for *ex situ* inspection of semiconductors that is essentially modulated photoreflectance mapping. The physical basis of this method may be the modulation in the index of refraction arising from the photomodulated free-carrier density, and not due to temperature changes or to the surface-band bending that is important in conventional photoreflectance (Section 9.6).

Photothermal methods have been applied to *in situ* monitoring of deposition. Geiler *et al.* (1992) used a thermal wave double-modulation technique for *in situ* monitoring of the evaporation of metallic layers using the appara-

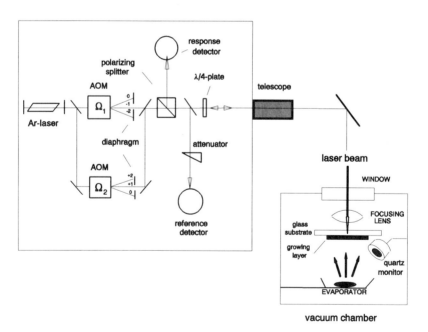

Figure 15.5 Thermal-wave analysis system that can be used for *in situ* probing. [From H. D. Geiler, N. Winkler, and D. Schiller, in *Springer Series in Optical Sciences* **69,** *In situ* measurement of thin film growth by photothermal induced frequency conversion, page 689, Figure 1, 1992. Copyright © Springer-Verlag.]

15.2 Thermal Wave Optical Spectroscopies

tus shown in Figure 15.5, with acousto-optic modulation. Schneider *et al.* (1993) and Schork *et al.* (1994) have also employed this method for *in situ* measurement of the thickness of metal films. In their investigation, two semiconductor lasers emitting at 785 nm, modulated between 100 kHz and 1.2 MHz—one shifted 10 kHz relative to the other—impinge the sample. These lasers both modulate the surface temperature and carrier concentration and sense the change in reflectance due to this modulation, which is detected at 10 kHz; this instrument has been described in more detail by Wagner and Geiler (1991). Figure 15.6 shows the TWA signal for an aluminum layer atop a 1-μm-thick SiO_2 layer as a function of Al thickness. The sensitivity increases as the metal film gets thinner and as the insulating film gets thicker. Schork *et al.* (1994) have shown that this method is potentially useful for *in situ* temperature analysis, as is detailed in Section 15.2.1. Schneider *et al.* (1993) have demonstrated how this compact thermal wave analysis system can be utilized to measure the thickness of thin metal films in the transport chamber of a cluster tool. Schlemm *et al.* (1994) used this method for on-line measurement of ion implantation; it has potential for *in situ* control.

Higashi *et al.* (1985) used pulsed optoacoustic spectroscopy to record the infrared absorption spectra of surface adsorbates by using a piezoelectric transducer that was bonded to the transparent sapphire substrate. A tunable dye laser (615 to 655 nm) was Raman-shifted in H_2 to the third Stokes line and provided the tunable infrared radiation from 2800 to 3800 cm^{-1} for absorption in the range of C—H and O—H stretches. This technique has submonolayer sensitivity and was used by Higashi and Rothberg (1985a,b) to follow *in situ* the nucleation and growth of Al by excimer-laser-assisted photodeposition with trimethylaluminum. Rothberg (1987) has surveyed the use of pulsed laser optoacoustic spectroscopy to study surface adsorbates.

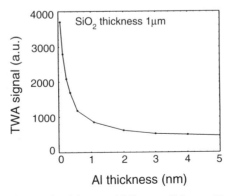

Figure 15.6 Thermal-wave signal from an Al film atop SiO_2 on a Si wafer (from Schneider *et al.*, 1993).

The sensitivity of these methods to transient processes has made them useful in real-time study of laser ablation for etching and pulsed laser deposition. Dyer and Srinivasan (1986) used piezoelectric transducers to measure the transient stress waves that are formed due to recoil effects during pulsed laser ablation of organic polymers. Photothermal beam deflection and photoacoustic spectroscopies have been used to probe the plume formed in pulsed laser deposition, by using an arrangement with a cw laser, often a He–Ne laser at 632.8 nm, passing parallel to the surface, as in Figure 15.4. Time-dependent nonuniformities in the refractive index due to the plume are easily monitored with a position-sensitive detector. Measurements can be performed as a function of distance between the probing laser and the surface. In early work, Koren (1987) found a shock wave and a slower cooling wave during CO_2 laser-produced plumes from Kapton films.

Chen and Yeung (1988) applied this technique to laser ablation from Si and noted that this application is technically different from photothermal deflection: even though there is a refractive index gradient, there is not enough time for collisions to establish a thermal gradient. Sell *et al.* (1989, 1991) and Ventzek *et al.* (1990) used this method to probe the plumes formed during UV laser ablation of polymers and $YBa_2Cu_3O_{7-x}$. Figure 15.7 shows the deflection signal Sell *et al.* (1989) obtained during the laser ablation of the polymer polyethyleneterephthalate (PET). Ventzek *et al.* (1991) used beam deflection to measure the mass of the expanding ablated material during UV laser ablation of PET and also to measure the depth of the ablation pit on the surface, via the acoustic wave that travels from the pit to the gas. In this latter technique, called photoacoustic beam deflection, the depth is determined by measuring the delay time of acoustic waves that travel from the laser-irradiated surface to the beam when ablation occurs in a gas or liquid background. Petzoldt *et al.* (1988) used this photoacoustic beam deflection method to determine the threshold for surface damage due to lasers incident on the surface.

Faris and Bergström (1991) developed a two-color beam thermal deflection method to measure electron density and expansion velocity in a laser-produced plasma. They used two wavelengths, 632.8 nm (He–Ne laser) and 441.6 nm (He–Cd laser), to separate the contributions from electrons from that of all other species, neutral atoms, molecules, ions, and particles. The contribution to the refractive index from free electrons decreases the refractive index and is very dependent on wavelength (from Equation 4.19 with $\nu_0 = 0$), while that due to the other species increases n and varies little with λ, except near resonance.

In analogy with thermal deflection of a beam in the gas phase, Spear and Russo (1991) examined transverse photobeam deflection in a solid, to monitor the refractive index gradients that accompany thermal gradients during beam processing.

15.2 Thermal Wave Optical Spectroscopies

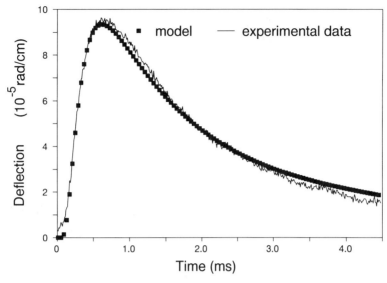

Figure 15.7 Single-shot laser beam deflection during excimer-laser ablation of polyethyleneterephthalate (PET), compared with the model (from Figure 1, J. A. Sell, D. A. Heffelfinger, P. Ventzek, and R. M. Gilgenbach, *Appl. Phys. Lett.* **55**, 2435 (1989).)

15.2.1 *Photoacoustic and Thermal Wave Thermometry*

The speed of acoustic waves in a solid medium is related to the elastic constants and the density, each of which depends on temperature (Victorov, 1967). Lee *et al.* (1989, 1990a,b) relied on this property to monitor the temperature of a silicon wafer by using lasers to both excite and detect the acoustic waves. They focused a nitrogen laser (337 nm) onto the wafer, which heated the wafer locally and caused elastic perturbations to propagate. A He–Ne laser (632.8 nm) was directed ~ 2–4 cm from the site of heating, and the deflection of the reflected beam from the initial position was monitored with a position-sensitive detector (Figure 15.8a). The speed of the acoustic mode is determined by the time it takes this perturbation to travel a given distance (Figure 15.8b), from which the (averaged) wafer temperature is determined. The fundamental plate mode (zeroth-order antisymmetric-mode Lamb wave) is the mode most sensitive to temperature changes. For propagation in the ⟨100⟩ direction on a (100) silicon wafer (thickness 0.5 mm), the group velocity linearly decreases from 2264 to 2229 m/sec as temperature increases from 0 to 1000°C (200 kHz) (Saraswat *et al.*, 1994). The temperature rise caused by the N_2 laser heating does not perturb the process because of the small duty cycle of the laser.

The same group has also examined two modified acoustic-based temperature sensors. In one, an air transducer launches a wave across the wafer,

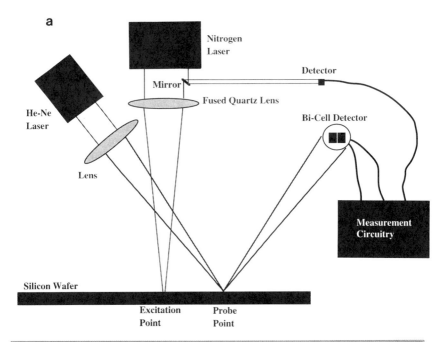

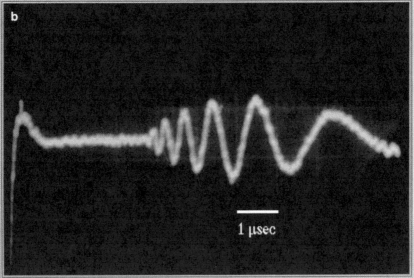

Figure 15.8 (a) Excitation and detection of acoustic waves and (b) observed fundamental plate mode (from Lee *et al.*, 1990b). [© 1990 IEEE.]

References

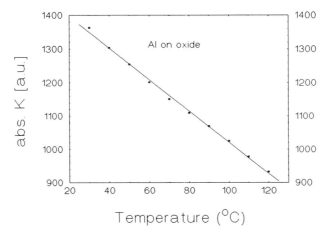

Figure 15.9 Temperature dependence of the thermal wave signal from 100 nm of evaporated Al on an oxidized silicon wafer (from Schork *et al.*, 1994).

and the deflection in the reflection of two lasers is used to measure the propagation time of the wave (Lee, 1990a,b). In a totally nonoptical variation, the wafer rests upon quartz pins. An extensional wave is launched in one quartz pin by a transducer and it is converted into a Lamb wave mode in the wafer. This wave is transmitted as an extensional wave into another pin, where it is detected by a transducer. Again, the propagation speed in the wafer, and consequently the temperature, is determined by the delay between wave launching and detection (Degertekin *et al.*, 1994).

Schork *et al.* (1994) have shown that the thermal wave signal of an Al layer on an insulator decreases nearly linearly with increasing temperature, as is shown in Figure 15.9. Because of its accuracy and repeatability (better than 1%) and rapid response (several msec), TWA shows some promise for *in situ* noncontact probing of temperature, at least at "lower" temperatures.

References

A. Aliouchouche, J. Boulmer, B. Bourguignon, J.-P. Budin, D. Débarre, and A. Desmur, *Appl. Surf. Sci.* **69,** 52 (1993).

P. Baeri and S. V. Campisano, in *Laser Annealing of Semiconductors* (J. M. Poate and J. M. Mayer, eds.), pp. 75–109. Academic Press, New York, 1982.

B. K. Bein and J. Pelzl, in *Plasma Diagnostics* (O. Auciello and D. L. Flamm, eds.), Vol. 2, Chapter 6, p. 211. Academic Press, Boston, 1989.

D. Bicanić, ed., *Photoacoustic and Photothermal Phenomena III,* Springer Ser. Opt. Sci., Vol. 69. Springer-Verlag, Berlin, 1991.

J. L. Brand and S. M. George, *Surf. Sci.* **167,** 341 (1986).

D. Burgess, Jr., P. C. Stair, and E. Weitz, *J. Vac. Sci. Technol. A* **4,** 1362 (1986).

G. Chen and E. S. Yeung, *Anal. Chem.* **60,** 864 (1988).
C. C. Cheng, K. V. Guinn, V. M. Donnelly, and I. P. Herman, *J. Vac. Sci. Technol. A* **12,** 2630 (1994).
C. C. Cheng, K. V. Guinn, I. P. Herman, and V. M. Donnelly, *J. Vac. Sci. Technol. A* **13,** 1970 (1995).
C. Christofides, *CRC Crit. Rev. Solid State Mater. Sci.* **18,** 113 (1993).
F. L. Degertekin, J. Pei, B. T. Khuri-Yakub, and K. C. Saraswat, *Appl. Phys. Lett.* **64,** 1338 (1994).
A. M. de Jong and J. W. Niemantsverdriet, *Surf. Sci.* **233,** 355 (1990).
R. de Jonge, J. Majoor, K. Benoist, and D. de Vries, *Europhys. Lett.* **2,** 843 (1986).
V. M. Donnelly, K. V. Guinn, C. C. Cheng, and I. P. Herman, *Mater. Res. Soc. Symp. Proc.* **334,** 425 (1994).
P. E. Dyer and R. Srinivasan, *Appl. Phys. Lett.* **48,** 445 (1986).
G. W. Faris and H. Bergström, *Appl. Opt.* **30,** 2212 (1991).
L. D. Favro, P.-K. Kuo, and R. L. Thomas, in *Photoacoustic and Thermal Wave Phenomena in Semiconductors* (A. Mandelis, ed.), Chapter 4, p. 69. North-Holland Publ., New York, 1987.
H. D. Geiler, N. Winkler, and D. Schiller, *Springer Ser. Opt. Sci.* **69,** 688 (1992).
S. M. George, in *Investigations of Surfaces and Interfaces* (B. W. Rossiter and R. C. Baetzold, eds.), Part A, Chapter 7, p. 453. Wiley, New York, 1993.
P. Gupta, P. A. Coon, B. G. Koehler, and S. M. George, *Surf. Sci.* **249,** 92 (1991).
I. P. Herman, V. M. Donnelly, K. V. Guinn, and C. C. Cheng, *Phys. Rev. Lett.* **72,** 2801 (1994).
G. S. Higashi and L. J. Rothberg, *J. Vac. Sci. Technol. B* **3,** 1460 (1985a).
G. S. Higashi and L. J. Rothberg, *Appl. Phys. Lett.* **47,** 1288 (1985b).
G. S. Higashi, L. J. Rothberg, and C. G. Fleming, *Chem. Phys. Lett.* **115,** 167 (1985).
G. Koren, *Appl. Phys. Lett.* **51,** 569 (1987).
Y. J. Lee, C. H. Chou, B. T. Khuri-Yakub, K. Saraswat, and M. Moslehi, *IEEE Ultrason. Symp., 1989,* p. 535 (1989).
Y. J. Lee, C. H. Chou, B. T. Khuri-Yakub, and K. Saraswat, *SPIE* **1393,** 366 (1990a).
Y. J. Lee, C. H. Chou, B. T. Khuri-Yakub, and K. Saraswat, *Symp. VLSI Technol., 1990,* p. 105 (1990b).
J.-T. Lin and T. F. George, *J. Appl. Phys.* **54,** 382 (1983).
A. Mandelis, ed., *Photoacoustic and Thermal Wave Phenomena in Semiconductors.* North-Holland Publ., New York, 1987.
L. Moenke-Blankenburg, *Laser Microanalysis.* Wiley, New York, 1989.
S. Petzoldt, A. P. Elg, M. Reichling, J. Reif, and E. Matthias, *Appl. Phys. Lett.* **53,** 2005 (1988).
J.-M. Philippoz, R. Zenobi, and R. N. Zare, *Chem. Phys. Lett.* **158,** 12 (1989).
J. F. Ready, *Effects of High-Power Laser Radiation.* Academic Press, New York, 1971.
A. Rosencwaig, *Photoacoustics and Photoacoustic Spectroscopy.* Wiley, New York, 1980.
A. Rosencwaig, in *Photoacoustic and Thermal Wave Phenomena in Semiconductors* (A. Mandelis, ed.), Chapter 5, p. 97. North-Holland Publ., New York, 1987.
L. Rothberg, *J. Phys. Chem.* **91,** 3467 (1987).
K. C. Saraswat, P. P. Apte, L. Booth, Y. Chen, P. C. P. Dankoski, F. L. Degertekin, G. F. Franklin, B. T. Khuri-Yakub, M. M. Moslehi, C. Schaper, P. J. Gyugyi,

References

Y. J. Lee, J. Pei, and S. C. Wood, *IEEE Trans. Semiconductor Manufact.* **7,** 159 (1994).

H. Schlemm, H. D. Geiler, and A. Kluge, *J. Phys. IV* (Supplément au J. de Physique III) **4,** C7-167 (1994).

C. Schneider, L. Pfitzner, and H. Ryssel, *Top. Conf. Manuf. Sci. 40th Nat. Symp. Am. Vac. Soc.,* Orlando, FL, *1993,* Presentation TC1-TuM10 (unpublished).

R. Schork, S. Krügel, C. Schneider, L. Pfitzner, and H. Ryssel, *J. Phys. IV* (Supplement au Journal de Physique III) **4,** C7-27 (1994).

J. A. Sell, ed., *Photothermal Investigations of Solids and Fluids.* Academic Press, Boston, 1989.

J. A. Sell, D. M. Heffelfinger, P. Ventzek, and R. M. Gilgenbach, *Appl. Phys. Lett.* **55,** 2435 (1989).

J. A. Sell, D. M. Heffelfinger, P. L. G. Ventzek, and R. M. Gilgenbach, *J. Appl. Phys.* **69,** 1330 (1991).

J. D. Spear and R. E. Russo, *J. Appl. Phys.* **70,** 580 (1991).

P. C. Stair and E. Weitz, *J. Opt. Soc. Am. B* **4,** 255 (1987).

P. G. Strupp, A. L. Alstrin, B. J. Korte, and S. R. Leone, *J. Vac. Sci. Technol. A* **10,** 508 (1992).

A. C. Tam, *Rev. Mod. Phys.* **58,** 381 (1986).

A. C. Tam, in *Photoacoustic and Thermal Wave Phenomena in Semiconductors* (A. Mandelis, ed.), Chapter 8, p. 175. North-Holland Publ., New York, 1987.

P. L. G. Ventzek, R. M. Gilgenbach, J. A. Sell, and D. M. Heffelfinger, *J. Appl. Phys.* **68,** 965 (1990).

P. L. G. Ventzek, R. M. Gilgenbach, D. M. Heffelfinger, and J. A. Sell, *J. Appl. Phys.* **70,** 587 (1991).

I. A. Victorov, *Rayleigh and Lamb Waves.* Plenum, New York, 1967.

M. Wagner and H. D. Geiler, *Meas. Sci. Technol.* **2,** 1088 (1991).

R. F. Wood and G. E. Giles, *Phys. Rev. B* **23,** 2923 (1981).

CHAPTER **16**

Nonlinear Optical Diagnostics

In linear optics, one photon interacts with the medium and one photon leaves the medium. Events occur at a rate proportional to the light intensity. In nonlinear optical processes, several photons, sometimes at different frequencies, simultaneously interact with the medium; sometimes incoherent or coherent light is emitted at a frequency that is different from the input frequencies. The rates of these events usually depend nonlinearly on laser intensity and become appreciably large only at very high laser intensities. Detailed descriptions of nonlinear optics have been presented by Shen (1984), Levenson and Kano (1988), Yariv (1989), and Boyd (1992). An overview of the physics of the nonlinear processes important in thin film diagnostics is presented in Section 4.4.

Because nonlinear optical processes are typically weaker, more complex, more expensive, less robust, and more intrusive than linear optical processes, they are not very suitable for real-time monitoring and control. However, they are still very useful for fundamental and process-development studies.

This chapter concentrates on the use of coherent anti-Stokes Raman scattering (CARS), second-harmonic generation (SHG), and third-harmonic generation (THG) in diagnostics. Several other nonlinear probes are closely linked to linear spectroscopies and are discussed in the chapter describing that linear diagnostic. Two-photon absorption has been used for LIF [laser-induced fluorescence] excitation in some atoms either because the wavelengths needed for one-photon excitation are shorter than those available or the most accessible transitions are not electric-dipole allowed. This two-photon LIF technique is presented in Section 7.2.1.2. Multiphoton ionization (MPI, REMPI), which is a multiple-photon absorption process where the final step ionizes a neutral atom or molecule to an ion and free electron, is described in Section 17.1.

Eckbreth (1996) has detailed the use of these gas-phase techniques in combustion studies, along with other nonlinear optical spectroscopies such as degenerate four-wave mixing (for detecting minority species) and polarization spectroscopy. Although these other techniques have not been widely used as thin film processing diagnostics, they show promise for process development studies. Of particular interest is that near saturation the use of degenerate four-wave mixing is relatively insensitive to the effects of collisions because it is a higher order (third order) process. As discussed in Section 7.1.1, this is not true for laser-induced fluorescence, which is a linear optical process.

Nonlinear optical methods, such as second-harmonic generation, sum frequency mixing, third harmonic generation, and stimulated Raman scattering, are also frequently used in generating light for linear and nonlinear optical diagnostics studies, as is described in Section 5.1.1.

Another potential application of nonlinear optical processes is in detection of infrared photons by frequency upconversion. In this process, the frequency of the infrared photon is summed with that of an input visible/ultraviolet photon from a high-intensity fixed-frequency laser. The resulting visible/ultraviolet photon is detected with a photomultiplier, which is faster, has higher quantum efficiency, and is less noisy than infrared detectors. While frequency upconversion has not been used much in thin film diagnostics, it has sometimes been used in the transient detection of gas-phase species by infrared diode lasers (Section 8.2.1.1).

16.1 Coherent Anti-Stokes Raman Scattering

16.1.1 *Theoretical and Experimental Considerations*

Coherent anti-Stokes Raman scattering (CARS), which is a powerful spectroscopy of gases, liquids, and solids, has been used to probe the gas-phase in thin film diagnostics. Two lasers at frequencies ω_1 and ω_2 ($<\omega_1$), which are typically high-power pulsed lasers, interact with the medium through the nonlinear susceptibility $\chi^{(3)}$ to produce a polarization $P^{(3)}$ at $\omega_3 = 2\omega_1 - \omega_2$, where $P^{(3)}(\omega_3) = \chi^{(3)}_{CARS} E_1^2(\omega_1) E_2(\omega_2)$. This process has been described in Section 4.4 (Equation 4.64, Figure 4.2b). The $\chi^{(3)}_{CARS}$ susceptibility is given by Equations 4.67–4.70. The induced polarization $P^{(3)}(\omega_3)$ generates coherent light at ω_3, which is then detected (Figure 16.1).

Coherent anti-Stokes Raman scattering has been reviewed by Tolles *et al.* (1977), Nibler and Knighten (1979), Harvey (1981), Eckbreth and Schreiber (1981), Druet and Taran (1981), Vogt (1982), Fleming (1983), Nibler and Yang (1987), Devonshire (1987), Levenson and Kano (1988), Eckbreth (1996), and Yariv (1989). The book by Eckbreth (1996) treats the theoretical and experimental aspects of CARS in detail, and describes the use of CARS

16.1 Coherent Anti-Stokes Raman Scattering

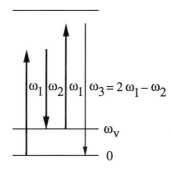

Figure 16.1 Energy-level diagram for CARS.

for diagnostics in combustion. Yuratich and Hanna (1977) have discussed the rotational structure and polarization properties of CARS.

In CARS, usually ω_1 is fixed (the pump), while ω_2 is tunable (the probe). Often a pulsed Nd:YAG laser or one of its harmonics is used as the pump, and a pulsed dye laser, pumped by a Nd:YAG laser harmonic, is used as the probe. The CARS signal $\propto |\chi^{(3)}_{CARS}|^2$ peaks when $\omega_1 - \omega_2$ is tuned near a resonance in the material, which is usually a vibrational resonance ω_v. Therefore, at resonance ω_2 is tuned to the Stokes frequency $\omega_1 - \omega_v$, and coherent emission at ω_3 is observed at the anti-Stokes frequency $\omega_1 + \omega_v$. Consequently, CARS is closely linked with spontaneous Raman scattering (Chapter 12). When the nonresonant background contribution to $\chi^{(3)}_{CARS}$ (χ^{NR}) is small ($|\chi^{res}_{CARS}| \gg \chi^{NR}$ in Equation 4.67), the lineshape of the CARS spectrum [power at ω_3 (P_3) versus $\omega_1 - \omega_2$] is very nearly a lorentzian centered at ω_v. However, when χ^{NR} is significant, the resonant and nonresonant terms interfere, leading to a lineshape that has a zero near ω_v.

Coherent anti-Stokes Raman scattering can be viewed as the scattering of ω_1 from the coherent oscillation produced by $\omega_1 - \omega_2 \approx \omega_v$ (Figure 16.1). In the closely related process called coherent Stokes Raman scattering (CSRS), the tunable frequency ω_2 is $<\omega_1$; at the $\omega_1 - \omega_2 \approx \omega_v$ resonance condition, coherent light (ω_3) is generated at the Stokes frequency $\omega_2 - \omega_v$, which is equal to $2\omega_2 - \omega_1$.

The power generated at ω_3 (P_3) by CARS is related to that at ω_2 (P_2) and the laser intensity at ω_1 (I_1) by (Druet and Taran, 1981; Eckbreth, 1996)

$$P_3 = \frac{16\pi^4 \omega_3^2}{n_1^2 n_2 n_3 c^4} |\chi^{(3)}_{CARS}|^2 I_1^2 P_2 L^2 \left(\frac{\sin(\Delta kL/2)}{\Delta kL/2}\right)^2 \quad (16.1)$$

where n_3 is the index of refraction at ω_3 and L is the distance that ω_1 and ω_2 overlap in the material. Δk is the magnitude of the momentum mismatch $\Delta \mathbf{k} = 2\mathbf{k}_1 - \mathbf{k}_2 - \mathbf{k}_3$, where $\mathbf{k}_i$ are the wavevectors of the respective beams and $k_i = n_i \omega_i / c$. (Note that because of the use of different definitions for the forms of the electric field and polarization in the literature, expressions such as Equations 16.1, 16.4, and 16.5 often have different coefficients in different sources.) Use of very high-intensity lasers (I_1, $\mathcal{P}_2$ large) that laterally overlap over a long distance L produces a large CARS signal, especially when conditions are near a $\chi_{CARS}^{(3)}$ resonance. $\mathcal{P}_3$ is maximized when the beams are phase-matched so that $\Delta \mathbf{k} = \mathbf{0}$ and the bracketed term becomes 1.

Several ways to achieve phase matching are shown in Figure 16.2, which depicts the paths of the lasers and the vector relationships of the $\mathbf{k}_i$. In the general two-beam scheme (Figure 16.2a), $\mathbf{k}_2$ intersects $\mathbf{k}_1$ at an angle θ. The $\mathbf{k}_3$ of the CARS signal is in the plane formed by $\mathbf{k}_1$ and $\mathbf{k}_2$, at an angle θ' on the other side of $\mathbf{k}_1$. Since ω_v/ω_1 is usually $\ll 1$, $\theta' \approx \theta$. θ is determined by dispersion of the index of refraction. When $\theta \gg 0°$ (solids and liquids), this arrangement spatially separates the CARS signal from the input beams; this leads to ready detection but limits the interaction distance. Because of the small dispersion in gases, collinearity ($\theta = 0°$) ensures phase matching in gases. In collinear CARS, ω_1 and ω_2 are combined by dichroic mirrors and the CARS signal is formed collinearly with them, so all $\mathbf{k}_i$ are parallel (Figure 16.2b). The interaction distance can be long in this scheme, but ω_3 must be separated from ω_1 and ω_2, perhaps by using a dichroic mirror. Spatial resolution is poor with collinearity because the CARS signal is integrated along a line, while with crossed beams the signal comes from a localized volume.

A crossed-beam arrangement called BOXCARS (Figure 16.2c) can be used to achieve phase matching within a localized region in gases. The ω_1 laser is split into two beams, each making an angle α with a given axis.

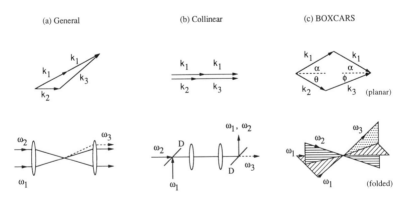

Figure 16.2 CARS phase-matching diagrams for (a) general, (b) collinear, and (c) BOXCARS configurations. The upper diagrams in (a) and (b) show the wavevectors of the beams, and the lower diagrams depict the beams in physical space. In (c), the upper diagram represents the beams in planar CARS, where all the beams are in one plane, while the lower diagram depicts the beam geometry (without lenses) for folded CARS, where $\mathbf{k}_2$ and $\mathbf{k}_3$ are not in the same plane as $\mathbf{k}_1$. D represents dichroic mirrors (from Eckbreth, 1996).

16.1 Coherent Anti-Stokes Raman Scattering

The ω_2 beam is introduced at angle θ and the anti-Stokes beam ω_3 appears at the phase-matched angle ϕ determined by

$$n_2\omega_2 \sin\theta = n_3\omega_3 \sin\phi \tag{16.2}$$

$$n_2\omega_2 \cos\theta + n_3\omega_3 \cos\phi = 2n_1\omega_1 \cos\alpha \tag{16.3}$$

In planar BOXCARS all four beams are in the same plane; but in folded BOXCARS the two ω_1 beams are in one plane and the ω_2 and ω_3 beams are in an orthogonal plane that intersects the common axis (Eckbreth, 1996).

In narrowband CARS, the lasers at ω_1 and ω_2 have narrow linewidths. The signal at ω_3 peaks when $\omega_1 - \omega_2$ is tuned near a resonance and is detected by a photodiode or photomultiplier, which may be preceded by a filter to remove residual light at ω_1 and ω_2. This is the most common CARS arrangement. The spectroscopic information, ω_v, etc., is obtained by tuning ω_2 in a controlled manner. This is fundamentally different from spontaneous Raman spectroscopy, in which spectroscopic information is obtained by measuring the wavelength of the weak scattered light.

In broadband CARS, the laser at ω_1 has a narrow linewidth, while that at ω_2 has a very broad bandwidth, say γ_2. The CARS light then has frequency components across this same width γ_2, but is sharply peaked only for those components near resonance. Spectroscopy occurs by dispersing the CARS signal in a spectrometer and analyzing the frequency components, as in spontaneous Raman scattering. This broadband version of CARS is needed when the process can be examined only in a limited time, perhaps when transient or fluctuating conditions make it impossible to probe the resonance by tuning the wavelength of the laser. An array detector is then needed to obtain all the spectral information from the single pair of laser pulses. Broadband CARS is rarely necessary for thin film processing diagnostics because most thin film processes can be probed over many laser pulses.

Because of the nonlinear dependence on laser intensity in Equation 16.1, the CARS signal can usually be increased by focusing the two lasers. If both input beams are gaussians with spot diameter d ($=2w$, Section 2.1), they can be focused by a lens of focal length f to a diameter $d_{focus} = 4\lambda f/\pi d$ (Equation 2.16). If the beams spatially overlap for several Rayleigh ranges (Section 2.1) and are phase-matched, then to a good approximation the signal becomes independent of the interaction length L (Druet and Taran, 1981; Eckbreth, 1996):

$$\mathcal{P}_3 = \left(\frac{2}{\lambda}\right)^2 \left(\frac{4\pi^2\omega_3}{c^2}\right)^2 \frac{1}{n_1^2 n_2 n_3} |\chi^{(3)}_{CARS}|^2 \mathcal{P}_1^2 \mathcal{P}_2 \tag{16.4}$$

This spatial overlap also defines the spatial resolution of CARS. If the lasers are not diffraction-limited, $\mathcal{P}_3$ is much smaller than predicted by this equation. Each of these expressions for P_3 is valid as long as very little light is converted to ω_3; that is, $P_3 \ll P_1, P_2$.

By using Equations 4.70 and 16.4, at peak conversion efficiency the anti-Stokes power can also be expressed as

$$P_3 = \left(\frac{2\lambda_2^4}{hc\lambda_1\lambda_3}\right)^2 \left(\frac{N_i - N_j}{\Gamma}\left(\frac{d\sigma}{d\Omega}\right)\right)^2 P_1^2 P_2 \qquad (16.5)$$

using $\lambda = \lambda_1$, $n_i = 1$, and $d\sigma/d\Omega$, the scattering cross-section for spontaneous Raman scattering. In practice, the observed CARS signals are an order of magnitude smaller than predicted by Equations 16.4 and 16.5, even with diffraction-limited lasers.

Most CARS studies are performed with vibrational–rotational scattering. However, pure rotational CARS, with $\Delta v = 0$, $\Delta J = \pm 2$, has the potential advantage of simultaneous multispecies detection by resolving overlapping rotational structure. Because of the small shifts in rotational scattering, the BOXCARS geometry must be used to separate the output and input beams. Also, the seeming advantage of the larger rotational Raman cross-section *vis-à-vis* vibrational cross-sections (~10×, Section 4.3.2.1; Eckbreth, 1996) is mostly canceled by the smaller population *differences* for rotational scattering, as seen in Equation 16.5. Since the $\omega_1 - \omega_2$ shift encountered in rotational CARS is smaller than the frequency shifts that occur when a dye laser is pumped, the laser system for pure rotational CARS can be more complicated than that needed in vibrational CARS, where ω_1 and ω_2 can be a fixed-frequency laser and the dye laser pumped by this laser.

16.1.2 *Attributes and Relative Strengths of CARS*

Because $P_3 \propto |\chi^{(3)}_{CARS}|^2 \propto (N_i - N_j)^2 \propto$ (partial pressure)^2, CARS is best suited for measuring majority species in relatively high-pressure reactors, although at times it is sensitive enough to measure the concentrations of minority species. The concentration of all molecules can be determined by CARS because all molecules have at least one Raman-active vibrational mode. It is also very useful in obtaining spatial profiles of the density. The CARS spatial resolution is determined by the laser geometry, while that in other diagnostics, such as LIF and spontaneous Raman scattering, depends on the laser geometry and on the collection system (Section 5.2.1.1). Temperature can be measured, and spatially profiled, by CARS using Equation 4.39.

The maximum power of the CARS signal (Equation 16.5) can be compared to that from spontaneous Raman scattering,

$$P_{Raman} = N_i L \, \Delta\Omega \left(\frac{d\sigma}{d\Omega}\right) P_1 \qquad (16.6)$$

where L is the length of the focal region and $\Delta\Omega$ is the solid angle collected. The CARS signal can be made many orders of magnitude larger than that for spontaneous Raman scattering by making $\mathcal{P}_1$ and $\mathcal{P}_2$ large. In spontaneous Raman scattering, only a fraction of the generated photons are collected, 0.7% for $f/3$ optics (Section 5.2.1.1), while in CARS all light is collected. Also, there are losses in dispersing the light in the spectrometer in spontaneous Raman scattering, while no such losses occur in narrowband CARS. CARS can be configured to achieve higher spatial resolution than in spontaneous Raman scattering; and at higher molecular densities and higher laser intensity, the sensitivity of CARS can be much higher.

Spontaneous Raman scattering (rotational and vibrational–rotational) is sensitive to the same transitions that can be probed by CARS, but it requires only one fixed-frequency laser. This is an advantage because CARS and many other diagnostics, such as LIF and IR-DLAS [IR diode laser absorption spectroscopy], usually require tunable lasers. Moreover, continuous-wave (cw) lasers can be used in spontaneous Raman scattering because probing with the high intensities attainable from pulsed lasers offers no advantage, as it does in CARS.

Infrared diode laser absorption and Fourier-transform infrared (FTIR) absorption are more sensitive to low concentrations than is CARS. However, infrared absorption can spatially profile only axially, while CARS can spatially resolve small volumes. Homonuclear diatomic molecules like N_2 and O_2 can be detected with CARS, but not by absorption. The time response of IR-DLAS is typically less than a millisecond, although with frequency upconversion it can be much faster. CARS takes a snapshot over the ~20-nsec length of the pulse and can get new information every several milliseconds, the separation between laser pulses. These time responses are usually more than sufficient for diagnostics.

CARS has several advantages over LIF. LIF requires tuning to an electronic resonance for excitation, while CARS does not, so species with highly-lying electronic states can be probed. Consequently, visible lasers can be used for CARS, while UV lasers are usually needed for LIF. Furthermore, the LIF signal must be collected by a lens, so close proximity to the probed region in the reactor is needed, and at best only a small fraction of all the fluorescence is collected and detected. Since the CARS signal is a coherent beam, virtually all of the signal can be collected, and the distance from the reaction window to the probed volume is of less concern. LIF can be performed in the excitation mode, as is usual with CARS, where a tunable laser is scanned through resonances and the signal is not spectrally dispersed but detected after passing through a bandpass filter.

16.1.3 *Density Measurements and Thermometry*

Devonshire (1987) has reviewed early work on using CARS to probe species during chemical-vapor deposition (CVD) and plasma-enhanced CVD, particularly during the deposition of amorphous silicon.

Chen *et al.* (1992) used CARS to probe the temperature and concentration distributions of H_2 in a hot-filament reactor used for diamond CVD by examining the Q-branch transitions of H_2 ($\omega_v \sim 4160$ cm^{-1}) from $v = 0$ to $v = 1$. The folded BOXCARS configuration was used with lasers intersecting at 6° and focused to ~30 μm, so the spatial resolution was ~30 × 30 × 300 μm^3. The CARS beam was spatially filtered through a 50-μm pinhole to reject the background due to the bright filament emission.

Figure 16.3a shows the CARS spectra for H_2 at 13°C, while Figure 16.3b shows the spectrum of pure H_2 in the reactor 0.5 mm from the filament. The data from both figures are plotted in Figure 16.4, using Equation 4.39, to give a Boltzmann plot for the rotational temperature. The $f(J)$ factor in

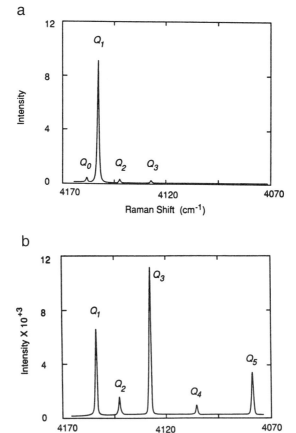

Figure 16.3 CARS spectrum of H_2 (a) near room temperature (13°C) and (b) 0.5 mm away from the substrate in a hot-filament CVD reactor (found to be at 1590 K in Figure 16.4) (from Chen *et al.*, 1992).

16.1 Coherent Anti-Stokes Raman Scattering

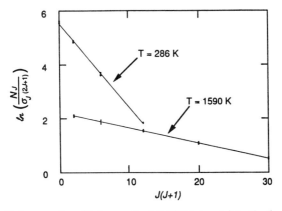

Figure 16.4 Boltzmann fits of ln(normalized CARS intensity) to the (normalized) rotational energy for the data in Figures 16.3a and b, to obtain the rotational temperature (from Chen *et al.*, 1992).

Equation 4.39 includes the usual $2J + 1$ population factor, the nuclear spin degeneracy factor for H_2 [1 (even J): 3 (odd J)], and the J-dependence of the Q-branch Raman scattering cross-section, as given in Equation 4.60. [The ratio $45a'^2/4\gamma'^2 = 6$ was used for H_2 (Section 4.3.2.1).] Figure 16.5 shows that the CARS signal increases quadratically with H_2 pressure (13°C, reactor off), as is expected from Equations 16.1, 16.4, and 16.5. Using CARS profiling, Chen *et al.* found that the temperature decreased with distance from the filament and increased with filament power for CVD mixtures of

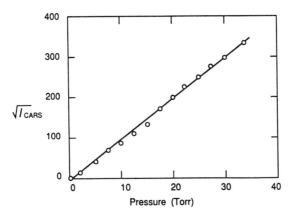

Figure 16.5 The pressure dependence of the CARS signal in H_2 (at room temperature with the CVD reactor filament turned off), which shows that the CARS intensity increases quadratically with density (from Chen *et al.*, 1992).

H_2/CH_4. They also deduced the H-atom concentration and profile from measurements of H_2 density and temperature.

In an earlier study of diamond deposition, Hay et al. (1990) used collinear CARS to measure temperature and the relative concentrations of CH_4 (ν_1, 2917 cm^{-1}), H_2, and C_2H_2 ($\nu_1 \sim 3373$ cm^{-1}) in hot-filament and RF CVD reactors operating with CH_4/H_2 mixtures. CARS of H_2 was used for thermometry. The concentration of CH_4, profiled axially in the filament reactor, decreased monotonically toward the filament, suggesting reactant depletion; depletion was also seen in the RF reactor. The intermediate C_2H_2 was identified in the hot-filament reactor, but not in the RF reactor.

Hata et al. (1983, 1987) used CARS, along with LIF and OES [optical emission spectroscopy], to investigate plasma-assisted CVD from silane ($\nu_1 \sim 2032$ cm^{-1}) and germane ($\nu_1 \sim 2110$ cm^{-1}). Figure 16.6a shows the CARS signal from GeH_4. The peak at 2111 cm^{-1} corresponds to the ν_1 Q-branch. The CARS signals from both reactants decreased with the same

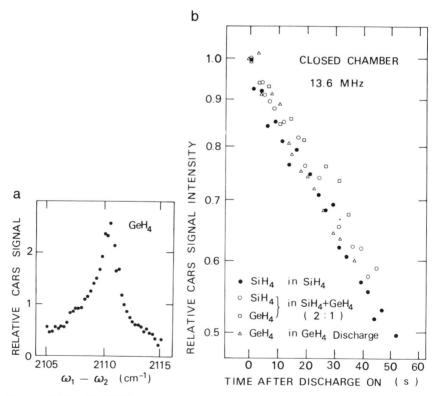

Figure 16.6 (a) CARS spectrum of GeH_4 as a function of $\omega_1 - \omega_2$, which peaks at 2111 cm^{-1} corresponding to the ν_1 Q branch. (b) Relative CARS signal versus time for SiH_4 and GeH_4 in a closed RF plasma-enhanced deposition chamber, operated with neat SiH_4, neat GeH_4, and a SiH_4/GeH_4 mixture (from Hata et al., 1987).

16.2 Surface Second-Harmonic Generation

exponential decay in closed reactors operating with SiH$_4$ only, GeH$_4$ only, and SiH$_4$ + GeH$_4$ reactants, as is seen in Figure 16.6b. Hanabusa and Kikuchi (1983) used CARS to follow silane (2186 cm^{-1}) in the CO$_2$-laser CVD of silicon, and in SiH$_4$/N$_2$ mixtures they spatially profiled temperature by CARS scattering in N$_2$. Kajiyama *et al.* (1985) employed CARS to follow both SiH$_4$ and NH$_3$ (3337 cm^{-1}) in the glow-discharge deposition of silicon nitride from SiH$_4$/NH$_3$ mixtures. Figure 16.7 shows that when the discharge is turned on, the CARS signal for NH$_3$ decreases (because the density of NH$_3$ decreases, while that of added C$_2$H$_2$ is unchanged). Lückerath *et al.* (1988) used collinear CARS to monitor the decomposition of AsH$_3$ (2114 cm^{-1}) and PH$_3$ (~2320 cm^{-1}) in a MOVPE [metalorganic vapor-phase epitaxy] reactor. These *in situ* measurements indicate that the thermal decomposition of AsH$_3$ over GaAs substrates and PH$_3$ over InP substrates occurs at lower temperature than had been reported by sampling and *ex situ* mass and IR spectroscopy.

CARS has been used to monitor many gas-phase species in combustion diagnostics, e.g., N$_2$, H$_2$, O$_2$, CO, H$_2$O, CO$_2$ (Nibler and Yang, 1987; Eckbreth, 1996), and in supersonic jets, e.g. N$_2$, O$_2$, H$_2$, D$_2$, C$_2$H$_4$ (Huber-Wälchli and Nibler, 1982).

16.2 Surface Second-Harmonic Generation

Second-harmonic generation (SHG) and the closely related technique of sum frequency generation (SFG) (Superfine *et al.*, 1988) are probes of

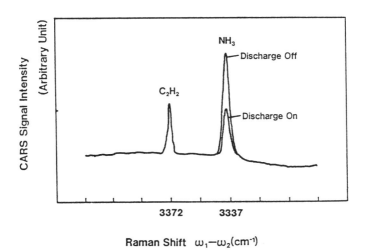

Figure 16.7 The CARS signals for NH$_3$ and C$_2$H$_2$ in a plasma reactor with an NH$_3$/SiH$_4$/C$_2$H$_2$ mixture, with the discharge turned on and off. [From K. Kajiyama, K. Saito, K. Usuda, S. S. Kano, and S. Maeda, *Appl. Phys. B* **38,** CARS study of SiH$_4$-NH$_3$ reaction process in glow discharge plasma, page 140, Figure 2, 1985. Copyright © Springer-Verlag.]

surfaces (gas–solid interfaces), buried solid–solid interfaces, and other interfaces, and have submonolayer sensitivity at the interfacial region. These probes have been of particular interest in studying surface processes on centrosymmetric materials, such as Si, because the electric-dipole contribution to second-harmonic generation from the bulk is zero due to symmetry considerations. Therefore, the dominant contributions to SHG in such materials come from the electric-dipole contribution of the surface and from lower-order bulk terms, such as those due to electric-quadrupole and magnetic-dipole interactions. In many cases, these latter bulk terms either are small compared to the surface term or their contribution can be separated from the surface term by judicious choice of polarizations and geometries.

Second-harmonic (SH) analysis of surfaces is often performed with a Nd:YAG laser (1.06 μm), which resonantly enhances SHG for Si, and in reflection geometry. The reflected SH radiation is separated from the pump beam by filters and a monochromator. Often SHG is analyzed as a function of sample azimuthal angle, as is demonstrated in Figure 16.8. [This is convenient in reactors where sample rotation is needed during processing, such as during molecular beam epitaxy (MBE).]

Shen (1989), Heinz (1991), Corn and Higgins (1994), and Liebsch (1994) have reviewed the nonlinear optics of surfaces and interfaces, while Reider and Heinz (1995) have reviewed these processes at semiconductor surfaces. Sipe *et al.* (1987) have published a detailed phenomenological theory of

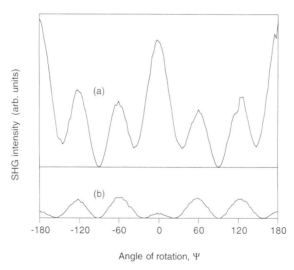

Figure 16.8 Comparison of surface second-harmonic generation from (a) O- and (b) H-terminated vicinal 5° Si(111) surfaces as a function of the angle of rotation (from Bjorkman *et al.*, 1993b).

16.2 Surface Second-Harmonic Generation

surface and bulk contributions to SHG. Many of the SH studies cited in these reviews were performed outside the processing chamber. Although the emphasis in this section is on the use of surface SHG as a real-time probe, several *ex situ* studies are cited here because they involve important systems and could have easily been performed *in situ*.

In one early study, Heinz *et al.* (1985) used SHG to study the symmetry of well-ordered Si(111)-7 × 7 surfaces and to investigate how these surfaces can become disordered during processing. One particularly useful feature of surface SHG is the distinctive angular pattern of the signal. Figure 16.9 shows the second-harmonic (SH) electric field on this well-ordered surface as a function of input polarization, with the SH signal resolved both parallel and perpendicular to the pump electric field. During *in situ*, real-time monitoring, they observed a decrease in SH intensity during exposure to oxygen (surface oxidation), which they attributed to the loss of long-range order. They also saw a decay in SH signal (versus ion fluence) during ion bombardment. Iyer *et al.* (1987) and Heinz *et al.* (1987) saw a similar decay during the evaporation of Si on a Si(111) substrate held at room temperature (Figure 16.10), because the film being deposited was amorphous. This figure also shows that they saw no change in the SH signal during deposition when the surface was held at 650°C, because the Si film being deposited was crystalline.

In another study of Si surfaces and interfaces, Tom *et al.* (1986) examined *in situ* oxidation and the effect of surface phosphorus on the Si(111)-7 × 7 surface by using SHG. Verheijen *et al.* (1991) demonstrated how to separate the effects of the surface and residual bulk terms in SHG by examining the rotational anisotropy of oxidized vicinal Si(111) surfaces, with well-

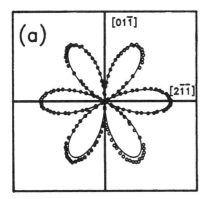
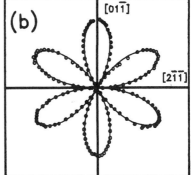

Figure 16.9 Second-harmonic electric field resolved along directions (a) parallel and (b) perpendicular to the pump electric-field vector, as a function of the angular orientation of the linearly polarized pump electric-field vector for a clean Si(111)-7 × 7 surface (from Heinz *et al.*, 1985).

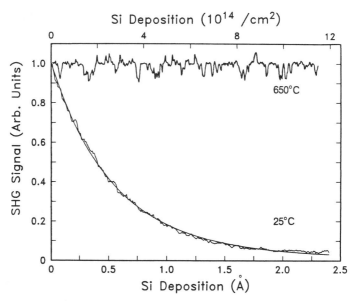

Figure 16.10 Real-time SH intensity from a Si(111)-7 × 7 surface during Si deposition by evaporation for substrate temperatures of 25°C (room temperature) and 650°C, which indicates the growth of amorphous and epitaxial material at the respective temperatures (from Heinz et al., 1987).

defined steps. Bjorkman et al. (1993a,b) and Emmerichs et al. (1994) used these results to study *ex situ* the interface of $SiO_2/Si(111)$ interfaces, which were formed by thermal oxidation, plasma-assisted deposition, and chemical oxidation. As seen in Figure 16.8, they found that the angular dependence of SH radiation is very different for oxygen- and hydrogen-terminated vicinal 5° Si(111). Ito and Hirayama (1994) found that the rotational asymmetry in the SHG at the $SiO_2/Si(111)$ interface disappeared upon hydrogen annealing, but not upon nitrogen annealing. This suggested that $\chi^{(2)}_{zxx}$, which causes this asymmetry, is due to interface dangling bonds (which are annealed by heating in hydrogen but not in nitrogen) and not to distorted bonds at the interface. Haraichi et al. (1994, 1995) found that the SHG intensity from the surface of Si(111) decreased when it was exposed to the etchants XeF_2 and Cl_2; for XeF_2, there was a concomitant loss of surface symmetry. The SHG intensity recovered after annealing.

In other studies of Si surfaces and interfaces, Power and McGilp examined SHG of Sb adsorbed on Si(111) (1993) and Si(100) (1994). Krüger et al. (1993) found that the real-time SH signal decreases during the electrodeposition of Ni on oxidized *n*-Si(111), probably due to a shadowing effect. Heinz et al. (1989) spectroscopically studied the epitaxial $CaF_2/Si(111)$ interface *ex situ*.

16.2 Surface Second-Harmonic Generation

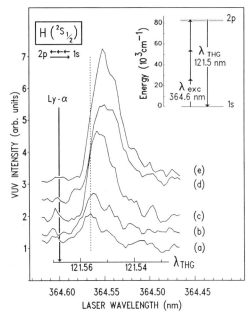

Figure 16.11 Use of THG to monitor ground-state H atoms in a H_2 (or CH_4/H_2) DC plasma, with excitation with a tunable dye laser near 364.6 nm and detection near 121.5 nm. The plasma power was (a) 190 W, (b) 370 W, (c) 530 W, (d) 630 W, and (e) 680 W; the estimated hydrogen density for curve (e) was $2 \pm 1 \times 10^{14}$ cm^{-3} (from Celii et al., 1990a).

Zhang et al. (1991) monitored *in situ* the deposition of Cu on polyimide films by using SHG. The SH signal increased rapidly and then decreased to a steady level. The initial increase indicated the formation of clusters on the surface that strongly enhance SHG via a plasma resonance, while the latter behavior was due to SHG on a smooth surface. The SH signal was followed during heating, during which cluster formation and diffusion into the polyimide were competing processes. The SH signal did not change with time when the copper was deposited on a thin layer of Ti atop the polyimide, which demonstrated that Ti is a barrier to diffusion.

Stehlin et al. (1988) have shown that, with the proper choice of polarizations, SHG can be used as an effective surface probe even for media without inversion symmetry, such as GaAs. Because the symmetry of the bulk of a material is generally different from that of the surface, the electric-dipole-allowed $\chi^{(2)}_{bulk}$ can be suppressed, without suppressing $\chi^{(2)}_{surface}$. They demonstrated this during the real-time evaporation of Sn on GaAs(001). More recently, Armstrong et al. (1993) has examined SHG from clean and oxide-covered c(4 × 4) and (2 × 4) reconstructed GaAs(001) surfaces. Yamada and Kimural (1995) used SHG to probe the surface during GaAs epitaxy.

16.3 Third-Harmonic Generation in Gases

Harmonic generation can also be used to study species in the gas phase. Because of inversion symmetry, the lowest-order contribution in gases is third-harmonic generation (THG). Detection of the ground state of light atoms, such as H, O, N, and F, is challenging because the lowest electronic transitions are in the vacuum ultraviolet. Celii *et al.* (1990a) used THG to monitor ground-state hydrogen atoms ($1s\,^2S_{1/2}$) in H_2 and CH_4/H_2 plasmas; the latter discharge is used for CVD of diamond films. They tuned an excimer-laser pumped dye laser to ~364.6 nm and generated third-harmonic light at 121.5 nm, near the $2p\,^2P_J^0 \rightarrow 1s\,^2S_{1/2}$ Lyman-α transition (Figure 16.11). The intensity and lineshift (relative to the Lyman-α transition) helped determine the H-atom density in the plasma. Celii *et al.* (1990b) have also detected O atoms via THG by using ~390-nm photons to generate 130 nm, which is nearly resonant with $3s\,^3S_1^0 - 2p^4\,^3P_J$, and Celii *et al.* (1991) have detected N atoms via THG by using ~360-nm photons to generate 120 nm, which is nearly resonant with $3s\,^4P_J - 2p^3\,^4S_{3/2}^0$. When possible, it is probably simpler to detect these atoms by using two-photon LIF (Section 7.2.1.2) than by THG.

References

S. R. Armstrong, R. D. Hoare, M. E. Pemble, I. M. Povey, A. Stafford, A. G. Taylor, S. A. Joyce, J. H. Neave, and J. Zhang, *Surf. Sci. Lett.* **291,** L751 (1993).

C. J. Bjorkman, C. E. Shearson, Jr., Y. Ma, T. Yasuda, G. Lucovsky, U. Emmerichs, C. Meyer, K. Leo, and H. Kurz, *J. Vac. Sci. Technol. A* **11,** 964 (1993a).

C. J. Bjorkman, T. Yasuda, C. E. Shearson, Jr., Y. Ma, G. Lucovsky, U. Emmerichs, C. Meyer, K. Leo, and H. Kurz, *J. Vac. Sci. Technol. B* **11,** 1521 (1993b).

R. W. Boyd, *Nonlinear Optics.* Academic Press, Boston, 1992.

F. G. Celii, H. R. Thorsheim, J. E. Butler, L. S. Plano, and J. M. Pinneo, *J. Appl. Phys.* **68,** 3814 (1990a).

F. G. Celii, H. R. Thorsheim, M. A. Hanratty, and J. E. Butler, *Appl. Opt.* **29,** 3135 (1990b).

F. G. Celii, H. R. Thorsheim, and J. E. Butler, *J. Chem. Phys.* **94,** 5248 (1991).

K.-H. Chen, M.-C. Chuang, C. M. Penney, and W. R. Banholzer, *J. Appl. Phys.* **71,** 1485 (1992).

R. M. Corn and D. A. Higgins, *Chem. Rev.* **94,** 107 (1994).

R. Devonshire, *Chemtronics* **2,** 183 (1987).

S. A. J. Druet and J. P. Taran, *Prog. Quantum Electron.* **7,** 1 (1981).

A. C. Eckbreth, *Laser Diagnostics for Combustion Temperature and Species,* 2nd ed. Gordon & Breach, Luxembourg, 1996.

A. C. Eckbreth and P. W. Schreiber, in *Chemical Applications of Nonlinear Raman Spectroscopy* (A. B. Harvey, ed.), Chapter 2, p. 27. Academic Press, New York, 1981.

References

U. Emmerichs, C. Meyer, H. J. Bakker, F. Wolter, H. Kurz, G. Lucovsky, C. E. Bjorkman, T. Yasuda, Y. Ma, Z. Jing, and J. L. Whitten, *J. Vac. Sci. Technol. B* **12,** 2484 (1994).

J. W. Fleming, *Opt. Eng.* **22,** 317 (1983).

M. Hanabusa and H. Kikuchi, *Jpn. J. Appl. Phys.* **22,** L712 (1983).

S. Haraichi, F. Sasaki, S. Kobayashi, M. Komuro, and T. Tani, *Jpn. J. Appl. Phys. Part 1* **33,** 7053 (1994).

S. Haraichi, F. Sasaki, S. Kobayashi, M. Komuro, and T. Tani, *J. Vac. Sci. Technol. A* **13,** 745 (1995).

A. B Harvey, ed., *Chemical Applications of Nonlinear Raman Spectroscopy.* Academic Press, New York, 1981.

N. Hata, A. Matsuda, and K. Tanaka, *J. Non-Cryst. Solids* **59/60,** 667 (1983).

N. Hata, A. Matsuda, and K. Tanaka, *J. Appl. Phys.* **61,** 3055 (1987).

S. O. Hay, W. C. Roman, and M. B. Colket, III, *J. Mater. Res.* **5,** 2387 (1990).

T. F. Heinz, in *Nonlinear Surface Electromagnetic Phenomena* (H.-E. Ponath and G. I. Stegeman, eds.), Chapter 5, p. 353. North-Holland Publ., Amsterdam, 1991.

T. F. Heinz, M. M. T. Loy, and W. A. Thompson, *J. Vac. Sci. Technol. B* **3,** 1467 (1985).

T. F. Heinz, M. M. T. Loy, and S. S. Iyer, *Mater. Res. Soc. Symp. Proc.* **75,** 697 (1987).

T. F. Heinz, F. J. Himpsel, E. Palange, and E. Burstein, *Phys. Rev. Lett.* **63,** 644 (1989).

P. Huber-Wälchli and J. N. Nibler, *J. Chem. Phys.* **76,** 273 (1982).

F. Ito and H. Hirayama, *Phys. Rev. B* **50,** 11208 (1994).

S. S. Iyer, T. F. Heinz, and M. M. T. Loy, *J. Vac. Sci. Technol. B* **5,** 709 (1987).

K. Kajiyama, K. Saito, K. Usuda, S. S. Kano, and S. Maeda, *Appl. Phys. B* **38,** 139 (1985).

J. Krüger, N. Sorg, J. Reif, and W. Kautek, *Appl. Surf. Sci.* **69,** 388 (1993).

M. D. Levenson and S. S. Kano, *Introduction to Nonlinear Laser Spectroscopy.* Academic Press, Boston, 1988.

A. Liebsch, *Surf. Sci.* **307-309,** 1007 (1994).

R. Lückerath, P. Tommack, A. Hertling, H. J. Koss, P. Balk, K. F. Jensen, and W. Richter, *J. Cryst. Growth* **93,** 151 (1988).

J. W. Nibler and G. V. Knighten, in *Raman Spectroscopy of Gases and Liquids* (A. Weber, ed.), Chapter 7. Springer-Verlag, Berlin, 1979.

J. W. Nibler and J. J. Yang, *Annu. Rev. Phys. Chem.* **38,** 349 (1987).

J. R. Power and J. F. McGilp, *Surf. Sci.* **287/288,** 708 (1993).

J. R. Power and J. F. McGilp, *Surf. Sci.* **307-309,** 1066 (1994).

G. A. Reider and T. F. Heinz, in *Photonic Probes of Surfaces; Electromagnetic Waves: Recent Developments* (P. Halevi, ed.), Vol. 2. Chapter 9. Elsevier, Amsterdam, 1995.

Y. R. Shen, *The Principles of Nonlinear Optics.* Wiley, New York, 1984.

Y. R. Shen, *Nature (London)* **337,** 519 (1989).

J. E. Sipe, D. J. Moss, and H. M. van Driel, *Phys. Rev. B* **35,** 1129 (1987).

T. Stehlin, M. Feller, P. Guyot-Sionnest, and Y. R. Shen, *Opt. Lett.* **13,** 389 (1988).

R. Superfine, P. Guyot-Sionnest, J. H. Hunt, C. T. Kao, and Y. R. Shen, *Surf. Sci.* **200,** L445 (1988).

W. M. Tolles, J. W. Nibler, J. R. McDonald, and A. B. Harvey, *Appl. Spectrosc.* **31,** 253 (1977).

H. W. K. Tom, X. D. Zhu, Y. R. Shen, and G. A. Somorjai, *Surf. Sci.* **167,** 167 (1986).
M. A. Verheijen, C. W. van Hasselt, and T. Rasing, *Surf. Sci.* **251/252,** 467 (1991).
H. Vogt, in *Light Scattering in Solids II* (M. Cardona and G. Guntherodt, eds.), Chapter 4. Springer-Verlag, Berlin, 1982.
C. Yamada and T. Kimura, *Jpn. J. Appl. Phys. (Part 1)* **34,** 1102 (1995).
A. Yariv, *Quantum Electronics.* Wiley, New York, 1989.
M. A. Yuratich and D. C. Hanna, *Mol. Phys.* **33,** 671 (1977).
J. Y. Zhang, Y. R. Shen, D. S. Soane, and S. C. Freilich, *Appl. Phys. Lett.* **59,** 1305 (1991).

CHAPTER **17**

Optical Electron/Ion Probes

Electrons can be ejected from a medium after it absorbs one or more photons that have sufficient energy. Electron emission from gas-phase species, which is called photoionization (PI), occurs when the energy of the photon(s) exceeds the ionization potential (see Figure 17.1). The analogous ejection of electrons from surfaces, called photoemission (PE), occurs when the photon energy exceeds the work function. With careful choice of wavelength, a single type of atom, molecule, or ion can be selectively identified by using photoionization, even for very small concentrations. Similarly, the nature of a film surface or adsorbates on this surface can be analyzed with photoemission.

For both photoionization and photoemission, the electron flux can be measured as a function of laser wavelength (near the work function or resonances) by tuning the wavelength, or the energy distribution of the emitted electrons can be resolved for a given wavelength. The maximum pressure in the chamber is limited (<1 mTorr) for PI and PE because of electron collection.

Optogalvanic spectroscopy probes the absorption of light in a plasma discharge by monitoring a change in a discharge characteristic, such as current, voltage, or electron density. Photoionization of the plasma and photoemission from electrodes are two types of optogalvanic spectroscopies. Absorption need not directly release electrons to lead to noticeable changes in the discharge.

Both PI and PE have been used frequently in fundamental studies. They have found limited use as process-development diagnostics and are rarely employed for real-time diagnostics. Photoionization and photoemission are discussed in Sections 17.1 and 17.2, respectively; however, the use of PI and PE in optogalvanic diagnostics is discussed separately in Section 17.3.

17.1 Photoionization

Photoionization can occur after the absorption of a single photon (Figure 17.1a), usually in the deep ultraviolet, or several photons (Figure 17.1b), often in the ultraviolet and/or visible. The latter method is a stepwise excitation process between different electronic states that can involve photons at one or more wavelengths, and is called multiphoton ionization (MPI). The MPI method has been used to measure very small concentrations of species and, in particular, to detect single atoms and to identify and measure gas-phase species in fundamental gas–surface studies; MPI has also been utilized to separate isotopes. Letokhov (1987) has reviewed these applications.

The goal of this diagnostic is to ionize a given species selectively and with high probability. The final step in single and multiphoton ionization is absorption from a discrete level to a continuum of free-electron states. Usually the cross-section of this last step is fairly independent of wavelength and is small, particularly when the energy of the free electron is large. However, in some instances, particularly in atoms, there are resonant enhancements in this cross-section at selected wavelengths due to autoionization; this occurs when the continuum state of the ion and free electron is isoenergetic with a "discrete" atomic state having two excited electrons. Also, ionization of highly excited Rydberg states can occur with a large cross-section by photoionization with an infrared laser, collisional excitation, or the application of a strong static electric field (Stark field). Some-

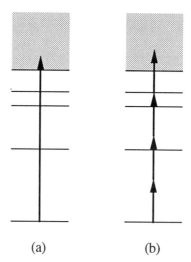

Figure 17.1 Energy levels for (a) single photon ionization (SPI) and (b) resonant-enhanced multiphoton ionization (REMPI) illustrating a 3 + 1 process.

17.1 Photoionization

times photoionization selectivity of a given species can occur with selective excitation of the ground-state species by the absorption of an infrared laser, followed by ultraviolet photoionization (Letokhov, 1987). These enhancement methods have rarely been applied to the optical diagnostics of thin film processing, although they have been used for other applications, such as isotope separation. Instead, the best ionization method for studying gas-phase species in film processing is resonant-enhanced multiphoton ionization, which is commonly called REMPI.

In REMPI, a molecule can be unambiguously identified and its density can be measured with high sensitivity, both because of selective resonant enhancement. The first step in an $n + m$ REMPI process is resonant (one-step) n-photon absorption from the ground electronic state (Section 4.4), which is followed by the nonresonant absorption of m photons that leads to ionization. The wavelength of the laser is tuned to the n-photon resonance, as is seen in Figure 17.1b which depicts $3 + 1$ REMPI. Although more than one laser wavelength can be used, use of a single laser is common. High-intensity pulsed lasers are employed in this highly nonlinear process. If no steps are saturated, the cross-section for REMPI $\propto I^{n+m}$, where I is the laser intensity. Electrons are collected by an electrode that is biased to a positive voltage, as is seen in Figure 17.2.

REMPI is actually a variant on laser-induced fluorescence (LIF; see Chapter 7) because the laser is tuned so the first step resonantly excites the species to an excited electronic state in both processes (Parker, 1983). Consequently, REMPI has the same spectral identification capability as does LIF because the rate of electron production is proportional to the absorption profile, albeit in a nonlinear manner. REMPI can have much higher sensitivity than LIF because all of the electrons can be collected with unity efficiency when a static electric field is applied to the collecting electrode. In LIF only a small fraction of emitted photons are detected because of the limited solid angle of collection, 0.7% for $f/3$ optics, and the limited quantum efficiency of detection (~10%). Moreover, larger species

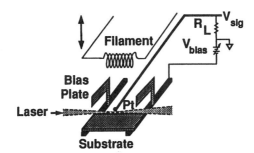

Figure 17.2 Schematic of REMPI experimental apparatus used during filament-assisted CVD of diamond, with movable filament (from Celii and Butler, 1992).

like CH_3, SiH_3, and GeH_3, which are important intermediates during CVD, do not fluoresce (as is necessary for LIF), yet they can be detected by REMPI. Since REMPI uses (invasive) electrodes for electron collection, it is useful for fundamental studies and process development, but not for real-time manufacturing control. Also, the production of electrons by REMPI can sometimes perturb the thin film process.

In MPI, electron collection by the ionization probe should be saturated. The collection efficiency must be independent of the process parameters that are varied. As in other gas-phase spectroscopies, signal analysis should reflect the Boltzmann distribution in vibrational and rotational states in the probed (ground) electronic state.

Okada *et al.* (1990) showed how REMPI can probe CH_3 in a laser chemical vapor deposition (CVD) environment via two-photon absorption at 333.4 nm. Overall, a three-photon process was suggested by the I^3 dependence of the REMPI signal, which assumes no laser saturation. Earlier studies by Hudgens *et al.* (1983) identified this CH_3 REMPI peak as the unresolved Q-branch of the resonant two-photon absorption from the $X\,^2A_2''(v=0)$ electronic state of CH_3 to the $3p^2\,A_2''(v=0)$ Rydberg state, followed by incoherent single-photon ionization. CH_3 can also be detected by UV absorption at 216 nm or IR absorption (Section 8.2.1.1.1), but not by LIF.

Celii *et al.* (1989), Celii and Butler (1992), and Corat and Goodwin (1993) detected CH_3, a dominant growth intermediate, during filament-assisted diamond growth with CH_4/H_2 mixtures. They used this 2 + 1 REMPI process, which they found to peak at 333.5 nm (Figure 17.3). Figure 17.3 also shows 2 + 1 REMPI of excited-state $C(^1D)$ atoms at 340.8 nm, where the two-photon resonance is $3p\,^1P_1 \leftarrow 2p\,^1D_2$; there is another REMPI resonance in C at 320.2 nm. These groups saw a rapid increase in the CH_3 concentration when the filament temperature increased above ~2150 K, as is seen in Figure 17.4. At this same temperature, the density of CH_4 decreased and that of C_2H_2 increased, as measured by sampling mass spectrometry. Corat and Goodwin (1993) used REMPI to show that the CH_3 density increases as the substrate temperature increases from 700 to 1000 K. This group verified their REMPI calibration procedure by examining the 3 + 1 REMPI signal at 314.4 nm in argon, with resonant three-photon absorption $3p^54s\,4s'[1/2]^0\,J = 1 \leftarrow 3p^6\,^1S$.

Celii *et al.* (1989) also monitored H atoms during diamond CVD, by using the 3 + 1 REMPI scheme described by Tjossem and Cool (1983). They saw a strong REMPI signal at 364.7 nm only in a standing-wave geometry, where a spherical mirror refocused the laser back along the optical axis in the reactor. In the usual traveling-wave geometry, with no backfocusing, only a weak REMPI feature was seen (364.54 nm) because of interference between odd harmonics of the laser field. Third-harmonic-generated light at 121.57 nm drives the excited state out of phase for

17.1 Photoionization

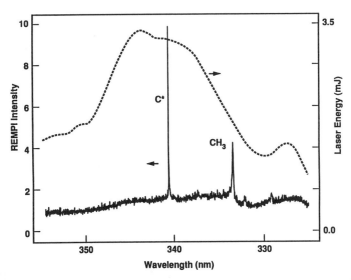

Figure 17.3 REMPI spectrum of C(^1D) and CH$_3$, along with dye laser power (dashed line) during filament-assisted diamond CVD (from Celii and Butler, 1992).

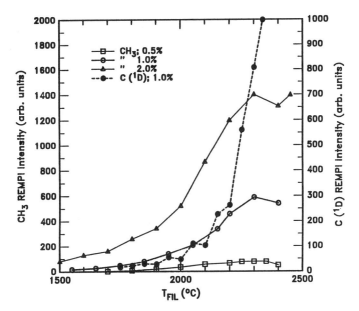

Figure 17.4 Dependence of CH$_3$ and C(^1D) REMPI signals on filament temperature during diamond CVD, with given CH$_4$/H$_2$ ratios (from Celii and Butler, 1992).

REMPI, so REMPI is suppressed. In the standing-wave configuration, third-harmonic generation is suppressed and a strong REMPI signal is seen. This is generally observed in odd harmonic REMPI processes (odd n), as for REMPI in Ar. Other ways to detect H atoms optically include multiphoton LIF (Section 7.2.1.2) and third-harmonic generation (Section 16.3).

Also of interest in thin film processing is the 2 + 1 REMPI detection of Cl by Arepalli *et al.* (1985). The first step is the $3p^44p\ ^4S^0 \leftarrow 3p^5\ ^2P^0$ two-photon transition, which uses 230–245-nm photons. Section 7.2.1.2 describes the use of this transition to detect Cl atoms by two-photon LIF by Selwyn *et al.* (1987). Within the theme of photoionization, that study also suggested that a photon could photodetach an electron from Cl$^-$ or XCl$^-$ to form Cl atoms (Equations 7.2 and 7.3), which would then be detected by two-photon LIF.

REMPI is also used extensively to investigate the fundamentals of gas–surface interactions, such as in probing the interactions of NO with surfaces (Misewich and Loy, 1986; Hamers *et al.,* 1988). Also, Zhang and Stuke (1988) have used a variation of this method, in which the photocreated ions are detected by mass spectrometry, to study the large-area photolysis of adsorbed and gas-phase aluminum alkyls. This technique is known as laser ionization mass spectroscopy (LIMS).

Goodwin and Otis (1989) used 2 + 1 REMPI to probe CO in the plume of laser-ablated polyimide. Instead of collecting electrons, they collected the ions, which they mass-resolved. Resonant two-photon absorption excited the CO $B\ ^1\Sigma^+ \leftarrow X\ ^1\Sigma^+$ transition using ~230-nm photons. Rotational and vibrational temperatures were determined from the REMPI signal (Figure 17.5) and analyzed as a function of delay time. Both the $(v', v'') = (0,0)$ and $(1,1)$ bands were excited, and the Q branches of these bands in this two-photon absorption were clearly resolved. The R and P branches of these bands overlap in one-photon absorption, which complicates rotational analysis.

Either electrons or ions can be detected in photoionization, although electrons are usually detected in REMPI. The laser intensities used in REMPI are often so high that the product ions continue to absorb photons and they fragment. This is a disadvantage if the ions must be probed by mass analysis, because the cracking pattern into these fragments can be very sensitive to the laser wavelength and intensity. One such example is in *in situ* monitoring of the flux of reactants and products during MBE [molecular-beam epitaxy] growth. Since there is very little fragmentation in single-photon photoionization (SPI; see Figure 17.1a) when the photon energy is barely higher than the ionization potential, SPI may be preferable to REMPI in such applications. While mass spectrometry (MS) can be used to monitor fluxes that reflect from the substrate, MS cannot measure the incoming flux without changing the configuration; SPI can be used to measure concentrations anywhere. Also, the ionizer in a mass spectrometer

17.1 Photoionization

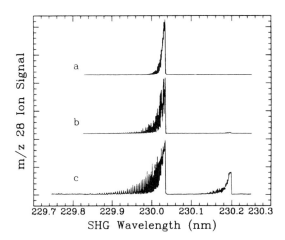

Figure 17.5 Mass-resolved ions from CO REMPI for (a) room temperature CO; (b) CO from low-fluence, submonolayer-per-pulse laser ablation from polyimide (6 mJ/cm^2); (c) CO from high-fluence laser ablation (150 mJ/cm^2, 100 Å removed per pulse) (from Goodwin and Otis, 1989).

cracks the species in addition to ionizing them, which can make analysis difficult.

Strupp *et al.* (1993) used SPI time-of-flight MS (SPI-TOFMS) to examine the arsenic species As, As$_2$, and As$_4$ in an MBE chamber, as is illustrated in Figure 17.6. They frequency-tripled a Nd:YAG laser (1.06 μm) in frequency summation crystals to produce 355 nm, and then tripled this output in Xe gas to produce 118-nm (10.5-eV) photons. Products were then detected by time-of-flight mass spectrometry (TOF-MS), without an ionizer. Table 17.1 shows that 10.5-eV photons can photoionize many atoms and molecules involved in film deposition.

When they operated the arsenic source for MBE at low temperature to produce only As$_4$, Strupp *et al.* (1993) saw large As$_4^+$ signals by SPI. With 118-nm SPI, Strupp *et al.* observed only very small As$_2^+$ signals (0.4%× that for As$_4^+$) and essentially no As$_3^+$; but with 266-nm MPI (at intensities high enough to give comparable signals), they saw much cracking, which gave larger signals for As$_2^+$ than for As$_4^+$. At higher temperatures, the source produces As$_2$ and As$_4$, which were detected as As$_2^+$ and As$_4^+$ with SPI; no fragmentation to As$^+$ was seen. Alstrin *et al.* (1993) used 118-nm SPI to detect both As and As$_2$ desorbing from silicon, with As$_4$ incident on the surface. Figure 17.7, from Alstrin *et al.* (1994), shows simultaneous monitoring by reflection high-energy electron diffraction (RHEED) and SPI-TOFMS. RHEED oscillations, indicating layer-by-layer growth, are seen only when both Ga and As$_4$ impinge the surface. A slow decay of the desorbing As$_2$ flux is seen when the As$_4$ flux is turned off. This technique has potential for real-time monitoring.

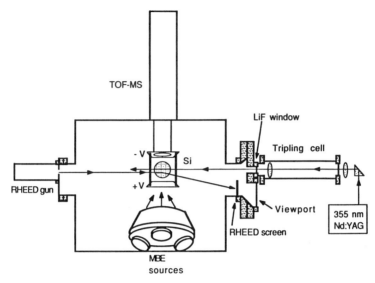

Figure 17.6 Schematic of the apparatus used for SPI-TOFMS during MBE (from Strupp *et al.*, 1993).

Photoionization at very high intensities (MPI and SPI) is also used in the detection of sputtered materials during *ex situ* characterization (Becker and Hovis, 1994).

Table 17.1 Species Ionizable at 118 nm (10.5 eV) ($\lambda_{Nd:YAG}/9$)*

Species	Ionization potential (eV)	Species	Ionization potential (eV)
Al	6.0	In	5.8
$Al(CH_3)_3$	9.1	Mn	7.4
As	9.8	P_4	9.7
As_2	9.7	PH_3	10.0
As_4	8.5	Se	9.8
AsH_3	10.0	Si	8.1
Cd	9.0	Sn	7.3
$Cd(CH_3)_2$	8.6	$Sn(CH_3)_2$	8.0
Ga	6.0	Te	9.0
GaAs	7.2	Zn	9.4
Hg	10.4	$Zn(CH_3)_2$	9.0

* From Strupp *et al.* (1993); Alstrin *et al.* (1994).

17.2 Photoemission

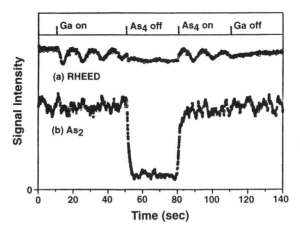

Figure 17.7 Simultaneous monitoring with (a) RHEED and (b) SPI-TOFMS of As_2 during the described GaAs MBE sequences at 800 K (from Alstrin et al., 1994).

17.2 Photoemission

Photoemission, based on the photoelectric effect, is a standard surface probe used in ultrahigh vacuum studies of surfaces. It is commonly known as XPS [x-ray photoelectron spectroscopy] or UPS [ultraviolet photoelectron spectroscopy], depending on the wavelength of the source used to eject electrons, or ESCA [electron spectroscopy for chemical analysis]. Feldman and Mayer (1986) have discussed the physics and applications of XPS.

XPS is frequently used to study surfaces after thin film processing. When processing is conducted at pressures too high for XPS analysis ($>10^{-6}$ Torr), the sample has to be transferred from the reactor to the XPS chamber (Oehrlein et al., 1991; Guinn and Donnelly, 1994). While such a measurement is neither in real-time nor *in situ*, sometimes the surface layers that form during processing, as during plasma etching, do not change appreciably during the pumpdown and transfer steps (Section 15.1).

Photoemission has also been used to probe surfaces during and after film processing by using ultraviolet lasers. The use of photoemission to probe electrodes during plasma processing is discussed in Section 17.3. Those studies are concerned with how the photoemission yield depends on surface modifications. Photoemission is also affected by doping. The work function of Si, and therefore the photoemission yield near threshold, depends on the type and level of doping (Allen and Gobeli, 1962). Quiniou et al. (1989) have used this property of PE to probe very thin doped line features in Si, with ~1-μm lateral resolution, by using tightly focused cw UV lasers (257 nm). Although this work was performed *ex situ*, it has some potential for *in situ* analysis during processing.

17.3 Optogalvanic Spectroscopy

Spectroscopic information is obtained in optogalvanic spectroscopy by tuning the laser to characteristic absorption features in the gas or the exposed surface in the plasma, and then monitoring changes in discharge current, voltage, or electron density. Figure 17.8 depicts one possible optogalvanic mechanism.

Dreyfus *et al.* (1985) have reviewed the use of optogalvanic spectroscopies to probe species in the plasma in low-pressure discharges of interest in thin film processing. Other references on optogalvanic effects are the conference proceedings edited by Camus (1983) and Stewart and Lawler (1991); the latter includes a review by Telle (1991). Absorption can lead to photoionization (PI) or photodetachment, which directly produce electrons, or other excitations that lead to the creation of electrons indirectly. For example, atoms and molecules excited to high-lying Rydberg states can ionize with high probability by collisions, which can be very sensitive to local electric fields (Nakajima *et al.,* 1983; Doughty and Lawler, 1984; Doughty *et al.,* 1984; Ganguly and Garscadden, 1985).

While the processes described so far are expected to increase electron densities and current, other optogalvanic events can decrease the discharge current (Figure 17.8). Metastably excited species, which are in states that decay radiatively only very slowly, constitute a source of excited species for electron impact ionization. If laser absorption excites a metastable species to a higher state that radiatively decays fast, the metastable density decreases, leading to decreases in electron production and discharge current (Doughty and Lawler, 1983).

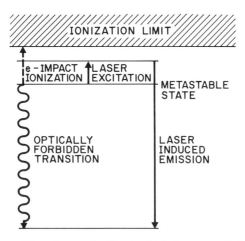

Figure 17.8 One possible optogalvanic effect, in which the more easily ionizable metastable species are depleted, causing a decrease in discharge current (from Donnelly, 1989).

17.3 Optogalvanic Spectroscopy

Photodetachment can determine the concentrations of negative ions. Greenberg *et al.* (1984) found large concentrations of negative ions in NF_3 discharges, with the predominant species being F^-, by photodetachment with a XeCl laser (308 nm). The main absorption process was probably $F^- + h\nu \rightarrow F + e^-$. The excess electron population was formed by photodetachment, which they measured by microwave interferometry. Gottscho and Gaebe (1986) found a transient current pulse in a Cl_2 RF discharge when a pulsed N_2 laser (337 nm) was passed near the momentary anode, which they attributed to photodetachment from Cl^-. Other negative ions detected by photodetachment include H^- and O^- (Bacal *et al.*, 1979) and I^- (Webster *et al.*, 1983). Walkup *et al.* (1983) used optogalvanic spectroscopy to probe positive ions in the cathode sheath region, where they are the main charge carriers. Figure 17.9 shows the laser optogalvanic spectrum of N_2^+, in which they tuned a dye laser across the rotational lines in the $B\ ^2\Sigma_u^+ \leftarrow X\ ^2\Sigma_g^+$ (0,0) band.

Downey *et al.* (1988) used photoemission optogalvanic spectroscopy (POGS) to characterize the surface of electrodes in a RF plasma. While in normal (gas-phase) optogalvanic spectroscopy, irradiation of the electrodes is avoided, in POGS electrons are purposely injected into the plasma from the surface. During plasma etching, the surface composition of the electrode is different from that in vacuum because of changes in work function and carrier mobility, and this affects the photoemission yield. They monitored the plasma current when a pulsed ultraviolet laser [248 nm (KrF excimer laser), 222 nm (KrCl), 193 nm (ArF)] irradiated the electrode surface to induce photoemission. Downey *et al.* also used a frequency-doubled dye laser to examine the wavelength dependence of PE. PE and

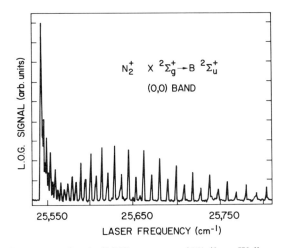

Figure 17.9 Laser optogalvanic (LOG) spectrum of N_2^+ (from Walkup *et al.*, 1983).

POGS of Al (work function of 4.06–4.41 eV, depending on surface conditions) and SiO_2 (8.9 eV, much greater than all excimer-laser photon energies) -covered Si (5.1 eV; 243 nm) were examined. The change in current with the plasma turned on (POGS) was usually much greater than that with it off and in vacuum (PE).

In vacuum and in Ar plasmas, the change in current was limited by laser power at lower intensities and was space-charge-limited at higher intensities. Figure 17.10 shows why this technique has potential as an endpoint detector and for *in situ* contamination monitoring. The POGS signal for 222 nm (5.6 eV) during CF_4/O_2 plasma etching of a SiO_2 film atop Si is plotted versus time, along with the laser power reflected from the electrode. The POGS signal was small when the SiO_2 layer was thick (time < 6 min) because electrons formed by photoemission from the Si surface could not escape through the thick oxide ($\gg$100 Å). The POGS signal increased when the oxide became sufficiently thin ($\sim$6 min), and decreased again when the bare silicon surface was exposed to the F-containing plasma (>7 min), which introduced electron traps that increased the Si work function. This decrease was not seen at higher photon energies (193 nm). The interference fringes observed in reflection were sometimes mirrored in the POGS signal.

In similar work, Selwyn *et al.* (1988) examined this plasma-enhanced photoemission (PEP) in several systems. In each case, an argon plasma increased the photoelectric signal greatly *vis-à-vis* the PE signal. This amplification results from cascade ionization in the plasma following electron acceleration in the cathode sheath. Addition of the reactive species oxygen to the plasma did not affect the PEP signal with a graphite electrode because

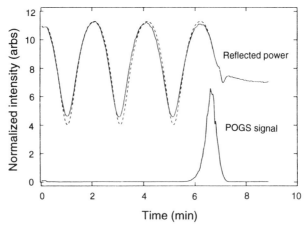

Figure 17.10 Photoemission optogalvanic spectroscopy (POGS) signal (lower curve) and reflectance (upper curve—solid/measured, dotted/calculated) during plasma etching of SiO_2/ *p*-type Si (Downey *et al.*, 1988).

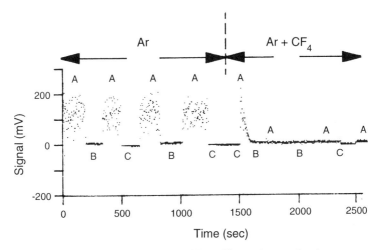

Figure 17.11 Averaged photoelectric signal for a silicon substrate showing sequences with the discharge on/laser on (A), discharge on/laser off (B), and discharge off/laser on (C) (Selwyn et al., 1988).

the CO and CO_2 etching products readily desorbed. In contrast, the addition of the reactive species CF_4 to the plasma with Al and Si surfaces effectively quenched the PEP signal, because a surface-passivating layer formed on the aluminum and fluorine penetrated into the silicon bulk. Figure 17.11 shows this for Si surfaces. Selwyn and Singh (1989a) found that the PEP signal in an Ar plasma is greater for p-type Si than for n-type by a factor of ~2.4–2.9 at lower laser powers, and that this ratio decreases with laser power for heavily doped samples, but is still >1. This doping effect is thought to be due to band bending at the surface and has been seen for vacuum PE (Allen and Gobeli, 1962).

Selwyn and Singh (1989b) have reviewed these two studies and have investigated the silicon doping effect in fluorinating plasmas. In these plasmas, the PEP signal increases initially, particularly for lightly doped n-type Si, and then rapidly decreases for all doped substrates because of the fluorinated surface layer. In general, PEP can be used either synchronized or unsynchronized with respect to the RF cycle, but better sensitivity is obtained by synchronizing the laser pulse to the maximum cathodic segment of the RF cycle.

References

F. G. Allen and G. W. Gobeli, *Phys. Rev.* **127**, 150 (1962).
A. L. Alstrin, P. G. Strupp, and S. R. Leone, *Appl. Phys. Lett.* **63**, 815 (1993).

A. L. Alstrin, A. K. Kunz, P. G. Strupp, and S. R. Leone, *Mater. Res. Soc. Symp. Proc.* **324,** 359 (1994).
S. Arepalli, N. Presser, D. Robie, and R. J. Gordon, *Chem. Phys. Lett.* **118,** 88 (1985).
M. Bacal, G. W. Hamilton, A. M. Bruneteau, H. J. Doucet, and J. Taillet, *J. Phys., Colloq. (Orsay, Fr.)* **40,** C7-491 (1979).
C. H. Becker and J. S. Hovis, *J. Vac. Sci. Technol. A* **12,** 2352 (1994).
P. Camus, *J. Phys. Colloque C7* (supplement au nº 11) **44** (1983).
F. G. Celli and J. E. Butler, *J. Appl. Phys.* **71,** 2877 (1992).
F. G. Celii, P. E. Pehrsson, H.-T. Wang, H. H. Nelson, and J. E. Butler, *AIP Conf. Proc.* **191,** 747 (1989).
E. J. Corat and D. G. Goodwin, *J. Appl. Phys.* **74,** 2021 (1993).
V. M. Donnelly, in *Plasma Diagnostics* (O. Auciello and D. L. Flamm, eds.), Vol. 1, Chapter 1, p. 1. Academic Press, Boston, 1989.
D. K. Doughty and J. E. Lawler, *Phys. Rev. A* **28,** 773 (1983).
D. K. Doughty and J. E. Lawler, *Appl. Phys. Lett.* **45,** 611 (1984).
D. K. Doughty, S. Salih, and J. E. Lawler, *Phys. Lett.* **103A,** 41 (1984).
S. W. Downey, A. Mitchell, and R. A. Gottscho, *J. Appl. Phys.* **63,** 5280 (1988).
R. W. Dreyfus, J. M. Jasinski, R. E. Walkup, and G. S. Selwyn, *Pure Appl. Chem.* **57,** 1265 (1985).
L. C. Feldman and J. W. Mayer, *Fundamentals of Surface and Thin Film Analysis,* Chapter 9, p. 213. North-Holland Publ., New York, 1986.
B. N. Ganguly and A. Garscadden, *Appl. Phys. Lett.* **46,** 540 (1985).
P. M. Goodwin and C. E. Otis, *Appl. Phys. Lett.* **55,** 2286 (1989).
R. A. Gottscho and C. E. Gaebe, *J. Vac. Sci. Technol. A* **4,** 1795 (1986).
K. E. Greenberg, G. A. Hebner, and J. T. Verdayen, *Appl. Phys. Lett.* **44,** 299 (1984).
K. V. Guinn and V. M. Donnelly, *J. Appl. Phys.* **75,** 2227 (1994).
R. J. Hamers, P. L. Houston, and R. P. Merrill, *J. Chem. Phys.* **88,** 6548 (1988).
J. W. Hudgens, T. G. DiGuiseppe, and M. C. Lin, *J. Chem. Phys.* **79,** 571 (1983).
V. S. Letokhov, *Laser Photoionization Spectroscopy.* Academic Press, Orlando, FL, 1987.
J. Misewich and M. M. T. Loy, *J. Chem. Phys.* **84,** 1939 (1986).
T. Nakajima, N. Uchitomi, Y. Adachi, S. Maeda, and C. Hirose, *J. Phys., Colloq. (Orsay, Fr.)* **44,** C7-497 (1983).
G. S. Oehrlein, G. M. W. Kroesen, E. de Frésart, Y. Zhang, and T. D. Bestwick, *J. Vac. Sci. Technol. A* **9,** 768 (1991).
T. Okada, H. Andou, Y. Moriyama, and M. Maeda, *Appl. Phys. Lett.* **56,** 1380 (1990).
D. H. Parker, in *Ultrasensitive Laser Spectroscopy* (D. S. Kluger, ed.), Chapter 4, p. 234. Academic Press, New York, 1983.
B. Quiniou, R. Scarmozzino, Z. Wu, and R. M. Osgood, *Appl. Phys. Lett.* **55,** 481 (1989).
G. S. Selwyn and J. Singh, *J. Vac. Sci. Technol. A* **7,** 982 (1989a).
G. S. Selwyn and J. Singh, *IEEE J. Quantum Electron.* **QE-25,** 1093 (1989b).
G. S. Selwyn, L. D. Baston, and H. H. Sawin, *Appl. Phys. Lett.* **51,** 898 (1987).
G. S. Selwyn, B. D. Ai, and J. Singh, *Appl. Phys. Lett.* **52,** 1953 (1988).
R. S. Stewart and J. E. Lawler, eds., *Optogalvanic Spectroscopy, Proceedings of the Second International Meeting on Optogalvanic Spectroscopy,* Inst. Phys. Conf. Ser. No. 113. Institute of Physics, Bristol, 1991.
P. G. Strupp, A. L. Alstrin, R. V. Smilgys, and S. R. Leone, *Appl. Opt.* **32,** 842 (1993).

References

H. H. Telle, in *Optogalvanic Spectroscopy, Proceedings of the Second International Meeting on Optogalvanic Spectroscopy* (R. S. Stewart and J. E. Lawler, eds.), Inst. Phys. Conf. Ser. No. 113, p. 1. Institute of Physics, Bristol, 1991.

P. J. H. Tjossem and T. A. Cool, *Chem. Phys. Lett.* **100,** 479 (1983).

R. Walkup, R. W. Dreyfus, and Ph. Avouris, *Phys. Rev. Lett.* **50,** 1846 (1983).

C. R. Webster, I. S. McDermid, and C. T. Rettner, *J. Chem. Phys.* **78,** 646 (1983).

Y. Zhang and M. Stuke, *J. Cryst. Growth* **93,** 143 (1988).

CHAPTER 18

Optical Thermometry

This chapter links and compares the optical probes of temperature that have already been described in Chapters 6–17. These probes are all useful for process development, and several are potentially useful for real-time monitoring during manufacturing.

This chapter attempts an overview of optical thermometry that avoids excessive repetition of the material covered in previous chapters. As such, the organization of this chapter differs from that of the preceding chapters. (1) First, the concepts of temperature and thermal equilibrium are presented. (2) Next, the need for temperature measurements in thin film processing is surveyed, along with nonoptical thermometry. (3) The fundamental basis of optical thermometry is then discussed. Although the potential use of each optical spectroscopy to measure temperature has been addressed in previous chapters, the fundamental physical principles underlying each technique have not always been emphasized; this is done here. (4) The applications of the various optical diagnostics to the thermometry of thin film processing are then described briefly, along with their strengths and weaknesses. Details about each technique can be found in the cited sections. (5) Finally, the appendix to this chapter lists many of the literature references cited in earlier chapters and serves as a link to the more detailed discussions in these chapters.

A critical comparison of optical probes for real-time measurements depends on many factors: the specific application, the current state of development of each probe, competitive nonoptical probes, probe adaptability to tools, and the cost of implementation and operation. Another factor is how the probe would be accepted by industry; although this is a somewhat nebulous consideration, it is vital to assess the likelihood of acceptance. Some of these factors are rapidly changing due to new breakthroughs in technology and newly recognized needs, while others are highly subjective. No critical comparison or absolute assessment of optical diagnostics of temperature is attempted here, but the limiting features of each probe are cited and their suitability for real-time control is discussed.

18.1 Temperature

When the temperature T of a system does not change in time, the system is said to be in thermal equilibrium. If, in addition, there are no mechanical or chemical changes, the system is in thermodynamic equilibrium (Sears and Salinger, 1975). During thin film processing, the system is surely not in thermodynamic equilibrium, even though it can be in thermal equilibrium.

If a small region of space is in thermal equilibrium, all properties of that medium can be described, at least locally, by a temperature T. This means that the distribution of particles in states i with energies $\mathcal{E}_i$ is described by one of three distribution functions (Reif, 1965; Sears and Salinger, 1975). Each of these has been used explicitly or implicitly in previous chapters.

Classical and distinguishable particles are described by the Maxwell–Boltzmann distribution function. The fractional population of atomic and molecular states is given by

$$f(\mathcal{E}_i) = g_i \exp(-\mathcal{E}_i/k_B T)/Z \qquad (18.1)$$

where g_i is the degeneracy of state i and Z is the partition function $\Sigma_j\, g_j \exp(-\mathcal{E}_j/k_B T)$. The population density of molecules in state i is $N(\mathcal{E}_i) = N f(\mathcal{E}_i)$, where N is the number density of molecules; this was used in Section 3.2.4.

The fractional population of states of fermions, such as electrons and holes in solids, is described by the Fermi–Dirac distribution function

$$f(\mathcal{E}_i) = \frac{g_i}{\exp[(\mathcal{E}_i - \mu_c)/k_B T] + 1} \qquad (18.2)$$

where μ_c is the chemical potential, which is closely related to the Fermi energy. This expression is important in analyzing free-carrier absorption (Equation 4.45).

The population of bosons, such as phonons in solids and photons, is given by the Bose–Einstein distribution function

$$f(\mathcal{E}_i) = \frac{g_i}{\exp[(\mathcal{E}_i - \mu_c)/k_B T] - 1} \qquad (18.3)$$

For phonons and photons, $\mu_c = 0$. This expression is used to obtain the blackbody distribution law, Equation 13.1, which describes photons in thermal equilibrium, and to determine the thermal phonon density, which is used in understanding phonon-assisted absorption (Equations 4.42–4.44) and Raman scattering thermometry (Section 12.4.2.1).

If a given medium is excited either transiently or in a steady-state manner, it may no longer be in thermal equilibrium. However, at some time the population distributions of different subsystems of that material may each be characterized by a "temperature," albeit a different temperature for each. One example of this limited thermal equilibrium (LTE) is a gas consisting of molecules whose translational, rotational, vibrational, and electronic degrees of freedom are characterized by different Boltzmann distributions, each at a different temperature (T_{trans}, T_{rot}, T_{vib}, T_{elect}). Other examples are transiently excited solids (where the electrons and phonons eventually "equilibrate"), the plasma plume in pulsed-laser deposition, and semiconductor lasers operating in steady state (where electrons and holes reach their own separate "equilibria" with their own quasi-Fermi levels). Limited thermal equilibrium occurs because relaxation within each of the subsystems, which can be caused by collisions, is faster than the coupling between the different subsystems. Although the system is not in thermal equilibrium, it may be in steady state. With transient excitation, such as with pulsed-laser heating, the system eventually reaches (total) thermal equilibrium. With steady-state excitation, as in a plasma, this limited thermal equilibrium is a steady-state condition. Griem (1964) and Konuma (1992) describe spatially local thermal equilibrium in plasmas.

Determining the temperature of each subsystem is quite useful in understanding film processes, particularly those occurring in discharges. Clark and De Lucia (1981) observed that $T_{rot} \approx T_{trans}$ in an OCS discharge; Farrow (1985) made a similar observation in a BCl_3 RF discharge. Rotational/translational equilibrium in glow discharges is reasonable because of rapid rotational/translational energy transfer. While these measurements were conducted at 50–400 mTorr, which is typical of "plasma etching" and reactive-ion etching (RIE), $T_{rot} \approx T_{trans}$ is also expected in high-charge-density, low-pressure discharges that operate near 1 mTorr if the mean free path is still much smaller than the chamber dimensions. (This may not be true for a species that is created very hot rotationally.)

Several researchers have found that $T_{rot} \ll T_{vib}$ in glow discharges, including Knights et al. (1982) for a SiH_4 discharge and Davis and Gottscho (1983) for a CCl_4 plasma. While Clark and De Lucia (1981) determined that $T_{rot} < T_{vib}$ for the vibrational mode of CS in an OCS discharge, they observed that $T_{rot} \approx T_{vib}$ for the vibrational modes of OCS in this discharge. In this report they also cited their earlier findings that in HCN and SO_2 discharges, T_{vib} of the stretching modes is higher than that for the bending modes and that it is also higher than the rotational/translational temperature. Most studies suggest that the measurement of T_{rot} gives a good measure of T_{trans}, while T_{vib} and T_{elect} do not. In these and virtually every other study, the population distributions of rotational and vibrational states in glow discharges have been found to be well described by thermal (Boltzmann) distributions.

Furnaces and other hot-walled reactors are systems in thermal equilibrium. The temperature is the same everywhere, on the walls and on the wafer, and it is easily measured with a thermocouple. In nonequilibrium environments, e.g., cold-walled reactors as in rapid thermal processing, the wafer is directly heated and the wafer temperature needs to be measured. Temperature uniformity is a potential problem in such reactors.

18.2 The Need for Thermometry in Thin Film Processing

Most thin film processes involve thermally activated steps. In many of these processes, the rate-limiting step is determined by kinetics rather than mass transport; it is therefore controlled by temperature. Since this dependence is usually exponential, as in the Arrhenius form $k_r = A \exp(-\mathcal{E}_{act}/k_B T)$, small changes in T can significantly affect the rate of the process. This is particularly true in deposition processes, such as CVD (chemical vapor deposition), and in rapid, high-temperature processes, such as RTP (rapid thermal processing). Using this Arrhenius form, the fractional change in layer thickness d deposited during CVD due to a change in temperature is $\delta d/d = (\mathcal{E}_{act}/k_B T)(\delta T/T)$. With $\mathcal{E}_{act} = 1.6$ eV, as in polysilicon CVD from silane, $\mathcal{E}_{act}/k_B T = 19$ at 700°C, which is typical of RTP–CVD conditions (Öztürk et al., 1991). At this temperature, the temperature must be controlled within 0.25% (±2.5°C) to maintain the thickness within 5% of the targeted value, as is seen from Figure 18.1. Moreover, such temperature uniformity is needed across the wafer. If the temperature is high by 40°C, as is not uncommon when using uncalibrated pyrometry for thermometry (Chapter 13), the deposited film can be twice the intended thickness. In discussing sensor needs during integrated circuit manufacturing, Moslehi et al. (1992) and Barna et al. (1994) have concluded that the control of wafer temperature is essential in virtually every fabrication step (see Table 1.1).

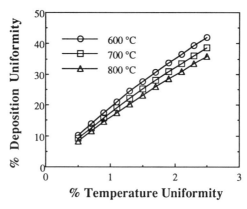

Figure 18.1 Effect of deviations from temperature uniformity on the uniformity of the thickness of polysilicon deposited at three target temperatures, assuming an activation energy of 1.6 eV (from Özturk et al., 1991). (© 1991 IEEE.)

18.2 The Need for Thermometry in Thin Film Processing

Anderson (1990) has discussed the ranges of temperature operation needed for each thin film process used in the semiconductor industry. He also critiqued temperature-measurement technologies for real-time monitoring, emphasizing conventional, nonoptical methods. The operating temperature ranges of semiconductor manufacturing processes are listed in Table 18.1. Anderson separately discussed the thermometric needs in low-temperature (-150 to $+60°C$), "room"-temperature ($10-80°C$), moderate-temperature ($80-200°C$), intermediate-temperature ($200-600°C$), and high-temperature ($550-1250°C$) processes. Though many potential applications for optical diagnostics of temperature are in the last two categories ($200-1250°C$), there is also interest in the second and third ranges ($10-200°C$). Thermocouples are commonly used in many applications, although the accuracy of thermocouples mounted in chucks depends on thermal contact and conduction. Because RF excitation is common in intermediate-temperature processing (except for annealing in tube furnaces), the use of noncontact methods, rather than thermocouples, is suggested because of potential RF pickup. More conventional methods, such as pyrometry and fluoroptic probes, have other problems in this temperature range (see below).

Table 18.1 Ranges of Wafer Temperature Measurements in Semiconductor Processing*

Process	Temperature range (°C)	Measurement specification (accuracy/resolution) (°C)
Resist development	18–25	±0.5/±0.1
Ion implanting	20–70	
Soft resist baking	80–120	±0.5/±0.1
Post-exposure harden (UV flood)	120–190	±1/±0.1
HMDS ovens (resist adhesion)	25–200	±1/±1
Plasma etching	~200	±2.5/±0.1
Alloying (and annealing)	375–450	±0.5/±0.25
Sputtering (metals, etc.)	25–500	±3/±3
Plasma-assisted deposition	200–500	
Tungsten CVD (cold wall)	300–600	±3/±0.5
MBE	300–1000	
CVD (furnace)	525–800	±0.5/±0.25
Oxidation (and annealing)	800–950	±1/±0.5
HIPOX (high pressure oxidation) tubes	750–1000	±1/±0.1
Oxidation furnace	750–1100	±0.5/±0.25
RTP	500–1250	±0.8/<±0.5

* Adapted from R. L. Anderson, Review of temperature measurements in the semiconductor industry, in *Advanced Techniques for Integrated Circuit Processing, SPIE* **1392**, 437 (1990), which also included lower temperature processes.

Extremely tight control of wafer temperature is needed in rapid thermal processing (RTP) to obtain reproducibility and to minimize slippage and warpage due to temperature nonuniformity (Peyton *et al.*, 1990). Rapid thermal processing temperature sensors must have high precision, ~1–2°C, and fast response, ~0.03 – 1 sec, for closed-loop control. Real-time measurements must not involve contact because contacting sensors rely on heat conduction, which is relatively slow, and are also sources of contamination. Peyton *et al.* (1990) surveyed noncontact temperature measurements for rapid thermal processing, and compared the strengths of each process. They emphasized the use of single-wavelength pyrometry, which is commonly used in RTP (even though the wafer emissivity is sometimes uncertain), and variations of conventional pyrometry, including dual-wavelength and ellipsometric pyrometries (Chapter 13). They also addressed the availability and cost of several of these sensors. Peters (1991) has surveyed the need for temperature measurements in RTP. Roozeboom and Parekh (1990) and Roozeboom (1991, 1993a,b) have detailed technological developments in pyrometry that can improve thermometry in RTP reactors.

Another potential problem in lamp-heated processes, such as MBE [molecular beam epitaxy] and RTP, is the changing absorption of lamp radiation by the wafer during the process; this can lead to undesired changes in the wafer temperature unless the lamp output is controlled to maintain constant temperature. For example, the wafer temperature can increase as a semiconductor film is deposited on a substrate when the film has a smaller band gap than the substrate. Moreover, the emissivity changes during such a deposition, which can greatly affect pyrometric measurements, as is detailed in Chapter 13. These effects must be monitored and controlled.

Temperature measurements can be made either on the front side or back surface of a wafer. Given the tool design, access to either side may be difficult. The "high cost of real estate" on the front side argues against dedicating a region on the front surface for measurements. The temperature is effectively the same on both wafer surfaces in virtually all RTP processes, and in all slower thin film processes. This can be understood by considering the characteristic thermal diffusion time in one dimension, across the thickness of the wafer, $\tau_d = h^2/D_t$, where h is the thickness of the wafer and D_t is the thermal diffusion constant. (See Section 4.5.) For Si, D_t monotonically decreases from 0.91 cm^2/sec at 300 K to 0.084 cm^2/sec at 1600 K (Moody and Hendel, 1982). Consequently, for silicon wafers with $h \sim 500$ μm, τ_d increases from ~3 to ~30 msec as temperature increases from 300 to 1600 K. When only one side of the wafer is heated, the other side will be at essentially the same temperature after several τ_d.

Thermal sensors that contact the wafer, such as thermocouples, pose a serious potential source of contamination in microelectronics processing. Furthermore, the thermal contact to the wafer may not be adequate or reproducible. Temperatures derived from noncontact methods, such as optically based temperature sensors, do not have this sensitivity. Still, the optical determination of temperature is indirect; physical or optical parameters are directly determined and temperature must be inferred from these parameters.

A temperature diagnostic may be suitable for process analysis and development, but unsuitable for real-time monitoring and control because of high cost or lack of robustness. Some temperature probes are useful only for specific purposes, i.e., for specific materials (e.g., for Si but not GaAs), in limited temperature ranges, for particular processes (e.g., processes that do not modify the surface by oxidizing, doping, roughening—or smoothing—it), or for specific manufacturing tools (e.g., those with a specific type of optical access).

Many of the temperature methods used in film processing are precise and reproducible, but not accurate. For example, pyrometry depends on emissivity, which depends on the wafer conditions, and can give readings that are off by 100°C. Such precise, yet inaccurate, sensors of wafer temperature have a definite, though limited, value. This is illustrated by the application of pyrometry in assessing temperature uniformity. With such sensors, processes are tuned for that one tool with test wafers. The temperature measurements are repeatable for future runs in that one tool, but cannot be transferred to other similar tools at that location or at other locations, or to even slightly different processes. These temperature sensors are not useful for flexible or adaptive manufacturing. Improved accuracy would reduce the need for test wafers and improve technology transfer (Anderson, 1990). Furthermore, while some temperature measurements determine an absolute value of temperature, others, such as reflection interferometry at one wavelength, measure temperature changes only; still, this suffices in many applications.

18.3 Nonoptical Probes of Temperature

"Conventional" temperature sensors have been reviewed by Anderson and Kollie (1976), Hudson (1980), Benedict (1984), Schooley (1986), McGee (1988), Norton (1989), and Anderson (1990). This includes most nonoptical methods and pyrometry.

Thermocouples measure the difference in temperature between two junctions, one in the processing unit and one maintained at a reference temperature. By choosing a particular thermoelement pair, thermocouples can be used from -269 to $3000°C$. This is accomplished by measuring the emf, which is related to the line integral of the temperature gradient between the two junctions

$$\text{emf} = -\int S_i(x) \nabla T(x) \cdot dx \qquad (18.4)$$

where $S_i(x)$ is the Seebeck coefficient of the thermoelements. The uncertainty in type-S (0–1500°C) and base-metal thermocouples (0–1100°C) is 0.2–3°C, depending on the temperature region. Resistance temperature detectors (RTDs) are also useful. Platinum RTDs can be used from 0 to 1064°C with <0.001–0.02°C uncertainty in different temperature regions. Fluidic thermometry is a noncontact method that uses the linear temperature dependence of the viscosity of He to determine wafer temperature (Anderson, 1990).

Some novel nonoptical temperature sensors are being developed. For example, Degertekin *et al.* (1994) have developed a nonoptical version of the laser-acoustic probe described in Section 15.2.1. The wafer temperature is still measured by the speed of acoustic waves in the wafer, but in this sensor the wave is directed to the wafer and retrieved from the wafer by quartz pins, upon which the wafer sits.

18.4 The Physical Basis of Optical Thermometry

18.4.1 *Thermometry of Gases*

In gases, optical spectroscopies determine temperature by measuring the relative distribution of states by using Equation 18.1. At lower pressures, the Doppler profile can be measured by absorption or emission to find T_{trans} (Equation 4.22). Although this can be done by using the probes described here, such as tunable infrared diode laser absorption (IR-DLAS) (Hershberger *et al.*, 1988; Park *et al.*, 1991), it is not often used in process environments. Still, there are important examples in plasmas and in examining the plume during pulsed laser deposition, including the high-resolution OES (optical emission spectroscopy) and LIF (laser-induced fluorescence) investigations cited in Sections 6.1 and 6.3 and in Section 7.2.1.1, respectively.

Much more common is the determination of T_{rot} by measuring the relative populations of rotational states $N(J)$, corrected by the rotational dependence of the transition matrix elements and other factors, as in Equation 4.39. Virtually every gas-phase spectroscopy can be used to measure T_{rot}: OES (Section 6.3.5), LIF (Section 7.2.1.1), infrared absorption (Section 8.2.1.3), rotational Raman scattering (Section 12.3), coherent anti-Stokes Raman scattering (CARS; Section 16.1.3), and resonant-enhanced multiphoton ionization (REMPI; Section 17.1). Each of these spectroscopies, with the exception of OES, is usually employed to measure T_{rot} in the ground electronic state; OES measures T_{rot} in an excited electronic state. The accuracy of these measurements is typically ±50 K. T_{vib} and T_{elect} can also be obtained by using Equation 18.1. The interferometric and photographic methods described in Chapter 10 can monitor temperature because the refractive index of a gas depends on density, and density and temperature are inversely related by the ideal-gas law. Electron temperatures in plasmas can be determined by Thomson scattering (Chapter 11).

18.4.2 Thermometry of Wafers

Measurement of wafer temperature requires much better accuracy and precision (Table 18.1) than those cited for the gas phase. Thermometry by pyrometry involves measuring the thermal distribution of photons, described by Equations 18.3 and 13.1. The emissivity of the radiating structure strongly affects the determination of temperature and is itself a function of temperature through the temperature dependence of the optical parameters and thermal expansion (Equation 13.5).

Most other optical probes of the temperature of solids relate either directly or indirectly to the thermal excitation of phonons (lattice vibrations) (Equation 18.3) and lattice anharmonicity. One exception is the intraband (free carrier) excitation of electrons and holes (Equation 18.2). With this latter mechanism, below-band gap wafer absorption can monitor the thermal density of intrinsic conduction-band electrons and valence-band holes to determine temperature (Equation 4.45, Section 8.4).

Thermal expansion is directly related to the phonon population because lattice vibrations cause the lattice constant to increase, since the interatomic forces that bind the lattice are anharmonic. The volume coefficient of thermal expansion β_T ($=3\tilde{\alpha}$) is related to other lattice parameters by

$$\beta_T = \frac{\gamma_{G\text{-th}} \kappa_T C_V}{V} = \frac{\gamma_{G\text{-th}} \kappa_S C_P}{V} \quad (18.5)$$

where $\gamma_{G\text{-th}}$ is the thermodynamic Grüneisen function, κ is the compressibility at constant temperature (T) or entropy (S), C is the heat capacity at constant volume (V) or pressure (P), and V is the volume (Barron et al., 1980; Mitra and Massa, 1982; Böttger, 1983). The connection of thermal expansion with anharmonicity is through the Grüneisen function $\gamma_{G\text{-th}}$, which is a weighted average of the mode Grüneisen parameters $\gamma_G = -d \ln \omega_p / d \ln V$, where ω_p is the phonon frequency for a given mode.

Several optical probes sense only the effect of thermal expansion, such as speckle and optical diffraction interferometries (Sections 10.1, 11.2.2.3, 11.3) and reflection microscopies (Section 9.4.1). Many other optical diagnostics probe both the effects of thermal expansion (directly) and the dependence of optical properties on temperature, which themselves often depend on the phonon population and thermal expansion.

In equilibrium, material properties are functions of T, V (volume), and P (pressure), two of which, say T and V, can be considered independent variables. Since most thin film processes occur at constant pressure, the temperature dependence of a property X (band gap energy, dielectric function, etc.) can be expressed as

$$\left(\frac{\partial X}{\partial T}\right)_P = \left(\frac{\partial X}{\partial T}\right)_V + \left(\frac{\partial X}{\partial V}\right)_T \left(\frac{\partial V}{\partial T}\right)_P \quad (18.6a)$$

$$= \left(\frac{\partial X}{\partial T}\right)_V + \beta_T V \left(\frac{\partial X}{\partial V}\right)_T = \left(\frac{\partial X}{\partial T}\right)_V + \beta_T \left(\frac{\partial X}{\partial \ln V}\right)_T \quad (18.6b)$$

where $\beta_T = (1/V)(\partial V/\partial T)_P$ is the volume coefficient of thermal expansion. So temperature can affect material parameters explicitly, with constant volume, and implicitly, by how the parameter is affected by volume changes through thermal expansion.

The optical properties of a material in the visible and ultraviolet are often characterized by the critical points (CP) in the dielectric function, including the fundamental band gap. These critical-point energies of a semiconductor depend on temperature because of lattice vibrations. Four interactions are important: (1) thermal expansion, which changes the band gap through its dependence on lattice constant (volume); (2) smearing out of the periodic potential (Debye–Waller factor); (3) electron–phonon coupling in second-order perturbation causing mutual repulsion of intra-band electronic states, i.e., the Fan terms for intraband coupling; and (4) the Fan terms for interband coupling (Heine and Van Vechten, 1976; Cohen and Chadi, 1980; Ridley, 1988). The effect of thermal expansion corresponds to the implicit dependence in Equation 18.6 (second term), and can be determined from the known dependence of the band gap on pressure (volume). It usually contributes to a relatively small fraction of the whole effect of the temperature change. The last three terms correspond to the explicit dependence (first term in Equation 18.6). In many semiconductors these four terms cause the fundamental band gap to decrease with temperature, and to decrease linearly with T at high temperature. [In some materials such, as PbTe, the gap increases with T (Keffer et al., 1968).]

This dependence of critical-point energy on temperature is often modeled by using one of two expressions. The Varshni expression (Varshni, 1967) is an empirical fit

$$\mathcal{E}(T) = \mathcal{E}(0) - \frac{\alpha_v T^2}{T + \beta_v} \qquad (18.7)$$

with the Varshni coefficients α_v and β_v, and the parameter $\mathcal{E}(0)$ for each critical point. Another expression linearly includes the Bose–Einstein factor for phonon population (Equation 18.3), and so explicitly includes the electron–phonon interaction (Lautenschlager et al., 1987a)

$$\mathcal{E}(T) = \mathcal{E}_b - a_b\left(1 + \frac{2}{e^{\theta_b/T} - 1}\right) \qquad (18.8)$$

with parameters $\mathcal{E}_b$, interaction strength a_b, and mean frequency of phonons involved (in temperature units) θ_b.

These parameters have been determined by several groups. For example, Cardona and co-workers used ellipsometry to determine the CP parameters in Equations 18.7 and 18.8 for critical points in Si (Lautenschlager et al., 1987c), Ge (Viña et al., 1984), GaAs (Lautenschlager et al., 1987a), InP (Lautenschlager et al., 1987b), InSb (Logothetidis et al., 1985), and GaP (Zollner et al., 1993). Shen et al. (1988) and Shen (1990) used photore-

18.4 The Physical Basis of Optical Thermometry

flectance to obtain these parameters for GaAs, GaAlAs, InP, and InGaAs. Table 18.2 lists these parameters for the band gap of several semiconductors.

Photoreflectance (Section 9.6) and ellipsometry (Section 9.11) are used to monitor critical energies to obtain the temperature. Other spectroscopies more specifically monitor features related to the fundamental band gap energies in direct-gap semiconductors. Spectral analysis of photoluminescence gives the band gap (Chapter 14), from which temperature can be determined. Also, the change in band-edge absorption in a wafer is a sensitive probe of temperature (Equation 4.41, Section 8.4).

Below-band-gap absorption is a temperature probe often used for indirect-gap semiconductors, such as Si (Section 8.4). Very near the gap, phonon-assisted absorption is very important (Equations 4.42–4.44). This is strongly affected by the thermal population of phonons (Equation 18.3) and by the thermal shifting of the fundamental band gap. At even lower photon energies, free-carrier absorption due to intrinsic and extrinsic carriers dominates. As mentioned earlier, the intrinsic carrier density is affected by temperature (Equation 4.45, Section 8.4).

Changes in temperature affect the complex dielectric function $\tilde{\varepsilon}(\omega)$. This function can be modeled near a CP by using Equation 4.40, which is parameterized by the critical-point energy (Equations 18.7 and 18.8), transition amplitude, and broadening parameter; these latter two parameters also depend on temperature. As in Equation 18.8, the broadening factor can be modeled by using a term that depends on phonon population (Equation 18.3)

$$\Gamma_L^{(E)}(T) = \Gamma_0 + \Gamma_1 \left(1 + \frac{2}{e^{\theta_b/T} - 1}\right) \tag{18.9}$$

Thermal expansion also affects the dielectric function because the density of oscillators changes with temperature. These temperature-dependent dielectric functions are used for thermometry by reflectometry from interfaces (Section 9.4.1), reflection interferometry (Section 9.5.2), and ellipsometry (Section 9.11), usually at wavelengths where the measured optical parame-

Table 18.2 Parameters from Equations 18.7 (the Varshni equation: $\mathcal{E}(0)$, α_v, β_v) and 18.8 (the Bose–Einstein expression: $\mathcal{E}_b$, a_b, θ_b) That Describe the Direct Gap ($\mathcal{E}_0 = \mathcal{E}_{bg}$) for Several Semiconductors*

Material	$\mathcal{E}(0)$ (meV)	α_v (0.1 meV/K)	β_v (K)	$\mathcal{E}_b$ (meV)	a_b (meV)	θ_b (K)
GaAs	1515 ± 5	5.1 ± 0.5	190 ± 82	1562 ± 10	50 ± 5	215 ± 20
$Ga_{0.82}Al_{0.18}As$	1771 ± 7	6.3 ± 0.5	236 ± 73	1885 ± 10	96 ± 5	332 ± 15
InP	1432 ± 7	4.1 ± 0.3	136 ± 60	1474 ± 10	51 ± 5	259 ± 20
$In_{0.16}Ga_{0.84}As$	1285 ± 6	5.1 ± 0.3	242 ± 50	1360 ± 10	82 ± 5	340 ± 20

* Obtained from Shen (1990).

ter is particularly sensitive to temperature changes. These latter two methods are also strongly affected by thermal expansion of any films atop the substrate, because the optical path length of a layer with thickness d is $n(\lambda, T)d(T)$. [The temperature dependence of these optical functions has been measured for many semiconductors (see Table 9.1) using ellipsometry, photoreflectance, etc.]

Using a more phenomenological model based on Equation 18.6, Thomas (1991) has modeled the temperature dependence of the complex index of refraction $\tilde{n}$ ($\tilde{\varepsilon} = \tilde{n}^2$) by using the Lorentz–Lorenz formula (Equation 4.4), finding

$$2\tilde{n}\frac{d\tilde{n}}{dT} = \frac{(\tilde{n}^2 + 2)(\tilde{n}^2 - 1)}{3\hat{\alpha}}\left(\frac{\partial \hat{\alpha}}{\partial T}\right)_V \quad (18.10)$$
$$- (\tilde{n}^2 + 2)(\tilde{n}^2 - 1)\bar{\alpha}\left(1 - \frac{V}{\hat{\alpha}}\left(\frac{\partial \hat{\alpha}}{\partial V}\right)_T\right)$$

where $\bar{\alpha}$ is the linear coefficient of thermal expansion and $\hat{\alpha}$ is the electric polarizability. The first term describes how the polarizability depends explicitly on temperature, while the second term describes the effect of volume change due to thermal expansion and how the polarizability and density both depend on volume (Equation 18.6).

Electronic relaxation also affects these temperature probes. The broadening parameter in Equations 4.40 and 18.9 is partly determined by relaxation processes. The temperature dependence of the decay rate of photoluminescence, which is affected by nonradiative relaxation processes, is monitored in fluorometric probing, and is described in Chapter 14.

The connection of the thermal excitation of lattice vibrations to temperature is direct in vibrational Raman scattering in solids (Section 12.4.2.1). Equation 12.4 shows that the phonon frequency of a given mode, which is the Raman shift, includes a term that depends on thermal expansion (Equation 12.5) and one that depends on the thermal population of phonons in that optical mode and in lower-energy, acoustic modes (Equation 12.6). This first term depends on anharmonicity within that mode, while the second term depends on anharmonicity that causes intermode coupling. The phonon linewidth (Equation 12.8) is derived from the phonon population and intermode coupling terms that are similar to those in Equation 12.6. The ratio of Stokes to anti-Stokes Raman scattering rates also depends on the phonon density, as is seen in Equation 12.10; it further depends on the ratio of the temperature-dependent optical parameters also given in this equation.

The change of acoustic velocity with temperature (Section 15.2.1) is closely tied to properties of phonons because both are determined by elastic constants. Furthermore, the acoustic velocity also depends on density, which varies inversely with volume and therefore decreases with increasing temperature.

18.5 A Comparison of Optical Thermometry Probes

Table 1.5 lists the different optical spectroscopies used for thermometry during thin film processing. This section briefly assesses and compares these diagnostics. The appendix to this chapter lists many of the citations on thermometry that are detailed in previous chapters. It is meant to be a representative and thorough compilation, but not a complete review of all published work.

18.5.1 *Gas Temperatures*

One common goal of measuring gas-phase temperatures in film processing is to determine the translational temperature of neutral species. Interferometric and photographic methods, such as interference holography, are sensitive to refractive-index changes and therefore determine translational temperature and density variations in flow reactors. This is relatively straightforward experimentally, but analysis can be relatively complicated. T_{trans} can also be measured directly and locally with Doppler spectroscopy; however, Doppler measurements are not always possible due to collisional broadening or because they require unduly complicated laser spectroscopic procedures. Laser-induced fluorescence and optical emission spectroscopy have proved to be very important in investigating high-charge-density plasmas. Still, spatially localized rotational temperatures are usually measured spectroscopically, since T_{rot} and T_{trans} are usually equal during most thin film processes. While determining T_{vib} can provide useful information, it rarely equals the translational temperature under processing conditions.

The relative merits of each gas-phase probe of T_{rot}—OES, LIF, absorption (IR-DLAS), spontaneous Raman scattering, CARS, and REMPI—are closely tied to the general application of that spectroscopy in that medium. For instance, OES can be used only with media with excited species. Infrared absorption cannot probe homonuclear diatomic molecules, but the other probes can. CARS is very useful when the gas densities are high enough, and it can map temperature with very good spatial resolution; however, CARS requires a more elaborate laser system than do the other techniques. Sometimes, temperature measurements are aided by adding a few percent of a new species, which is nonreactive and nonperturbative, to the reaction mixture, and then measuring T_{rot} of this species optically. When measuring complicated rotational profiles with much band overlap, temperatures can be determined within ~50°C; precision is better when the rotational structure is more clearly resolved.

Measuring the temperature, and more generally the energy-level distribution, of gas-phase species is very important in fundamental studies and in process development; however, it is rarely tracked for real-time control during manufacturing. Tracking wafer temperature is preferred because it can be controlled more directly and because it is often significant in the rate-limiting step of the film process.

18.5.2 Wafer Temperatures

Many optical diagnostics can be used to measure wafer temperature. Most involve no contact and are noninvasive. Their relative merits for use in feedback and control depend on many criteria, which are listed here (in no particular order):

(1) Required accuracy, precision. The required precision and repeatability is often several degrees Celsius or less (Table 18.1).

(2) Measurement speed. Most are fast enough for steady-state processing, although several are not fast enough for rapid thermal processing, for which measurements must be made in $\sim$30 msec $-$ 1 sec.

(3) Ability to measure absolute temperature or only changes in temperature.

(4) Ability to probe uniformity of temperature profile.

(5) Sensitivity to changes that occur on the wafer as the process ensues.

(6) Sensitivity to the model that is used to obtain temperature.

(7) Sensitivity to instrument alignment, calibration, and drifts.

(8) Nature of the measurement, i.e., determining the value of a cross-section or the position of a spectral peak. The measurement of a spectral peak sometimes becomes more difficult at high temperatures because of thermal broadening, such as that of electronic critical points (in ellipsometry, photoluminescence, photoreflectance) and phonon energies (in Raman scattering). The minimum detectable temperature change is related to $\Gamma_L^{(\mathcal{E})}/(d\mathcal{E}_x/dT)$, where $\mathcal{E}_x$ is the energy that is being measured (e.g., band gap, phonon energy) and $\Gamma_L^{(\mathcal{E})}$ is the linewidth.

(9) Sensitivity to external light. This is significant when gas-phase species emit light, as in plasma processing, and when devices that heat the wafer also emit light, such as lamps and resistive heaters, as in pyrometry. Reflections from the chamber walls can also be a problem.

(10) Need for special preparation of the wafer, such as the fabrication of line or grating structures. This also means the potential loss of wafer real estate.

(11) Ability to probe the back side of the wafer, which can be more complicated when the back side is rough, as is usual, or coated with films.

(12) Cost of ownership and upkeep, and overall savings. Ability to multiplex to many tools to save costs may be important.

(13) State of development of technology.

(14) Ease of implementation in current and new-generation tools (access).

(15) Small footprint (size).

18.5 A Comparison of Optical Thermometry Probes

(16) Simplicity. The diagnostic must not increase the complexity of the overall process and must be simple enough to permit turnkey operation.

(17) Robustness.

(18) Nonperturbative nature.

(19) Versatility of the technique to other processes, materials, and temperature ranges.

Narrow-band pyrometry (radiation thermometry) is widely used, especially in RTP reactors, and is straightforward to implement (Chapter 13). Measurement accuracy critically depends on knowing the correct value of the emissivity, which depends on the properties of the wafer, overlying films, and surface conditions. Supporting real-time optical measurements and accurate models of the surface structure can improve the accuracy greatly. Light guides internal to the reactor can be used for collection and calibration. Pyrometry does not have sufficient accuracy in the lower temperature ranges for silicon wafer thermometry because the wafer is transparent in the infrared regions that need to be measured.

Absorption measurements are gaining acceptance because of their low cost, high precision, and sometimes also high accuracy (Section 8.4). Band-edge measurements are made on direct-gap semiconductor wafers, and below-gap phonon-assisted and free-carrier absorption measurements are performed for indirect-gap materials. In the transmission geometry, access to the wafer on both sides is needed, while in "reflection" versions, access to only one side is necessary. Corrections for wafer roughness may be needed with this probe when data from wafers polished on both sides are used for calibration.

Reflection interferometry is a simple, straightforward, and inexpensive monitor of wafer temperature (Section 9.5.2). It is also applicable at low temperatures, and can even be used with wafers that are rough on one side. Using single-wavelength monitoring, fringes must be counted relative to a reference temperature. Temperature increases can be distinguished from decreases only by using special procedures. Interference effects in other optical probes may be useful in determining temperature, as in pyrometric interferometry.

Reflectometry at the interface of a bare wafer (Section 9.4.1) is also a simple, yet sensitive, optical thermometer. It is practical when used in the comparison mode (optical bridge), but is also sensitive to process-related changes on the surface. Reflection from an edge of the wafer that moves with thermal expansion, which is a form of interferometry, is another relatively inexpensive diagnostic that determines (only) the average wafer temperature; it also shows promise.

Use of reflection ellipsometry to measure temperature can be based on changes in the optical dielectric function (thick films, bare wafer) and/or interference effects (films atop a wafer) (Section 9.11). It is wiser to make measurements based on how the dielectric function changes with temperature than to directly track the temperature-dependent shifts of the critical-point energies, because the peaks are broad and the determination of the exact peak involves a careful fitting procedure (second derivatives, variations in the phase angle, etc.). Ellipsometry is very sensitive to model assumptions, such as the model of the surface and overlayers, and to instrument alignment and calibration (points 6 and 7 above). Although it is relatively expensive, it is being utilized increasingly for real-time measurements of film thickness and composition and should be used for thermometry if it is already implemented on a tool for these applications. Signal analysis can be fast enough for applications in RTP.

Photoreflectance (Section 9.6) tracks critical points (derivatives of the dielectric function), while photoluminescence (Section 14.3) tracks the fundamental band gap of direct-gap semiconductors. Both methods are greatly limited by strong thermal broadening at high temperatures.

Optical thermometers based on thermal expansion make measurements relative to a reference temperature. They are relatively inexpensive and accurate, but may need special surface preparation or use valuable wafer real estate [optical diffraction (moiré) interferometry, Section 11.2.2.3; interferometry across a wafer, Section 10.1]. Use of speckle interferometry (Section 11.3) is promising because it does not require special wafer preparation, has the advantage of being a back-side measurement, and can be employed to map the temperature across the wafer. The above-cited probe of reflection from the edge of the wafer has the first two advantages, but measures only the average wafer temperature.

Raman scattering has potentially high spatial resolution and can be used without special surface preparation (Section 12.4.2.1). It is relatively insensitive to transparent surface overlayers. However, since Raman signals are weak, efficient collection is needed, along with the rejection of background light. It is often difficult to obtain the large solid angles needed for efficient light collection within production tools. Raman measurements of phonon frequencies are straightforward, but require determination the peak frequency, while measurements of Stokes/anti-Stokes intensity ratios have a limited temperature range and require careful correction involving optical parameters (near a resonance). Raman scattering is generally limited to semiconductors and insulators (not metals). Since the lasers and detection systems commonly employed in laboratory Raman scattering experiments are too expensive for real-time control, specialized systems must be designed to lower the cost of Raman thermometry.

Optical techniques can also be used to generate acoustic waves at one place on the wafer and detect them elsewhere on the wafer, thereby determining the temperature-dependent propagation time (Section 15.2.1).

Thermal wave spectroscopies also sense acoustic waves optically (Section 15.2.1).

These cited methods probe the optical properties of the wafer. In contrast, optical emission from optically excited theromographic phosphors placed at the tip of a probe is analyzed in fluoroptic measurements (Section 14.3). In this thermometry, either the probe is in contact with the wafer—hence the temperature measurement is accurate but invasive (and can lead to contamination)—or the probe does not make contact with the wafer—hence the method is less accurate. Temperature measurement errors are reduced if the phosphor is painted on the surface, but this is not possible during production because it would produce contamination. Commercial probes can be operated up to 450°C; however, research suggests with proper lightguiding these probes can be used up to 1200°C.

APPENDIX
Representative Citations in Real-Time Optical Thermometry in Thin Film Processing

18A.1 Reviews of Optical Thermometry

Reviews That Include Optical Thermometry in Thin Film Processing

R. L. Anderson, *SPIE* **1392,** 437 (1990).
W. G. Breiland and P. Ho, in *Chemical Vapor Deposition* (M. L. Hitchman and K. F. Jensen, eds.), Chapter 3, p. 91. Academic Press, Boston, 1993.
V. M. Donnelly, in *Plasma Diagnostics* (O. Auciello and D. L. Flamm, eds.), Vol. 1, Chapter 1, p. 1. Academic Press, Boston, 1989.
L. Peters, *Semicond. Int.* **14**(9), 56 (1991).
D. Peyton, H. Kinoshita, G. Q. Lo, and D. L. Kwong, *SPIE* **1393,** 295 (1990).
E. A. Rohlfing, *J. Chem. Phys.* **89,** 6103 (1988).
F. Roozeboom, *Mater. Res. Soc. Symp. Proc.* **224,** 9 (1991).
F. Roozeboom, *Mater. Res. Soc. Symp. Proc.* **303,** 149 (1993).
F. Roozeboom, in *Rapid Thermal Processing Science and Technology* (R. B. Fair, ed.), Chapter 9, p. 349. Academic Press, Boston, 1993.
F. Roozeboom and N. Parekh, *J. Vac. Sci. Technol. B* **8,** 1249 (1990).

Reviews That Include the Use of Optical Thermometry in Related Fields

M. C. Drake, *Mater. Res. Soc. Symp. Proc.* **117,** 203 (1988).
A. C. Eckbreth, *Laser Diagnostics for Combustion Temperature and Species,* 2nd ed. Gordon & Breach, Luxembourg, 1996.

18A.2 Citations for Optical Thermometry in Gases

Interferometry (Section 10.1)

S. Benet, R. Bergé, S. Brunet, S. Charar, B. Armas, and C. Combescure, in *Proceedings of the Eighth International Conference on Chemical Vapor Deposition* (J. M. Blocher, Jr., G. E. Vuillard, and G. Wahl, eds.), p. 188. Electrochemical Society, Pennington, NJ, 1981.

H. Chehouani, S. Bénet, S. Brunet, and B. Armas, *J. Phys. IV* [Colloque C3, Supplément au Journal de Physique II] **3**, 83 (1993).

L. J. Giling, *J. Electrochem. Soc.* **129**, 634 (1982).

L. J. Giling, *J. Phys., Colloq. (Orsay, Fr.)* **43**, C5-235 (1982).

J. E. Williams and R. W. Peterson, *J. Cryst. Growth* **77**, 128 (1986).

Laser-Induced Fluorescence (LIF) (Chapter 7)

J. P. Booth, G. Hancock, and N. D. Perry, *Appl. Phys. Lett.* **50**, 318 (1987).

G. P. Davis and R. A. Gottscho, *J. Appl. Phys.* **54**, 3080 (1983).

V. M. Donnelly, D. L. Flamm, and G. Collins, *J. Vac. Sci. Technol.* **21**, 817 (1982).

R. W. Dreyfus, R. Kelly, and R. E. Walkup, *Appl. Phys. Lett.* **49**, 1478 (1986).

K. E. Greenberg and P. J. Hargis, Jr., *J. Appl. Phys.* **68**, 505 (1990).

T. Nakano, N. Sadeghi, D. J. Trevor, R. A. Gottscho, and R. W. Boswell, *J. Appl. Phys.* **72**, 3384 (1992).

N. Sadeghi, T. Nakano, D. J. Trevor, and R. A. Gottscho, *J. Appl. Phys.* **70**, 2552 (1991).

R. V. Smilgys and S. R. Leone, *J. Vac. Sci. Technol. B* **8**, 416 (1990).

P. Van de Weijer and B. H. Zwerver, *Chem. Phys. Lett.* **163**, 48 (1989).

R. Walkup, P. Avouris, R. W. Dreyfus, J. M. Jasinski, and G. S. Selwyn, *Appl. Phys. Lett.* **45**, 372 (1984).

R. C. Woods, R. L. McClain, L. J. Mahoney, E. A. Den Hartog, H. Persing, and J. S. Hamers, *SPIE* **1594**, 366 (1991).

Optical Emission Spectroscopy (OES) (Section 6.3.5)

R. d'Agostino, F. Cramarossa, S. De Benedictis, and G. Ferraro, *J. Appl. Phys.* **52**, 1259 (1981).

G. P. Davis and R. A. Gottscho, *J. Appl. Phys.* **54**, 3080 (1983).

R. A. Gottscho and V. M. Donnelly, *J. Appl. Phys.* **56**, 245 (1984).

J. C. Knights, J. P. M. Schmitt, J. Perrin, and G. Guelachvili, *J. Chem. Phys.* **76**, 3414 (1982).

J. T. Knudtson, W. B. Green, and D. G. Sutton, *J. Appl. Phys.* **61**, 4771 (1987).

M. Oshima, *Jpn. J. Appl. Phys.* **17**, 1157 (1978).

D. M. Phillips, *J. Phys. D* **8**, 507 (1975).

R. A. Porter and W. R. Harshbarger, *J. Electrochem. Soc.* **126**, 460 (1979).

S. W. Reeve and W. A. Weimer, *J. Vac. Sci. Technol. A* **13**, 359 (1995).

E. A. Rohlfing, *J. Chem. Phys.* **89**, 6103 (1988).

D. V. Tsu, R. T. Young, S. R. Ovshinsky, C. C. Klepper, and L. A. Berry, *J. Vac. Sci. Technol. A* **13**, 935 (1995).

F. Zhang, Y. Zhang, Y. Yang, G. Chen, and X. Jiang, *Appl. Phys. Lett.* **57**, 1467 (1990).

Raman Scattering

Spontaneous: Mostly Rotational, Some Vibrational (Section 12.3)

W. G. Breiland and G. H. Evans, *J. Electrochem. Soc.* **138**, 1806 (1991).
W. G. Breiland and P. Ho, in *Proceedings of the Ninth International Conference on Chemical Vapor Deposition* (McD. Robinson, C. H. J. van den Brekel, G. W. Cullen, J. M. Blocher, and P. Rai-Choudhury, eds.), p. 44. Electrochemical Society, Pennington, NJ, 1984.
W. G. Breiland, M. E. Coltrin, and P. Ho, *J. Appl. Phys.* **59**, 3267 (1986).
J. Doppelbauer, G. Leyendecker, and D. Bäuerle, *Appl. Phys. B* **33**, 141 (1984).
D. I. Fotiadis, M. Boekholt, K. F. Jensen, and W. Richter, *J. Cryst. Growth* **100**, 577 (1990).
M. Koppitz, O. Vestavik, W. Pletschen, A. Mircea, M. Heyen, and W. Richter, *J. Cryst. Growth* **68**, 136 (1984).
M. Koppitz, W. Richter, R. Bahnen, and M. Heyen, in *Laser Processing and Diagnostics* (D. Bäuerle, ed.), p. 530. Springer-Verlag, Berlin, 1984.
G. Leyendecker, J. Doppelbauer, D. Bäuerle, P. Geittner, and H. Lydtin, *Appl. Phys. A* **30**, 237 (1983).
Y. Monteil, M. P. Berthet, R. Favre, A. Hariss, J. Bouix, M. Vaille, and P. Gibart, *J. Cryst. Growth* **77**, 172 (1986).
T. O. Sedgwick and J. E. Smith, Jr., *J. Electrochem. Soc.* **123**, 254 (1976).
T. O. Sedgwick, J. E. Smith, Jr., R. Ghez, and M. E. Cowher, *J. Cryst. Growth* **31**, 264 (1975).

CARS (Section 16.1)

K.-H. Chen, M.-C. Chuang, C. M. Penney, and W. R. Banholzer, *J. Appl. Phys.* **71**, 1485 (1992).
M. Hanabusa and H. Kikuchi, *Jpn. J. Appl. Phys.* **22**, L712 (1983).
S. O. Hay, W. C. Roman, and M. B. Colket, III, *J. Mater. Res.* **5**, 2387 (1990).

REMPI (Section 17.1)

P. M. Goodwin and C. E. Otis, *Appl. Phys. Lett.* **55**, 2286 (1989).

Thomson Scattering (Electrons) (Chapter 11)

T. Hori, M. D. Bowden, K. Uchino, K. Muraoka, and M. Maeda, *J. Vac. Sci. Technol. A* **14**, 144 (1996).

Transmission (Absorption) (IR) (Section 8.2.1.3)

W. W. Clark, III and F. C. De Lucia, *J. Chem. Phys.* **74**, 3139 (1981).
T. A. Cleland and D. W. Hess, *J. Appl. Phys.* **64**, 1068 (1988).

L. A. Farrow, *J. Chem. Phys.* **82,** 3625 (1985).
M. Haverlag, E. Stoffels, W. W. Stoffels, G. M. W. Kroesen, and F. J. de Hoog, *J. Vac. Sci. Technol. A* **12,** 3102 (1994).
J. C. Knights, J. P. M. Schmitt, J. Perrin, and G. Guelachvili, *J. Chem. Phys.* **76,** 3414 (1982).
J. Wormhoudt, A. C. Stanton, A. D. Richards, and H. H. Sawin, *J. Appl. Phys.* **61,** 142 (1987).

18A.3 Citations for Optical Thermometry of Wafers

Ellipsometry (Section 9.11.4)

K. A. Conrad, R. K. Sampson, E. A. Irene, and H. Z. Massoud, *J. Vac. Sci. Technol. B* **11,** 2096 (1993).
G. M. W. Kroesen, G. S. Oehrlein, and T. D. Bestwick, *J. Appl. Phys.* **69,** 3390 (1991).
G. N. Maracas, J. L. Edwards, K. Shiralagi, K. Y. Choi, R. Droopad, B. Johs, and J. A. Woolam, *J. Vac. Sci. Technol. A* **10,** 1832 (1992).
R. K. Sampson and H. Z. Massoud, *J. Electrochem. Soc.* **140,** 2673 (1993).
R. K. Sampson, K. A. Conrad, E. A. Irene, and H. Z. Massoud, *J. Electrochem. Soc.* **140,** 1734 (1993).
R. K. Sampson, K. A. Conrad, H. Z. Massoud, and E. A. Irene, *J. Electrochem. Soc.* **141,** 539, 737 (1994).
T. Tomita, T. Kinosada, T. Yamashita, M. Shiota, and T. Sakurai, *Jpn. J. Appl. Phys.* **25,** L925 (1986).
(Also see ellipsometric pyrometry.)

Interferometry (Not in Films or Substrates)

On Surfaces

Lateral Expansion of the Wafer (Section 10.1)

D. Bollmann and K. Haberger, *Microelectron. Eng.* **19,** 383 (1992).

Optical Diffraction Thermometry/Projection Moiré Interferometry (Section 11.2.2.3)

S. R. J. Brueck, S. H. Zaidi, and M. K. Lang, *Mater. Res. Soc. Symp. Proc.* **303,** 117 (1993).
A. Durandet, O. Joubert, J. Pelletier, and M. Pichot, *J. Appl. Phys.* **67,** 3862 (1990).
M. K. Lang, G. W. Donohoe, S. H. Zaidi, and S. R. J. Brueck, *Opt. Eng.* **33,** 3465 (1994).
M. Pichot and M. Guillaume, U. S. Patent 4,525,066 (1982).
S. H. Zaidi, S. R. J. Brueck, and J. R. McNeil, *J. Vac. Sci. Technol. B* **10,** 166 (1992).

Speckle Interferometry (Section 11.3)

D. Burckel, S. H. Zaidi, M. K. Lang, A. Frauenglass, and S. R. J. Brueck, *Mater. Res. Soc. Symp. Proc.* **342,** 17 (1994).

D. Burckel, S. H. Zaidi, A. Frauenglass, M. Lang, and S. R. J. Brueck, *Opt. Lett.* **20,** 315 (1995).
D. W. Voorhes and D. M. Hall, *SPIE* **1595,** 61 (1991).

Photoacoustic and Thermal Wave Thermometry (Section 15.2.1)

Y. J. Lee, C. H. Chou, B. T. Khuri-Yakub, K. Saraswat, and M. Moslehi, *IEEE Ultrason. Symp., 1989,* p. 535 (1989).
Y. J. Lee, C. H. Chou, B. T. Khuri-Yakub, and K. Saraswat, *SPIE* **1393,** 366 (1990).
Y. J. Lee, C. H. Chou, B. T. Khuri-Yakub, and K. Saraswat, *Symp. VLSI Technol., 1990,* p. 105 (1990).
R. Schork, S. Krügel, C. Schneider, L. Pfitzner, and H. Ryssel, *J. Phys. IV* [Supplement au Journal de Physique III] **4,** C7–27 (1994).

Photoluminescence/Fluorescence (including the Fluoroptic probe) (Section 14.3)

D. J. As and L. Palmetshofer, *J. Appl. Phys.* **62,** 369 (1987).
A. R. Bugos, S. W. Allison, and M. R. Cates, *IEEE Proc. Southeastcon, 1989,* p. 361 (1989).
E. J. Egerton, A. Nef, W. Millikin, W. Cook, and D. Baril, *Solid State Technol.* **25,** 84 (August, 1982)
K. T. V. Grattan and Z. Y. Zhang, *Fiber Optic Fluorescence Thermometry.* Chapman & Hall, London, 1995.
I. Hussla, K. Enke, H. Grünwald, G. Lorenz, and H. Stoll, *J. Phys. D* **20,** 889 (1987).
D. Kirillov and J. L. Merz, *J. Appl. Phys.* **54,** 4104 (1983).
P. Kolodner and J. A. Tyson, *Appl. Phys. Lett.* **40,** 782 (1982).
P. Kolodner and J. A. Tyson, *Appl. Phys. Lett.* **42,** 117 (1983).
P. Kolodner, A. Katzir, and N. Hartsough, *Appl. Phys. Lett.* **42,** 749 (1983).
Luxtron Corp., Product Information (1995).
J. S. McCormack, *Electron. Lett.* **17,** 630 (1981).
R. P. Salathé, H. H. Gilgen, and Y. Rytz-Froidevaux, *IEEE J. Quantum Electron.* **QE-17,** 1989 (1981).
C. J. Sandroff, F. S. Turco-Sandroff, L. T. Florez, and J. P. Harbison, *Appl. Phys. Lett.* **59,** 1215 (1991).
M. Takai, H. Nakai, J. Tsuchimoto, K. Gamo, and S. Namba, *Jpn. J. Appl. Phys.* **24,** L705 (1985).

Photoreflectance (Section 9.6)

K. Capuder, P. E. Norris, H. Shen, Z. Hang, and F. H. Pollak, *J. Electron. Mater.* **19,** 295 (1990).

Pyrometry (Thermal Radiometry) (Chapter 13)

General and Single-Wavelength Pyrometry (Sections 13.2 and 13.5)

R. L. Anderson, *SPIE* **1392,** 437 (1990).
F. G. Boebel and H. Möller, *IEEE Trans. Semicond. Manuf.* **SM-6,** 112 (1993).
M. Delfino and D. T. Hodul, *IEEE Trans. Electron Devices* **ED-39,** 89 (1992).
J.-M. Dilhac, C. Ganibal, and N. Nolhier, *Mater. Res. Soc. Symp. Proc.* **224,** 3 (1991).
A. T. Fiory, C. Schietinger, B. Adams, and F. G. Tinsley, *Mater. Res. Soc. Symp. Proc.* **303,** 139 (1993).
E. Fujita, K. Furusawa, H. Kataoka, N. Tsumita, T. Yonekawa, and N. Shige, *J. Vac. Sci. Technol. A* **10,** 1657 (1992).
D. B. Geohegan, *Appl. Phys. Lett.* **62,** 1463 (1993).
D. S. Katzer and B. V. Shanabrook, *J. Vac. Sci. Technol. B* **11,** 1003 (1993).
B. Linnander, *IEEE Circuits Devices* **9**(4), 35 (1993).
T. Loarer, J.-J. Greffet, and M. Huetz-Aubert, *Appl. Opt.* **29,** 979 (1990).
D. Mazzarese, K. A. Jones, and W. C. Conner, *J. Electron. Mater.* **21,** 329 (1992).
M. M. Moslehi, R. A. Chapman, M. Wong, A. Paranjpe, H. N. Najm, J. Kuehne, R. L. Yeakley, and C. J. Davis, *IEEE Trans. Electron Devices* **ED-39,** 4 (1992).
T. M. Niemczyk, J. E. Franke, B. Wangmaneerat, C. S. Chen, and D. M. Haaland, in *Computer-Enhanced Analytical Spectroscopy* (C. L. Wilkins, ed.), Vol. 4, p. 165. Plenum, New York, 1993.
T. M. Niemczyk, J. E. Franke, S. Zhang, and D. M. Haaland, *Mater. Res. Soc. Symp. Proc.* **341,** 119 (1994).
M. Nouaoura, L. Lassabatere, N. Bertru, J. Bonnet, and A. Isamail, *J. Vac. Sci. Technol. B* **13,** 83 (1995).
J. Nulman, S. Antonio, and W. Blonigan, *Appl. Phys. Lett.* **56,** 2513 (1990).
M. C. Öztürk, F. Y. Sorrell, J. J. Wortman, F. S. Johnson, and D. T. Grider, *IEEE Trans. Semicond. Manuf.* **SM-4,** 155 (1991).
V. Patel, M. Patel, S. Ayyagari, W. F. Kosonocky, D. Misra, and B. Singh, *Appl. Phys. Lett.* **59,** 1299 (1991).
D. W. Pettibone, J. R. Suarez, and A. Gat, *Mater. Res. Soc. Symp. Proc.* **52,** 209 (1986).
D. Peyton, H. Kinoshita, G. Q. Lo, and D. L. Kwong, *SPIE* **1393,** 295 (1990).
E. A. Rohlfing, *J. Chem. Phys.* **89,** 6103 (1988).
F. Roozeboom, *Mater. Res. Soc. Symp. Proc.* **224,** 9 (1991).
F. Roozeboom, *Mater. Res. Soc. Symp. Proc.* **303,** 149 (1993).
F. Roozeboom and N. Parekh, *J. Vac. Sci. Technol. B* **8,** 1249 (1990).
C. Schietinger, B. Adams, and C. Yarling, *Mater. Res. Soc. Symp. Proc.* **224,** 23 (1991).
T. O. Sedgwick, *Appl. Phys. Lett.* **39,** 254 (1981).
B. V. Shanabrook, J. R. Waterman, J. L. Davis, R. J. Wagner, and D. S. Katzer, *J. Vac. Sci. Technol. B* **11,** 994 (1993).
P. Vandenabeele and K. Maex, *J. Appl. Phys.* **72,** 5867 (1992).
P. Vandenabeele, R. J. Schreutelkamp, K. Maex, C. Vermeiren, and W. Coppeye, *Mater. Res. Soc. Symp. Proc.* **260,** 653 (1992).
B. Wangmaneerat, J. A. McGuire, T. M. Niemczyk, D. M. Haaland, and J. H. Linn, *Appl. Spectrosc.* **46,** 340 (1992).

B. Wangmaneerat, T. M. Niemczyk, and D. M. Haaland, *Appl. Spectrosc.* **46,** 1447 (1992).
S. Wood, P. Apte, T.-J. King, M. Moslehi, and K. Saraswat, *SPIE* **1393,** 337 (1990).

Ellipsometric Pyrometry (Section 13.2)

G. P. Hansen, S. Krishnan, R. H. Hauge, and J. L. Margrave, *Metall. Trans. A* **19A,** 1889 (1988).
G. P. Hansen, S. Krishnan, R. H. Hauge, and J. L. Margrave, *Appl. Opt.* **28,** 1885 (1989).

Dual-Wavelength Pyrometry (Section 13.3)

D. Mordo, Y. Wasserman, and A. Gat, *SPIE* **1595,** 52 (1991).
D. Mordo, Y. Wasserman, and A. Gat, *Semicond. Int.* **14**(11), 86 (Oct. 1991).
P. C. Nordine, *High. Temp. Sci.* **21,** 97 (1986).

Pyrometric Interferometry (Section 13.4)

F. G. Boebel and H. Möller, *IEEE Trans. Semicond. Manuf.* **SM-6,** 112 (1993).
F. G. Böbel, H. Möller, A. Wowchak, B. Hertl, J. Van Hove, L. A. Chow, and P. P. Chow, *J. Vac. Sci. Technol. B* **12,** 1207 (1994).
F. G. Böbel, H. Möller, B. Hertel, G. Ritter, and P. Chow, *Solid State Technol.* **37**(8), 55 (1994).
F. G. Böbel, A. Wowchak, P. P. Chow, J. Van Hove, and L. A. Chow, *Mater. Res. Soc. Symp. Proc.* **324,** 105 (1994).
K. A. Snail and C. M. Marks, *Appl. Phys. Lett.* **60,** 3135 (1992).

Raman Scattering (Spontaneous, Vibrational) (Section 12.4.2.1)

B. J. Kip and R. J. Meier, *Appl. Spectrosc.* **44,** 707 (1990).
G. D. Pazionis, H. Tang, and I. P. Herman, *IEEE J. Quantum Electron.* **QE-25,** 976 (1989).
M. Yamada, K. Nambu, Y. Itoh, and K. Yamamoto, *J. Appl. Phys.* **59,** 1350 (1986).
(see other references in Section 12.4.2.1)

Reflection Interferometry (in Wafers, Films) (Section 9.5.2)

R. A. Bond, S. Dzioba, and H. M. Naguib, *J. Vac. Sci. Technol.* **18,** 335 (1981).
V. M. Donnelly, *J. Vac. Sci. Technol. A* **11,** 2393 (1993).
V. M. Donnelly, *Appl. Phys. Lett.* **63,** 1396 (1993).
V. M. Donnelly and J. A. McCaulley, *J. Vac. Sci. Technol. A* **8,** 84 (1990).
V. M. Donnelly, D. E. Ibbotson, and C.-P. Chang, *J. Vac. Sci. Technol. A* **10,** 1060 (1992).

D. Hacman, *Optik* **28,** 115 (1968).
T. R. Hayes, P. A. Heimann, V. M. Donnelly, and K. E. Strege, *Appl. Phys. Lett.* **57,** 2817 (1990).
J. Kikuchi, R. Kurosaki, S. Fujimura, and H. Yano, *SPIE* **2336,** 111 (1994).
R. Kurosaki, J. Kikuchi, Y. Kobayashi, Y. Chinzei, S. Fujimura, and Y. Horiike, *SPIE* **2635,** 224 (1995).
J. A. McCaulley, V. M. Donnelly, M. Vernon, and I. Taha, *Phys. Rev. B* **49,** 7408 (1994).
V. R. McCrary, V. M. Donnelly, S. G. Napholtz, T. R. Hayes, P. S. Davisson, and D. C. Bruno, *J. Cryst. Growth* **125,** 320 (1992).
K. Murakami, Y. Tohmiya, K. Takita, and K. Masuda, *Appl. Phys. Lett.* **45,** 659 (1984).
K. L. Saenger, *J. Appl. Phys.* **63,** 2522 (1988).
K. L. Saenger and J. Gupta, *Appl. Opt.* **30,** 1221 (1991).
K. L. Saenger, F. Tong, J. S. Logan, and W. M. Holber, *Rev. Sci. Instrum.* **63,** 3862 (1992).
H. Sankur and W. Gunning, *Appl. Phys. Lett.* **56,** 2651 (1990).
P. J. Timans, R. A. McMahon, and H. Ahmed, *Appl. Phys. Lett.* **53,** 1844 (1988).

Reflectometry at an Interface (Section 9.4.1)

J. M. C. England, N. Zissis, P. J. Timans, and H. Ahmed, *J. Appl. Phys.* **70,** 389 (1991).
D. Guidotti and J. G. Wilman, *Appl. Phys. Lett.* **60,** 524 (1992).
D. Guidotti and J. G. Wilman, *J. Vac. Sci. Technol. A* **10,** 3184 (1992).
H. Kempkens, W. W. Byszewski, P. D. Gregor, and W. P. Lapatovich, *J. Appl. Phys.* **67,** 3618 (1990).
B. Peuse and A. Rosekrans, *Mater. Res. Soc. Symp. Proc.* **303,** 125 (1993).
B. Peuse and A. Rosekrans, *SPIE* **2091,** 301 (1994).
B. W. Peuse, A. Rosekrans, and K. A. Snow, *SPIE* **1804,** 45 (1993).

Thermal Expansion–Based Methods

D. Bollmann and K. Haberger, *Microelectron. Eng.* **19,** 383 (1992). (Section 10.1)
S. R. J. Brueck, S. H. Zaidi, and M. K. Lang, *Mater. Res. Soc. Symp. Proc.* **303,** 117 (1993). (Section 11.2.2.3)
D. Burckel, S. H. Zaidi, M. K. Lang, A. Frauenglass, and S. R. J. Brueck, *Mater. Res. Soc. Symp. Proc.* **342,** 17 (1994). (Section 11.3)
D. Burckel, S. H. Zaidi, A. Frauenglass, M. Lang, and S. R. J. Brueck, *Opt. Lett.* **20,** 315 (1995). (Section 11.3)
A. Durandet, O. Joubert, J. Pelletier, and M. Pichot, *J. Appl. Phys.* **67,** 3862 (1990). (Section 11.2.2.3)
M. K. Lang, G. W. Donohoe, S. H. Zaidi, and S. R. J. Brueck, *Opt. Eng.* **33,** 3465 (1994).
B. Peuse and A. Rosekrans, *Mater. Res. Soc. Symp. Proc.* **303,** 125 (1993). (Section 9.4.1)
B. Peuse and A. Rosekrans, *SPIE* **2091,** 301 (1994). (Section 9.4.1)
B. W. Peuse, A. Rosekrans, and K. A. Snow, *SPIE* **1804,** 45 (1993). (Section 9.4.1)
M. Pichot and M. Guillaume, U. S. Patent 4,525,066 (1982). (Section 11.2.2.3)

D. W. Voorhes and D. M. Hall, *SPIE* **1595,** 61 (1991). (Section 11.3)
S. H. Zaidi, S. R. J. Brueck, and J. R. McNeil, *J. Vac. Sci. Technol. B* **10,** 166 (1992). (Section 11.2.2.3)
[also see Reflection Interferometry (in wafers, films) (Section 9.5.2)]

Transmission (Absorption) (Section 8.4)

M. E. Adel, Y. Ish-Shalom, and H. Gilboa, *SPIE* **1803,** 290 (1992).
R. Duek, N. Vofsi, S. Mangan, and M. Adel, *Semicond. Int.* July, p. 208, 1993.
E. S. Hellman and J. S. Harris, Jr., *J. Cryst. Growth* **81,** 38 (1987).
S. R. Johnson, C. Lavoie, T. Tiedje, and J. A. Mackenzie, *J. Vac. Sci. Technol. B* **11,** 1007 (1993).
S. R. Johnson, C. Lavoie, E. Nodwell, M. K. Nissen, T. Tiedje, and J. A. Mackenzie, *J. Vac. Sci. Technol. B* **12,** 1225 (1994).
D. S. Katzer and B. V. Shanabrook, *J. Vac. Sci. Technol. B* **11,** 1003 (1993).
K. Kyuma, S. Tai, T. Sawada, and M. Nunoshita, *IEEE J. Quantum Electron.* **QE-18,** 676 (1982).
W. S. Lee, G. W. Yoffe, D. G. Schlom, and J. S. Harris, Jr., *J. Cryst. Growth* **111,** 131 (1991).
J. A. Roth, T. J. de Lyon, and M. E. Adel, *Mater. Res. Soc. Symp. Proc.* **324,** 353 (1994).
B. V. Shanabrook, J. R. Waterman, J. L. Davis, R. J. Wagner, and D. S. Katzer, *J. Vac. Sci. Technol. B* **11,** 994 (1993).
J. C. Sturm and C. M. Reaves, *IEEE Trans. Electron Devices* **ED-39,** 81 (1992).
J. C. Sturm, P. V. Schwartz, and P. M. Garone, *Appl. Phys. Lett.* **56,** 961 (1990).
J. C. Sturm, P. M. Garone, and P. V. Schwartz, *J. Appl. Phys.* **69,** 542 (1991).

References

R. L. Anderson, *SPIE* **1392,** 437 (1990).
R. L. Anderson and T. G. Kollie, *CRC Crit. Rev. Anal. Chem.* **6,** 171 (1976).
G. G. Barna, M. M. Moslehi, and Y. J. Lee, *Solid State Technol.* **37**(4), 57 (1994).
T. H. K. Barron, J. G. Collins, and G. K. White, *Adv. Phys.* **29,** 609 (1980).
R. P. Benedict, *Fundamentals of Temperature, Pressure, and Flow Measurements.* Wiley, New York, 1984.
H. Böttger, *Principles of the Theory of Lattice Dynamics,* p. 47. Physik-Verlag, Berlin, 1983.
W. W. Clark, III and F. C. De Lucia, *J. Chem. Phys.* **74,** 3139 (1981).
M. L. Cohen and D. J. Chadi, in *Handbook of Semiconductors* (M. Balkanski, ed.), Vol. 2, Chapter 4B, p. 155. North-Holland Publ., Amsterdam, 1980.
G. P. Davis and R. A. Gottscho, *J. Appl. Phys.* **54,** 3080 (1983).
F. L. Degertekin, J. Pei, B. T. Khuri-Yakub, and K. C. Saraswat, *Appl. Phys. Lett.* **64,** 1338 (1994).
L. A. Farrow, *J. Chem. Phys.* **82,** 3625 (1985).
H. R. Griem, *Plasma Spectroscopy.* McGraw-Hill, New York, 1964.
V. Heine and J. A. Van Vechten, *Phys. Rev. B* **13,** 1622 (1976).

J. F. Hershberger, J. Z. Chou, G. W. Flynn, and R. E. Weston, Jr., *Chem. Phys. Lett.* **149,** 51 (1988).
R. P. Hudson, *Rev. Sci. Instrum.* **51,** 871 (1980).
C. Keffer, T. M. Hayes, and A. Bienenstock, *Phys. Rev. Lett.* **21,** 1676 (1968).
J. C. Knights, J. P. M. Schmitt, J. Perrin, and G. Guelachvili, *J. Chem. Phys.* **76,** 3414 (1982).
M. Konuma, *Film Deposition by Plasma Techniques.* Springer-Verlag, Berlin, 1992.
P. Lautenschlager, M. Garriga, S. Logothetidis, and M. Cardona, *Phys. Rev. B* **35,** 9174 (1987a).
P. Lautenschlager, M. Garriga, and M. Cardona, *Phys. Rev. B* **36,** 4813 (1987b).
P. Lautenschlager, M. Garriga, L. Viña, and M. Cardona, *Phys. Rev. B* **36,** 4821 (1987c).
S. Logothetidis, L. Viña, and M. Cardona, *Phys. Rev. B* **31,** 947 (1985).
T. D. McGee, *Principles and Methods of Temperature Measurement.* Wiley, New York, 1988.
S. S. Mitra and N. E. Massa, in *Handbook of Semiconductors* (W. Paul, ed.), Vol. 1, Chapter 3, p. 81. North-Holland Publ., Amsterdam, 1982.
J. E. Moody and R. H. Hendel, *J. Appl. Phys.* **53,** 4364 (1982).
M. M. Moslehi, R. A. Chapman, M. Wong, A. Paranjpe, H. N. Najm, J. Kuehne, R. L. Yeakley, and C. J. Davis, *IEEE Trans. Electron Devices* **ED-39,** 4 (1992).
H. N. Norton, *Handbook of Transducers.* Prentice-Hall, Englewood Cliffs, NJ, 1989.
M. C. Öztürk, F. Y. Sorrell, J. J. Wortman, F. S. Johnson, and D. T. Grider, *IEEE Trans. Semicond. Manuf.* **SM-4,** 155 (1991).
J. Park, Y. Lee and G. W. Flynn, *Chem. Phys. Lett.* **186,** 441 (1991).
L. Peters, *Semicond. Int.* **14**(9), 56 (1991).
D. Peyton, H. Kinoshita, G. Q. Lo, and D. L. Kwong, *SPIE* **1393,** 295 (1990).
F. Reif, *Fundamentals of Statistical and Thermal Physics.* McGraw-Hill, New York, 1965.
B. K. Ridley, *Quantum Processes in Semiconductors,* 2nd ed., p. 35. Oxford Univ. Press, New York, 1988.
F. Roozeboom, *Mater. Res. Soc. Symp. Proc.* **224,** 9 (1991).
F. Roozeboom, *Mater. Res. Soc. Symp. Proc.* **303,** 149 (1993a).
F. Roozeboom, in *Rapid Thermal Processing Science and Technology* (R. B. Fair, ed.), Chapter 9, p. 349. Academic Press, Boston, 1993b.
F. Roozeboom and N. Parekh, *J. Vac. Sci. Technol. B* **8,** 1249 (1990).
J. F. Schooley, *Thermometry.* CRC, Boca Raton, FL, 1986.
F. W. Sears and G. L. Salinger, *Thermodynamics, Kinetic Theory and Statistical Thermodynamics.* Addison-Wesley, Reading, MA, 1975.
H. Shen, *SPIE* **1286,** 125 (1990).
H. Shen, S. H. Pan, Z. Hang, J. Leng, F. H. Pollak, J. N. Woodall, and R. N. Sacks, *Appl. Phys. Lett.* **53,** 1080 (1988).
M. E. Thomas, in *Handbook of Optical Constants of Solids II* (E. D. Palik, ed.), Chapter 8, p. 177. Academic Press, Boston, 1991.
Y. P. Varshni, *Physica* **34,** 149 (1967).
L. Viña, S. Logothetidis, and M. Cardona, *Phys. Rev. B* **30,** 1979 (1984).
S. Zollner, M. Garriga, J. Kircher, J. Humlicek, M. Cardona, and G. Neuhold, *Thin Solid Films* **233,** 185 (1993).

CHAPTER **19**

Data Analysis and Process Control

In their presentation of optical diagnostics that can be used to obtain information during thin film processing, Chapters 6–18 describe the connection between the raw data obtained from these probes and physical quantities of interest, such as temperature, film thickness, and species density in the gas phase. This chapter focuses on how to analyze and interpret this experimental information, and how to make decisions based on it. This involves data analysis, model design, and process control strategies.

The perspective here is somewhat broader than that in the rest of this volume because much of the discussion in this chapter is pertinent to all sensors, not just to optical diagnostics. Moreover, for illustration, several examples of models and process control are presented that do not use information from real-time sensors at all. Still, the interpretation of data from optical diagnostics and the use of optical probes as sensors for process control are emphasized. This chapter extends the overview given in Section 1.4.

Data analysis, model design, and control strategy are all important in utilizing the information from process sensors. Analysis of the raw data is relatively straightforward in fundamental and process-development studies, but can become overwhelming during manufacturing when continuous readings from the multiple sensors in each tool start streaming in for real-time analysis (Section 19.1).

Models are needed to transform these data into an understanding of the process (from a fundamental or process-development viewpoint) or at least a characterization that interrelates the state of the equipment, the process, and the wafer (Moslehi *et al.*, 1992; Barna *et al.*, 1994a) from a manufacturing perspective (Section 19.2). These models are mathematical constructs that need not necessarily be based on the physical and chemical steps, and they can be important both in process development and control. Some models utilize sensor readings of intermediate process conditions (*in situ* measurements), while others relate input process variables (equipment-state inputs) to process results (final wafer state), as is discussed in Section 19.2.

These data and models are needed for process control during manufacturing (Section 19.3). Real-time process control requires rapid, efficient analysis and interpretation of data from all sensors, followed by feedback to the equipment parameters by means of a control algorithm. An integral part of this procedure is developing a model in which experimentally controllable parameters are related to process results and the outcome of the process is characterized, optimized, and controlled.

The specific wafer-state parameters that need to be tracked and controlled depend on the specific process. Examples are listed in Tables 1.1 and 1.9 and are also given by Moslehi *et al.* (1992) and Barna *et al.* (1994a). They can include wafer temperature, film thickness, film lateral structure, composition, and electrical properties. In every case, it is important to achieve both the desired set point and uniformity across the wafer. For some parameters *in situ*, real-time tracking is absolutely necessary, while for others *in situ*, post-process analysis suffices. Furthermore, some cases require only in-line or *ex situ* post-process analysis.

19.1 Data Acquisition and Analysis

Usually, multivariate statistics are needed to analyze thin film experiments because of the interplay of the (1) multidimensional input (equipment) and process-state parameters, such as gas and wafer temperatures, total pressure, and the fractional composition of each gas; (2) multiple *in situ* probes, which can provide continuous readings as a function of time and, in some cases, continuous readings in multiple dimensions (e.g., wavelength for OES—an optical probe—or mass, for mass spectrometry—a nonoptical probe); and (3) various process results (wafer-state parameters). Efficient data analysis is essential because of the continuous stream of readings from many sensors and the need for rapid data interpretation during real-time process control. The term "rapid" is relative to the speed of the process. For example, annealing by rapid thermal processing (RTP) has very different analysis requirements ($\sim 0.03 - 1$ sec) than do laser annealing with excimer lasers (many nanoseconds) and furnace annealing (many minutes).

The data provided by multichannel detectors (Section 5.1.2), such as diode arrays and CCD [charge-coupled device] cameras for OES, spectroscopic reflection experiments, and Raman scattering, can improve the understanding of a process and can affect each of these three points. The flow of such data from these detectors is rapid enough for many real-time control applications. However, unless these data streams are processed by mathe-

19.1 Data Acquisition and Analysis

matical routines that can analyze and interpret these data very rapidly, this data flow is useless for process control.

Chemometrics is the term commonly used in the chemical community for the application of statistical and mathematical methods for efficient data analysis and interpretation and for the design and optimization of chemical experiments. The data analysis and design features of chemometrics are also relevant to thin film processing, and have been detailed in several books, including those by Sharaf *et al.* (1986), Miller and Miller (1988), Brereton (1990), Meloun *et al.* (1992), and Meier and Zünd (1993). Important methods in multivariate statistics, such as multiple linear regression, discriminant analysis, and principal-component analysis, are described in the texts by Harris (1985), Flury and Riedwyl (1988), and Bernstein *et al.* (1988).

Haaland and Thomas (1988a,b), Haaland (1988), and Thomas and Haaland (1990) have compared spectral analysis by four types of multivariate analysis: inverse least-squares (ILS), classical least-squares (CLS), partial least-squares (PLS), and principal-component regression (PCR) (Martens and Naes, 1989). They found that the full-spectrum methods, CLS, PLS, and PCR, usually outperformed ILS, which is frequency-limited. While PCR and PLS gave similar information, PLS was usually either optimal or very nearly so.

The use of one of these multivariate analysis routines, principal-component regression analysis, is now outlined. The goal of this technique is to reduce a complex data set of n elements, such as that from a multichannel spectrum (of n wavelengths or masses), to a smaller set $m \ll n$, to give a compressed view of the distinctive features of the original data set (Jolliffe, 1986; Shadmehr *et al.*, 1992). Since n is typically ~500 or 1000 in multichannel optical detectors, rapid compression down to a small number of components, say $m = 4$, is very desirable. With the measurement of property X, such as the emission intensity, in channel k during run j called X_{jk}, the sampled covariance matrix is

$$A_{jk} = \frac{1}{p-1} \sum_{i=1}^{p} (X_{ij} - \overline{X}_j)(X_{ik} - \overline{X}_k) \qquad (19.1)$$

where p is the number of runs and

$$\overline{X}_j = \frac{1}{p} \sum_{i=1}^{p} X_{ij} \qquad (19.2)$$

The n eigenvectors of matrix A are the principal components, and only those eigenvectors with the largest eigenvalues are of interest. X is then projected onto the principal components. The new data set is less complex than the original data set because it has many fewer dimensions, and yet it retains essentially the same information content. As detailed in Section

19.2.3, Shadmehr et al. (1992) used principal-component analysis to reduce the 1024 OES data points and 99 RGA [residual gas analyzer] points to four points per run for each.

Multivariate analysis has also been used in several other *ex situ* and *in situ* analyses of thin films and thin film processes. Haaland (1990) has discussed how multivariate calibration methods can be applied to Fourier-transform infrared (FTIR) analysis. Haaland (1988) used CLS, PLS, and PCR methods for quantitative IR analysis of borophosphosilicate films. Niemczyk, Haaland, Franke and co-workers applied these techniques to the investigation of dielectric film composition and temperature by infrared absorption (Section 8.3), reflection (Section 9.5), and emission (Section 13.2; Figure 13.8) spectroscopies. Wangmaneerat et al. (1992a) used multivariate PLS to analyze the infrared emission spectra of phosphosilicate glass taken with an FTIR spectrometer, while Wangmaneerat et al. (1992b) employed this method to analyze the optical emission spectrum obtained during plasma etching by using a photodiode array. Splichal and Anderson (1991) used multivariate statistics to correlate OES spectra taken during plasma etching with etch rates and electrical probe measurements (Figure 6.11).

Other analysis methods have also been employed. Krukar et al. (1993) measured the scattergrams (over 180°) from Si wafers with photolithographically defined lines and spaces produced by reactive-ion etching (RIE). A range of etch depths and sidewall profiles, generated by varying etching conditions, was examined (Figures 11.26 and 11.27; Section 11.2.2.1). They used discriminant analysis and neural network training (Section 19.2.2) to catalog these scattergrams according to the etch depth and profile of the microstructures. With this method, these etch features were classified with >95% accuracy, which bodes well for the use of scatterometry as a diagnostic for real-time monitoring and control during etching. In earlier work, Giapis et al. (1991) used principal-component analysis to classify the scattergrams according to etch profile and depth.

Such methods do a fine job of handling the flow of data when analysis of data obtained at a time t is unaffected by the results of the analysis at a previous time $t - dt$. However, optical monitoring of the growth or etching of films often depends on conclusions drawn earlier during the process. One such example is monitoring epitaxial growth, or in fact any deposition, by ellipsometry to control the composition with submonolayer accuracy. As described in Section 9.11.1, the interpretation of ellipsometric data depends on the model and on the data set used for the optical parameters of the system. This interpretation must be performed with minimal uncertainty for real-time control of composition with submonolayer control. Section 9.11.1 details several algorithms that can be used for rapid analysis of the near-surface dielectric function, and consequently composition, and of the layer thickness (Aspnes, 1996, also private communication, 1995). One

19.2 Process Modeling

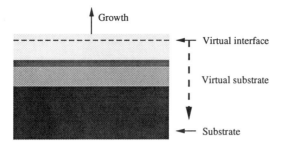

Figure 19.1 Virtual interface model.

widely employed method is the virtual interface model (Figure 19.1), which updates the interpretation based only on new data.

19.2 Process Modeling

There are four goals of process modeling: (1) process understanding and development; (2) process optimization for manufacturing, including adaptive manufacturing, and for making the process robust, i.e., insensitive to small disturbances (open-loop control); (3) error diagnosis during manufacturing; and (4) error correction during manufacturing for feedback to the current run (real-time, closed-loop control) or at least for subsequent runs (Run-to-Run control).

Models interrelate equipment, process, and wafer parameters, as diagrammed in Figure 19.2. They can be physics- and chemistry-based, using

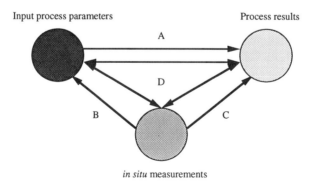

Figure 19.2 Relation of input process parameters and process results, and how *in situ* measurements can be used to analyze both. Input process parameters (equipment-state parameters) directly affect process results (wafer-state parameters) (A). *In situ* measurements can be used to check equipment- (and process-) state parameters (B) and to monitor process results (C). In real-time process control *in situ* measurements can be used to interrelate all process-related parameters (D).

first principles, or empirical ("black-box"-based), such as those using statistical and neural network approaches. Sometimes models are developed for a given process in a specific type of tool.

When models are based solely on physical and chemical processes, they can be extrapolated outside the range of available data to give good predictions of process sensitivities. They can also be used to understand the same process in different reactors and in other thin film processes having similar chemistry. In contrast, empirical models, which are included within the realm of chemometrics, correlate process parameters to results without input about physical processes. These models are easy to apply and minimize the need for detailed fundamental studies. However, empirical models can be interpolated with good accuracy only within the range of experimental data. Heuristically less appealing than physical models and limited to the parameter space studied, empirical models are still often found to be useful for process modeling and real-time control. Sometimes such chemometrics models can also give important physical and chemical insight into the process. Hybrid, semiempirical modeling attempts to adapt the better features of physical-based and empirical modeling (Sachs *et al.*, 1992).

Models can relate experimental equipment parameters directly to the process goals (wafer-state parameters), as in path A in Figure 19.2. For example, in plasma processing, the input conditions set by the operator can include RF power, pressure, electrode gap, and the flow rates of each gas, and the process goals or specifications include the etch or deposition rates of each film (for etching this gives information about selectivity), uniformity, and (for etching) edge profile. This mapping gives the recipe for operation. However, the conditions of the reactor may change with repeated use, and the same input conditions may not yield the same process results; i.e., the recipe may change. This may occur because the set input condition cannot be achieved due to malfunctioning equipment or to more subtle, though equally devastating, changes (drifts), such as those caused by changing wall conditions.

Models can also involve the internal conditions of the reactor (process parameters and current wafer parameters), which can be monitored with appropriate diagnostics. Typical internal conditions are film thickness and composition in CVD and the plasma characteristics in RIE. Plasma characteristics can be monitored by using OES and/or RGA. These models can try to (1) determine the input parameters from these internal conditions (path B in Figure 19.2), because the latter are known to correlate with the process inputs; (2) relate the internal conditions to process results (C); or (3) interrelate input, internal, and final conditions (D).

19.2.1 *Statistical Models*

Several different approaches can be used in constructing empirical models, including statistical and neural network models. In statistical models, the

f process variables I_j are correlated to m process specifications O_i, either linearly or nonlinearly. If there are k levels or settings for each of the f factors, k^f trials or experiments in a "full factorial" model design are needed for each process specification. A selected subset of such trials is called a *fractional factorial design.*

In one statistical modeling technique, a response surface is created (Box and Draper, 1987; Jenkins *et al.,* 1986; Myers *et al.,* 1989) in which process specifications or results, also called responses, are expressed as sums of linear and often bilinear terms of input process variables, also called factors, such as

$$O_i = O_i^0 + \sum_j a_{ij}I_j + \sum_{jj'} b_{ijj'}I_jI_{j'} \tag{19.3}$$

where a_{ij} and $b_{ijj'}$ are parameters that characterize the surface for the ith process output. Response surface methodology (RSM) requires at least as many experimental runs as parameters, O_i^0, a_{ij}, and $b_{ijj'}$. Also, there must be at least three levels ($k \geq 3$) for the quadratic design of Equation 19.3, and at least four levels for a cubic design. The algorithms for choosing the trial subset in this fractional factorial design, such as the Box–Behnken, Face-Centered Cube (FCC), and central composite rotatable designs, differ in the number of trials they require and their sensitivity to small variations in process parameters from their "usual" settings. Least-squares regression analysis is used to examine the RSM "data." A broad range of input variables should be chosen if the outer regions of the factor space must be examined, because, unlike physical models, empirical models cannot be extrapolated outside the given parameter space. Statistical models become exceedingly complex and time-consuming with more than six process variables. Less significant variables can be set at a typical level when there are many parameters (>6). Use of response surface methods in integrated circuit manufacturing has been reviewed by Lombardi (1991). Statistically designed experiments using factorial design for robust operation, which are also called matrix experiments, have been reviewed by Hendrix (1979) and Phadke (1989).

Fractional factorial designs have also been developed for process optimization. In orthogonal design (Yin and Jillie, 1987), a matrix is constructed in which the row elements are the independent process inputs and the column elements are the trials. The subset of trials to be performed is given by the "orthogonal" matrix, which is often expressed as an orthogonal table. These orthogonal designs were further developed by Taguchi, resulting in the Taguchi quality-control methods for the design of experiments (orthogonal arrays, factor design) (Logothetis and Wynn, 1989; Roy, 1990; Tjantelé, 1991; Taguchi, 1993; Barker, 1994).

19.2.2 *Neural Network Training*

Neural network modeling is a qualitatively different approach in which the model "adaptively learns" from training experiments (McClelland *et al.,*

1986; Himmel and May, 1993; Zupan and Gasteiger, 1993). Neural networks are highly nonlinear algorithms, in which learning occurs in many parallel processing units, called neurons. Each neuron is a simple processor in which incoming connections from the previous layer are weighted and then "squashed" by an exponential sigmoid function to give the weighted output for the next layer, as is illustrated in Figure 19.3. There are several parallel neurons in each layer of neurons, and the interconnections between layers are highly branched. The first layer (input layer) receives the n input conditions, which are fed into the first "hidden" layer of neurons. The output of this layer is fed into the next hidden layer of neurons, which is fed into the next level, etc. The output of the last hidden layer feeds the last neuron layer (output layer), whose outputs are the m process results. This is illustrated in Figure 19.4. The hidden layers "represent" the physical and chemical steps in the process. The number of neurons per hidden layer and the number of hidden layers are both adjustable in the model. Since neural networks that use the feed-forward error back-propagation algorithm (described below) need only two hidden layers, the number of neurons (k) in each layer is varied to optimize performance. In an "n-q-m" structure, n input neurons feed hidden layers with q neurons each, which results in m output neurons.

The input to a given neuron (Figure 19.3), say the neuron i in layer k, is the weighted sum

$$I_{i,k} = \sum_j w_{i,j,k} \, O_{j,k-1} \tag{19.4}$$

where $w_{i,j,k}$ is the weighting function between neuron i in layer k and neuron j in layer $k-1$ and $O_{j,k-1}$ is the output of this latter neuron. The output of this neuron is

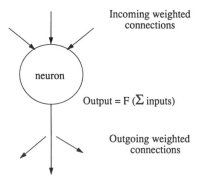

Figure 19.3 Schematic of a single neuron, as is used in neural network analysis. The output of a neuron is F(the weighted sum of the inputs), where F is the exponential sigmoid function, as is seen in Equations 19.4 and 19.5 (from Himmel and May, 1993). (© 1993 IEEE.)

19.2 Process Modeling

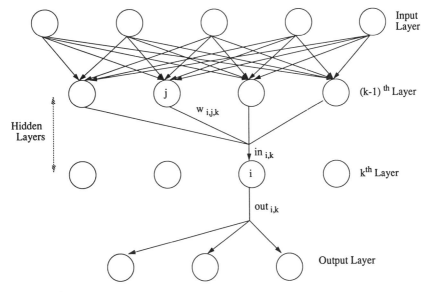

Figure 19.4 Feed-forward error back-propagation (FFEBP) neural network (from Himmel and May, 1993). (© 1993 IEEE.)

$$O_{i,k} = \frac{1}{1 + \exp(-I_{i,k})} \tag{19.5}$$

which is the exponential sigmoid function.

The feed-forward error back-propagation (FFEBP) method is one algorithm that efficiently "learns" the relationships between noisy input and output data; this amounts to optimizing the weighting functions $w_{i,j,k}$. After an initial random choice of $w_{i,j,k}$, the algorithm is run forward and the error is determined by comparing the output of the network with the experimental output. The error is then propagated backward, and the error is minimized by using a gradient-descent approach that optimizes the weighting functions one layer at a time. The layer-to-layer propagation of the algorithm depends on the gradient in the error, and it is followed until the model converges (Himmel and May, 1993). For testing, the data set is often subdivided into a training set, to "teach" the neural network, and a set to test network predictions.

Sometimes a process is subdivided into subprocesses, each of which is analyzed by individual FFEBP networks. These subprocesses, linked by "influence diagrams," are then integrated by a single associative-memory (AM) network that looks at all process parameters by using a "learning" algorithm (Nadi *et al.*, 1991).

19.2.3 *Applications to Thin Film Processing Development*

References to physical models of CVD can be found in Sherman (1988) and Hitchman and Jensen (1993). One example is the detailed model of Si chemical vapor deposition developed by Coltrin *et al.* (1984), which incorporates both flow dynamics and chemical kinetics. Physical models of plasma etching are cited in Manos and Flamm (1989) and Himmel and May (1993). Many of these models are large area, tool-scale models that are focused on the net flow of species to and from the surface. Physical models and simulations on the scale of features are also very important (Cale *et al.*, 1991; Singh and Shaqfeh, 1993; Singh *et al.*, 1994), as are models that link feature- and tool-scale simulations (Cale *et al.*, 1993), and atomic-scale simulations. Empirical models have also been increasingly applied to optimizing and controlling CVD and plasma etching.

Lin and Spanos (1990) developed an equipment-specific linearized model of polysilicon deposition by combining a physical model with statistical experimental design and standard regression analysis. Sachs *et al.* (1992) constructed a model for the low-pressure chemical vapor deposition (LPCVD) of polysilicon in a horizontal-tube furnace for batch processing, which accounts for wafer-to-wafer variations in thickness. This is a semiempirical model that combines physical modeling, based on the work of Roenigk and Jensen (1985), and statistical experimental design to create a "smart" response surface. Guo and Sachs (1993) developed the method of multiple response surfaces to model, optimize, and control the spatial uniformity of film processes, including etching and CVD operating in either the batch-processing or single-wafer mode.

DePinto (1993) used Taguchi methods to characterize and optimize a LPCVD tool for a constant-temperature polysilicon process. Aceves *et al.* (1992) characterized aluminum evaporation by using the Taguchi technique.

Nadi *et al.* (1991) compared statistical and neural network models of LPCVD of polysilicon, and found that the neural network with influence diagrams is more than three times as accurate in predicting the deposition rate and more than twice as accurate in predicting film stress, even when given half as much experimental information.

Jenkins *et al.* (1986) and Allen *et al.* (1986a) used response-surface methodology to model plasma etching, and Allen *et al.* (1986b) compared these predictions with those from a physical/chemical model. Yin and Jillie (1987) used orthogonal design to optimize plasma etching of polysilicon and silicon nitride by C_2F_6. May *et al.* (1991) followed a fractional factorial approach with an empirical response surface to model the etching of n^+-type doped polysilicon in a $CCl_4/He/O_2$ plasma. Himmel and May (1993) compared empirical models of plasma etching of polysilicon, specifically the response surface and neural network models. There were six input conditions and

19.2 Process Modeling

four process outputs, with six neurons per hidden layer in the neural network model. They concluded that neural network modeling is superior to statistical modeling using response surfaces in that it is more accurate when both models utilize equivalent experimental data, and it can be as accurate as statistical modeling even when it is fed relatively less experimental data. Kim and May (1994) optimized the strategies used in modeling plasma etching with a neural network, using polysilicon etching by CCl_4 as a example. Rietman and Lory (1993) also used a neural network to model plasma etching.

Shadmehr *et al.* (1992) used a neural network to determine the effect of process parameters on RIE etching of Si by CHF_3/O_2 by using *in situ* measurements with OES, an optical probe with information at 1024 wavelengths, and RGA, a nonoptical probe with information at 99 masses. They employed principal-component analysis to reduce the 1024 OES data points and 99 RGA points each to four points per run in the 72 experiments they conducted. This reduced data "input" from OES, RGA, or OES + RGA was then related to the experimental input parameters, RF power, O_2 fraction, pressure, and initial Teflon-like thin film present on inside walls of the chamber, which constituted the "output" of the model. The "input" was related to the "output" (path B in Figure 19.2) by using either a linear statistical model or a neural network. Figure 19.5 shows the correlation of the first and second principal components projected from the OES data set for different experimental parameters. Both OES and RGA made good "predictions" of the input parameters, based on root-mean-square errors. OES data determined RF power better, while RGA determined pressure better. Both predicted thin film content and mixture fraction fairly well. Analysis was vastly better when both OES and RGA data sets were used than when either set was used alone. Neural network analysis was usually superior to the linear regression method, but it was sometimes decidedly inferior in pressure analysis.

Modeling of temperature distributions in cold-wall reactors can be very important. It is especially so in rapid thermal processing (RTP), where it is essential to design the lamp system and control the lamp output to achieve uniformity (Gyurcsik *et al.,* 1991; Apte and Saraswat, 1992; Saraswat *et al.,* 1994; Schaper *et al.,* 1994). The lamps can also be controlled to account for emissivity changes in the wafer, which can affect the temperature (and its measurement by pyrometry) (Sorrell and Gyurcsik, 1993; also see Chapter 13). Cho and Kailath (1993) have tested physics-based and black-box statistical models of RTP processing.

Smadi *et al.* (1992) have modeled particle deposition and etch depth after plasma etching of silicon by using statistical experimental design. They found that second-order response-surface methodology uniquely related particulate density on the surface, measured by laser light scattering, and

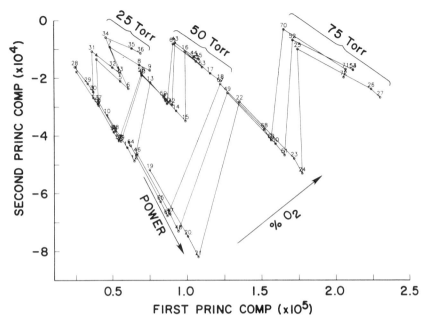

Figure 19.5 Plot of the projection of the optical emission spectrum from a CHF$_3$/O$_2$ plasma on its first and second principal components. Each point represents a given input condition, with points at the same pressure connected and trends for RF power and O$_2$ fraction shown by arrows (from Shadmehr *et al.*, 1992).

etch depth to plasma process parameters, such as RF power, total pressure, flow rate, and etch time.

In more general applications, Boning *et al.* (1992) developed a versatile framework for modeling semiconductor processing, Boning and Mozumder (1994) used response-surface modeling for process optimization of film processing, and Gaston and Walton (1994) integrated RSM with process and device simulation tools to help reduce the time needed to optimize the design/analysis cycle time, which is important in adaptive manufacturing.

19.3 Process Control

The goal of a control system is to achieve the desired system response. The control system can be used to compensate for deviations from the desired response or to improve manufacturing capabilities (Butler, 1995). For thin film processing, the desired response is the structure on the wafer: its composition, thickness, lateral patterning, electrical and structural properties, etc. There are two general system strategies to achieve control. In the

19.3 Process Control

open-loop control system, as depicted in Figure 19.6a, a controller sets the input process conditions that are supposed to give the desired output conditions. In the *closed-loop control system,* shown in Figure 19.6b, there is feedback control. The actual output is compared to the desired output response, and if there is a difference (an error), the controller corrects the input variables into the process. The error can be due to disturbances acting on the system or to parameter variations in the system. The algorithm that corrects for errors is usually based on linear system theory, and can rely on statistical process models or more physical models of the system. The fundamentals of feedback-control systems are covered in the textbooks by Seborg *et al.* (1989), Van de Vegte (1990) and Dorf (1992).

A more sophisticated view of control possibilities includes regulation with different time scales and control structures (Butler, 1995). The time

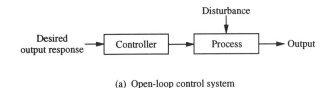

(a) Open-loop control system

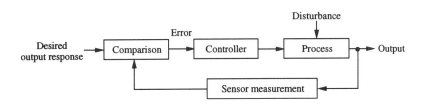

(b) Closed-loop feedback control system

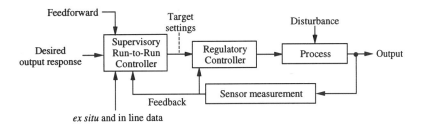

(c) Regulatory and supervisory control system

Figure 19.6 (a) Open- and (b) closed-loop feedback control systems, and (c) more complex systems that include supervisory and regulatory control (adapted from Butler, 1995).

scale can be real-time, in which settings are changed during a run, or Run-to-Run (RtR), in which target settings are determined for the next run. Fast and dynamic processes require real-time control, while slow and unchanging processes may require only RtR control. The control structure can be regulatory, in which the regulatory controller changes settings to achieve the target values, or supervisory, in which the supervisory controller provides target settings for the regulatory controllers. Figure 19.6c depicts a control system with a RtR supervisory controller, which can use *in situ* feedback, feedforward, in line or *ex situ* data, and a real-time regulatory controller, which uses *in situ* feedback data. The feedforward information can come from a previous run or from a previous step in a multistep run.

More representative than the single-input (to the process)/single-output (SISO)/single-sensor measurement closed-loop control system depicted in Figure 19.6b is a *multivariable control system*, which can have several inputs, outputs, and sensor readings, and not necessarily the same number of each of these. (See Figures 1.2 and 1.4.) A single multiple input–multiple output (MIMO) controller can be more stable than a corresponding set of SISO controllers.

The goal in designing open-loop systems is to choose the input process parameters in such a way that disturbances and parameter variations do not significantly affect the output. One way to do this is by statistical design of the experiment (or process), such as the response-surface and Taguchi methods described in Section 19.2.1, which permits efficient exploration of parameter space to determine the best and most robust operating point. Making the optimal choice of equipment-state parameters is part of the widely employed control approach called *statistical process control*. Traditional *ex situ* statistical process control (SPC) has been overviewed by Spanos *et al.* (1992).

In SPC, critical process parameters are measured after the process step is completed, *ex situ* either on-line or off-line, and are monitored by using control charts. When a system is in statistical process control, "chance causes" lead to natural variations or background noise. "Assignable causes" due to changes in input or operating conditions cause the system to lose control. SPC routines decide whether a system is under control by using charts based on statistics; if it is not in control, an alarm can be generated. It is assumed that data for each measured parameter follow identical normal distributions and are independent of each other; this is known as the IIND assumption (identically, independently, and normally distributed). Processes out of statistical process control are retuned as needed by the operator. There is often a very long delay in detecting errors with these methods, which is unsatisfactory in advanced manufacturing environments. SPC also makes extensive use of pilot wafers for calibration.

These traditional methods have certainly improved manufacturing reliability in many areas, including batch-processing integrated-circuit fabrica-

19.3 Process Control

tion, and yields can be very high with SPC. However, SPC is not expected to work as well when the demands in semiconductor processing get tighter. In contrast, improved control strategies such as *real-time process control*, which use sensors for closed-loop control, may improve performance as process demands increase. In addition to the possibility of improved process yields during large-scale manufacturing and single-wafer processing, real-time process control has definite advantages for flexible and specialty fabrication. Moreover, it is often difficult to acquire the conventional on-line, post-processing information needed for traditional SPC in modern reactors, such as multichamber cluster tools.

There are several reasons why this type of active control is necessary (Butler, 1995). Disturbances, such as those due to reactor aging, cleaning, and component malfunctioning, can alter the process. There might be dynamic process targets, as in specialty fabrication, which cannot be achieved without control. The setpoint conditions of a process can also change for other reasons. If two different thin film processes are run in the same tool, control strategies are needed to set the target variables for each process on the basis of potential interactions with the other, which can include changes of the coatings on the reactor walls; this is known as the "two process problem." In addition to using *in situ* diagnostics for such feedback control to the same process step, these diagnostics can be used for feedforward control to subsequent process steps to help correct for random errors or long-term drift incurred in the current run (Figure 1.2).

Modern process control strategies are sometimes collectively called *advanced process control* (APC) (Butler, 1995), which includes the aforementioned regulatory and supervisory control. The capabilities of APC include fault detection, fault classification and diagnosis, fault prognosis, and process-control correction. APC can improve process capability (e.g., minimize variability), increase equipment availability and productivity (minimize failure, accelerate repair, reduce the use of test wafers, assess equipment aging), decrease the manufacturing cycle time (reduce setup and inspection times), and increase manufacturing flexibility. In particular, real-time control with *in situ* diagnostics for feedback permits control of dynamic processes, accelerates decision making, removes deadtime, and reduces the need for pilot wafers. Feedback to the same process step and feedforward to the next process step are both important in Run-to-Run control. Good control strategies track changes in the process with minimum lag, are resistant to disturbances (unexpected real changes in the process), and reject noise (spurious fluctuations in the process). There is a tradeoff in these strategies because a system that responds rapidly to disturbances is also sensitive to noise.

Four potential applications of *in situ* sensors are described here in the context of process control. Goals 1–3 constitute a hierarchy of error detection and correction that culminates with real-time closed-loop control. The fourth goal is control for adaptive manufacturing, including *in situ* control of the process with submonolayer accuracy.

(1) The least ambitious goal is sounding an "alarm" to denote a problem. However, the alert comes only after the process step has been completed—perhaps because of slow data analysis and interpretation. This is quite valuable, nevertheless, and is more ambitious than the goal of traditional SPC in two ways: (a) Errors are likely to be detected more quickly; and (b) runs in such multistep systems as multifunctional reactors or cluster tools can be aborted after a defective step—and before subsequent steps—when *in situ* sensors are used. Subsequent processing steps can bury the problem, i.e., hide the "defect," until final testing and thus represent a waste of time and money. (These *in situ* measurements can be conducted either in real time during the process step or after it.) Moreover, usual SPC methods work best by detecting deviations among many wafers, which is suitable for batch-processing systems. Statistical control strategies are less applicable to single-wafer systems, while *in situ* sensing is fully compatible with single-wafer systems and is useful even when detection is slow. This goal is consistent with Run-to-Run control. A more mudane, though still important, aim of real-time sensing is merely archiving real-time diagnostic data for evaluation only if an error is detected in a later process step.

(2) Real-time analysis is fast and an alarm can be sounded during the step in which the problem occurs. While the problem cannot be alleviated, the benefits of goal 1 still apply. Corrective action can be taken prior to subsequent runs (supervisory Run-to-Run control) before much money and time are wasted, such as by altering the run recipe (menu) to compensate for drifts in the equipment and the process, conducting preventive maintenance on the equipment, and correcting equipment malfunctioning. In a multistep run corrective action could be taken before the next process step (feedforward control).

(3) Real-time analysis detects errors so fast that they can be corrected by using feedback to a regulatory controller to modify the recipe of the current run. This also requires a satisfactory process model. In addition to error detection and correction, this fast process monitoring also permits endpoint control of a process (Table 1.2), which can stop an etching, deposition, resist curing, oxidation, or doping step when the desired wafer parameter is attained. These two capabilities are the ultimate goals of closed-loop control with real-time sensors.

(4) The techniques in goal 3 are so refined and process simulation tools are so advanced that adaptive and flexible manufacturing can occur, including prototyping. For example, submonolayer control of composition during epitaxy would then be possible, so heterostructures with programmable composition profiles could be grown.

In each case, process control entails comparing the measured equipment-, process-, and wafer-state parameters with the expected values (Figure 1.4).

19.3 Process Control

This involves analysis of the data and deciding what the appropriate action should be: (a) continuing the process, (b) taking corrective action after error detection, or (c) stopping the process because the endpoint has been reached. The analysis and decision making steps may involve simple procedures or the use of elaborate models, as presented in Section 19.2.

Moslehi *et al.* (1992) categorized feedback control signals that change the equipment parameters into indirect measurement signals, from equipment sensors, and direct measurement signals, from process and wafer sensors, as is illustrated in Figure 1.4. The controllers adjust the equipment parameters continuously to minimize the difference between the observed parameters and the setpoints. The direct signals should lead to more robust control and less spread in process and wafer parameters than indirect control.

Relatively unsophisticated models are needed for endpoint detection. For example, monitoring of plasma etching by OES (Section 6.3.1.2) often involves tracking the derivative of the optical emission signal. The process is stopped when appropriate. Sometimes, it is better to track the intensity of emission from one species, normalized by that of another. (This is also common in sputter deposition.) In some cases, the etching process recipe may be changed just before the endpoint is achieved so that the endpoint is reached more gently, which may cause less damage to the underlying layers. Section 6.3.1.3 describes one way to achieve such a flexible endpoint by using optical emission interferometry.

In statistical process control, the operating point for input variables is chosen so that the output is relatively insensitive to small changes in these variables, as is illustrated by the "sweet spot" D in Figure 19.7. This choice as the operating point is actually poor for a system with feedback control because the output cannot be changed by varying the process variable (Butler, 1995). Several other operating points in Figure 19.7 are also unsatisfactory, either because the output varies too quickly (A) or too slowly (C, G) with the input variable or because the gain [d(output variable)/d(input variable)] changes slope (B, E). The best operating point is F because it has the optimum slope for control. In a real process the optimum region in a multidimensional input-variable/output-variable space must be determined.

Some sensors directly monitor the wafer-state parameter of interest or monitor a parameter that is closely linked with that parameter, while other sensors monitor features that may correlate with that wafer parameter, but only indirectly. For example, reflection interferometry monitors film thickness and can therefore give the etch rate during plasma etching. Plasma-induced optical emission (OE) of a given species may correlate with the measured etch rate, even though the connection between them may not be clear. In this case, OES may be called a "virtual sensor" (Butler, 1995).

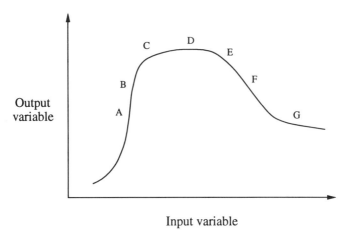

Figure 19.7 Choice of an operating point for a system under real-time control. Operation at point F may be best for closed-loop control because it has the optimum slope for control, whereas the optimal point for open-loop control, point D, is unsatisfactory. Other potential operating points are also unsatisfactory, either because the output varies too quickly (A) or too slowly (C, G) with the input variable or because the gain [d(output variable)/d(input variable)] changes slope (B, E) (from Butler, 1995).

Some process- and equipment-state parameters are relatively easy to control. For example, direct control of resistive heaters or the heating lamps changes wafer temperature; heating models are used to help achieve the temperature setpoint and uniformity. Other parameters, such as etching rate, depend on many parameters and require more elaborate control strategies.

Saraswat *et al.* (1994) described real-time control of wafer temperature for uniformity and repeatability in a rapid thermal processing reactor that was integrated with sensors to permit adaptive manufacturing. Control was achieved by using either an empirically determined model and a linear quadratic control strategy or a physical model with an internal model control design procedure. The linear quadratic control strategy minimizes a quadratic cost function subject to linear constraints and the effect of Gaussian noise. Schaper *et al.* (1994) detailed this latter real-time multivariable strategy, which uses input from pyrometric temperature readings across the wafer.

Numerous examples of the use of individual optical diagnostics to achieve real-time control can be found throughout this book. In one comprehensive application of optical probes, Barna *et al.* (1994b) integrated several *in situ* sensors for machine control for monitoring, diagnosing, and controlling plasma etching and rapid thermal processing reactors. They included several optical sensors—a single-wavelength ellipsometer (SWE), spectroscopic ellipsometers (SE), monochromators (for OES), scatterometry-based

19.3 Process Control

critical-dimension sensors, and pyrometers—and two nonoptical sensors—a two-point probe and eddy current sensors.

As part of that effort, Maung *et al.* (1994) detailed the use of *in situ* spectroscopic ellipsometry for real-time machine control. They selected SE rather than SWE because SE has sensitivity for a large range of thicknesses and avoids nonunique solutions that are common in SWE, in addition to its ability to identify material properties (which is important in deposition). They solved the inverse problem by using a nonlinear, unconstrained optimization scheme and applied it to the control and endpoint monitoring of remote plasma etching and the control of rapid thermal oxidation. This group also used single-wavelength ellipsometry for Run-to-Run supervision with a process control strategy employing RSM-statistically designed experiments (Butler and Stefani, 1994).

One control strategy used in ellipsometry employs the virtual interface method described by Aspnes (1993a,b, 1994, 1995, also private communication, 1995) and Urban and Tabet (1994). This approach is illustrated in Figure 19.1 and is discussed in Sections 9.11.1 and 19.1. The goal of this technique is to achieve submonolayer knowledge of the composition near the surface so that submonolayer control of composition can be achieved by varying equipment parameters. Urban and Tabet (1994) have shown that data analysis using the virtual interface method can be rapid enough for real-time feedback with neural network analysis.

Sometimes the advantages of real-time (kinetic) spectroscopic reflectometry and ellipsometry for process monitoring *vis-à-vis* real-time single-wavelength monitoring can be retained by simply analyzing data obtained from a few optimally selected wavelengths (Chapter 9). This can accelerate analysis and lead to more rapid process control. However, rapid interpretation of spectroscopic experiments is also possible with fast multivariate analysis routines.

The more sophisticated models described in the previous section are being increasingly used for control. Ramamurthi (1994) has noted that control by using physical, first-principles models is often not robust, because these models do not easily adapt to process drifts. Statistical models are useful within the window of experimental design, but also have difficulty in tuning in response to process drifts. However, modifications of these models are more robust, as is described below.

Sachs *et al.* (1991) have devised a process control system for VLSI fabrication that optimizes and controls run conditions and can be used for flexible manufacturing. It consists of three components: a flexible recipe generator that determines reasonable initial conditions, a Run-by-Run controller that tunes the recipe between runs on the basis of post-process and *in situ* measurements from previous runs, and a real-time controller that controls the current run by using *in situ* measurements. They used this procedure to optimize and control LPCVD of polysilicon. This improves the use of statistically designed experiments.

Chang and Spanos (1991) have developed a system to conduct real-time malfunction diagnosis based on evidential reasoning techniques and numerical models using sensor information and expected results; they demonstrated this technique in a low-pressure CVD (LPCVD) reactor used for depositing polysilicon. May and Spanos (1993) applied these automated malfunction methods to plasma etching. Saxena and Unruh (1994) evaluated several methods of rapid diagnosis of equipment malfunctioning and misprocessing using models that relate equipment inputs to observable wafer- and equipment-state parameters. One method uses the response surface model (RSM), which gives quantitative predictions for small deviations from the operating point, but which becomes inaccurate with equipment modifications and aging. The others involve a combination of this approach with more qualitative features.

One form of real-time process monitoring employs statistical process control methods. Traditional statistical process control methods cannot utilize data from process tool sensors because they violate the IIND assumption; they can be nonstationary, autocorrelated, and cross-correlated (Spanos *et al.*, 1992). Furthermore, these methods are not designed to use the multiple sensor readings needed in high-volume and complex fabrication. Spanos *et al.* (1992) have modified traditional SPC methods so they can use tool data. By applying time-series filtering and multivariate statistical process control methods, they were able to use real-time sensor information to generate alarms due to malfunctioning while the wafer was still in the processing chamber.

Mozumder and Barna (1994) also developed a real-time statistical feedback control procedure for plasma etching that does not have a large out-of-control false alarm rate and is protected against drifts in the state of the equipment. For etching silicon nitride (atop polysilicon atop silicon dioxide) in a $CHF_3/CF_4/O_2$ plasma, the only *in situ* process sensor they used was the detection of the optical emission line of CN at 388.3 nm. They also employed equipment monitors of chamber pressure, delivered RF power, gas flow rates, and substrate temperature. Barna (1992) has described automatic problem detection in a plasma etching reactor, which involves coupling information from the optical emission endpoint monitoring system (that detects the endpoint *and* determines the quality of the etch) with statistical process control capability. Elta *et al.* (1994) explored the application of advanced feedback control methods, including multivariable feedback control, to regulate a reactive ion etching chamber. Many of the strategies used for plasma processing control, which can involve either optical or nonoptical probes, have been reviewed by Hershkowitz and Maynard (1993).

The volume edited by White and Sofge (1992) discusses novel intelligent control strategies, including adaptive control, fuzzy logic control, neural network control, and fault diagnosis. Ramamurthi (1994) has described a self-learning fuzzy logic system to diagnose drifts in a mass flow controller, which is readily adaptable to variations in equipment design.

References

M. Aceves, J. A. Hernández, and R. Murphy, *IEEE Trans. Semicond. Manuf.* **SM-5,** 165 (1992).
K. D. Allen, H. H. Sawin, M. T. Mocella, and M. W. Jenkins, *J. Electrochem. Soc.* **133,** 2315 (1986a).
K. D. Allen, H. H. Sawin, and A. Yokozeki, *J. Electrochem. Soc.* **133,** 2331 (1986b).
P. P. Apte and K. C. Saraswat, *IEEE Trans. Semicond. Manuf.* **SM-5,** 180 (1992).
D. E. Aspnes, *Appl. Phys. Lett.* **62,** 343 (1993a).
D. E. Aspnes, *J. Opt. Soc. Am. A* **10,** 974 (1993b).
D. E. Aspnes, *Surf. Sci.* **307–309,** 1017 (1994).
D. E. Aspnes, *IEEE J. Sel. Top. Quantum Electron.* **1,** 1054 (1995).
T. B. Barker, *Quality by Experimental Design,* 2nd ed. Dekker, New York, 1994.
G. G. Barna, *IEEE Trans. Semicond. Manuf.* **SM-5,** 56 (1992).
G. G. Barna, M. M. Moslehi, and Y. J. Lee, *Solid State Technol.* 37(4), 57 (1994a).
G. G. Barna, L. M. Loewenstein, R. Robbins, S. O'Brien, A. Lane, D. D. White, Jr., M. Hanratty, J. Hosch, G. B. Shinn, K. Taylor, and K. Brankner, *IEEE Trans. Semicond. Manuf.* **SM-7,** 149 (1994b).
I. H. Bernstein, with C. P. Garbin and G. K. Teng, *Applied Multivariate Statistics.* Springer, New York, 1988.
D. S. Boning and P. K. Mozumder, *IEEE Trans. Semicond. Manuf.* **SM-7,** 233 (1994).
D. S. Boning, M. B. McIlrath, P. Penfield, Jr., and E. M. Sachs, *IEEE Trans. Semicond. Manuf.* **SM-5,** 266 (1992).
G. Box and N. Draper, *Empirical Model-Building and Response Surfaces.* Wiley, New York, 1987.
R. G. Brereton, *Chemometrics: Applications of Mathematics and Statistics to Laboratory Systems.* Ellis Horwood, New York, 1990.
S. W. Butler, *J. Vac. Sci. Technol. B* **13,** 1917 (1995).
S. W. Butler and J. A. Stefani, *IEEE Trans. Semicond. Manuf.* **SM-7,** 193 (1994).
T. S. Cale, T. H. Gandy, and G. B. Raupp, *J. Vac. Sci. Technol. A* **9,** 524 (1991).
T. S. Cale, J.-H. Park, T. H. Gandy, G. B. Raupp, and M. K. Jain, *Chem. Eng. Commun.* **119,** 197 (1993).
N. H. Chang and C. J. Spanos, *IEEE Trans. Semicond. Manuf.* **SM-4,** 43 (1991).
Y. M. Cho and T. Kailath, *IEEE Trans. Semicond. Manuf.* **SM-6,** 233 (1993).
M. E. Coltrin, R. J. Kee, and J. A. Miller, *J. Electrochem. Soc.* **131,** 425 (1984).
G. DePinto, *IEEE Trans. Semicond. Manuf.* **SM-6,** 332 (1993).
R. C. Dorf, *Modern Control Systems,* 6th ed. Addison-Wesley, Reading, MA, 1992.
M. E. Elta, J. P. Fournier, J. S. Freudenberg, M. D. Giles, J. W. Grizzle, P. P. Khargonekar, B. A. Rashap, F. L. Terry, Jr. and T. Vincent, *SPIE* **2091,** 438 (1994).

B. Flury and H. Riedwyl, *Multivariate Statistics: A Practical Approach*. Chapman & Hall, London, 1988.

G. J. Gaston and A. J. Walton, *IEEE Trans. Semicond. Manuf.* **SM-7**, 22 (1994).

K. P. Giapis, R. A. Gottscho, L. A. Clark, J. B. Kruskal, D. Lambert, A. Kornblit, and D. Sinatore, *J. Vac. Sci. Technol. A* **9**, 664 (1991).

R.-S. Guo and E. Sachs, *IEEE Trans. Semicond. Manuf.* **SM-6**, 41 (1993).

R. S. Gyurcsik, T. J. Riley, and F. Y. Sorrell, *IEEE Trans. Semicond. Manuf.* **SM-4**, 9 (1991).

D. M. Haaland, *Anal. Chem.* **60**, 1208 (1988).

D. M. Haaland, in *Practical Fourier Transform Infrared Spectroscopy: Industrial and Laboratory Chemical Analysis* (J. R. Ferraro and K. Krishnan, eds.), Chapter 8, p. 395. Academic Press, San Diego, CA, 1990.

D. M. Haaland and E. V. Thomas, *Anal. Chem.* **60**, 1193 (1988a).

D. M. Haaland and E. V. Thomas, *Anal. Chem.* **60**, 1202 (1988b).

R. J. Harris, *A Primer of Multivariate Statistics,* 2nd ed. Academic Press, Orlando, FL, 1985.

C. D. Hendrix, CHEMTECH, March, p. 167 (1979).

N. Hershkowitz and H. L. Maynard, *J. Vac. Sci. Technol. A* **11**, 1172 (1993).

C. D. Himmel and G. S. May, *IEEE Trans. Semicond. Manuf.* **SM-6**, 103 (1993).

M. L. Hitchman and K. F. Jensen, eds., *Chemical Vapor Deposition: Principles and Applications.* Academic Press, Boston, 1993.

M. W. Jenkins, M. T. Mocella, K. D. Allen, and H. H. Sawin, *Solid State Technol.* **29** (4), 175 (Apr. 1986).

I. T. Jolliffe, *Principal Component Analysis.* Springer-Verlag, New York, 1986.

B. Kim and G. S. May, *IEEE Trans. Semicond. Manuf.* **SM-7**, 12 (1994).

R. Krukar, A. Kornblit, L. A. Clark, J. Kruskal, D. Lambert, E. A. Reitman, and R. Gottscho, *J. Appl. Phys.* **74**, 3698 (1993).

K.-K. Lin and C. J. Spanos, *IEEE Trans. Semicond. Manuf.* **SM-3**, 216 (1990).

N. Logothetis and H. P. Wynn, *Quality Through Design.* Oxford Univ. Press (Clarendon), Oxford, 1989.

C. Lombardi, *Microelectron. Eng.* **10**, 287 (1991).

D. M. Manos and D. L. Flamm, *Plasma Etching: An Introduction.* Academic Press, Boston, 1989.

H. Martens and T. Naes, *Multivariate Calibration.* Wiley, New York, 1989.

S. Maung, S. Banerjee, D. Draheim, S. Henck, and S. W. Butler, *IEEE Trans. Semicond. Manuf.* **SM-7**, 184 (1994).

G. S. May and C. J. Spanos, *IEEE Trans. Semicond. Manuf.* **SM-6**, 28 (1993).

G. S. May, J. Huang, and C. J. Spanos, *IEEE Trans. Semicond. Manuf.* **SM-4**, 83 (1991).

J. L. McClelland, D. E. Rumelhart, and the PDP Research Group, *Parallel Distributed Processing.* MIT Press, Cambridge, MA, 1986.

P. C. Meier and R. E. Zünd, *Statistical Methods in Analytical Chemistry.* Wiley, New York, 1993.

M. Meloun, J. Militký, and M. Forina, *Chemometrics for Analytical Chemistry,* Vol. 1. Ellis Horwood, New York, 1992.

J. C. Miller and J. N. Miller, *Statistics for Analytical Chemistry,* 2nd ed. Ellis Horwood, New York, 1988.

References

M. M. Moslehi, R. A. Chapman, M. Wong, A. Paranjpe, H. N. Najm, J. Kuehne, R. L. Yeakley, and C. J. Davis, *IEEE Trans. Electron Devices* **ED-39,** 4 (1992).

P. K. Mozumder and G. G. Barna, *IEEE Trans. Semicond. Manuf.* **SM-7,** 1 (1994).

R. H. Myers, A. I. Khuri, and W. H. Carter, Jr., *Technometrics* **31**(2), 137 (1989).

F. Nadi, A. M. Agogino, and D. A. Hodges, *IEEE Trans. Semicond. Manuf.* **SM-4,** 52 (1991).

M. S. Phadke, *Quality Engineering using Robust Design.* Prentice-Hall, Englewood Cliffs, NJ, 1989.

R. K. Ramamurthi, *IEEE Trans. Semicond. Manuf.* **SM-7,** 42 (1994).

E. A. Rietman and E. R. Lory, *IEEE Trans. Semicond. Manuf.* **SM-6,** 343 (1993).

K. F. Roenigk and K. F. Jensen, *J. Electrochem. Soc.* **132,** 448 (1985).

R. K. Roy, *A Primer on the Taguchi Method.* Van Nostrand-Reinhold, New York, 1990.

E. Sachs, R.-S. Guo, S. Ha, and A. Hu, *IEEE Trans. Semicond. Manuf.* **SM-4,** 134 (1991).

E. Sachs, G. H. Prueger, and R. Guerrieri, *IEEE Trans. Semicond. Manuf.* **SM-5,** 3 (1992).

K. C. Saraswat, P. P. Apte, L. Booth, Y. Chen, P. C. P. Dankoski, F. L. Degertekin, G. F. Franklin, B. T. Khuri-Yakub, M. M. Moslehi, C. Schaper, P. J. Gyugyi, Y. J. Lee, J. Pei, and S. C. Wood, *IEEE Trans. Semicond. Manuf.* **SM-7,** 159 (1994).

S. Saxena and A. Unruh, *IEEE Trans. Semicond. Manuf.* **SM-7,** 220 (1994).

C. Schaper, M. Moslehi, K. Saraswat, and T. Kailath, *IEEE Trans. Semicond. Manuf.* **SM-7,** 202 (1994).

D. E. Seborg, T. F. Edgar, and D. A. Mellichamp, *Process Dynamics and Control.* Wiley, New York, 1989.

R. Shadmehr, D. Angell, P. B. Chou, G. S. Oehrlein, and R. S. Jaffe, *J. Electrochem. Soc.* **139,** 907 (1992).

M. A. Sharaf, D. L. Illman, and B. R. Kowalski, *Chemometrics.* Wiley, New York, 1986.

A. Sherman, *J. Electron. Mater.* **17,** 413 (1988).

V. K. Singh and E. S. G. Shaqfeh, *J. Vac. Sci. Technol. A* **11,** 557 (1993).

V. K. Singh, E. S. G. Shaqfeh, and J. P. McVittie, *J. Vac. Sci. Technol. B* **12,** 2952 (1994).

M. M. Smadi, G. Y. Kong, R. N. Carlile, and S. E. Beck, *J. Electrochem. Soc.* **139,** 3356 (1992).

F. Y. Sorrell and R. S. Gyurcsik, *IEEE Trans. Semicond. Manuf.* **SM-6,** 273 (1993).

C. J. Spanos, H.-F. Guo, A. Miller, and J. Levine-Parrill, *IEEE Trans. Semicond. Manuf.* **SM-5,** 308 (1992).

M. P. Splichal and H. M. Anderson, *SPIE* **1594,** 189 (1991).

G. Taguchi, *Taguchi Methods: Design of Experiments.* ASI Press, Dearborn, MI, 1993.

E. V. Thomas and D. M. Haaland, *Anal. Chem.* **62,** 1091 (1990).

M. Tjantelé, *Microelectron. Eng.* **10,** 277 (1991).

F. K. Urban, III and M. F. Tabet, *J. Vac. Sci. Technol. A* **12,** 1952 (1994).

J. Van de Vegte, *Feedback Control Systems,* 2nd ed. Prentice-Hall, Englewood Cliffs, NJ, 1990.

B. Wangmaneerat, J. A. McGuire, T. M. Niemczyk, D. M. Haaland, and J. H. Linn, *Appl. Spectrosc.* **46,** 340 (1992a).

B. Wangmaneerat, T. M. Niemczyk, G. Barna, and D. M. Haaland, in *Plasma Processing* (G. S. Mathad and D. V. Hess, eds.). p. 115. Electrochem Soc., Pennington, NJ, 1992b.

D. A. White and D. A. Sofge, eds., *Handbook of Intelligent Control.* Van Nostrand–Reinhold, New York, 1992.

G. Z. Yin and D. W. Jillie, *Solid State Technol.* **30**(5), 127 (1987).

J. Zupan and J. Gasteiger, *Neural Networks for Chemists: An Introduction.* VCH, Weinheim, 1993.

Index

A

Abelès matrix methods, 340
Abel transform, 150–151, 203–206
  see also tomographic reconstruction
Absorbance (absorbed fraction), 264–265, 267–268, 270–272, 275, 279, 286, 290, 295–296, 298, 301, 304, 306–307, 310–312, 466
Absorption (transmission), 14, 21–22, 26, 38, 46–47, 50, 60–61, 81, 83, 85, 90, 94–100, 120–121, 123, 139, 141, 148, 151–152, 193, 195, 263–319, 367, 495, 638
  coefficient [absorption length =1/ absorption coefficient], 7, 60, 82, 84, 86–90, 97–100, 112–113, 115, 146, 149, 203, 205, 215, 217–218, 220–222, 232, 238, 264–265, 268, 279, 287, 290, 295, 309, 313, 319, 338, 342, 564, 621, 623, 639, 673
  comparison with other methods, 160, 174, 216, 561, 701
  cross section, 84, 218, 220, 222, 234, 264–265, 290, 296
  for particles, 499
  depth, 7, 264, 309, 564
  endpoint, 14, 283, 286–287
  for calibration of other methods, 174–176, 218
  for gas-phase thermometry, 27, 295–296, 696, 701, 707–708
  for monitoring flow and flux, 29, 38, 280, 282, 292–294, 298, 300, 305–308
  in adsorbates, 309
  in films, 309–312
  in the gas phase, 41, 45, 272–309, 696, 701, 707–708
  in the infrared, 26, 125, 137, 199, 399, 676, 718
    for gas-phase thermometry, 27, 295–296, 696, 701, 707–708
    for particle detection, 496
    for wafer thermometry, 27, 312–319, 703, 712–713
    in gases, 273–296
    of surfaces, 381–399, 647
    see also diode lasers (IR-DLAS), Fourier transform infrared spectroscopy
  in the ultraviolet/visible, 26, 676
    in gases, 296–309
  in wafers, 312–319, 703, 712–713
  profile, 265, 267–268
  three-photon, 249, 676
  two-photon (TPA), 85–86, 109, 111–112, 216, 244–250, 257, 655, 676, 678
  see also laser-induced fluorescence (two-photon LIF), band-edge absorption, optogalvanic spectroscopy, photoionization, resonant-enhanced multiphoton ionization, self absorption, transmission
Absorption edge sensing, *see* band edge-absorption
Acoustic
  probes, waves, 31, 649–651, 696, 704–705

739

Acoustic (*continued*)
  velocity, 649, 700
  see also thermal wave analysis, ultrasonic detection
Acousto-optic modulator, 153, 644, 646
Actinometry, 174–176, 198, 204–206, 247, 284
Activation energy, 20, 253–254, 363, 602, 638, 640, 643, 692
Adaptive
  control, 735
    manufacturing, *see* flexible manufacturing
  multilayer control model, 450
Adducts, 296
Adlayers, *see* adsorbates
Adsorbates, 5, 31, 252, 254, 309, 312, 327, 331, 333–334, 336–340, 343, 381–400, 403–412, 463–465, 576–578, 626, 637–643, 647, 666–668, 673
  see also differential reflectometry, overlayers, reflectance-difference spectroscopy, surface composition, surface photoabsorption, very thin films
Advanced process control (APC), 14, 20, 729
  see also closed-loop control, feedback control, feedforward control, process control, regulatory control, supervisory control
Ag, 194
  deposition and monitoring, 444, 580
  dielectric function, 344
Al
  atom monitoring, 166, 172, 177, 181, 189–190, 208, 226, 232, 234, 254–256, 302, 304, 307, 483–484, 489
  deposition (including pulsed laser deposition and laser ablation), 189–190, 194, 199–201, 234, 256, 293, 302, 304, 396, 489, 724
    and monitoring, 357–360, 445, 647
    and nucleation, 647
  dielectric function, 344
  electrode (etching), 170, 172, 208, 234, 247–248, 507, 509
  etching, 177, 181, 183–184, 290
  film, 243
    thermometry, 647, 651

  thickness, 651
  in eutectic, 350
  ionization potential, 680
  particulates, 234, 507, 509–510
  substrate for polymers, 383, 385
  surface monitoring, work function, 684
Al^+ [Al(II)] monitoring, 166, 199–201
AlAs
  deposition and monitoring, 307–308, 313, 418, 421–423, 451–452, 463, 609–612
  dielectric function, optical properties, 344, 620
  etching and monitoring, 360
  phonon energies, 108
  see also AlGaAs, heterostructures
AlCl monitoring, 166, 177, 181–183
$AlCl_3$, 290
Al_2Cl_6, 290
ALE, *see* atomic layer epitaxy
Alexandrite, *see* $BeAl_2O_4:Cr^{3+}$
AlF monitoring, 166, 170
AlGaAs
  deposition and monitoring, 315, 354, 357–360, 380, 422, 430, 432, 434–435, 444, 449–452, 463, 609, 611–612, 626
  dielectric function, optical properties, critical points, 343–345, 449, 621
  temperature variations, 344, 699
  diode laser, *see* diode laser
  etching, 181–183
    and monitoring, 361–362, 458, 625
  Raman scattering temperature effects, 585
  temperature, 380, 699
  see also AlAs, GaAs, heterostructures
Allowed transitions, 94–96, 100–105, 242, 655
  see also forbidden transitions, selection rules
Alloy
  semiconductor (composition, monitoring, control), 343–345, 432, 443, 446, 449–453, 582, 621, 627
    see also AlGaAs, GeSi, heterostructures, indium tin oxide (ITO), InGaAs, InGaAsP, TiN
  metal (sputtering of), 187–190
Alloying, 693
AlO monitoring, 96, 226, 243, 254–256

Index

Al_2O_3
  etching, 243
  deposition (including pulsed laser deposition and laser ablation), 195, 254–255, 293, 483–484, 724
  and monitoring, 453–454
  atomic layer, 312
  dielectric function, 344, 371
  spherical particles, 517–520
  thermal properties, 371
  see also alumina
$Al_2O_3:Cr^{3+}$ (ruby), 633–634
AlON dielectric function, 344
AlSb
  dielectric function, 344
  Raman scattering temperature effects, 585
AlSi eutectic, 350
Alternating line intensities, see nuclear spin statistics
Alumina (porous), 312
  see also Al_2O_3
Aluminum alkyl reactants, 678
  see also triethylaluminum, trimethylaluminum, tritertiarybutylaluminum
Ammonium fluoride, see NH_4F
Amorphous materials, 106, 351, 458, 562
  see also crystallization, plasma-enhanced chemical vapor deposition
Amorphous silicon, see Si
Amplitude scattering matrix, 498–500, 515–520
  see also scattering matrix
Anemometry, see laser velocimetry
Angle of incidence, 328–343
  see also reflection
Angle-resolved scatter, 535
Angular dissymmetry, 507, 509
Anharmonicity of
  molecular vibrations, 68, 71, 95, 104
  phonons, 68, 583–585, 697, 700
  rotation, 70
Anisotropy of
  dielectric tensor, 82
  etching, 11, 43
  (bulk) media, 335, 338, 666–668, 670
  polarizability, 92, 100–101, 338
    see also polarizability
  surface, 338, 413–425, 666–668
  see also reflectance-difference spectroscopy, surface morphology (roughness, anisotropic)

Annealing, 46, 49, 348, 376, 459, 531, 582, 668, 693, 716
  see also laser annealing
Anomalous dispersion, 485
Antireflection (AR) coating, 355, 450
  see also dielectric mirrors
Anti-Stokes scattering, lines, processes, 23, 93, 101, 103–104, 123, 248–249, 559–560, 567, 583, 657
  see also coherent anti-Stokes Raman scattering (CARS), Stokes/anti-Stokes scattering intensity ratio
Aperture, 143–147, 149, 165, 489
  see also numerical aperture
Ar
  arc lamps, 597
    see also lamps
  flow, 282, 455, 486
  leak, 172
  monitoring, 162–163, 166–167, 170–171, 174–176, 178–179, 184–185, 187–189, 192–193, 199, 204–208, 225, 247, 507, 678
  for calibration, 676
    see also actinometry
  plasma (or sputtering in), 162–163, 167, 169–171, 174–176, 178–180, 187–193, 204–208, 234, 237–238, 246, 281, 360–361, 381, 410, 444, 506–510, 548–549, 684–685
Ar^+ [Ar(II)]
  bombardment (for etching, etc.), 184–185, 237, 257, 458, 580
  monitoring, 166–167, 184–185, 188, 192, 199, 206, 208, 226, 238–239
Ar-ion laser, 120–121, 224, 238, 287, 367, 405, 409, 506–507, 513, 529, 536, 563, 567, 570, 574–577, 578–579, 581–583, 586, 604, 622–623, 627–628, 646, 681
ArF lasers, 120, 195, 224, 229, 235, 237, 247, 288, 398, 683–684
  see also excimer lasers, laser ablation, pulsed laser deposition
Array detection, see multichannel detection
Arrhenius activation energies, form for rate equation, 20, 363, 602, 640, 692
Arsenosilicate glass films, 312
As monitoring
  gas phase, 166, 177, 181–183, 226, 237, 679–680
  ionization potential (for), 680

As monitoring (*continued*)
 on surface, 254, 405–410, 414, 418–423, 450, 460
As_2
 ionization potential, 680
 monitoring, 96, 226, 229–233, 254, 297, 679–681
  nonoptical, 409–410
 reactant, 39, 679
As_4
 ionization potential, 680
 monitoring, 297, 679–680
 reactant, 232–233, 254, 679, 681
AsH, 308
As-H vibration at surface, 266, 393–395, 626–627
AsH_2, 229, 308
AsH_3, 96
 ionization potential, 680
 reactant, 229–231, 282, 405–407, 409, 420–424, 449, 507, 574
  and monitoring, 229, 279–280, 288–289, 292–293, 297–300, 570, 665
Ashing (plasma stripping), 43, 50, 179–181, 204–206, 365, 497
 see also photoresist
As-O vibration at surface, 391–392, 626–627
As_2S_3 dielectric function, 344
As_2Se_3 dielectric function, 344
Associated ellipsometry parameters, 428–429, 438–439
Atomic
 density measurements, 302, 304–305
 flux monitoring, 232, 305–308
 see also density, laser-induced fluorescence, optical emission spectroscopy
Atomic force microscopy (AFM), 31–32, 523, 529, 531
Atomic layer epitaxy (ALE), 40, 42, 307, 355–357, 405–409, 411–412, 422–423, 578–580
 see also layer-by layer growth
Atomic (emission) resonance lamps, 89, 120–121, 124, 129, 232, 267, 269, 273, 301–302, 304–308
Atomic spectroscopy, *see* spectroscopy
Attenuated total internal reflection (ATR, ATIR) spectroscopy, 7, 22, 25–26, 31, 48, 263, 266, 329, 381–383, 385–399, 578, 626–627

Au
 deposition, 455
 dielectric function, 344
 in eutectic, 350
Auger electron spectroscopy (AES), 31–33, 582
Au/Si eutectic, 350
Autoionization, 674
Azimuth (incident), 405–408, 530–534

B

B atom monitoring, 301
Ba
 atom monitoring, 166, 194, 202, 226, 256, 302–303
 dielectric function, 344
 laser ablation, 256
Ba^+ monitoring, 166, 194, 202, 302, 304
Backscattering collection configuration, 105–108, 142–147, 222, 243, 564–565, 578
$BaClF:Sm^{2+}$ phosphor, 632
Baking, *see* photoresist
$BaMgF_4$, 36–37
Band bending, 685
Band-edge (and below-band-gap) absorption, 312–319, 403, 535, 697, 699, 703, 713
Band gap, 76–77, 99–100, 312–319, 368–369, 371, 379, 403, 429–433, 567, 578, 586, 599–600, 622, 628, 694
 temperature dependence, 379–380, 621, 628–631, 698–699, 704
 see also critical point energies
Bandhead, 96, 104
Band offsets, 578
Band shrinkage, 629
Band-to-band transitions, 89, 99–100, 619–622
BaO, 166, 194, 226, 256
BaO_2, 194
Batch processing, 15–18, 728–730
$BaTiO_3$, 36–37, 194–195
 dielectric function, 344
BBO, *see* beta-barium borate
BCl monitoring, 166, 170, 226, 242
BCl_2 monitoring, 166, 170
BCl_3 plasma, 178, 241–242, 624–625, 691
 and monitoring, 170, 295
Be dielectric function, 344
$BeAl_2O_4:Cr^{3+}$ (alexandrite), 633–634

Index   743

Beam
   divergence, 61–62
   propagation, 57–65
Beer's law, 60, 84, 86, 89–90, 149, 264–265,
      268, 279, 307, 315
   deviations from, 89–90, 268, 307
BeO dielectric function, 344
Beta(β)-barium borate (BBO), 123, 224,
      232, 249
BH_2 monitoring, 301
B_2H_6 plasma, 301
Bi
   atom monitoring, 307
   dielectric function, 344
Biacetyl, 244
Bi-cell detectors, see position-sensitive
      detectors
Bidirectional reflection distribution
      function (BRDF), 525
Blackbody, 592–595, 608, 613–614
   radiation, 21, 157, 165, 193, 195, 387,
      466, 591–614
   see also pyrometry
Black-box models, see empirical models,
      neural network models, statistical
      models
BN laser ablation, 613–614
B-O vibration monitoring, 266, 446
Boltzmann factor, distribution function, 73,
      75, 88, 98, 234–235, 237, 254–257, 276,
      662, 676
   see also Maxwell-Boltzmann distribution
      function
Born-Oppenheimer approximation,
      67–68
Borophosphosilicate glass (BPSG), 266,
      293, 312, 604, 718
   see also phosphosilicate glass
Bose-Einstein distribution function,
      690–691, 698–699
   see also Planck blackbody law
Boxcar integration, 126, 153, 170–171, 223,
      232, 640
   see also gated analysis
BOXCARS, 658–659, 662
BPSG, see borophosphosilicate glass
Br, 166, 175
Br_2 plasma and monitoring, 175,
      296–297
Bragg reflectors, see distributed Bragg
      reflectors
Bremsstrahlung, 193

Brewster's (and pseudo-Brewster) angle,
      140, 185–186, 331–332, 337, 355, 382,
      389, 403, 405, 413, 439, 441
   (Brewster angle) reflection spectroscopy
      (BARS), 25, 403, 413
   see also p-polarized reflectance
      spectroscopy, surface
      photoabsorption
Bright field analysis, 520–521
Brillouin scattering, 23, 496, 559–560
Broadening, see line broadening
Bromination (of surface), 253, 640–641
Bruggerman expression, 430
Bubbles, 515, 517–520
Buried
   layers, 342, 426
   metal layers, 382, 384, 389
Burstein-Moss shift, 629

C

C
   ablation, 194–195, 302, 614
   amorphous (a-C), 446
   atom monitoring, 166, 169–170, 186, 308,
      676–677
   deposition, 193, 567–569
      and monitoring, 446, 455
   dielectric function, 344, 505–506,
      517–520
   particles, 505–506, 508, 511, 517–520,
      614
   see also diamond, graphite
C_2 monitoring, 166, 187, 195, 199, 203, 226,
      230, 255–256, 483
Ca
   atom monitoring, 302, 307
   dielectric function, 344
Ca^+ monitoring, 302
CaF monitoring, 302
CaF_2
   deposition and monitoring, 454–455
   dielectric function, 344
   laser ablation, 302
   /Si interface, 668
Carriers, see free carriers
CARS, see coherent anti-Stokes Raman
      scattering
CCD array, see charge-coupled device
      array detector
CCl monitoring, 166, 181, 197–198, 226,
      235–236
CCl_2 monitoring, 166, 170

CCl$_4$
  monitoring, 291
  plasma, 169–170, 181, 197–198, 235–236, 290–291, 510, 691, 724–725
Cd
  alkyl monitoring, 298
    see also dimethylcadmium
  atom ionization potential, 680
  dielectric function, 344
CD, *see* critical dimension
CdS
  deposition and monitoring, 574–575
  dielectric function, 344
CdSe dielectric function, 344
CdTe
  deposition, 231
    and monitoring, 354–357, 452, 454, 575–576
  dielectric function, 344
  laser heating, thermometry, 628–629
  phonon energies, 108
  surface modification and monitoring, 402
CdZnTe thermometry, 318
Ce dielectric function, 344
CF monitoring, 226, 235, 299, 308
CF$_2$, 73
  monitoring, 166, 169–170, 176–177, 226, 235, 254, 281, 283, 287, 295, 299
  reactivity with surfaces, 254
CF$_2^+$ monitoring, 235
CF$_3$ monitoring, 226, 281, 287, 290
CF$_3^+$ monitoring, 226
CF$_4$
  monitoring, 226, 290–291, 295, 571
  plasma, 160, 162, 167–169, 172–176, 178–179, 197, 235, 237, 247, 281, 283, 287–288, 290–291, 299, 455–456, 458, 510, 684–685, 734
  reactant, 571–572
CF$_4^+$ monitoring, 226
CF$_x$ monitoring
  gas phase, 172–173
  overlayer, 456
C$_2$F$_4$ plasma, 290
C$_2$F$_6$
  monitoring, 281, 291
  plasma, 199, 235, 283, 287–288, 299, 724
n-C$_3$F$_6$ plasma, 290
C$_3$F$_8$
  monitoring, 291
  plasma, 287–288, 290

CFCl$_3$
  monitoring, 291
  plasma, 245, 290–291, 510
CF$_2$Cl$_2$
  monitoring, 291
  plasma, 170, 181, 245–246, 283, 290–291, 361, 510
CF$_3$Cl
  monitoring, 291
  plasma, 284, 290–291, 510
CF$_2$O monitoring, 283
CGS (gaussian units), 57, 59, 81, 113–116
CH, 166, 170, 187, 193, 195, 199, 308
CH$^+$, 187
CH$_2$, 166, 187, 280, 308
CH$_3$
  monitoring, 27, 279–282, 284, 309, 676–677
  reactant, 280
CH$_4$, 100
  monitoring, 27, 280–281, 284, 289, 291–292, 567–570, 664, 676
  plasma, 169–170, 178, 180, 182, 184, 187, 281, 284, 360–361, 397, 508, 511, 669–670
  reactant, 230, 248, 280, 567–569, 664, 676–677
CH$_x$ surface vibrations, 266, 391–392, 394–395, 647
C$_2$H monitoring, 226, 230
C$_2$H$_2$
  monitoring, 27, 280, 283, 568, 664–665, 676
  plasma, 187, 665
  reactant, 195, 230, 568, 611
C$_2$H$_4$ monitoring, 280–281, 284, 568, 665
(C$_2$H$_4$)$_n$ (polyethylene) dielectric function, 344
C$_2$H$_6$ monitoring, 280
C$_3$H$_4$ (allene and propyne) monitoring, 280
C$_3$H$_6$ (propene and cyclopropane) monitoring, 280
Chalcogenide optical fibers, *see* optical fibers (infrared)
Charge-coupled device (CCD) array detectors (cameras), 126–129, 131, 135, 141–142, 148, 153, 164–165, 225, 437, 484, 506–507, 513, 545–546, 566, 573, 586, 603, 628, 634
  intensified (ICCD array), 128–129, 148, 151, 195–198, 491, 613–614, 716
  see also multichannel detection

Index

CH$_3$AsH$_2$ (methylarsine) monitoring, 292
(CH$_3$)$_2$AsH (methylarsine) monitoring, 292
C$_2$H$_5$AsH$_2$ (ethylarsine) monitoring, 292
CH$_3$CHO (acetaldehyde), 294
CH$_3$COOH (acetic acid), 294
CH$_3$D monitoring, 291
Chemical beam epitaxy (CBE), 40
Chemically-amplified resists, 543
Chemical vapor deposition (CVD), 11, 17, 26–33, 36–43, 45, 48, 218–219, 227–231, 244, 250–251, 256, 270, 275, 279–282, 287, 290–302, 304–305, 351, 383–384, 387, 404–412, 443–445, 453, 534–536, 561, 567–572, 580, 601–602, 613–614, 661–665, 692–693, 720, 724, 734
  diagnostics locator, 28, 38
  model, 227–229, 568–571, 724
  see also physical modeling, process modeling
  particle monitoring during, 496–497, 507, 512–514
  reactor, 39–40, 141, 358, 481–482, 486–488
  rotating disk, 228, 357–360, 513, 569, 571
  see also laser deposition, plasma-enhanced chemical vapor deposition, metalorganic chemical vapor deposition, reactant delivery
Chemiluminescence, 158–160, 184, 230–231
Chemometrics, 19–20, 166, 173, 717, 720
  see also multivariate analysis
CHF$_3$ plasma, 170, 172–173, 178–179, 185–186, 235, 283, 725–726, 734
CH$_2$F monitoring, 308
C$_2$H$_2$F$_4$ plasma, 299
CH$_3$I reactant, 280, 282
Chlorination (of a surface), 171–172, 252–253, 581–582, 640–642
CHO, 187
Chopper (mechanical), 153, 223, 270–272, 276, 279, 284, 296, 298, 300, 305, 345, 599, 612, 622, 631, 635, 644
  see also lock-in detection
Cl, 72
  adsorption, 574
  generated, 184
  monitoring, 94, 160, 162–163, 166, 169–172, 175, 177, 184, 205, 216, 226, 244–246, 274, 279, 283–285, 308, 508, 678
Cl$^+$ monitoring, 166–167, 169, 171, 226, 238–240
Cl$^-$ monitoring, 683
Cl$_2$, 70–71
  plasma, 160–163, 170–172, 175, 178, 180–182, 205, 236, 238, 240–241, 245–246, 252–253, 283–285, 290, 360–363, 455–456, 508, 624–625, 640–642, 683
  and monitoring, 166, 169–172, 181
  reactant, 161, 361, 581–582, 641–642, 668
Cl$_2^+$, 166, 169–171, 226, 236, 241
Classical free electron model, *see* free electron model
Classical least-squares (CLS) analysis, 717
Cleaning surfaces, *see* surface cleaning
Closed-loop (feedback) control, 10, 13, 20, 49–50, 340, 353–354, 434, 449–452, 454, 458, 462, 543, 719, 727–732
  see also endpoint detection, feedback control, process control, real-time monitoring
Cluster tools, 13, 16–17, 43, 45, 391, 647, 729–730
CN, 166, 168, 177, 187, 226, 256, 734
Co, 37
  dielectric function, 344
Co-alloy/Cr films, 603
CO, 73
  laser, 601
  monitoring, 27, 166–169, 172, 176–181, 184–187, 197, 199, 279, 665, 678–679
  for wavelength calibration, 278
  plasma, 287, 446, 448
  products, 684
CO$^+$, 166, 169, 187
CO$_2$, 76, 95, 103
  laser, 295, 310, 648, 665
  monitoring, 166, 176, 279, 294, 665
  for wavelength calibration, 278
  plasma, 187
  products, 684
CO$_2^+$, 166, 169
Coaxial collection configuration, 142–147
  see also backscattering collection configuration
COF$_2$ plasma, 290
Coherence length, 357

Coherent anti-Stokes Raman scattering (CARS), 23, 26–27, 98, 109–111, 116, 123, 147, 561, 655–665
  comparison with other methods, 701
  for gas-phase thermometry, 27, 661–665, 696, 701, 706
  vibrational-rotational vs pure rotational, 660
Coherent Stokes Raman scattering (CSRS), 110, 657
Cold wall reactors, 725
  see also molecular beam epitaxy, rapid thermal processing
Collected solid angle, *see* solid angle (collected)
Collection efficiency of
  electrons by electrodes, 675–676
  photons, 141–147, 163–165, 222, 562, 565
Collisional
  broadening, 88–89, 267, 276, 701
  decay, relaxation, 161–162, 218, 220–221, 230
  excitation, 160–162, 174–176, 255
  see also collisions, lorentzian lineshape
Collisions, 157, 160–162, 174, 193, 195, 220–221, 223, 656, 674, 682, 691, 701
  see also collisional broadening, collisional decay, collisional excitation
Combination bands, 95
Computer-assisted tomography (CAT), *see* tomographic reconstruction
Computer integrated manufacturing (CIM), 15, 17
Concentration, gas phase, *see* density
Confocal microscopy, 7, 243
Conformal coverage, 293–294
Connes-type interferometer, 199, 295
Constant-angle reflection interference spectroscopy (CARIS), 353
Contact holes, *see* vias
Contamination (of wafers), 16, 496, 684, 694
Continuous wave (cw) lasers, *see* lasers
Control, *see* advanced process control, closed-loop control, feedback control, feedforward control, open loop control, process control, real-time monitoring
Controllers, 727–728
CoSi, CoSi$_2$ transformation, 603

Coupled wave (electromagnetic diffraction) theory, 537, 542–543
Cr
  atom monitoring, 166, 189–190
  dielectric function, 344
  films, 603
  substrate for polymers, 383
Cr^{3+}-doped phosphors, 633–634
Critical angle, 329, 385
Critical dimension (CD), 5, 11, 36, 38, 46, 49, 365, 538–539, 542–544, 733
  diagnostics locator, 36
  see also metrology
Critical point (CP) (energies), 76–77, 93–94, 99, 343, 377–379, 430, 443, 463, 698, 704
  dependence on temperature, 698–699, 702, 704
  see also band gap
Critical thickness (of films), 531, 533–534, 575, 578
  see also incommensurate growth
Cross section
  collisional, 160–162, 174–176
  ionization, 674–675
  see also absorption cross section, absorption coefficient, Raman scattering, scattering (light)
Crystal originated particles (COPs), 518
Crystallinity, 667–668
Crystallization, 349–351, 363–364, 376, 443
Cs dielectric function, 344
CsI dielectric function, 344
CSRS, *see* coherent Stokes Raman scattering
Cu
  atom monitoring, 166, 169, 172, 188–190, 194, 202, 226, 231, 255–256, 302, 304, 307
  chlorination and etching, 581–582
  deposition, 231, 293, 535
  dielectric function, 344
  electrode, 169, 172
  in Al films, 189–190
  oxidation and monitoring, 367, 459
  pulsed laser deposition, laser ablation, 194, 255, 485
  substrate for polymers, 383
  target, sputter deposition, 189, 302, 304
Cu$^+$ monitoring, 166, 194, 226, 255
Cu$_2$ monitoring, 226, 255
CuCl, CuCl$_2$ formation and monitoring, 581–582

Index

CuF monitoring, 166, 169
Cu(hfac)$_2$ reactant, 231
CuO
 /Cu$_2$O dielectric function, 344
 molecule monitoring, 166, 194, 226, 256
 pulsed laser deposition, 194
Cu$_2$O
 formation and monitoring, 367
 monitoring, 581–582
Curvature
 laser beam, 60–61
 wafer, 350–351

D

D$_2$, 224, 247
 lamps, *see* lamps
 monitoring, 665
 plasma, 391
Damage (in crystals), 402, 427, 456, 578, 580, 582–583, 624–627
 see also defects
Damping factor, 87, 92
 see also line broadening, lineshape, relaxation rates
Dark field analysis, 515, 520–521
Data acquisition and analysis, 19–20, 715–719
DCO$^+$ monitoring, 287
Debris, *see* walls (flaking from)
Defects
 in manufacturing processes, 730
  see also disturbances, error detection, noise (in a process run), process rifts
 monitoring in semiconductors, 413, 537, 621, 623–624, 643–644
  see also damage, dislocations
 on surface, 623–627
Deformation potential, 93, 105–106, 576
Degeneracy factor (of level), 70, 73, 75, 86–87, 97, 200, 268, 690
Degenerate four-wave mixing, 110, 656
Density (profile), 141, 149, 217–218, 221–223, 244, 264, 481–491, 567–574, 660–665, 690
 atomic, 302–308
 calibration, 174–176, 234, 236–237, 247–248, 272–273, 281, 283, 304
  see also actinometry
 measurement, locator, 26–27
  see also population distribution function

Deoxidation, 424–425
 see also surface cleaning
Depletion layer, *see* surface depletion layer
Depolarization ratio, 563
Depolarized scattering, 101, 105–107, 562–563
 see also polarized scattering
Deposition, 34–42, 389, 438, 444
 see also chemical vapor deposition, metalorganic chemical vapor deposition, molecular beam epitaxy, photolysis, plasma-enhanced chemical vapor deposition, rapid thermal chemical vapor deposition, sputter deposition
Depth of focus, 145–146
Desorption, *see* temperature programmed desorption, laser-induced thermal desorption
DETe, *see* diethyltellurium
Detection limit, 152–153, 216, 229, 237, 242, 247, 270–272, 279, 287, 290, 295–296, 300, 464, 571–572
 see also absorbance, submonolayer control
Detectivity, 124–125, 152, 597
Detectors (photodetectors), 124–129, 152–154, 597
 noise, 264, 270–271, 273
 quantum efficiency (and detectivity), 124–128, 136, 142–143, 145–146, 152–154, 165, 222, 264, 562, 565, 597
 see also charge-coupled device array detector, multichannel detection, noise, photoconductive detectors, photodiode array, photomultipliers, position-sensitive detectors, quantum efficiency, signal-to-noise ratio
dEuFOD, 634–635
Deuterated triglycine sulfate (DTGS) detector, 124, 135
Development, *see* photoresist
DEZn, *see* diethylzinc
DF-treated surfaces, 389–390
Diagnostics
 locators, 14, 25–28, 36, 38, 41
 needs of thin film processing, 34–50
Diamond
 adsorbate vibrations, 266
 deposition, 26–28, 30, 35, 186–187, 199, 230, 232, 248–249, 255, 280, 283, 662–664, 669–670, 675–677

Diamond (*continued*)
  and monitoring, 355, 446–448, 580, 609, 611
  dielectric function, 344
  Raman properties, 564
    phonon energies, 105, 504
    thermometry, 584–586
Dielectric films, 11, 36–37, 334, 337–338, 718
  see also borophosphosilicate glass, phosphosilicate glass, silicon nitride, SiO_2, SiO_xN_y
Dielectric function (constant, tensor), 59, 82–85, 99, 114–115, 330–331, 337–338, 342–346, 369–372, 378, 382, 385–386, 399–400, 404, 414–415, 417–418, 426, 443–444
  temperature variation, 343–346, 461–463, 699–700, 704
  values, 343–346, 446–447, 449–451, 463–464
  see also ellipsometry, index of refraction, near-surface dielectric function spectroscopy, pseudodielectric function
Dielectric mirrors (deposition of coating), 340, 355, 450
  see also distributed Bragg reflectors
Diethyltellurium (DETe) reactant, 231
Diethylzinc (DEZn)
  monitoring of
    adsorbate, 309
    gas, 298–299
    reactant, 298–299
Difference frequency mixing, 109, 287
Differential reflectometry, 25–26, 336–338, 345, 377, 399–402, 464
  see also photoreflectance, reflectance-difference spectroscopy, surface photoabsorption
Differential scattering cross section, *see* cross section, Raman scattering, scattering (light)
Diffraction, 57–58, 61–62, 495–552
  angle, 524
  gratings, 129, 130–134, 154, 483–485, 537–549
    see also grating equation, laser light scattering, spectrometers (grating), surface morphology (periodic features)
  order, 131–134, 541–544
  theory, 58, 495
  vector theory, 522, 524–525
Diffractometry, 522
Diffuse reflection, *see* laser light scattering
Diffusion, *see* interdiffusion
Dimers (on surfaces), 394–395, 414, 418–421
Dimethylamine, 397
Dimethylarsine $(CH_3)_2AsH$, 292
Dimethylcadmium $[Cd(CH_3)_2$, DMCd]
  adsorbate monitoring, 309, 397–398
  ionization potential, 680
  see also Cd alkyl monitoring
Dimethylethylamine alane, 396
Dimethyltin $[Sn(CH_3)_2$, DMSn] ionization potential, 680
Dimethylzinc $[Zn(CH_3)_2$, DMZn]
  ionization potential, 680
Diode arrays, *see* photodiode arrays
Diode lasers, 89, 120–121, 123, 137, 317–319, 512, 550, 647
  AlGaAs, 286
  lead salt, 276–278, 286
  modulated, pulsed (infrared), 374–376
  tunable infrared, 121, 123, 270–272, 274–287, 399, 656
  visible, 355, 361–362, 631, 647
  see also infrared diode laser absorption spectroscopy, InGaAsP diode lasers
Dipole moment (electric, operator), 81–83, 85–87, 90–91, 94–95, 103, 108–109, 111, 114–115
  matrix element, 86–87, 89–90, 92, 95–98, 105, 114–115, 220–221, 265, 284, 386
Direct gap transition (semiconductors), 99, 619–631
Direct laser writing, 243, 491, 580–582
  see also laser-assisted processing, laser deposition, laser etching
Discriminant analysis, *see* linear discriminant analysis
Disilane, *see* Si_2H_6
Dislocations, 33, 531–534
Disorder, 578, 580, 582–583
Distributed Bragg reflectors (DBRs), 359, 307–308, 610–612
  see also dielectric mirrors, heterostructures, vertical-cavity surface-emitting lasers
Distributed feedback reflectors, 537–538

Index

Disturbances (in manufacturing processes), 729
   see also defects, error detection, noise, process drifts
Divergence, *see* beam divergence
DMAAs, *see* tris(dimethylamino)arsine
DMAP, *see* tris(dimethylamino)phosphine
DMASb, *see* tris(dimethylamino)stibine
DMCd, *see* dimethylcadmium
DN_2^+ monitoring, 287
D_2O decomposition and monitoring, 312
Doped
   crystals (hosts, insulators), 619–620, 629–635
   semiconductors, 99–100, 237–238, 317–319, 343, 346, 387–388, 462, 613–614, 681, 685
   see also doping
Doping (and implantation), 11, 35, 46, 582–583
   see also doped semiconductors
Doppler
   profile (lineshape), broadening (linewidth), 68, 88–89, 162, 167, 176, 199, 223, 238, 256, 267–268, 273, 275–276, 284, 304, 307, 690, 701
   see also temperature (translational, T_{trans})
   shift (and spectroscopy), 68, 88, 238–240, 257, 287, 306
   shifted one-photon LIF, 238–240
   shifted two-photon LIF, 257
   velocimetry, 41
   see also elastic scattering, laser velocimetry
Drifts, *see* process drifts
Dry etching and processing, 34–35
   see also ion assisted etching, plasma-enhanced chemical vapor deposition, plasma etching, plasma processing
DTGS detector, *see* deuterated triglycine sulfate detector
Dual-wavelength pyrometry, 608–609
Dye lasers, 90, 120–123, 224, 227, 229–232, 236–238, 240–241, 244, 249–251, 254, 257, 267, 270–271, 273, 287, 300–302, 304, 490–491, 507, 563, 570, 624–627, 647, 660, 669–670, 677, 683
Dynamic random access memory (DRAM), 520

E

ECR plasmas, *see* electron cyclotron resonance plasmas
Effective
   charge, 88
   mass, 88, 99–100
   medium approximation (EMA), 430
Einstein A coefficient, 86–88, 115, 200, 268
   see also radiative decay rate
Elastic constants, 106–107, 700
Elastic light scattering, 90–92, 121, 140, 243, 495–552
   for particle monitoring, 22–23, 36, 41, 48, 496–522
   see also laser light scattering, Rayleigh scattering, scatterometry, Thomson scattering
Elastic waves, 643–644
   see also acoustic waves, thermal waves
Electrical probes, 11, 14, 30, 718, 733
   see also Langmuir probes, nonoptical probes
Electric-dipole allowed transitions, *see* allowed transitions
Electric dipole moment, *see* dipole moment
Electric displacement vector, 82, 84, 114
Electric field(s)
   effects (Stark), 674, 682
   induced Raman scattering, 574
   of light, 57–65, 81–85, 89, 108–110, 112, 114–116, 135, 328–330, 332, 340–341, 385–386, 413, 427–428, 498, 562–563
   local field correction, 82
   of scattered light, 498
   static, 93, 241–242, 377–380
   see also Langmuir probes, Stark broadening
Electric polarizability, *see* polarizability
Electric susceptibility, *see* susceptibility
Electrodeposition, 668
Electrodes
   collection by, 675–676
   photoemission of, 673, 683–685
Electroluminescent polymers, 385
Electron beam
   exposed resists, 367
   heating, 346–348, 376
Electron cyclotron resonance (ECR) plasmas, 44, 88, 167, 170, 176, 178, 180, 199, 238–240, 245–246, 281, 283, 290, 383, 410, 458–459, 548–549

Electron detection (using photon excitation), 673–685
Electronic
　energy levels, 68–69
　　in atoms, 71–72
　　in molecules, 69, 72–74
　　in solids, 76–77
　　partition function, population, 73, 75–76
　Raman scattering (in molecules), 91, 559, 578
　-speckle interferometry (ESPI), 550–551
　transitions, 69, 95–98, 485, 619–635
　　see also absorption (in the ultraviolet/visible), laser-induced fluorescence, optical emission spectroscopy
　　see also temperature (electronic, T_{elect})
Electron paramagnetic resonance (EPR), 308
Electron-phonon coupling, 93, 698
Electrons (in plasmas), 160–162, 174–176
　density, 483–484, 496, 648
　temperature, 167, 176, 496, 696
　see also Langmuir probes
Electrons and holes (semiconductors), 619–621
Electron spectroscopy for chemical analysis (ESCA), 32, 681
　see also x-ray photoelectron spectroscopy
Electro-optic modulator, 153, 300, 644
Electroreflectance, 377
Electrostatic particle traps, see particle traps
Ellipsometers (ellipsometry), 435–442
　automatic (dynamic), 435–440, 443ff
　null, 435, 437–438
　photometric, 435
　polarization-modulated (PME) [or phase-modulated], 415, 436–438, 444, 446
　rotating-
　　analyzer, 435–441, 450, 460, 463, 601
　　compensator, 435, 439, 455, 458
　　element, 435, 438–441
　　polarizer, 435, 437–439, 449, 458
Ellipsometric angles, 341–342, 418, 428, 432, 438–441, 445–446, 448, 453, 456, 459–460
Ellipsometric pyrometry, see pyrometry (ellipsometric)

Ellipsometry (single wavelength (SWE) and spectroscopic (SE)/multiwavelength), 17, 22, 25–26, 36, 38, 46, 48, 50, 57, 64, 120, 126, 139, 327–328, 336, 352–353, 425–465, 530, 535–536, 718
　comparison with other methods, 341–343, 359, 379, 400–401, 418, 524
　for wafer thermometry, 27, 443–444, 450, 458, 460–463, 698–700, 702–704, 708
　imaging, 441
　immersion, 459
　infrared, 439, 441, 446–448
　kinetic, 433, 442–443, 447
　SWE vs SE, 426–427, 443–444, 732–733
　variable angle, 418, 438, 442, 459
　see also associated ellipsometry parameters, pseudodielectric function
Emerald:Cr^{3+}, 634
Emission, 94–98, 696, 718
　see also laser-induced fluorescence, optical emission spectroscopy, pyrometry, radiative decay
Emissivity, 16, 139, 315, 463, 466, 592–609, 694–695, 697, 703, 725
　see also pyrometry
Emittance, 466, 592–594
Empirical (black-box) model (of processing), 20, 720–725, 732
　see also neural network model, statistical models
Endpoint detection, control, 4, 10–14, 20–21, 29, 35, 38, 45–46, 50, 139, 238, 350, 354, 684, 730–731, 733–734
　diagnostics locator, 14
　flexible, 185–186, 731
　nonoptical, 14
　photoresist, 14, 49–50, 354, 365–368, 542–543
　plasma processing, 14, 38, 45–46, 158, 160–161, 163, 166, 168, 172, 168, 176–186, 238, 283, 286–287, 354, 360–361, 456–458, 603, 684, 734
Energy
　distribution, see population distribution function
　levels, 67–78
Epioptics, 24–26

Index

Epitaxial growth, 28, 667–668
  see also metalorganic chemical vapor deposition, molecular beam epitaxy, vapor phase eptiaxy
Equipment-state parameters (inputs), 10–12, 715, 719, 730–734
Error (fault) detection and correction (in manufacturing), 9, 14–15, 20, 458, 719, 729–731, 735
  see also defects, disturbances, noise, process drifts
Etalon, 238, 267, 278, 280–281, 283–285, 483
  see also Fabry Perot interferometer
Etch
  depth, profiles, 537–542, 718
  rates, 731–732
  see also reflection interferometry (for film thickness)
Etching, 42–48, 354, 389, 580–582
  neutral beam, 361, 409–410
  patterned, 163, 178–179, 287, 456–458, 537–542
  see also dry etching, plasma etching
Étendue (throughput), 130, 145
Ethylarsine ($C_2H_5)AsH_2$), 292
Eu-doped phosphors, 630–635
Europium thenoyltrifluoroacetonate phosphor (EuTTA), 634–635
Eutectic temperature, 350
Evaporation (for deposition), 36–39, 302, 305–308, 646–647, 724
  monitoring, 444, 667–669
  see also molecular beam epitaxy, pulsed laser deposition
Excimer laser(s), 35, 62, 120, 122–123, 224, 231, 304, 309–310, 349–350, 442, 507, 570, 645, 647, 670, 716
  heating, 369
  see also ArF laser, dye lasers, KrCl laser, KrF laser, laser ablation, pulsed laser deposition, XeCl lasers
Excitation
  laser-induced fluorescence, 23, 215, 217, 224–225
  photoluminescence, 23, 623
Excitons, 621
Experimental methods (overview), 119–154, 223–225, 578, 621, 626–628, 638, 643, 646, 680, 716, 727–728
Explosive crystallization, 349–350, 363

Exponential
  sigmoid function, 722–723
  spiral approximation, 339–340, 432–435
Ex situ analysis (materials characterization, post-process analysis), 2–3, 23, 33–34, 465, 728–729
Extensometry (of wafer diameter, for wafer temperature), 347, 349, 483–484, 545–552
  see also interferometry across wafer (for wafer thermometry)
Extinction
  coefficient, 84, 265
  see also absorption coefficient, index of refraction
  cross section, 499–501
Extractive batch mode, 515

F

F monitoring, 94, 96, 160, 162, 166–170, 174–177, 181, 183, 205, 235, 274, 284–285, 670
F^+ monitoring, 166, 169
F^- monitoring, 683
F_2 monitoring, 226
F_2^+ monitoring, 226
$f_\#$, 61–62, 132–133, 139, 144–147, 163
  see also numerical aperture
Fabrication methods (integrated circuits for microelectronics and optoelectronics), 11, 34–50, 496, 692–694, 728–729
  see also batch processing, flexible manufacturing
Fabry Perot interferometer, 129, 162–163, 167, 175–176, 267, 278, 482–483, 560
  see also etalon
Factorial design (fractional, full), 721, 724
Factors (in statistical design of experiments), 721
Faraday cell, 438
Fast Fourier analysis, transform, 341, 354, 546
Fault detection and correction (in manufacturing), see error detection and correction
Fe
  atom monitoring, 166, 188–190, 226, 257
  dielectric function, 344
  sputtering of, 37, 257

Fe-18Cr-3Mo oxidation, 577
Feedback control, 4, 10, 12–13, 15, 305–308, 340, 449–452, 454, 458, 462, 612, 702, 716, 727–731
  see also closed-loop control, process control, real-time monitoring
Feedforward control, 4, 14, 727–730
  see also process control
Feed-forward error back-propagation (FFEBP) algorithm, 722–723
Fellgett advantage, 128, 130, 137
Fe_2O_3, 194
Fermi-Dirac distribution function, 620, 628, 690
Fermi level, energy, 690–691
  pinning/unpinning, 380, 623–625
Ferroelectric films, 36–37, 455
Fiber optic fluorescence thermometry, 629–633
Fiber optics, see optical fibers
Film absorption, 309–312
Film characterization (ex situ), 33
Film composition, 188–191, 353–354, 582
  diagnostics locator, 38
  see also alloy (semiconductor), ellipsometry, photoluminescence
Film thickness, 11, 31, 36, 38, 279, 294–295, 304–308, 311, 327, 460, 483, 609–614, 718–719
  diagnostics locator, 36
  see also ellipsometry, metrology, reflection interferometry
Films, optics of, 333–341
  multiple films, 338–341
Filters (bandpass, colored glass, holographic notch, interference (dielectric)), 120, 129, 134, 148, 165–166, 232, 269, 276, 294, 306, 365, 566, 586, 597, 599–602, 622, 628, 631, 661, 666
  see also quarter-wave filters
Final wafer state parameters, see wafer state parameters
Fine structure, 246–247
Flaking from reactor walls, see walls
Flashlamps (Xe), 298, 301, 304
  see also lamps
Flexible endpoint, see endpoint detection
Flexible (adaptive, specialty) manufacturing, 16–18, 695, 719, 726, 729–730, 732–733

Flow monitoring, 29, 31, 38, 292–294, 449–452
  see also flux monitoring; reactant delivery; time-resolved (kinetic) measurements; transient turn-on, turn-off behavior
Flow rate-modulation epitaxy (FME), see atomic layer epitaxy
Flow velocity, velocimetry, see laser velocimetry
Flow visualization, 38, 40–41, 150, 243–244, 481–482, 486–490, 496, 512–514
  diagnostics locator, 41
Fluorescence, 41, 215, 619, 638
  see also laser-induced fluorescence, photoluminescence
Fluoride fibers, see optical fibers (infrared)
Fluorinated overlayers on Si, 455–456, 685
Fluorocarbon overlayers, 459, 465
Fluorophors, see phosphors
Fluoroptic probe, 374, 376, 460, 629–633, 693, 700, 705, 709
  see also photoluminescence
Flux monitoring (to and from the surface), 29, 31, 38, 42, 250–252, 305–308
  see also flow monitoring; reactant delivery; time-resolved (kinetic) measurements; transient turn-on, turn-off behavior
Focusing light, 61–62
  see also imaging
Forbidden transitions, 94–96, 108, 242, 578, 582–583
  by spin, 95–96, 230, 245–246, 255, 619
  see also allowed transitions, selection rules
Foucault knife edge test, 347, 349
  see also schlieren photography
Fourier transform
  analysis, 129–130, 383, 435, 438, 440, 526–527
    see also fast Fourier analysis
  infrared spectroscopy (FTIRS), 26, 30, 121, 135–137, 268–269, 273–276, 280, 287, 290–295, 312, 383–384, 386–399, 483, 567, 604, 613–614, 626–627, 696, 718
    comparison with other methods, 275, 661
  lens, 521
  mass spectrometry, 638
  Raman spectroscopy, 137, 560, 567
  spectrometer, 124, 135–137, 290, 386–399, 439, 483, 560, 718

Index

Fractal-like surfaces, 529
Fractional factorial design, 721, 724
Franck-Condon factor, 96–97, 161, 219–221, 273
Franz-Keldysh effect, oscillations, 379–380
Free carriers, 312–313, 317–319, 376–373, 387, 578, 613, 621, 628, 646–647, 697
  absorption by, 100, 312–313, 317–319, 387, 599–600, 690, 699, 703
Free electron model, 87–88, 99, 114
  see also Lorentz oscillator model
Frequency (of light), 58
Frequency doubling, *see* second harmonic generation
Frequency modulation spectroscopy, 271–272, 278–279, 283–287, 300–301
Frequency summation, 109, 120, 123, 249, 656, 665
Frequency upconversion (of infrared light), 276, 656, 661
Fresnel relations, 330, 340, 364, 433, 450
Fringe(s), 335–336, 352–377, 481–489, 514, 703
  visibility (contrast), 361, 365–366, 372–373
  see also interferogram, interferometry
Fröhlich interaction, 93, 108, 574–577
FTIR, *see* Fourier transform infrared spectroscopy
Fundamental scattering, 92
Fundamental thin film science, 2, 8–9, 18–20
Furnaces, 39, 48, 306, 613, 691, 716
Fused silica, *see* SiO$_2$
Fuzzy logic control, 735

G

Ga
  atom
    ionization potential, 680
    monitoring, 166, 170, 177–178, 181–183, 226, 230, 232, 254, 304–307, 642
    reactant, 298, 679, 681
  dielectric function, 344
  monitoring on surface, 254, 405–410, 414, 418–423, 464
GaAlAs, *see* AlGaAs
GaAs, 76–78, 313
  cleaning (including deoxidation), 391–392, 410, 424, 460, 530–531
  deposition, 40, 229, 288–289, 291–293, 306, 396, 507, 529–531, 533, 669, 679, 681
    and monitoring, 307–308, 315, 357–360, 380, 405–409, 418–424, 450–451, 463, 609–612, 626
  dielectric function, critical points, optical properties, 99, 266, 331–332, 344–345, 387, 578–580, 619, 623
    temperature variation, 344, 372, 387, 463, 698–699
  doping, 237–238, 387–388
  emissivity, 599
  etching, 170, 178, 181–183, 189, 229, 237, 583, 628
    and monitoring, 360–361, 381, 409–410, 458, 624–626
  ionization potential (of molecule), 680
  oxidation, 459–460
  passivation, 391–392, 624–627
  phonon energies, 108
  photowashing (surface treatment), *see* photowashing
  Raman scattering temperature effects, 585
  substrate, 249, 280, 354, 422, 424, 432, 452, 454, 529, 531, 533–534, 578–580, 665, 669
  surface, 387–388, 391–397, 399, 414, 418–424
    and monitoring, 576–577, 669
    see also surface photoabsorption, reflectance-difference spectroscopy
  thermometry, 313–318, 368, 371–372, 379–380, 628–630
    laser heating, 628
  see also heterostructures
GaAsP deposition and monitoring, 423–424
GaCl
  desorption, 409–410
  monitoring, 166, 170, 226, 230, 297–298
  reactant, 409
GaCl$_3$ reactant, 288
Ga-H vibration on surface, 266, 393–395
GaInAs, *see* InGaAs
GaInP
  deposition and monitoring, 423–424
  dielectric function, 344
  Raman scattering temperature effects, 585

GaN deposition, 167, 290–291, 396, 604
GaP
    deposition, 530
        and monitoring, 355, 357, 411–412, 423–424, 580
    dielectric function, critical points, 344
        temperature variation, 698
    phonon energies, 108
    substrate, 411–412
GaSb
    deposition and monitoring, 315
    dielectric function, 344
    phonon energies, 108
    thermometry, 315
Gas chromatography, 30
Gas delivery monitoring, *see* flow monitoring; flux monitoring; reactant delivery; time-resolved (kinetic) measurements; transient turn-on, turn-off behavior
Ga_2Se_3 monitoring, 576–577
Gaseous electronics conference (GEC) cell, 171–173, 205–206, 225, 283
Gas phase diagnostics locators
    flow visualization, 41
    species identification, density measurements, 26–27
    temperature, 27
Gas source MBE, *see* molecular beam epitaxy
Gated analysis, detection, 45, 123, 240–241, 580, 613–614, 622, 624–627
    see also boxcar integration
Gaussian beam
    beam waist, 60–62
    propagation, 60–62, 639, 659
    radius of curvature, 60–61
Gaussian CGS units, 57, 81, 113–116
Gaussian profile (of spectral line), 88, 267–268, 515
    see also Doppler profile
GdCo, 37
$(Gd_{0.99}Eu_{0.01})_2O_2S$ phosphor, 630, 633
Ge
    ablation, 195
    atom monitoring, 159, 166, 182, 226, 304
    deposition, 159, 228, 304, 310, 664–665
        and monitoring, 410–411
    detector, 125, 129, 314, 374–375, 597
    dielectric function, critical points, 344
        temperature variation, 346, 698
    etching and monitoring, 582
    films, 388
    melting, 350
        and monitoring, 582
    optical properties, spectroscopy (solid), 99–100, 313, 388–389
    Raman scattering (solid), 564
        phonon energies, 105, 108, 564
        thermometry, 584–585
    substrate, 580–581
    surface, 395–396, 398
        and monitoring, 418, 423
Ge^+, 166
GEC cell, *see* gaseous electronics conference cell
GeH monitoring, 159, 308
Ge-H vibration, 395–396
GeH_3 monitoring, 309, 676
GeH_4 (germane)
    plasma, 187, 228
        and monitoring, 664–665
    reactant, 159, 224, 304, 310
Ge_2H_6 (digermane) reactant, 395–396, 410–411
GeO deposition, 195
Geometric optics, 57–58, 61, 141, 499, 546
Germane, *see* GeH_4
GeSi alloys
    amorphous (a-GeSi), 582
    annealing, 582
    etching, 182
        and monitoring, 455
    deposition, 534–536, 664–665
        and monitoring, 453, 582
        and thermometry, 318
    dielectric function, 343–344
        temperature variation, 344
    Raman scattering thermometry, 584–585
    see also heterostructures
Glan-Taylor polarizer, 435
Glan-Thompson polarizer, 140, 413
Glass
    reflow, 11
    spherical particles, 505–506, 517–520
    thermometry, 369
    see also arsenosilicate glass films, borophosphosilicate glass, phosphosilicate glass, SiO_2
Glowbar, 119–120, 136, 269, 274, 276, 294, 439
Glow discharge, *see* plasmas
Graphite
    dielectric function, 344

Index

electrodes, 247–248, 508–509, 684–685
laser ablation/pulsed laser deposition,
194–195, 203, 255, 614
see also C
Grating equation, 131, 524, 529, 545
Gratings, *see* diffraction gratings
Graybody, 593, 595, 597, 599, 608
Growth: layer-by-layer, 2D vs. 3D, 31–32
see also deposition
Grüneisen parameter, function, 584, 697

H

H
 adsorbed on surface, 48, 389–399,
  577–578, 666
 atom
  cleaning, 531
  monitoring, 27, 166, 170, 177–179,
   186–188, 216, 226, 230, 244,
   247–249, 676, 678, 669–670, 676,
   678
  reactant, 250, 280
H^- monitoring, 683
H_2, 70–71, 76, 101–102, 104
 flow, 282, 288–289, 292, 393–395, 419,
  486–487, 514, 668
 for stimulated Raman scattering, 224,
  249, 647
 in plasma (and sputtering in), 160, 170,
  178, 180, 182, 184, 187, 193, 249,
  360–361, 391–392, 410, 444, 446,
  448, 624–626, 669–670
 monitoring, 27, 166, 187, 563, 567–571,
  662–665
 reactant, 227, 230, 243, 248, 280,
  408–409, 424–425, 567–572, 665,
  676–677
 thermometry, 662–664
Hadamard summation, 440
Hamiltonian, 85–86, 95
Harmonic generation
 see also second harmonic generation,
  third harmonic generation
Harmonic oscillator approximation, 71, 95,
 97, 583–584
 see also anharmonicity
Hayfield and White matrix method,
 340–341
HBr plasma, 253, 455–456, 458, 640–641
HCl, 70–71, 297–298
 monitoring, 181, 279, 288, 297
 plasma, 171, 458

reactant, 243, 409–410
 wet, 454
HCN
 for wavelength calibration, 278
 plasma, 691
HCO^+ monitoring, 287
H_2CO for wavelength calibration, 278
HCS^+ monitoring, 287
He, 227, 293, 568–570
 flow, 486
 monitoring, 171
 plasma, 171, 181–183, 238, 287, 361, 724
Heat capacity, *see* specific heat
Heat flow equation, 112–113, 639, 644
He-Cd laser, 120–121, 404–406, 529, 542,
 648
Helical resonators (HR), 44, 171–172,
 252–253, 456, 640–642
Helicon-wave excited plasma, 44, 300,
 455–456
Hemispherical spectral radiant intensity,
 593
He-Ne laser, 120–121, 123–124, 286, 310,
 349–350, 354, 356–357, 360–363,
 367–369, 376, 405, 410, 435, 444, 455,
 483–484, 505, 507–508, 510–511, 516,
 529, 533, 540, 542–544, 546, 548, 550,
 586, 609, 623, 645, 648–650
Heterobipolar junction transistor (HBT),
 182
Heterostructures, 181–184, 328, 340, 621,
 730
 AlAs/AlGaAs, 611–612
 GaAs/AlAs, 307–308, 313, 360, 610–612
 GaAs/AlGaAs, 181–183, 315, 361, 458,
  463, 625–626, 629
 GaAs/InGaAs, 529, 531–533
 HgTe/CdTe, 354–357, 452
 InP/InGaAs, 182, 184, 423–424, 458
 InP/InGaAsP, 360–361
 Si/GeSi, 182, 453, 534–536
 see also distributed Bragg reflectors,
  multiphase models
HF
 monitoring, 295, 572
 wet etching, passivation, treated
  surfaces, 48, 389–390, 392–393,
  398–399, 459
HfN sputter deposition, 190
Hg
 atom ionization potential, 680
 dielectric function, 344

Hg (*continued*)
  lamp radiation, spectral lines, 444, 454, 543
  see also lamps
HgCdTe
  deposition and monitoring, 452–453
  detector, 125, 129, 279, 294, 388, 391, 597, 645
  dielectric function, 343–344
  formed after interdiffusion, 354–357
  substrate, 383
$HgNH_3$, 159
HgTe deposition, 231
  and monitoring, 354–357, 452
High charge density plasmas, 43–44, 691, 701
  see also electron cyclotron resonance plasmas, helical resonators, helicon-wave excited plasmas
High electron mobility transistor (HEMT), 315
  see also modulation doped field-effect transistor (MODFET)
High reflection (HR) coatings, 355
  see also dielectric mirrors
High-resolution electron energy loss spectroscopy (HREELS), 31–32
High T_c superconductors, 37, 194–198, 202–203, 256, 302–304, 307
  see also $YBa_2Cu_3O_{7-x}$ compounds
HN_2^+ monitoring, 287
H_2O, 48, 172, 178, 294
  decomposition, 312
  dielectric function, 344
  monitoring
    as a gas, 279, 665
    for wavelength calibration, 278
    on surface, 312, 382
      of vibrations, 266, 391–392
  reactant, 312
Hollow cathode atomic lamps, *see* atomic resonance lamps
Holographic
  double exposure, pulsed, 488–489
  interferometer, 482
  interferometry, *see* interference holography
  notch filer, 134, 566, 586
Homoepitaxy monitoring, 612–613
Homogeneous broadening, linewidth, 88–89, 238
  see also collisional broadening, radiative decay

Hönl-London factor, 88–90, 97–98, 196, 221, 223
Hook spectroscopy, 485
Hot band, 568
H_2S
  plasma, 287
  reactant, 422
HSiCl monitoring, 215, 217, 226–227, 243
Hund's rules, 96
Hydrocarbon surface contamination, 48

I
I^- monitoring, 683
ICCD array, *see* charge-coupled device array detectors (intensified)
Imaged volume, 143–147
Imaging, 6–7, 61–62, 137–138, 141–151, 264, 273, 481–491
  of radicals interacting with surfaces (IRIS), 250–252
  three-dimensional (3D), 141–147, 163–164, 169, 193–195, 202–208, 218–219, 225, 232, 234–236, 241–245, 247–248, 497, 505–520, 567–573, 660
  two-dimensional (2D), 141, 148, 193–198, 203–208, 225–226, 441, 465, 516–522, 613–614, 626–628
  see also laser induced fluorescence (planar)
  see also photography, tomographic reconstruction (one-dimensional, 1D imaging)
Immersion ellipsometry, 459
Impurity detection, 29, 48, 172, 178–180, 187, 189–190
In
  atom
    ionization potential, 680
    monitoring, 166, 170, 177–178, 180, 182, 184, 189, 191–193, 226, 230, 232, 254, 304–305, 642–643
  dielectric function, 344
  -free mounting, 317
  on surface, 254
  reactant, 297
  sputtering, 189
InAlAs dielectric function, 344
InAlGaAs dielectric function, 344
InAs
  deposition, 531

Index

757

and monitoring, 315, 407, 409,
423–425, 451–452
detector, 125, 597
dielectric function, 344
phonon energies, 108
surface, 574
thermometry, 315
InAsP deposition and monitoring, 423–424
InCl
monitoring, 170, 226, 229–230, 297–298
reactant, 409
Incoherent light sources, 59–60, 89,
119–121, 123–124, 152, 269–270
see also lamps
Incommensurate growth, 424–425, 531,
533–534, 575
see also critical thickness
Inconel 718, 189–190
Indenecarboxylic acid, 310–311
Index of refraction, 11, 58–59, 82–86,
99–100, 114–115, 263, 265, 327–340,
343, 352, 364, 368–369, 385–386, 404,
481–483, 499, 502, 515, 620–621, 657
changes, 643, 645–646, 648
continuously varying, gradients, 336,
364–365, 549
control of, 355, 461–463
inhomogeneities, gradients, 481,
485–491, 642–649, 696, 701, 706
ratio, relative, 499, 515
temperature variation, 343–345,
369–372, 376, 461–463, 699–700
see also dielectric function
India ink, 517–520
Indirect transitions, gap semiconductor,
99–100, 312, 317–319, 567
Indium tin oxide (ITO) deposition,
191–193, 455
Inelastic scattering, 22–23, 537
see also Brillouin scattering, Raman
scattering
Influence diagrams, 723–724
Infrared absorption, *see* absorption (in the
infrared)
Infrared (IR) active/inactive transitions,
92–93, 103, 105
Infrared diode laser, *see* diode laser
Infrared (IR) diode laser absorption
spectroscopy (IR-DLAS), 26, 267,
274–287, 309, 696
comparison with other methods, 661,
701

Infrared emission, 165, 167, 199, 312, 718
see also pyrometry
Infrared modulation spectroscopy,
270–272, 278–287
Infrared reflection-absorption spectroscopy
(IRAS, IRRAS, RAIRS), 23, 31, 263,
290, 381–385, 389
Infrared thermography, 603
see also pyrometry
InGaAs
deposition, 529, 531, 533–534
and monitoring, 423–424, 451–452
dielectric function, critical points,
344–345
temperature variation, 699
etching, 182, 184
and monitoring, 360–361, 458
see also heterostructures
InGaAsP
deposition, 229, 296–298, 375
diode laser, 372–373, 375
etching, 229
and monitoring, 360–361
InGaP, *see* GaInP
Inhomogeneous broadening, linewidth,
88–89, 238
see also Doppler profile
In-line analysis, 3, 13, 16, 32, 515, 642, 681,
716
In-line scattering, 566
see also coaxial collection configuration
InP, 313
deposition, 229, 292, 296–298, 532–535
and monitoring, 407, 409, 423–424
desorption from, 254
dielectric function, critical points (and
optical properties), 344, 619
temperature variation, 344, 346,
698–699
etching, 170, 178, 180, 182, 184, 229
and monitoring, 360–361, 375–376,
458
phonon energies, 108
substrate, 424–425, 665
surface, 423–425, 450, 574–576
cleaning, deoxidation, and monitoring,
424–425, 464
thermometry, 318, 368, 371, 375, 377
laser heating, 628, 631–632
In_2S_3 monitoring, 575
InSb
deposition and monitoring, 315

InSb (*continued*)
  detector, 295, 603
  dielectric function, critical points, 344
    temperature variation, 346, 698
  melting, 350
  phonon energies, 108
  surface and monitoring, 575–576
  thermometry, 315
  see also heterostructures
In situ
  monitoring, control, and diagnostics (sensors), 1–50, 139, 495–497, 515, 522, 537, 559, 716, 727–734
    see also closed-loop control, real-time monitoring
  processing, 13, 16–17
  scanning coherent lidar (ISICL), 510–512
InSn target, 191–193
Instrumentation (overview), 119–154
In$_2$Te$_3$ monitoring, 575–576
Integrated circuit
  inspection, 519–522, 643
  processing, *see* fabrication methods
Intelligent control strategies, 735
  see also advanced process control, closed-loop control, neural network control
Intensified detectors, *see* charge-coupled device array detectors, photodiode arrays
Intensity (of light beam, also power), 59–64, 84, 86, 89–90, 98, 108, 112, 115, 124, 135–136, 146, 149, 220, 222–224, 245, 247, 249, 264, 329, 341, 385–386, 499, 592, 639, 655, 657–660, 675–676, 678
  see also scattered intensity
Interdiffusion, 355–357, 452
Interface(s), 463–465
  model, 331
  modes, 572
  monitoring, 572–578, 665–669
    see also surface-sensitive spectroscopy
  state density, 623–624
  see also adsorbates, overlayers
Interfacial
  reactions, 383, 385
  regions, 561, 572–578
Interference-enhanced Raman scattering (IERS), 577–580

Interference holography, 26, 41, 244, 482, 485–489, 701
Interferogram, 352–376, 614
  see also fringes
Interferometers, 481–489
Interferometric profilometry, 523
Interferometry, 21, 41, 46–47, 411, 481–489, 512, 586, 644
  for film thickness, 14, 21, 36
    ellipsometry, 36, 427
    laser light scattering, 36, 530, 534
    optical emission spectroscopy, 14, 36, 184–186
    pyrometric, 14, 36, 609–614
    Raman, 14, 36, 579–580
    reflection, 14, 36, 352–368
      see also reflection interferometry
  for gas-phase thermometry, 27, 486–489, 696, 706
  for wafer thermometry,
    across wafer, lateral distances, 27, 36, 483–485, 543–552, 703–704, 708–709, 712
    reflection, 368–377, 703, 711–713
      see also reflection interferometry
  Nomarski, polarization, 489
  see also Fourier transform infrared spectroscopy, Fourier transform spectrometer
Intracavity effects, 301, 566
Inverse least-squares (ILS) analysis, 717
Ion
  assisted (sputter) etching, 42–46, 184–185, 257
    see also plasma etching
  bombardment, 237, 667
  flux to surface, 641
  impact, 160
  implantation, 402, 578, 580, 582–583, 603, 647, 693
    particle monitoring, 496
  monitoring of, 287
    see also optical emission spectroscopy
Ion detection (using photon excitation), 673–683
Ionization, *see* multiphoton ionization, photoionization, plasmas, resonant-enhanced multiphoton ionization, single photon ionization
Ionization potential, 673, 678–680
Ir dielectric function, 344
Isotope effects, 238, 280, 290–291, 312, 389–391

Index

J
Jacquinot advantage, 130, 137
Jones vector, 62–63, 65

K
K dielectric function, 344
Kapton film, 247, 648
KBr dielectric function, 344
KCl
  aqueous solutions, 422–423
  dielectric function, 344
KDP, *see* potassium dihydrogen phosphate
Kerosene (for particle production), 513
Kinetic measurements, *see* time-resolved (kinetic) measurements
Kirchoff's law of radiation, 593, 609
Knife edges, *see* Foucault knife edge test, schlieren photography
KOH etch, 483
Kramers Kronig relations, 84–85, 342
KrCl laser, 683–684
  see also excimer lasers, laser ablation, pulsed laser deposition
KrF laser, 120, 122, 159, 194–198, 202, 224, 229, 231–232, 235–236, 254–255, 280, 288, 302, 304, 398, 402, 484, 489, 642, 683
  see also excimer lasers, laser ablation, pulsed laser deposition
Kr-ion laser, 120–121, 376, 506, 563, 613

L
La dielectric function, 344
Lambert cosine law, 593
Lamb wave modes, 649–651
Lamp(s), 347
  focusing, 62
  for spectroscopy, 89, 119–121, 123–124, 134, 267, 269, 296–298, 301, 304, 308, 314, 316, 357–360, 365, 400–401, 404–405, 415–416, 435–437, 535, 597, 631, 644
  pulsed, 298, 301, 304
  for wafer heating, 314, 596–599, 603, 606–608, 694, 702, 725, 732
  see also atomic resonance lamps, flashlamps, rapid thermal processing
Langmuir probes, 14, 30, 44, 46, 509, 641
LaO_2S:Eu phosphor, 632
$LaPO_4$:Eu^{3+} phosphor, 634

Large-scale manufacturing, 10, 17–18, 729
  see also batch processing
Laser ablation
  of films, substrates, 35, 37, 47, 137, 143, 193–201, 234, 254–256, 483–485, 488–491, 645, 648–649
  of particulates, 234
  see also pulsed laser deposition
Laser annealing, 348, 582, 716
Laser-assisted processing, 47
  see also direct laser writing, laser ablation, laser deposition, laser etching, photolysis, pulsed laser deposition
Laser deposition, 47, 159, 195, 231–232, 243, 302, 309–310, 354, 580–581, 647, 665
  see also photolysis, pulsed laser deposition
Laser-desorption
  laser-induced fluorescence (LD-LIF), 171–172, 252–253, 639–642
  plasma induced emission (LD-PIE), 171–172, 639–640
  see also laser-induced thermal desorption
Laser desorption spectroscopy (LDS), *see* laser-induced thermal desorption
Laser diodes, *see* diode lasers
Laser direct write, *see* direct laser writing
Laser Doppler velocimetry (LDV), *see* laser velocimetry
Laser etching, 47, 581–583, 628, 642
Laser heating, 112–113, 140, 369, 564, 580–583, 586, 604, 628–629, 631–632, 637–651, 691
Laser-induced fluorescence (LIF), 7, 22–24, 26–27, 29, 32, 38, 41, 44, 47, 85, 90, 98, 121, 123, 125, 129, 134, 139, 141, 145–148, 153, 171–172, 176, 195, 215–257, 263–264, 301, 567–568, 619, 638–642, 656, 664, 696
  after laser-induced thermal desorption, *see* laser desorption
  calibration of, 218, 221–223, 247, 304
  comparison with other methods, 160, 162, 174, 263–264, 273, 296, 327, 660–661, 675–676, 701
  Doppler shifted, 238–240, 257
  excitation LIF, 23, 215, 217, 224–225
  for gas-phase thermometry, 27, 197–198, 221, 231, 233–237, 696, 701, 706

Laser-induced fluorescence (LIF) (*continued*)
  for particle monitoring, 234, 507–508
  high-spectral resolution, 235, 238–240, 483, 696
  microLIF, 242–244
  planar LIF (light sheet technique, two-dimensional LIF), 148, 225–226, 230, 255–256, 489
  time-resolved, 230–231, 240–242, 252–257
  two-photon LIF, 24, 27, 109, 111–112, 176, 216, 226, 244–250, 257, 508, 655, 670, 678
Laser-induced photofragment emission (LIPE), 218, 228–229
Laser-induced thermal desorption (LITD), 24–25, 123, 151, 171–172, 252–253, 456, 637–643
Laser ionization mass spectroscopy (LIMS), 30, 678
Laser light scattering (LLS), 23, 25–26, 33, 38, 45, 121, 140, 495–552, 725–726
  for strain relaxation, 38, 529, 531, 533–534
  from rough/smooth surfaces (diffuse reflection), 25–26, 33, 315–317, 327, 411–412, 432, 489, 522–536
  see also elastic light scattering, interferometry, particles, Rayleigh scattering, scatterometry
Laser linewidth, 89, 121, 267–268, 275, 563
Laser magnetic resonance (LMR), 308–309
Laser melting, 347–350, 582–583
Laser microanalysis, microvaporization, 638
Laser optogalvanic spectroscopy (LOGS), 683
  see also optogalvanic spectroscopy
Laser radar, *see* lidar
Lasers, 60–62, 119–124, 152, 218
  continuous wave (cw), 60–62, 89–90, 112–113, 120–124, 224, 238–240, 563, 628–629
  chopped, 622
  intensity of, *see* intensity
  pulsed, 59, 61–62, 89–90, 112–113, 120–124, 151, 223–225, 238, 240–241, 244, 563, 622, 624–627, 638, 644, 647, 656, 675
  pulse length, 59, 89, 121, 151, 639
  tunable, 120–123
  see also diode lasers
  see also diode lasers, excimer lasers, vertical-cavity surface-emitting lasers
Laser velocimetry (anemometry, flow or laser (Doppler) velocimetry), 41, 512–514, 549, 552
  see also Doppler shift
Latent image evaluation, 11, 14, 46, 50, 537–538, 542–544
  see also photoresist, scatterometry
Lateral distances on wafers, surfaces diagnostics locator, 36
Latex paint spheres, *see* white latex paint spheres
Lattice mismatched growth, *see* incommensurate growth
Lattice vibrations in solids, *see* phonons
Layer-by-layer (2D) growth, 307, 312, 405–409, 411–412, 420, 444–445, 578
  see also atomic layer epitaxy
Layer thickness, *see* film thickness
Lead-salt diode lasers, *see* diode lasers
Least-squares (regression) analysis, 717, 721, 724–725
Level mixing, 241–242
LiBaAlF$_6$(LIBAF):Cr^{3+}, 634
LiCaAlF$_6$(LICAF):Cr^{3+}, 634
Lidar, 510–512
Li dielectric function, 344
LiF dielectric function, 344
Lifetime (radiative), *see* radiative decay
Light collection, *see* imaging
Light detection and ranging, *see* lidar
Light emitting diodes (LEDs), 313, 361, 612, 631
Light-matter interactions, 81–116
Light pipes (guides), 297–298, 596, 604–607, 703–704
  see also optical fibers
Light scattering, 7, 22, 90–94
  see also elastic light scattering, laser light scattering, Raman scattering, Thomson scattering
Light sheet technique, *see* laser-induced fluorescence (planar LIF)
Light sources, *see* incoherent light sources, lamps, lasers
Limited thermal equilibrium, *see* thermal equilibrium

Index

LiNbO$_3$, 36–37, 287
  dielectric function, 344
Linear discriminant analysis, 542, 717–718
Linear optical spectroscopies (overview), 21–23, 81–108, 113–116
Line broadening (width), 87–89, 93, 99, 193, 265, 267–268, 271, 275, 302–304, 306–307, 699
  homogeneous, 88–89, 267
  inhomogeneous, 88–89, 267
  also see collisional broadening, Doppler profile, lineshape, plasma broadening, relaxation rates, Stark broadening
Lineshape, linewidth (spectral), 86–89, 93, 110–111, 265, 379, 562–563, 657, 700, 702
  see also Doppler profile, lorentzian profile, line broadening
Linewidth (of pattern) control, 539–544
  see also critical dimension
Liquid precursors, 292–294
LiSrAlF$_6$(LISAF):Cr^{3+}, 634
LiTaO$_3$ electro-optic modulator, 300
Lithography, *see* photolithography
Lock-in detection, 126, 153, 223, 270–271, 276, 279, 284, 287, 294, 296, 298, 301, 306–307, 317, 374, 400–401, 631, 644
  see also chopper
Lorentzian profile, lineshape, 87–88
Lorentz-Lorenz formula, 82, 700
Lorentz oscillator model, 87–88, 99, 115, 430, 463
  see also free electron model
Lorenz field, 82
Low energy electron diffraction (LEED), 31–32
L-S (Russel-Saunders) coupling, 95–96
Luminescence, 215, 620
  see also laser-induced fluorescence, phosphorescence, photoluminescence, resonance fluorescence
LuPO$_4$:Eu^{3+} phosphor, 634

M

Mach-Zehnder interferometry, 41, 482, 485
Magnesium fluorogermanate phosphor (activated with tetravalent manganese), 632
Magnetic
  dipole transitions, 94
  field, 238–239
  films, recording, 37–38, 603
  permeability, 83
Magnetron, 207–208
  etching, 44, 245–246
  sputtering, 37–38, 191–193, 355, 455
Malfunctioning, *see* defects, disturbances, error detection, noise (in a process run), process drifts
Manufacturing (of integrated circuits), *see* batch processing, fabrication methods, flexible manufacturing, large-scale manufacturing
Markov process (of random surface defect scatterers), 528
Masks, 34–35, 42, 49–50, 543
  in endpoint detection, 178–179
  in etching, 360
  patterned (micro-LIF), 243–244
  see also photoresist
Mass spectrometry (spectroscopy), 14, 29–30, 45, 47, 532, 638, 642, 676, 678–681, 716–717
  laser ionization (LIMS), 30
  reflection, 30
  time-of-flight (TOFMS), 30, 679–681
  also see residual gas analysis
Mass transport models, *see* physical modeling (of a process)
Materials characterization, 2, 23, 33–34, 559, 621, 626–628, 643
  see also ex situ analysis
Matrices in optics, 62–65, 430
  Abelès method, 340
  coherency, 65
  for Jones vectors, 62–63, 65
  Hayfield and White method, 340–341
  Mueller, for Stokes parameters, 65
Matrix (transition) elements, 86–87, 89, 93, 96–98, 161, 220–222, 265, 284, 620–621, 696
  see also dipole moment
Matrix experiments, 721
Maxwell-Boltzmann distribution function; speed, velocity distributions, 88, 202, 254–257, 690, 697
  see also Boltzmann distribution, Doppler broadening
Maxwell's equations, 57–58, 499
MBE, *see* molecular beam epitaxy
Mechanical chopper, *see* chopper
Melting, 347–350, 582–583
Menu, *see* run recipe

Mepsicrons, 126–127, 566
Metal
  film deposition, 11, 16, 36–37
    thickness, 646–647
  oxidation, 465
  surface monitoring, 328, 333, 337, 465, 537
Metal alkyl monitoring, 298
  see also specific individual alkyls
Metalorganic chemical vapor deposition (MOCVD), 31, 37, 40, 151, 232, 244, 279–282, 302, 354–360, 375, 377, 379–380, 396–397, 404–409, 411–412, 414, 419–424, 444, 452, 454, 507, 514, 529–530, 532–535, 569–574, 603–604, 665
  see also chemical vapor deposition
Metalorganic molecular beam epitaxy (MOMBE), see molecular beam epitaxy
Metalorganic vapor phase epitaxy (MOVPE), see metalorganic chemical vapor deposition
Methyl, see CH_3
Methylarsine $(CH_3)AsH_2$, 292
Methyleneimine, 292
Methylmethyleneimine, 292
Methylsilazane, 287
Metrology, 5, 17, 35–36
  across wafer, 347, 349, 483–484, 543–552
  of films, 354–368
  see also critical dimension, film thickness, interferometry, particles
Mg dielectric function, 344
$MgAl_2O_4$ (spinel) dielectric function, 344, 371
MgF_2 dielectric function, 344
MgO dielectric function, 344, 371
Michelson interferometer, 129, 135–137, 482–485, 510–512, 567, 613–614
Microcalorimetry, 645–646
MicroLIF, 242–244
Microscopy (optical), 7, 33, 142, 144, 491, 523, 536, 543, 638
  see also Raman microprobe analysis
Microwave interferometry, 30, 683
Microwave plasmas, 388, 446
  see also electron cyclotron resonance plasmas
Mie scattering, 499
MIR, see multiple internal reflection

Mirage effect, 645
  internal, 646
Mirau microscopy, 7
Mirrors, see dielectric mirrors
Misfit dislocations, 33, 531–534
  see also defects
MKSA (Rationalized, SI) units, 57, 81, 113–116
Mn
  atom ionization potential, 680
  dielectric function, 344
Mo
  deposition and monitoring, 455
  dielectric function, 344
  substrate, 445–446
MOCVD, see metalorganic chemical vapor deposition
Modeling (of a material), see multiphase models
Modeling (of a process), see process modeling
Modulated reflectance spectroscopy, 646
  see also differential reflectometry, infrared modulation spectroscopy, photoreflectance
Modulation doped field-effect transistor (MODFET), 378
  see also high electron mobility transistor (HEMT)
Modulation methods
  in absorption, 271–272, 278–287, 300–301
  in pyrometry, 605–608
  in reflection, see differential reflectometry, photoreflectance
Moiré interferometry, see projection moiré interferometry
Molecular beam epitaxy (MBE), 29–33, 36–42, 151, 167, 232, 253–254, 302, 305–308, 414, 419, 421–423, 452–453, 463–464, 530–535, 561, 574–576, 623, 626, 666–668, 679–680
  atomic beam flux, 29, 302, 305–308
  diagnostics locator, 38
  ellipsometer installation, 442, 444
  gas source MBE (GSMBE), 40, 254
  growth rate during, 609–612
  see also ellipsometry
  metalorganic MBE (MOMBE), 40, 354, 397, 422–423, 432, 443–444, 449–453, 463
  thermometry during, 312–318, 350, 359–360, 463, 596, 603, 609–612, 623, 628–630, 693–694

Index

Molecular beams, 232, 250–252
  pulsed, 409–411
  see also molecular beam epitaxy
Molecular spectroscopy, see spectroscopy
MOMBE, see molecular beam epitaxy
Monel K-500 alloy, 188–189
Monitor (pilot, test) wafers, 12, 497, 543, 695, 728–729
Monochromator, see spectrometers
Monolayer control, see submonolayer control
MOVPE, see metalorganic chemical vapor deposition
Mueller matrices, 65
Multichannel (array, parallel) detection, 124, 126–129, 131, 134, 141, 151, 164–166, 224–225, 269, 302, 304, 353–354, 437, 440, 445, 447, 566, 573, 623, 659, 716–717
  see also charge-coupled device array detectors, mepsicrons, photodiode arrays, vidicons
Multielement detection, see multichannel detection, position-sensitive detectors
Multilayers, see dielectric mirrors, heterostructures
Multimedia models, see multiphase models
Multipass configuration, 274–275, 279–280, 283–284, 287, 565–566
  see also White cell
Multiphase models, 311, 331–341, 429–435, 440, 459, 718–719
  see also nucleation (during deposition)
Multiphoton
  absorption, 90
    see also absorption (three-photon), absorption (two-photon), laser-induced fluorescence (two-photon LIF)
  processes, 23
  ionization (MPI), 23–24, 147, 255, 263, 638, 655, 674, 676, 679–680
    see also resonant-enhanced multiphoton ionization
Multiple internal reflection, element (MIR), 386–399
  see also attenuated total internal reflection spectroscopy
Multiple linear regression, 717, 725
Multiplexing, 130
  see also multichannel detection

Multistep run, 730
  see also cluster tools
Multivariate analysis, correlation, and statistics, 19–20, 166, 172–173, 183, 312, 604–606, 716–718, 733–734
  see also chemometrics
Multi- vs single variable (input, ouput) control, 728, 731–734
Multiwavelength
  pyrometry, 596, 608–609
  spectroscopy, 353–354, 357–360, 365–366, 732–733
  see also spectroscopic analysis

N

N monitoring, 167, 177, 183, 216, 226, 244–245, 247, 270, 670
N_2, 37, 73–74, 76, 95, 101–104, 194, 288
  flow (added gas), 486–487, 514, 665, 668
  laser, 120, 649–650, 683
  monitoring, 96, 159, 166–168, 172, 176–177, 179–180, 187–188, 190–191, 197–200, 235–236, 569–572, 661, 665
  plasma, 159, 167, 182, 199–200, 235–238, 240–241, 369–370
  seeding in, 197–198, 284
  reactant, 256, 571–572
  thermometry, 197–200, 235–236, 569–570, 572, 665
N_2^+, 73–74, 96
  monitoring, 166–167, 190–191, 197–198, 226, 236, 238, 240–241, 683
NA, see numerical aperture
Na dielectric function, 344
NaCl dielectric function, 344
NaF dielectric function, 344
Naphthoquinone diazide, 310–311
Na_2S passivation, 399, 422, 624
  see also $(NH_4)_2S$ treatment, photowashing, sulfide layers
Nb
  atom monitoring, 166, 193
  dielectric function, 344
NbN sputter deposition, 193
Nb_2O_5 deposition, etching, and monitoring, 454
Nd dielectric function, 344
ND_3
  monitoring and reactant, 290–291, 312
  plasma, 391

Nd:glass laser, 348
Nd:YAG laser, and harmonics, 120, 123, 137, 224, 235, 237, 249, 506, 567, 569, 580, a642, 657, 666, 679–680
Ne
 monitoring, 189–190
 sputtering plasma, 189–190, 193
Near-surface dielectric function spectroscopy (NSDFS), 427, 433–435, 718
Neural network models, training, 20, 170, 434, 542, 718, 720–725, 733, 735
Neurons (in neural network), 722–723, 725
NF_3 plasma, 174–175, 683
NH
 monitoring, 166, 187–188, 226, 250–252, 308
 reactivity with surfaces, 250–252
NH_3
 for wavelength calibration, 278
 plasmas, 187–188, 252, 391–393, 510, 626–627
 and monitoring, 665
 reactant, 159, 396, 604
 and monitoring, 250, 290–291, 312
NH_4F, 389, 398
$(NH_4)_2S$ treatment, 399
 see also Na_2S passivation, photowashing, sulfide layers
Ni, 37
 atom monitoring, 166, 188–190
 deposition, 310, 668
 dielectric function, optical properties, 344, 382, 384
Nitridation, 11, 48
 see also silicon nitride
N-methyl-2-pyrrolidone (NMP), 365
NO
 monitoring, 166, 177, 179–180, 250, 678
 for wavelength calibration, 278
 surface interactions, 678
N_2O
 monitoring, 167, 279, 295
 for wavelength calibration, 278, 280–281
 plasma, 167, 187, 290, 295
Noise (in a detector), 152–153, 270–272, 306, 345, 403
 cancellation, 510–512, 516
 see also signal-to-noise ratio
Noise (in a process run), 728–729
 see also defects, disturbances, error detection, process drifts

Nomarski
 interferometry, 489, 515–520
 microscopy, 523
Nonlinear optical spectroscopies, 23–24, 81, 84, 90, 108–112, 116
 for light generation, 120, 123
 for monitoring, 123, 139, 222–223, 655–670
 see also coherent anti-Stokes Raman scattering (CARS), laser-induced fluorescence (two-photon LIF), multiphoton ionization, resonant-enhanced multiphoton ionization, second harmonic generation, third harmonic generation
Nonoptical probes, sensors, 14, 27–34, 695–696, 733–734
 see also electrical probes, Langmuir probes, mass spectrometry
Nonradiative decay, relaxation, 220, 264, 619, 622–623, 628–629, 642–643, 700
 see also collisional decay
Normal coordinates, 91
Nuclear spin statistics (alternating line intensities), 30, 76, 98, 102, 105, 233, 569, 571–572, 683
Nucleation
 during deposition
  in the film, 42, 444–448, 647
  in the gas phase, 497
 during recrystallization, 363
Numerical aperture (NA), 132, 138–139, 144
 see also $f_\#$

O
O
 adsorbed on Si, 666
 atom monitoring, 96, 166–169, 176–177, 192, 204–206, 216, 226, 244–245, 247–248, 270, 308, 670
O^+, 166, 169
O^- monitoring, 683
O_2, 37, 48, 76, 242
 adsorbed, 537, 574
 monitoring, 661, 665
 (in) plasma (or sputtering in), 43, 160, 162, 167–170, 174–176, 184–187, 191–193, 197, 199, 204–206, 235–237, 247, 250, 283, 290, 455, 640, 684–685, 724–726, 734
 see also ashing, plasma stripping
 reactant, 159, 230, 256, 383, 611

Index

O_2^+, 184
O-branch lines, 103–105
O_3 reactant, 392–393
  and monitoring, 293–294
OCS
  monitoring, 295
    for wavelength calibration, 278
  plasma, 296, 691
Off-line, 3, 728
  see also ex situ analysis, materials characterization
OH
  monitoring, 27, 159, 166, 172, 177, 180–181, 187, 226, 230, 250–252, 308
  reactivity with surfaces, 250–252
O-H vibrations, 266, 647
OMCVD, see metalorganic chemical vapor deposition
OMVPE, see metalorganic chemical vapor deposition
One-color pyrometry, 596–608
On-line analysis, 3, 515, 728–729
Open-loop control, 10, 12–13, 49, 719, 721, 728, 732
Optical beam deflection (OBD) spectroscopy, 645
Optical coatings, 28, 37
Optical density, 264
  see also optical thickness
Optical detectors, see detectors
Optical diffraction thermometry (for wafers), 27, 36, 543–549, 697, 704, 708
Optical emission
  as a light source, 269–270, 302, 304
  background, 580, 622, 632
Optical emission spectroscopy (OES) [plasma-induced emission (PIE)], 17, 21–23, 26–27, 29–30, 38, 44–47, 50, 98, 125–126, 129, 133–134, 140–141, 143–145, 147–151, 153–154, 157–208, 217, 220, 235–236, 240–241, 247, 267, 273, 284, 299, 302–304, 361, 483–484, 489, 507, 510, 622, 638, 664, 716–717, 720, 725–726, 731–732, 734
  after laser-induced thermal desorption, see laser desorption
  calibration of, 160, 302,
    see also actinometry
  comparison with other methods, 217, 327, 381, 701
  endpoint control, 11, 14, 158, 160–161, 163, 166, 168, 176–186, 734
  excitation mechanisms, 160–162
  for gas-phase thermometry, 27, 195–203, 235–236, 284, 696, 701, 706–707
  high resolution, 162–163, 167, 483, 696
  interferometry for film thickness, 14, 36, 184–186, 731
  time-resolved, 162, 170–172, 193–198, 240–241, 302–303, 483–484
  see also actinometry, emission, infrared emission, self absorption, tomographic reconstruction
Optical factor, 526
Optical fiber(s), 128, 132, 135, 137–139, 141–144, 148, 163–164, 173, 178–179, 206, 269, 275, 294–295, 313–314, 357–358, 365–366, 519, 566, 586, 596, 604, 608, 630–633
  -based sensor, 313, 629–633
  infrared (chalcogenide, silver halide), 137, 139, 275, 290–292, 294–295, 399, 601
  thermometry, 604–608, 629–633
  ultraviolet (silica), 137, 144
  visible, 137
  see also light pipes
Optical micrometry (of wafer diameter), see extensometry
Optical multichannel analyzers (OMAs), 124, 126–129, 304
  see also charge-coupled device array detectors, mepsicrons, multichannel detection, photodiode arrays
Optical parametric oscillators (OPOs), 120, 224
Optical particle counters, see particles
Optical path length, 41, 58, 352, 364, 368, 427, 481–489, 512, 609, 700
Optical sources, 119–124, 265–272, 306–307
  see also lamps, lasers, noise, signal-to-noise ratio
Optical thickness/thick media, 89, 264–265, 268, 335, 342, 357, 564, 594
Optics
  of reflection, 328–341
  overview, 57–65
Optoacoustic detection, spectroscopy, 263, 301, 644, 647
Optogalvanic spectroscopy, 38, 46, 263, 673, 682–685

Organometallic chemical vapor deposition (OMCVD), *see* metalorganic chemical vapor deposition
Organometallic vapor phase epitaxy (OMVPE), *see* metalorganic chemical vapor deposition
Orthogonal design, 721, 724
Oscillator strength, 87, 102, 485
Os dielectric function, 344
Oven temperature, 306
  see also furnaces
Overlayers, 331, 336–340, 343, 347, 430–431, 433–435, 443, 455–456, 459, 463–465, 581–582, 586, 599–601, 683–685, 704
  see also adsorbates, very thin films
Overtones, 95, 574–577
Oxidation, 11, 17, 35, 46, 48, 367, 389, 443, 459, 465, 577, 581, 693
  anodic, 422–423
  see also Si (oxidation), SiO_2
Oxide removal, *see* surface cleaning
Oxides, 48
  see also SiO_2
Oxyacetylene flame, 230
Oxynitride films, 181, 355, 439
Ozonator, 293–294
Ozone, *see* O_3

P

P adsorbed on Si, 667
P_2 monitoring, 226, 229, 297
P_4
  ionization potential, 680
  monitoring, 297–298
Packaging, 34
PACVD (plasma-assisted chemical vapor deposition), *see* plasma-enhanced chemical vapor deposition
Parabolic (quantum) wells, 450, 452
Parallel detection, *see* multichannel detection
Paraxial equation, 57–58, 645
Partial least-squares (PLS), 604, 717–718
Particle counters, *see* particles
Particle temperature after laser ablation, 613–614
Particle trajectory, 510–516
Particle traps, 505, 508–509
Particles (particulates) [detection, monitoring, and sizing of], 9, 11, 35–36, 38, 41, 45, 174, 193, 234, 495–522
  in a gas, 41, 234, 243, 495–496, 505–514
    see also flow visualization
  in a liquid, 495–496, 514–520
  on repetitively patterned substrates, 519–522
  on surfaces, 48, 496, 516–522, 725–726
    see also crystal originated particles (COPs), single particle detection, surface morphology
Partition function (Z), 73, 75, 97–98
Passivation, 48, 389–392, 398–399, 577, 624–627
  see also photowashing, Si (wafer)
Patterned films, 163, 178–179, 287, 365–366, 519–522
Patterning, 9, 49–50, 243–244, 310–311, 456–458, 537–544, 580
  see also diffraction gratings, laser light scattering, masks, surface morphology
Pb dielectric function, 344
P-branch lines, 95, 97, 103
PbS
  camera, detector, 125, 597, 628
  dielectric function, 344
PbSe
  detector, 125, 597
  dielectric function, 344
PbSnTe dielectric function, 344
PbTe dielectric function, 344
$PbZr_{0.53}Ti_{0.43}O_3$ (PZT), 194
Pd
  dielectric function, 344
  /Si interface, 577
  surface treatment, 465
PdO reduction and monitoring, 582
Pd silicide, 577
perdeutero(*tris*-6,6,7,7,8,8,8-heptafluoro-2,2-dimethyl-3,5-octanedionato)europium phosphor (dEuFOD), 634–635
Periodic features on surfaces, *see* diffraction gratings, laser light scattering, surface morphology
Perpendicular collection configuration, *see* right angle collection configuration
Perturbation theory, *see* quantum mechanics
PET, *see* polyethyleneterephthalate
PH monitoring, 229
PH_2 monitoring, 229

Index

PH₃
  ionization potential, 680
  reactant, 297–298, 407, 423–425
  and monitoring, 229, 297, 665
Phase change, *see* evaporation, melting
Phase-lock, -shift detection, 632
Phase matching, 123, 658–659
Phase-modulated ellipsometer, *see* ellipsometer (polarization-modulated)
Phase-sensitive detection, 438
  see also lock-in detection
Phase shift, 499, 515–516
  interferometry (PSI), 550
Phonon(s) (optical), 68, 77–78, 93, 105–108, 312–313, 317–319, 380, 388, 559, 561–562, 564, 572–586, 690–691, 697–699
  -assisted absorption, 99–100, 312–313, 317–319, 690, 699, 703
  longitudinal optical (LO), 77–78, 105–108, 574–580, 583
  population (density), occupation number, 93, 584, 586, 690, 697–700
  temperature variations of, 582–586, 700, 702
  transverse optical (TO), 77–78, 105–108, 574–578, 580
  see also Brillouin scattering (for acoustic phonons), Raman scattering, vibrations
Phosphorescence, 215–216, 244, 514, 619
  see also laser-induced fluorescence, photoluminescence
Phosphors (fluorophors), 619, 629–635, 704
Phosphosilicate glass (PSG), 266, 293, 353–354, 604–607, 718
  see also borophosphosilicate glass, glass
Photoabsorption, *see* surface photoabsorption
Photoactive compound (PAC), 310–311, 367, 543
Photoacoustic
  beam deflection, 648
  spectroscopy (PAS), microscopy (PAM), 24, 642–651
  see also thermal wave analysis
  thermometry, 649–651, 709
Photo-assisted growth and processing, 39, 47
  see also laser-assisted processing, photolysis
Photoconductive detectors (photodiode, photovoltaic modes), 124–125, 152, 306, 314, 597, 622
Photodeflection, 47, 645–646, 648–651
Photodetachment, 245, 678, 682–683
Photodetectors, *see* detectors
Photodiode arrays (PDAs), 126, 131, 148, 165–166, 171, 206–207, 244, 250–252, 297–298, 302, 357–360, 365–366, 437, 458, 716, 718
  gated, 126, 171, 194
  intensified, 126, 165, 171, 250–252, 566
  see also charge-coupled device array detectors, multichannel detection
Photodiodes, *see* photoconductive detectors
Photodissociation cross section, 247
Photoelastic modulators, 415–417
  see also ellipsometers, reflectance-difference spectroscopy
Photoelectric effect, *see* photoemission
Photoemission (PE), 24, 263, 642, 673, 681, 683–685
  endpoint detection, 14, 684
  optogalvanic spectroscopy (POGS), 14, 46, 683–685
  yield, 683
Photography, 23, 27, 47, 481, 489–491, 506–508, 513
  high-speed (fast), 151–152, 193–198, 256, 302, 489–491
  see also time-resolved measurements
Photoionization, 29, 263, 673–682
  see also multiphoton ionization, resonant-enhanced multiphoton ionization, single photon ionization
Photolithography, 34–35, 43, 49–50, 539–544
  see also photoresist
Photoluminescence (PL), 23, 25–26, 33, 35, 38, 45, 85, 147, 263–264, 391–392, 489, 567, 619–635
  comparison with other methods, 264, 327, 379–380
  excitation, 23, 623
  for wafer thermometry, 27, 628–635, 699–700, 704, 709
Photolysis, photolytic processes (including deposition), 47, 159, 195, 229, 231–232, 280, 288, 298, 304, 309–310, 354, 398, 582, 647, 678
Photomultipliers (PMTs), 125–126, 142, 152–153, 165–166, 170, 224, 232, 298, 301, 306, 400–401, 415–416, 436–437, 566, 586, 622, 656, 659

Photon-assisted growth and processing, 39, 47
  see also laser-assisted processing, photolysis
Photon counting, 125
Photons, 57–59, 690
Photopyroelectric characterization, 646
Photoreflectance, 120, 328, 376–381, 646
  for wafer thermometry, 27, 379–380, 698–700, 702, 704, 709
Photoresist, 9, 11, 14, 34–35, 46, 49–50, 172, 310–312, 364–368, 399, 439, 497, 543–544
  baking, 49–50, 311–312, 365–366, 543, 693
  coating, 49–50, 311, 365–367
  development, 354, 365–368, 543, 693
  dissolution, 459
  (as mask during) etching, 178–179
  exposure, 49–50, 310–311, 365–366, 537, 542–544
  monitoring, 46, 49–50, 310–312, 364–368, 399, 439, 543–544
    see also optical emission monitoring
  patterned, 537, 542–544
  planarization, 181
  stripping (ashing), 49–50, 179–181, 204–206
    particle monitoring 496–497
  see also latent image evaluation
Photothermal
  beam deflection (PBD) spectroscopy, 126, 645, 648–649
  deflection spectroscopy (PDS), 645
  displacement spectroscopy, 644–645
  effects, 599
  radiometry, 645–646
  spectroscopy, 24, 599, 642–651, 705
  see also thermal wave analysis
Photowashing, 380, 399, 410, 624–625
  see also Na_2S passivation, $(NH_4)_2S$ treatment, sulfide layers
Physical (and chemical) modeling (of a process), 20, 227–229, 568–571, 719–722, 724–725, 727, 732–733
Physical vapor deposition (PVD), 36–39, 41–42
  see also evaporation, molecular beam epitaxy, pulsed laser deposition, sputtering
PIE (plasma-induced emission), see optical emission spectroscopy (OES)

Pilot wafers, see monitor wafers
Pits on surfaces, 517–518
Placzek polarizability theory, 104
Placzek-Teller coefficients, 102, 105
Planarization, 181, 542
Planar LIF, see laser-induced fluorescence (planar LIF)
Planck blackbody law, spectra, 592–594, 614, 690, 697
  see also Bose-Einstein distribution function
Plasma(s), 21, 38, 42–46, 217, 223, 234, 691, 693, 701
  cleaning, 388, 410
  diagnostics, 27, 30–31, 38
  -enhanced chemical vapor deposition (PECVD) [assisted CVD (PACVD)], 39, 42–43, 157, 166, 186–188, 228, 250–252, 302, 355, 383, 397, 413, 437, 444–448, 453–454, 459, 580, 661–665, 668–670, 693
  -enhanced photoemission (PEP), 683–685
  etching, 10–11, 14, 17, 27, 38, 42–46, 157, 166–186, 235–237, 245–246, 250–253, 283–291, 296–297, 299, 360–363, 365–377, 381, 410, 443, 455–458, 460–461, 537–542, 548–549, 578, 603, 623–625, 633, 635, 639–642, 683–685, 691, 693, 718, 724–726 731–734
  diagnostics locator, 38
  induced damage, 624
  model, 724
  see also reactive ion etching
  heating, 369–370, 460–461
  -induced broadening, 89, 267, 302
  -induced emission (PIE), 141, 149–151, 157–208
    see also optical emission spectroscopy (OES)
  laser produced (plume), 482–482, 648, 691
    see also laser ablation, pulsed laser deposition
  oxidation, 459–460
  particle formation and monitoring in, 496–497, 505–512
  processing, treatment (etching, deposition, and other), 10–11, 27, 42–46, 48, 157–158, 166, 218, 220,

Index

224, 247–252, 270, 275, 309, 369–370, 465, 496, 563, 639, 673, 682–685, 691, 702, 720
pulse-modulated, pulsed, 151, 171, 374–375, 510
remote, 388, 391–392, 413, 444, 626–627, 733
stripping (ashing), 49–50, 179–181, 204–206, 365, 496–497
see also high charge density plasmas, optogalvanic spectroscopy, tomographic reconstruction
Plasmoid, 361–363
Plasmon, plasma resonances, *see* surface plasmon resonances
Plumes, *see* laser ablation, pulsed laser deposition
PMMA, *see* polymethylmethacrylate
Pockels cell, 438
Poisson's ratio, 350
Polarimetry, 341–343
Polarizability (electric), 82–84, 87, 90–92, 100–105, 109, 114–116, 562–563, 700
anisotropic, 100–105, 563
derived, dynamic, 82–83, 100–101, 104–105
mean (isotropic), 100–101, 103–105, 563
static, equilibrium, 82–83, 101–103
see also susceptibility
Polarization (electric), 81–83, 90–91, 109–110, 114–116, 656
Polarization configuration, 139–140, 415–417, 435–440, 572, 574–577
Polarization-modulated ellipsometer,
Polarization of light, 21, 58, 62–65, 91, 101, 105–108, 131, 154, 222, 345–347, 352–353, 415, 495, 497–504
analysis, 415–417, 435–440, 510–512
see also polarization configuration, polarization optics
circular, 63–64, 515–516
effects, 21, 355, 413, 526, 529, 562–563, 572, 574, 575–578, 582–583, 601, 657, 666–667, 669
on windows, 415, 417, 441–442
see also ellipsometry, reflectance-difference spectroscopy
elliptical, 63–64, 435, 437–438
linear, 58, 62–64
s, σ, TE, 63–64, 328–343, 385–386, 390, 392–395, 405–406, 427–428, 515–516, 522, 531, 537

p, π, TM, 63–64, 186, 328–343, 382, 385–386, 389–390, 392–395, 403–413, 427–428, 515–516, 522, 531, 537
see also ellipsometry, laser light scattering, reflectance-difference spectroscopy, surface photoabsorption
Polarization-modulated ellipsometer, *see* ellipsometers
Polarization optics (polarizers), 139–140, 415–417, 435–440
analyzers, 415–417, 435–440
compensators, 415–417, 435–440
modulators, 415–417, 436–438
polarizers, 140, 415–417, 435–400
Polarization selection rules, *see* polarization effects, selection rules
Polarization state, 341, 428, 440
Polarized
interferometry, 489
scattering, 101–107, 562–563, 574–577
see also depolarized scattering
Polyamic acid, 365
Polyethylene dielectric function, 344
Polyethyleneterephthalate (PET), 490, 648–649
Polyimide, 383, 385
as substrate for deposition, 669
curing, 365, 383, 385
laser ablation, 195, 256, 678–679
see also photoresist, polymers
Polymers, 50, 235, 364–368, 385, 578
ablation, pulsed laser deposition, 194, 302, 483, 490, 648–649
degradation, 383, 385
etching (and dissolution), 177, 179–181, 204–206, 365, 455, 465
on reactor walls, 184, 299
(reactor) wall cleaning, 184
see also photoresist
Polymethylmethacrylate (PMMA), 490–491
reactivity of, 251–252
see also photoresist, polymers
Polysilicon, *see* silicon
Polystyrene latex spheres (PLS), 515, 517–520
Population density, *see* density
Population distribution function, 5, 73, 75–76
p-polarized reflectance spectroscopy (PRS), 185–186, 332, 355, 357, 403–413

Porosity, 343, 443
  see also voids
Porous materials (for monitoring adsorbates)
  alumina, 312
  Si, 312
  SiO_2, 312
Position-sensitive detectors (bi-cell, quadrant), 126, 374–375, 645, 648–650
  see also charge-coupled device array detectors, mepsicrons, multichannel detection, photodiode arrays
Post-process analysis, *see ex situ* analysis, in-line analysis, on-line analysis
Potassium dihydrogen phosphate (KDP)
  dielectric function, 344
  for frequency mixing, 123, 224
Powder formation in plasmas, 510, 512
  see also particles
Power (of light beam), *see* intensity
Power broadening, 89
  see also saturation effects
Pressure broadening, *see* collisional broadening
Principal-component analysis, 542, 717–718, 725–726
Printed circuit boards, 603
Process control, 4, 9–18, 715–716, 726–735
  multiple-input multiple-output (MIMO), 728
  single-input single-output (SISO), 728
  see also advanced process control, closed-loop control, feedback control, feedforward control, open-loop control, process modeling, real-time monitoring, regulatory control, Run-to-Run control, statistical process control, supervisory control
Process development, 2, 5, 9–10, 17–21, 29, 41, 50, 276, 715, 719, 724–726
Process drifts, 729, 733–734
  see also defects, disturbances, error detection, noise (in a process run)
Processing (semiconductor), *see* fabrication methods
Process malfunctioning, *see* defects, disturbances, error detection, noise (in a process run), process drifts
Process modeling, 19–21, 715–716, 719–726, 730–735
  see also empirical models, neural network models, physical modeling, statistical models

Process reactors, *see* walls (flaking from), tools
Process-state parameters (conditions, results), 10–12, 716, 719–720, 730–732
Production reactors, *see* tools
Profilometry, *see* interferometric profilometry
Projection moiré interferometry, 544–550, 704, 708
Prototyping, 730
  see also flexible manufacturing
Pseudo-Brewster angle, *see* Brewster's angle
Pseudodielectric function, 429, 431–433, 445, 447
Pt
  /a-Si interface, 577
  dielectric function, 344
  monitoring, 349–350
PtSi detector, 129
Pt silicide (PtSi, Pt_2Si) formation and monitoring, 349–350, 577
Pu dielectric function, 344
Pulsed laser deposition (PLD), 6, 28, 30, 37, 47, 88, 129, 137, 148, 151–152, 193–201, 254–256, 301–304, 490–491, 613–614, 648–649, 691, 696
  see also laser ablation
Pulsed lasers, *see* lasers
Pulsed (pulse-modulated) plasmas, *see* laser ablation, plasmas (pulse-modulated, pulsed), pulsed laser deposition
PYRITE (method), 612
Pyroelectric detectors, 124, 135, 645
Pyrometers (types of), 591, 596–599, 608
  see also pyrometry
Pyrometric interferometry, 361, 592, 609–614, 703, 711
Pyrometry (infrared, optical), 157, 591–614, 692–695, 697, 702–703, 710–711, 718, 725, 732–733
  dual-wavelength (two-color), 608–609, 694, 711
  ellipsometric, 463, 601, 694, 711
  film measurements (by interferometry), 14, 36, 592, 609–614
  micro-, 604
  signals, 596
  single-wavelength (one-color), 596–608, 694, 710–711
  wafer thermometry, 14, 17–18, 21, 27, 47, 125, 129, 140, 153, 312, 315, 358, 591–614
  see also blackbody, graybody

Index

Q

Q-branch lines, 95, 102–105
Quadrant detectors, *see* position-sensitive detectors
Quantum efficiency, 125, 127–128, 136, 143, 145–146, 165, 675
  see also detectors
Quantum mechanics, of light-matter interaction, 85–94, 101
Quantum wells (or superlattices), 36, 77, 378, 578, 627
  see also heterostructures
Quarter-wave
  coatings, 355
  filters, 313
  see also dielectric mirrors, distributed Bragg reflectors, heterostructures
Quartz fibers, *see* optical fibers (ultraviolet)

R

Radiation constants (first, second, and third), 592
Radiation thermometry, *see* pyrometry
Radiation trapping, *see* self absorption
Radiative decay, 81, 85–87, 98, 157, 161–162, 215, 264, 619, 642
  rate (lifetime), 86–88, 97, 115, 145–146, 157, 161, 193, 200, 220, 225, 264, 267–268, 273, 302, 622–623, 629
  see also optical emission spectroscopy
Radiometry nomenclature, units, 59–60, 593
Radius of curvature, *see* curvature
Radon transform, 149–150, 203, 207–208
Raman active, inactive mode, 92, 103–108, 660
Raman microprobe analysis, 7, 244, 564, 580–582
Raman scattering (inelastic, spontaneous), 7, 22–23, 25–26, 35, 38, 41, 45, 92, 121, 125, 129, 132, 134, 137, 139, 141, 145–147, 154, 227, 496, 559–586, 716
  comparison with other methods, 659–661, 701
  cross-section, 90–94, 101–105, 562–564, 566, 660
  efficiency, 564
  for
    gas-phase thermometry, 27, 567–573, 696, 707
    particle analysis, 496
    strain analysis, 35, 38, 106–107, 578, 580, 586
    surface and interface sensitivity, 25, 572–578
    thin film analysis, 578–583
    wafer thermometry (phonon shifts, phonon lifetimes, Stokes/anti-Stokes scattering intensity ratio), 27, 583–586, 690, 700, 702, 704, 711
  fundamentals, 81–83, 85–86, 90–94, 98–108, 110–111, 561–563
  in molecules (gases), 26, 91–93, 100–105, 567–574
  in solids, 93–94, 105–108, 572–586
  interferometry for film thickness, 14, 36, 578–580
  lineshape, linewidth, 92, 101, 563, 583–586
  resonance, 92–94, 108
  tensor, 91, 564
  see also coherent anti-Stokes Raman scattering (CARS), electronic (Raman scattering), phonons, resonant Raman scattering, rotational (Raman scattering), scattering cross-section, Stokes/anti-Stokes scattering intensity ratio, vibrational (Raman scattering), vibrations in solids
Raman scattering (stimulated), for light generation, 110, 120, 123, 224, 247–249, 647, 656
Rapid thermal (RT-)
  annealing (RTA), 46, 49, 402
  chemical vapor deposition (RTCVD, RTP-CVD), 39, 46, 592, 602–603, 692
  oxidation (RTO), 46, 462, 733
  processing (RTP), 6, 15, 39, 46, 151, 350, 436, 716, 725
    lamps, 35, 317, 596–598
    reactors, 596–598, 691–694, 702–704, 732–733
    thermometry for, 31, 46, 312, 317–319, 350, 462, 550, 591–592, 596, 601–609, 691–694, 702–704
    see also pyrometry
Raster scanning, 506–509, 519, 627–628
Rationalized MKSA units, 57, 81, 113–116
Rayleigh
  –Gans scattering, 499

Rayleigh (*continued*)
  range, 61, 659
  scattering, 23, 41, 90–92, 101, 103, 496–502, 560
    see also elastic light scattering, laser light scattering
Rb dielectric function, 344
R-branch lines, 95, 97, 103
Re dielectric function, 344
Reabsorption, *see* self absorption
Reactant delivery (control of flow), 231, 244, 292–295, 297–298, 300
  see also flow monitoring; flux monitoring; time-resolved (kinetic) measurements; transient turn-on, turn-off behavior
Reactive ion
  beam etching, 44, 167
  etching (RIE), 38, 43–46
    diagnostics locator, 38
    see also plasma etching
Reactive sputtering, 158, 190–193
  see also sputter deposition
Reactors, *see* tools
Real-time monitoring, feedback, and (process) control, 2–6, 10–20, 27–29, 31, 42, 157–158, 166, 170, 174, 178, 181, 187, 189, 194, 218, 253, 264, 276, 286–287, 290, 293–295, 297–298, 304–308, 310–311, 316, 342, 353, 360, 364–368, 413–414, 426, 435, 440–441, 443, 446, 495, 529, 539–544, 547, 578–582, 609–614, 640–642, 645–651, 667, 676, 679, 689, 694–695, 701–705, 715–716, 718–720, 727–734
  see also closed-loop control, ellipsometry, endpoint detection, process control
Recipe, *see* run recipe
Recrystallization, *see* crystallization
Recursion relations for monitoring, 339–340, 430
Reflectance, 112–113, 329, 331–342, 345–350, 352–377, 378–413, 415, 417–418, 465–466, 526, 564, 593, 609
Reflectance-difference (anisotropy) spectroscopy (RDS/RAS), 7, 22, 25, 31–32, 84, 139, 328, 338, 413–425
  comparison with other methods, 403–405, 417–418, 442
  rotating-element systems, 417

Reflection, 21–22, 35, 38, 46, 129, 134, 139, 310–311, 327–465, 601
  coefficient, 186, 329–342, 415, 417, 419, 428, 465–466, 609
    virtual, 339
  coefficient ratio, 341–342, 418, 428–430, 435, 437
  endpoint, 14, 181, 183, 286, 350, 354, 360, 365–368, 456–458
  infrared, 353–354, 718
    see also attenuated total internal reflection spectroscopy, reflection (interferometry)
  interferometry, 45, 50, 121, 327–329, 335–336, 338–341, 352–377
    for film thickness, 14, 36, 181–183, 286, 311, 313, 352–368, 411–412, 609, 612, 641, 684, 731
    for wafer thermometry, 27, 126, 313, 352–354, 368–377, 695, 699, 703, 711–712
    see also ellipsometry, reflectometry
Reflection high-energy electron diffraction (RHEED), 31–33, 39, 405, 420–422, 451, 679–681
Reflectivity, 328, 339, 348, 465–466
  see also time-resolved reflectivity
Reflectometry, 120, 341–343
  at one interface, 328–335, 345–351
    for wafer thermometry, 27, 316–318, 345–347, 688, 712
  compared to other methods, 442–443
  spectroscopic, 353–354, 357–360, 365–366, 716, 733
  see also reflection (interferometry)
Refractive index, *see* index of refraction
Regression analysis, 440–441
  see least-squares regression analysis, multiple linear regression
Regulated flow, 449–452
Regulatory control (of a process), 13–14, 727–730
Relaxation rates, 98, 100, 145–146, 153, 264, 619, 622–623, 700
  see also collisional decay, nonradiative decay, radiative decay rate
Remote plasmas, *see* plasmas
Residual gas analysis (RGA), 29, 170, 717, 720, 725
  see also mass spectrometry
Resist, *see* photoresist
  see also laser-induced fluorescence

Index 773

Resonance fluorescence, 215, 232, 305
Resonant-enhanced multiphoton ionization (REMPI), 24, 26–27, 30, 32, 109, 112, 123, 218, 263, 638, 655, 674–679
  comparison with other methods, 701
  for gas-phase thermometry, 27, 678–679, 696, 701, 707
  see also multiphoton ionization
Resonant Raman scattering, 92–94, 108, 566, 572–579, 582, 586
Resonant tunnelling diodes (RTDs), 15, 451–452
Response surface, response surface methodology (RSM), 721, 724–726, 728, 733–734
Reticle inspection, 522
RF discharges, see plasmas
Rh dielectric function, 344
Right angle collection configuration, 143–147, 243, 565–566
Rigid rotor approximation and deviations from, 70, 568–569
Ripple-amplitude method for pyrometry, 605–607
Rochon polarizer, 435
Root-mean-square (RMS) roughness, 527–528
Rotating disk reactors, 228, 327, 357–360, 442, 513, 569, 571, 666
  see also wafer (rotation)
Rotating-element (analyzer, compensator, polarizer) systems, ellipsometers, see ellipsometers
  see also reflectance-difference spectroscopy
Rotational
  alternating intensities, 76
    see also nuclear spin statistics
  energy levels, 68–71
  partition function, population, 73, 75–76
  Raman scattering (pure rotational), 90–92, 95, 98, 100–103, 561, 563, 567–573, 696, 707
  transitions, 69, 94–98, 101–105
    see also temperature (T_{rot}), vibrational-rotational Raman scattering
Rough surfaces, see laser light scattering, surface morphology
Rovibronic states (bands), 69, 96
Ru atom monitoring, 307
Ruby, see $Al_2O_3:Cr^{3+}$

Run recipe, menu, 10, 13–14, 720, 730–731, 733
Run-to-Run (RtR) control, 13, 458, 719, 727–730, 733
  see also process control
Russel-Saunders coupling, 95–96, 307
Rydberg states (highly excited), 674, 676, 682

S

S, 177
S_2, 166, 176
S alkyl monitoring, 298
Sapphire, 376, 604, 609, 647
  see also silicon-on-sapphire
Saturation effects, intensity, and spectroscopy, 23, 89–90, 112, 222–223, 231, 238, 249, 304, 656, 675–676
Sb
  adsorbates, 423, 464, 576, 668
  dielectric function, 344
S-branch lines, 101–105
Scanning electron microscopy (SEM), 31, 33
  with energy dispersive x-ray analysis (EDAX), 31, 33
Scanning tunneling microscopy (STM), 31–32
Scattergrams, 495, 537–544, 718
Scattering (of light), 22–23, 489
  angle, 498–506, 523–552, 561–562, 564
  coefficients, 500–501
  cross section (total and differential), 90–91, 101–105, 110–111, 115, 146, 499–506, 524–529
    see also Raman scattering
  efficiency, 500–506
  elastic, see elastic light scattering, laser light scattering, particles, scatterometry
  inelastic, see Raman scattering
  (scattered) intensity, 498–506, 524–529, 562–566, 562–563
  matrix, 498
    see also amplitude scattering matrix
    see also Brillouin scattering, Raman scattering, Rayleigh scattering, Thomson scattering
Scatterometers, 537–539
  two theta (2Θ), 538–539
Scatterometry, 23, 495–552, 718, 732–733
  from periodic surfaces

Scatterometry (*continued*)
  for endpoint detection, 14
  for etch profile widths, depths, and profiles, 11, 14, 23, 36, 46, 537–544
  see also latent image evaluation
  from particles, *see* particles, elastic scattering
  from rough (smooth) surfaces, *see* laser light scattering
Schlieren photography, 41, 151, 244, 481, 489–490
  see also Foucault knife edge test
Schottky barrier heights, 623
Se
  adsorbates, 626
  alkyl monitoring, 298
  atom ionization potential, 680
  dielectric function, 344
Second harmonic generation (SHG), 23, 109–110
  for light generation, 120, 123, 224, 236–237, 249, 656
  surface, 23, 25, 655, 665–669
Selection rules, 86, 90–97, 100–108, 161, 221, 265, 572, 574, 575–578
Self absorption (radiation trapping), 89, 161, 234, 267, 269–270, 300, 302, 304, 623
Semiconductor lasers, *see* diode lasers
Semiconductor photodetectors, *see* detectors, photoconductive detectors
Semiempirical modeling, 720, 724
Sensors, 3–13, 18, 29–31, 119, 139, 715–716, 727–734
  see also virtual sensor
Seraphim coefficients, 378
SF_6 plasma, 169–170, 176, 205, 236, 286, 455–456, 548–549
Shadowgraphy, 41, 151–152, 481, 489–491
Shot noise, *see* noise
Si
  amorphous (a-Si)
    crystallization of, 349–351, 363–364
    deposition of, 187, 228, 250, 280, 300, 355, 397, 444–447, 577, 661, 667–668
      see also plasma-enhanced chemical vapor deposition
    properties of, 344, 430–431
  atom
    ionization potential, 680

  monitoring, 96, 166, 169–172, 177, 182, 187–188, 193, 195, 226–229, 304, 308, 640
  deposition, 11, 17, 187, 193, 215, 217, 227–229, 234, 243, 250, 280, 288, 300–301, 304, 567–570, 602, 609, 613–614, 661, 664–665, 692, 724, 734
    and monitoring, 355, 397, 409, 437, 444–447, 453, 455, 458, 534–536, 580–581, 667–668
    crystallinity, 667–668
    of doped (Si) [homoepitaxy], 613–614
  detectors, 125–129, 314, 597
    see also charge coupled device (CCD) detectors, photodiode arrays
  dielectric function, critical points (and optical properties), 313, 344, 346, 348, 369, 388–389, 430–431, 444, 464, 620
    doping, 462
    temperature variation, 344, 346, 460–463, 698
  doped, 237, 317–319, 346, 462, 613–614, 681, 685
  (monitoring) during ion bombardment, 667
  emissivity, 597, 599–603, 608
  etching, 161, 167–172, 177, 179–181, 205–206, 235, 237, 243, 252–253, 283, 286–287, 290–291, 508–510, 540, 578, 603, 640–642, 724–726
    and monitoring, 360–363, 455–456, 458–459, 582–583
    of As-doped, *n*-type, 237
  heating, 604
  in eutectic, 350
  interface with, *see* CaF_2, SiO_2
  ion implantation, 580
  laser ablation, 648
  melting, 348–350
    and monitoring, 582–583
  -on-insulator (SOI), 376, 586, 600, 603
  -on-sapphire (SOS), 376, 586
  oxidation, 17, 242, 351, 392–393, 611, 666–668
    see also SiO_2
  particles, 234, 506
  phonon energies, 77–78, 105, 108, 564
  porous, 312
  scatterometry of, 540–542, 718

Index 775

(as a) substrate, 355, 382–383, 388, 401–402, 410–412, 455, 530, 534–536, 580
surface, 382, 384, 388–393, 396–399, 401–402, 423, 577–578
  and monitoring, 576, 666–668
  adsorption on, 577–578, 626, 668
  bromination, 253, 640–641
  chlorination, 171–172, 252–253, 640–642
  desorption from, 171–172, 252–254, 640–643, 679
  orientation, 402
  oxidized, 396, 398
  photoemission, 681, 684–685
  reactivity of, 250
  treatment, 464
thermophysical properties, 369, 694
vibrations in, 266
  see also Si (phonon energies)
wafer (and solid), 237, 249, 310, 312, 359–350, 355, 580
  absorption in, 99–100, 313, 317–319, 699
  cleaning/passivation, 17, 48, 389–392, 453, 463–464, 530
  electrodes, 509
  emissivity, 139, 597, 599–602
  n-type doped, 317–319, 613–614
    As-doped, 237
  Raman scattering in, 105, 108, 564, 580, 585–586
  thermometry, 317–319, 345–349, 368–376, 460–463, 483–484, 543–552, 584–586, 599–603, 632, 649–651, 703
    see also temperature (of the wafer)
  work function, 684
    effect of doping, 681, 685
    see also heterostructures, SiO_2 (formation by oxidation)
Si_2 monitoring, 96, 226–227
SiBr monitoring, 166, 226, 253, 641
SiC
  amorphous (a-SiC), 445–446
  deposition, 195, 397
    and monitoring, 445–446
  dielectric function, 344
  Raman scattering temperature effects, 585
SiCl monitoring, 161, 166, 169–172, 177, 181, 226, 252–253, 640–642

$SiCl_2$ monitoring, 227, 288
  on surface, 409
$SiCl_3$ monitoring, 288
$SiCl_4$
  monitoring, 288, 290
  plasma, 181–183, 187, 250, 361
  reactant, 243, 288, 628
SiD_4 reactant, 280
Sidewall angles, profiles, 537–542, 718
SiF monitoring, 177, 226, 237
SiF_2 monitoring, 226, 235
SiF_3 monitoring, 161, 166
SiF_3^+ monitoring, 226
SiF_4
  monitoring, 226, 286–287, 290–291
  plasma, 290
SiF_4^+ monitoring, 226
SiF_x overlayer monitoring, 455
SiGe alloys, *see* GeSi alloys
Sigmoid function, 722–723
Signal
  averaging, 125–126
    see also boxcar integration, lock-in detection
  calibration, 153–154, 178–179
  collection and analysis (overview), 139–154
  processing, 152
  strength, 145–147, 152–154, 415, 440, 562, 565–566
Signal-to-noise (S/N) ratio, 130, 152–153, 183, 270–272, 279, 306, 566
  see also detectivity, noise
SiH
  monitoring, 166, 187–188, 195, 199, 226, 250, 280, 308
  reactivity with surfaces, 250
SiH_2 monitoring, 300–301
SiH_3
  monitoring, 280, 309, 676
  reactant, 280
SiH_4 (silane)
  monitoring, 199, 568–570
  plasma, 187–188, 199, 228, 234, 247–248, 250, 280, 295, 300–301, 304, 397, 444, 506–507, 510, 512, 691
    and monitoring, 664–665
  reactant, 159, 227–230, 234, 290, 383–384, 465, 568–570, 580–581, 602, 692
Si-H_x vibrations on surfaces, 266, 389–390, 392–393, 395–399

Si$_2$H$_6$ (disilane)
  plasma, 300
  reactant, 195, 228, 280, 395–396
SiHCl$_3$ monitoring, 288
SiH$_2$Cl$_2$
  monitoring, 288
  reactant, 215, 217, 227–228, 288, 409
SiH$_3$Cl reactant, 227
SiHF$_3$ monitoring, 290
Silane, *see* SiH$_4$
Silicides, 603
  see also Pd silicide, Pt silicide
Silicon nitride (Si$_3$N$_4$, SiN$_x$)
  deposition, 159, 187–188, 197, 665
    and monitoring, 413, 439, 446
  dielectric function, 344
  etching, 166, 168, 170, 177, 183, 541, 724, 734
    and monitoring, 360, 439
  reactivity of surface, 251–252
  substrate, 444
  vibrations in, 266
SiN (molecule) monitoring, 96, 226, 237
Single particle detection, 510–512, 515–520
Single photon ionization (SPI), 638, 674, 678–681
Single wafer processing (SWP), 15–16, 729–730
Single-wavelength pyrometry, 596–608
SiO
  dielectric function, 344
  monitoring (molecule), 159, 226, 230–231, 237, 242, 250
  (molecule) reactivity with surfaces, 250
SiO$_2$ (including fused silica)
  dielectric function, 344, 369, 371
  etching, 170, 172–173, 177–179, 184–186, 235, 237, 283, 286–287, 299, 541
    and monitoring, 360, 401–402, 439, 455, 458, 460–461
  formation, 48
    by deposition, 48, 159, 187, 230–231, 250, 290, 293–294, 312, 668
      and monitoring, 383, 413, 439, 450, 453–454
    by oxidation, 48, 242, 351, 459, 462, 611, 666–668
    see also Si oxidation
  interface with Si, 459, 666–668
  on Si, 599–602, 647, 666–668, 684
  particles
    formed during sputtering, 510
    spherical, 517–520
  porous, 312
  reactivity of surface, 250
  substrate, 309–310, 444
  thickness, 462
  vibrations in, 266
  work function (and photoemission), 684
  see also glass, optical fibers (ultraviolet)
SiO$_x$N$_y$ (silicon oxynitride) deposition and monitoring, 181, 355, 439
Sn
  atom ionization potential, 680
  deposition, 669
  dielectric function, 344
  see also indium tin oxide (ITO)
Sn(CH$_3$)$_2$, *see* dimethyltin
Snell's law, 328–331
SnTe dielectric function, 344
SO monitoring, 226, 236–237
SO$_2$
  monitoring, 226, 236
    for wavelength calibration, 278
  plasma, 691
Solid angle (collected), 144–147, 149, 163, 205, 222, 232, 264, 562, 565, 660–661, 675, 704
Solids, spectroscopy, 89, 99–100
Spatial filter, 144–145, 520–521
Spatial resolution (imaging, mapping, and analysis), *see* imaging
Specialty manufacturing, *see* flexible manufacturing
Specific heat, 112–113, 639–640, 644, 697
Speckle
  metrology/thermometry (interferometry and photography), 27, 36, 41, 495, 548–552, 697, 704, 708–709
  velocimetry, 549
Spectral lineshape, *see* lineshape
Spectral linewidth, *see* line broadening
Spectral radiance, radiant flux, 593
Spectrometers, 119–120, 129–137, 142–144, 148
  dispersive (mostly grating), 129–135, 139, 144, 154, 163–167, 175, 193, 215, 219–220, 224–225, 250, 267, 269, 274–275, 279, 296–298, 301, 304, 308, 314, 353, 357, 365, 400–401, 415–416, 435–437, 439, 485, 560, 564, 566, 586, 622–623, 640, 661–666, 732
    astigmatic (imaging), 134–135, 142, 148, 164

Index

stigmatic, 485
(dispersive) prism, 129, 134, 274, 287–290, 437
  see also Fourier transform spectrometer
Spectroscopic
  analysis, 21, 120, 151, 716, 732–733
    see also ellipsometry, multiwavelength spectroscopy, reflectometry
  instrumentation, 129–137
Spectroscopy, overview, 2, 21–28, 67–78, 81–116
Spikes on surfaces, 517–518
Spin-forbidden transitions, see forbidden transitions (by spin)
Spontaneous emission (rate), see radiative decay (rate)
Sputtering (including reactive sputtering), 36–38, 158, 161, 187–193, 234, 257, 302, 304, 355, 444, 455, 535, 603, 693, 731
  particles formed during, 507, 510
Sr
  atom monitoring, 307
  dielectric function, 344
$SrRuO_3$ deposition, 307
$SrTiO_3$ dielectric function, 344
Stainless steel electrodes, 509
Standard regression analysis, 724
Stark broadening, 89, 162, 267, 302
  see also electric field (effects, static)
Statistical design of experiments, 20, 166, 720–721, 724–726, 728, 733–734
  see also chemometrics
Statistical models (of processing), 20, 720–721, 724–725, 727
  see also chemometrics
Statistical process control (SPC), 12–13, 20, 365, 368, 728–731, 734
Stefan-Boltzmann constant, law, 593
Stimulated emission, 86, 246
Stimulated Raman scattering, see Raman scattering (stimulated)
Stokes/anti-Stokes scattering intensity ratio, 93, 567, 583, 586, 700, 704
Stokes parameters, 64–65, 341–342, 438, 498
Stokes scattering, lines, processes, 23, 93, 102–104, 123, 247, 559–560, 567, 583, 647, 657
  see also coherent Stokes Raman scattering, Stokes/anti-Stokes scattering intensity ratio

Stoney's equation, 350
Strain, 35–36, 38, 106–107, 350–351, 379, 486, 550–551, 619, 644
  biaxial, 106–107, 578, 580, 586
  in windows, 415, 417, 441–442
  relaxation, 38, 531, 533–534
    see also critical thickness, incommensurate growth
Streak photography, 195
  see also photography (high-speed)
Streamlines, see flow visualization
Stress, 11, 35, 106–107, 350–351, 379, 580, 586, 644, 648
  see also strain
Structure of matter, 67–78
Subfeature speckle interferometry (SUFSI), 550–551
Submonolayer control, monitoring, and sensitivity, 6–7, 15, 24–26, 31–33, 42–43, 46, 48, 327, 427, 443, 463–465, 561, 564–567, 572–577, 637–643, 647, 665–669, 718–719, 730, 733
  see also adsorbates, near-surface dielectric function spectroscopy, surface composition
Sulfide layers
  on InP, 465
  see also Na_2S passivation, $(NH_4)_2S$ treatment, photowashing, reflectance-difference spectroscopy
Sum frequency generation, 109, 120, 123, 249, 656
  surface, 665
Superconductors, see high T_c superconductors
Supervisory control (of a process), 14, 727–730
Surface cleaning, 43, 48, 389–392, 398–399, 424–425, 460, 463–464, 517–518, 529–531, 535
  see also passivation
Surface composition, monitoring, 25, 38, 250–257, 381–425, 572–578, 665–669, 673, 681, 683–685, 718–719
  diagnostics locator, 25
  see also adsorbates, ellipsometry, near-surface dielectric function spectroscopy, submonolayer control
Surface contamination, 16, 496, 684, 694
Surface coverage, 637–643
  see also submonolayer control, surface composition

Surface damage, defects, 516–517, 623–627
Surface deformation, 643
Surface depletion layer, 578
Surface desorption, *see* laser-induced thermal desorption, temperature programmed desorption
Surface differential spectroscopy, 328
  see also differential reflectometry, photoreflectance, reflectance-difference spectroscopy, surface photoabsorption
Surface-enhanced Raman scattering (SERS), 566–567, 580
Surface factor, 525–529
Surface/gas interactions, 678
Surface infrared
  reflectometry (SIRR), 25–26, 337, 381–399
  spectroscopy (SIRS), 25, 31, 137, 328, 381–399
Surface morphology
  periodic features, 495, 522, 537–549
  roughness (smoothness), 11, 25, 38, 343, 427, 430, 432, 453, 459, 462–463, 517, 522–537, 548–552, 593, 600–601, 669
  anisotropic, 424–425, 529, 531–534
Surface passivation, *see* passivation, photowashing, surface cleaning
Surface photoabsorption (SPA), 22, 25–26, 328, 332, 336–338, 403–413
  comparison with other methods, 417–418, 442
Surface plasmon resonances, waves, 537, 669
Surface power spectral density (PSD), 525–529
Surface processes, 250–251
  see also laser ablation, laser-induced thermal desorption, pulsed laser deposition
Surface recombination, 623–627
Surface reconstruction, 392–395, 414, 418–423, 626, 667–669
Surface roughness, *see* surface morphology (roughness)
Surface second harmonic generation, 665–669
Surface-sensitive spectroscopy, diagnostics
  optical, 24–26, 48, 100
    diagnostics locator, 25
    see also submonolayer control, surface composition, surface morphology
  nonoptical, 31–33

Surface-state density, 623–624
Surface topography, profiles, 522–552
  see also surface morphology
Surface treatment, *see* photowashing
Susceptibility (electric), 82, 84, 87–88, 90–91, 93, 100, 105, 109–111, 114–116, 656–660, 669
  see also polarizability
Synchronous detection, 162–163, 170–171, 224, 240–242, 631–632, 635, 685
  see also lock-in detection, time-resolved (kinetic) measurements

T
Ta
  dielectric function, 344
  etching, 360
Taguchi quality control, 721, 724, 728
Ta_2O_3 depostion, etching, and monitoring, 454
$TaSi_x$ deposition and stress, 351
Te
  adsorbate, 626
  alkyl monitoring, 298
  atom
    monitoring, 226, 231
    ionization potential, 680
  dielectric function, 344
TEGa, *see* triethylgallium
TEM_{00} mode, 60–62
Temperature (thermometry), 9–10, 16, 27, 97–99, 153–154, 306, 689–713
  -dependent dielectric functions, 343–346
    see also dielectric function, index of refraction
  nature of, 690–692
  of electrons, 496, 696
  of films, 604–606, 718
  of the gas phase (atoms, molecules, ions), 27, 38, 41, 162, 195–203, 221, 224, 276, 295–296, 481–483, 485–489, 567–573, 661–665, 696, 701–702, 706–708, 716, 725
    electronic (T_{elect}), 99, 199–201, 304, 691, 696
    rotational (T_{rot}), 99, 102, 105, 196–200, 203, 221, 231, 233, 235–237, 255, 276, 284, 290, 295–296, 567–573, 642, 662–663, 678, 691, 696, 701

Index 779

translational (T_{trans}), 99, 199, 202–203, 223, 238, 254–255, 276, 284, 296, 691, 696, 701
 see also time of flight
vibrational (T_{vib}), 99, 197, 199, 203, 221, 231, 233–237, 255, 276, 642, 678, 691, 696, 701
of the wafer, 11, 16, 27, 38, 46, 264, 312–319, 345–349, 352, 368–377, 443, 450, 460–463, 483–485, 535, 543–549, 551–552, 583–586, 591–614, 621, 628–635, 649–651, 692, 695, 697–700, 702–705, 708–713, 716, 725
 see also thermometry
Temperature-dependent optical parameters, 697, 700
 see also band gap, critical point, dielectric function, index of refraction
Temperature measurements, monitoring (thermometry), *see* temperature
Temperature programmed desorption (TPD), 31–32, 252–254, 637–638, 679–681
 see also laser-induced thermal desorption (LITD)
Temporal resolution, *see* time-resolved measurements
TEOS, *see* tetraethoxysilane
tert butylarsine reactant, 420, 422
tert butylphosphine (TBP) reactant, 409
TESb, *see* triethylantimony
Test wafers, *see* monitor wafers
Tetraethoxysilane (TEOS)
 monitoring, 293
 reactant, 293–294, 312
Tetraethylorthosilicate (TEOS), *see* tetraethoxysilane
TGS detector, *see* deuterated triglycine sulfate detector
Th dielectric function, 344
Thermal conductivity, 112–113, 639–640, 644
Thermal desorption spectroscopy (TDS), *see* temperature programmed desorption, laser-induced thermal desorption
Thermal diffusion, 112–113, 639, 644
 distance, 113, 644
 time, 694
 see also thermal wave analysis

Thermal distributions, distribution function, *see* Boltzmann factor, Bose-Einstein distribution function, Fermi-Dirac distribution function, Maxwell-Boltzmann distribution function
Thermal emission, *see* pyrometry
Thermal equilibrium, 88, 97–99, 591–592, 628, 690–692
 limited (deviations from), 99, 200, 296, 690–691
Thermal expansion, 317, 351
 coefficient, 347, 369, 371–372, 376, 544, 546, 551, 580, 584, 698, 700
 of the wafer (for thermometry), 347, 349, 368–377, 483–484, 543–552, 580, 583–584, 697, 703–704, 708–709, 712–713
Thermal lensing probing, 646
Thermal radiation, *see* pyrometry
Thermal wave analysis (TWA), spectroscopy, 24, 263, 642–651
 double-modulation technique, 646–647
 mapping, 646
 number, 644
 thermometry, 649–651, 705, 709
Thermal waves, 643–644
Thermocouples, 29, 124, 312, 315, 350, 371, 462, 545, 551, 569, 571, 646, 691, 693–696
Thermodynamic equilibrium, 690
 see also thermal equilibrium
Thermoelastic response sensing, 644
Thermometry, 5, 10–11, 27–28, 35, 38–39, 97–99, 153–154, 306, 689–713
 diagnostics locators, 27
 need for, requirements of, 692–695
 of electrons, 496, 696
 of the gas phase, *see* temperature (of the gas phase)
 of the wafer, *see* temperature (of the wafer)
 see also temperature
ThF_4 dielectric function, 344
Thickness monitoring, *see* film thickness, metrology
Thin film processing, overview, 34–50
Third harmonic generation (THG) [of gases], 110
 for light generation, 120, 123, 656, 679–680
 for monitoring, 26–27, 655, 669–670
 suppression during REMPI, 676, 678

Thomson scattering (of electrons), 38, 96, 696, 707
Three-dimensional (3D, island) growth, 424–425, 578
Three-photon absorption, *see* absorption (three-photon)
Ti
  ablation, 194, 256
  atom monitoring, 166, 190–191
  dielectric function, 344
  (as) diffusion barrier, 669
$TiCl_4$, 41, 513
Time of flight, 193–198, 202–203, 254–257, 638, 679–681
  see also mass spectrometry
Time-resolved (kinetic) measurements, 151–152, 162–163, 170–172, 193–198, 202–203, 235, 240–241, 256, 302–303, 342, 345, 347, 349, 433, 442–443, 447, 631–635, 637–644, 656, 661, 733
  see also boxcar integration; endpoint detection; flow monitoring; flux monitoring; gated detection; photography (high-speed); reactant delivery; synchronous detection; time of flight; time-resolved reflectivity; transient, turn-on, turn-off behavior
Time resolved reflectivity (TRR), 345–350, 354, 362–364, 376, 403, 410–411
  see also surface photoabsorption
TiN (TiN_x)
  etching, 177
  deposition, 190–191, 194, 256, 293
  and monitoring, 455
TiO_2
  depostion, etching, and monitoring, 454
  dielectric function, 344
  particles, 41, 513, 517–520
$TiSi_2$ (C49) formation and stress, 351
Titanium sapphire lasers, 120–121
Titration, 248–249
TMAl, *see* trimethylaluminum
TMB, *see* trimethylborane
TMGa, *see* trimethylgallium
TMIn, *see* trimethylindium
TMP, *see* trimethylphosphine
TMSb, *see* trimethylantimony
Tomographic reconstruction, 7, 141, 148–151, 203–208, 264, 270, 361
Tool
  cleaning, maintenance, 184, 496–497
  control, 495
  design, access, data from, 509, 694–695, 702, 715
  malfunctioning, 729–730, 734
  performance, 729
  -scale models, 724
  see also cluster tools, rotating disk reactors, walls
Total integrated scatter (TIS), 527–528
Total internal reflection, 385, 399
  microscopy (TIRM), 536–537
  spectroscopy, *see* attenuated total internal reflection (ATR) spectroscopy
  see also light pipes
Tracers, 41, 512–514
  see also flow visualization
Transformer (inductively) coupled plasmas [TCP (ICP)], 44
Transient, turn-on, turn-off behavior, 230–231, 244, 276, 280, 282, 294–295, 298, 300, 302–304, 306
  see also flow monitoring; flux monitoring; reactant delivery; time-resolved (kinetic) measurements
Transition moment, *see* dipole moment, matrix elements
Transition strengths (atoms, molecules), 96–99
Translational energy, 167, 195–198, 202–203, 254–257
  see also temperature (translational, T_{trans})
Transmission, 25, 263–319
  coefficient, 329–330, 465–466
  wafer thermometry, 27, 312–319
  see also absorption
Transmission electron microscopy (TEM), 31, 33, 534
Transmittance, 264, 329, 465–466, 593
  see also absorbance
Triethylaluminum (TEAl) reactant, 449
Triethylantimony (TESb)
  monitoring, 295, 298–299
Triethylgallium (TEGa)
  monitoring, 292, 295, 304–305
  reactant, 292, 449, 405–408
Triethylindium (TEIn) reactant, 407
Triethylphosphene (TEP) adducts, 296
Trimethylaluminum (TMAl)
  ionization potential, 680
  reactant, 312
  and monitoring, 279–280, 298–300

Index   781

Trimethylantimony (TMSb)
  monitoring and reactant, 298–299
Trimethylborane (TMB)
  monitoring and reactant, 293
Trimethylgallium (TMGa)
  reactant, 40, 232, 394–396, 405, 408,
    420–423, 507, 604
  and monitoring, 279–282, 288–292,
    295–296, 298–300
Trimethylindium (TMIn)
  adducts, 296
  monitoring, 304–305
  reactant, 232, 407
  and monitoring, 279–280, 282, 295,
    298–299, 570, 574
Trimethylphosphene (TMP)
  adducts, 296
  monitoring and reactant, 293
tris(dimethylamino)arsine (DMAAs)
  monitoring and reactant, 291–292
tris(dimethylamino)phosphine (DMAP)
  monitoring and reactant, 291–292, 397
tris(dimethylamino)stibine (DMASb)
  monitoring and reactant, 291–292
Tritertiarybutylaluminum (TTBAl)
  monitoring, 295
TRR, see time-resolved reflectivity
TTBAl, see tritertiarybutylaluminum
Tunable lasers, 120–123, 215, 224
  see also dye lasers, diode lasers, titanium
    sapphire lasers
Two-color pyrometry, 608–609
Two-dimensional (2D) growth, see layer-
  by-layer growth
Two-phonon Raman scattering
  (overtones), 574–577
Two-photon
  absorption, see absorption (two-photon)
  laser-induced fluorescence, see laser-
    induced fluorescence (two photon
    LIF)
Two theta (2Θ) scatterometers, 538–539

U
Ultrahigh vacuum (UHV), 6, 31–32, 38,
  418–421, 582, 613
  see also molecular beam epitaxy
Ultrasonic detection, 31, 295
Ultraviolet (UV)
  exposure, 392–393
    of photoresist, 49–50, 310–311,
      365–366, 537, 542–544

see also photolysis
photoelectron spectroscopy (UPS), 681
  see also excimer lasers, vacuum
    ultraviolet light
Units and units conversion, 57, 81,
  113–116
Univariate analysis (correlation), 172–173

V
V dielectric function, 344
Vacuum leaks (detection of), 46, 172
Vacuum ultraviolet (VUV) light, 244, 247,
  255, 670
Vapor phase epitaxy (VPE), 39–40
  see also chemical vapor deposition,
    metalorganic chemical vapor
    deposition
Variable angle ellipsometry, 418, 438, 442,
  459
Variable-angle monochromatic fringe
  observation (VAMFO), 353
Varshni equation, parameters, 313, 315,
  380, 698–699
  see also band gap
Vector potential, 85, 93
Velocity-modulated infrared laser
  spectroscopy, 287
Velocity (speed) profiles, see Doppler,
  electron cyclotron resonance plasmas,
  laser ablation, laser velocimetry,
  Maxwell-Boltzmann distribution
  function, pulsed laser deposition, time
  of flight
Vertical-cavity surface-emitting lasers
  (VCSEL), 355, 359–360, 611
  see also distributed Bragg reflectors
Very large scale integration (VLSI), 34
Very thin films, 336–340, 572–578, 626
  see also adsorbates, differential
    reflectometry, ellipsometry,
    overlayers, reflectance-difference
    spectroscopy, surface
    photoabsorption
Vias, 178–180, 360
Vibrational (in molecules)
  (-rotational) absorption transitions, 69,
    94–97
  energies, energy levels, 68–69, 71, 74,
    266
  mode energies (typical), 71, 266
  motion, 68
  partition function, population, 73, 75–76

Vibrational (in molecules) (*continued*)
  Raman scattering (and vibrational-rotational), 90–93, 98, 100–101, 103–105, 237, 559–563, 567–572, 574
    see also coherent anti-Stokes Raman scattering (CARS), temperature (vibrational, T_{vib})
Vibrations in solids, 77–78
  Raman scattering, 90–91, 93–94, 105–108, 559–563, 578–586, 690, 700, 702, 704, 711
    see also phonons
Vibrations on surfaces, 266, 381–399
Vibronic states (bands), 69, 96
Vicinal (miscut) substrates, 532–535
Vidicons, 628
  see also multichannel detection
Virtual
  interface model, theory, 339–340, 433–434, 440, 450, 719, 733
  reflection coefficient, 339
  sensor, 731
  substrate approximation, 434, 450, 719
Voids (in materials), 430–431, 454
  see also porosity
Voigt profile, 88
Volumetric instruments (for particle monitoring), 515

W

W
  atom monitoring, 195
  clusters, 195
  deposition, 195, 290, 310, 571–572, 693
    and monitoring, 455
    and stress, 351
  dielectric function, 344
  etching, 169–170, 177
  /halogen lamps, 597
    see also lamps
  surface, 383
W^+ monitoring, 195
Wafer
  absorption, 312–319
  contamination, 16, 29, 496, 684, 694
  curvature, 350–351
  diameter(s), 16
    see also extensometry, thermal expansion (of the wafer)
  rotation, 33, 184–185, 228, 327, 357–360, 442, 449, 451, 513, 531–532, 534, 569, 571, 666
  single wafer processing (SWP), 15–16, 729–730
  -state parameters, 10–12, 715–716, 719–720, 730–732, 734
  temperature (thermometry)
    diagnostics locators, 27
    uniformity (across the wafer), 692, 695, 704, 716, 725, 732
    see also temperature (of the wafer)
  warpage, slippage, 693–694
  see also monitor wafers
Walls (in a reactor)
  coatings on, 299, 720, 725, 729
  flaking from, 184, 497, 510–512
Wavelength, 57–58
Wavelength-modulation spectroscopy, 271–272, 278–279, 283–287, 301
  see also frequency-modulation spectroscopy
Wave vector, 58
Wet chemical
  etching, 47–48, 361–362, 459, 541
  processing, 34–35, 47–48
WF_6 reactant, 195, 290, 351, 383–384
  and monitoring, 571–572
White (multipass) cell, 274–275, 279, 284
  see also multipass configuration
White latex paint spheres, 517–520
White light sources, *see* lamps
Wien's
  approximation, 594, 608
  displacement law, 592, 596
Work function, 14, 46, 642, 673, 681, 683–684

X

Xe
  arc lamps, 597
    see also lamps
  plasma, 175
XeCl laser, 120, 122–123, 172, 224, 227, 229, 252–253, 491, 640–642
XeF laser, 120, 224, 255
  see also excimer lasers, laser ablation, pulsed laser deposition
XeF_2 reactant, 668
X-ray
  fluorescence, 31
  optics, 455

Index

photoelectron spectroscopy (XPS), 17, 31–32, 455–456, 582, 642, 681
scattering, 523

Y

Y atom monitoring, 166, 194, 202, 226, 256, 302–303
Y^+ monitoring, 166, 194
YAG laser, *see* Nd:YAG laser
YAG:Tb phosphor, 633
$YBa_2Cu_3O_{7-x}$ compounds, 194–198, 202–203, 256, 302–304, 307, 490, 613–614, 648
YO monitoring, 166, 194, 226, 256, 302–303
Y_2O_3
   dielectric function, 344
   laser ablation, 194
Y_2O_3:Eu phosphor, 633
Young's modulus, 350
$Y_2O_3ZrO_2$ dielectric function, 371
YVO_4:Eu phosphor, 633

Z

Zeeman effect, 238, 308
Zn
   adsorbate, 626
   alkyl monitoring, 298
      see also dimethylzinc

atom
   ionization potential, 680
   monitoring, 182, 226, 250, 257
dielectric function, 344
sputtering, 257
ZnO
   Raman scattering temperature effects, 585
   sputtering, 193
ZnS
   ablation, 257
   dielectric function, 344
   etching, 182
   phonon energies, 108
   Raman scattering temperature effects, 585
ZnSe
   deposition, 534
      and monitoring, 576–580
   dielectric function, 344
   phonon energies, 108
   spectroscopy, 99
ZnTe
   deposition, 534
   dielectric function, 344
   phonon energies, 108
   thermometry, 318
Zr
   atom monitoring, 16, 191
   dielectric function, 344
ZrN sputter deposition, 190–191
ZrO_2
   deposition, 195
   film, 536–537